Advanced Oxidation Processes for Wastewater Purification

Advanced Oxidation Processes for Wastewater Purification

Topic Editors

Ying Zhang

Fangke Yu

MDPI

Basel • Beijing • Wuhan • Barcelona • Belgrade • Novi Sad • Cluj • Manchester

Topic Editors
Ying Zhang
College of Environmental Science and Engineering
Nankai University
Tianjin
China

Fangke Yu
School of Environmental Science and Engineering
Shaanxi University of Science and Technology
Xi'an
China

Editorial Office
MDPI AG
Grosspeteranlage 5
4052 Basel, Switzerland

This is a reprint of the Topic, published open access by the journals *Catalysts* (ISSN 2073-4344), *Molecules* (ISSN 1420-3049), *Nanomaterials* (ISSN 2079-4991), *Processes* (ISSN 2227-9717), *Separations* (ISSN 2297-8739), *Sustainability* (ISSN 2071-1050) and *Water* (ISSN 2073-4441), freely accessible at: https://www.mdpi.com/topics/1SA560VE8F.

For citation purposes, cite each article independently as indicated on the article page online and as indicated below:

Lastname, A.A.; Lastname, B.B. Article Title. *Journal Name* **Year**, *Volume Number*, Page Range.

ISBN 978-3-7258-7813-0 (Hbk)
ISBN 978-3-7258-7814-7 (PDF)
https://doi.org/10.3390/books978-3-7258-7814-7

Contents

About the Editors

Ying Zhang

Ying Zhang is a Professor and Doctoral Supervisor at the College of Environmental Science and Engineering, Nankai University. She received her bachelor's, master's, and doctoral degrees from Shandong Agricultural University, Tongji University, and the University of Alberta in Canada, respectively, and previously conducted postdoctoral research at the University of Toronto. Her research is primarily focused on water pollution control, environmental interfacial catalysis, advanced oxidation technologies, and the development of environmental nanomaterials. Moreover, she has led multiple research projects funded by the National Natural Science Foundation of China and the Tianjin Natural Science Foundation, and has published 80 high-quality SCI papers. In addition, she serves as a reviewer and young Editorial Board member for several core journals, as well as teaching undergraduate courses, demonstrating a strong commitment to both scientific research and talent development.

Fangke Yu

Fangke Yu is an Associate Professor, PhD, and Master's Supervisor at the School of Environmental Science and Engineering, Shaanxi University of Science and Technology. A Fellow of the Royal Society of Chemistry, he graduated from the School of Environmental Science and Engineering, Nankai University. He was selected as one of the first batch of outstanding young talents at the university and a young scholar for ecological environmental protection in the Qinling Mountains of Shaanxi Province. He also serves as an expert reviewer for national and provincial natural science foundations and a reviewer for several authoritative environmental SCI journals. His main research areas are electrochemical advanced oxidation water treatment, functional electrode material development, new materials for water purification, and battery recycling technology. He has presided over several provincial and national-level research projects, including those funded by the National Natural Science Foundation of China, the Postdoctoral Science Foundation, and the Shaanxi Provincial Natural Science Foundation. He has published over forty SCI papers, applied for several national invention patents, and was shortlisted for the national finals of the National Postdoctoral Innovation and Entrepreneurship Competition. He also teaches courses such as "Principles of Environmental Engineering" and "Physical Pollution Control Engineering," and actively guides students in scientific and technological innovation competitions and postgraduate training.

catalysts

MDPI

Article

PVP-Assisted Synthesis of Fe-TiO_2 for Efficient Tetracycline Degradation via Peroxymonosulfate Activation

Xin Wang [1,2], Longxue Yu [1], Shuo Wang [1], Lingyun Xue [1], Yi Liu [1,2,*], Zhuhui Qiao [2,*] and Xunyong Liu [1,2,*]

1 Shandong Laboratory of Yantai Advanced Materials and Green Manufacturing, Yantai 264006, China; wangxin1230518@163.com (X.W.); yulongxue0303@163.com (L.Y.); 2024120762@m.ldu.edu.cn (S.W.); lingyun302a@126.com (L.X.)

2 School of Chemistry and Chemical Engineering, Ludong University, Yantai 264025, China

* Correspondence: liuyi@ldu.edu.cn (Y.L.); zhqiao@licp.cas.cn (Z.Q.); xunyongliu@ldu.edu.cn (X.L.); Tel./Fax: +86-535-6672176 (X.L.)

Abstract

Tetracycline (TC) is chemically stable and recalcitrant to natural degradation. Peroxymonosulfate (PMS)-based advanced oxidation processes offer an effective removal strategy, the efficacy of which relies on high-performance heterogeneous catalysts. Titanium dioxide (TiO_2) is an ideal material due to its stability and environmental compatibility, yet its practical application is hindered by inadequate PMS activation capacity, particle agglomeration, and difficult recovery. To address these limitations, a heterogeneous Fe/TiO_2 catalyst was constructed via Fe^{3+} doping, innovatively utilizing polyvinylpyrrolidone (PVP) as a structure-directing agent. PVP's steric hindrance effectively suppressed nanoparticle agglomeration and enabled high dispersion of Fe active sites, simultaneously enhancing catalytic activity and stability. Under optimized conditions, the Fe/TiO_2/PMS system achieved 94.3% TC degradation, following pseudo-first-order kinetics and significantly outperforming pure TiO_2 used in this experimental system. Radical quenching verified sulfate radicals ($SO_4{\bullet}^-$) as the dominant species. The catalyst demonstrated excellent recyclability, retaining over 80% degradation efficiency after six cycles and enabling convenient magnetic separation. Moreover, in complex water matrices (tap water and seawater), it sustained high removal efficiency (>90% initially, >70% after six cycles), highlighting its superior anti-interference capability and practical potential. This work offers a strategic material design strategy for efficient and robust TC removal in challenging water environments.

Keywords: Fe-doped TiO_2; peroxymonosulfate activation; tetracycline degradation; heterogeneous catalysis

1. Introduction

With the rapid development of the economy and the continuous deepening of industrialization, water pollution issues have become increasingly severe. Among them, the widespread presence of antibiotic contaminants in water bodies and their potential ecological risks have attracted extensive attention [1]. Tetracycline (TC), a typical broad-spectrum antibiotic, is extensively used in human medicine and livestock farming. However, its persistence and tendency to accumulate in aquatic environments pose a serious threat to ecological balance and public health [2]. Moreover, the accumulation of TC in organisms may induce adverse physiological effects, including nephrotoxicity, mutagenicity, and teratogenicity [3]. Consequently, the development of efficient TC treatment technologies holds significant importance.

Currently, techniques such as adsorption, biodegradation, membrane separation, and catalytic degradation have been employed for TC removal [4–6]. Among them, advanced oxidation processes (AOPs) are regarded as one of the most promising approaches for TC elimination due to their high efficiency, low cost, and mild reaction conditions [7]. In particular, the peroxymonosulfate (PMS)-activated AOP has garnered widespread attention owing to its exceptional degradation capability for emerging contaminants and broad pH adaptability [8]. Conventional PMS activation strategies primarily rely on homogeneous Fenton-like catalytic systems (e.g., using soluble iron salts). However, these systems face significant limitations: the continuous leaching of transition metal ions not only leads to catalyst deactivation but also causes secondary pollution, severely restricting their practical application [9]. To address these issues, the development of stable heterogeneous Fenton-like catalysts has become a research focus [10–12]. Among various heterogeneous catalysts, titanium dioxide (TiO_2) stands out due to its comprehensive advantages including non-toxicity, high chemical stability, abundance, and easily modifiable surface [13,14]. However, TiO_2-based materials generally suffer from limitations such as restricted PMS activation capability and difficult recovery [15].

To address these challenges, researchers have explored modification strategies including crystal phase control, elemental doping, and surface sensitization [16–18]. Among these, metal ion doping has been demonstrated as an effective approach to enhance the overall performance of TiO_2-based catalysts [19–21]. Fe^{3+}-modified TiO_2 composite materials have attracted significant attention due to their high stability and tunable surface properties. The Fe^{3+} ion, with its similar radius to Ti^{4+}, can be firmly incorporated into the TiO_2 lattice through lattice site substitution or surface anchoring [22]. For instance, Diamandescu et al. confirmed through extended X-ray absorption fine structure (EXAFS) analysis that Fe^{3+} can successfully substitute Ti^{4+} sites in the TiO_2 lattice, inducing local structural distortion characterized by an increase in Ti-O bond length and a decrease in Ti-Ti bond length. This structural alteration provides a microscopic basis for explaining the enhancement of its catalytic performance [23]. This process not only establishes efficient and stable heterogeneous active centers within the material but also introduces ferromagnetic components through successful Fe^{3+} doping, endowing the catalyst with responsiveness to external magnetic fields. The resulting magnetic properties enable rapid catalyst recovery through magnetic separation after reaction completion, eliminating dependence on traditional solid–liquid separation techniques such as centrifugation or filtration. This approach significantly enhances recovery efficiency and operational convenience. For example, Sun et al. [24] fabricated a magnetic Fe_2O_3/mesoporous black TiO_2 hollow-sphere heterojunction (M-Fe_2O_3/b-TiO_2) via wet impregnation and surface hydrogenation strategies, which exhibits broad-spectrum response and magnetic separation capability. Furthermore, the Fe active centers can directly and efficiently activate PMS by triggering homolytic cleavage of its O-O bond through interfacial electron transfer, continuously generating highly reactive oxygen species dominated by sulfate radicals ($SO_4\bullet^-$). Simultaneously, a unique synergistic mechanism may form between potential Ti^{3+} defects and Fe species in the material: Ti^{3+} acts as an endogenous electron donor, facilitating the reduction in surface Fe^{3+} to Fe^{2+}, thereby significantly accelerating the Fe^{2+}/Fe^{3+} interfacial catalytic cycle and enhancing both PMS activation efficiency and sustained radical generation capacity [25]. For instance, Xu et al. [26] activated the originally redox-inert TiO_2 by inducing a surface Fe^{2+}/Fe^{3+} cycle through Fe doping in TiO_2. This modification not only confers exceptional Fenton-like activity to TiO_2 but also substantially addresses the technical challenges associated with the recovery and reuse of TiO_2 nanoparticles. Although the Fe^{3+} doping strategy can enhance the catalytic activity of TiO_2 and improve its recovery performance, such materials still face a critical challenge in practical Fenton-like applications: nanoparticles are prone to agglomeration during synthesis, leading to reduced active sites and diminished mass transfer efficiency, which consequently causes significant attenuation of catalytic and stability performance [27,28].

In response to the issues outlined above, several strategies have been explored in existing studies. For instance, constructing hollow or mesoporous structures to increase specific surface area or introducing carbon materials as dispersion carriers has been proposed. However, these approaches often involve complex synthesis procedures, and the introduction of additional components may lead to unfavorable interfacial charge recombination [29–31]. On the other hand, the use of conventional polymers or surfactants as dispersing agents represents an alternative approach. However, their insufficient stability under Fenton-like reaction conditions often leads to desorption from the catalyst surface, resulting in non-persistent dispersion effects [32,33]. It is noteworthy that even polyvinylpyrrolidone (PVP), a polymer capable of effectively suppressing random nanoparticle aggregation through coordination by its polar groups and steric hindrance from its hydrophobic chains, has been predominantly utilized in existing studies merely as a physical barrier [34]. This limited utilization hinders the full exploitation of its potential as a "structure-directing agent" for synergistically optimizing the overall material performance. Consequently, existing strategies often focus predominantly on addressing isolated challenges, such as agglomeration during synthesis or recovery difficulties, while lacking an integrated design capable of simultaneously optimizing activity, stability, and operational convenience. Crucially, the deep integration of the interfacial modulation capability of PVP with the magnetic and catalytic properties of Fe^{3+} doping remains a significant challenge. The fundamental goal of constructing a catalytic material that concurrently possesses high-density active sites, superior operational stability, and convenient magnetic separation is thus a critically underexplored scientific problem.

Based on the above analysis, we propose a PVP-assisted Fe^{3+} doping strategy to construct a high-performance Fe/TiO_2 catalyst via a facile one-step hydrothermal method. In this design, PVP molecules not only suppress particle agglomeration but also, through their coordination effect, guide and stabilize the highly dispersed state of Fe species while they are incorporated into the TiO_2 lattice, thereby establishing a robust Fe-O-Ti interface at the nanoscale. This synergistic design effectively inhibits nanoparticle aggregation, thereby enhancing the initial catalytic activity while laying a structural foundation for long-term cycling stability. Furthermore, the successfully incorporated Fe^{3+} serves as highly efficient active sites while simultaneously endowing the material with convenient magnetic separability. Performance tests demonstrate that the material maintains over 90% TC removal efficiency under complex water matrix conditions, along with favorable cycling stability. These results strongly validate the effectiveness of our proposed integrated design strategy, using PVP as a structure-directing agent, in simultaneously enhancing catalytic activity, stability, and operational convenience, offering a new perspective for the treatment of TC-containing wastewater in practical environments.

2. Results and Discussion

As illustrated in Scheme 1, a modified TiO_2 catalyst (Fe/TiO_2) was developed via a facile PVP-assisted hydrothermal method. The resulting catalyst effectively activated PMS through a synergistic action between TiO_2 and Fe species, generating highly oxidative radicals that subsequently attacked and degraded TC.

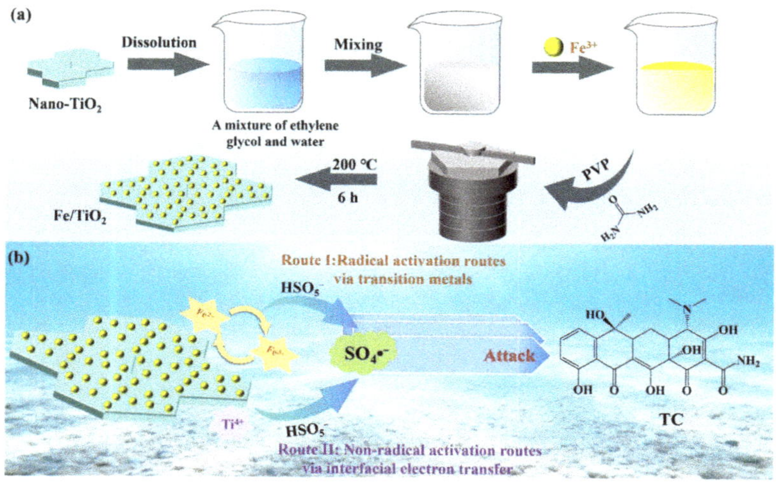

Scheme 1. Preparation of Fe/TiO_2 (**a**) and its mechanism for TC degradation (**b**).

2.1. Characterization of Fe/TiO_2

The crystal structure and chemical environment of Fe/TiO_2 were systematically characterized by X-ray diffraction (XRD) and Fourier transform infrared (FT-IR) spectroscopy. The XRD pattern (Figure 1a) exhibits characteristic diffraction peaks at 25.3°, 37.8°, 48.0°, and 53.9°, which are indexed to the (101), (004), (200), and (105) crystal planes of anatase TiO_2 (JCPDS No. 21-1272) [35], respectively, indicating the formation of a well-crystallized anatase matrix. Moreover, no diffraction peaks corresponding to free iron oxide phases were detected, suggesting that the Fe^{3+} precursor did not form separate crystalline phases but was highly dispersed within the TiO_2 support. The chemical bonding states were further investigated by FT-IR spectroscopy (Figure 1b). The absorption band at 484.71 cm^{-1} is attributed to the Ti-O-Ti skeletal vibration, confirming the preservation of the anatase structure as determined by XRD. The observed shoulder peak at 554.62 cm^{-1}, assigned to Fe-O bond vibration, directly demonstrates the successful incorporation of Fe species, forming stable configurations such as Fe-O-Fe or Ti-O-Fe, thereby constructing efficient heterogeneous active sites for PMS activation. Additionally, the appearance of the C-N stretching vibration peak at 1244 cm^{-1} indicates the residual presence of PVP in the material. After the catalytic reaction, these key spectral features remained largely unchanged, demonstrating the excellent chemical bond stability of the catalyst under strong oxidizing conditions. N_2 adsorption–desorption analysis revealed fundamental differences in the pore characteristics between pristine TiO_2 (Figure 1c) and Fe/TiO_2 (Figure 1d). Fe/TiO_2 exhibited an order-of-magnitude enhancement in both specific surface area (237.2 m^2/g) and total pore volume (0.275 cm^3/g) compared to pristine TiO_2 (11.99 m^2/g and 0.021 cm^3/g, respectively), creating favorable conditions for high-density loading and full exposure of Fe active sites. The pore size distribution provided further structural insight: pristine TiO_2 showed a concentration of pores around 3.7 nm, indicative of tight nanoparticle packing, whereas Fe/TiO_2 displayed a dominant peak at 35.2 nm. This prominent shift confirms that PVP acted as a structure-directing agent, guiding the assembly of nanoparticles into a three-dimensional network with open mesoporous channels. Crucially, this mesoporous structure is highly advantageous over a microporous one, as it provides sufficiently large channels for the rapid diffusion of bulky PMS and TC molecules. Consequently, this optimized pore architecture significantly enhanced mass transfer of PMS and TC molecules, ensuring their efficient access to internal active sites [36].

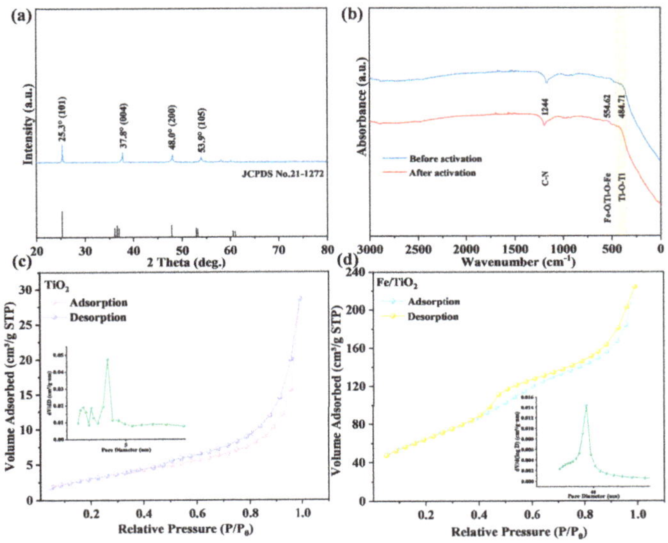

Figure 1. (**a**) XRD pattern and (**b**) FT-IR spectra before and after the reaction of the Fe/TiO_2 catalyst; N_2 adsorption–desorption isothermsand the corresponding pore size distribution curves of pristine TiO_2 (**c**) and Fe/TiO_2 (**d**).

To further investigate the morphological and structural characteristics of the catalysts, TiO_2 and Fe/TiO_2 were characterized by transmission electron microscopy (TEM). As shown in Figure 2a, pure TiO_2 exhibits a sheet-like morphology with poor crystallinity, as evidenced by vague lattice fringes. After Fe modification, Fe/TiO_2 (Figure 2b) maintained a uniformly dispersed nanosheet morphology without significant agglomeration, confirming that PVP effectively functioned as both a steric hindrance agent and a dispersant during synthesis. The high-resolution TEM image (Figure 2c) further reveals that the Fe/TiO_2 nanosheets possess clear and continuous lattice fringes with an interplanar spacing of approximately 0.35 nm, which corresponds well to the (101) plane of the anatase phase. This result is consistent with XRD analysis and indicates significantly enhanced crystallinity due to Fe doping and the hydrothermal process, providing a solid foundation for charge transfer and structural stability. Furthermore, the elemental mapping images (Figure 2d–f) clearly demonstrate a highly uniform distribution of Fe throughout the TiO_2 matrix with no observable local aggregation. This provides direct microscopic evidence that Fe species were successfully doped and uniformly anchored onto the TiO_2 surface.

X-ray photoelectron spectroscopy (XPS) analysis was conducted to further investigate the surface elemental composition and chemical states of the Fe/TiO_2 catalyst. The survey spectrum (Figure 3a) confirms the presence of Fe, Ti, and O elements. In the high-resolution Fe 2p spectrum (Figure 3b), the characteristic peaks observed at 711.2 eV and 709.8 eV are assigned to Fe^{3+} and Fe^{2+}, respectively. The coexistence of the Fe^{3+}/Fe^{2+} redox couple provides key sites for the electron transfer process in PMS activation during the Fenton-like reaction [37]. In the O 1s spectrum (Figure 3c), the peak located at 529.8 eV is attributed to contributions from both Fe-O and Ti-O bonds. The characteristic peak at 458.7 eV in the Ti 2p spectrum (Figure 3d) verifies the presence of Ti^{4+}, corresponding to anatase TiO_2. Collectively, the XPS spectral analysis reveals strong interfacial interactions between the Fe and Ti species, which stabilize their stable and dispersed coexistence.

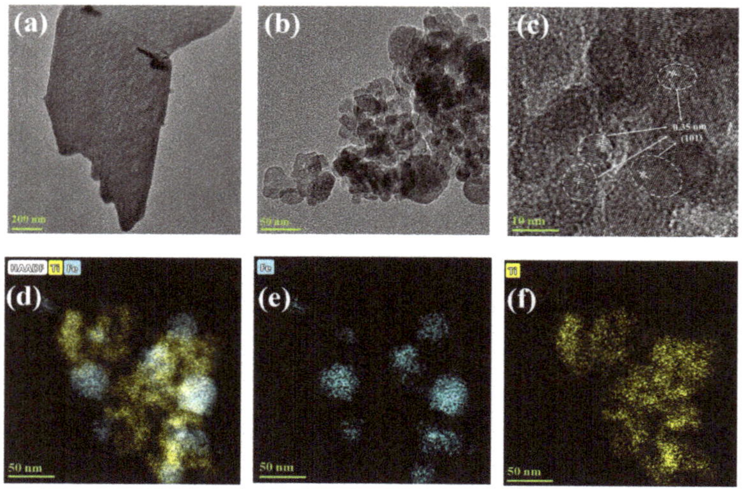

Figure 2. TEM image of TiO_2 (**a**); TEM images of Fe/TiO_2 (**b**,**c**); Elemental mapping images of Fe and Ti (**d**–**f**).

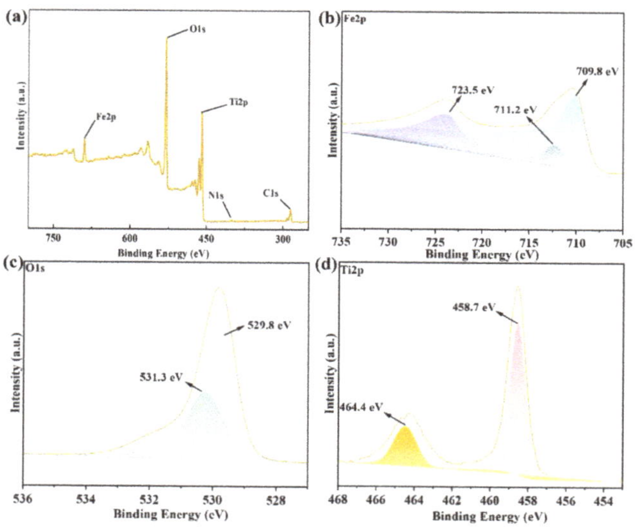

Figure 3. (**a**) XPS survey spectrum and high-resolution spectra of (**b**) Fe 2p, (**c**) O 1s, and (**d**) Ti 2p for the Fe/TiO_2 catalyst.

2.2. Optimization of Catalytic Condition

The optimal reaction conditions for TC degradation in the Fe/TiO_2/PMS system were determined by systematically investigating key parameters, including PMS concentration, initial TC concentration, and catalyst dosage. As the oxidant precursor, the PMS concentration directly governs the yield of active radicals in the system. As shown in Figure 4a, the degradation efficiency of TC initially increased and then decreased with increasing PMS concentration, reaching a maximum at 0.6 mM. In the low-concentration range (0.2–0.6 mM), the enhancement in degradation efficiency can be attributed to the increased activation of PMS molecules by the active sites on the Fe/TiO_2 catalyst surface, generating sufficient amounts of $SO_4\bullet^-$ and $\bullet OH$ radicals, thereby accelerating the oxidative decomposition of TC. However, when the PMS concentration exceeded 0.6 mM, the TC degradation efficiency decreased. This decline can be attributed to two primary reasons: first, excess PMS may scavenge $SO_4\bullet^-$ radicals in the system, consuming the highly reactive radical species; second, high PMS concentrations may lead to the occupation of catalyst active sites by its decomposition products or unreacted molecules, thereby reducing TC adsorption and inhibiting the reaction progress. Subsequently, the effect of initial TC concentration on the degradation performance was investigated. The effect of the initial TC concentration was also examined (Figure 4b). As expected, the degradation efficiency decreased with increasing TC concentration, which is consistent with expectations. This trend occurs because, at fixed catalyst and PMS loadings, the finite radical population becomes insufficient to degrade the higher quantity of TC molecules. Furthermore, TC and its degradation intermediates compete for the available radicals, ultimately leading to reduced apparent degradation efficiency.

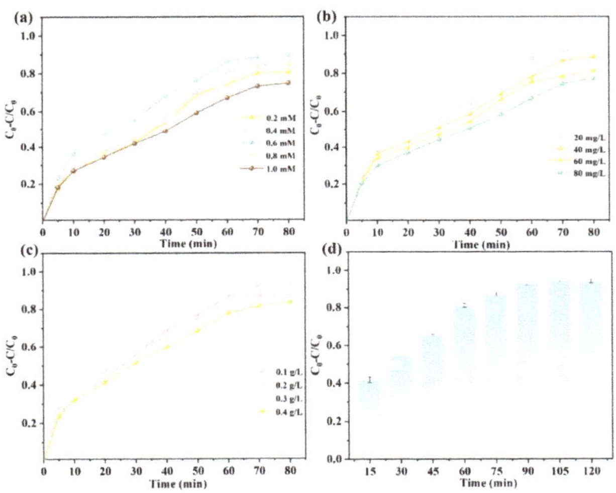

Figure 4. Univariate optimization experiments. The degradation performance was evaluated under the following conditions: (**a**) Effect of PMS concentration with fixed reaction time (80 min), (**b**) initial TC concentration (40 mg/L), and Fe/TiO_2 dosage (0.2 g/L); (**c**) Effect of initial TC concentration with fixed PMS concentration (0.6 mM), Fe/TiO_2 dosage (0.2 g/L), and reaction time (80 min); (**d**) Effect of reaction time with fixed PMS concentration (0.6 mM), initial TC concentration (20 mg/L), and Fe/TiO_2 dosage (0.2 g/L).

Following the investigation of reactant concentrations, the role of catalyst dosage was evaluated. As the core component for PMS activation and radical generation, the catalyst dosage significantly influenced the process. As shown in Figure 4c, when the catalyst dosage increased from 0.1 g/L to 0.2 g/L, the degradation efficiency improved markedly and reached an optimum at 0.2 g/L, owing to the increased availability of active sites promoting PMS activation and sustained radical generation. However, a further increase to 0.4 g/L led to a decrease in efficiency, likely due to particle aggregation and increased mass transfer resistance at higher loadings, which reduced the accessibility of active sites. The reaction kinetics are shown in Figure 4d. The degradation of TC primarily occurred within the first 80 min, reaching 94.3%, after which the curve plateaued. This indicates that the catalytic oxidation reaction approached equilibrium by 80 min.

In summary, through systematic univariate optimization experiments, the optimal conditions for TC degradation were determined as follows: PMS concentration 0.6 mM, initial TC concentration 20 mg/L, Fe/TiO_2 dosage 0.2 g/L, and reaction time 80 min.

The reaction kinetics of the system and the enhancement effect of Fe modification on TiO_2 were further investigated. The degradation kinetics of TC under optimal conditions were fitted, indicating that the process follows a pseudo-first-order kinetic model (Figure 5a). This behavior demonstrates that the degradation rate of TC is proportional to its instantaneous concentration when the catalyst and PMS concentrations are held constant, a characteristic of typical heterogeneous catalytic oxidation processes. Furthermore, by comparing the catalytic performance of Fe/TiO_2 and pure TiO_2 under identical conditions, the significant effect of Fe modification was directly verified. As shown in Figure 5b, under natural light conditions, Fe/TiO_2 achieved a TC degradation rate three times that of pure TiO_2 used in this experimental system within the 80 min reaction period. This significant enhancement is primarily attributed to the successful introduction of Fe species, which creates highly dispersed active sites (e.g., Fe^{2+}/Fe^{3+}) on the TiO_2 surface. These sites are markedly more efficient than pure TiO_2 in activating PMS to generate highly oxidative $SO_4\bullet^-$ and •OH radicals.

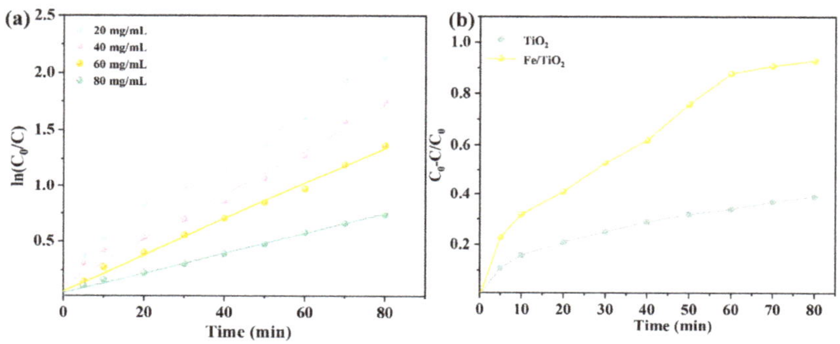

Figure 5. (**a**) Pseudo–first–order kinetic fitting curve for TC degradation; (**b**) Comparison of degradation performance between Fe/TiO_2 and pure TiO_2.

2.3. *Cycling Stability and Mechanism Investigation*

The cycling stability and reaction mechanism are crucial for assessing the practical application potential of a catalyst and understanding its underlying chemical kinetics. The cycling performance of the Fe/TiO_2 catalyst was first evaluated. As shown in Figure 6a, the degradation rate of TC remained at approximately 80% after six consecutive cycles, indicating good structural stability and reusability. The slight decrease in the degradation rate could be attributed to the minor physical loss of the catalyst during each cycle. Furthermore, partial deactivation likely occurred because some active sites were occupied by intermediate products that were difficult to desorb during the reaction. To elucidate the dominant reaction mechanism, radical scavenging experiments were conducted. The inhibitory effects of different scavengers on the TC degradation rate showed significant differences (Figure 6b): Compared to the blank group, the reaction was strongly suppressed upon the addition of methanol, whereas the addition of TBA resulted in a relatively limited inhibitory effect. This contrast provides key evidence for identifying the dominant radical species. Methanol can efficiently quench both $SO_4\bullet^-$ and •OH, while TBA primarily quenches •OH, with its quenching rate constant for $SO_4\bullet^-$ being several orders of magnitude lower. Consequently, the experimental results clearly demonstrate that $SO_4\bullet^-$ plays a dominant role in this degradation system, while •OH plays a secondary role. Moreover, the addition of L-histidine and BQ only slightly inhibited the TC degradation process, indicating that singlet oxygen ($^1O_2$) and superoxide radicals ($O_2\bullet^-$) contributed minimally to the catalytic oxidation reaction. The possible degradation mechanism of TC in the Fe/TiO_2/PMS system is proposed to proceed via synergistic activation pathways. Specifically, Ti^{4+} sites form a surface complex with $HSO_5\bullet^-$, which undergoes interfacial electron transfer (likely from adsorbed TC molecules or the TiO_2 substrate itself) to generate $SO_4\bullet^-$ without altering the Ti oxidation state, which represents a non-radical activation route. Concurrently, the Fe species drive the radical-based activation through the Fe^{2+}/Fe^{3+} redox cycle: Fe^{2+} reacts with $HSO_5\bullet^-$ to yield $SO_4\bullet^-$ and Fe^{3+}, and the subsequent reduction of Fe^{3+} back to Fe^{2+} (facilitated by electron donors or TiO_2-mediated electron transfer) is crucial for sustaining the catalytic activity. Notably, the two pathways reinforce each other: the TiO_2 substrate not only contributes to direct PMS activation but also promotes the Fe^{3+}/Fe^{2+} cycle, thereby collectively enhancing $SO_4\bullet^-$ production and enabling efficient TC mineralization.

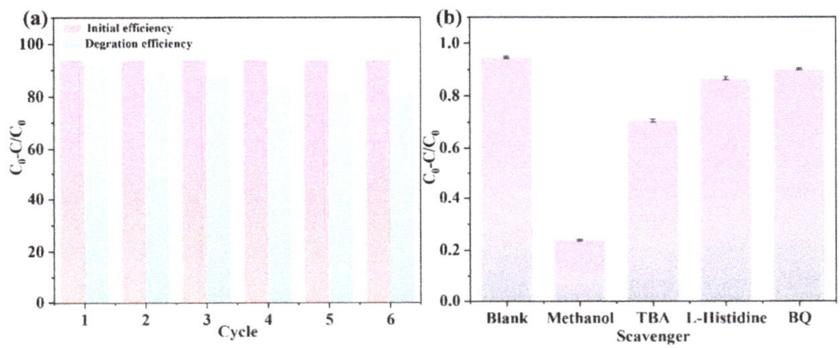

Figure 6. Investigation of catalyst cycling stability and reaction mechanism. (**a**) Degradation performance of the Fe/TiO_2 catalyst over six consecutive cycles of reuse; (**b**) Effects of different radical scavengers on the TC degradation rate. Testing conditions: Methanol (50 mM), TBA (50 mM), L-histidine (20 mM), benzoquinone (10 mM).

2.4. TC Degradation in Real Water Samples

To evaluate the practical application potential of the catalyst, its performance and cycling stability for TC degradation were investigated in complex water matrices (tap water and seawater). As shown in Figure 7a,c, the TC degradation efficiency exceeded 90% in both real water matrices, with only a slight decrease compared to the ultrapure water system. This result indicates that the Fe/TiO_2/PMS system maintained excellent intrinsic catalytic activity despite the presence of common inorganic ions (e.g., Cl^-, HCO_3^-) in tap water or high salinity and complex organic matter in seawater. This robustness is likely due to the strong PMS activation capability of the stable Fe active sites on the catalyst surface, which were not readily completely poisoned by background constituents in the water. Furthermore, the degradation profiles over successive cycles (Figure 7a,c) and the corresponding histograms of degradation efficiency (Figure 7b,d) clearly show that after six consecutive cycles, the efficiency decreased to approximately 75% in tap water and 70% in seawater. This performance attenuation can be ascribed to competitive anions scavenging radicals, thereby reducing degradation efficiency and accelerating irreversible catalyst passivation. Additionally, natural organic matter (e.g., humic acid) or metal cations in seawater may form a surface coating on the catalyst, physically shielding active sites and hindering the diffusion and contact of PMS and TC. Nevertheless, the catalyst retained over 70% degradation efficiency after six cycles in these complex aqueous environments. This performance, superior to most reported results (Table S5), demonstrates its stable catalytic performance and potential for practical application.

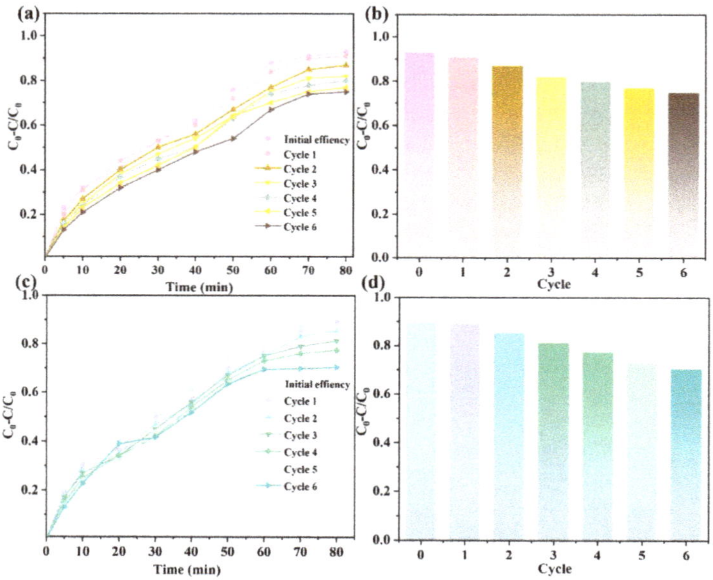

Figure 7. Performance and stability of Fe/TiO_2 for TC degradation in tap water and seawater. Degradation performance and stability in tap water matrix (**a**) and corresponding degradation efficiency histogram (**b**); Degradation performance and stability in seawater matrix (**c**) and corresponding degradation efficiency histogram (**d**).

3. Materials and Methods

3.1. Materials

Ferric chloride hexahydrate ($FeCl_3{\cdot}6H_2O$) was purchased from Tianjin Opulent Chemical Co., Ltd., Shanghai, China. Other reagents including nano-titanium dioxide (TiO_2), urea (CH_4N_2O), polyvinylpyrrolidone ($(C_6H_9NO)_n$, PVP), potassium peroxymonosulfate ($2KHSO_5{\cdot}KHSO_4{\cdot}K_2SO_4$, PMS), tert-butanol ($C_4H_{10}O$, TBA), p-benzoquinone ($C_6H_4O_2$, BQ), and tetracycline ($C_{22}H_{24}N_2O_8$, TC) were obtained from Aladdin Chemical Reagent Co., Shanghai, China. All chemicals were used as received without further purification.

3.2. Characterization

The morphological features of TiO_2 and Fe/TiO_2 were investigated using transmission electron microscopy (TEM, Libra 200, Zeiss, Carl Zeiss AG, Jena, Germany). Crystal structure analysis was performed by X-ray powder diffraction (XRD, Rigaku, Tokyo, Japan) with 2θ scanning from 10° to 80° at a rate of 2°/min. The N_2 adsorption–desorption isotherms were measured using a Micromeritics ASAP 2020 surface area and porosity analyzer (Micromeritics, Norcross, GA, USA) to determine the specific surface area and pore structure. The elemental composition and surface chemical states of the materials were analyzed by X-ray photoelectron spectroscopy (XPS, Kratos AXIS ULTRA DLD, Kratos Analytical Ltd., Manchester, UK). Surface functional groups of the catalysts were characterized using Fourier transform infrared spectroscopy (FT-IR, JASCO FT-IR-420, JASCO Corporation, Tokyo, Japan). Additionally, ultraviolet-visible absorption spectra were recorded on a UV–Vis spectrophotometer (Purkinje General T700, Beijing Purkinje General Instrument Co., Ltd., Beijing, China).

3.3. Synthesis of Fe/TiO$_2$

1.0 g of nano-titanium dioxide (TiO_2) was dispersed in 50 mL of a solvent mixture of ethylene glycol and deionized water (1:1, v/v) and treated with ultrasonication to form a homogeneous suspension. Then, 0.56 g of ferric chloride hexahydrate ($FeCl_3 \cdot 6H_2O$) was added to the suspension, and the mixture was stirred for 1 h to achieve sufficient adsorption of Fe^{3+} ions onto the TiO_2 support. Subsequently, 0.3 g of urea (CH_4N_2O) and 0.2 g of polyvinylpyrrolidone ($(C_6H_9NO)_n$, PVP) were added sequentially under continuous stirring, until a uniform mixture was obtained. The mixture was transferred into a Teflon-lined stainless-steel autoclave and reacted at 200 °C for 6 h. After the autoclave was naturally cooled to room temperature, the product was collected by centrifugation and washed three times with deionized water and twice with anhydrous ethanol. The solid was finally dried at 60 °C for 8 h under vacuum to obtain the powder sample.

3.4. Optimization of Catalytic Conditions

The effects of key operational parameters, including peroxymonosulfate ($2KHSO_5 \cdot KHSO_4 \cdot K_2SO_4$, PMS) concentration, tetracycline ($C_{22}H_{24}N_2O_8$, TC) initial concentration, catalyst dosage, and reaction time, on the degradation performance of Fe/TiO_2 were systematically investigated using a one-factor-at-a-time approach. The specific procedure was as follows: a predetermined mass of Fe/TiO_2 catalyst was placed in a beaker, and a TC solution of specified concentration was added. The mixture was sealed and then stirred at 500 rpm for 60 min in the dark to achieve adsorption–desorption equilibrium. Immediately after equilibrium, a sample was taken to measure the initial absorbance of TC at 355 nm. Subsequently, a PMS solution of given concentration was introduced to initiate the catalytic oxidation reaction. At predetermined time intervals, samples were continuously collected, and the variation in the characteristic absorption peak intensity of TC was monitored using a UV–Vis spectrophotometer. The degradation efficiency was calculated according to Equation (1):

$$Degradation\ (\%) = [(C_0 - C)/C_0] \times 100\% \quad (1)$$

where C_0 is the initial absorbance of the pollutant after adsorption equilibrium and before the catalytic oxidation reaction starts, and C is the absorbance of TC at a specific time during the catalytic reaction.

The parameter optimization procedure was as follows. First, with the catalyst dosage fixed at 10 mg and the initial TC concentration fixed at 40 mg/L, the influence of PMS concentration (0.2, 0.4, 0.6, 0.8, and 1 mM) was systematically investigated. Based on these optimization results, the optimal PMS concentration was applied to subsequently investigate the effects of the initial TC concentration (20, 40, 60, and 80 mg/L) and then the catalyst dosage (0.1, 0.2, 0.3, and 0.4 g/L) on the degradation performance. Finally, the reaction kinetics were evaluated over a time period of 15–120 min (Tables S1–S4).

3.5. Cycling Stability Test

The cycling stability of the Fe/TiO_2 catalyst was evaluated through six consecutive rounds of recovery and reuse. After each reaction cycle, the catalyst was recovered by magnetic separation, followed by ultrasonic cleaning sequentially with 20 mL of anhydrous ethanol and 20 mL of ultrapure water. This washing procedure was repeated three times. After vacuum drying at 35 °C for 8 h, the catalyst was reintroduced into the next catalytic reaction under optimal conditions. The stability was assessed by monitoring the variation in degradation efficiency across the cycles.

3.6. TC Degradation in Real Water Samples

To evaluate the applicability and anti-interference capability of the as-prepared Fe/TiO_2 in practical aquatic environments, its degradation behavior toward TC was systematically assessed in two complex aqueous matrices: tap water and seawater. All environmental water samples were filtered through 0.45 μm microporous membranes after collection to remove suspended particulate matter. Subsequently, TC solutions were prepared using these real water samples as the solvent, and degradation experiments were conducted under the optimized catalytic conditions. The performance was compared with that observed in an ultrapure water system.

4. Conclusions

This study demonstrates that PVP serves as a highly effective structure-directing agent for fabricating a high-performance Fe/TiO_2 catalyst via a facile one-step hydrothermal route. Beyond suppressing nanoparticle agglomeration, PVP guided the formation of a robust Fe–O–Ti interface, which was crucial for creating highly dispersed and stable Fe active sites. This synergistic design resulted in a catalyst that integrates exceptional PMS activation capability (achieving 94.3% TC degradation under optimum conditions), remarkable recyclability (>80% efficiency after 6 cycles), and convenient magnetic separation. The dominance of sulfate radicals in the system and its robust performance in complex water matrices underscore its potential for practical wastewater treatment.

Supplementary Materials: The following supporting information can be downloaded at: https://www.mdpi.com/article/10.3390/catal15121105/s1, Table S1: Optimization of PMS concentration; Table S2: Optimization of initial TC concentration; Table S3: Optimization of Fe/TiO2 dosage; Table S4: Optimization of reaction time; Table S5: Comparison of degradation and cycling performance among some reported catalysts. References [38–43] are cited in the Supplementary Materials.

Author Contributions: Conceptualization, X.W.; methodology, L.Y.; validation, S.W. and L.X.; formal analysis, X.W. and L.Y.; investigation, Y.L. and Z.Q.; resources, Y.L., Z.Q., and X.L.; data curation, S.W.; writing—original draft preparation, X.W.; writing—review and editing, X.W.; visualization, L.Y.; supervision, X.L. and Z.Q.; project administration, X.L.; funding acquisition, Z.Q. All authors have read and agreed to the published version of the manuscript.

Funding: This research received no external funding.

Data Availability Statement: The data supporting the findings of this study are available from the corresponding author upon reasonable request. All data have been processed and analyzed as presented in the manuscript.

Acknowledgments: This work was financially supported by the Natural Science Foundation of Shandong Province (ZR2023MB024), the Yantai High-End Talent Introduction and Cultivation "Double Hundred Plan" Scholars and Young Experts, the Yantai Basic Research Project for Science and Technology Innovation Development Plan (2024JCYJ064), the Key R&D Program of Shandong Province (Major Innovative and Technological Engineering) (2024CXGC010316, 2024CXGC010406, 2025CXGC010506), the Science Fund of Shandong Laboratory of Yantai Advanced Materials and Green Manufacturing (AMGM2023A02, AMGM2024A06), and the National Natural Science Foundation of China (52073134).

Conflicts of Interest: The authors declare no conflicts of interest.

References

1. Li, X.; Hu, J.; Deng, Y.; Li, T.; Liu, Z.-Q.; Wang, Z. High stable photo-Fenton-like catalyst of FeP/Fe single atom-graphene oxide for long-term antibiotic tetracycline removal. *Appl. Catal. B Environ.* **2023**, *324*, 122243. [CrossRef]

2. Wang, A.; Zheng, Z.; Wang, H.; Chen, Y.; Luo, C.; Liang, D.; Hu, B.; Qiu, R.; Yan, K. 3D hierarchical H_2-reduced Mn-doped CeO_2 microflowers assembled from nanotubes as a high-performance Fenton-like photocatalyst for tetracycline antibiotics degradation. *Appl. Catal. B Environ.* **2020**, *277*, 119171. [CrossRef]
3. Wang, C.; Lin, C.-Y.; Liao, G.-Y. Degradation of antibiotic tetracycline by ultrafine-bubble ozonation process. *J. Water Process Eng.* **2020**, *37*, 101463. [CrossRef]
4. Zhang, X.; Hou, J.; Zhang, S.; Cai, T.; Liu, S.; Hu, W.; Zhang, Q. Standardization and micromechanistic study of tetracycline adsorption by biochar. *Biochar* **2024**, *6*, 12. [CrossRef]
5. Chen, X.; Yang, Y.; Ke, Y.; Chen, C.; Xie, S. A comprehensive review on biodegradation of tetracyclines: Current research progress and prospect. *Sci. Total Environ.* **2022**, *814*, 152852. [CrossRef]
6. Gan, J.; Wang, H.; Li, J.; Song, X.; Liu, X.; Jiang, J.; Huo, P. g-C_3N_4/$BaTiO_3$/PVDF membrane photocatalytic degradation of tetracycline. *J. Ind. Eng. Chem.* **2025**, *152*, 461–473. [CrossRef]
7. Sivanesan, J.; Sivaprakash, B.; Rajamohan, N.; Phanindra, V.S.S.; Sonne, C.; Liew, R.K.; Lam, S.S. Remediation of tetracycline pollution using microplastics, green materials, membranes and sonocatalysts: A review. *Environ. Chem. Lett.* **2024**, *22*, 2943–2975. [CrossRef]
8. Hua, B.; Zheng, L.; Adeboye, A.; Li, F. Defect- and nitrogen-rich porous carbon embedded with Co NPs derived from self-assembled Co-ZIF-8 @ anionic polyacrylamide network as PMS activator for highly efficient removal of tetracycline hydrochloride from water. *Chem. Eng. J.* **2022**, *443*, 136439. [CrossRef]
9. He, W.; Huang, L.; Wang, X.; Zhang, J. Molybdenum nitride(γ-Mo_2N) as a novel co-catalyst to enhance Fe(III)/Fe(II) cycle for homogeneous and heterogeneous peroxymonosulfate activation: Performance and mechanism. *J. Environ. Chem. Eng.* **2024**, *12*, 112404. [CrossRef]
10. Fu, W.; Huo, S.; Zhang, M.; Song, L.; Zhao, Q.; Wu, X.; Gao, M. Efficient degradation of oxytetracycline by glucose modified $CuFeO_2$ in visible-light-assisted heterogeneous activation of peroxymonosulfate system: Performance, mechanism and DFT calculation. *J. Environ. Chem. Eng.* **2023**, *11*, 111225. [CrossRef]
11. Wang, Y.; Lv, M.; Dong, H.; Hassan, M.; Wang, D.; Gong, Z. Electro-assisted heterogeneous peroxymonosulfate activation on $NiFe_2O_4$ coated nickel foam cathode for tetracycline degradation: Synergistic efficacy and mechanism. *J. Taiwan Inst. Chem. Eng.* **2024**, *163*, 105656. [CrossRef]
12. Qiu, X.; Zhao, Y.; Jia, Z.; Li, C.; Jin, R.; Mutabazi, E. Fe and Zn co-doped carbon nanoparticles as peroxymonosulfate activator for efficient 2,4-dichorophenol degradation. *Environ. Res.* **2024**, *240*, 117313. [CrossRef] [PubMed]
13. Azami, M.S.; Jalil, A.A.; Hassan, N.S.; Hussain, I.; Fauzi, A.A.; Aziz, M.A.A. Green carbonaceous material–fibrous silica-titania composite photocatalysts for enhanced degradation of toxic 2-chlorophenol. *J. Hazard. Mater.* **2021**, *414*, 125524. [CrossRef] [PubMed]
14. Lim, C.; An, H.-R.; Lee, H.; Lee, R.; Choi, Y.; Park, J.-I.; Yoon, J.; Lee, H.U.; Lee, Y.-S. Carbon-titanium dioxide heterogeneous (photo)catalysts (C–TiO_2) for highly efficient visible light photocatalytic application. *Compos. Part B Eng.* **2022**, *241*, 109997. [CrossRef]
15. Ayappan, C.; Xing, R.; Zhang, X.; Luo, D.; Liu, S.; Fujishima, A. TiO_2-based photocatalysts for emerging gaseous pollutants removal: From photocatalysts to reactors design. *Coord. Chem. Rev.* **2024**, *515*, 215960. [CrossRef]
16. Marszewski, M.; Marszewska, J.; Pylypenko, S.; Jaroniec, M. Synthesis of porous crystalline doped titania photocatalysts using modified rrecursor strategy. *Chem. Mater.* **2016**, *28*, 7878–7888. [CrossRef]
17. Chen, F.; Li, W.; Sahoo, B.; Kreyenschulte, C.; Agostini, G.; Lund, H.; Junge, K.; Beller, M. Hydrogenation of pyridines using a nitrogen-modified titania-supported cobalt catalyst. *Angew. Chem.-Int. Edit.* **2018**, *57*, 14488–14492. [CrossRef]
18. Fang, W.; Xing, M.; Zhang, J. Modifications on reduced titanium dioxide photocatalysts: A review. *J. Photochem. Photobiol. C Photochem. Rev.* **2017**, *32*, 21–39. [CrossRef]
19. Chen, L.; Lei, Y.; Yang, Y.; Huang, J.; Zhang, W.; Hoong Ng, K.; Lai, Y. Metal organic framework-assisted copper-modified titania (Cu/TiO_2) with abundant exposed active sites and highly accessible pore channels for an enhanced photo-generation of hydrogen. *J. Colloid Interface Sci.* **2025**, *677*, 647–656. [CrossRef]
20. Siemer, N.; Muñoz-Santiburcio, D.; Marx, D. Solvation-enhanced oxygen activation at gold/titania nanocatalysts. *ACS Catal.* **2020**, *10*, 8530–8534. [CrossRef]
21. Shen, X.-Z.; Guo, J.; Liu, Z.-C.; Xie, S.-M. Visible-light-driven titania photocatalyst co-doped with nitrogen and ferrum. *Appl. Surf. Sci.* **2008**, *254*, 4726–4731. [CrossRef]
22. Ahmad, I.; Zou, Y.; Yan, J.; Liu, Y.; Shukrullah, S.; Naz, M.Y.; Hussain, H.; Khan, W.Q.; Khalid, N.R. Semiconductor photocatalysts: A critical review highlighting the various strategies to boost the photocatalytic performances for diverse applications. *Adv. Colloid Interface Sci.* **2023**, *311*, 102830. [CrossRef]
23. Diamandescu, L.; Vasiliu, F.; Tarabasanu-Mihaila, D.; Feder, M.; Vlaicu, A.M.; Teodorescu, C.M.; Macovei, D.; Enculescu, I.; Parvulescu, V.; Vasile, E. Structural and photocatalytic properties of iron-and europium-doped TiO2 nanoparticles obtained under hydrothermal conditions. *Mater. Chem. Phys.* **2008**, *112*, 146–153. [CrossRef]

24. Sun, B.; Zhou, W.; Li, H.; Ren, L.; Qiao, P.; Xiao, F.; Wang, L.; Jiang, B.; Fu, H. Magnetic Fe_2O_3/mesoporous black TiO_2 hollow sphere heterojunctions with wide-spectrum response and magnetic separation. *Appl. Catal. B Environ.* **2018**, *221*, 235–242. [CrossRef]
25. Sood, S.; Umar, A.; Mehta, S.K.; Kansal, S.K. Highly effective Fe-doped TiO_2 nanoparticles photocatalysts for visible-light driven photocatalytic degradation of toxic organic compounds. *J. Colloid Interface Sci.* **2015**, *450*, 213–223. [CrossRef]
26. Xu, Q.-Q.; Cheng, X.-L.; Zhang, B.-Y.; Zhang, F.; Wang, X.; Li, S.-S.; Zhang, Y.-X. Fe-doping induced surface Fe^{2+}/Fe^{3+} cycle and activated redox-inert TiO_2 for enhanced Hg(II) electrochemical sensing: An efficient strategy to strengthen the redox activity. *Anal. Chim. Acta.* **2022**, *1232*, 340472. [CrossRef]
27. Ji, Z.; Cai, R.; Ye, W.; Lu, P.; Dong, C.-L.; Huang, Y.-C.; She, X.; Butenko, D.S.; Liu, Y.; Zhu, Y.; et al. Confined Fe single atomic sites on (100) plane of anatase TiO_2 nanofibers boost white LED driven Fenton-like norfloxacin degradation. *J. Clean. Prod.* **2023**, *382*, 135161. [CrossRef]
28. Wei, Z.; Janczarek, M.; Endo, M.; Wang, K.; Balčytis, A.; Nitta, A.; Méndez-Medrano, M.G.; Colbeau-Justin, C.; Juodkazis, S.; Ohtani, B.; et al. Noble metal-modified faceted anatase titania photocatalysts: Octahedron versus decahedron. *Appl. Catal. B Environ.* **2018**, *237*, 574–587. [CrossRef]
29. Kausar, F.; Varghese, A.; Pinheiro, D.; Devi K R, S. Recent trends in photocatalytic water splitting using titania based ternary photocatalysts—A review. *Int. J. Hydrogen Energy* **2022**, *47*, 22371–22402. [CrossRef]
30. Li, Y.; Wei, Y.; He, W.; Tang, Z.; Xiong, J.; Zhao, Z. Ordered macroporous structured TiO_2-based photocatalysts for CO_2 reduction: A review. *Chin. Chem. Lett.* **2023**, *34*, 108417. [CrossRef]
31. Singh, S.; Mahalingam, H.; Singh, P.K. Polymer-supported titanium dioxide photocatalysts for environmental remediation: A review. *Appl. Catal. A-Gen.* **2013**, *462–463*, 178–195. [CrossRef]
32. Doustdar, F.; Ghorbani, M. ZIF-8 enriched electrospun ethyl cellulose/polyvinylpyrrolidone scaffolds: The key role of polyvinylpyrrolidone molecular weight. *Carbohydr. Polym.* **2022**, *291*, 119620. [CrossRef] [PubMed]
33. Safira, A.R.; Alluhayb, A.H.; Aadil, M.; Alkaseem, M.; Fattah-alhosseini, A.; Kaseem, M. Enhanced photocatalytic reduction of p-nitrophenol by polyvinylpyrrolidone-modified MOF/porous MgO composite heterostructures. *Compos. Part B Eng.* **2024**, *284*, 111710. [CrossRef]
34. Liu, L.; Song, D.; Kaiser, M.E.; Liu, J.; De Yoreo, J.J.; Sushko, M.L. Epitaxial growth of 2D core-crown $SnS_2/SnSe_2$ heterostructure through interfacial modification with polyvinylpyrrolidone. *Small* **2025**, *21*, 2410526. [CrossRef] [PubMed]
35. Liu, W.; Yang, Y.; Li, Y.; Zhou, Y.; Wang, C.; Zhou, Y.; Shang, J.; Cheng, X. Oxygen vacancies enhanced natural manganese sand activation by PMS for CBZ degradation: Intermediate toxicity and DFT calculations. *Sep. Purif. Technol.* **2024**, *329*, 125015. [CrossRef]
36. Wang, Y.; Wu, L.; Zhou, Y.; Zhang, Y.; Sun, S.-P.; Duo Wu, W.; Wang, X.; Wu, Z. Ternary FeS/γ-Fe_2O_3@N/S-doped carbon nanohybrids dispersed in an ordered mesoporous silica for efficient peroxymonosulfate activation. *Chem. Eng. J.* **2022**, *435*, 135124. [CrossRef]
37. Cui, B.; Tian, T.; Duan, L.; Rong, H.; Chen, Z.; Luo, S.; Guo, D.; Naidu, R. Towards advanced removal of organics in persulfate solution by heterogeneous iron-based catalyst: A review. *J. Environ. Sci.* **2024**, *146*, 163–175. [CrossRef]
38. Wei, Y.; Chen, M.; Guo, S.; Li, Y.; Zhang, Z.; Xie, Y.; Cai, C.; Wei, Q. Hierarchically porous TiO_2 photocatalyst prepared by additive manufacturing and dealloying for highly efficient photocatalytic degradation of tetracycline. *ACS Environ. Au* **2024**, *4*, 1590–1600. [CrossRef]
39. Moharrami, E.; Keshipour, S. Photocatalytic degradation of tetracycline antibiotic using nitrogen-doped reduced graphene oxide-supported titania/platinum nanoparticles. *npj Mater. Degrad.* **2025**, *9*, 57. [CrossRef]
40. Liu, X.; Yang, Y.; Li, H.; Yang, Z.; Fang, Y. Visible light degradation of tetracycline using oxygen-rich titanium dioxide nanosheets decorated by carbon quantum dots. *Chem. Eng. J.* **2021**, *408*, 127259. [CrossRef]
41. Luo, X.; Zhang, L.; Gong, X.; Liu, Y.; Tian, J. Defect-enriched ilmenite-type catalyst derived from titanium dioxide slag for peroxymonosulfate activation in efficient tetracycline removal. *Chem. Eng. Sci.* **2023**, *280*, 119068. [CrossRef]
42. Liu, X.; Yang, Z.; Yang, Y.; Li, H. Construction of carbon quantum dots sensitized porous carbon nitride/titanium dioxide nanosheets for enhancing visible light photocatalytic degradation of tetracycline. *J. Environ. Chem. Eng.* **2022**, *10*, 108083. [CrossRef]
43. Sun, S.; Ming, W.; Zhang, H.; Li, A.; Xu, G. Frustrated defect pairs of oxygen vacancies and doped nitrogen in amorphous titanium dioxide for enhanced photodegradation of tetracycline. *New J. Chem.* **2025**, *49*, 12716–12734. [CrossRef]

molecules

Article

Titanium Dioxide Solar Photocatalytic Microbial Inactivation Assessment Utilizing Viability Tests and a Novel Triplex qPCR Assay for Nucleic Acid Degradation Determination

Ioannis Paspaltsis [1], Eirini Kanata [1], Sotirios Sotiriadis [1], Chrysanthi Berberidou [2], Sophia Tsoumachidou [2], Athanasios Arsenakis [3], Konstantinos Xanthopoulos [1], Dimitra Dafou [4], Ioannis Poulios [2] and Theodoros Sklaviadis [1,*]

1 Laboratory of Pharmacology, School of Pharmacy, Faculty of Health Sciences, Aristotle University of Thessaloniki, 54124 Thessaloniki, Greece; ipaspalt@pharm.auth.gr (I.P.); ekanata@pharm.auth.gr (E.K.); sotosotiriadis96@gmail.com (S.S.); xantho@pharm.auth.gr (K.X.)
2 Laboratory of Physical Chemistry, School of Chemistry, Faculty of Sciences, Aristotle University of Thessaloniki, 54124 Thessaloniki, Greece; cberber@chem.auth.gr (C.B.); stsoum@gmail.com (S.T.); poulios@chem.auth.gr (I.P.)
3 Sterimed SA, G' Fassi, Building Block 52b, Sindos Industrial Area, Thessaloniki 57022, Greece; arsenakis@sterimed.gr
4 Department of Genetics, Development and Molecular Biology, School of Biology, Faculty of Sciences, Aristotle University of Thessaloniki, 54124 Thessaloniki, Greece; dafoud@bio.auth.gr
* Correspondence: sklaviad@pharm.auth.gr; Tel.: +30-2310997720

Abstract

Photocatalytic oxidation of microorganisms is a powerful alternative to established disinfection approaches, applicable to a variety of water matrices. Bacterial vegetative cells, spores, fungi, and viruses, represent potential biopathogens and photocatalysis targets. Inactivation efficiency is usually evaluated by assessing viability through culture. However, additional inactivation assessment approaches are needed, as some microbes, despite being unculturable, remain metabolically active and pathogenic. Nucleic acid quantification approaches (qPCR) can assess nucleic acid release and degradation during photocatalysis. We developed a novel multiplex qPCR assay for simultaneous detection/quantification of genomic DNA from different bacterial and fungal species and of MS2 bacteriophage load. Following small-scale solar titanium dioxide photocatalysis on a microbial suspension mixture containing different biopathogen classes, we assessed photocatalytic efficiency by conventional microbiological assays (culture) and our novel molecular assay. Microbiological assays show a significant reduction in microbe viability within one hour of processing, following previously reported patterns of microbial species resistance. Molecular analysis data show that nucleic acids released in solution due to microbial oxidative damage were significantly reduced due to oxidative degradation within six hours. Through targeting different biopathogen classes, our assay could be a useful tool for assessment of photocatalytic microbe inactivation both in laboratory and real-wastewater applications.

Keywords: solar; titanium dioxide; photocatalysis; multiplex quantitative PCR; microorganism inactivation efficiency; *Escherichia coli*; *Staphylococcus aureus*; *Candida albicans*; MS2 bacteriophage; *Geobacillus stearothermophilus* spores

1. Introduction

Photocatalysis, an advanced oxidation process (AOP), attracts increasing interest in diverse research areas, including the oxidation of hazardous chemicals and the inactivation

 https://doi.org/10.3390/molecules30224333

of microbes. Artificial or solar-emitted photons are utilized by photocatalysts for the generation of reactive oxygen species (ROS) such as $O^{\bullet -2}$ and H_2O_2 [1]. The most prominent ROS are the short-lived •OH radicals [2] with a lifetime of 10^{-9} s [3] and a high redox potential of 2.8 eV [4]. Produced ROS oxidize nonspecifically organic molecules (chemical pollutants or microbial components) in proximity to the catalyst. As a result, the toxicity of chemicals or microbial viability/pathogenicity is decreased.

Photocatalysis constitutes a powerful alternative to established detoxification and disinfection technologies in terms of environmental protection and remediation. Different photocatalysts in diverse experimental configurations have shown great potential in reducing toxicity conferred by pesticides, pharmaceuticals, pigments, detergents, organic diluents, and plastics [5–10]. Moreover, photocatalytic inactivation of a variety of potential biopathogens has been reported, including different bacterial species, either in the form of vegetative cells or spores, fungi, viruses, or prions [11,12].

TiO_2 and several TiO_2-based, performance-enhancing composites [13] are often used for the treatment of chemicals, microbes, or their combination in water [14]. Application of TiO_2 and TiO_2 doping for photocatalytic purposes gained popularity because of its important advantages: it is inexpensive, powerful in terms of ROS generation potential, and can be reused without significant loss of action; at the same time, it is environmentally friendly, as it is biologically inert, displaying no known toxicity. Importantly, it can be powered by both artificial and solar irradiation.

Regarding the antimicrobial activity of TiO_2, substantial research has focused on bacterial inactivation. The initial inactivation step includes adsorption of the bacterial cell on the surface of the catalyst. Irradiation of the catalytic molecule leads to ROS generation, which oxidizes molecules present at proximal bacterial structures. As a result, cell wall or outer cell membrane decomposition is initiated in Gram-positive or Gram-negative bacterial cells, respectively [2,15,16]. Disruption of the outer layer surrounding the bacterium may lead rapidly to bacterial inactivation, resulting in the reduction or complete loss of proliferation ability [17]. A similar microbial particle decomposition pattern, from outside to inside, during photocatalytic treatment, is proposed for other potential biological pathogens, such as fungi and viruses [18–20].

Oxidative damage of the cell wall, cell membrane, or viral capsid eventually causes the release of the cellular or viral organic content, which accounts for approximately 96% of microbial dry weight [2], to the surrounding solution. Released biopolymers, such as nucleic acids (DNA and RNA), proteins, lipids, and carbohydrates, as well as organic monomers such as nucleotides, amino acids, and sugar monomers, may be further oxidized as the reaction proceeds. Complete oxidation of organic matter results in photocatalytic mineralization, i.e., the complete decomposition of the organic substances into inorganic compounds [17].

In photocatalytic-treated suspensions, microbial inactivation is usually estimated by classical microbiological methods based on culturing, to assess microbial viability. However, this approach cannot provide precise information related to microbial viability; it was reported four decades ago that microbes under environmental stress, such as the oxidative environment during photocatalytic treatment, may lose their ability to grow on culture media, becoming viable but not culturable [21]. This is a survival strategy, associated with metabolic and genetic changes, resulting in metabolically, transcriptionally, and translationally active, but unculturable cells. Crucially, microbes in this state can retain their pathogenicity, as they express virulence genes; when favorable conditions are restored, they may resuscitate, regaining culturability [22]. Thus, relying exclusively on culture to quantify bacterial density may lead to inaccurate results, posing a significant public health risk, especially when culture methods are used for assessing pathogen inactivation in medical liquid waste before disposal. In this regard, molecular techniques such as quantitative PCR (qPCR), which allow detection/quantification of nucleic acids released from damaged cells and nucleic acid integrity assessment, provide a more detailed and reliable assessment of photocatalysis performance [14].

Thanks to its sensitivity, the qPCR assay may provide accurate data on the degradation state of microbial macromolecules in treated water matrices. This is especially significant when photocatalytic approaches are applied for water remediation or when treated water will be released to the environment. In these cases, any naked DNA derived from dead bacteria may act as a substrate for horizontal gene transfer [23], leading to undesirable gene recombination. Furthermore, through the detection of DNA degradation, the application of qPCR to monitor photocatalytic treatment may provide data on the degradation course of microbes [17]. Since ROS are produced throughout photocatalytic treatment, they oxidize nonspecifically the existing organic matter. However, it is important to note that through qPCR, only DNA fragments larger than approximately 100 bp can be detected; thus, qPCR cannot provide evidence of complete mineralization.

Aiming at providing an accurate molecular tool for assessing photocatalytic efficiency on microbial inactivation, we have developed a novel, hydrolysis probe-based multiplex qPCR assay for the estimation of nucleic acid leakage/damage level during photocatalytic inactivation of microbes. The assay allows the simultaneous detection of bacterial, fungal, and viral genomes. It has been developed for the detection of *Geobacillus stearothermophilus*, *Staphylococcus aureus*, *Escherichia coli*, *Candida albicans*, and the *Emesvirus zinderi* (MS2) bacteriophage. Importantly, the bacterial- and fungal-specific primer–probe sets may detect a wide range of bacterial and fungal species commonly present in urban or medical wastewater, highlighting the potential use of the assay in both laboratory and real-life settings.

Application of the assay on a small-scale solar photocatalytic setting, using TiO_2/H_2O_2 for inactivation of a microbial mixture, provided additional data regarding microbial inactivation. Our data show that microbial viability is markedly reduced within the first hour of photocatalysis, and significant microbial nucleic acid degradation occurs within six hours of treatment. This data indicates that safe estimation of microbial inactivation requires a combination of both culture and molecular assays. The assay developed and evaluated in this study could serve as a molecular tool for the assessment of microbial inactivation under different photocatalysis settings, allowing easier comparison of experimental data obtained under different experimental setups.

2. Results and Discussion

Photocatalytic approaches are gaining increasing attention as an efficient, environmentally friendly, and sustainable alternative to traditional disinfection approaches, aiming at the inactivation of various microorganisms, identified as potential biopathogens in water matrices [16,24–30].

Accurate evaluation of photocatalytic disinfection efficiency is highly needed to ensure safe disposal of treated materials [21,31]. The traditional culture method for the determination of microbial viability is usually applied for assessing photocatalytic efficiency. Molecular assays targeting nucleic acids released from the microorganisms and further degraded by free radicals during photocatalytic treatment, provide an additional and more precise approach for the assessment of photocatalytic efficiency [14,21,32–34] and are urgently needed [17].

Aiming to provide a molecular tool for assessing the disinfection efficiency of photocatalytic approaches applied in aqueous matrices, and considering that different water matrices display a wide microbial diversity, we developed and utilized a triplex qPCR assay for the simultaneous detection of representatives from different potential biopathogen classes. Our assay utilizes three primer–probe sets and enables the detection of at least three different bacterial species (set 1), a fungus (set 2), and a bacteriophage (set 3). The assay targets model microorganisms (*Escherichia coli*, *Geobacillus stearothermophilus*, *Staphylococcus aureus*, *Candida albicans*, and MS2), which are widely used for optimization of small-scale photocatalytic approaches in the laboratory [35–37] and are also present in real wastewater. Thus, our assay can be utilized in both experimental and real-life settings.

The efficiency of each primer–probe set against the targeted microbes was determined based on standard curves generated by serial dilution of commercial DNA/RNA standards (*Escherichia coli*, *Candida albicans*, MS2). DNA extracted from bacterial cultures of *S. aureus* and *G. stearothermophilus* was also utilized for standard curve preparation. Assay efficiencies were calculated based on curve slopes, using the equation Efficiency = $10^{(-1/\text{slope})} - 1$. Table 1 summarizes assay efficiencies and detection limits in terms of genome copy numbers per microorganism. Figure S1 displays pertinent standard curves.

Table 1. Amplification efficiencies and detection limits determined for each of the targeted microorganisms by the newly developed multiplex qPCR assay.

Target Microorganism	Genome Size (kbp)/Target Gene Copies in Genome	Standard Curve Equation/R2	Assay Efficiency **	Standard Deviation of Linear Regression	Genome Copy Number Detection Limit ¶
MS2 #	3569/ 1	Y = −3.303X + 40.55 (R2 = 0.9974)	100.8	0.3551	1 (0.4)
Candida albicans #	16,000/ 21–176 ¥	Y = −3.821X + 33.69 (R2 = 0.9972)	82.7	0.2951	1 (0.56)
Escherichia coli #	~5000/ 7 **	Y = −3.308X + 33.86 (R2 = 0.996)	100.6	0.4394	1 (0.08)
Staphylococcus aureus (ATCC 6538) $	2,800,485/ 5	Y = −3.539X + 38.30 (R2 = 0.9994)	91.675	0.1811	2 (1.47)
Geobacillus stearothermophilus (ATCC 7953) $	2,787,229/ 10	Y = −3.476X + 40.04 (R2 = 0.9988)	93.949	0.2432	2 (1.70)

Assay efficiency determined by construction of standard curves using commercially available nucleic acid standards. $ Assay efficiency determined by construction of standard curves using DNA corresponding to predetermined genome copy numbers. ** Assay efficiency was determined based on the slope of the corresponding standard curve using the equation Efficiency = $10^{(-1/\text{slope})} - 1$; ¶ Genome copy number detection limit was determined using the equation LOD = (3.3 × standard deviation of linear regression)/slope of the regression line, standard deviation of linear regression = Sy.x; ¥ Michael L Pendrak (DOI: 10.1261/rna.028050.111); ECOCYS database, https://www.biocyc.org/gene?orgid=ECOLI&id=RRSA-RRNA, accessed on 25 August 2025.

By targeting the highly conserved 16S rRNA and utilizing a degenerate primer (Bact91F, Table 2, Figure 1), our bacteria-specific primer–probe set not only efficiently identifies all three targeted microbes (*Geobacillus stearothermophilus*, *Staphylococcus aureus*, *Escherichia coli*) but also, as indicated by BLASTn analysis, has the potential to identify additional bacteria from other genera, such as *Salmonella*, *Pseudomonas*, *Vibrio*, *Klebsiella*, *Listeria*, *Streptococcus*, and *Enterococcus*, commonly present in water matrices [38,39] (Table S1). Similarly, the fungi-specific primer–probe set targeting 18S rRNA, apart from *Candida*, has the potential to identify additional fungal genera, such as *Fusarium* and *Penicillium*, commonly detected in both urban wastewater [40] and medical liquid waste [41]. These observations highlight the potential versatility of our assay. However, in cases where the targeted microorganisms are other than the ones validated in this study (i.e., *Escherichia coli*, *Candida albicans*, MS2, *S. aureus*, *G. stearothermophilus*), we recommend determining the efficiencies of the corresponding assay primer–probe sets for the microorganisms of interest.

Table 2. Oligonucleotide sequences of primers and probes. FAM, 6-carboxyfluorescein; IBFQ, 3′ Iowa Black Fluorescent Quencher; IBRQ, 3′ Iowa Black Red Quencher; Cy5, Cyanine5. R = G or A, Y = C or T.

Target	Target Gene	Oligo Name	Sequence (5′—>3′)	Binding Site	Length (bp)
Bacterial gDNA	16S rRNA	Bact91F	RCCGCAAGGYTRAAACTCAA	880–899 [1] 901–920 [2] 806–825 [3]	20
		Bact91R	GGTAAGGTTCTTCGCGTTGC	951–970 [1] 974–993 [2] 877–896 [3]	20
		Bact91probe	5-FAM/CGC ACA AGC/ZEN/ GGT GGA GCA T/IBFQ	917–935 [1] 940–938 [2] 843–861 [3]	19
Fungal gDNA	18S rRNA	18S C.al 110F	TAGTTGAACCTTGGGCTTGG	522–541	20
		18S C.al 110R	CAAAGTAAAAGTCCTGGTTCGC	610–631	22
		18S C.al 110probe	ATTO550N/CTGGACCCAGCCGAGCCTTT/IBRQ	569–588	20
MS2 cDNA	Capsid protein	MS2 86F	AATCAGGCAACGGCTCTCTA	1287–1306	20
		MS2 86R	TTGTCGACGAGAACGAACTG	1353–1372	20
		MS2 86pr	Cy5/AGA GCC CTC AAC CGG AGT TTG AAG/IBRQ	1309–1333	24

Binding sites of oligos at reference sequences mentioned above: [1] *Geobacillus stearothermophilus*, [2] *Staphylococcus aureus*, [3] *Escherichia coli*.

```
Escherichia coli                AGTACGGCCGCAAGGTTAAAACTCAAATGAATTGACGGGGGCCCGCACAAGCGGTGGAGC
Geobacillus stearothermophilus  AGTACGGCCGCAAGGCTGAAACTCAAAGGAATTGACGGGGGCCCGCACAAGCGGTGGAGC
Staphylococcus aureus           AGTACGACCGCAAGGTTGAAACTCAAAGGAATTGACGGGGACCCGCACAAGCGGTGGAGC
                                ****** ******** * ********* ************ *******************

Escherichia coli                ATGTGGTTTAATTCGATGCAACGCGAAGAACCTTACCTGGTCTTGACATCCACGGAAGTT
Geobacillus stearothermophilus  ATGTGGTTTAATTCGAAGCAACGCGAAGAACCTTACCAGGTCTTGACATCCCCTGACAAC
Staphylococcus aureus           ATGTGGTTTAATTCGAAGCAACGCGAAGAACCTTACCAAATCTTGACATCCTTTGACAAC
                                **************** ********************   ***********   **
```

Figure 1. Partial nucleotide alignment sequences from *Geobacillus stearothermophilus* (GenBank AB680491.1), *Staphylococcus aureus* (GenBank NR_118997.2), and *Escherichia coli* (GenBank MW349588.1). The sequences correspond to the 16S rRNA gene and were retrieved from the NCBI nucleotide database. Primer annealing regions are marked in red, the probe region is marked in blue, and mixed bases corresponding to sequence variability among the species are shown in green. The forward primer was designed as a degenerate oligo. Asterisks (*) indicate conserved nucleotide positions shared by all three sequences. Multiple alignment was performed with Clustal omega software (online available at https://www.ebi.ac.uk/jdispatcher/msa/clustalo, accessed on 15 June 2023).

We utilized our multiplex qPCR assay on photocatalytically treated microbial mixture suspensions. We selected TiO_2 since it is a well-established catalyst, previously used for the photocatalytic inactivation of various microorganisms, including the highly resistant-to-inactivation prion agents [42–46]. Its non-toxicity, high photochemical stability, and potential for reuse further contribute to its wide use as a photocatalyst [47]. TiO_2 excitation is achieved by irradiation at a wavelength below 375 nm, equivalent to the UV-A range of the spectrum [48], allowing the use of both artificial (UV-A lamps) and natural (solar light) sources of illumination. Considering that electricity and maintenance costs are the major operational costs in artificial UV-A photocatalysis [49], exploitation of solar UV-A light for TiO_2-mediated photocatalytic disinfection has gained much attention [27,33,42,48,50–52]. Targeting a cost-effective and sustainable disinfection approach, we focused on the solar-driven photocatalytic inactivation of microorganisms.

We utilized natural sunlight to treat a mixture of different microbial suspensions (*Geobacillus stearothermophilus* spores, *Staphylococcus aureus*, *Escherichia coli*, MS2, and *Candida albicans*) in the presence of 0.5 g/L TiO_2 and 500 ppm H_2O_2 for 6 h. H_2O_2 was added to enhance disinfection efficiency [26,49]. All microorganisms were added at a final concentration of 10^6 cfu/mL, except for MS2, which was added at 10^8 pfu/mL. H_2O_2 levels were assessed at 60 min, and additional H_2O_2 was added to ensure the desired concentration (500 ppm) was maintained throughout the reaction.

Experiments were performed in July, a month of high insolation and high temperature in Greece. Solar UV-A irradiation intensity and temperature were monitored during the experimental process. UV-A irradiation and temperature ranged between 4.1 and 0.31 mW/cm^2 and 44–30.4 °C, respectively, peaking at the beginning of each experiment and then gradually decreasing.

At predefined time points, samples were collected and used either for microbe viability testing through culture or for nucleic acid isolation for application of our newly developed qPCR assay. Similar reactions, performed under dark conditions, were included in each experiment and served as the negative control.

For the microbial viability tests, samples taken at 0, 15, 30, and 60 min were plated on appropriate agar media using the top agar overlay method. The number of colonies or plaques was determined following 20 h incubation at 37 °C or at 60 °C for *Geobacillus stearothermophilus* spores.

We observed a significant reduction in microbial viability within the first hour of treatment. Figure 2 depicts microbial viability data acquired from two independent experiments (sampling of each time point in triplicate for each experiment) in relation to accumulated solar energy (kJ/L). Accumulated energy was determined as previously described.

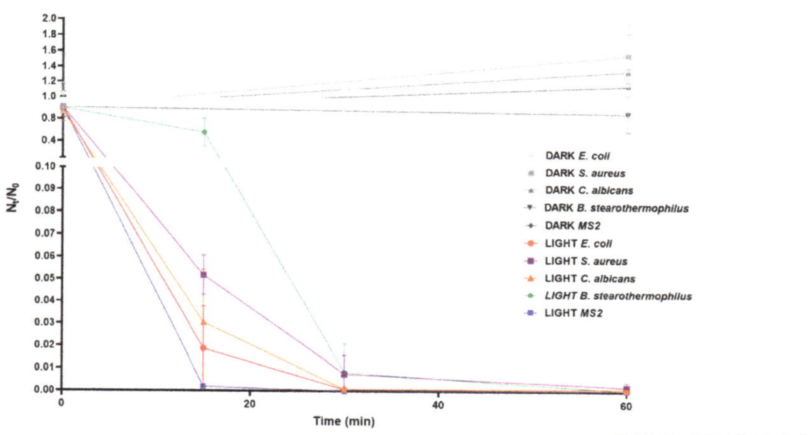

Figure 2. Microbial viability reduction by solar TiO_2 photocatalysis supplemented with H_2O_2 in the tested microbial suspension mixture. The graph depicts viability reduction (N_t/N_0) of treated samples and corresponding controls (gray symbols and lines) in relation to processing time and the calculated accumulated energy after 60 min of treatment. The graph summarizes data acquired from two independent experiments (sampling of each time point in triplicate for each experiment). Error bars correspond to standard errors. The legend shows experimental conditions (dark, light) and the tested microorganisms.

As shown in Figure 2, we observed microorganism-related differences in terms of inactivation rates; *Geobacillus stearothermophilus* spores displayed the highest resistance to inactivation, followed by *Staphylococcus aureus*, *Candida albicans*, *Escherichia coli*, and

MS2 (Figure 2). This pattern is in line with previously described patterns of microbe-specific resistance to inactivation [28], reflecting the impact of individual microorganism characteristics, such as size, structure, and chemical composition. Indeed, spores are the most resistant, due to their compact structure [15], while the MS2 bacteriophage appears to be highly susceptible, possibly due to the lack of genetic material repair mechanisms, which results in higher vulnerability to nucleic acid damage through ROS [29].

Parameters affecting the efficiency of photocatalytic disinfection include catalyst concentration, microorganism load and diversity, composition of treated material, irradiation intensity, and temperature [32]. High catalyst concentrations negatively impact photocatalytic efficiency, resulting in increased turbidity and light scattering [53]. Similarly, microbial load affects the processing time required for disinfection. In addition, disinfection efficiency is affected by the microbe species present in the treated material, and the overall photocatalytic disinfection efficiency of microbial mixtures differs from the inactivation rates of individually tested microorganisms [54]. Moreover, the presence of salts and/or suspended residues, corresponding to components of the treated material, may affect disinfection efficiency [55]. Finally, light intensity and temperature have a considerable impact on disinfection efficiency, with higher light intensities and lower temperatures promoting photocatalysis [49]. Heterogeneity in terms of experimental parameters, including catalyst configuration and concentration, photocatalytic setup, microorganism strains and their initial load, further diversified by fluctuations in light intensity and temperature that characterize outdoor experimentation, hampers a direct comparison of our results with similar published studies.

Nevertheless, we observed the previously reported impact of light intensity and temperature on photocatalytic efficiency. This is depicted in Figure S2, which presents individual data from two experiments conducted under different temperature and solar intensity ranges. Accumulated energy values determined for each experiment are shown. Even though the same pattern of differential microbial resistance to inactivation was observed, the rate of microbial inactivation was altered (Figure S2A,B). Higher inactivation rates were determined under increased accumulated energy and lower temperature (~35 °C), in line with previous reports [49]. Even though not the focus of this study, the effects of photolysis and direct peroxygen chemistry were assessed in a set of experiments through the use of appropriate controls (Supplementary Figure S3).

To gain a better understanding of microbial inactivation during photocatalysis, we next focused on nucleic acid release and degradation, utilizing our newly developed multiplex qPCR method. Tested samples were collected at timepoints 0, 1h, 3h, and 6h, centrifuged to pellet intact bacteria, and the supernatants, harboring DNA released in the solution due to bacterial damage and RNA from both intact and damaged MS2 virions, were processed for DNA/RNA extraction, followed by qPCR. Similar processing was performed for samples under 'Dark' conditions. qPCR data were converted to genome equivalents based on corresponding standard curves (Table 1). In the case of bacteria, genome equivalents of *E. coli* were calculated. Table S2 provides detailed qPCR data of two independent experiments for which viability culture assays have been performed (culture assay results shown in Figure 2). Figure 3 depicts results from the corresponding qPCR analyses, expressed as a percentage of nucleic acid signal at each time point relative to an arbitrarily selected reference sample per condition (photocatalytically treated, "Light", or the dark control, "Dark") and microbe class (Bacteria, Candida, and MS2).

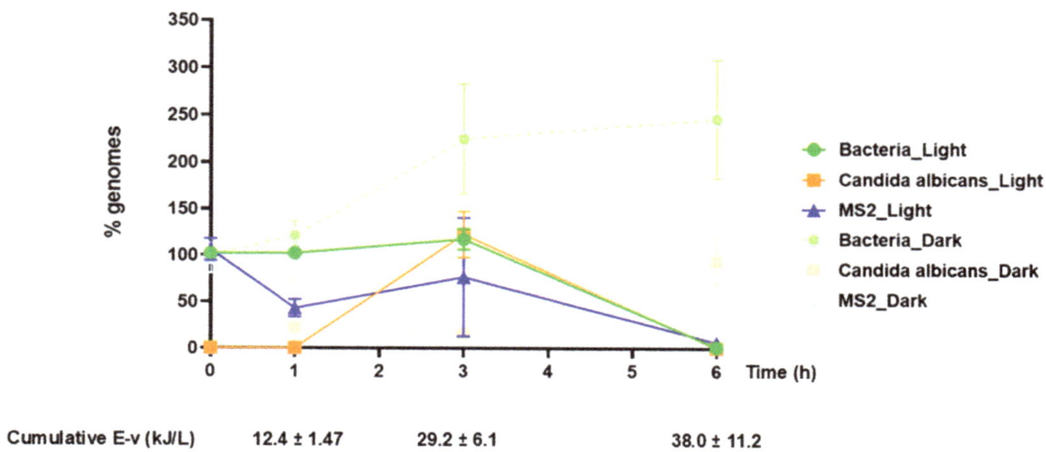

Figure 3. qPCR analysis results of two independent photocatalysis experiments (Light) and their corresponding controls (Dark). The graph shows the percentage of detected genomes at the indicated time points, 0, 1, 3, and 6 h, and in relation to the accumulated energy. For bacteria, values refer to *E. coli* genome equivalents. As a 100% genome was considered the detected microbial genome number at time point 0 h of each condition, except for *C. albicans*, for which the 3 h (Light) and 6 h (Dark) values were assumed to be 100%. Samples of each experiment were analyzed in triplicate.

During photocatalysis, microbial structural elements are oxidized and damaged, leading to nucleic acid release in the solution. This is followed by further oxidation of biological macromolecules by ROS, resulting in nucleic acid degradation. Our molecular assay enables the detection of nucleic acid release and estimation of their integrity through amplification of short nucleic acid fragments (86–110 bp, depending on the assay). In this regard, we expect no detectable bacterial and fungal DNA in the tested supernatant at time point 0, as intact cells are pelleted and not processed. In contrast, the MS2 nucleic acid signal, corresponding to intact virions present in the tested supernatant, is expected to be present at time point 0. As photocatalytic treatment proceeds, compromised microbes and virions release macromolecules, leading to a higher nucleic acid signal. This peak in macromolecule concentration, reflected in the higher nucleic acid level readings, is followed by signal reduction due to nucleic acid fragmentation to the extent that target nucleic acid regions are not intact and the primers can no longer recognize them, initiating their amplification.

Our molecular assay covers a wide range of bacterial species and provides a generalized detection/quantification of bacterial DNA. In our experimentation, this corresponds to DNA from *Geobacillus stearothermophilus*, *Staphylococcus aureus*, and *Escherichia coli*. Our data (Figure 3) showed that in the photocatalytically treated samples, total bacterial DNA decreases significantly after 6 h of treatment (~1.5% of the initial load). A relatively high bacterial DNA signal at time point 0 was observed, which was unexpected, considering that bacterial cells should be intact at this time point. This high initial signal, which is indicative of cellular damage, could be attributed to centrifugation and resuspension of bacterial pellets during the preparation steps. The integrity of *Escherichia coli* and *Staphylococcus aureus* is expected to be affected more severely by this processing than *Geobacillus stearothermophilus* spores, which are compact and thus more resistant to mechanical stress. Bacterial DNA levels, representing DNA released from *Geobacillus stearothermophilus*, *Escherichia coli*, and *Staphylococcus aureus*, remained virtually unaffected for the first 3 h of treatment. The kinetics of the reaction are considered to reflect a dual process, entailing both the oxidation of already released DNA, resulting in signal reduction, and the release of additional DNA from newly damaged bacterial cells during treatment, resulting in signal increases. Considering the results acquired from the corresponding culture assays, we speculate that bacterial DNA release at earlier photocatalysis time points is mostly contributed by the most susceptible bacterial species (*Escherichia coli*), followed by intermediate (*Staphylococcus aureus*) and the most resistant (*Geobacillus stearothermophilus* spores) bacteria. In the corresponding "Dark" samples, the qPCR signal increased with time, probably due to damage related to oxidation caused by the added H_2O_2 and mechanical stress related to constant stirring.

Candida albicans DNA signal was detected after 3 h of solar illumination, suggesting that a higher treatment time is required for the initial cellular damage of yeast cells, compared to vegetative bacteria (*E.coli*, *S. aureus*). This is in line with previous reports [56] that correlate structural effects of *Candida albicans* to inactivation resistance. Of note, our culture data showed that after 1 h of solar illumination, *Candida albicans* showed practically no growth ability on Sabouraud plates; considering that *Candida albicans* DNA was not detectable until 3 h of treatment, it could be suggested that after 1 h of treatment, *Candida albicans* cells lose their ability to propagate, but remain intact; however, they may retain their metabolic activity and pathogenicity, as previously reported for both *Candida albicans* and other microbes under stress conditions [21,31,57]. Following an additional 3 h of treatment (time point 6 h), *Candida albicans* DNA in the processed supernatants was below the detection limit of our assay (<1 genome equivalent, Table 1), reflecting a highly significant degradation of fungal DNA, indicative of fungal cell damage and thus reduced pathogenicity. Taken together, culture and molecular assay data indicate that a 6 h photocatalytic treatment is preferable to evaluate inactivation of *Candida albicans* and reduced pathogenicity, further highlighting the value of molecular assays for the assessment of microbial photocatalytic inactivation. The "Dark" samples showed, as expected, no *Candida* DNA signal at time point 0, which gradually increased to reach a maximum at 6 h.

RNA from the MS2 virus was isolated directly from the supernatant, corresponding to either RNA from intact or from damaged virions. Considering that the MS2 molecular target assessed by our assay represents one copy/genome, in contrast to the other microbe targets (e.g., *E. coli* target: 7 copies/genome, *G. stearothermophilus* target: 10 copies/genome, *S. aureus* target: 6 copies/genome, fungal target: 21–176 copies/genome, Table 1), we used $\times 100$ more concentrated MS2 (10^8 pfu/mL) compared to other microbes to ensure efficient signal detection. A significant decrease in MS2 nucleic acid load (down to 10%) was achieved after 6 h of solar irradiation, suggesting significant virion and nucleic acid damage at this time point. On the contrary, no significant signal variations were observed, in the dark control samples.

In summary, we have developed a multiplex qPCR assay allowing the simultaneous detection/quantification of different microbial classes, including bacteria, fungi, and viruses, and utilized it for assessing the inactivation efficiency of a microbe mixture following small, laboratory-scale TiO_2 solar photocatalysis. Our molecular data provides additional information on the photocatalytic inactivation of different microbial species under solar irradiation, allowing the determination of an efficient processing time in terms of microbe inactivation, based on the estimation of nucleic acid integrity.

Considering that molecular assays contribute to a better estimation of photocatalytic microbial inactivation, our assay, which allows the detection of different microorganism classes, covering a wide range of species commonly detected in urban wastewater and medical liquid waste, could serve as a valuable tool for assessing photocatalytic microbial inactivation in both laboratory and real-life settings. Moreover, widespread assessment of microbial inactivation using the same assay would also allow more efficient comparisons among the results acquired under different experimental settings. In this context, we next plan to use the newly developed multiplex qPCR assay to assess the operation of a novel, much larger (reaction volume 50 L) solar photocatalytic reactor recently developed by our research team.

3. Materials and Methods

3.1. Multiplex qPCR Assay Development

3.1.1. Primers and Probes

The multiplex assay developed includes three primer-probe sets that target (a) bacteria, (b) fungi, and (c) the MS2 bacteriophage, thus allowing simultaneous detection/quantification of different microbe classes. The bacteria-specific primer–probe set targets the highly conserved region of 16S rRNA, present in all bacterial species used in this study (*Escherichia coli, Staphylococcus aureus*, and *Geobacillus stearothermophilus*). The forward bacteria-specific primer was designed as a degenerate oligo to ensure amplification of all the bacterial species of interest (Figure 1). The fungi-specific primer-probe set targets the 18S rRNA of *Candida albicans* (designed based on the reference sequence GenBank AF114470.1), while the MS2 primer–probe set was designed (based on the reference sequence GenBank NC_001417.2) to target an 86 bp region within the gene encoding for the viral capsid protein.

All oligonucleotides were custom-designed by the Primer3web online tool, version 4.1.0, https://primer3.ut.ee/ (accessed on 13 June 2023) [58] and synthesized by Integrated DNA Technologies (IDT). Primers and probes were analyzed with the BLASTn tool [https://blast.ncbi.nlm.nih.gov/Blast.cgi, [59]] (accessed on 15 July 2023) to ensure specificity to the target genome. Possible homo- and heterodimer formations, GC content, and secondary structure were evaluated using the online OligoAnalyzer tool (Integrated DNA Technologies, IDT), https://www.idtdna.com/calc/analyzer (accessed on 8 June 2023). All probes were HPLC-purified and labeled at the 5′ end with a specific fluorophore: 6-FAM for bacterial targets (with double quenching using ZEN as an internal quencher), ATTO550N for yeast, and Cy-5 for MS2. Primer and probe sequences are listed in Table 1.

A BLASTn analysis of the bacterial primer–probe sets showed a 100% sequence identity with various bacterial species due to the conservation of the targeted region. Similarly, several fungal species are recognized by the fungi-targeting primer-probe set. A summary of selected bacterial and fungal species targeted by the designed primer-probe sets is shown in Table S1.

3.1.2. qPCR

Reactions were set in a final volume of 20 µL, containing appropriate DNA/cDNA amounts (serial dilutions of standards for standard curve preparation, isolated nucleic acids at indicated time points after photocatalytic treatment, or the corresponding dark control samples). Primer concentration for *C. albicans* and MS2 genome detection was 300 nM, while the concentration of the bacteria-specific Bac91F and Bac91R primers was 450 nM. Each hydrolysis probe was used at a concentration of 200 nM. Each reaction contained 10 µL of master mix (final concentration in reaction 1×), supplemented with the reference dye ROX (2× Luna Universal Probe qPCR Master Mix, New England Biolabs, M3004). Thermal cycling was performed in an Applied Biosystems 7500Fast real-time PCR system (Applied Biosystems, Waltham, MA, USA). Cycling conditions included an initial denaturation step at 95 °C for 2 min followed by 40 cycles of 95 °C for 15 s, 60 °C for 30 s, and 68 °C for 20 s.

All samples were tested in triplicate, and appropriate negative controls (no template reactions) were included in each run.

3.1.3. DNA/RNA Standards, Standard Curves and Reaction Efficiency Estimation

The following commercial nucleic acid standards were purchased from Merck & Co., Inc. (Rahway, NJ, USA):: (a) MS2 phage RNA standard (10165948001, 800 ng/μL), (b) fungal DNA standard from *Candida albicans* (MBD0044-0.3UG, 10 ng/μL), (c) microbial DNA standard from *E. coli* (MBD0013-0.3UG, 10 ng/μL), and used for standard curve preparation for assay efficiency estimation and subsequent microbe quantification. The MS2 RNA standard was reverse transcribed to cDNA using the Takara PrimeScript RT Reagent Kit (TAKARA Bio Inc, Kusatsu, Shiga, Japan) (RR037A) according to the manufacturer's instructions.

To assess the primer-probe efficiencies, standards were serially diluted (1:20 and then four additional 1:10 dilutions) and used for the preparation of standard curves. Reaction efficiencies were determined from the acquired curves using the equation Efficiency = $10^{(-1/\text{slope})} - 1$. For the determination of the limit of detection, the following equation was used: LOD = (3.3 $\times$ standard deviation of linear regression)/slope of the regression line (standard deviation of linear regression = Sy.x).

3.2. Solar Titanium Dioxide Photocatalytic Inactivation of Microbial Mixture Suspensions

3.2.1. Microbial Strains

The microbial strains used in this study were selected to include representatives of all potential biopathogen classes, namely bacterial spores, fungi, viruses, and Gram-positive and Gram-negative bacteria. Model organisms, including *Geobacillus stearothermophilus* (ATCC 7953) endospores, *Staphylococcus aureus* (ATCC 6538), and MS2 bacteriophage (ATCC 15597-B1), were used. In addition, a clinical isolate of the yeast *Candida albicans* and an ampicillin-resistant laboratory strain of *Escherichia coli* (TOP10 transformed with a plasmid vector carrying ampicillin resistance) were utilized.

Bacillus spores were used as a representative of species highly resistant to inactivation. *Staphylococcus aureus* and *Escherichia coli* were used as representatives of Gram-positive and Gram-negative bacteria, respectively, and *Candida albicans* as a fungal representative. The MS2 bacteriophage, displaying morphological and genomic similarities to human enteric viruses (RNA genome, size, and shape), was used as a non-pathogenic viral surrogate.

Microbial strains were kindly provided by Emeritus Professor Minas Arsenakis (School of Biology, Aristotle University of Thessaloniki, Thessaloniki, Greece).

3.2.2. Microbial Propagation and Culture Assays

Microorganisms were propagated in appropriate or selective nutrient media (broth or agar). Tryptic Soy Broth Agar-TSB (Applichem, 413820) was used at 60 °C for *Geobacillus stearothermophilus*, Sabouraud Dextrose Agar—SDA (dextrose 40 g/L, peptone 10 g/L, agar 15 g/L, pH 5.6) was used at 37 °C for *Candida albicans*; Luria–Bertani-LB (10 g/L tryptone, 5 g/L yeast extract, and 5 g/L NaCl) supplemented with 100 μg/mL ampicillin was used at 37 °C for *Escherichia coli* (TOP10 transformed with an ampicillin resistance conferring plasmid); and Mannitol Salt Agar—MSA (peptone 10 gr/L, sodium chloride 75 gr/L, D-mannitol 10 g/L, meat extract 1g/L, agar 15 g/L)—was used at 37 °C as a selective for *Staphylococcus* species medium due to its high salt concentration. The MS2 host strain (*E. coli* Top10F') was propagated in Luria–Bertani-LB.

Culture assays for microbe viability estimation were performed on appropriate agar plates by overlaying corresponding top agar media (7 g/L of agar) to ensure homogeneous distribution of microbes. *Geobacillus stearothermophilus* endospores were processed as previously described [42].

Plates were incubated at 37 °C (*S. aureus, E. coli, C. albicans*, MS2) or 60 °C (*G. stearothermophilus*) for 20 h, and colonies/plaques were counted to determine colony/plaque-forming units.

3.2.3. Solar Photocatalytic Treatment of Microbial Mixture Suspensions

Photocatalysis was conducted using titanium dioxide (TiO_2) P25 (Aeroxide, Evonik, (Essen, Germany); 70% anatase–30% rutile) and solar light as the irradiation source. Treatment was performed in Phosphate-Buffered Saline (PBS) under continuous stirring (400 rpm), in 6-well plates (Greiner, 657185), at a final volume of 10 mL per well, containing 0.5 g/L TiO_2 and 500 ppm H_2O_2 per well. Freshly prepared cultures of *E. coli* and *S. aureus* at the exponential phase (OD600 = 0.5) and overnight *C. albicans* cultures (OD600~ 1.3) were used. Cultures at the desired growth phase were left on ice for 10 min to stop microbial growth, centrifuged (2500× *g*/20 min/4 °C), and resuspended in PBS. Previously prepared and titrated stocks of MS2 (10^{11} pfu/mL) and *G. stearothermophilus* spores (2×10^7 cfu/mL) were used. All microorganisms were added at a final concentration of 10^6 cfu/mL except MS2, which was added at a higher concentration (10^8 pfu/mL). We used ×100 more concentrated MS2 (10^8 pfu/mL) compared to other microbes to ensure efficient signal detection, taking into account that the MS2 molecular target assessed by our assay represents one copy/genome, in contrast to the other microbe targets (e.g., bacterial target: 7 copies/genome, fungal target: 21–176 copies/genome, Table 1) and considering the genome copy numbers detection limits per microbial class (Table 1).

At different time points (0, 15, 30, 60, 180, and 360 min), samples were collected and used either for determining residual viability by plating, utilizing the top agar overlay method (timepoints 0, 15, 30, and 60 min), or for extracting nucleic acids for use as templates in subsequent multiplex qPCR assays (timepoints 0, 60, 180, and 360 min). H_2O_2 was added at a final concentration of 500 ppm immediately after sampling at t = 0.

Collected samples were plated on appropriate nutrient media, incubated overnight, and the colonies/plaques were counted to determine residual viability. For *G. stearothermophilus* endospores, samples were boiled for 10 min to stimulate germination, mixed with 3 mL of TSB soft agar, overlayed onto TSB agar plates, and incubated overnight at 60 °C. All samples were plated in triplicate.

Solar UV-A intensity (mW/cm^2) was measured with a photometer/radiometer (UV-1700, PharmaSpec, Shimadzu, Japan), and temperature was monitored during the experiments. H_2O_2 concentration was evaluated at t = 60 min using indicator strips (MQuant, Merck,1.10337.0001) and supplemented to 500 ppm. Accumulated solar energy was calculated as previously described [60].

In parallel, control reactions were conducted, containing the catalyst, H_2O_2, and the microbial mix suspension, in the dark. Even though not the focus of this study, the effects of photolysis and direct peroxygen chemistry were assessed in a set of experiments through the use of appropriate controls (Supplementary Figure S3).

3.2.4. Microbial Viability Estimation Through Culture Assays

Microbial culture assays were used to quantify the inactivation of microbes during the photocatalytic treatment. Viability reduction over photocatalytic time was quantified using the ratio N_t/N_0, where N_t represents the number of viable and culturable cells/virus-like particles (cfu/mL or pfu/mL) at specific time points, and N_0 corresponds to initial content at time point zero. The inactivation efficiency was estimated for samples (s) and time points based on the colony or plaque counts relative to the zero time point (N_{ts}/N_{0s}).

3.2.5. Nucleic Acid Purification from Photocatalytically Treated Samples

Samples were collected at defined time points (0, 1, 3, 6 h), centrifuged for 15 min, 3000× *g* at 4 °C to remove TiO_2, cells, and cellular debris, and the supernatant, containing free DNA/RNA and MS2 viral particles, was transferred to new tubes and stored at −20 °C until further processing for nucleic acid purification.

A total of 400 μL of the supernatant was used for DNA recovery using linear polyacrylamide (LPA) as a carrier [61] (France). Briefly, 8 μL of 0.25% LPA was added to each sample. Following the addition of 2.5 volumes of absolute ethanol, samples were incubated for 15 min at −80 °C, centrifuged for 15 min at 20.000× *g* at 4 °C, washed with 70 % ethanol, and centrifuged again under the same conditions. Pellets were air-dried and resuspended in 30 μL of water for injection.

For RNA extraction, 400 μL of supernatant was used. MS2 RNA was extracted using the NucleoSpin RNA virus kit (Macherey-Nagel, Düren, Germany, 740956.50) according to the manufacturer's instructions. Isolated RNA was finally eluted in 30 μL of elution buffer. Viral RNA was reverse transcribed using the LunaScript RT SuperMix (New England Biolabs, Ipswich, MA, USA, E3010). First strand cDNA synthesis reaction contained 8 μL of RNA and 2 μL of the LunaScript RT SuperMix

1 μL of the isolated DNA and 1.25 μL of the prepared cDNA were applied as template in each single PCR reaction as described in Section 3.1.2.

Supplementary Materials: The following supporting information can be downloaded at: https://www.mdpi.com/article/10.3390/molecules30224333/s1, Figure S1: Assay performance against the targeted microbes. Graphs depict standard curves constructed using serial dilutions of commercially available nucleic acid standards. DNA extracted from bacterial cultures, corresponding to predetermined genome copy numbers, was also used for standard curve construction in the case of *S. aureus* and *G. stearothermophilus*. Assay efficiency in each case was determined from the slope of the corresponding standard curve. Figure S2: The effect of solar UV-A intensity and temperature on photocatalytic inactivation of the tested microbial mixture. The graphs depict viability reduction (N_t/N_0) of treated samples and corresponding controls (gray symbols and lines) in relation to accumulated energy and processing time for two independent experiments (A, B, sampling of each time point in triplicate for each experiment). Error bars correspond to standard errors. The legends show experimental conditions (dark, light) and tested microorganisms. Higher accumulated energy was associated with higher inactivation rates (A). Figure S3: Culture and molecular data corresponding to the tested (A) bacteria (*E.coli, S. aureus, G. stearothermophilus*), (B) fungi (*C. albicans*), and (C) virus (*MS2* bacteriophage). TiO_2 solar light driven photacatalysis in the presence of H_2O_2 was applied to a mixture of all tested microorganisms; controls corresponding to solar light-only (microorganisms under the solar light without addition of TiO_2 and H_2O_2), TiO_2-only (microorganisms in the presence of TiO_2, without H_2O_2, in dark) and H_2O_2-only (microorganisms in the presence of H_2O_2, without TiO_2, in dark) were included to evaluate the contribution of photolysis (solar light), TiO_2 effects (TiO_2-only) and peroxygen chemistry (H_2O_2-only). Samples were collected at defined time points (0, 60, 180, 270 min) and used for plating on appropriate culture media for evaluation of viable microorganisms (cfu/mL, pfu/mL, microbiological assay, left side panels) and for DNA/RNA extraction applied on molecular analysis utilizing our newly developed qPCR assay (right side panels). The graphs on the left present the % ratio of culturable microorganisms at each time point relative to corresponding initial counts (% N_t/N_0). The graphs on the right present relative genome copy number changes during the experiment, expressed as % genome copy number change relative to the highest signal acquired in each case. The *x*-axis presents accumulative Energy (kJ/L), which was determined considering (i) UV-A measurements during experimentation, (ii) treated volume, and (iii) irradiated surface [61]. Sampling time points (min) are also depicted below the cumulative energy values. The relation to the Energy *X*-axis is not applicable to the TiO_2 and H_2O_2 controls, as they were performed in the dark. TiO_2 and H_2O_2 controls are shown in dashed lines to highlight that for these, only time associations are applicable. Since our molecular assay does not discriminate among the tested bacterial species (A, right panel), culture data in panel A left correspond to cumulative data. Data presented correspond to triplicate, and error bars represent standard errors. Table S1: List of selected microbial species with 100% sequence identity to the designed primers and probes, based on BLASTn analyses. Table S2: qPCR data from two independent photocatalytic experiments. The table shows data from photocatalytically treated ("Light") microbes at four time points (0, 1, 3 and 6h) and corresponding controls ("Dark") which were not illuminated. Genome equivalents were calculated according to the equations derived from the standard curves using commercially obtained nucleic acid standards. In the case of bacteria, data refers to genome equivalents of *E. coli*. StDev: standard deviation.

Author Contributions: Conceptualization, K.X., D.D., T.S., A.A. and I.P. (Ioannis Poulios); funding acquisition, K.X., D.D., A.A., T.S. and I.P. (Ioannis Poulios); investigation, I.P. (Ioannis Paspaltsis), S.S., E.K., C.B., S.T. and K.X.; methodology, I.P. (Ioannis Paspaltsis), S.S., E.K., K.X., D.D., I.P. (Ioannis Poulios) and T.S.; project administration, K.X.; resources, K.X., D.D., I.P. (Ioannis Poulios) and T.S.; supervision, K.X., D.D., I.P. (Ioannis Poulios) and T.S.; writing—original draft, I.P. (Ioannis Paspaltsis), S.S., E.K., K.X., S.T. and C.B.; writing—review and editing, I.P. (Ioannis Paspaltsis), S.S., E.K., D.D., K.X., S.T., C.B., I.P. (Ioannis Poulios) and T.S. All authors have read and agreed to the published version of the manuscript.

Funding: This research was implemented in the framework of the Action Investment Plans of Innovation of the Program Central Macedonia, PARD 2021–2027, which is co-funded by the European Union and Greece (Project code: KMP6-0281571).

Institutional Review Board Statement: Not applicable.

Informed Consent Statement: Not applicable.

Data Availability Statement: The original contributions presented in this study are included in the article/Supplementary Material. Further inquiries can be directed to the corresponding author.

Acknowledgments: The authors would like to acknowledge Minas Arsenakis, School of Biology, Aristotle University of Thessaloniki, Thessaloniki, Greece, for providing microbial strains used in this study.

Conflicts of Interest: Author Athanasios Arsenakis was employed by the company STERIMED SA. The remaining authors declare that the research was conducted in the absence of any commercial or financial relationships that could be construed as a potential conflict of interest.

Abbreviations

The following abbreviations are used in this manuscript:

qPCR	Quantitative Polymerase Chain Reaction
cfu	Colony-Forming Units
pfu	Plaque-Forming Units
AOP	Advanced Oxidation Process
ROS	Reactive Oxygen Species
TiO_2	Titanium Dioxide
HPLC	High-Performance Liquid Chromatography
TSB	Tryptic Soy Broth
SDA	Sabouraud Dextrose Agar
LB	Luria–Bertani
MSA	Mannitol Salt Agar
PBS	Phosphate-Buffered Saline
LPA	Linear Polyacrylamide

References

1. Nosaka, Y.; Nosaka, A.Y. Generation and Detection of Reactive Oxygen Species in Photocatalysis. *Chem. Rev.* **2017**, *117*, 11302–11336. [CrossRef]
2. Dalrymple, O.K.; Stefanakos, E.; Trotz, M.A.; Goswami, D.Y. A Review of the Mechanisms and Modeling of Photocatalytic Disinfection. *Appl. Catal. B Environ.* **2010**, *98*, 27–38. [CrossRef]
3. Gogniat, G.; Thyssen, M.; Denis, M.; Pulgarin, C.; Dukan, S. The Bactericidal Effect of TiO_2 Photocatalysis Involves Adsorption onto Catalyst and the Loss of Membrane Integrity. *FEMS Microbiol. Lett.* **2006**, *258*, 18–24. [CrossRef] [PubMed]

4. Liu, H.; Wang, C.; Wang, G. Photocatalytic Advanced Oxidation Processes for Water Treatment: Recent Advances and Perspective. *Chem. An Asian J.* **2020**, *15*, 3239–3253. [CrossRef] [PubMed]
5. Chu, T.T.H.; Nguyen, T.B.H.; Huong, P.T. An Approach for the Treatment of Chlorpyrifos and Atrazine Pesticides Using Graphitic Carbon Nitride Photocatalyst. *J. Environ. Sci. Health Part B Pestic. Food Contam. Agric. Wastes* **2025**, *60*, 200–207. [CrossRef]
6. Qiu, Y.; Lu, J.; Yan, Y.; Niu, J. Enhanced Visible-Light-Driven Photocatalytic Degradation of Tetracycline by 16% Er^{3+}-Bi_2WO_6 Photocatalyst. *J. Hazard. Mater.* **2022**, *422*, 126920. [CrossRef]
7. Mubeen, K.; Safeen, K.; Irshad, A.; Safeen, A.; Ghani, T.; Shah, W.H.; Khan, R.; Ahmad, K.S.; Casin, R.; Rashwan, M.A.; et al. ZnO/CuSe Composite-Mediated Bandgap Modulation for Enhanced Photocatalytic Performance against Methyl Blue Dye. *Sci. Rep.* **2023**, *13*, 19580. [CrossRef]
8. Wysokowska, K.; Cupiał, Z.; Staszak, M.; Zgoła-Grześkowiak, A.; Koziolek, J.; Ławniczak, Ł.; Wysokowski, M.; Wyrwas, B. Photocatalytic Degradation of Non-Ionic, Anionic, and Cationic Surfactants: From Batch Experiments through Equilibrium/Kinetic Study to Ecotoxicology Analysis. *Chem. Pap.* **2024**, *78*, 761–777. [CrossRef]
9. Zhang, J.; Nosaka, Y. Photocatalytic Oxidation Mechanism of Methanol and the Other Reactants in Irradiated TiO_2 Aqueous Suspension Investigated by OH Radical Detection. *Appl. Catal. B Environ.* **2015**, *166–167*, 32–36. [CrossRef]
10. Díez, A.M.; Licciardello, N.; Kolen'ko, Y.V. Photocatalytic Processes as a Potential Solution for Plastic Waste Management. *Polym. Degrad. Stab.* **2023**, *215*, 110459. [CrossRef]
11. Mohamadpour, F.; Amani, A.M. Photocatalytic Systems: Reactions, Mechanism, and Applications. *RSC Adv.* **2024**, *14*, 20609–20645. [CrossRef]
12. Paspaltsis, I.; Kotta, K.; Lagoudaki, R.; Grigoriadis, N.; Poulios, I.; Sklaviadis, T. Titanium Dioxide Photocatalytic Inactivation of Prions. *J. Gen. Virol.* **2006**, *87*, 3125–3130. [CrossRef]
13. Al Zoubi, W.; Salih Al-Hamdani, A.A.; Sunghun, B.; Ko, Y.G. A Review on TiO_2-Based Composites for Superior Photocatalytic Activity. *Rev. Inorg. Chem.* **2021**, *41*, 213–222. [CrossRef]
14. Venieri, D.; Chatzisymeon, E.; Gonzalo, M.S.; Rosal, R.; Mantzavinos, D. Inactivation of Enterococcus Faecalis by TiO_2-Mediated UV and Solar Irradiation in Water and Wastewater: Culture Techniques Never Say the Whole Truth. *Photochem. Photobiol. Sci.* **2011**, *10*, 1744–1750. [CrossRef]
15. Foster, H.A.; Ditta, I.B.; Varghese, S.; Steele, A. Photocatalytic Disinfection Using Titanium Dioxide: Spectrum and Mechanism of Antimicrobial Activity. *Appl. Microbiol. Biotechnol.* **2011**, *90*, 1847–1868. [CrossRef] [PubMed]
16. Laxma Reddy, P.V.; Kavitha, B.; Kumar Reddy, P.A.; Kim, K.H. TiO_2-Based Photocatalytic Disinfection of Microbes in Aqueous Media: A Review. *Environ. Res.* **2017**, *154*, 296–303. [CrossRef] [PubMed]
17. Rokicka-Konieczna, P.; Morawski, A.W. Photocatalytic Bacterial Destruction and Mineralization by TiO_2-Based Photocatalysts: A Mini Review. *Molecules* **2024**, *29*, 2221. [CrossRef]
18. Ishiguro, H.; Nakano, R.; Yao, Y.; Kajioka, J.; Fujishima, A.; Sunada, K.; Minoshima, M.; Hashimoto, K.; Kubota, Y. Photocatalytic Inactivation of Bacteriophages by TiO_2-Coated Glass Plates under Low-Intensity, Long-Wavelength UV Irradiation. *Photochem. Photobiol. Sci.* **2011**, *10*, 1825–1829. [CrossRef] [PubMed]
19. Thabet, S.; Simonet, F.; Lemaire, M.; Guillard, C.; Cotton, P. Impact of Photocatalysis on Fungal Cells: Depiction of Cellular and Molecular Effects on Saccharomyces Cerevisiae. *Appl. Environ. Microbiol.* **2014**, *80*, 7527–7535. [CrossRef]
20. Regmi, C.; Joshi, B.; Ray, S.K.; Gyawali, G.; Pandey, R.P. Understanding Mechanism of Photocatalytic Microbial Decontamination of Environmental Wastewater. *Front. Chem.* **2018**, *6*. [CrossRef]
21. Kacem, M.; Bru-Adan, V.; Goetz, V.; Steyer, J.P.; Plantard, G.; Sacco, D.; Wery, N. Inactivation of *Escherichia coli* by TiO_2-Mediated Photocatalysis Evaluated by a Culture Method and Viability-QPCR. *J. Photochem. Photobiol. A Chem.* **2016**, *317*, 81–87. [CrossRef]
22. Liu, J.; Yang, L.; Kjellerup, B.V.; Xu, Z. Viable but Nonculturable (VBNC) State, an Underestimated and Controversial Microbial Survival Strategy. *Trends Microbiol.* **2023**, *31*, 1013–1023. [CrossRef]
23. Hasegawa, H.; Suzuki, E.; Maeda, S. Horizontal Plasmid Transfer by Transformation in *Escherichia coli*: Environmental Factors and Possible Mechanisms. *Front. Microbiol.* **2018**, *9*, 2365. [CrossRef]
24. Ajiboye, T.O.; Babalola, S.O.; Onwudiwe, D.C. Photocatalytic Inactivation as a Method of Elimination of *E. coli* from Drinking Water. *Appl. Sci.* **2021**, *11*, 1313. [CrossRef]
25. Ren, G.; Han, H.; Wang, Y.; Liu, S.; Zhao, J.; Meng, X.; Li, Z. Recent Advances of Photocatalytic Application in Water Treatment: A Review. *Nanomaterials* **2021**, *11*, 1804. [CrossRef]
26. Hooshmand, S.; Kargozar, S.; Ghorbani, A.; Darroudi, M.; Keshavarz, M.; Baino, F.; Kim, H.W. Biomedical Waste Management by Using Nanophotocatalysts: The Need for New Options. *Materials* **2020**, *13*, 3511. [CrossRef]
27. Moreira, N.F.F.; Narciso-da-Rocha, C.; Polo-López, M.I.; Pastrana-Martínez, L.M.; Faria, J.L.; Manaia, C.M.; Fernández-Ibáñez, P.; Nunes, O.C.; Silva, A.M.T. Solar Treatment (H_2O_2, TiO_2-P25 and GO-TiO_2 Photocatalysis, Photo-Fenton) of Organic Micropollutants, Human Pathogen Indicators, Antibiotic Resistant Bacteria and Related Genes in Urban Wastewater. *Water Res.* **2018**, *135*, 195–206. [CrossRef] [PubMed]

28. Lonnen, J.; Kilvington, S.; Kehoe, S.C.; Al-Touati, F.; McGuigan, K.G. Solar and Photocatalytic Disinfection of Protozoan, Fungal and Bacterial Microbes in Drinking Water. *Water Res.* **2005**, *39*, 877–883. [CrossRef]
29. Kokkinos, P.; Venieri, D.; Mantzavinos, D. Advanced Oxidation Processes for Water and Wastewater Viral Disinfection. A Systematic Review. *Food Environ. Virol.* **2021**, *13*, 283–302. [CrossRef]
30. Aranciaga Pajuelo, R.B.; Vargas López, J.P.; Castañeda Olivera, C.A.; Jave Nakayo, J.L.; Benites Alfaro, E.G.; Cabrera Carranza, C.F. Inactivation of Antibiotic Resistant Bacteria in Hospital Wastewater by TiO_2/H_2O_2 Photocatalysis. *Chem. Eng. Trans.* **2021**, *86*, 853–858. [CrossRef]
31. Ji, H.; Cai, Y.; Wang, Z.; Li, G.; An, T. Sub-Lethal Photocatalysis Promotes Horizontal Transfer of Antibiotic Resistance Genes by Conjugation and Transformability. *Water Res.* **2022**, *221*, 118808. [CrossRef]
32. Venieri, D.; Chatzisymeon, E.; Politi, E.; Sofianos, S.S.; Katsaounis, A.; Mantzavinos, D. Photoelectrocatalytic Disinfection of Water and Wastewater: Performance Evaluation by QPCR and Culture Techniques. *J. Water Health* **2013**, *11*, 21–29. [CrossRef]
33. Polo-López, M.I.; Castro-Alférez, M.; Nahim-Granados, S.; Malato, S.; Fernández-Ibáñez, P. Legionella Jordanis Inactivation in Water by Solar Driven Processes: EMA-QPCR versus Culture-Based Analyses for New Mechanistic Insights. *Catal. Today* **2017**, *287*, 15–21. [CrossRef]
34. Guo, B.; Snow, S.D.; Starr, B.J.; Xagoraraki, I.; Tarabara, V.V. Photocatalytic Inactivation of Human Adenovirus 40: Effect of Dissolved Organic Matter and Prefiltration. *Sep. Purif. Technol.* **2018**, *193*, 193–201. [CrossRef]
35. Seven, O.; Dindar, B.; Aydemir, S.; Metin, D.; Ozinel, M.A.; Icli, S. Solar Photocalytic Disinfection of a Group of Bacteria and Fungi Aqueous Suspensions with TiO_2, ZnO and Sahara Desert Dust. *J. Photochem. Photobiol. A Chem.* **2004**, *165*, 103–107. [CrossRef]
36. Krýsa, J.; Musilová, E.; Zita, J. Critical Assessment of Suitable Methods Used for Determination of Antibacterial Properties at Photocatalytic Surfaces. *J. Hazard. Mater.* **2011**, *195*, 100–106. [CrossRef]
37. Matsuura, R.; Aida, Y. Purification of Living Environments Using Photocatalysts: Inactivation of Microorganisms and Decomposition of Allergens. *J. Vet. Med. Sci.* **2024**, *86*, 689–699. [CrossRef] [PubMed]
38. Kaur, R.; Yadav, B.; Tyagi, R.D. Microbiology of Hospital Wastewater. In *Current Developments in Biotechnology and Bioengineering*; Tyagi, R.D., Sellamuthu, B., Eds.; Elsevier: Amsterdam, Netherlands, 2020; pp. 103–148.
39. Folorunso, O.C. Microbial Contamination in Urban Wastewater Systems: Emerging Health Threats and Mitigation Strategies. *Int. J. Sci. Res. Arch.* **2025**, *14*, 1449–1463. [CrossRef]
40. Ren, B.; Shi, X.; Chi, Y.; Ren, T.; Jin, X.; Wang, X.C.; Jin, P. A Comprehensive Assessment of Fungi in Urban Sewer Biofilms: Community Structure, Environmental Factors, and Symbiosis Patterns. *Sci. Total Environ.* **2022**, *806*, 150728. [CrossRef] [PubMed]
41. Arroyo, M.G.; Ferreira, A.M.; Frota, O.P.; Brizzotti-Mazuchi, N.S.; Peresi, J.T.M.; Rigotti, M.A.; Macedo, C.E.; Sousa, A.F.L.D.; Andrade, D.D.; Almeida, M.T.G.D. Broad Diversity of Fungi in Hospital Water. *Sci. World J.* **2020**, *2020*, 9358542. [CrossRef]
42. Berberidou, C.; Paspaltsis, I.; Pavlidou, E.; Sklaviadis, T.; Poulios, I. Heterogenous Photocatalytic Inactivation of *B. stearothermophilus* Endospores in Aqueous Suspensions under Artificial and Solar Irradiation. *Appl. Catal. B Environ.* **2012**, *125*, 375–382. [CrossRef]
43. Ripolles-Avila, C.; Martinez-Garcia, M.; Hascoët, A.S.; Rodríguez-Jerez, J.J. Bactericidal Efficacy of UV Activated TiO_2 Nanoparticles against Gram-Positive and Gram-Negative Bacteria on Suspension. *CYTA J. Food* **2019**, *17*, 408–418. [CrossRef]
44. Cho, M.; Chung, H.; Choi, W.; Yoon, J. Different Inactivation Behaviors of MS-2 Phage and *Escherichia coli* in TiO_2 Photocatalytic Disinfection. *Appl. Environ. Microbiol.* **2005**, *71*, 270–275. [CrossRef] [PubMed]
45. Gupta, K.; Singh, R.P.; Pandey, A.; Pandey, A. Photocatalytic Antibacterial Performance of TiO_2 and Ag-Doped TiO_2 against *S. aureus. P. aeruginosa* and *E. coli*. *Beilstein J. Nanotechnol.* **2013**, *4*, 345–351. [CrossRef] [PubMed]
46. Paspaltsis, I.; Kanata, E.; Sotiriadis, S.; Correia, S.S.; Schmitz, M.; Zerr, I.; Dafou, D.; Xanthopoulos, K.; Sklaviadis, T. A Comparison of RML Prion Inactivation Efficiency by Heterogeneous and Homogeneous Photocatalysis. *Pathogens* **2024**, *13*, 420. [CrossRef]
47. Hashimoto, K.; Irie, H.; Fujishima, A. TiO_2 Photocatalysis: A Historical Overview and Future Prospects. *Japanese J. Appl. Physics, Part 1 Regul. Pap. Short Notes Rev. Pap.* **2005**, *44*, 8269–8285. [CrossRef]
48. Dufner, L.; Hofmann, P.; Dobslaw, D.; Kern, F. Degradation of Bacteria for Water Purification in a TiO_2-Coated Photocatalytic Reactor Illuminated by Solar Light. *Appl. Water Sci.* **2025**, *15*, 101. [CrossRef]
49. Villar-Navarro, E.; Levchuk, I.; Rueda-Márquez, J.J.; Homola, T.; Moriñigo, M.Á.; Vahala, R.; Manzano, M. Inactivation of Simulated Aquaculture Stream Bacteria at Low Temperature Using Advanced UVA- and Solar-Based Oxidation Methods. *Sol. Energy* **2021**, *227*, 477–489. [CrossRef]
50. Rodrigues, C.P.; Ziolli, R.L.; Guimarães, J.R. Inactivation of *Escherichia coli* in Water by TiO_2-Assisted Disinfection Using Solar Light. *J. Braz. Chem. Soc.* **2007**, *18*, 126–134. [CrossRef]
51. Fagan, R.; McCormack, D.E.; Dionysiou, D.D.; Pillai, S.C. A Review of Solar and Visible Light Active TiO_2 Photocatalysis for Treating Bacteria, Cyanotoxins and Contaminants of Emerging Concern. *Mater. Sci. Semicond. Process.* **2016**, *42*, 2–14. [CrossRef]
52. Venieri, D.; Gounaki, I.; Binas, V.; Zachopoulos, A.; Kiriakidis, G.; Mantzavinos, D. Inactivation of MS2 Coliphage in Sewage by Solar Photocatalysis Using Metal-Doped TiO_2. *Appl. Catal. B Environ.* **2015**, *178*, 54–64. [CrossRef]

53. Adán, C.; Magnet, A.; Fenoy, S.; Pablos, C.; del Águila, C.; Marugán, J. Concomitant Inactivation of *Acanthamoeba* Spp. and *Escherichia coli* Using Suspended and Immobilized TiO_2. *Water Res.* **2018**, *144*, 512–521. [CrossRef]
54. Kuliesiene, N.; Sakalauskaite, S.; Tuckute, S.; Urbonavicius, M.; Varnagiris, S.; Daugelavicius, R.; Lelis, M. TiO_2 Application for the Photocatalytical Inactivation of *S. enterica*, *E. coli* and *M. luteus* Bacteria Mixtures. *Environ. Clim. Technol.* **2020**, *24*, 418–429. [CrossRef]
55. Blanchon, C.; Toulza, E.; Calvayrac, C.; Eichendorff, S.; Travers, M.A.; Vidal-Dupiol, J.; Montagnani, C.; Escoubas, J.M.; Stavrakakis, C.; Plantard, G. Inactivation of Two Oyster Pathogens by Photocatalysis and Monitoring of Changes in the Microbiota of Seawater: A Case Study on Ostreid Herpes Virus 1 MVar and Vibrio Harveyi. *Chemosphere* **2024**, *346*, 140565. [CrossRef] [PubMed]
56. Tzeng, J.H.; Weng, C.H.; Yen, L.T.; Gaybullaev, G.; Chang, C.J.; de Luna, M.D.G.; Lin, Y.T. Inactivation of Pathogens by Visible Light Photocatalysis with Nitrogen-Doped TiO_2 and Tourmaline-Nitrogen Co-Doped TiO_2. *Sep. Purif. Technol.* **2021**, *274*, 118979. [CrossRef]
57. Xiao, Y.; Wang, J.; Sun, P.; Ding, T.; Li, J.; Deng, Y. Formation and Resuscitation of Viable but Non-Culturable (VBNC) Yeast in the Food Industry: A Review. *Int. J. Food Microbiol.* **2025**, *426*, 110901. [CrossRef] [PubMed]
58. Untergasser, A.; Cutcutache, I.; Koressaar, T.; Ye, J.; Faircloth, B.C.; Remm, M.; Rozen, S.G. Primer3-New Capabilities and Interfaces. *Nucleic Acids Res.* **2012**, *40*, e115. [CrossRef]
59. Altschup, S.F.; Gish, W.; Miller, W.; Myers, E.W.; Lipman, D.J. Basic Local Alignment Search Tool. *J. Mol. Biol.* **1990**, *215*, 403–410. [CrossRef]
60. Kositzi, M.; Poulios, I.; Malato, S.; Caceres, J.; Campos, A. Solar Photocatalytic Treatment of Synthetic Municipal Wastewater. *Water Res.* **2004**, *38*, 1147–1154. [CrossRef]
61. Gaillard, C.; Strauss, F. Ethanol precipitation of DNA with linear polyacrylamide as carrier. *Nucleic Acids Res.* **1990**, *18*, 378. [CrossRef]

sustainability

MDPI

Article

Removal of 2-Methyl-4-Isothiazolin-3-One by VUV/UV/Persulfate for Sustainable Wastewater Reclamation: Effects of Inorganic Anions on the Concentrations and Contributions of Hydroxyl Radicals and Sulfate Radicals

Yi-Fan Zhang [1], Hong-Wei Xu [2] and Nan Huang [1,*]

[1] National Engineering Laboratory for Advanced Municipal Wastewater Treatment and Reuse Technology, College of Environmental Science and Engineering, Beijing University of Technology, Beijing 100124, China; zhangyifan-huangnan@emails.bjut.edu.cn

[2] Environmental Simulation and Pollution Control State Key Joint Laboratory, State Environmental Protection Key Laboratory of Microorganism Application and Risk Control (SMARC), School of Environment, Tsinghua University, Beijing 100084, China; xuhw22@tsinghua.org.cn

* Correspondence: huangnan@bjut.edu.cn

Abstract

The non-oxidizing antimicrobial 2-Methyl-4-Isothiazolin-3-one (MIT) poses a significant environmental risk given its frequent detection in municipal wastewater. This study showed that the combination of Vacuum UV/UV (VUV/UV) and persulfate (PDS) efficiently achieved the rapid transformation and removal of 10 μM MIT within 90 s, which is much faster than UV, UV/PDS, and VUV/UV. Increasing the PDS dosage improved MIT degradation, whereas changes in pH between 4 and 10 had little effect. Radical quenching experiments showed that 93% of the MIT oxidation was attributable to the hydroxyl radical (•OH) and the sulfate radical ($SO_4\bullet^-$). $SO_4\bullet^-$ and •OH at concentrations of 8.6×10^{-12} M and 1.5×10^{-12} M accounted for 32% and 61% of the MIT degradation, respectively, and the greater contribution of •OH was attributed to its higher reaction rate constant with MIT compared to $SO_4\bullet^-$. Sulfate had a negligible impact on the radical concentrations. Chloride (1 mM) reduced the $SO_4\bullet^-$ and •OH concentrations by 61% and 27%, respectively. And the $SO_4\bullet^-$ contribution to MIT degradation fell to 19%. Nitrate (5 mM) readily quenched •OH but minimally affected $SO_4\bullet^-$. The •OH concentration decreased by 79%, reducing its contribution to 27%. Bicarbonate/carbonate (5 mM) simultaneously reduced the $SO_4\bullet^-$ and •OH by 26–30% and had little effect on their contributions. Because of the quenching effect of organic matter and inorganic anions on radicals, secondary effluent inhibited the degradation of MIT. After a 120 s treatment, the total organic carbon, UV_{254}, and fluorescence regional integration were reduced by 5%, 8%, and 17–24%, respectively. This study provides a quantitative analysis of how inorganic ions alter the concentrations and contributions of •OH and $SO_4\bullet^-$, elucidating the MIT removal mechanisms in VUV/UV/PDS for sustainable wastewater reclamation.

Keywords: VUV/UV/persulfate; 2-methyl-4-Isothiazolin-3-one; hydroxyl and sulfate radicals; inorganic anions; sustainable wastewater reclamation

1. Introduction

2-Methyl-4-Isothiazolin-3-one (MIT), a non-oxidizing antimicrobial, is widely used across industries—including papermaking, leather, dye coatings, cosmetics, and reverse osmosis systems—to inhibit the growth of bacteria, fungi, and algae [1,2]. The antimicrobial mechanism of MIT involves reducing the sulfur atom to form a disulfide bond with cysteine, thereby inactivating the protein. Studies have shown that MIT has adverse effects on rat lung macrophages [3], cortical neurons [4], and zebrafish embryos [5] and can also cause skin contact allergies [6]. The environmental risks posed by MIT have become a subject of concern in recent years. Water resource shortages are driving an increased demand for municipal wastewater reclamation. In water-deficient areas, treated wastewater that contains MIT (such as secondary effluent from a wastewater treatment plant) is often used to replenish rivers and lakes [7]. The use of reclaimed water in many cities for street cleaning and landscaping directly increases public exposure to MIT [8]. Since MIT biodegrades slowly in the environment, continuous discharges will inevitably increase its environmental load [9,10]. Therefore, MIT removal is necessary during the wastewater reclamation process to reduce potential environmental and health hazards [11].

A variety of technologies have been studied and applied in MIT processing. The soil filtration method is low-cost but requires a long hydraulic retention time to achieve a high removal rate [12]. Due to its antimicrobial activity, MIT at mg/L levels can inhibit the biological treatment systems [13]. Advanced oxidation processes, such as photocatalytic oxidation and ozonation, can efficiently remove MIT. However, photocatalytic oxidation faces the challenge of catalyst separation and recovery [14]. Ozonation risks forming harmful by-products such as bromate [15].

Compared with established processes, the emerging vacuum UV/UV (VUV/UV) technology shows promise for pollutant removal [16,17]. 185 nm and 254 nm photons can be emitted by the conventional low-pressure mercury lamp. However, the traditional glass tube wall will block 185 nm photons. In contrast, the VUV/UV light source uses a high-permeability quartz tube wall, which can utilize both wavelengths at the same power consumption [18,19]. The 185-nm VUV can also activate H_2O molecules to form hydroxyl radicals (•OH). VUV/UV enables the rapid degradation of a broader spectrum of organic pollutants, including many that are refractory to UV alone [20,21].

The VUV/UV/persulfate (VUV/UV/PDS) technology uses VUV, UV, and persulfate to drive superior contaminant removal, further enhancing the removal efficiency. The simultaneous activation of persulfate and H_2O by VUV/UV produces large amounts of sulfate radicals ($SO_4{\bullet}^-$) and •OH radicals. The high operational stability and strong adaptability of VUV/UV/PDS make it a very promising candidate for full-scale water treatment applications [22].

Although MIT degradation by 254-nm UV combined with oxidants or catalysts has been reported [23], applications of the VUV/UV light source for MIT removal remain limited [23], and MIT removal by VUV/UV/PDS has not been reported. The technical advantages of VUV/UV/PDS for removing MIT and treating real reclaimed water must be investigated. While inorganic anions in real water matrices are known to reduce MIT removal efficiency, their effects on radical concentrations have not been quantified.

The objective of the study was to explore MIT removal by VUV/UV/PDS under different oxidation conditions and water matrices. The steady-state concentrations and MIT degradation contributions of $SO_4{\bullet}^-$ and •OH in the presence of different inorganic anions were quantified. In addition, the efficacy of VUV/UV/PDS for MIT removal from actual secondary effluent was evaluated, along with subsequent changes in total organic carbon, UV_{254}, and three-dimensional fluorescence spectrum.

2. Materials and Methods

2.1. Chemicals and Materials

2-Methyl-4-Isothiazolin-3-one (MIT, 98.0%) was sourced from Xushuo Biotechnology (Shanghai, China). Sodium persulfate ($Na_2S_2O_8$, 99%) was purchased from SennChem (Shanghai, China). Sodium thiosulfate ($Na_2S_2O_3 \cdot 5H_2O$, 99%) was purchased from Innochem (Beijing, China). Disodium hydrogen phosphate dodecahydrate ($Na_2HPO_4 \cdot 12H_2O$, 99%) was purchased from Jingchun (Beijing, China). Formic acid (99%) was acquired from Alicheem Chemical Technology Co., Ltd. (Shanghai, China). Methanol (chromatographic grade) was obtained from MREDA (Beijing China). Sodium chloride (NaCl) was provided by Sinopharm Chemical Reagent Co., Ltd. (Shanghai, China), sodium nitrate ($NaNO_3$,99.0%) was provided by Alfa Aesar (Shanghai, China), sodium sulfate (Na_2SO_4, 99%) was provided by Macklin (Shanghai, China), sodium dihydrogen phosphate dihydrate ($NaH_2PO_4 \cdot 2H_2O$), ethanol (chromatographic grade, 99.5%), sodium bicarbonate ($NaHCO_3$, 99.8%), p-nitrobenzoic acid (pNBA), p-chlorobenzoic acid (pCBA) were provided by Aladdin (Shanghai, China). The ultrapure water was prepared by Milli-Q ultrapure water device. The actual reclaimed water used was the secondary effluent from a Beijing sewage treatment plant.

2.2. Experimental Procedure

A quasi-collimated VUV/UV system was used. The employed light sources included 254 nm UV lamps and 185 nm + 254 nm VUV/UV lamps, both purchased from Foshan Comwin (Foshan, China). Nitrogen gas is required throughout the experiment to prevent ozone production by VUV reactions with air. A 30-min stabilization period was applied to the device's VUV and UV irradiance before the start of the experiment. The reaction solution was continuously stirred with a magnetic stirrer throughout the reaction.

A 50-mL solution containing 10 μM of MIT was introduced into the reaction dish, after which the PDS solution at the required concentration was added dropwise. The initial pH was regulated by sodium hydroxide and sulfuric acid. The samples were tested when adjusted. To block 185-nm UV light, filters must be installed in advance when conducting experiments on separate UV and UV-PDS treatment technologies. Before the start of the experiment, the irradiance (3.3×10^3 μEinstein/(cm^2·s) and 1.3×10^3 μEinstein/(cm^2·s)) of the ultraviolet lamp was gauged using an ILT2400 ultraviolet radiometer (International Light Technologies, Peabody, MA, USA). The secondary effluent was subjected to a 0.45-μm membrane filtration and was amended with MIT to a final concentration of 10 μM. All experiments were performed in at least duplicate.

2.3. Determination of Radical Concentration

Reaction rate constants that associated MIT, pCBA, and pNBA with the •OH radicals were 7.5×10^9 M^{-1} s^{-1}, 5×10^9 M^{-1} s^{-1}, and 2.6×10^9 M^{-1} s^{-1}, respectively. The pNBA did not react with the $SO_4 \bullet^-$ radicals, and the reaction rate of pCBA with the $SO_4 \bullet^-$ radicals was 3.6×10^8 M^{-1} s^{-1} [24]. In the mixed solution of MIT, pCBA, and pNBA, the respective degradation kinetics were shown in Equations (1)–(3). Details of the calculation methods were shown in Text S1. The synergistic coefficient for combining VUV/UV and PDS was calculated using Equation (4).

$$-\ln\frac{[\mathrm{MIT}]}{[\mathrm{MIT}]_0} = k_{\mathrm{photo,MIT}}t + k_{\bullet\mathrm{OH,MIT}}[\bullet\mathrm{OH}]_{\mathrm{ss}}t + k_{\bullet\mathrm{SO}_4^-,\mathrm{MIT}}\left[\bullet\mathrm{SO}_4^-\right]_{\mathrm{ss}}t \quad (1)$$

$$-\ln\frac{[\mathrm{pNBA}]}{[\mathrm{pNBA}]_0} = k_{\mathrm{photo,pNBA}}t + k_{\bullet\mathrm{OH,pNBA}}[\bullet\mathrm{OH}]_{\mathrm{ss}}t \quad (2)$$

$$-\ln\frac{[\text{pCBA}]}{[\text{pCBA}]_0} = k_{\text{photo,pCBA}}t + k_{\bullet\text{OH,pCBA}}[\bullet\text{OH}]_{ss}t + k_{\bullet\text{SO}_4^-,\text{pCBA}}[\bullet\text{SO}_4^-]_{ss}t \tag{3}$$

$$Synergistic\ coefficient = \frac{Removal\ Effieciency_{VUV/UV/PDS}}{Removal\ efficiency_{VUV/UV} + Removal\ efficiency_{PDS}} \tag{4}$$

2.4. Analysis Method

The MIT was analyzed by HPLC (Shimadzu, Kyoto, Japan) fitted with a C8 column (250 × 4.6 mm, 5 μm) from J & K Scientific (Beijing, Shanghai). A 1:1 mixture of organic and inorganic phases served as the mobile phase. The organic phase was methanol, and the inorganic phase was a 10 mM phosphate buffer at pH 3 with 0.1% formic acid.

The TOC in the samples before and after the reactions was determined with a TOC analyzer (Sievers 5310 C, Wilmington, DE, USA). The UV_{254} absorbances of the sample before and after the reactions were measured using an ultraviolet spectrophotometer. The three-dimensional fluorescence spectra of the samples before and after the reactions were measured by the fluorescence spectrophotometer (F-7000, Hitachi, Tokyo, Japan).

3. Results and Discussion

3.1. MIT Removal by VUV/UV/PDS

The curves depicting MIT degradation by different technologies are manifested in Figure S1. The MIT degradation by PDS alone and UV alone was extremely slow, with less than 10% MIT degradation achieved within 90 s. The VUV/UV/PDS, VUV/UV, and UV/PDS processes reached removal efficiencies of 100%, 71%, and 25%, respectively, at 90 s. The synergistic coefficient was >1 throughout the degradation process, indicating a synergistic effect between VUV/UV and PDS.

It is worth noting that the UV and VUV/UV experiments have different intensities. Under the same time condition, the fluence of VUV/UV exceeds that of UV. The time unit was converted to fluence based on the light intensity, as described by Xue et al. [25], to eliminate the effect of different fluences and to ensure that the efficiencies of the techniques could be compared under the same fluence. Following this adjustment (Figure 1), VUV/UV/PDS still showed the best performance on MIT removal [26].

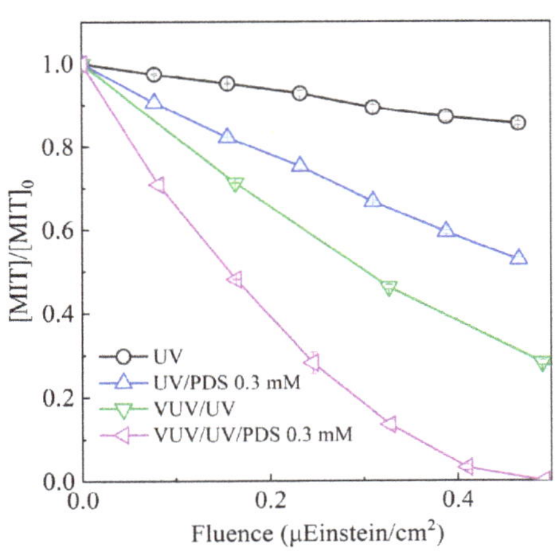

Figure 1. Degradation of MIT by different technologies.

The MIT degradation achieved by a combined UV/PDS technique showed a synergistic effect, with a degradation rate significantly higher than the sum of the degradation rates from the individual UV and PDS processes under the same conditions. During interactions between UV and PDS, UV can activate PDS and produce strong oxidizing radicals that degrade MIT rapidly [7,26]. The VUV/UV and PDS combination also had a synergistic effect on MIT removal, through PDS activation and the generation of sulfate radicals. Similar findings have also been reported during sulfamethoxazole degradation by schwertmannite with the UV/PDS system [27,28].

The conditions and efficiencies for MIT removal by other treatment technologies are summarized in Table S1. VUV/UV/PDS outperforms UV-based AOPs [7,9], ozonation [15], electro-oxidation [8,29], and microalgal treatment [30], in terms of reaction time and removal efficiency, and its performance is comparable to that of O_3/PMS [31]. It should be noted that the experimental conditions across studies vary and may influence treatment outcomes. Future work should conduct more comprehensive technology comparisons from the perspectives of cost-effectiveness and product toxicity.

3.2. Influence of PDS Concentration and pH

3.2.1. Influence of Initial PDS Concentration

The MIT degradation by VUV/UV/PDS at different PDS dosages is portrayed in Figure 2. As the PDS concentration increased, the MIT degradation efficiency by VUV/UV/PDS gradually increased. For 0.1 mM PDS, the removal efficiency surpassed 90% within 90 s. For 0.5 mM PDS, the MIT removal was 100% in 75 s. The MIT degradation adhered to the pseudo-first-order kinetics. The apparent rate rose from 0.023 s^{-1} to 0.038 s^{-1} when the initial PDS concentration increased from 0.1 mM to 0.5 mM. Similar Effects of PDS dosage have been mentioned during carbamazepine degradation by VUV/UV/PDS [22].

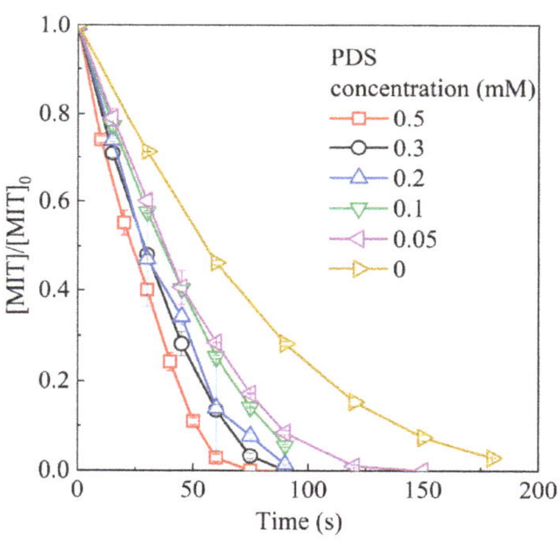

Figure 2. Effects of PDS concentration on MIT removal by VUV/UV/PDS.

3.2.2. Influence of Initial pH Values

A buffer was not used in this study. The pH values were set using H_2SO_4 or NaOH before reactions. However, Wu et al. [32] noticed shifts in the solution pH throughout the VUV/UV/PDS process. We therefore evaluated the effects of pH changes during the process. As shown in Figure 3, the MIT degradation remained stable over an initial solution pH range of 4–10. Therefore, the pH had little effect on MIT degradation. MIT and PDS maintain consistent speciation from pH 4 to 9. pH does not affect either the reaction rate between MIT and radicals or the photolysis of PDS [22]. As pH rises, $SO_4{\bullet}^-$ may be converted to •OH [7]. Based on the findings in Section 3.3.2, the oxidation contribution of •OH is dominant, so the overall impact of pH remains limited. Comparable findings have been documented in studies on MIT degradation via UV/persulfate [7] and ozonation [15], in which pH showed little influence on degradation efficiency.

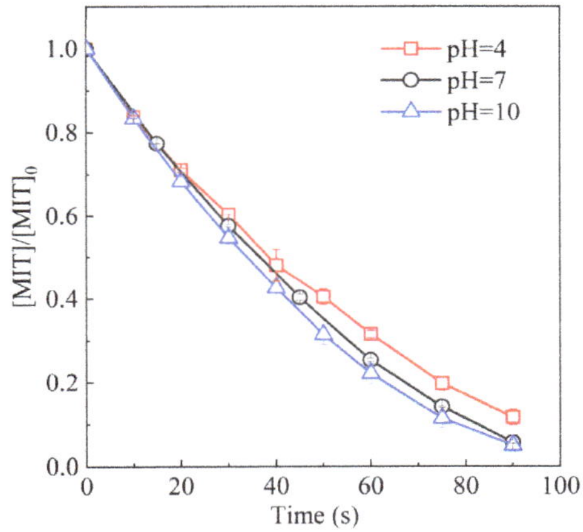

Figure 3. Effects of pH on MIT removal by VUV/UV/PDS.

3.3. Concentration and Contribution of Radicals

3.3.1. Effects of Radical Quencher

Ethanol can quench •OH and SO_4•$^-$ radicals in the reaction system and was used to assess their contribution to MIT removal. As the ethanol concentration increased, the apparent rate dropped gradually from 0.023 s^{-1} to 0.00196 s^{-1}. When the ethanol concentration exceeded 10 mM, the degradation rate of MIT did not decrease further, which suggests that the radicals were completely quenched (Figure 4). Based on these results, the contribution ratios of the radical oxidation and photolysis to the MIT degradation were 94% and 6%, respectively.

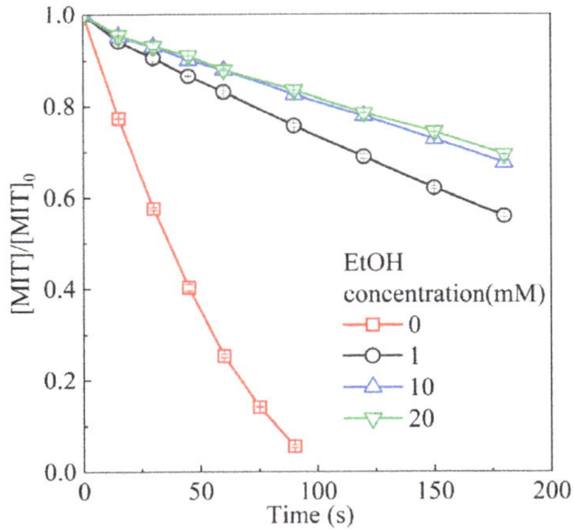

Figure 4. Effects of ethanol on MIT removal during VUV/UV/PDS.

3.3.2. Concentration of •OH and SO_4•$^-$

The •OH and SO_4•$^-$ radical concentrations during the MIT degradation by VUV/UV /PDS were determined using the pCBA and pNBA probe methods and competitive kinetics. Details of the calculation of concentration and contribution of •OH and SO_4•$^-$, and the reaction rate constant of $SO_4\bullet^-$ to MIT ($k_{MIT, SO4\bullet-}$) were shown in Figure S2 and Text S1. The $k_{MIT, SO4\bullet-}$ value was computed to be 6.9 × 10^8 M^{-1} s^{-1} (Text S1). In ultrapure water, the VUV/UV/PDS system contained •OH and $SO_4\bullet^-$ at concentrations of 1.5 × 10^{-12} M and 8.6 × 10^{-12} M, respectively. Their respective contributions to degradation accounted for 61% and 32%.

3.4. Effects of Inorganic Anions on Radical Concentration and Degradation Contribution

The presence of inorganic anions in actual water might affect the •OH and $SO_4\bullet^-$ concentrations, thereby influencing MIT removal. The •OH and $SO_4\bullet^-$ concentrations in the existence of Cl^-, $SO_4{}^{2-}$, $NO_3{}^-$, and $HCO_3{}^-/CO_3{}^{2-}$ at different concentrations were determined using the pCBA and pNBA probe methods. The contribution ratios of the two radicals to MIT degradation were also calculated.

3.4.1. Chloride

The effect of Cl^- on MIT degradation is shown in Figure S4a. The MIT degradation by VUV/UV/PDS was inhibited by Cl^-, and the inhibition effect progressively rose with increasing Cl^- concentration. At Cl^- concentrations of 0 mM, 1 mM, and 10 mM, the 90-s degradation rates of MIT were 100%, 72%, and 48%, respectively. Despite the inhibitory effect of Cl^-, it is worth noting that MIT degradation by VUV/UV/PDS with 10 mM Cl^- was still higher than that by UV/PDS without Cl^- (25%).

$$Cl^- + SO_4\bullet^- \rightarrow Cl\bullet + SO_4^{2-} \left(k = 2.7 \times 10^8 M^{-1} s^{-1}\right) \tag{5}$$

$$Cl^- + HO\bullet \rightarrow Cl\bullet + OH^- \left(k = 4.3 \times 10^9 M^{-1} s^{-1}\right) \tag{6}$$

$$Cl\bullet + OH^- \rightarrow Cl^- + \bullet OH \left(k = 6.1 \times 10^9 M^{-1} s^{-1}\right) \tag{7}$$

When Cl^- concentration rose to 1 mM, the •OH concentration dropped by 26% to 1.1 × 10^{-12} M, while the $SO_4\bullet^-$ concentration fell by 60% to 3.4 × 10^{-12} M. The $SO_4\bullet^-$ concentration was always higher (between 3 and 5.6 times higher) than that of •OH.

As the Cl^- concentration rose, the apparent reaction rate constant declined (Figure 5a). Specifically, when the Cl^- concentration augmented from 0 mM to 1 mM, the apparent degradation rate dropped from 0.0189 s^{-1} to 0.0126 s^{-1}.

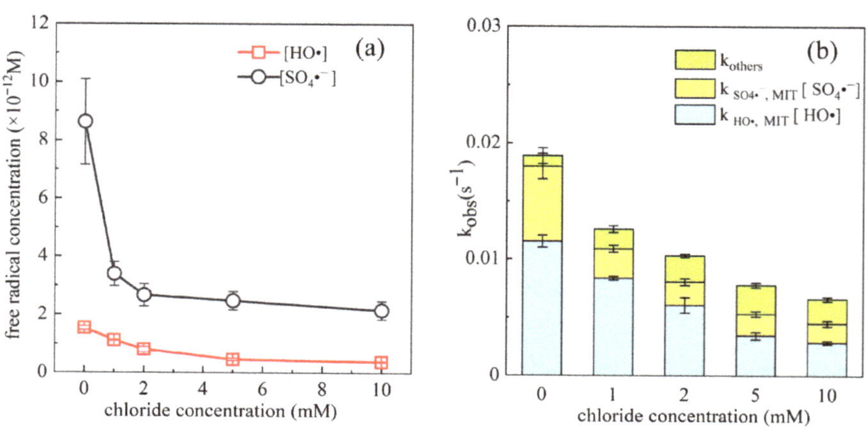

Figure 5. Effect of chloride on the concentration of sulfate radicals ($SO_4\bullet^-$) and hydroxyl radicals (•OH) (**a**), and their contribution to MIT degradation (**b**).

Though the concentration of •OH was lower than that of $SO_4•^-$, the •OH degradation contribution was higher than that of $SO_4•^-$ because •OH had a higher reaction rate constant with MIT (7.5×10^9 M^{-1} s^{-1}) than $SO_4•^-$ (6.9×10^8 M^{-1} s^{-1}). When the Cl^- concentration increased from 0 mM to 1 mM, the $SO_4•^-$ contribution dropped from 32% to 19%, while the •OH contribution increased from 62% to 66% (Figure 5b).

With the Cl^- addition, the $SO_4•^-$ reacted with the Cl^- (Equation (5)), and the $SO_4•^-$ concentration decreased sharply, resulting in a decrease of the $SO_4•^-$ degradation contribution. In contrast, the reaction between Cl^- and •OH was reversible (Equations (6) and (7)) and had little effect on the •OH concentration.

VUV/UV/PDS oxidation of methylene blue showed similar effects, where the Cl^- consumed most of the $SO_4•^-$, reaching 67.5–82.2%, followed by MB (11.4–21.6%) [24]. It can be inferred that, during the MIT treatment with VUV/UV/PDS, chlorides mainly affected the transformation of $SO_4•^-$ and inhibited the MIT degradation by $SO_4•^-$.

3.4.2. Sulfate

The presence of SO_4^{2-} had little impact on MIT degradation, which is presented in Figure S4b. Similar results were reported during VUV/UV/PDS oxidation of methylene blue [24].

The effects of SO_4^{2-} on the •OH and $SO_4•^-$ concentrations are shown in Figure 6a. The addition of SO_4^{2-} (0–10 mM) to the reaction system led to the conversion of a small amount of •OH to $SO_4•^-$, and a slight fluctuation in the apparent rate constant. In the presence of SO_4^{2-}, the •OH, $SO_4•^-$, and photolysis contributions were 36–61%, 32–43%, and 6–28%, respectively (Figure 6b).

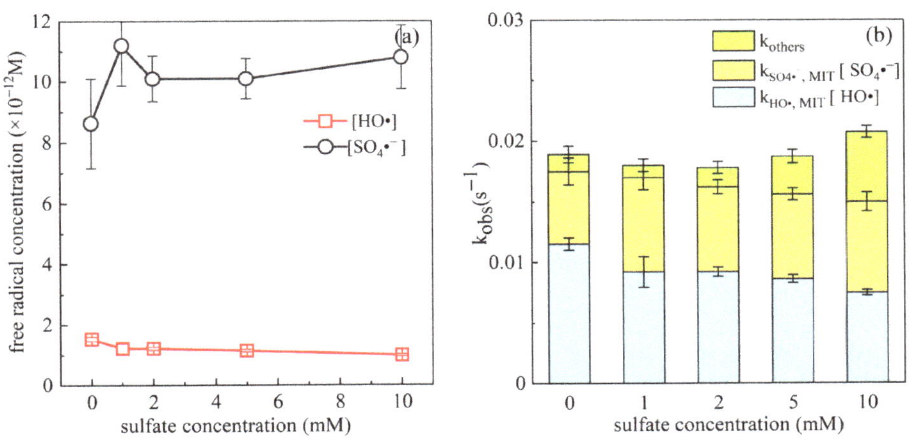

Figure 6. Effect of sulfate on the concentration of $SO_4•^-$ and •OH (**a**), and their contribution to MIT degradation (**b**).

3.4.3. Nitrate

The effect of NO_3^- on MIT degradation is shown in Figure S4c. Yang et al. [33] observed a drop in MIT degradation after NO_3^- addition. When the NO_3^- concentration rose from 0 mM to 1 mM, the degradation rate fell from 90% to 60% in 90 s, and, when the NO_3^- concentration was 10 mM, the degradation rate was 51% at 90 s.

The effects of NO_3^- on the •OH and $SO_4•^-$ concentrations in the reaction system are shown in Figure 7a. When NO_3^- (0–5 mM) was added to the reaction system, there was no noticeable change in the $SO_4•^-$ concentration, but the •OH concentration decreased significantly. A rise in NO_3^- concentration from 0 mM to 1 mM led to a 62% reduction in •OH concentration, which fell from 1.5 × 10^{-12} M to 5.7 × 10^{-13} M. When the NO_3^- concentration increased from 0 mM to 5 mM, the SO_4•$^-$ degradation contribution changed from 32% to 71%, and the •OH degradation contribution decreased from 61% to 27% (Figure 7b).

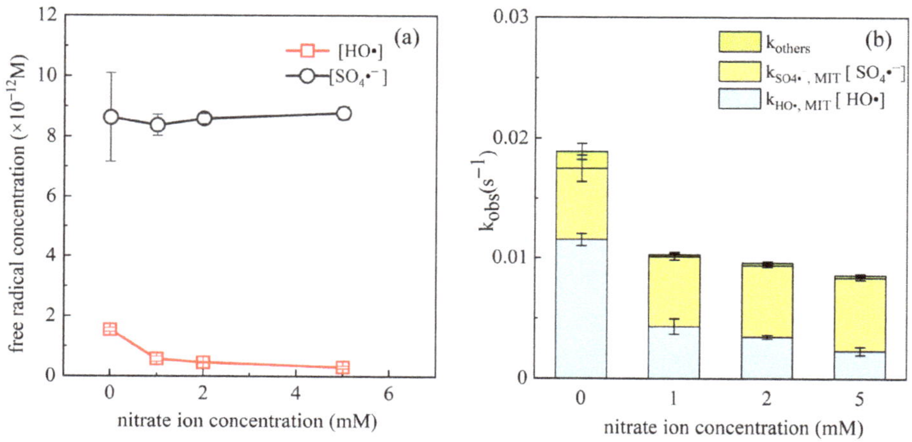

Figure 7. Effects of nitrate on the concentration of SO_4•$^-$ and •OH (**a**), and their contribution to MIT degradation (**b**).

Studies have shown that NO_3^- can inhibit the degradation of pollutants in UV photolysis and UV-based AOPs. It has been reported that NO_3^- can compete with pollutants to consume • OH, hydrogen atoms, and electrons [34,35], resulting in a decrease in the •OH concentration and degradation contribution, and little change in the SO_4•$^-$ concentration.

3.4.4. Bicarbonate/Carbonate

The effects of bicarbonate/carbonate (HCO_3^-/CO_3^{2-}), common anions in wastewater and natural water, on MIT degradation by VUV/UV/PDS are presented in Figure S4d. As the HCO_3^-/CO_3^{2-} concentration grew, the MIT degradation rate gradually fell. At a concentration of 10 mM, the degradation rate at 90 s was 32%.

The addition of HCO_3^-/CO_3^{2-} significantly reduced the SO_4•$^-$ and •OH concentrations (Figure 8a). When HCO_3^-/CO_3^{2-} concentrations rose to 10 mM, the SO_4•$^-$ concentrations dropped by 79% to 1.8 × 10^{-12} M, and the •OH concentrations dropped by 82% to 2.7 × 10^{-13} M. The apparent reaction rate constant of MIT degradation decreased, and there were no obvious patterns in the degradation contributions of each part. The SO_4•$^-$, •OH, and photolysis contributions were 30–37%, 53–61%, and 7–15%, respectively, as shown in Figure 8b.

HCO_3^-/CO_3^{2-} can react rapidly with •OH and SO_4•$^-$ [36,37]. Wen et al. [24] documented that the HCO_3^-/CO_3^{2-} reaction rate constants with •OH and SO_4•$^-$ ranged from 8.5 × 10^6–3.9 × 10^8 M^{-1} s^{-1}, and 1.6 × 10^6–6.1 × 10^6 M^{-1} s^{-1}, respectively. Therefore, HCO_3^-/CO_3^{2-}, when present, inhibited MIT removal during VUV/UV/PDS oxidation.

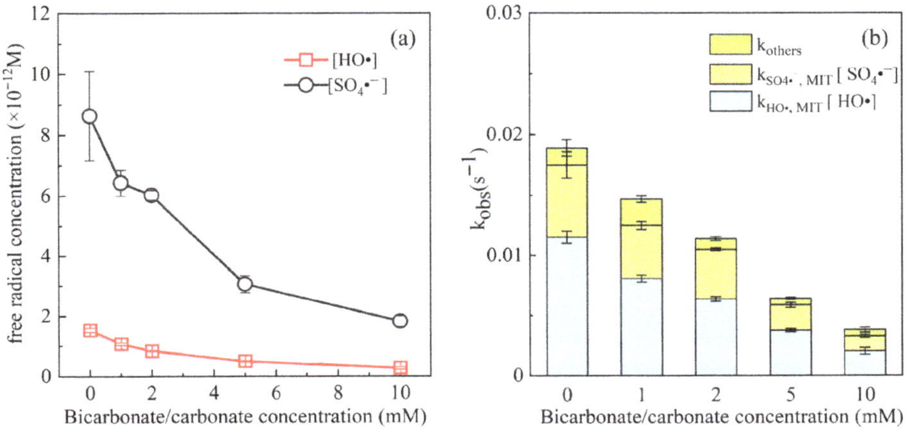

Figure 8. Effects of bicarbonate/carbonate on the concentration of $SO_4\bullet^-$ and $\bullet OH$ (**a**) and their contribution to MIT degradation (**b**).

3.5. MIT Removal in Actual Secondary Effluent

3.5.1. Removal of MIT, TOC and UV_{254}

In actual secondary effluent, the MIT degradation by VUV/UV/PDS was inhibited. The degradation rate was 16% at 90 s (Figure 9). The low rate may have been due to various anions in the secondary effluent, which competitively absorbed UV light and quenched radicals, thereby decreasing the degradation rate of MIT.

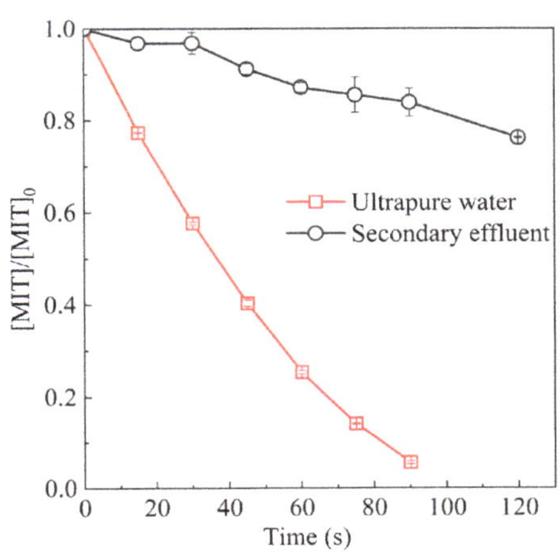

Figure 9. MIT removal by VUV/UV/PDS in actual secondary effluent.

The TOC and UV_{254} were determined before and after the VUV/UV/PDS oxidation. The TOC concentration was 5% lower after the reaction than before. The slight difference may reflect the short reaction time, during which there was minimal mineralization of pollutants by VUV/UV/PDS. The UV_{254} was 8% lower after the reaction than before, indicating that some aromatic C=C and C=O bonds were removed in the secondary effluent.

3.5.2. The Three-Dimensional Fluorescence Spectrum

A five-point method was used to divide the three-dimensional fluorescence region [38]. The fluorescence of the samples was detected before and after 120 s of treatment (Figure 10). The fulvic acid, humic acid, and soluble microbial product concentrations in the secondary effluent were higher than those in the spiked MIT solution. The fluorescence intensity in II, III, IV, and V decreased after the VUV/UV/PDS treatment.

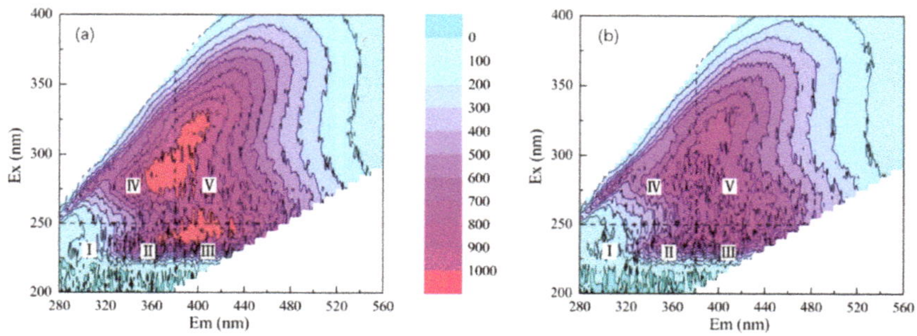

Figure 10. The three-dimensional fluorescence spectrum without treatment (**a**) and after 120 s of VUV/UV/PDS treatment (**b**). I and II: aromatic proteins; III: fulvic acid substances; IV: soluble microbial products; V: humic acid substances.

The fluorescence intensity of the five regions was integrated separately. The changes in the fluorescence regional integration are shown in Figure S3. After 120 s of VUV/UV/PDS treatment, the fluorescence regional integrations of I, II, III, IV, and V were 24%, 23%, 18%, 21%, and 17% lower than at the start of the reaction, respectively. The reduction differences in each region were similar, indicating that the VUV/UV/PDS treatment technology had a general ability to degrade most organic matter in secondary effluent.

4. Conclusions

VUV/UV/PDS showed synergistic performance on MIT removal. The degradation reaction was dominated by radical oxidation. Enhancing PDS dosage promoted MIT degradation, while the effects of pH values (4–10) are slight.

Under UPW conditions, the $SO_4\bullet^-$ concentration during MIT degradation by VUV/UV/PDS is 5.7-fold that of •OH. Because •OH reacts with MIT 10.9 times faster than $SO_4\bullet^-$ does, •OH governs MIT removal. Actual secondary effluents contain various inorganic ions that scavenge radicals, altering radical concentrations and thereby inhibiting MIT degradation. This study provides a quantitative analysis of how inorganic ions alter both the concentration and contribution of •OH and $SO_4\bullet^-$.

Sulfate had a negligible effect on radical concentration and MIT degradation. Chloride significantly reduced $SO_4\bullet^-$ concentration and moderately reduced the •OH concentration. Nitrate significantly reduced the •OH concentration but had little effect on $SO_4\bullet^-$. In the presence of 5 mM nitrate, MIT degradation switched to $SO_4\bullet^-$ domination. Bicarbonate/carbonate decreased both •OH and $SO_4\bullet^-$ concentrations and had little effect on their contributions.

Removing MIT using VUV/UV/PDS during wastewater reclamation processes can reduce risks associated with the ecological use of reclaimed water, address the drying up of rivers and lakes in water-scarce regions, and directly support the long-term goals of water recycling and sustainable urban development. Although this study has shown advantages in MIT removal using VUV/UV/PDS, several limitations remain to be addressed. Firstly, the operating costs have not been systematically evaluated to clarify the economic feasibility of large-scale application. It is necessary to construct a pilot continuous-flow system to simulate performance and conduct a cost-effectiveness analysis. Research on the degradation products and changes in toxicity still needs to be deepened, and the whole life cycle and carbon footprint assessments need to be carried out.

Supplementary Materials: The following supporting information can be downloaded at: https://www.mdpi.com/article/10.3390/su17219882/s1, Figure S1: The degradation performances of MIT by UV alone, PDS alone, UV/PDS, VUV/UV and VUV/UV/PDS; Figure S2: MIT logarithmic degradation curves of pCBA and pNBA as radical probes; Figure S3: The change of fluorescence intensity in each region before and after degradation; Figure S4: Effect of anions on degradation of MIT by VUV/UV/PDS. (a) Cl^-, (b) SO_4^{2-}, (c) NO_3^-, and (d) HCO_3^-/CO_3^{2-}; Table S1: Removal of MIT by different treatment processes; Text S1: Calculation of the steady-state concentration of sulfate radicals ($SO_4\bullet^-$), hydroxyl radicals (•OH), the second order rate constant between MIT and $SO_4\bullet^-$ ($k_{MIT, SO4\bullet-}$), and the contributions of radicals.

Author Contributions: Conceptualization, N.H.; Formal analysis, Y.-F.Z. and H.-W.X.; Funding acquisition, N.H.; Investigation, Y.-F.Z. and H.-W.X.; Methodology, H.-W.X.; Supervision, N.H.; Writing—original draft, Y.-F.Z.; Writing—review & editing, N.H. All authors have read and agreed to the published version of the manuscript.

Funding: This study was funded by the Major Program of National Natural Science Foundation of China (No. 52293440, No. 52293442) and the National Natural Science Foundation of China (No. 52100051).

Institutional Review Board Statement: Not applicable.

Informed Consent Statement: Not applicable.

Data Availability Statement: The original contributions presented in this study are included in the article/Supplementary Material. Further inquiries can be directed to the corresponding author.

Conflicts of Interest: The authors declare no conflicts of interest. The funders had no role in the design of the study; in the collection, analyses, or interpretation of data; in the writing of the manuscript; or in the decision to publish the results.

References

1. Bollmann, U.E.; Vollertsen, J.; Carmeliet, J.; Bester, K. Dynamics of biocide emissions from buildings in a suburban stormwater catchment-Concentrations, mass loads and emission processes. *Water Res.* **2014**, *56*, 66–76. [CrossRef]
2. Carbajo, J.B.; Perdigón-Melón, J.A.; Petre, A.L.; Rosal, R.; Letón, P.; García-Calvo, E. Personal care product preservatives: Risk assessment and mixture toxicities with an industrial wastewater. *Water Res.* **2015**, *72*, 174–185. [CrossRef] [PubMed]
3. Poon, R.; Rigden, M.; Edmonds, N.; Charman, N.; Lamy, S. EVects of 5-chloro-2-methyl-4-isothiazolin-3-one and other candidate biodiesel biocides on rat alveolar macrophages and NR8383 cells. *Arch. Toxicol.* **2011**, *85*, 1419–1427. [CrossRef] [PubMed]
4. He, K.; Huang, J.; Lagenaur, C.F.; Aizenman, E. Methylisothiazolinone, a neurotoxic biocide, disrupts the association of Src family tyrosine kinases with focal adhesion kinase in developing cortical neurons. *J. Pharmacol. Exp. Ther.* **2006**, *317*, 1320–1329. [CrossRef]
5. Li, A.; Chen, Z.; Wu, Q.Y.; Huang, M.H.; Liu, Z.Y.; Chen, P.; Mei, L.C.; Hu, H.Y. Study on the removal of benzisothiazolinone biocide and its toxicity: The effectiveness of ozonation. *Chem. Eng. J.* **2016**, *300*, 376–383. [CrossRef]
6. Marrero-Alemán, G.; Santana, P.S.; Liuti, F.; Hernández, N.; López-Jiménez, E.; Borrego, L. The Role of Cleaning Products in Epidemic Allergic Contact Dermatitis to Methylchloroisothiazolinone/Methylisothiazolinone. *Dermatitis* **2018**, *29*, 77–80. [CrossRef]
7. Cui, J.Z.; Cai, S.K.; Zhang, S.R.; Wang, G.Q.; Gao, C.Z. Degradation of a non-oxidizing biocide in circulating cooling water using UV/persulfate: Kinetics, pathways, and cytotoxicity. *Chemosphere* **2022**, *289*, 133064. [CrossRef]
8. Wang, Y.C.; Chen, M.; Wang, C.; Meng, X.Y.; Zhang, W.Q.; Chen, Z.F.; Crittenden, J. Electrochemical degradation of methylisothiazolinone by using Ti/SnO_2-Sb_2O_3/α, β-PbO_2 electrode: Kinetics, energy efficiency, oxidation mechanism and degradation pathway. *Chem. Eng. J.* **2019**, *374*, 626–636. [CrossRef]
9. Chai, Z.Z.; Wang, J.Q.; Dai, Y.X.; Du, E.R.; Guo, H.G. Synergy between UV light and trichloroisocyanuric acid on methylisothiazolinone degradation: Performance, kinetics and degradation pathway. *Environ. Res.* **2023**, *236*, 116693. [CrossRef]

10. Gómez-Rodríguez, P.; Calza, P.; Fabbri, D.; Medana, C.; van-Grieken, R.; López-Muñoz, M.J. Photocatalytic degradation of methylisothiazolinone in water by TiO_2 and TiO_2/persulfate systems with simulated solar radiation. *Catal. Today* **2023**, *413*, 113942. [CrossRef]
11. López-Timoner, R.; Duarte-Alvarado, V.; Castillo, M.A.; Santos-Juanes, L.; Arques, A.; Amat, A.M. Parabens and Methylisotiazolinone (MIT): Preservatives with Different Behaviors When Subjected to Ozone and Ultraviolet Light Treatments. *Water* **2023**, *15*, 3837. [CrossRef]
12. Bester, K.; Banzhaf, S.; Burkhardt, M.; Janzen, N.; Niederstrasser, B.; Scheytt, T. Activated soil filters for removal of biocides from contaminated run-off and waste-waters. *Chemosphere* **2011**, *85*, 1233–1240. [CrossRef]
13. Laopaiboon, L.; Laopaiboon, P. Biodegradation of isothiazolone biocide (Kathon WT) and treatment efficiency in a laboratory-scale rotating biological contactor. *J. Biotechnol.* **2010**, *150*, S227. [CrossRef]
14. Kandavelu, V.; Kastien, H.; Thampi, K.R. Photocatalytic degradation of isothiazolin-3-ones in water and emulsion paints containing nanocrystalline TiO_2 and ZnO catalysts. *Appl. Catal. B-Environ. Energy* **2004**, *48*, 101–111. [CrossRef]
15. Li, A.; Wu, Q.Y.; Tian, G.P.; Hu, H.Y. Effective degradation of methylisothiazolone biocide using ozone: Kinetics, mechanisms, and decreases in toxicity. *J. Environ. Manag.* **2016**, *183*, 1064–1071. [CrossRef]
16. Zheng, X.; Yang, Q.; Xia, J.; Cheng, R.; Li, H.; Zheng, J. Resistance differences of representative model microorganisms in different disinfection processes: ZIF-67, UVC and ozone. *Water Cycle* **2024**, *5*, 259–265. [CrossRef]
17. Li, M.K.; Wang, C.; Yau, M.L.; Bolton, J.R.; Qiang, Z.M. Sulfamethazine degradation in water by the VUV/UV process: Kinetics, mechanism and antibacterial activity determination based on a mini-fluidic VUV/UV photoreaction system. *Water Res.* **2017**, *108*, 348–355. [CrossRef] [PubMed]
18. Huang, N.; Shao, W.-T.; Wang, Q.; Wang, W.-L.; Wu, Q.-Y.; Hu, H.-Y. Degradation of chloromethylisothiazolinone antimicrobial by Vacuum-Ultraviolet/Ultraviolet irradiation: Reactive species, degradation pathway and toxicity evaluation. *Chemosphere* **2022**, *302*, 134821. [CrossRef] [PubMed]
19. Liu, J.; Bai, X.; Bai, Y. Exploration and case analysis of treatment processes and reuse pathways for industrial brine wastewater in China. *Water Cycle* **2024**, *5*, 278–285. [CrossRef]
20. Li, M.K.; Qiang, Z.M.; Hou, P.; Bolton, J.R.; Qu, J.H.; Li, P.; Wang, C. VUV/UV/Chlorine as an Enhanced Advanced Oxidation Process for Organic Pollutant Removal from Water: Assessment with a Novel Mini-Fluidic VUV/UV Photoreaction System (MVPS). *Environ. Sci. Technol.* **2016**, *50*, 5849–5856. [CrossRef]
21. Xiao, Z.-Y.; Huang, N.; Wang, Q.; Wang, W.-L.; Wu, Q.-Y.; Hu, H.-Y. Advanced oxidation of dodecyl dimethyl benzyl ammonium chloride by VUV/UV/chlorine: Synergistic effect, radicals, and degradation pathway. *Sep. Purif. Technol.* **2022**, *292*, 121012. [CrossRef]
22. Nong, Y.J.; Zhang, Y.L.; Hübner, U.; Wang, W.L.; Wu, Q.Y.; Huang, N.; Drewes, J.E.; Hu, H.Y. Roles of radical species in vacuum-UV/UV/peroxydisulfate advanced oxidation processes and contributions of the species to contaminant degradation at different water depths. *J. Hazard. Mater.* **2023**, *446*, 130660. [CrossRef] [PubMed]
23. Huang, N.; Shao, W.-T.; Wang, W.-L.; Wang, Q.; Chen, Z.-Q.; Wu, Q.-Y.; Hu, H.-Y. Removal of methylisothiazolinone biocide from wastewater by VUV/UV advanced oxidation process: Kinetics, mechanisms and toxicity. *J. Environ. Manag.* **2022**, *315*, 115107. [CrossRef] [PubMed]
24. Wen, D.; Li, W.T.; Lv, J.R.; Qiang, Z.M.; Li, M.K. Methylene blue degradation by the VUV/UV/persulfate process: Effect of pH on the roles of photolysis and oxidation. *J. Hazard. Mater.* **2020**, *391*, 121855. [CrossRef]
25. Xue, H.H.; Gao, S.Y.; Li, M.; Wang, Y.; Liu, B.S. Performance of ultraviolet/persulfate process in degrading artificial sweetener acesulfame. *Environ. Res.* **2020**, *188*, 109804. [CrossRef]
26. Lou, F.; Qiang, Z.M.; Zou, X.; Lv, J.R.; Li, M.K. Organic pollutant degradation by UV/peroxydisulfate process: Impacts of UV light source and phosphate buffer. *Chemosphere* **2022**, *292*, 133387. [CrossRef]
27. Li, G.; Li, H.; Zhang, S.; Mi, X.; Guo, Y.; Song, Z.; Liu, Y.; Pan, H.; Wang, B.; Liu, Y.; et al. Effect of confined Mn oxides on regulating capability of activated coke for persulfate-based oxidation of a sweetener acesulfame. *Water Cycle* **2024**, *5*, 247–258. [CrossRef]
28. Yan, S.; Zhan, L.L.; Meng, X.Q.; Wang, D.A.Z.; Wang, X.M.; Zheng, G.Y.; Lu, J.H.; Zhou, L.X. Role of schwertmannite or jarosite in photocatalytic degradation of sulfamethoxazole in ultraviolet/peroxydisulfate system. *Sep. Purif. Technol.* **2021**, *274*, 118991. [CrossRef]
29. Li, C.F.; Peng, S.Q.; Li, H.Q.; Zhang, M.Y.; Yang, P. Methylisothiazolinone (MIT) degradation performance of an air-lift moving bed electro-Fenton reactor: Homogeneous and heterogeneous phases. *Sep. Purif. Technol.* **2025**, *378*, 134768. [CrossRef]
30. Wang, X.X.; Wang, W.L.; Dao, G.H.; Xu, Z.B.; Zhang, T.Y.; Wu, Y.H.; Hu, H.Y. Mechanism and kinetics of methylisothiazolinone removal by cultivation of Scenedesmus sp. LX1. *J. Hazard. Mater.* **2020**, *386*, 121959. [CrossRef]
31. Yang, Z.W.; Wang, W.L.; Lee, M.Y.; Wu, Q.Y.; Guan, Y.T. Synergistic effects of ozone/peroxymonosulfate for isothiazolinone biocides degradation: Kinetics, synergistic performance and influencing factors*. *Environ. Pollut.* **2022**, *294*, 118626. [CrossRef] [PubMed]

32. Wu, Z.D.; Yang, L.X.; Tang, Y.B.; Qiang, Z.M.; Li, M.K. Dimethoate degradation by VUV/UV process: Kinetics, mechanism and economic feasibility. *Chemosphere* **2021**, *273*, 129724. [CrossRef] [PubMed]
33. Yang, L.X.; Zhang, Z.H.; Chen, Z.M. Formation of nitrite and ammonium during the irradiation of nitrate-containing water by VUV/UV. *J. Water Process Eng.* **2021**, *40*, 101801. [CrossRef]
34. Mack, J.; Bolton, J.R. Photochemistry of nitrite and nitrate in aqueous solution: A review. *J. Photochem. Photobiol. A-Chem.* **1999**, *128*, 1–13. [CrossRef]
35. Zoschke, K.; Dietrich, N.; Börnick, H.; Worth, E. UV-based advanced oxidation processes for the treatment of odour compounds: Efficiency and by-product formation. *Water Res.* **2012**, *46*, 5365–5373. [CrossRef] [PubMed]
36. Li, J.Y.; Jing, B.Q.; Tian, Y.; Li, G.H.; Li, X.K. Electrochemical activation of peroxymonosulfate with commercial electrodes in the mixed reaction chamber: A cost-effective way for sulfamethazine degradation. *Desalination Water Treat.* **2025**, *323*, 101279. [CrossRef]
37. Lee, J.; von Gunten, U.; Kim, J.-H. Persulfate-Based Advanced Oxidation: Critical Assessment of Opportunities and Roadblocks. *Environ. Sci. Technol.* **2020**, *54*, 3064–3081. [CrossRef]
38. Zhang, W.C.; Zhang, Y.J.; Yu, L.J.; Gu, L.K.; Ji, J.T.; Zhang, G.Y.; Peng, Z.X. The effect of nano-Fe3O4 addition on the nitrogen transfer pathways and biomass morphology in floc-granule coexistence system. *J. Environ. Chem. Eng.* **2024**, *12*, 114831. [CrossRef]

sustainability

MDPI

Article

Activation of Persulfate by Sulfide-Modified Nanoscale Zero-Valent Iron Supported on Biochar for 2,4-Dichlorophenol Degradation: Efficiency, Sustainability, and Mechanism Investigation

Mu Wang [1,2,3], Yan Zhao [1], Zongsheng An [1] and Changming Dou [1,*]

1 Anhui Provincial Academy of Eco-Environmental Science Research, Hefei 230026, China; wangmucjl@163.com (M.W.); yanzhao.ahhky@gmail.com (Y.Z.); anzongsheng100@sina.com (Z.A.)
2 College of Civil and Hydraulic Engineering, Bengbu University, Bengbu 233030, China
3 School of Earth and Space Sciences, University of Science and Technology of China, Hefei 230026, China
* Correspondence: doucm@sina.com

Abstract

The activation of persulfate (PS) to oxidize and degrade 2,4-dichlorophenol (2,4-DCP) in aqueous solution represents a prevalent advanced oxidation technology. This study established a PS activation system using sulfide-modified nanoscale zero-valent iron supported on biochar (S-nZVI@BC). The optimal conditions included a PS:2,4-DCP mass ratio of 70:1 and S-nZVI@BC:PS of 1.5:1. The activator had excellent stability after being reused five times, which lead to high cost-effectiveness and sustainable usability. This system exhibited broad pH adaptability (3–11), with enhanced efficiency under acidic/neutral conditions. Chloride ion, nitrate, and carbonate had effects during the degradation. During the initial degradation phase, S-nZVI@BC played a primary role, with a greater contribution rate of adsorption than reduction. Fe^0 played a dominant role in the PS activation process; reactive species—including HO•, $SO_4•^-$, and $O_2•^-$—were identified as key agents in subsequent degradation stages. The overall degradation processes comprised three distinct stages: dechlorination, ring-opening, and mineralization.

Keywords: 2,4-dichlorophenol; sulfide modification; nanoscale zero-valent iron; persulfate; sustainable usability

1. Introduction

Chlorophenols (CPs) are a class of chlorinated aromatic compounds composed of benzene rings, hydroxyl groups, and chlorine atoms. Owing to their antibacterial properties and unique reactivity, CPs have found extensive application in agricultural and industrial sectors [1,2]. However, their high chemical stability and resistance to biodegradation lead to persistence in the natural environment, rendering them significant and widespread environmental pollutants [3]. Especially, 2,4-dichlorophenol (2,4-DCP) is a common persistent pollutant detected in agricultural production, chlorination-disinfected water, and paper mill wastewater [4,5]. Conventional methods for CPs' removal include adsorption, membrane separation, and biodegradation [6]. Nevertheless, these approaches exhibit limitations: adsorbents, upon saturation, require incineration or disposal, posing risks as secondary pollution sources [7]; membrane separation proves inefficient for treating wastewater with high CP concentrations; and the biodegradation processing cycle is long [1].

Advanced oxidation processes (AOPs) currently represent more effective technologies for treating toxic organic pollutants [8]. They utilize catalytic methods—such as light, ultrasound, electricity, and catalysts—to activate various oxidants, generating active substances with high redox potential. These enable the efficient degradation, even mineralization of recalcitrant organic pollutants. Common oxidants incorporate ozone (O_3), hydrogen peroxide (H_2O_2), and persulfate (PS) [8–10]. Among them, the PS-based advanced oxidation process (PS-AOP) has garnered significant attention in recent years [11]. The PS oxidation technology was proposed after experimental verification [12]. During the electrolysis process, sulfate can generate PS, which has oxidizing ability. Compared to O_3 and H_2O_2, PS offers distinct advantages: a wider variety of activation modes [13], simultaneously producing multiple reactive oxygen species (ROS) during the activation process, including sulfate radical ($SO_4\bullet^-$), hydroxyl radical (HO•), and superoxide radical ($O_2\bullet^-$) [14], higher free radical yields [15], lower dependence on environmental conditions [16], and existing in solid state makes it more conducive for storage and transportation [17].

PS has certain oxidation ability, but the degradation efficiency is relatively low. Generally, it needs to be activated through different methods for generating more active substances with higher redox potential [18]. Iron-based materials have been extensively investigated as activators for PS due to their environmentally friendly nature, relative low toxicity, and cost-effectiveness [19,20]. However, nZVI exhibits low efficiency as a kind of PS activator, primarily attributed to particle aggregation, surface passivation, and inefficient electron transfer [21]. To address these deficiencies, various nZVI modification techniques have been explored. Among them, biochar (BC) loading and sulfide modification have been proven particularly effective [21–24].

BC is the thermal decomposition product of waste biomass under the anoxic condition and is a stable compound rich in carbon [25]. The BC-supported nZVI composite has gained significant attention owing to its high contaminant adsorption capacity, good electrical conductivity, and enhanced nZVI reactivity. The properties and preparation conditions of BC (such as pore structure, functional group, raw material composition, and pyrolysis temperature) are of crucial importance in determining the properties of BC-supported nZVI [21,22].

On the other hand, sulfide-modified iron-based materials are easy to prepare. Depending on the form of sulfur, sulfur-iron compounds such as FeS, Fe_3S_4, FeS_2, and Fe_9S_{11} can be obtained. Among them, FeS is more common [26]. Sulfide modification enhances the hydrophobicity and salt resistance of nZVI, reduces electron transfer resistance to contaminants, blocks atomic hydrogen adsorption sites, and inhibits the water reduction reaction and consequent H_2 evolution [23,24].

Despite possessing these advantages, research concerning activation methods and reaction mechanisms in combination utilizing sulfide-modified and BC-supported iron-based materials remains relatively scarce. For instance, Chen et al. [27] demonstrated the feasibility and mechanism of S-nZVI@BC for trichloroethylene (TCE) removal in groundwater remediation, and their results indicated that S-nZVI@BC, synergistically combining the high adsorption capacity of BC with the strong reductive capacity of S-nZVI, outperformed either component alone. Similarly, Xie et al. [28] synthesized S-nZVI@BC and specifically investigated the influencing factors with activation mechanisms involved in the PS degradation of 2-chlorophenol; they found that S-nZVI@BC had excellent stability and sustainability, could be reused multiple times, and was worthy of in-depth research.

When using degradable materials for pollutants, it is very important to conduct an analysis regarding their environmental friendliness and sustainability. Kayode et al. [29] employed a life-cycle assessment (LCA) to evaluate the environmental impact of coupling hydrothermal carbonization (HTC) with anaerobic digestion (AD) compared to conventional AD treatment. HTC degraded wastewater sludge in an aqueous medium, producing carbon-dense hydrochar while reducing sludge volumes. Elif et al. [30] presented a foundational overview of LCA principles and described a systematic, step-by-step procedure for its effective application. Additionally, this article revisited the fundamental concepts of carbon footprint (CF) analysis as a complementary tool for quantifying greenhouse gas emissions associated with products and activities.

In this study, S-nZVI@BC composites were synthesized using different composition ratios. The effects of key operational parameters on degradation efficiency were systematically investigated by using the method of controlling variables, including the following: the mass ratio of PS:2,4-DCP and S-nZVI@BC:PS, initial pH, reaction temperature, and the concentrations of specific background ions. Component contribution analysis quantified the adsorption and reduction in S-nZVI@BC according to the 2,4-DCP removal rate at the initial stage of the reaction. Furthermore, the degradation mechanism and the relative contributions of various reactive species were elucidated using oxidant scavenger experiments and electron spin resonance (ESR) spectroscopy.

2. Materials and Methods

2.1. Chemicals and Reagents

All chemical reagents were of analytical grade or higher and used as received without further purification. Iron nitrate nonahydrate ($Fe(NO_3)_3{\bullet}9H_2O \geq 98.5\%$), sodium sulfide ($Na_2S{\bullet}9H_2O \geq 98.5\%$), 1,10-phenanthroline ($C_{12}H_8N_2{\bullet}H_2O \geq 98.0\%$), and tert-butyl alcohol (TBA, $C_4H_{10}O \geq 98.0\%$) were supplied by Sinopharm Chemical Reagent Co., Ltd. (Shanghai, China). Sodium persulfate (PS, $Na_2S_2O_8 \geq 99.0\%$), 2,4-dichlorophenol (2,4-DCP, $C_6H_4Cl_2O \geq 99.0\%$), potassium borohydride (PBH, $KBH_4 \geq 99.0\%$), 5, 5-Dimethyl-1-pyrroline-N-oxide (DMPO, $C_6H_{11}NO \geq 99.0\%$), iron sulfate heptahydrate ($FeSO_4{\bullet}7H_2O \geq 99.0\%$), and p-benzoquinone (BQ, $C_6H_4O_2 \geq 99.0\%$) were obtained from Shanghai Macklin Biochemical Co., Ltd. (Shanghai, China). Hydrochloric acid (HCl, 36.0–38.0%), sulfuric acid (H_2SO_4 95.0%–98.0%), sodium hydroxide ($NaOH \geq 99.0\%$), sodium chloride ($NaCl \geq 99.5\%$), sodium carbonate ($Na_2CO_3 \geq 99.5\%$), and sodium nitrate ($NaNO_3 \geq 99.5\%$) were sourced from Tianjin Kermel Chemical Reagent Co., Ltd. (Tianjin, China). Methanol (MeOH, $CH_4O \geq 99.9\%$) was a guaranteed reagent provided by Shanghai Weston Trading Co., Ltd. (Shanghai, China). Rice husk was purchased from a green plant base in Hefei, China. The ultrapure water used throughout the experiments was produced by a Synergy UV water purification system (Merck Millipore, Grenoble, France).

2.2. Synthesis of S-nZVI@BC

The synthesis of S-nZVI@BC included BC preparation, a reaction in the liquid phase to obtain nZVI and sulfide-modification. Rice husk was chosen as the material for preparing BC as a common agricultural waste. It was collected and dried under 105 °C in the air-dry oven for 24 h before being used. The dried material was then placed in an OTF-1200X type tube-heating furnace. The pyrolysis process was carried out under a constant nitrogen purge. The temperature was first ramped from ambient to 600 °C at a rate of 10 °C/min, held at 600 °C for 2 h, and dropped to room temperature in four hours. The final BC product was ground into a powder and sieved (50-mesh) for subsequent use.

S-nZVI@BC was synthesized in a two-step method. Quantitative BC was added to a quantitative $Fe(NO_3)_3$ solution of 200 mL, stirring constantly for 30 min with constant nitrogen purge, and then the suspension was subsequently treated by the dropwise addition of 200 mL KBH_4 solution under stirring. After 10 min of reaction, 110 mL of Na_2S solution was introduced, followed by 30 min of ultrasonication. The resulting solid product was then collected by filtration and thoroughly washed with ultrapure water until the filtrate conductivity fell below 50 μS/cm. Lastly, the material was dried in a freeze dryer (the temperature was set to −50 °C) and the S-nZVI@BC was obtained after 24 h. The mass ratio of PBH and $Fe(NO_3)_3 \bullet 9H_2O$ was 1.3:1 when preparing the reaction reagents, and the additive amounts of BC, $Fe(NO_3)_3 \bullet 9H_2O$, and $Na_2S \bullet 9H_2O$ were regulated to make the mass ratio of Fe to C approximately 1:2 and S to Fe approximately 1:4, respectively.

2.3. Experimental Processes

All reactions were conducted in 500 mL flasks containing 400 mL of 2,4-DCP solution. S-nZVI@BC and PS were added at specified dosages, with the total volume adjusted to 500 mL using ultrapure water. The degradation experiments were conducted in a thermostatic reciprocating shaker (SHA-B, Shanghai, China) at 180 rpm and set temperatures (10, 30, and 50 °C). Liquid samples were withdrawn at predetermined time intervals (10, 30 min, 1, 2, 4, 8, 16, 24, 48, 72 h) and immediately filtered through 0.22 mm membranes. Then, 0.5 mL of solution was transferred into the special sample vial and capped for determination after adding 0.5 mL MeOH to stop the reaction.

The material of BC was rice husk, while the pyrolysis temperature value was 600 °C, the mass ratio of S:Fe:C = 1:4:8, and the initial 2,4-DCP concentration was 20 mg/L. A series of batch experiments were conducted to determine the optimal reaction conditions and evaluate the influence of key parameters. The investigated variables and their ranges were as follows: the mass ratio of PS to 2,4-DCP (0, 10, 20, 30, 40, 50, 60, 70, and 80), the mass ratio of S-nZVI@BC to PS (0, 0.2, 0.4, 0.6, 0.8, 1.0, 1.2, 1.5, and 2.0), the initial pH (3.0, 5.0, 7.0, 9.0, and 11.0), the reaction temperature (10 °C, 30 °C, and 50 °C), and the concentration of background anions (chloride ion, nitrate ion, carbonate ion; 0, 50, 100, 200, and 500 mg/L). All experiments were performed in duplicate, and the reported data represent the average of the measurements, and were analyzed mainly by Excel 2016 and Origin 2021.

Component contribution analysis was applied to analyze the actions of adsorption and reduction according to the mass ratio of BC, S-nZVI, and nZVI in the S-nZVI@BC. Four ingredients were added to the 2,4-DCP solution with 20 mg/L of 400 mL, which were S-nZVI@BC 0.84 g, BC 0.517 g, S-nZVI 0.177 g, and nZVI 0.146 g (the dosages of ingredients were calculated by the mass ratio of S:Fe:C = 1:4:8, and FeS was regarded as the main component of S-nZVI). The actions of adsorption reduction were quantified by determinating the degradation of 2,4-DCP at predetermined time intervals of 10, 30 min, 1, 2, 4, 8, 16, 24, 48, 72 h. Then, to desorb the 2,4-DCP from the ingredients, the pH of the solution was adjusted to 12 by adding 1 mM NaOH. After 5 min of violent shaking, 2,4-DCP was measured in the solution after immediate filtration, and the difference value before and after desorption could be used to quantify the adsorption capacity of materials [31].

2.4. Analytical Methods

The concentration of 2,4-DCP was determined by high performance liquid chromatograph (HPLC) (Thermo Scientific U3000, Thermo Scientific, Waltham, MA, USA) with the Athena C18-WP column (250 mm × 4.6 mm, 5 μm). The following operating conditions were employed: column temperature of 40 °C, detection wavelength of 254 nm, a mobile phase ratio of V(water):V(MeOH) = 3:7, a flow rate of 1.0 mL/min, and a sample injection volume of 10 μL. The pH and conductivity of aqueous solutions were measured using a PHBJ-260 portable pH meter and a DDS-307 conductivity meter, respectively.

The physicochemical properties of S-nZVI@BC before and after reactions were characterized as follows. The Brunauer–Emmett–Teller (BET) surface areas were measured by N_2 adsorption using a speci-tellfic surface area and mesoporous distribution analyzer (Micromeritics Tristar II 3020, Micromeritics, Atlanta, GA, USA). The surface features used a scanning electron microscope (SEM) with Energy Disperse Spectroscopy (EDS) (GeminiSEM 500, ZEISS, Oberkochen, Germany). X-ray photoelectron spectroscopy (XPS) data were obtained with an X-ray photoelectron spectrometer (Thermo Scientific ESCALAB 250Xi, Thermo Scientific, Waltham, MA, USA). X-ray diffraction (XRD) analysis was conducted with a multifunctional rotating target X-ray diffractometer (Rigaku SmartLab, Tokyo, Japan). The functional group analysis was obtained with a fourier transform infrared spectrometer (FTIR) (Nicolet 8700, Thermo Scientific, Waltham, MA, USA).

Electron spin resonance (ESR) spectroscopy (JES-FA200, JEOL Ltd., Tokyo, Japan) was employed to identify reactive oxygen species (HO•, $SO_4 \bullet^-$ and $O_2 \bullet^-$) generated in the reaction systems. ESR measurements were conducted at the University of Science and Technology of China under the following conditions: temperature maintained at 25 °C using a circulating water system, a microwave frequency set at 9.84 GHz (X-band), and 5,5-dimethyl-1-pyrroline-N-oxide (DMPO) used as the spin-trapping agent. HO• and $SO_4 \bullet^-$ were detected in pure water solution, and $O_2 \bullet^-$ was detected with MeOH reagent (v:v = 1:1). Spectrometer parameters were as follows: a microwave frequency of 8.75–9.65 GHz (X-band), central magnetic field of 3245.0 G, microwave energy of 2 mW, modulation frequency of 100 kHz, scanning width of 50 G, time constant of 0.03 s, and scanning frequency of once every 30 s.

3. Results and Discussion

3.1. Characterization of Prepared Materials

The basic structures of BC, S-nZVI, and S-nZVI@BC before and after reaction were analyzed using SEM, the element compositions of them were obtained by EDS, and the results are presented in Figure 1a–d and Figure S1, respectively. As shown in Figure 1a, BC exhibited a smooth surface and provided a honeycomb structure for S-nZVI, with carbon and oxygen being the predominant elements (Figure S1a). Figure 1b reveals that S-nZVI possessed a flocculent agglomeration structure with numerous pores, primarily composed of iron, oxygen, carbon, and sulfur (Figure S1b). Figure 1c demonstrates that S-nZVI nanoparticles were densely packed onto the BC surface, nearly covering it completely. Compared to BC, the increased proportions of iron and sulfur elements (Figure S1c) indicate the successfully loading of these elements on the BC surface during S-nZVI@BC preparation. Following the reaction, Figure 1d shows a reappearance of a smooth carbon skeleton on the surface of S-nZVI@BC. This observation, coupled with a reduction in the proportions of iron and sulfur (Figure S1d), suggests the consumption of these elements during the reaction. Collectively, these results demonstrate that BC provided abundant binding sites, effectively mitigating S-nZVI agglomeration, and consequently enhancing the activation of PS by S-nZVI to generate free radicals.

The crystal structures of S-nZVI@BC materials before and after reaction were analyzed by XRD. Figure 1e presents the obtained XRD patterns within the 2θ range of 10° to 80°. The spectra for both fresh and reacted S-nZVI@BC exhibit a diffraction peak at approximately 2θ = 22.5° and 22.6°, respectively, corresponding to the characteristic peak of rice husk BC. A distinct characteristic peak of Fe^0 is observed at 2θ = 44.92° in both spectra, confirming the successful reduction in Fe^{3+} to Fe^0 during material synthesis. Characteristic peaks indicative of FeS are also evident at 2θ = 35.46°(fresh) and 35.32°(reacted), demonstrating the presence of FeS and successfully sulfidation modification of nZVI [32]. Finally, weak diffraction peaks appeared at 2θ = 61.92°(fresh) and 62.12°(reacted), suggesting that an iron oxide protective film was formed on the material surface. This observation is consistent with the increased oxygen content detected in Figure S1c,d.

Functional group analysis of S-nZVI@BC before and after reaction was conducted using FT-IR spectroscopy. Figure 1f displays the obtained spectra within the wavenumber range of 400–4000 cm^{-1}. The most prominent absorption band at ~3412 cm^{-1} corresponds to the O–H stretching vibrations of hydroxyl groups on the BC surface. A distinct band at ~1385 cm^{-1} arises from the stretching vibrations of other BC surface functional groups. Post-reaction, both bands exhibited reduced intensity relative to fresh S-nZVI@BC, indicating hydroxyl group participation in degradation reactions and possible coverage by precipitated iron hydroxides. The band at ~1627 cm^{-1}, due to C = O stretching in carboxyl groups (–COOH), maintained consistent intensity before and after reaction, suggesting minimal involvement of carboxyl groups in 2,4-DCP degradation. The C–O vibration band at ~1104 cm^{-1} [33] intensified following the reaction, implying increased carbon–oxygen bonding. This likely results from oxygen incorporation and iron leaching during degradation, consistent with EDS elemental analysis. A new band emerged at ~568 cm^{-1} in reacted samples, potentially assignable to either Fe–O vibrations (from surface Fe^0 oxidation) or sulfide-related bonds, confirming successful material sulfidation.

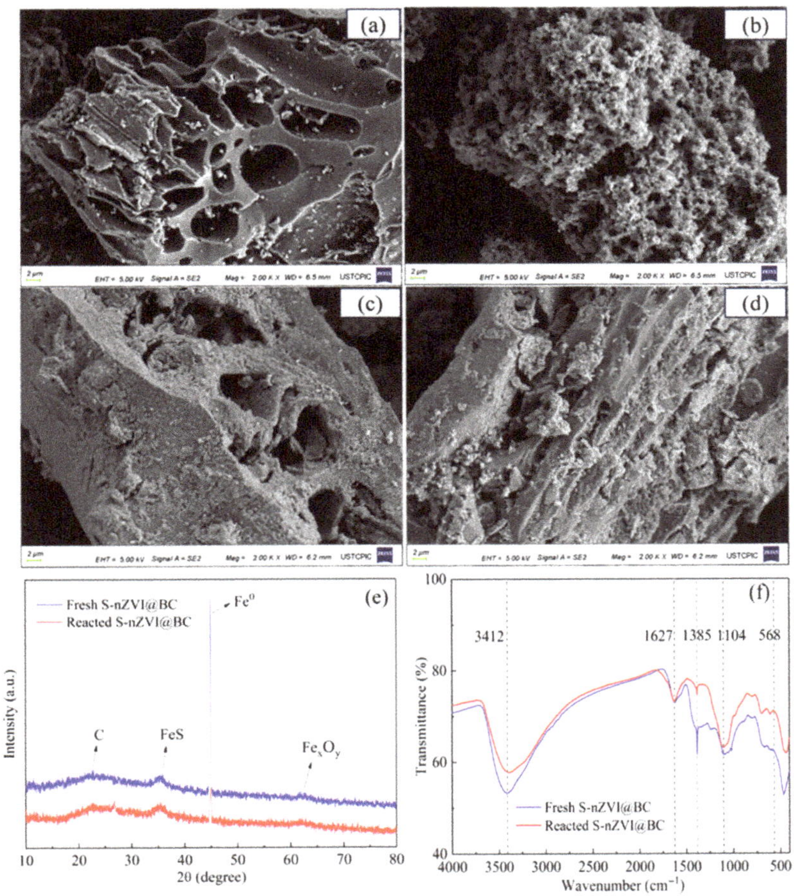

Figure 1. SEM images of (**a**) BC, (**b**) S-nZVI, (**c**) fresh S-nZVI@BC, and (**d**) reacted S-nZVI@BC; the magnifications of the SEM images are all 2.00 k×. (**e**) XRD analysis and (**f**) infrared spectra of fresh S-nZVI@BC and reacted S-nZVI@BC. Reaction condition: PS = 1400 mg/L, [2,4-DCP] = 20 mg/L of 400 mL, BC@S-nZVI = 2100 mg/L, initial pH = 9 ± 0.2, T = 30 °C.

To investigate the adsorption properties of materials through their specific surface area and pore structure characteristics, BET analysis was conducted on dehydrated S-nZVI, BC, and S-nZVI@BC composites before and after reaction. The results are shown in Table S1. BC exhibited the largest BET surface area (351.76 m^2/g) and total pore volume (0.29 cm^3/g) among all samples. These values exceeded those of pristine S-nZVI, fresh S-nZVI@BC, and reacted S-nZVI@BC. The reduction in both parameters for S-nZVI@BC is attributed to pore blockage by S-nZVI particles during composite formation. Post-reaction, the partial consumption of S-nZVI liberated occupied pore spaces, resulting in an increased surface area and pore volume relative to fresh composite [34].

3.2. *Effects of Different Degradation Systems*

Determining optimal oxidant and activator dosages constitutes fundamental research in organic pollutant degradation systems [28,35]. The effects of reagent dosage were evaluated by varying the mass ratios of PS to 2,4-DCP and S-nZVI@BC to PS, with the corresponding degradation efficiencies shown in Figure 2a,b. Figure 2a demonstrates the influence of the PS:2,4-DCP mass ratio (where C_t and C_0 represent 2,4-DCP concentrations at time t and initial state, respectively). At a fixed activator concentration, increasing the PS:2,4-DCP ratio from 0 to 70 enhanced 2,4-DCP removal from 10.9% to 97.4% within 72 h. This improvement is attributed to enhanced radical generation via S-nZVI@BC-mediated PS activation (Equations (1)–(6)) [36], where a higher PS dosage promotes oxidative species production. However, further increasing the ratio to 80 reduced the removal efficiency to 91.7%, likely due to radical scavenging through $SO_4\bullet^-$ self-quenching or reaction with excess PS (Equations (7) and (8)). Consequently, a PS:2,4-DCP of 70 was the optimum mass ratio for subsequent experiments.

$$Fe^0 + S_2O_8^{2-} \rightarrow Fe^{2+} + 2SO_4^{2-} \tag{1}$$

$$2Fe^0 + O_2 + 2H_2O \rightarrow 2Fe^{2+} + 4OH^- \tag{2}$$

$$Fe^{2+} + S_2O_8^{2-} \rightarrow Fe^{3+} + SO_4^{2-} + SO_4\bullet^- \tag{3}$$

$$Fe^0 + S_2O_8^{2-} + 2H_2O \rightarrow Fe^{2+} + 2HO\bullet + 2SO_4^{2-} + 2H^+ \tag{4}$$

$$Fe^0 + 2S_2O_8^{2-} \rightarrow Fe^{2+} + 2SO_4\bullet^- + 2SO_4^{2-} \tag{5}$$

$$2H_2O + 2S_2O_8^{2-} \rightarrow O_2\bullet^- + 4H^+ + 3SO_4^{2-} + SO_4\bullet^- \tag{6}$$

$$SO_4\bullet^- + SO_4\bullet^- \rightarrow S_2O_8^{2-} \tag{7}$$

$$SO_4\bullet^- + S_2O_8^{2-} \rightarrow SO_4^{2-} + S_2O_8\bullet^- \tag{8}$$

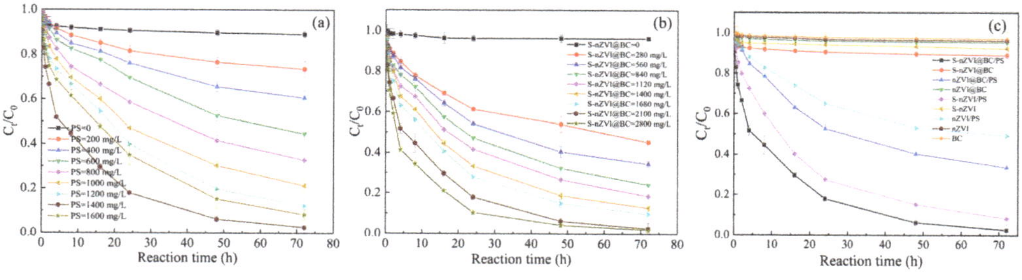

Figure 2. Effect of different activation systems on the degradation of 2,4-DCP: (**a**) different mass concentration of PS with fixed value [S-nZVI@BC] = 2100 mg/L, (**b**) different mass concentration of S-nZVI@BC with fixed value [PS] = 1400 mg/L, (**c**) different reaction systems with mass ratio of PS to 2,4-DCP was 70, S-nZVI@BC, nZVI@BC, S-nZVI, and nZVI to PS were 1.5. Reaction condition: [2,4-DCP] = 20 mg/L of 400 mL, initial pH = 9 ± 0.2, T = 30 °C.

Figure 2b illustrates the mass ratio of S-nZVI@BC:PS's influence on 2,4-DCP degradation. At a fixed PS concentration, increasing the S-nZVI@BC:PS ratio from 0 to 1.5 enhanced 2,4-DCP removal from 3.8% to 97.4% within 72 h. A further increase to 2.0 with the degradation rate only improved slightly (98.0%), indicating that the activation of PS by S-nZVI@BC reached a saturated state. This plateau effect likely results from the excessive Fe^{2+} release at a high activator dosage, which scavenges $SO_4\bullet^-$ via oxidative consumption Equation (9) [28], thereby reducing radical availability for pollutant degradation. Consequently, the optimal S-nZVI@BC:PS mass ratio was determined as 1.5 for subsequent studies.

$$Fe^{2+} + SO_4\bullet^- \rightarrow Fe^{3+} + SO_4^{2-} \quad (9)$$

Figure 2c compares 2,4-DCP degradation efficiencies across nine relevant systems under identical conditions. After 72 h, the removal rates followed the order: S-nZVI@BC/PS (97.4%) > S-nZVI/PS (91.7%) > nZVI@BC/PS (66.3%) > nZVI/PS (50.5%) ≫ S-nZVI@BC (10.9%) ≈ S-nZVI (7.6%) ≈ nZVI@BC (4.6%) ≈ nZVI (3.8%) ≈ BC (2.8%).

Notably, systems lacking PS activation exhibited minimal removal (<11%), while nZVI@BC/PS and nZVI/PS achieved only 50.5–66.3% efficiency. In contrast, both sulfide modification systems (S-nZVI@BC/PS and S-nZVI/PS) demonstrated significantly enhanced degradation, with S-nZVI@BC/PS achieving near-complete pollutant removal (97.4%). These results confirm that sulfide modification and BC support synergistically activate PS, where there is a higher degradation rate after sulfide modification due to the improved nZVI hydrophobicity and electron transfer capacity [27,28]. Consequently, the S-nZVI@BC/PS system represents an efficient PS-based AOP for 2,4-DCP wastewater remediation.

3.3. Effects of Different Influencing Factors

The reusability of S-nZVI@BC in PS activation was evaluated through five consecutive 2,4-DCP degradation cycles (72 h each). Post-reaction, the catalyst underwent triple washing with deionized water/methanol, followed by freeze-drying for reuse. As depicted in Figure 3a, 2,4-DCP removal decreased from 97.4% (cycle 1) to 74.0% (cycle 5), confirming retained catalytic activity after multiple cycles.

Notably, efficiency declined rapidly during initial cycles (cycles 1–3) but stabilized subsequently. This biphasic deactivation pattern is attributed to the following: (i) Irreversible adsorption of degradation intermediates on active sites, persisting despite solvent washing [36]. (ii) Progressive iron precipitation reducing accessible Fe^{2+} species and active component content [37]. Despite these mechanisms, S-nZVI@BC demonstrates robust stability in PS activation for sustained 2,4-DCP degradation.

The energy required for recycling materials was lower than that needed for reproducing, and it also reduced the usage of related raw materials and lead to widespread environmental and economic benefits.

According to the calculation, oxidation degradation one ton 2,4-DCP with concentration of 20 mg/L, the cost of using H_2O_2 is CNY 8640. Not including the equipment cost for preparing O_3, the cost is CNY 2716, and the cost of the S-nZVI@BC/PS system is CNY 4860. The cost of an industrial O_3 generator is approximately several tens of thousands of yuan. Moreover, the preparation of O_3 requires a large amount of electricity. After comprehensive analysis, the S-nZVI@BC/PS system is still more energy-efficient and environmentally friendly.

Under optimized conditions, the key parameters on 2,4-DCP degradation effects are presented in Figure 3b–f. Batch experiments evaluating pH influence (Figure 3b) revealed distinct kinetic patterns: complete removal (100%) occurred at 4 h (pH = 3), 48 h (pH = 5), and 72 h (pH = 7), and efficiency declined to 91.1% at pH = 11.

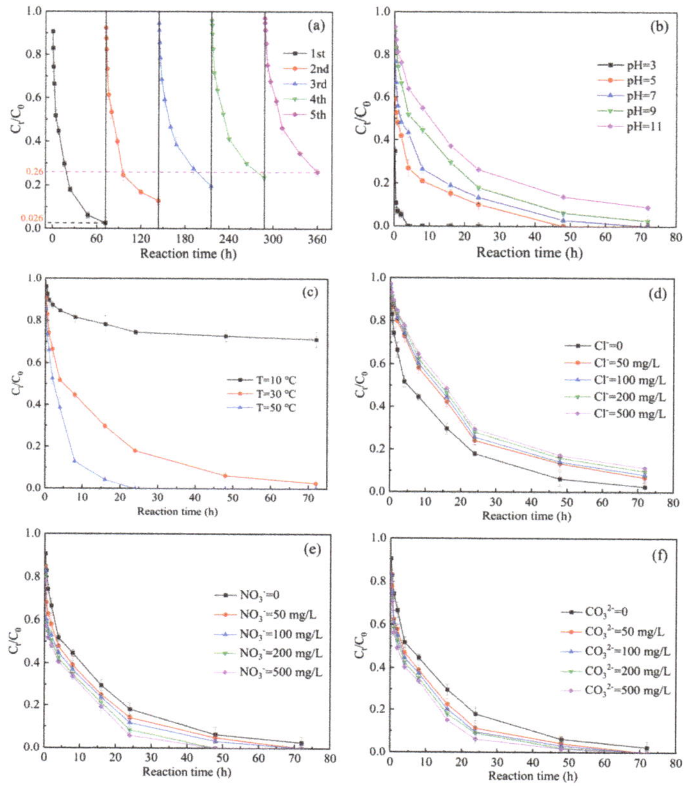

Figure 3. Effect of different influence factors on the degradation of 2,4-DCP (**a**) recycle tests of S-nZVI@BC, (**b**) pH, (**c**) temperature, (**d**) Cl^-, (**e**) NO_3^-, (**f**) CO_3^{2-}. Reaction condition: PS = 1400 mg/L, [2,4-DCP] = 20 mg/L of 400 mL, BC@S-nZVI = 2100 mg/L, initial pH = 9 ± 0.2, T = 30 °C.

This demonstrates effective PS activation by S-nZVI@BC across acidic to neutral pH (3–7), outperforming alkaline conditions. The enhanced degradation effects at a lower pH are attributed to the following: (i) Accelerated $SO_4\bullet^-$ generation via facilitated PS decomposition (Equation (10)). (ii) Reduced activation energy for radical production. (iii) Suppressed formation of passivating iron hydroxide layers. Conversely, alkaline conditions (pH = 11) promote the following: active site occlusion by $Fe(OH)_x$ precipitates, $SO_4\bullet^-$ conversion to less oxidative $\bullet OH$ (E^0 = 1.9–2.7 V vs. 2.5–3.1 V), and Equations (11) and (12) collectively diminishing degradation efficiency [38]. Furthermore, acidic conditions lower the activation energy barrier for PS activation, accelerating radical generation kinetics and thereby enhancing 2,4-DCP degradation efficiency [39].

$$S_2O_8^{2-} + H^+ \rightarrow SO_4\bullet^- + SO_4^{2-} + H^+ \tag{10}$$

$$Fe^{3+} + 3OH^- \rightarrow Fe(OH)_3 \downarrow \tag{11}$$

$$SO_4\bullet^- + OH^- \rightarrow SO_4^{2-} + HO\bullet \tag{12}$$

Figure 3c demonstrates the temperature dependence of 2,4-DCP degradation. A dramatic reduction in removal efficiency occurred with decreasing temperature (50 °C → 10 °C), evidenced by a complete pollutant elimination within 16 h at 50 °C versus merely 23.7%

removal after 72 h at 10 °C. This aligns with the established literature confirming enhanced persulfate activation kinetics and contaminant degradation at elevated temperatures [40].

The influence of anions on 2,4-DCP degradation presented in the S-nZVI@BC/PS system. Chloride ion (Cl^-) exhibited inhibitory effects (Figure 3d), reducing the removal efficiency from 97.4% (no Cl^-) to 88.6% at 500 mg/L within 72 h. This attenuation is attributed to radical scavenging mechanisms: Cl^- carries out a reaction with $SO_4\bullet^-$ to form less reactive chlorine radicals (Cl•, Equation (13)) [41], which exhibit a lower oxidative capacity toward phenolic compounds.

Conversely, both nitrate (NO_3^-) and carbonate (CO_3^{2-}) enhanced degradation kinetics (Figure 3e,f). At 500 mg/L concentration, NO_3^- achieved complete removal (100%) within 48 h (vs. 97.4% control) and CO_3^{2-} reached 100% removal within 72 h. This acceleration aligns with reports [42,43] on secondary radical generation ($NO_3\bullet/CO_3\bullet^-$ via Equations (14) and (15)), which rapidly oxidize phenolic contaminants at elevated concentrations.

$$SO_4\bullet^- + Cl^- \rightarrow SO_4^{2-} + Cl\bullet \quad (13)$$

$$SO_4\bullet^- + NO_3^- \rightarrow SO_4^{2-} + NO_3\bullet \quad (14)$$

$$SO_4\bullet^- + CO_3^{2-} \rightarrow SO_4^{2-} + CO_3\bullet^- \quad (15)$$

3.4. Analysis of XPS

XPS was employed to probe iron speciation and binding energy evolution in S-nZVI@BC before and after degradation (Figure 4). The Fe2p 3/2 high-resolution spectrum of fresh material (Figure 4a) revealed three deconvoluted peaks at 706.1 eV (Fe^0), 710.9 eV [Fe(II)], and 713.3 eV [Fe(III)], confirming the successful reduction in Fe^{3+} to Fe^0 during synthesis.

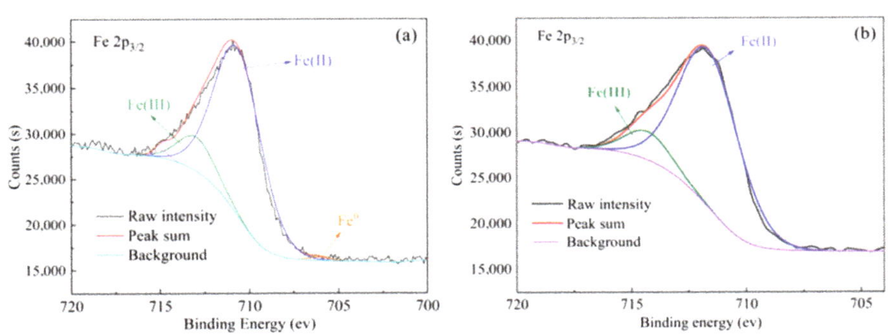

Figure 4. XPS analysis of (**a**) fresh S-nZVI@BC and (**b**) reacted S-nZVI@BC. Reaction condition: PS = 1400 mg/L, [2,4-DCP] = 20 mg/L of 400 mL, BC@S-nZVI = 2100 mg/L, initial pH = 9 ± 0.2, T = 30 °C.

Post-reaction (Figure 4b), the Fe2p 3/2 spectrum exhibited two dominant components at 711.9 eV [Fe(II)] and 714.6 eV [Fe(III)], indicating the surface oxidation and depletion of Fe^0 during 2,4-DCP degradation. Survey spectra (Figure S2) further verified the coexistence of Fe, O, C, and S elements through characteristic Fe 2p, O 1s, C 1s, and S 2p photoelectron peaks [44].

3.5. Degradation Mechanisms Analysis

The mechanisms of S-nZVI@BC on 2,4-DCP includes the adsorption of BC and reductive de-chlorination of S-nZVI and nZVI [27,37]. Component contribution analysis

(Figure 5a) quantified these mechanisms according to the 2,4-DCP removal rate. The sum of BC adsorption, S-nZVI reduction, and nZVI reduction was greater than the actions of S-nZVI@BC, which simultaneously includes adsorption and reduction; it meant that the preparation of the S-nZVI@BC was not a simple mixture of several components, but S-nZVI loaded on the BC as a whole. The reactive sites and pores of BC were filled; therefore, the overall performance of the S-nZVI@BC was lower than the sum of the removal effects of each raw material. Furthermore, the removal rate of S-nZVI@BC on 2,4-DCP increased from 4.5% of 10 min to 10.9% of 72 h, in which the action of adsorption was more dominant than the action of reduction, increasing from 2.9% after 10 min to 7.5% after 72 h.

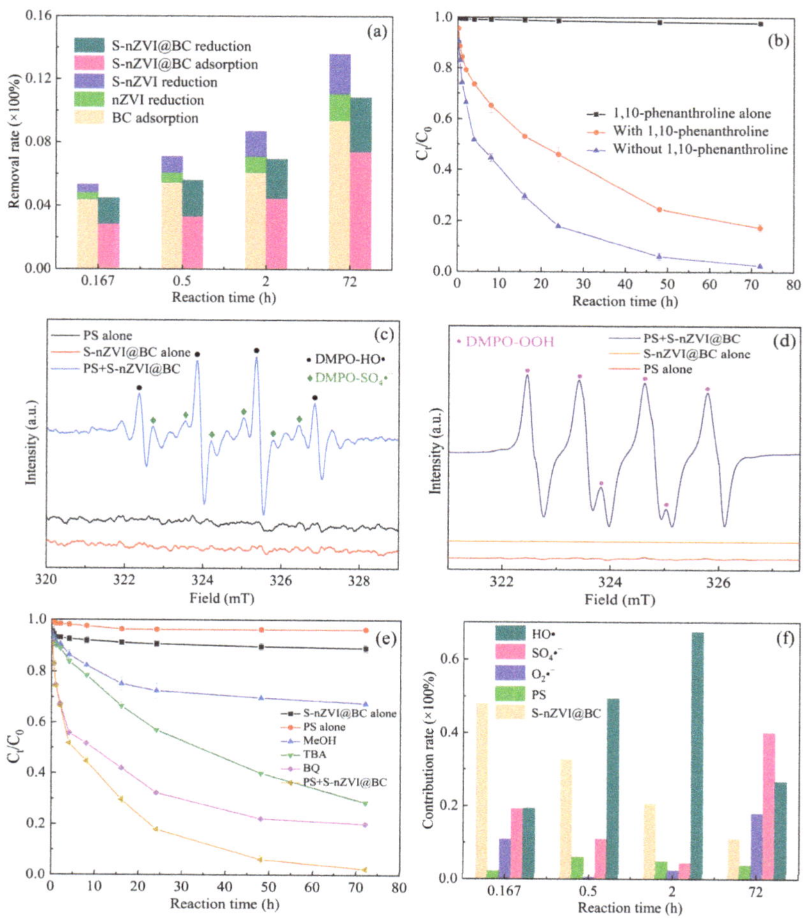

Figure 5. (**a**) Analysis of S-nZVI@BC ingredients' adsorption and reduction; (**b**) effect of 1,10-phenanthroline on inactivation of Fe^{2+}; (**c**) detection of HO• and $SO_4•^-$, (**d**) detection of $O_2•^-$; (**e**) removal of 2,4-DCP in different reaction systems; (**f**) contribution rates of five chemical factors at discrete time intervals. Reaction condition: PS = 1400 mg/L, [2,4-DCP] = 20 mg/L of 400 mL, BC@S-nZVI = 2100 mg/L, initial pH = 9 ± 0.2, T = 30 °C.

The persulfate activation mechanism in S-nZVI@BC/PS systems involves three iron species: (i) solid-phase Fe^0, (ii) dissolved Fe^{2+}, and (iii) surface-bound Fe^{2+}. To delineate their respective contributions, 1,10-phenanthroline (0.258 g, equimolar to theoretical iron content) was employed as a Fe^{2+} scavenger [45]. Figure 5b reveals the following: Control:

97.4% 2,4-DCP removal (72 h), Scavenger in solution: 82.5% removal (14.9% decrease), Scavenger alone: 2.2% removal (negligible adsorption). This demonstrates that Fe^0-mediated activation dominates (>80% contribution), while homogeneous/heterogeneous Fe^{2+} accounts for only 14.9% of the activation capacity. These findings align with Jin et al.'s report establishing solid Fe^0 as the primary PS activation pathway [32].

ESR spectroscopy with DMPO spin trapping was employed to identify reactive oxygen species (ROS) in the S-nZVI@BC/PS reaction system. As shown in Figure 5c, the characteristic signals of DMPO-HO• (1:2:2:1 quartet) and DMPO-$SO_4{\bullet}^-$ were exclusively detected in the complete system, absent in PS-only and S-nZVI@BC-only controls. The attenuated DMPO-$SO_4{\bullet}^-$ signal intensity is attributable to its transient nature and lower trapping efficiency [46]. Similarly, Figure 5d confirms $O_2{\bullet}^-$ generation through DMPO-OOH adduct signatures [47]. These results collectively demonstrate the co-generation of HO•, $SO_4{\bullet}^-$, and $O_2{\bullet}^-$ radicals during 2,4-DCP degradation, consistent with the PS activation mechanisms reported for iron-based systems [36,48].

Radical quenching experiments were performed to quantify the contributions of key species (S-nZVI@BC, PS, HO•, $SO_4{\bullet}^-$ and $O_2{\bullet}^-$) to 2,4-DCP degradation. MeOH, TBA, and BQ served as scavengers for HO•/$SO_4{\bullet}^-$, HO•, and $O_2{\bullet}^-$, respectively, dosed at a 10:1 mass ratio relative to PS. Compared to the S-nZVI@BC/PS reaction system as shown in Figure 5e, the degradation efficiency at 72 h decreases from 97.4% to 10.9% in the presence of S-nZVI@BC alone, to 3.8% in the presence of PS alone, to 32.3% in the presence of MeOH, to 71.4% in the presence of TBA, and to 79.9% in the presence of BQ. Correspondingly, the contribution of various factors to the degradation of 2,4-DCP were different.

The relative contributions of five chemical factors were quantified by comparing 2,4-DCP removal efficiencies across experimental systems at discrete time intervals. Taking 72 h as a representative time point: the residual concentration ratio C_t/C_0 was 0.026 in the complete S-nZVI@BC/PS system, corresponding to 97.4% total removal. As the control systems showed, C_t/C_0 was 0.891 (10.9% removal) for S-nZVI@BC alone and 0.962 (3.8% removal) for PS alone. Radical quenching revealed partial contributions: C_t/C_0 was 0.677 with MeOH ($SO_4{\bullet}^-$/HO• quenching) in the system, corresponding to 32.3% residual removal. C_t/C_0 was 0.286 with TBA (HO• quenching) in the system, corresponding to 71.4% residual removal. C_t/C_0 was 0.201 with BQ ($O_2{\bullet}^-$ quenching) in the system, corresponding to 79.9% residual removal. Radical-specific contributions were derived by differential inhibition: $SO_4{\bullet}^-$ contribution 39.1% (71.4–32.3%), HO• contribution 26.0% (97.4–71.4%), and $O_2{\bullet}^-$ contribution 17.5% (97.4–79.9%). The final contribution rates were calculated by the ratio between 10.9%, 3.8%, 39.1%, 26.0%, and 17.5% to 97.4%, respectively, and the results were as follows: S-nZVI@BC 11.2%, PS 3.9%, $SO_4{\bullet}^-$ 40.2%, HO• 26.7%, $O_2{\bullet}^-$ 18.0%. The time-resolved contributions calculated analogously are summarized in Figure 5f.

The relative contribution of S-nZVI@BC (incorporating both adsorption and reductive dechlorination of 2,4-DCP) was highest (48.0%) at the initial reaction stage (0.167 h), which can be attributed to the abundant accessible sites on the BC surface. Subsequently, the contribution of HO• increased and exceeded other factors after 0.5 h, reaching a maximum of 67.3% at 2 h. This could be caused by the rapid activation of PS by Fe^0, favoring the generation of HO• (Equation (4)) over $SO_4{\bullet}^-$ or $O_2{\bullet}^-$ (Equations (3), (5) and (6)) at this stage. In the later reaction phase, the contribution of HO• declined, while that of $SO_4{\bullet}^-$ became dominant (40.2% at 72 h), coinciding with the precipitation of $Fe(OH)_3$ resulting from OH^- consumption (Equation (11)). Meanwhile, the contribution of $O_2{\bullet}^-$ stabilized at 18.1% after initial fluctuations.

3.6. Analysis of Degradation Pathway

The products of 2,4-DCP degradation were analyzed by an ultra high-resolution mass spectrometer, and the schematic pathways for 2,4-DCP degradation in the S-nZVI@BC/PS system are presented based on the obtained intermediates in Figure 6. The entire reaction was mainly divided into dechlorination, ring-opening, and mineralization. First, ortho-chloride on an aromatic ring of 2,4-DCP was directly reductive dechlorination by S-nZVI@BC, and the 4-chlorophenol (m/z = 127.16) was produced. Then, para-chloride on the intermediate product was further attacked and phenol (m/z = 93.03) was produced. Second, the aromatic ring was broken by ROS and PS, and low molecular weight products were generated, such as acetic acid (m/z = 60.95), propionic acid (undetected, predicted product based on Equation (16)), oxalic acid (m/z = 90.67), and maleic acid (undetected, predicted product based on Equation (17) [36]. Finally, the products with smaller molecular weights undergo further mineralization to CO_2 and H_2O. The biological toxicity intensities of the above several small-molecule organic substances in water are as follows: oxalic acid > maleic acid > propionic acid > acetic acid. However, their toxicity is much lower than that of 2,4-DCP at the low concentration condition.

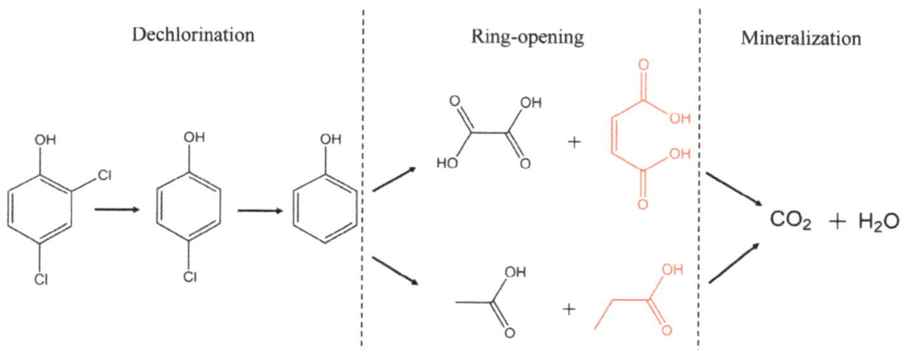

Figure 6. Degradation pathways of 2,4-DCP in the S-nZVI@BC/PS system. Reaction condition: PS = 1400 mg/L, [2,4-DCP] = 20 mg/L of 400 mL, BC@S-nZVI = 2100 mg/L, initial pH = 9 ± 0.2, T = 30 °C.

Furthermore, through total organic carbon (TOC) analysis, it was determined that the initial TOC of the 2,4-DCP solution was 8.7 mg/L, and it decreased to 2.1 mg/L after degradation for 72 h. This indicated that the TOC removal rate of the solution was 75.9%, while the removal rate of 2,4-DCP was 97.4% at this time. This confirmed that the 2,4-DCP degradation process involved the conversion into smaller organic compounds, and most of them were completely mineralized.

$$SO_4\bullet^- + HO\bullet + O_2\bullet^- + C_6H_6O \rightarrow SO_4^{2-} + C_3H_6O_2 + C_2H_4O_2 \quad (16)$$

$$SO_4\bullet^- + HO\bullet + O_2\bullet^- + C_6H_6O \rightarrow SO_4^{2-} + C_2H_2O_4 + C_4H_4O_4 \quad (17)$$

3.7. Summary and Reflection

Overall, comparing the findings of this article to the established literature, the results are summarized in Table 1. The basic research thoughts, analysis methods, and research results were consistent. However, this article focused more on the adsorption and reduction effects of S-nZVI@BC on 2,4-DCP, and conducted a detailed study on this. This article lacks electrochemical-related tests, and quantitative analysis electron transfer in a reaction system is part of the study on the degradation mechanism. Meanwhile, the lack of quantifying PS generation to directly link catalyst activity to oxidant production is also a drawback of this article. The application of this system to real wastewater containing two or more contaminants is also worthy of deep research. These contents also represent the research directions in the near future.

Table 1. The finding comparisons of this research to the established literature.

Contents	This Research	The Established Literatures
To prepare S-nZVI@BC and characterize its microscopic shape	S-nZVI@BC was synthesized with two-step method in liquid phase; analyzed the microscopic structures, functional groups and chemical bonds of it	The preparation methods were mature, morphological characterization more comprehensive, and test equipment more advanced
To establish an S-nZVI@BC/PS reaction system for 2,4-DCP degradation	The optimal PS:2,4-DCP mass ratio was 70:1 and S-nZVI@BC:PS of 1.5:1; the degradation effect was better than that of either one or more components of the S-nZVI@BC/PS system	Due to the differences in material composition and degradable pollutants, the optimal ratios were not the same. However, the highest degradation rate of the entire system was consistent
To analyze different influencing factors of the system	The reusability of S-nZVI@BC, pH, temperature, and anions in solution was evaluated	The conventional influencing factors studied were the same
To quantify the contribution of S-nZVI@BC	Component contribution analysis quantified the adsorption and reduction in S-nZVI@BC; this was a notable highlight in this research	Some studies focused on the adsorption properties of S-nZVI@BC, but very little literature has paid attention to and quantitatively analyzed other chemical effects
To explore the PS activation mechanism	Fe^0 played a dominant role in the PS activation process, HO•, $SO_4•^-$, and $O_2•^-$ were key agents in subsequent degradation stages	The methods for determining the activation mode and identifying the active substances were basically the same
To analyze the degradation pathway	The degradation products and pathways were analyzed, but propionic acid and maleic acid were predicted products based on chemical equations	Every degradation product could be detected, and determined the clear degradation pathway

4. Conclusions

This study established an S-nZVI@BC-activating PS reaction system, and demonstrated exceptional 2,4-DCP degradation efficiency with varied influencing factors under the optimal conditions. The S-nZVI was stably loaded on the surface and in the pores of BC. The dominant activation was derived by Fe^0 in S-nZVI@BC, while adsorption predominated in the initial stage. Degradation proceeded via dechlorination, ring-opening, and mineralization, primarily mediated by HO•, $SO_4•^-$, and $O_2•^-$. In addition, the S-nZVI@BC had excellent stability and high reuse efficiency, the energy required for recycling it was lower than that needed for reproducing, and it also reduced the usage of related raw materials, leaded to high cost-effectiveness and sustainable usability. This work provides foundational insights for scaling up environmental remediation technologies, and will help advance the field of sustainability.

Supplementary Materials: The following supporting information can be downloaded at: https://www.mdpi.com/article/10.3390/su17198721/s1, Table S1: The total volume in pores and BET surface area statistics of materials; Figure S1: EDS images; Figure S2.: XPS analysis.

Author Contributions: Conceptualization, M.W. and C.D.; methodology, Y.Z. and Z.A.; software, M.W.; validation, M.W., Y.Z., Z.A. and C.D.; formal analysis, Z.A.; investigation, M.W. and Y.Z.; resources, Y.Z.; data curation, M.W.; writing—original draft preparation, M.W.; writing—review and editing, Y.Z.; visualization, M.W.; supervision, C.D.; project administration, Y.Z.; funding acquisition, C.D. All authors have read and agreed to the published version of the manuscript.

Funding: This research was funded by the Anhui Provincial Academy of Eco-Environmental Science Research grant number 52400198. And the APC was funded by the Anhui Provincial Academy of Eco-Environmental Science Research.

Institutional Review Board Statement: Not applicable.

Informed Consent Statement: Not applicable.

Data Availability Statement: The data are unavailable due to the related research topic still being under investigation.

Acknowledgments: This study is thankful for the support of Hefei Key Technology Major Research and Development Projects (2021GJ063) and National Natural Science Foundation of China for Young Scientists (52400198).

Conflicts of Interest: The authors declare no conflicts of interest.

References

1. Yadav, S.; Kumar, S.; Haritash, A.K. A comprehensive review of chlorophenols: Fate, toxicology and its treatment. *J. Environ. Manag.* **2023**, *342*, 118254. [CrossRef] [PubMed]
2. Yang, C.F.; Lee, C.M. Pentachlorophenol contaminated groundwater bioremediation using immobilized Sphingomonas cells inoculation in the bioreactor system. *J. Hazard. Mater.* **2008**, *152*, 159–165. [CrossRef]
3. Igbinosa, E.O.; Odjadjare, E.E.; Chigor, V.N.; Igbinosa, I.H.; Emoghene, A.O.; Ekhaise, F.O.; Igiehon, N.O.; Idemudia, O.G. Toxicological profile of chlorophenols and their derivatives in the environment: The public health perspective. *Sci. World J.* **2013**, *2013*, 460215. [CrossRef] [PubMed]
4. Zada, A.; Khan, M.; Khan, M.A.; Khan, Q.; Habibi-Yangjeh, A.; Dang, A.; Maqbool, M. Review on the hazardous applications and photodegradation mechanisms of chlorophenols over different photocatalysts. *Environ. Res.* **2021**, *195*, 110742. [CrossRef] [PubMed]
5. Li, R.C.; Jin, X.Y.; Megharaj, M.; Naidu, R.; Chen, Z.L. Heterogeneous Fenton oxidation of 2,4-dichlorophenol using iron-based nanoparticles and persulfate system. *Chem. Eng. J.* **2015**, *264*, 587–594. [CrossRef]
6. Garba, Z.N.; Zhou, W.M.; Lawan, I.; Xiao, W.; Zhang, M.; Wang, L.; Chen, L.; Yuan, Z. An overview of chlorophenols as contaminants and their removal from wastewater by adsorption: A review. *J. Environ. Manag.* **2019**, *241*, 59–75. [CrossRef]
7. Ma, T.G. Efficiency and Mechanism of Persulfate Activation by Polyaniline-Derived Carbon Materials for Chlorophenols Degradation in Water. Ph.D. Thesis, Jilin University, Changchun, China, November 2023.
8. Wang, J.L.; Chen, H. Catalytic ozonation for water and wastewater treatment: Recent advances and perspective. *Sci. Total Environ.* **2020**, *704*, 135249. [CrossRef]
9. Anjali, R.; Shanthakumar, S. Insights on the current status of occurrence and removal of antibiotics in wastewater by advanced oxidation processes. *J. Environ. Manag.* **2019**, *246*, 51–62. [CrossRef]
10. Zhou, Y.B.; Zhang, Y.L.; Hu, X.M. Enhanced activation of peroxymonosulfate using oxygen vacancy-enriched $FeCo_2O_{4-x}$ spinel for 2,4-dichlorophenol removal: Singlet oxygen-dominated nonradical process. *Colloids Surf. A* **2020**, *597*, 124568. [CrossRef]
11. Matzek, L.W.; Carter, K.E. Activated persulfate for organic chemical degradation: A review. *Chemosphere* **2016**, *151*, 178–188. [CrossRef]
12. Tsitonaki, A.; Petri, B.; Crimi, M.; Mosbaek, H.; Siegrist, R.L.; Bjerg, P.L. In situ chemical oxidation of contaminated soil and groundwater using persulfate: A review. *Crit. Rev. Environ. Sci. Technol.* **2010**, *40*, 55–91. [CrossRef]
13. Ike, I.A.; Linden, K.; Orbell, J.D.; Duke, M. Critical review of the science and sustainability of persulphate advanced oxidation processes. *Chem. Eng. J.* **2018**, *338*, 651–669. [CrossRef]
14. Wang, J.L.; Wang, S.Z. Reactive species in advanced oxidation processes: Formation, identification and reaction mechanism. *Chem. Eng. J.* **2020**, *401*, 126158. [CrossRef]
15. Ling, S.K.; Wang, S.; Peng, Y. Oxidative degradation of dyes in water using Co^{2+}/H_2O_2 and Co^{2+}/peroxymonosulfate. *J. Hazard. Mater.* **2010**, *178*, 385–389. [CrossRef] [PubMed]

16. Zhang, Y.Q.; Zhang, J.; Xiao, Y.; Chang, W.C.; Lim, T.T. Kinetic and mechanistic investigation of azathioprine degradation in water by UV, UV/H_2O_2 and UV/persulfate. *Chem. Eng. J.* **2016**, *302*, 526–534. [CrossRef]
17. Lee, J.; Von, G.U.; Kim, J.H. Persulfate-Based Advanced Oxidation: Critical Assessment of Opportunities and Roadblocks. *Environ. Sci. Technol.* **2020**, *54*, 3064–3081. [CrossRef] [PubMed]
18. Chen, G.Y.; Wu, G.Y.; Li, N.; Lu, X.K.; Zhao, J.H.; He, M.T.; Yan, B.B.; Zhang, H.Q.; Duan, X.G.; Wang, S.B. Landfill leachate treatment by persulphate related advanced oxidation technologies. *J. Hazard. Mater.* **2021**, *418*, 126355. [CrossRef]
19. Chen, H.; Zhang, Z.L.; Feng, M.B.; Liu, W.; Wang, W.; Yang, Q.; Hu, Y. Degradation of 2,4-dichlorophenoxyacetic acid in water by persulfate activated with FeS (mackinawite). *Chem. Eng. J.* **2017**, *313*, 498–507. [CrossRef]
20. Kim, C.L.; Ahn, J.Y.; Kim, T.Y.; Shin, W.S.; Hwang, I. Activation of Persulfate by Nanosized Zero-Valent Iron (NZVI): Mechanisms and Transformation Products of NZVI. *Environ. Sci. Technol.* **2018**, *52*, 3625–3633. [CrossRef]
21. Wang, S.S.; Zhao, M.Y.; Zhou, M.; Li, Y.C.; Wang, J.; Gao, B.; Sato, S.; Feng, K.; Yin, W.Q.; Igalavithana, A.D.; et al. Biochar-supported nZVI (nZVI/BC) for contaminant removal from soil and water: A critical review. *J. Hazard. Mater.* **2019**, *373*, 820–834. [CrossRef]
22. Jiang, X.Y.; Ouyang, Z.Z.; Zhang, Z.F.; Yang, C.; Li, X.Q.; Dang, Z.; Wu, P.X. Mechanism of glyphosate removal by biochar supported nano-zero-valent iron in aqueous solutions. *Colloids Surf. A* **2018**, *547*, 64–72. [CrossRef]
23. Cai, J.; Zhang, Y. Enhanced degradation of bisphenol S by persulfate activated with sulfide-modified nanoscale zero-valent iron. *Environ. Sci. Pollut. Res.* **2022**, *29*, 8281–8293. [CrossRef]
24. Wu, G.C.; Kong, W.J.; Gao, Y.; Kong, Y.; Dai, Z.G.; Dan, H.B.; Shang, Y.N.; Wang, S.Q.; Yin, F.J.; Yue, Q.Y.; et al. Removal of chloramphenicol by sulfide-modified nanoscale zero-valent iron activated persulfate: Performance, salt resistance, and reaction mechanisms. *Chemosphere* **2021**, *286*, 131876. [CrossRef] [PubMed]
25. Li, Y.F.; Hu, S.D.; Chen, J.H.; Muller, K.; Li, Y.C.; Fu, W.J.; Lin, Z.W.; Wang, H.L. Effects of biochar application in forest ecosystems on soil properties and greenhouse gas emissions: A review. *J. Soils Sediments* **2018**, *18*, 546–563. [CrossRef]
26. Xu, J.; Avellan, A.; Li, H.; Liu, X.T.; Noël, V.; Lou, Z.; Wang, Y.; Kaegi, R.; Henkelman, G.; Lowry, G.V. Sulfur Loading and Speciation Control the Hydrophobicity, Electron Transfer, Reactivity, and Selectivity of Sulfidized Nanoscale Zerovalent Iron. *Adv. Mater.* **2020**, *32*, e1906910. [CrossRef]
27. Chen, J.; Dong, H.R.; Tian, R.; Li, R.; Xie, Q.Q. Remediation of trichloroethylene-contaminated groundwater by sulfide-modified nanoscale zero-valent iron supported on biochar: Investigation of critical factors. *Water Air Soil Pollut.* **2020**, *231*, 432. [CrossRef]
28. Xie, R.H.; Wang, M.; Li, W.P.; Song, J.J. Degradation of 2-Chlorophenol in aqueous solutions using persulfate activated by biochar supported sulfide-modified nanoscale zero-valent iron: Performance and mechanisms. *Water* **2023**, *15*, 2805. [CrossRef]
29. Taiwo, K.J.; Oancea, A.V.; Kotha, N.S.; Usack, J.G. Enhancing wastewater treatment sustainability through integrated anaerobic digestion and hydrothermal carbonization: A life-cycle perspective. *Sustainability* **2025**, *17*, 7545. [CrossRef]
30. Kaynak, E.; Piri, I.S.; Das, O. Revisiting the Basics of Life Cycle Assessment and Lifecycle Thinking. *Sustainability* **2025**, *17*, 7444. [CrossRef]
31. Su, Y.M.; Jassby, D.; Song, S.K.; Zhou, X.F.; Zhao, H.Y.; Filip, J.; Petala, E.; Zhang, Y.L. Enhanced oxidative and adsorptive removal of diclofenac in heterogeneous Fenton-like reaction with sulfide modified nanoscale zerovalent iron. *Environ. Sci. Technol.* **2018**, *52*, 6466–6475. [CrossRef]
32. Jin, H.; Cang, Z.Z.; Ding, W.; Wu, W.T.; Ma, H.K.; Wang, C.X.; Qi, Z.W.; Li, Z.F.; Zhang, L.L. Oxidative removal of antibiotic resistant *E. coli* by sulfidated zero-valent iron: Homogeneous vs heterogeneous activation. *J. Hazard. Mater.* **2021**, *408*, 124411. [CrossRef]
33. Wang, X.Q.; Guo, Z.Z.; Hu, Z.; Ngo, H.H.; Liang, S.; Zhang, J. Adsorption of phenanthrene from aqueous solutions by biochar derived from anammoniation-hydrothermal method. *Sci. Total Environ.* **2020**, *733*, 139267. [CrossRef]
34. Zhao, H.Y.; Wang, Q.N.; Chen, Y.; Tian, Q.L.; Zhao, G.H. Efficient removal of dimethyl phthalate with activated iron-doped carbon aerogel through an integrated adsorption and electro-Fenton oxidation process. *Carbon* **2017**, *124*, 111–122. [CrossRef]
35. Zhang, P.; Song, D.; Xu, X.; Hao, Y.; Shang, X.; Wang, C.; Tang, J.; Sun, H. Sulfidated zero valent iron as a persulfate activator for oxidizing organo-phosphorus pesticides (OPPs) in aqueous solution and aged contaminated soil columns. *Chemosphere* **2021**, *281*, 130760. [CrossRef] [PubMed]
36. Wang, Y.Y.; Wang, L.; Zhang, Y.L.; Mao, X.H.; Tan, W.B.; Zhang, Y.L.; Wang, X.S.; Chang, M.; Guo, R.N.; Xi, B.D. Perdisulfate-assisted advanced oxidation of 2,4-dichlorophenol by bio-inspired iron encapsulated biochar catalyst. *J. Colloid Interface Sci.* **2021**, *592*, 358–370. [CrossRef] [PubMed]
37. Luo, X.W.; Shen, M.X.; Liu, J.H.; Ma, Y.J.; Gong, B.N.; Liu, H.Q.; Huang, Z.J. Resource utilization of piggery sludge to prepare recyclable magnetic biochar for highly efficient degradation of tetracycline through peroxymonosulfate activation. *J. Clean. Prod.* **2021**, *294*, 126372. [CrossRef]
38. Song, S.K.; Su, Y.M.; Adeleye, A.S.; Zhang, Y.L.; Zhou, X.F. Optimal design and characterization of sulfide-modified nanoscale zerovalent iron for diclofenac removal. *Appl. Catal. B-Environ.* **2017**, *201*, 211–220. [CrossRef]

39. Manz, K.E.; Carter, K.E. Investigating the effects of heat activated persulfate on the degradation of furfural, a component of hydraulic fracturing fluid chemical additives. *Chem. Eng. J.* **2017**, *327*, 1021–1032. [CrossRef]
40. Sun, Y.; Zhao, J.; Zhang, B.-T.; Li, J.; Shi, Y.; Zhang, Y. Oxidative degradation of chloroxylenol in aqueous solution by thermally activated persulfate: Kinetics, mechanisms and toxicities. *Chem. Eng. J.* **2019**, *368*, 553–563. [CrossRef]
41. Wang, Y.R.; Tian, D.F.; Chu, W.; Li, M.R.; Lu, X.W. Nanoscaled magnetic $CuFe_2O_4$ as an activator of peroxymonosulfate for the degradation of antibiotics norfloxacin. *Sep. Purif. Technol.* **2019**, *212*, 536–544. [CrossRef]
42. Gao, Y.Q.; Gao, N.Y.; Wang, W.; Kang, S.F.; Xu, J.H.; Xiang, H.M.; Yin, D.Q. Ultrasound-assisted heterogeneous activation of persulfate by nano zero-valent iron (nZVI) for the propranolol degradation in water. *Ultrason. Sonochem.* **2018**, *49*, 33–40. [CrossRef] [PubMed]
43. Wang, X.H.; Lin, Y.C. Phototransformation of cephalosporin antibiotics in an aqueous environment results in higher toxicity. *Environ. Sci. Technol.* **2012**, *46*, 12417–12426. [CrossRef] [PubMed]
44. Chen, J.S.; Luo, H.Y.; Luo, D.Y.; Chen, Y.; Tang, J.; Ma, H.; Pu, S.Y. New insights into the degradation of nitrobenzene by activated persulfate with sulfidated nanoscale zero-valent iron: Synergistic effects of reduction and reactive oxygen species oxidation. *Sep. Purif. Technol.* **2023**, *322*, 124252. [CrossRef]
45. Du, J.; Bao, J.; Lu, C.; Werner, D. Reductive sequestration of chromate by hierarchical $FeS@Fe^0$ particles. *Water Res.* **2016**, *102*, 73–81. [CrossRef]
46. Wang, Q.; Wang, B.B.; Ma, Y.; Xing, S.T. Enhanced superoxide radical production for ofloxacin removal via persulfate activation with Cu-Fe oxide. *Chem. Eng. J.* **2018**, *354*, 473–480. [CrossRef]
47. Webster, R.D. Electrochemistry combined with electron paramagnetic resonance (EPR) spectroscopy for studying catalytic and energy storage processes. *Curr. Opin. Electrochem.* **2023**, *40*, 101308. [CrossRef]
48. Liu, Y.; Zhao, Y.; Wang, J. Activation of peroxydisulfate by a novel Cu^0-Cu_2O@CNTs composite for 2, 4-dichlorophenol degradation. *Sci. Total Environ.* **2021**, *754*, 141883. [CrossRef]

sustainability

MDPI

Article

Persulfate Activation of Iron-Based Battery Catalytic Material (LFP) Modified on Polymeric Membrane (LFP@PVDF) for the Treatment of Textile Dye Wastewater

Ali Kemal Topaloğlu [1,*], Bekir Fatih Kahraman [1] and Semih Engün [2]

1 Department of Environmental Engineering, Zonguldak Bülent Ecevit University, Zonguldak 67100, Turkey; bfatihkahraman@beun.edu.tr
2 Department of Metallurgical and Materials Engineering, Zonguldak Bülent Ecevit University, Zonguldak 67100, Turkey; semihengun@beun.edu.tr
* Correspondence: alikemal.topaloglu@beun.edu.tr

Abstract

In this study, a novel LFP–catalytic microfiltration membrane (LFP@PVDF) was fabricated by loading a lithium-ion battery material $LiFePO_4$ (LFP) onto polymeric micro-porous polyvinylidene fluoride (PVDF) using a filter press coating method. The successful loading of LFP material onto the LFP@PVDF catalytic membrane was confirmed by the characterization of the material using FTIR, SEM, EDX, and XRD analysis. To evaluate the catalytic performance of the LFP@PVDF membrane, the reactive black 5 (RB5) dye-containing solution was used with or without the peroxymonosulfate (PMS) activator in a dead-end filtration under room conditions. The influence of parameters such as LFP loading, initial RB5 dye concentration, persulfate dosage, and solution pH on the performance of the persulfate oxidation process was comprehensively examined. It was found that the LFP@PVDF membrane/persulfate activation system can effectively remove RB5 dye with an efficiency of 97.3%. The RB5 dye removal by LFP@PVDF membranes with varying experimental conditions was found to fit the pseudo-second-order kinetic model. Quenching experiments showed that the reactive species $HO^{\bullet}$, $SO_4^{\bullet-}$ and $^1O_2$ were responsible for the dye removal. The LFP@PVDF membrane/persulfate activation system appeared to be a promising approach for the removal of organic contaminants.

Keywords: lithium-ion battery; advanced oxidation; catalytic degradation; catalytic membrane; peroxymonosulfate; $LiFePO_4$

1. Introduction

Fluctuating market prices of metals and metalloids complicate lithium-ion batteries' (LIBs) production and affect their overall demand, especially with the growing demand for electronics and electric vehicles [1,2]. Despite the many benefits that can be provided by LIBs, their relatively short life cycle means that environmental disposal and recycling plans must be employed. Spent LIBs contain dangerous metal elements such as cobalt, nickel, and lithium, which could find their way into the surrounding soil and water and endanger the lives of living organisms [3,4]. In addition, spent LIBs consist of critical raw materials such as lithium, cobalt, nickel, and iron, which are used for manufacturing new batteries and in other technologies [1]. The disposal of LIBs leads to pollution and depletion of the available materials; therefore, it is important to recover them for optimal recycling [5].

Given that resource recycling and environmental management are two major concerns, research on problems associated with spent LIBs remains a critical area of interest to establish efficient techniques for their effective recycling [2,6]. It should be noted that the traditional methods of recycling, including pyrometallurgical and hydrometallurgical recycling, require circuitous processes, a substantial amount of energy, and may result in secondary pollution [7]. Therefore, potential approaches to new and environmentally sound recycling methods should be investigated.

Cathode-active materials (CAM) used in LIBs attract great attention with their promising catalytic properties in environmental remediation applications. These materials can be evaluated in catalytic decontamination processes prior to disposal or recycling thanks to their redox-active transition metals such as nickel, cobalt, manganese, and iron. Transition metal oxides play a critical role in catalytic processes thanks to their regular crystal structures, providing high surface area and high surface energies of transition metal atoms [7,8]. Therefore, using expired LIB CAMs as a catalyst in water treatment applications finds its application in the management of LIB wastes and the cost-effective method of controlling wastewater.

One type of LIBs containing valuable CAMs is $LiFePO_4$ (LFP) batteries. Among various cathode chemistries, LFP has attracted considerable attention in the field of energy storage due to its long cycle life and excellent thermal stability, resisting temperatures as high as 340 °C without any adverse reactions, hence improving safety during operation [9]. Moreover, LFP batteries are nontoxic and environmentally friendly, possessing minimal environmental hazards, hence meeting the global sustainability goals [10]. Therefore, the use of LFP batteries in the industry is becoming increasingly widespread. According to Global EV Outlook 2024 data, LFP's share in the electric vehicle market has doubled compared to 2020 and reached 40% [11]. This rapid increase in demand for LFP in the battery market brings with it the question of how to recycle battery components that have reached the end of their service life. Using LFP-based cathode active material (CAM) wastes in wastewater treatment is a more economical and easier process than $LiCoO_2$ (LCO) and $LiNi_xCo_yMn_{1-x-y}O_2$ (NMC) based CAMs. The main reason for this is that the main components of LFP, iron (Fe) and phosphorus (P), are more abundant in nature and cheaper compared to the components of LCO and NMC, nickel (Ni), cobalt (Co), and manganese (Mn). Although recycling of LCO and NMC components for reuse in the battery industry is costly and requires complex processes, it is necessary for the efficient use of critical waste CAM components for battery production. On the other hand, the recovery of LFP CAM components to the battery industry does not carry the same level of necessity. Therefore, LFP wastes stand out as a suitable material source that can be evaluated in alternative applications such as water treatment [10]. In addition, the presence of iron in LFP makes it a suitable activator for peroxymonosulfate (PMS), a strong oxidant used in advanced oxidation processes (AOPs) [12].

The textile industry uses various synthetic dyes to color textile products, resulting in wastewaters containing high amounts of dyes [13,14]. When the dye wastewaters are discharged into water environments without treatment, they cause serious environmental pollution [13,15]. In addition, they threaten human health and other living organisms due to their carcinogenicity and toxicity [16,17]. Therefore, such wastewaters must be effectively treated before discharge. Advanced oxidation processes (AOPs) are one of the most effective methods for removing such pollutants [6]. PMS-based AOPs depend on the activation of PMS to produce reactive oxygen species (ROS), including sulfate radicals ($SO_4^{\bullet-}$) and hydroxyl radicals ($HO^{\bullet}$), which are highly effective in organic pollutant degradation [12,18]. Recent studies have shown that LFP is a potential catalyst for PMS activation in the degradation of organic pollutants [7,12,19]. LFP showed excellent catalytic activity, attributed to lithium defects and oxygen vacancies produced during battery cycling [20]. These defects could enhance the adsorption and activation of PMS, leading to efficient ROS generation and subsequent pollutant degradation [12]. However, it is well known that the application of nanomaterials without immobilization on a suitable supported material is limited by drawbacks such as aggregation, difficulties in separation, environmental risks, and technical challenges [21–24]. To eliminate these drawbacks, as polymeric membranes can be easily modified with catalytic materials using simple methods, and these membranes are generally resistant to acids, bases, and oxidants, they have been used as a support material [25–27]. Wastewater treatment applications of the PMS-based AOPs using composite catalytic membranes modified with LFP have not been carried out.

This work presents the preparation of a novel LFP-modified polyvinylidene fluoride microfiltration catalytic membrane (LFP@PVDF) to remove the organic dyes in wastewater. The opinion is to load LFP on the surface of the PVDF membrane to fabricate a catalytic membrane system of high efficiency. The proposed LFP@PVDF membrane offers several advantages over conventional powder-based catalysts: enhanced stability, reduced catalyst loss, and easier separation from the treated effluent. The activation of persulfate on the LFP@PVDF membrane and its effectiveness in the degradation of organic dyes were examined. This study investigated the impact of operational parameters, such as LFP loading, PMS concentration, solution pH, and dye concentration, on degradation efficiency. In addition, the reactive radicals involved and their contribution to the dye removal process were studied, along with the reusability of the material. Results obtained from this study will contribute to advancing appropriate wastewater treatment technologies and the circular economy by demonstrating the potential of LIB active materials for environmental remediation.

2. Materials and Methods

2.1. Materials

Commercial polymeric polyvinylidene difluoride (PVDF) membrane with 0.2 μm average pore diameter was purchased from Microdyn-Nadir membranes, Wiesbaden, Germany. The LFP was purchased from MTI Corporation (Salt Lake City, UT, USA), having a particle size of D_{50} ~3.5 μm. Tert-butyl alcohol (TBA $\geq$ 99.7) and ethanol (emsure) were purchased from Merck (Darmstadt, Germany). Potassium peroxymonosulfate, commercially named oxone ($KHSO_5 \cdot 0.5KHSO_4 \cdot 0.5K_2SO_4$), was purchased from Merck. Reactive Black 5 dye (RB5, $C_{26}H_{21}N_5Na_4O_{19}S_6$) was obtained from DyStar (Türkiye).

2.2. Modification of the LFP@PVDF Catalytic Membrane

The iron-based LFP catalytic membrane (LFP@PVDF) was fabricated using a straightforward and convenient filter press coating method. In the method, a 100 mg/L LFP stock solution was first prepared by mixing the LFP catalytic material with deionized water. Solutions with varying concentrations were prepared by diluting the stock solution, and then produced LFP@PVDF membranes with different loading amounts of LFP using these solutions under a vacuum filtration system. In this process, a flat sheet PVDF membrane with an active surface area of 11.94 cm^2 was placed in a cup filter device, and the LFP catalytic material was uniformly loaded onto the PVDF membrane surface under a vacuum pressure of −250 mm Hg at room temperature (25 ± 1 °C). The PVDF membranes loaded with LFP catalytic material were subsequently dried at 40 °C in a vacuum oven for at least 30 min. Following this procedure, LFP@PVDF membranes with varying loading amounts of LFP catalytic material were fabricated (0.2, 0.3, and 0.4 mg/cm^2).

2.3. Characterization

To confirm that the LFP catalytic materials were loaded on the surface of the PVDF membrane, infrared spectrophotometer (FT-IR), scanning electron microscopy (SEM), and energy dispersive X-ray (EDX) analyses were conducted. The LFP catalytic material, the unmodified membrane, and the LFP@PVDF membrane were analyzed by a Perkin Elmer Pyris infrared spectrophotometer (FT-IR) between 550 and 4000 cm^{-1} wavenumber at room temperature (PerkinElmer, Shelton, CT, USA). The structure of the unmodified PVDF membrane and the LFP@PVDF membrane was analyzed by a field emission scanning electron microscope (SEM, FEI Quanta 400F, FEI, Houston, TX, USA) at an operating voltage of 5 kV. Energy dispersive X-ray (EDX) analysis was performed in the attachment in SEM-Ametek EDAX Genesis at the operating voltage of 15 kV (Edax Inc., Berwyn, PA, USA). The crystal structures of samples were characterized by an X-ray diffractometer (XRD, Panalytical Empyrean, Almelo, The Netherlands), equipped with 2θ range = 10–90° Cu-Kα radiation (λ = 1.54 Å).

2.4. RB5 Dye Removal Performance Tests

The catalytic performance of the LFP@PVDF membrane with/without the PMS activator on the RB5 dye wastewater was investigated using different experimental tests. The experiments were conducted using a vacuum filtration system (dead-end filtration system) (Table 1, Figure S1) under −250 mm Hg vacuum pressure at room temperature (25 ± 1 °C). A stock RB5 dye wastewater of 500 mg/L was prepared with deionized water and then diluted to the desired concentrations for the experiments. In the RB5 dye removal experiments, LFP@PVDF membranes loaded with the target LFP dosages were placed in the dead-end filtration system. After adding a given concentration of PMS to the 50 mL RB5 dye solution and stirring it with a magnetic stirrer for 60 s to fully dissolve the PMS, the dye solution was poured into the filtration system cup. To determine the absorbance of the residual RB5 dye after the filtration, samples were taken from the filtrate every 5 min for 30 min, and the absorbance was measured immediately using a UV-visible spectrophotometer (Shimadzu-1800, Shimadzu, Duisburg, Germany) at the maximum absorption wavelength (λmax = 597 nm). The percentage of RB5 dye removal efficiency was described in Equation (1).

$$\text{The percentage of RB5 dye removal efficiency} = \frac{(A_0 - A_t)}{A_0} \times 100 \quad (1)$$

where at 597 nm of the reaction solution, A_0 is the initial absorbance value of RB5, and A_t is the absorbance value of RB5 at a time, t.

Table 1. Explanation of the terminology used to designate the treatments evaluated in the RB5 dye removal experiments.

Expression	Membrane	$LiFePO_4$ (LFP) Loading	PMS Application
0.2LFP@PVDFwPMS	+	0.2 mg/cm^2; 9.5 mg/g	+
0.3LFP@PVDFwPMS	+	0.3 mg/cm^2; 14.3 mg/g	+
0.4LFP@PVDFwPMS	+	0.4 mg/cm^2; 19.0 mg/g	+
0.2LFP@PVDF	+	0.2 mg/cm^2; 9.5 mg/g	−
0.3LFP@PVDF	+	0.3 mg/cm^2; 14.3 mg/g	−
0.4LFP@PVDF	+	0.4 mg/cm^2; 19.0 mg/g	−
PVDF	+	-	−
PVDFwPMS	+	-	+

+: Used, −: Not used.

The impacts of various parameters on RB5 dye removal were investigated. To find the optimum loading dosage, the dosage of LFP on the PVDF membrane surface was applied in different amounts from 0 to 0.4 mg/cm^2. The effects of PMS dosage (0–500 mg/L), RB5 dye concentration (20–80 mg/L), and pH level (3, 7, 11) were also investigated to determine the optimal conditions. Adjustment of pH was made using HCl and NaOH solutions. To identify the reactive species, experiments were carried out during the catalytic reaction by adding various quenchers, such as EtOH, TBA, and NaN_3, at a concentration of 1500 mM. Additionally, the recyclability of the LFP@PVDF membrane was studied to assess its potential for reuse. In detail, the contaminated catalytic membrane was cleaned using distilled water and then dried at 40 °C in the vacuum oven before reuse.

2.5. Decolorization Rate

In the study, Pseudo-first-order and Pseudo-second-order kinetic models were used to determine the decolorization kinetic rate of RB5 dye by the oxidation process. The pseudo-first and second-order reaction kinetic models were described in Equations (2) and (3), respectively [28,29]. The calibration curve of RB5 dye was prepared to determine the value of RB5 dye concentration using the RB5 dye solutions from 0 to 80 mg/L, and the calibration curve was found to be linear (R^2 values > 0.99).

$$\ln \frac{C_t}{C_0} = -K_1 t \tag{2}$$

$$-\frac{dC_t}{dt} = K_2 C_t^2 \tag{3}$$

where C_0 (mg/L) and C_t (mg/L) are concentration values of the RB5 dye at the initial and at a certain time t, respectively; K_1 (1/min) and K_2 (L/mg·min) are kinetic rate constants of pseudo-first-order reaction and pseudo-second-order reaction, respectively. After integrating Equation (3), Equation (4) can be obtained [28,29].

$$\frac{1}{C_t} - \frac{1}{C_0} = K_2 t \tag{4}$$

2.6. Statistical Analysis

All of the experimental results were subjected to one-way analysis of variance (ANOVA) with Tukey post hoc tests. SPSS software (v.20.0) was used, and the significance was attested by $p < 0.05$. ANOVA results are provided as tables in the Supplementary Material (Tables S2–S13).

3. Results and Discussion

The microstructures of unmodified PVDF and LFP-loaded PVDF (LFP@PVDF) membranes were examined by scanning electron microscopy (SEM); the chemical analyses of the elements forming the PVDF and LFP materials were examined by energy dispersive X-ray spectrum (EDX), and the obtained results are shown in Figure 1. When the unmodified PVDF micrograph is examined, it is seen that it consists of homogeneously distributed pores with a size distribution ranging from 100 to 300 nm (Figure 1a). The homogeneity of the pore distribution indicates that the solvent/PVDF balance is at an appropriate ratio, and it is seen that the PVDF membrane is in the microfiltration membrane range in terms of the pore size of the membrane [30,31]. The SEM image obtained after loading the PVDF membrane with LFP catalytic material (Figure 1b) shows that the LFP particles are distributed on the membrane surface. It is seen that the LFP particles have a wide particle size range of 400 nm–3 µm and are held on the PVDF surface in the form of agglomerations consisting of secondary particles (Figure S2). As seen in the SEM images (Figure 1c), the membrane maintains its porous structure. Partial contraction due to LFP loading did not negatively affect target pollutant removal. On the contrary, it increased the contact time between the wastewater and the LFP catalytic material, contributing positively to PMS activation and catalytic efficiency.

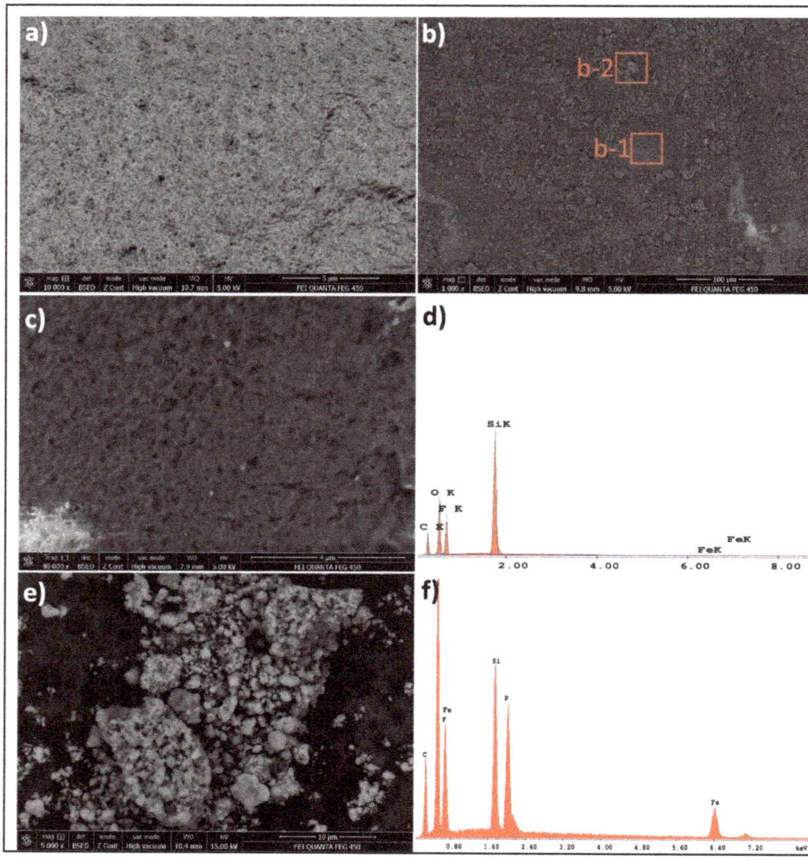

Figure 1. SEM images of (**a**) bare PVDF membrane surface, (**b**) LFP@PVDF membrane surface, (**c**) magnified view, and (**d**) EDX spectrum of LFP@PVDF membrane surface showing unloaded part (b-1), (**e**) magnified view, and (**f**) EDX spectrum of LFP@PVDF membrane surface showing LFP particles (b-2).

The EDX area analysis of the particle-loaded membrane surface (Figure 1f), together with the detailed spectra shown in Supplementary Materials (Figure S3), confirmed the presence of Fe and P atoms originating from LFP. The Fe^{2+}/Fe^{3+} redox couples in LFP are therefore expected to act as catalytic activators of PMS, thereby enabling the membrane/persulfate activation process.

The chemical structures of unmodified PVDF, LFP-loaded PVDF (LFP@PVDF), and LFP materials were investigated by Fourier Transform Infrared Spectroscopy (FT-IR) analysis, and the obtained spectra are presented in Figure 2. The FTIR spectra of the unmodified PVDF material indicate the characteristic PVDF properties. That is because the peaks obtained at 3023.6 cm^{-1} and 1402.7 cm^{-1} are related to C–H vibrations and indicate the presence of CH_2 groups in PVDF [32]. Similarly, the peak at 1180.4 cm^{-1} reflects the basic properties of C-F stretching vibrations in the 1400–1000 cm^{-1} range in the literature, while the peaks in the 830–520 cm^{-1} range are a clear indicator of C-F deformation vibration modes [33]. Turning now to the FT-IR spectrums of LFP-loaded PVDF membrane and pristine LFP, the LFP@PVDF membrane has the characteristic peaks of the pristine LFP, which were at 575.8 and 635.5 cm^{-1} associated with the vibrations of the (P-O-P) [34] and (Fe–O) [35] and at 1137.4 cm^{-1} related to the $(PO_4)^{3-}$ vibrations [34]. Furthermore, the 972.7 and 1050.7 cm^{-1} wavenumbers should be associated with the vibrations of the $(PO_4)^{3-}$ [34]. This is because the two peaks of the LFP@PVDF membrane shifted and changed significantly in terms of shape and wavenumber when compared with the two characteristic peaks of 975.0 and 1066.9 cm^{-1} in the PVDF membrane (Figure 2).

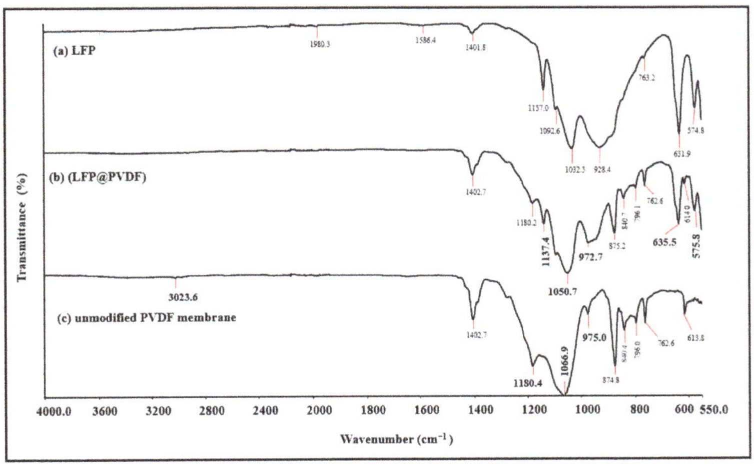

Figure 2. FTIR results of (**a**) LFP material, (**b**) LFP@PVDF membrane, and (**c**) unmodified PVDF membrane.

XRD analysis was used to characterize the loading of LFP catalytic material onto the PVDF membrane surface, and the obtained diffraction patterns are presented in Figure 3. In the XRD analysis of commercial LFP powder, all diffraction peaks were found to be compatible with the orthorhombic Pnma (62) crystal structure (JCPDS 83-2092). The absence of any impurities indicates that the commercial LFP has high purity, which is necessary for providing the required catalytic properties during water treatment. On the other hand, the sharp and broad diffraction patterns observed in the unmodified PVDF membrane reveal that the structure contains both crystalline and amorphous phases. Among these structures, a peak observed at approximately 28 degrees corresponds to monoclinic SiO_2 (PDF 98-016-2627) with the C12/c1 space group. Modification of PVDF membranes with silica nanoparticles improves mechanical properties, permeability, and degradation resistance compared to standard PVDF membranes [36]. Finally, XRD analyses of LFP-loaded PVDF membranes show that the major peaks of LFP are present, indicating that LFP is successfully loaded onto the PVDF surface. The preservation of the LFP crystal structure suggests that no chemical reaction occurs between PVDF and LFP. However, in the LFP-modified PVDF membrane, the SiO_2 peak observed at approximately 28 degrees in the unmodified membrane is no longer visible. This suggests that the diffraction signals from nano-sized SiO_2 are masked by the micro-sized LFP signals. EDX analyses confirm the presence of Si on the LFP-modified membrane surface, supporting this interpretation. As a result, the FT-IR, SEM, SEM-EDX, and XRD analyses confirmed that the LFP catalytic material was successfully loaded onto the surface of the PVDF membrane.

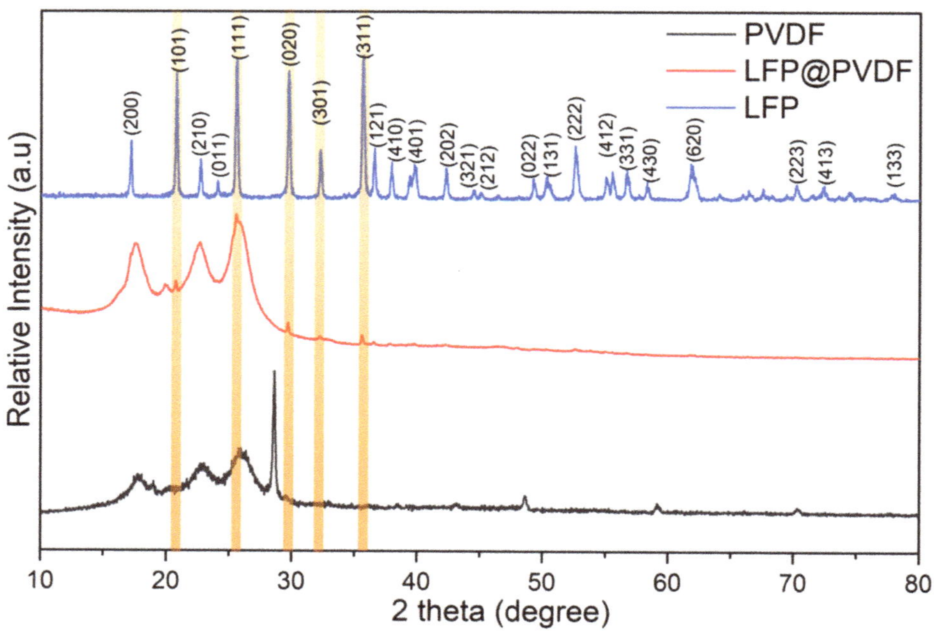

Figure 3. XRD results of LFP material, PVDF membrane, and LFP@PVDF membrane (orange highlights indicate the major peaks of LFP).

3.1. The Performance of the RB5 Dye Removal Using PMS Activator Agent with LFP@PVDF Catalytic Membranes

3.1.1. Effect of LFP Catalyst Dosage on the RB5 Dye Removal

Figure 4a shows RB5 dye removal efficiencies obtained using various loading amounts of LFP from 0 to 0.4 mg/cm^2 on the PVDF membrane surface and 300 mg/L PMS activator agent. When the unmodified PVDF membrane was used, the final RB5 dye removal efficiency increased from 7.25% to 33.62% after PMS was added to the RB5 dye solution. It was demonstrated that PMS has the potential to oxidize a range of organic pollutants to some extent, including antibiotics and pharmaceuticals, without the need for activation [37]. The degradation process primarily involves the generation of singlet oxygen ($^1O_2$) through PMS self-decomposition, although its role is limited due to rapid quenching by water [37]. Modified membranes (0.2LFP@PVDF, 0.3LFP@PVDF, and 0.4LFP@PVDF) resulted in limited RB5 dye removal efficiencies around 10% without PMS application. This suggests that it might be due to the limited adsorption to the lithium-ion battery material [38,39].

When PMS was added to the dye solutions, RB5 dye removal efficiency for 0.2LFP@PVDF, 0.3LFP@PVDF, and 0.4LFP@PVDF membranes was strongly improved to 93.8%, 97.3%, and 96.8%, respectively, because the LFP could activate the PMS effectively, which enhances the removal of dye [40]. In the first ten minutes, removal efficiencies of 0.3LFP@PVDF and 0.4LFP@PVDF were significantly higher ($p < 0.05$) than 0.2LFP@PVDF. This continued in the following phases of the experiment, but at the end of the experiment, there was no significant difference between the three treatments ($p > 0.05$). As the LFP dosage increased, it offered more active sites for the reaction with PMS [40]. However, when the amount of catalyst exceeds a certain level, the removal efficiency might decrease due to a self-quenching mechanism. This is the case when an excessively high catalytic dosage causes the catalyst to react with sulfate radicals, which increases the rate of self-quenching and decreases the degradation of organic substances [7,41,42].

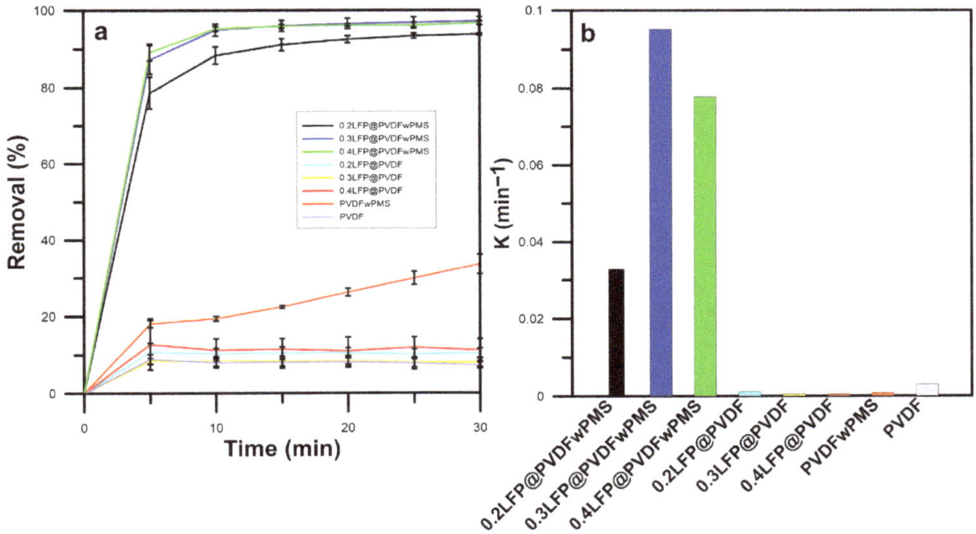

Figure 4. (**a**) Removal efficiencies of different processes; (**b**) pseudo-second-order kinetic model. (Experimental conditions: different concentrations of loadings of LFP catalyst, 20 mg/L RB5, 300 mg/L PMS, pH = 7). Error bars represent the standard error from triplicate experiments.

Moreover, the RB5 dye removal by LFP@PVDF membranes with varying LFP concentrations was found to fit the pseudo-second-order kinetic model well (Figure 4b and Table S1). For the different LFP loading amounts of 0.2, 0.3, and 0.4 mg/cm^2 on the PVDF membrane surface, the rate constant K_2 values of RB5 dye removal were 0.033, 0.095, and 0.078 L/mg·min, respectively. Notably, the kinetic rate of RB5 dye removal was highest for the 0.3 mg/cm^2 LFP loading, which aligns with the trend observed in the dye removal efficiencies. As a result, the loading amount of LFP with 0.3 mg/cm^2 on the PVDF membrane surface was selected for further experiments, as it has the highest kinetic constant as well as a high removal efficiency and thus provides a balance between economic and practical considerations compared to the other options.

Examples of advanced oxidation studies adopting persulfates and various catalysts together with their experimental conditions and performances were presented in Supplementary Materials (Table S14). Removal efficiencies reported are under conditions optimized for each catalyst. Despite variations in experimental parameters, iron-based catalysts activating PMS consistently demonstrate high dye removal efficiencies above 90%. LFP@PVDF catalytic membrane achieved a dye removal efficiency of approximately 97% for RB5 dye, which is comparable to or better than many reported catalytic AOP systems. These results indicate that the LFP@PVDF membrane is an effective catalyst for dye degradation.

3.1.2. Effect of Initial RB5 Dye Concentrations on the RB5 Dye Removal

The effect of RB5 dye removal at various initial RB5 concentrations was studied using LFP@PVDF membrane with a loading capacity of 0.3 mg/cm^2 and 300 mg/L of PMS. As shown in Figure 5, by increasing the initial concentration of RB5 dye from 20 to 80 mg/L, the efficiency of RB5 dye removal decreased from 97.1% to 69.1%. This trend was further proved by the pseudo-second-order kinetic model, in which the associated rate constant was calculated to drop dramatically from 0.095 to 0.0011 L/mg·min (Figure 5a and Table S1). This can be ascribed to the insufficient supply of radicals produced, which were insufficient for the efficient degradation of dye solutions with very high concentrations of RB5 dye [12,41]. The removal efficiencies of 20 and 40 mg/L were 97.3 and 89.7%, respectively, and they were not significantly different ($p > 0.05$). However, the rate constants were 0.095 to 0.006 L mg^{-1} min^{-1}, respectively. Therefore, it was concluded that the optimal RB5 dye concentration for further experiments was 20 mg/L.

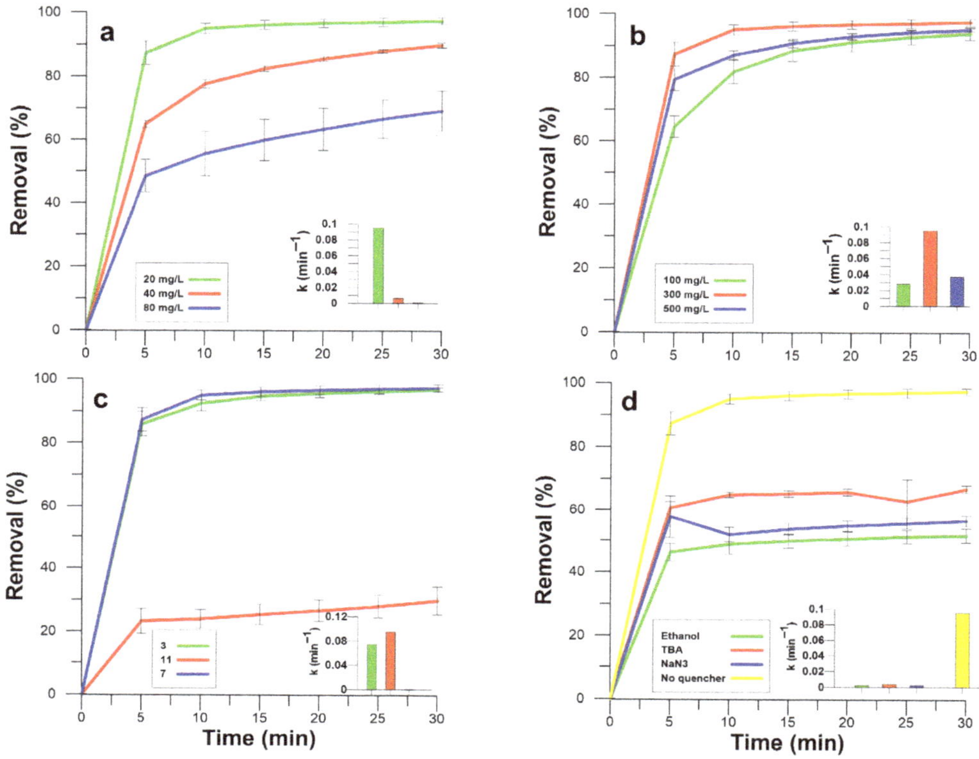

Figure 5. Effects of (**a**) initial RB5 dye concentrations (0.3 mg/cm^2 LFP, 300 mg/L PMS, pH = 7) (**b**) PMS dosages (20 mg/L RB5, 0.3 mg/cm^2 LFP, pH = 7) (**c**) initial pH values (20 mg/L RB5, 0.3 mg/cm^2 LFP, 300 mg/L PMS) and (**d**) quenchers (20 mg/L RB5, 0.3 mg/cm^2 LFP, 300 mg/L PMS, pH = 7) on removal and kinetic constants. Error bars represent the standard error for triplicate reactors.

It was observed that increasing the initial RB5 dye concentration from 20 mg/L to 80 mg/L led to a decrease in removal efficiency from 97.1% to 69.1%. Simultaneously, the associated pseudo-second-order kinetic rate constant dropped dramatically. On the other hand, 20, 40, and 80 mg/L initial concentrations led to 0.97, 1.79, and 2.76 mg dye removal at the end of the experimental period. When we calculate the absolute degraded amounts, we expect to illustrate that while the percentage efficiency declines due to the system being overwhelmed, the total mass of dye removed may still increase, up to a certain point. This phenomenon is, as previously stated, primarily ascribed to the insufficient supply of radicals produced, which becomes inadequate for the efficient degradation of dye solutions at very high concentrations.

3.1.3. Effect of the PMS Dosage on the RB5 Dye Removal

The amount of the persulfate dosage affects the efficiency of the treatment process and operating costs [43]. To determine the effect of PMS dosage on the RB5 dye removal, different initial PMS concentrations from 100 to 500 mg/L were tested (Figure 5b). Increasing the initial PMS concentration from 100 to 300 mg/L, the RB5 dye removal efficiency increased from 91.4 to 97.3% with a significant difference ($p < 0.05$) and similarly the Pseudo-second-order kinetic rate constant K_2 values of RB5 dye removal rose from 0.0281 to 0.0952 L/mg·min. This phenomenon may be due to the fact that when PMS concentration increases, the contact opportunities for catalytic material and PMS increase more, which generates more radicals [40]. However, when the initial PMS concentration increased to 500 mg/L, the RB5 dye removal efficiency decreased to 94.8% and there was no difference between 300 and 500 mg/L PMS statistically ($p > 0.05$). The result was consistent with the Pseudo-second-order kinetic model (Figure 5b and Table S1), in which the corresponding kinetic constant value declined to 0.0372 L/mg·min. The reason can be attributed to the self-quenching of excess ROS because of the excessive PMS concentration (Equations (5) and (6)) [38]. Additionally, using excessive PMS concentration can lead to a high sulfate concentration in the wastewater, potentially resulting in undesirable secondary contamination [44]. According to these results, and minimize operating costs, a 300 mg/L PMS concentration was chosen for the following experiments.

$$HSO_5{}^- + SO_4{}^{\bullet -} \rightarrow SO_5{}^{\bullet -} + HSO_4{}^- \quad (5)$$

$$HSO_5{}^- + HO^\bullet \rightarrow SO_4^- + H_2O + 0.5O_2 \quad (6)$$

3.1.4. Effect of Initial pH on the RB5 Dye Removal

The pH level of the solution had a significant impact on the removal of the RB5 dye from the wastewater. To investigate the effect of initial pH, the RB5 dye removal was studied across a range of pH values from 3 to 11 (Figure 5c). As the initial pH increased from 3 to 7, the RB5 dye removal efficiency remained relatively consistent at around 97% with no significant difference ($p > 0.05$). This shows the ability of the material to operate under neutral pH levels without the need for pH adjustment. However, the efficiency decreased dramatically to 29.8% when the pH reached 11. These results were confirmed by the pseudo-second-order kinetic model, which showed the corresponding rate constant values increasing from 0.0747 to 0.0952 L/mg·min as the pH increased from 3 to 7, before dropping to 0.0009 L/mg·min at a pH of 11 (Figure 5c, Table S1).

Persulfates are more effectively activated at lower pH values; the formed sulfate radicals, which have a high reactivity for organic dye decomposition, depend strongly on this parameter. The literature states that if an initial pH value is less than 3, there will be a higher rate of reaction due to a high amount of hydrogen ions that may easily help generate reactive sulfate radicals [40]. When the pH is increased from neutral values, the activation efficiency decreases. It was suggested that when the initial pH value is above 3.5, it may rise very fast as a result of the increase in OH^- ions that come from the dissociation of impurities like Li_3PO_4 present in solution. This can be a consequence of the stability of PMS. Persulfate has lower stability at alkaline conditions, which causes it to break down automatically and hence is less available for the next reactions [45]. Moreover, another factor is that when pH becomes higher, there is a tendency for the radicals generated (e.g., $SO_4{}^{\bullet -}$ and $HO^\bullet$) to recombine rather than reacting with organic pollutants, thus further decreasing the degradation rate [46].

3.2. *Identification of Reactive Oxygen Species*

To examine the effect of the dominant reactive oxygen species, such as $SO_4{}^{\bullet -}$, $HO^\bullet$ and ${}^1O_2$ generated during the reaction on the RB5 dye removal, the scavenging experiments were carried out by using various quenchers. The TBA, EtOH, and NaN3 were added to the RB5 dye solution to inhibit the reactive oxygen species during the removal of the RB5 dye. The RB5 dye removal efficiency was 97.3% without the addition of any quenchers, but decreased to 66.7, 56.3, and 51.5% after the introduction of 1500 mM TBA, NaN_3, and EtOH, respectively (Figure 5d). All of the quenchers led to significant differences ($p < 0.05$) in removal efficiencies, confirming the contribution of reactive species in treatments. Also, the corresponding kinetic constant value was 0.0952 L mg^{-1} min^{-1} without the addition of any quenchers, but decreased to 0.0043, 0.0031, and 0.0026 L mg^{-1} min^{-1} after the introduction of the TBA, NaN_3, and EtOH, respectively (Figure 5d, Table S1). Based on these results, it was concluded that $HO^{\bullet}$ (approximately 30.3% contribution) appeared to be more dominant than $SO_4^{\bullet-}$ (approximately 15.2% contribution). Previous studies show that either $SO_4^{\bullet-}$ or $HO^{\bullet}$ could contribute more to the oxidation process upon activation of PMS by $LiFePO_4$ [12,40]. This difference might be related to other experimental factors. Moreover, the results of the NaN_3 quenching suggest the presence of $^1O_2$, which reveals the contribution of a non-radical pathway. Several other studies in which $LiFePO_4$ and other cathode materials have been used as PMS activator have reported the presence of $^1O_2$ [6,39,47]. For instance, Xu et al. (2023) [39] found $^1O_2$ to be the primary reactive oxygen species after using $LiNixCoyMn_{1-x-y}O_2$, NCM as an activator for PMS. The results in this study indicated that reactive oxygen species. $SO_4^{\bullet-}$, $HO^{\bullet}$ and $^1O_2$ were all responsible for the RB5 dye removal.

3.3. Catalytic Mechanisms of LFP@PVDF/PMS System

The LFP@PVDF/PMS system's potential reaction pathways have been demonstrated (Equations (7)–(12)). It can be inferred that reactive radicals ($HO^{\bullet}$ and $SO_4^{\bullet-}$) are originated from the activation of PMS by LFP surface ≡Fe(II) (Equations (7) and (8)) [12,48]. Furthermore, the $\equiv Fe^{2+}$ sites on LFP are able to form a $\equiv Fe^{2+}$-OH complex in aqueous environments. This complex subsequently reacts with PMS, yielding the $\equiv Fe^{2+}\text{-OH-OSO}_3^-$ complex (Equation (9)). This complex might generate $\equiv Fe^{3+}$-OH and $SO_4^{\bullet-}$ (Equation (10)) [49] which leads to the reaction of $SO_4^{\bullet-}$ with OH or H_2O in solution to form $HO^{\bullet}$ (Equations (11) and (12)) [12,50,51]. In conclusion, $HO^{\bullet}$ and $SO_4^{\bullet-}$ radicals generated during PMS activation of the LFP membrane catalyst played an important role in RB5 dye removal.

$$\equiv \mathrm{Fe(II)} + \mathrm{HSO_5}^- \rightarrow \equiv \mathrm{Fe(III)} + \mathrm{HO}^{\bullet} + \mathrm{SO_4^{2-}} \quad (7)$$

$$\equiv \mathrm{Fe(II)} + \mathrm{HSO_5}^- \rightarrow \equiv \mathrm{Fe(III)} + \mathrm{SO_4}^{\bullet-} + \mathrm{OH}^- \quad (8)$$

$$\equiv \mathrm{Fe^{2+}\text{-}OH^-HSO_5}^- \rightarrow \equiv \mathrm{Fe^{2+}\text{-}(HO)\text{-}OSO_3^-} + \mathrm{OH}^- \quad (9)$$

$$\mathrm{Fe^{2+}\text{-}(HO)\text{-}OSO_3^-} \rightarrow \equiv \mathrm{Fe^{3+}\text{-}OH^-} + \mathrm{SO_4}^{\bullet-} \quad (10)$$

$$\mathrm{SO_4}^{\bullet-} + \mathrm{OH}^- \rightarrow \mathrm{SO_4^{2-}} + \mathrm{HO}^{\bullet} \quad (11)$$

$$\mathrm{SO_4}^{\bullet-} + \mathrm{H_2O} \rightarrow \mathrm{SO_4^{2-}} + \mathrm{HO}^{\bullet} + \mathrm{H}^+ \quad (12)$$

3.4. Reusability of the Catalytic Membrane and Cost Analysis

Reusability tests were conducted for three successive rounds to determine the reusability of the LFP@PVDF catalytic membrane. The RB5 dye removal efficiency was 97.3% for the first cycle; then, it decreased dramatically to 41.9% for the second one and further reduced to 37.3% in the third cycle (Figure 6). With increasing number of cycles, the efficiency of dye removal decreases for two possible main reasons: One might be that the active sites within the catalytic material are occupied after repeated cycles [40]. Also, the by-products generated during the reaction of dye removal could deposit on the surface of the catalytic material, thus blocking its active sites [6]. After the first cycle of the Rb5 dye experiment, FT-IR analysis of the LFP@PVDF membrane revealed a new peak at 682.5 cm^{-1}

(Supplementary Materials, Figure S5). This peak may be due to SO_4 formed as a result of RB5/LFP/PMS interactions [52,53], which might have accumulated on the catalyst surface and reduced its efficiency. The LFP catalytic material, with PMS as an activator, showed a good result in the removal of dye in the first cycle, but could not be reused efficiently to remove dye after that. A similar result was obtained with the LFP catalytic material using persulfate as an activating agent in the breakdown of orange G dye [40]. Lin et al. [40] also reported that the primary reason for the decreased removal efficiency of the LFP material is the partial degradation of the catalyst's active sites as the reaction is repeated and the dissolution and leaching of small amounts of Fe, PO_4^{3-}, and Li ions from the surface. This leaching, even at low levels, weakens the structural stability of the catalyst and partially limits the heterogeneous activation process. Moreover, regeneration using an acid (HCl) or a base (NaOH) also did not improve the reusability of the membrane. Further research is needed to improve the stability and reusability of LFP-based materials. Alternative regeneration methods that may be more effective in membrane reuse should be investigated.

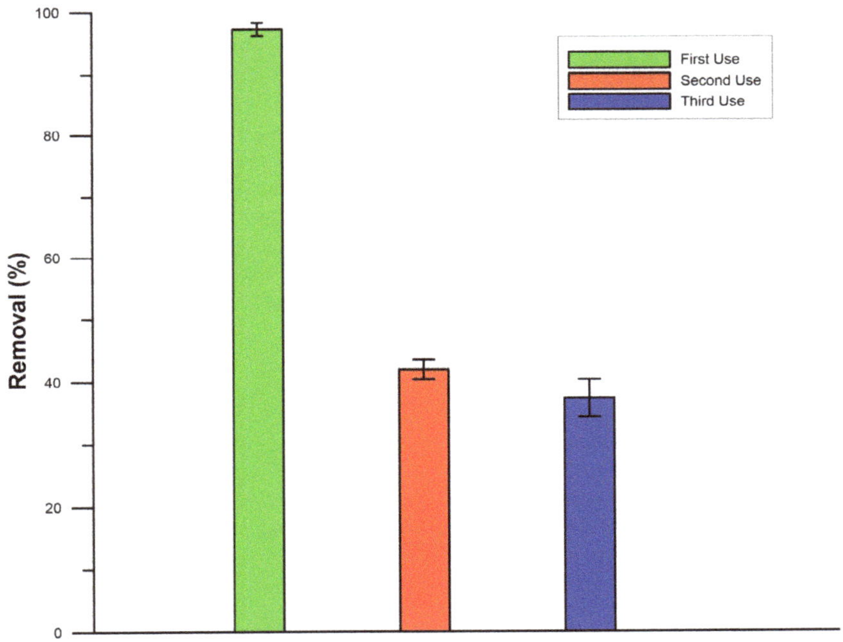

Figure 6. Reusability of the LFP@PVDF catalytic membrane (20 mg/L RB5, 0.3 mg/cm^2 LFP loading, 300 mg/L PMS, pH = 7). Error bars represent the standard error for triplicate reactors.

Table 2 shows the cost calculation of the materials used in catalytic membrane production and the amount of PMS used in the experiments. For example, in the treatment of a 20 mg/L RB5 solution in a volume of 50 mL using a 0.3 mg/cm^2 LFP@PVDF catalytic membrane and 300 mg/L PMS, the chemical cost is approximately 0.029 €. These experiments were conducted under low vacuum pressure (−250 mm Hg; equivalent to −0.33 bar), and the removal efficiency of the system was 97.3%. Similar efficiencies can also be achieved with nanofiltration (NF) or reverse osmosis (RO) membrane systems. However, NF membranes generally require high pressures in the range of 5–20 bar, while RO membranes require high pressures in the range of 7–60 bar [54]. Therefore, these membranes require higher initial investment and operating costs compared to MF membranes.

The high removal efficiency obtained in this study suggests that this method could be considered for industrial-scale applications in the future.

Table 2. Cost analysis of Materials used in catalytic membrane preparation and PMS usage.

Material	Supplier Price	Amount Used (Exp. Conc.)	Cost of Amount Used * (€)
PMS	1 Kg = 154.00 € (Merck)	150 mg (300 mg/L)	0.0231
LFP	150 g = 225.40 USD (MTI Corporation)	3582 mg/11.94 cm^2 (0.3 mg/cm^2)	0.0063
PVDF	200 × 300 mm = 97.67 USD (Sterlitech, Washington, DC, USA)	11.94 cm^2 (active membrane surface area)	0.0002
Total Cost			0.0296

* Note: In cost calculations, the Euro/Dollar ratio was taken as 1.17.

4. Conclusions

The present work investigated the removal of RB5 dye by an LFP@PVDF membrane/persulfate activation system using LFP as an activator and PMS as an oxidant. It was found that the LFP@PVDF membrane can efficiently mineralize RB5 dye in the presence of PMS on a dead-end filtration system. The removal of RB5 by the LFP@PVDF membrane/persulfate activation system follows pseudo-second-order kinetics. Higher LFP loading and PMS concentration enhance removal efficiencies in the experimental ranges tested in this study. In contrast, higher initial dye concentration or higher pH inhibits contaminant removal. The dominant species responsible for the contaminant removal were $SO_4^{\bullet-}$, $HO^{\bullet}$, and $^1O_2$. The LFP@PVDF membrane/persulfate activation system offers an efficient and environmentally friendly approach for degrading organic pollutants and provides a new perspective on the resource utilization of lithium-ion battery materials. Further research can focus on improving the reusability of the LFP@PVDF membrane and upscaling the process for large-scale applications.

Supplementary Materials: The following supporting information can be downloaded at: https://www.mdpi.com/article/10.3390/su17188469/s1, Figure S1: Vacuum filtration system used for RB5 dye de-colorization experiments; Figure S2: SEM image of LFP particles on the PVDF membrane surface; Figure S3: EDS analysis result of the LFP@PVDF membrane; Figure S4: EDS analysis result of the PVDF membrane; Figure S5: FTIR spectra: of LFP@PVDF membrane; Table S1: The values of the pseudo-first- and -second-order rate constants and the corresponding correlation coefficients for removal of RB5 by different systems; Table S2: ANOVA results of treatments tested with different LFP loadings and with/without LFP or PMS; Table S3: Tukey test results of treatments tested with different LFP loadings and with/without LFP or PMS; Table S4: ANOVA results of treatments tested with different RB5 concentrations; Table S5: Tukey test results of treatments tested with different RB5 concentrations; Table S6: ANOVA results of treatments tested with different PMS concentra-tions; Table S7: Tukey test results of treatments tested with different PMS concentrations; Table S8: ANOVA results of treatments tested with different initial pH values; Table S9: Tukey test results of treatments tested with different initial pH values; Table S10: ANOVA results of treatments tested with different quenchers; Table S11: Tukey test results of treatments tested with different quenchers; Table S12: ANOVA results of three cycles tested in reusability experiments; Table S13: Tukey test results of three cycles tested in reusability experiments; Table S14: Examples of advanced oxidation studies adopting persulfates and various catalysts.

Author Contributions: A.K.T.: Project administration, Conceptualization, Methodology, Formal Analysis, Investigation, Writing. B.F.K.: Conceptualization: Methodology, Formal Analysis, Investigation, Writing—Review and Editing. S.E.: Formal Analysis, Investigation, Writing—Review and Editing. All authors have read and agreed to the published version of the manuscript.

Funding: This research received no external funding.

Institutional Review Board Statement: Not applicable.

Informed Consent Statement: Not applicable.

Data Availability Statement: The relevant data from this study are available from the corresponding author upon request.

Acknowledgments: The authors would like to thank the support provided by the Zonguldak Bulent Ecevit University.

Conflicts of Interest: The authors declare no conflicts of interest.

References

1. Leal, V.M.; Ribeiro, J.S.; Coelho, E.L.D.; Freitas, M.B.J.G. Recycling of spent lithium-ion batteries as a sustainable solution to obtain raw materials for different applications. *J. Energy Chem.* **2023**, *79*, 118–134. [CrossRef]
2. Guo, H.; Sun, L.; Qiu, J.; Wu, J.; Wang, X.; Li, X.; Zhao, Y.; Wang, S.; Zeng, G. Sustainable recycling of $LiCoO_2$ cathode scrap on the basis of successive peroxymonosulfate activation and recovery of valuable metals. *Sci. Total Environ.* **2021**, *759*, 143478. [CrossRef] [PubMed]
3. He, T.; Zhao, J.; Chen, D.; Gu, S.; Yu, J. Self-actuated leaching and integrated separation of spent lithium-ion batteries cathode and anode sheets. *Sep. Purif. Technol.* **2024**, *345*, 127396. [CrossRef]
4. Zheng, S.; Zhu, L.; Lin, Z.; Kong, X.; Liu, Y.; Cao, S.; Ding, K.; Zhao, G. A review of cathode and electrolyte recovery from spent lithium-ion batteries: Recent technologies, processes and policies. *Resour. Chem. Mater.* **2024**, *3*, 188–229. [CrossRef]
5. Pavithra, S.; Sakunthala, A. Recovery Approaches for Spent Batteries: A Review. *Prabha Mater. Sci. Lett.* **2024**, *3*, 298–317. [CrossRef]
6. Zhao, Y.; Yuan, X.; Jiang, L.; Li, X.; Zhang, J.; Wang, H. Reutilization of cathode material from spent batteries as a heterogeneous catalyst to remove antibiotics in wastewater via peroxymonosulfate activation. *Chem. Eng. J.* **2020**, *400*, 125903. [CrossRef]
7. Wang, P.; Guo, Y.; Guan, J.; Wang, Z. Applications of Spent Lithium Battery Electrode Materials in Catalytic Decontamination: A Review. *Catalysts* **2023**, *13*, 189. [CrossRef]
8. Bell, A.T. The Impact of Nanoscience on Heterogeneous Catalysis. *Science* **2003**, *299*, 1688–1691. [CrossRef]
9. Balakrishnan, N.T.M.; Paul, A.; Krishnan, M.A.; Das, A.; Raphaez, L.R.; Ahn, J.H.; Fatima, M.J.J.; Prasanth, R. Lithium Iron Phosphate ($LiFePO_4$) as High-Performance Cathode Material for Lithium Ion Batteries. In *Metal, Metal-Oxides and Metal Sulfides for Batteries, Fuel Cells, Solar Cells, Photocatalysis and Health Sensors*; Rajendran, S., Karimi-Maleh, H., Qin, J., Lichtfouse, E., Eds.; Springer International Publishing: Cham, Switzerland, 2021; pp. 35–73. [CrossRef]
10. Chen, C.; Ma, H.; Wang, Y. The Progress and Future Prospects of Lithium Iron Phosphate Cathode Materials. *Highlights Sci. Eng. Technol.* **2022**, *3*, 31–42. [CrossRef]
11. The International Energy Agency. Global EV Outlook 2024. Available online: https://www.iea.org/reports/global-ev-outlook-2024 (accessed on 8 July 2025).
12. Wang, P.; Lou, X.; Chen, Q.; Liu, Y.; Sun, X.; Guo, Y.; Zhang, X.; Wang, R.; Wang, Z.; Chen, S.; et al. Spent $LiFePO_4$: An old but vigorous peroxymonosulfate activator for degradation of organic pollutants in water. *Environ. Res.* **2022**, *214*, 113780. [CrossRef]
13. Arica, T.A.; Balci, F.M.; Balci, S.; Arica, M.Y. Highly Porous Poly(o-phenylenediamine) Loaded Magnetic Carboxymethyl Cellulose Hybrid Beads for Removal of Two Model Textile Dyes. *Fibers Polym.* **2022**, *23*, 2838–2854. [CrossRef]
14. Forgacs, E.; Cserháti, T.; Oros, G. Removal of synthetic dyes from wastewaters: A review. *Environ. Int.* **2004**, *30*, 953–971. [CrossRef]
15. Liu, Y.; Chen, J.; Duan, D.; Zhang, Z.; Liu, C.; Cai, W.; Zhao, Z. Environmental Impacts and Biological Technologies Toward Sustainable Treatment of Textile Dyeing Wastewater: A Review. *Sustainability* **2024**, *16*, 10867. [CrossRef]
16. Yu, M.; Li, H.; Li, K.; Li, Y.; Liu, F.; Zhang, G.; Xiao, T.; Zhang, P.; Zhang, H.; Long, J. Magnetite-Based Biochar Coupled with Binary Oxidants for the Effective Removal of Mixed Dye from Wastewater. *Fibers Polym.* **2022**, *23*, 450–462. [CrossRef]
17. Zhang, D.; Sun, H.; Yu, B.; Xie, Y.; Li, F. Preparation of MnO_2@PP MB for Organic Dyes Removal. *Fibers Polym.* **2025**, *26*, 1251–1264. [CrossRef]
18. Kader, N.; Değermenci, G.D. Magnetic biochar production from agricultural waste and reactive blue 19 removal by peroxymonosulfate activation. *Fibers Polym.* **2024**, *26*, 1197–1208. [CrossRef]
19. Han, S.-J.; Xu, L.; Chen, C.; Wang, Z.-Y.; Fu, M.-L.; Yuan, B. Recovery of graphite from spent lithium-ion batteries and its wastewater treatment application: A review. *Sep. Purif. Technol.* **2024**, *330*, 125289. [CrossRef]

20. Wang, P.; Lou, X.; Sun, X.; Chen, Q.; Liu, Y.; Guo, Y.; Zhang, X.; Guan, J.; Wang, R.; Zhang, R.-Q.; et al. Spent rather than pristine $LiFePO_4$ cathode materials can catalytically activate sulfite for organic pollutants decontamination. *Chem. Eng. J.* **2022**, *446*, 137123. [CrossRef]
21. Zhang, Y.; Wu, B.; Xu, H.; Liu, H.; Wang, M.; He, Y.; Pan, B. Nanomaterials-enabled water and wastewater treatment. *NanoImpact* **2016**, *3–4*, 22–39. [CrossRef]
22. Zhao, W.; Chen, I.W.; Huang, F. Toward large-scale water treatment using nanomaterials. *Nano Today* **2019**, *27*, 11–27. [CrossRef]
23. Cai, Z.; Sun, Y.; Liu, W.; Pan, F.; Sun, P.; Fu, J. An overview of nanomaterials applied for removing dyes from wastewater. *Environ. Sci. Pollut. Res.* **2017**, *24*, 15882–15904. [CrossRef]
24. Nasrollahzadeh, M.; Sajjadi, M.; Iravani, S.; Varma, R.S. Green-synthesized nanocatalysts and nanomaterials for water treatment: Current challenges and future perspectives. *J. Hazard. Mater.* **2021**, *401*, 123401. [CrossRef]
25. Lalia, B.S.; Kochkodan, V.; Hashaikeh, R.; Hilal, N. A review on membrane fabrication: Structure, properties and performance relationship. *Desalination* **2013**, *326*, 77–95. [CrossRef]
26. Warsinger, D.M.; Chakraborty, S.; Tow, E.W.; Plumlee, M.H.; Loutatidou, S.; Karimi, L.; Mikelonis, A.M.; Achilli, A.; Ghassemi, A.; Padhye, L.P.; et al. A review of polymeric membranes and processes for potable water reuse. *Prog. Polym. Sci.* **2018**, *81*, 209–237. [CrossRef]
27. Yao, L.; Zhang, L.Z.; Wang, R.; Loh, C.H.; Dong, Z.L. Fabrication of catalytic membrane contactors based on polyoxometalates and polyvinylidene fluoride intended for degrading phenol in wastewater under mild conditions. *Sep. Purif. Technol.* **2013**, *118*, 162–169. [CrossRef]
28. Santana, C.S.; Ramos, M.D.N.; Velloso, C.C.V.; Aguiar, A. Kinetic evaluation of dye decolorization by Fenton processes in the presence of 3-hydroxyanthranilic acid. *Int. J. Environ. Res. Public Health* **2019**, *16*, 1602. [CrossRef]
29. Sun, S.P.; Li, C.J.; Sun, J.H.; Shi, S.H.; Fan, M.H.; Zhou, Q. Decolorization of an azo dye Orange G in aqueous solution by Fenton oxidation process: Effect of system parameters and kinetic study. *J. Hazard. Mater.* **2009**, *161*, 1052–1057. [CrossRef]
30. Topaloğlu, A.K.; Yildirim, Y. Removal of reactive black 5 dye by using polyoxometalate-membrane. *Membr. Water Treat.* **2021**, *12*, 23–35. [CrossRef]
31. Awad, A.M.; de Lannoy, C.-F. Electrochemical Poly(vinylidene fluoride) (PVDF) Membranes Using Polyethylenimine Cross-Linked Polydopamine-Bound Carbon Nanotubes. *ACS Appl. Polym. Mater.* **2024**, *6*, 4985–4998. [CrossRef]
32. Bai, H.; Wang, X.; Zhou, Y.; Zhang, L. Preparation and characterization of poly(vinylidene fluoride) composite membranes blended with nano-crystalline cellulose. *Prog. Nat. Sci. Mater. Int.* **2012**, *22*, 250–257. [CrossRef]
33. Obaid, M.; Ghouri, Z.K.; Fadali, O.A.; Khalil, K.A.; Almajid, A.A.; Barakat, N.A.M. Amorphous SiO_2 NP-Incorporated Poly(vinylidene fluoride) Electrospun Nanofiber Membrane for High Flux Forward Osmosis Desalination. *ACS Appl. Mater. Interfaces* **2016**, *8*, 4561–4574. [CrossRef] [PubMed]
34. Julien, C.M.; Mauger, A. Nano Aspect of Vibration Spectra Methods in Lithium-Ion Batteries. In *Nanoscale Technology for Advanced Lithium Batteries*; Springer: New York, NY, USA, 2014; pp. 167–206. [CrossRef]
35. Muruganantham, R.; Sivakumar, M.; Subadevi, R. Synthesis and electrochemical characterization of olivine-type lithium iron phosphate cathode materials via different techniques. *Ionics* **2016**, *22*, 1557–1565. [CrossRef]
36. Lin, F. Study on the performance of nanoparticle-modified PVDF membrane in delaying membrane aging. *Open Chem.* **2024**, *22*, 20240124. [CrossRef]
37. Yang, Y.; Banerjee, G.; Brudvig, G.W.; Kim, J.-H.; Pignatello, J.J. Oxidation of Organic Compounds in Water by Unactivated Peroxymonosulfate. *Environ. Sci. Technol.* **2018**, *52*, 5911–5919. [CrossRef]
38. Zhao, Y.; Wang, H.; Ji, J.; Li, X.; Yuan, X.; Duan, A.; Guan, X.; Jiang, L.; Li, Y. Recycling of waste power lithium-ion batteries to prepare nickel/cobalt/manganese-containing catalysts with inter-valence cobalt/manganese synergistic effect for peroxymonosulfate activation. *J. Colloid. Interface Sci.* **2022**, *626*, 564–580. [CrossRef] [PubMed]
39. Xu, L.; Liang, L.; Chen, C.; Chen, Z.-H.; Lv, Z.-B.; Fu, M.-L.; Yuan, B. Upcycling spent lithium battery cathodes into efficient PMS catalysts for organic contaminants degradation. *J. Environ. Chem. Eng.* **2023**, *11*, 111605. [CrossRef]
40. Lin, X.M.; Ma, Y.W.; Wang, Y.; Wan, J.Q.; Guan, Z.Y. Lithium iron phosphate ($LiFePO_4$) as an effective activator for degradation of organic dyes in water in the presence of persulfate. *RSC Adv.* **2015**, *5*, 94694–94701. [CrossRef]
41. Yu, B.; Li, Z.; Zhang, S. Zero-Valent Copper-Mediated Peroxymonosulfate Activation for Efficient Degradation of Azo Dye Orange G. *Catalysts* **2022**, *12*, 700. [CrossRef]
42. Topaloğlu, A.K.; Kahraman, B.F. Textile dye removal in wastewater by peroxymonosulfate (PMS) activation on a zero-valent iron nanoparticle–modified ultrafiltration catalytic membrane (nZVI@PES). *Environ. Sci. Pollut. Res.* **2023**, *30*, 94779–94789. [CrossRef]
43. El-monem, H.A.; Mahanna, H.; El-Halwany, M.; Samy, M. Photo-thermal activation of persulfate for the efficient degradation of synthetic and real industrial wastewaters: System optimization and cost estimation. *Environ. Sci. Pollut. Res.* **2024**, *31*, 24153–24162. [CrossRef]
44. Oh, W.-D.; Dong, Z.; Lim, T.-T. Generation of sulfate radical through heterogeneous catalysis for organic contaminants removal: Current development, challenges and prospects. *Appl. Catal. B* **2016**, *194*, 169–201. [CrossRef]

45. Li, J.; Shi, Q.; Sun, M.; Liu, J.; Zhao, R.; Chen, J.; Wang, X.; Liu, Y.; Gong, W.; Liu, P.; et al. Peroxymonosulfate Activation by Facile Fabrication of α-MnO_2 for Rhodamine B Degradation: Reaction Kinetics and Mechanism. *Molecules* **2023**, *28*, 4388.
46. Zhang, Y.; Kang, W.; Yu, H.; Chen, S.; Quan, X. Electrochemical activation of peroxymonosulfate in cathodic micro-channels for effective degradation of organic pollutants in wastewater. *J. Hazard. Mater.* **2020**, *398*, 122879. [CrossRef]
47. Liang, J.; Xue, Y.; Gu, J.-N.; Li, J.; Shi, F.; Guo, X.; Guo, M.; Min, X.; Lia, K.; Sun, T.; et al. Sustainably recycling spent lithium-ion batteries to prepare magnetically separable cobalt ferrite for catalytic degradation of bisphenol A via peroxymonosulfate activation. *J. Hazard. Mater.* **2022**, *427*, 127910. [CrossRef] [PubMed]
48. Brandt, C.; Fabian, I.; van Eldik, R. Kinetics and Mechanism of the Iron(III)-catalyzed Autoxidation of Sulfur(IV) Oxides in Aqueous Solution. Evidence for the Redox Cycling of Iron in the Presence of Oxygen and Modeling of the Overall Reaction Mechanism. *Inorg. Chem.* **1994**, *33*, 687–701. [CrossRef]
49. Tan, C.; Gao, N.; Fu, D.; Deng, J.; Deng, L. Efficient degradation of paracetamol with nanoscaled magnetic $CoFe_2O_4$ and $MnFe_2O_4$ as a heterogeneous catalyst of peroxymonosulfate. *Sep. Purif. Technol.* **2017**, *175*, 47–57. [CrossRef]
50. Kuo, D.T.F.; Kirk, D.W.; Jia, C.Q. The chemistry of aqueous S(IV)-Fe-O_2 system: State of the art. *J. Sulfur. Chem.* **2006**, *27*, 461–530. [CrossRef]
51. Hong, Y.; Zhou, H.; Xiong, Z.; Liu, Y.; Yao, G.; Lai, B. Heterogeneous activation of peroxymonosulfate by CoMgFe-LDO for degradation of carbamazepine: Efficiency, mechanism and degradation pathways. *Chem. Eng. J.* **2020**, *391*, 123604. [CrossRef]
52. Makreski, P.; Jovanovski, G.; Dimitrovska, S. Minerals from Macedonia: XIV. Identification of some sulfate minerals by vibrational (infrared and Raman) spectroscopy. *Vib. Spectrosc.* **2005**, *39*, 229–239. [CrossRef]
53. Yusuf, M.O. Characterization in Cementitious Material Binders Using Fourier-Transform Infrared Spectroscopy. *Appl. Sci.* **2023**, *13*, 3353. [CrossRef]
54. Anim-Mensah, A.R. Pressure-driven polymeric membrane performance prediction, new membrane dimensionless number, and considerations for effective membrane design, selection, testing, and operation. *Front. Membr. Sci. Technol.* **2025**, *3*, 1454589. [CrossRef]

water

MDPI

Article

Regenerable Biochar Catalyst from Biogas Residue for Peroxymonosulfate Activation in Bisphenol A-Containing Wastewater Treatment

Yating Pan [1], Xue Yang [2], Haijuan Wei [3], Xiang Liu [1,*], Pan Wang [2], Nina Duan [2] and Miao Lin [4,*]

1 Department of Environmental Science and Engineering, Fudan University, Shanghai 200438, China; 22210740064@m.fudan.edu.cn
2 Shanghai Municipal Engineering Design Institute (Group) Co., Ltd., Shanghai 200092, China; yangxue@smedi.com (X.Y.); wangpan@smedi.com (P.W.); duannina@smedi.com (N.D.)
3 Shanghai Chengtou Wastewater Treatment Co., Ltd., Shanghai 201203, China; weihaijuan@shanghaiwater.com
4 Shanghai Academy of Agricultural Sciences, Shanghai 201403, China
* Correspondence: liuxiang@fudan.edu.cn (X.L.); linmiao@saas.sh.cn (M.L.)

Abstract: The biogas residue (BR) from the anaerobic digestion of sludge poses a threat to the environment due to the presence of toxic and hazardous substances. Furthermore, emerging contaminants, such as bisphenol A (BPA), are widespread in domestic and industrial wastewater, requiring efficient and sustainable treatment technologies. In this study, the BR-based biochar was pyrolyzed from urea-modified BR and employed as a catalyst to activate peroxymonosulfate (PMS) for BPA degradation. With BR-based biochar pyrolyzed at 750 °C as a catalyst, 20 mg/L of BPA was completely removed. Free radical detection confirmed that hydroxyl radical (•OH) and sulfate radical ($\bullet SO_4^-$) generation decreased with the increase in catalyst reuse times. The X-ray photoelectron spectra showed that the catalyst deactivation mainly resulted from -COOH and C-OH group loss, which acted as active sites for generating •OH and $\bullet SO_4^-$, and HCl or NaOH regeneration for catalysts could recover oxygen-containing functional groups, boosting BPA removal from 54.7% to 91.5% and 100%, respectively. Thermal regeneration could only enlarge the catalyst's specific surface area (SSA) to recover adsorption capacity, but might not restore the free radical generation capability. This research offered a theoretical basis for the sustainable utilization of BR and provided a reference for reusing catalysts in wastewater treatment.

Keywords: biogas residue; biochar; wastewater; peroxymonosulfate; advanced oxidation process; catalyst regeneration

1. Introduction

The rapid pace of urbanization has led to a continuous rise in the quantity of sludge generated by municipal wastewater treatment plants (WWTPs) [1]. Anaerobic digestion (AD) has become one of the most widely used methods for stabilizing sludge [2]. This process involves the decomposition of organic matter in the sludge by microorganisms in an anoxic environment, resulting in the production of biogas, with biogas residue (BR) remaining. The BR contains an array of toxic and harmful substances, such as heavy metals, pathogens, and persistent organic pollutants [3]. The improper disposal of BR could result in the leaching of hazardous substances into aquatic systems, posing significant risks to water quality and ecosystem health [4]. Among the disposal methods for BR, pyrolysis could produce cost-effective and functional biochar and avoid the problem of secondary pollution caused by incineration and landfill disposal [5]. Additionally, the solid product, biochar, characterized by its high specific surface area (SSA) and abundant catalytic groups, is extensively utilized for the adsorption and catalytic removal of pollutants in wastewater [6].

In recent years, increasing concerns have been raised regarding the accumulation of endocrine-disrupting chemicals (EDCs) in the water environment and their associated adverse effects [7]. Among these, bisphenol A (BPA), a representative EDC, is widely used in industrial applications and can be released into aquatic environments through product leaching and wastewater discharge [8]. Its presence poses significant risks to ecosystems and biodiversity, while also threatening human reproductive health and developmental processes [9]. Sulfate-radical-based advanced oxidation processes (SR-AOPs) have emerged as an effective strategy for eliminating these emerging contaminants in water environments [10]. In recent years, a variety of materials, including carbon-based [11], aluminum-based [12], and perovskites materials [13], have been extensively investigated as catalysts for peroxymonosulfate (PMS) activation, demonstrating excellent catalytic performance. Due to the simple preparation methods and low cost, BR-based biochar has been utilized as a catalyst in PMS activation to degrade contaminants [14].

However, current research primarily focuses on exploring the mechanisms of activating PMS with new catalysts [15], with comparatively limited attention given to the deactivation and regeneration of catalysts and the underlying mechanisms involved. In practical wastewater treatment, the decrease in pollutant removal efficiency caused by catalyst deactivation was inevitable, thereby limiting the long-term sustainability of heterogeneous advanced oxidation processes (HAOPs) [16,17]. Although some studies have identified some reasons for catalyst deactivation, such as the adsorption of intermediates of contaminants, leading to the coverage of catalytic active sites and inactivation [18], the systematic and comprehensive investigation of the factors and mechanisms contributing to catalyst deactivation still needs to be explored.

Therefore, ensuring the stability and efficiency of HAOPs in wastewater treatment requires effective catalyst regeneration strategies. It is worth noting that physical and chemical modification methods, including acid, alkali, and thermal treatments, have been explored to restore catalytic activity [19,20]. Current research on catalyst deactivation and regeneration has primarily focused on Al_2O_3-based catalysts [21]. For carbon-based catalysts, studies have shown that porous carbon could effectively regain its catalytic performance through alkali and thermal treatment [22]. However, research on regenerating sludge-derived biochar catalysts remains limited, despite their potential for viable wastewater treatment [23]. Most importantly, the regeneration outcomes vary with the type of catalyst and regeneration conditions [19]. A comprehensive comparison of different regeneration strategies is essential to optimize catalyst reusability and enhance its practical applicability in water treatment.

This study first synthesized BR-based biochar with urea modification through pyrolysis. Subsequently, the prepared BR-based biochar was utilized to catalytically degrade BPA by activating PMS, and the stability of the catalysts with the increase in catalyst reuse times was evaluated. Then, three methods (HCl, NaOH, and thermal) were used to regenerate the used BR-based biochar catalysts, and the structural and morphological features, surface functional groups, elemental composition, and electron transfer capability of the catalysts before and after use and regeneration were comprehensively investigated and assessed. Combined with the catalytic performance of BPA removal, BR-based biochar's deactivation and regeneration mechanisms for PMS activation were revealed. This study aimed to provide a viable strategy for the environmental disposal of BR and a reference for the efficient use of catalyst recycling in activating PMS.

2. Materials and Methods

2.1. Materials

In this study, the BR was obtained from a water treatment plant in Shanghai, China. Other chemicals were analytical-grade (Text S1) and purchased from Sinopharm Chemical Reagent Co., Ltd. (Shanghai, China). All aqueous solutions were prepared with deionized water (Milli-Q/18.2 MΩ cm).

2.2. BR-Based Biochar Preparation

Firstly, BR was subjected to dehydration and freeze-drying, ground into a uniform powder, and stored at ambient temperature. The BR was mixed with urea at a mass ratio of 1:1 and meticulously ground in an agate mortar for 15 min to ensure thorough blending [24,25]. Then, three ceramic crucibles, each containing 15 g of the prepared samples, were positioned at the center of a quartz tube (length: 1000 mm, inner diameter: 50 mm) inside the tubular furnace to ensure uniform heating. The furnace stopper was secured at the end of the tube, and the system's seal was checked for integrity. Prior to the pyrolysis experiment, N_2 was introduced into the pyrolysis system at a flow rate of 100 mL/min for 10 min to purge the air from the system [26]. Then, the samples were heated gradually in N_2 at a rate of 10 °C/min to 550 °C, 650 °C, and 750 °C, respectively, and each maintained at the target temperature for 1 h [27]. The prepared N-doped biochar samples were named NBC550, NBC650, and NBC750. Contrastingly, the samples without N-doping were prepared under identical thermal conditions and labeled BC550, BC650, and BC750. All the prepared biochar was used as the catalysts for the activation of PMS.

2.3. BPA Degradation by Peroxymonosulfate Activation Test and Catalyst Stability Evaluation

In this study, the performance of the synthesized catalysts was assessed by activating PMS and degrading BPA. Experiments were conducted at room temperature (25.0 ± 1.0 °C) in 100 mL conical flasks using a magnetic stirrer (SN-MS-H280D, Sunne, Shanghai, China) at a rate of 450 rpm. Initially, the catalyst (0.3 g/L) was dispersed in a 100 mL solution containing 20 mg/L BPA and stirred for 30 min to establish adsorption–desorption equilibrium. Subsequently, PMS (0.1 g/L) was added to initiate the reaction. A 1 mL sample was periodically extracted using a pipette, followed by immediate filtration through a 0.45 μm filter membrane for analysis to assess the degradation of the pollutant.

To assess the stability of the catalyst, the NBC750 was subjected to five repeated use tests, and the catalyst after repeated uses was named NBC750-5th. The catalysts were vacuum-filtered, rinsed with ultrapure water, and dried at 60 °C. Furthermore, three methods (HCl, NaOH, and thermal regeneration) were used to regenerate the catalysts sequentially, named as NBC750-HCl, NBC750-NaOH, and NBC750-500 °C, respectively. Specifically, 1 g of deactivated catalyst was added to 100 mL of either 1 M NaOH or 1 M HCl solution and stirred at room temperature for 3 h. After filtration, catalysts were rinsed with deionized water to a neutral pH and dried at 60 °C, yielding regenerated HCl or NaOH regeneration catalysts. Moreover, 1 g of deactivated catalyst was thermally treated at 500 °C under a N_2 atmosphere for 30 min to obtain thermally regenerated catalysts. The reusability of the regenerated catalysts was assessed by reapplying them in PMS activation for BPA degradation, with repeated cycles of regeneration and application. The specific experimental procedures were identical to those used in the stability tests of fresh catalysts.

Reactive oxygen species (ROSs) generated during catalysis on BPA degradation were analyzed using quenching assays and electron paramagnetic resonance (EPR) technology [28]. For quenching experiments, some quenchers were selected based on the literature [29,30], and their concentrations were determined by experimental validation. In this study, 2000 mM of tert-butyl alcohol (TBA) was used to quench free hydroxyl radicals ($\bullet OH_{free}$), while the same dose of ethyl alcohol (EtOH) was applied to quench both $\bullet OH_{free}$ and free sulfate radicals ($\bullet SO_4{}^-{}_{free}$), and 30 mM of dimethyl sulfoxide (DMSO) was employed to quench the surface-bound $\bullet OH$ and $\bullet SO_4{}^-$ ($\bullet OH_{ads}$ and $\bullet SO_4{}^-{}_{ads}$). Moreover, 10 mM of trichloromethane ($CHCl_3$) was applied as a quencher for superoxide anions ($\bullet O_2{}^-$). Each experimental condition was conducted in triplicate.

2.4. Structural Characterizations

The structural and morphological features of the biochar were investigated by a scanning electron microscope (SEM, Sigma 300, Zeiss, Oberkochen, Germany) equipped with an energy dispersive X-ray detector (EDS). The X-ray photoelectron spectroscopy (XPS, Nexsa, Thermo Fisher, Waltham, MA, USA) was employed to identify the elemental composition and content of the catalysts. The N_2 adsorption–desorption isotherms were determined using a fully automated surface and porosity analyzer (Autosorb-iQ, Quantachrome, Boynton Beach, FL, USA). The SSA and pore size distribution of catalysts were determined by the Brunauer–Emmett–Teller (BET) method and the Density Functional Theory (DFT) model. X-ray diffraction (XRD, Panalytical Empyrean, Malvern Panalytical, Almelo, NL, USA) was used to examine the crystal structure of the catalysts. Raman spectra of the materials were obtained at an excitation wavelength of 532 nm by a Raman spectrometer (XploRA, Horiba, Longjumeau, France). Fourier-transform infrared spectroscopy (FT-IR, Nicolet 6700, Thermo Fisher, Waltham, MA, USA) was accepted to detect the surface functional groups of the catalysts. The elemental composition of BR (C, H, N, S) was analyzed by an elemental analyzer (EA3000, Euro Vector, Pavia, Italy), and the O content was calculated by subtracting the ash mass (from drying BR at 600 °C for 1 h in a muffle furnace) and the masses of C, H, N, and S from the total sample weight. Additionally, the electron transfer capability of the catalysts was assessed using an electrochemical workstation (CHI 600C, CH Instruments, Shanghai, China), where a Ag/AgCl electrode served as the reference electrode, a Pt wire electrode acted as the counter electrode, and a glassy carbon electrode (GCE) acted as the working electrode. To prepare the working electrodes, 5 mg of biochar powder was mixed with 5 μL of Nafion solution (5.0 wt%), 200 μL of ethanol, and 400 μL of deionized water. The mixture was then subjected to ultrasonic dispersion for 1 h to form a uniform suspension. Afterward, 6 μL of the suspension was dropped onto the surface of the glassy carbon electrode, and the catalyst-loaded electrode was dried at 60 °C for 20 min.

2.5. Analytical Methods

A high-performance liquid chromatograph (HPLC, Agilent 1260 infinity, Agilent, Santa Clara, CA, USA) was used for the determination of BPA concentration. The mobile phase comprised acetonitrile and water (60/40, v/v) with a 1.0 mL/min flow rate. PMS concentration was measured at 352 nm using a UV spectrophotometer (DR3900, Hach, Loveland, CO, USA). Total organic carbon (TOC) was tested by a total organic carbon analyzer (TOC-L, Shimadu, Kyoto, Japan). EPR spectroscopy (EMX nano, Bruker, Rheinstetten, Germany) with 5,5-dimethyl-1-pyrroline N-oxide (DMPO) as a spin-trapping agent was employed to detect the generated ROSs. The pH of the solution was measured using a pH meter (WTW PH/Oxi340i, Shanghai Jinbi Trading Co., Ltd., Shanghai, China), and the electrode was calibrated before each test. Each experimental condition was conducted in triplicate.

3. Results and Discussion

3.1. Catalytic Performance of BR-Based Biochar Catalysts

The catalytic activity of BR-based biochar catalysts was systematically evaluated by activating PMS and the degradation of BPA in aqueous solutions. Initially, catalysts were stirred in BPA solutions for 30 min to achieve adsorption–desorption equilibrium, and the adsorption ability of the prepared catalysts for BPA was similar at about 27–35%. After the adsorption, it was observable in Figure 1A that the removal rate of BPA within 60 min was only 4.5% without catalysts. In the systems with the addition of the catalysts, removal rates of 94.0% and 88.0% were obtained in NBC650 and NBC550 systems, revealing that a higher preparation temperature for biochar was also beneficial for catalytic activity. In BR-based biochar catalyst systems, N-doping enhanced the catalytic performance of catalysts, with NBC750 achieving 100% BPA removal in 60 min and the highest PMS consumption (Figure 1A,B). As modeled by first-order kinetic fits in various systems, the rate constants (k) for BPA degradation were calculated (Figure 1A), supporting the above conclusion. Moreover, the 69.3% TOC removal within 60 min in the NBC750 system indicated nearly 70% BPA mineralization (Figure S1).

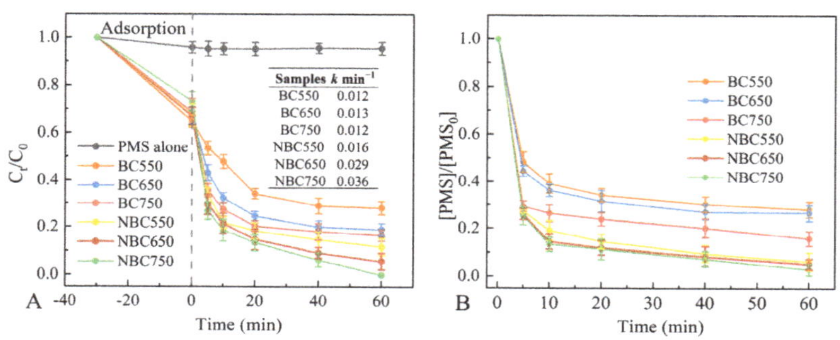

Figure 1. (**A**) Removal performance of BPA catalyzed by different BR-based biochar and corresponding k ($-\ln(C_t/C_0)$ vs. time) values. (**B**) Consumption of PMS in catalyst/PMS/BPA solution.

To investigate the impact factors of BPA degradation, the pollutant degradation tests were carried out under different initial conditions with NBC750 catalyst (Figure S2) [10,31]. The results showed that BPA degradation efficiency nearly peaked at an NBC750 dosage of 0.3 g/L, with no significant improvement at 0.4 g/L. For PMS, increasing the dosage from 0.1 g/L to 0.2 g/L only minimally enhanced BPA degradation. Additionally, NBC750 effectively activated PMS for BPA degradation in the pH range of 3–9, with the highest efficiency at pH 7. Therefore, for subsequent experiments, NBC750 and PMS dosages of 0.3 g/L and 0.1 g/L were selected, with a pH of 7, under the 20 mg/L BPA condition.

3.2. Stability Tests and Reactivation of BR-Based Biochar Catalysts

The stability and reusability of BR-based biochar catalysts are critical for their practical engineering application [32]. As illustrated in Figure 2A, compared with the first run (100% removal in 60 min) of the NBC750 system, the BPA removal decreased to 76.6%, 66.0%, 55.8%, and 54.7% in its continuous use of the 2nd–5th, respectively, which denoted that NBC750 gradually lost its catalytic activity with the increase in catalyst reuse times.

After five consecutive applications, the BR-based biochar catalysts were regenerated through HCl, NaOH, and thermal regeneration, respectively, and then re-employed to activate PMS for BPA degradation. From Figure 2B, the NBC750-500 °C system achieved 50.7% BPA removal, with 23.8% due to adsorption, but showed no sustained catalytic activity after 10 min. However, NBC750-HCl or NBC750-NaOH achieved a 91.5% or 100% BPA removal, demonstrating that acid or alkali methods were more effective than thermal treatment in catalyst regeneration. Consequently, no further experiments were conducted on the thermally regenerated catalysts. In the subsequent four runs, the BPA removal rates with NBC750-HCl were 63.4%, 59.2%, 55.2%, and 53.5%, while those with NBC750-NaOH were 65.5%, 62.3%, 57.3%, and 56.3%, revealing that the HCl and NaOH groups exhibited similar BPA removal rates.

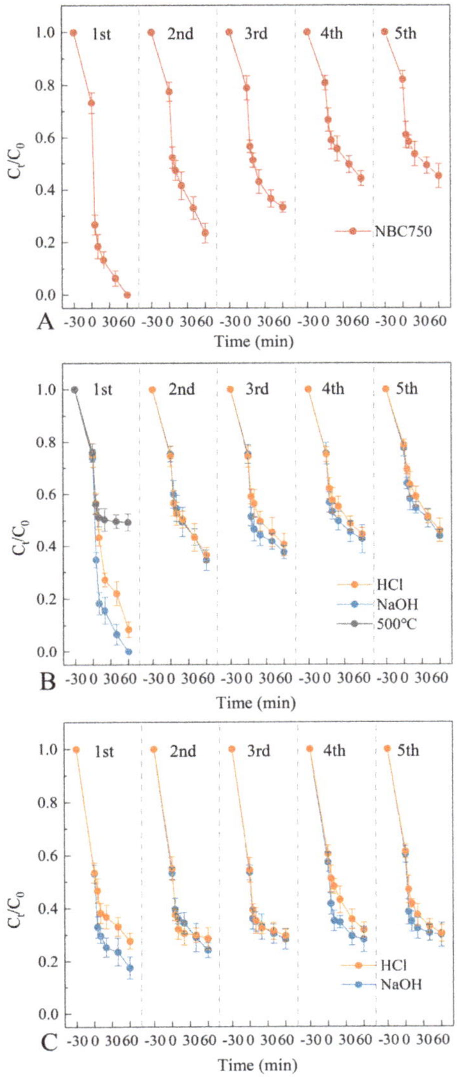

Figure 2. (**A**) Reusability tests of NBC750. (**B**) First and (**C**) second cycle tests of NBC750 catalysts regenerated by different reactivation methods (HCl, NaOH, and thermal regeneration).

Then, the BR-based biochar catalysts were regenerated a second time and used again, with the results shown in Figure 2C. The BPA removal rates with NBC750-HCl and NBC750-

NaOH increased from 53.5% and 56.3% to 72.3% and 82.4%, respectively, indicating that the BR-based biochar catalysts could be regenerated again.

3.3. Role of Radical and Non-Radical Process in BR-Based Biochar Catalysts in PMS System

The ROSs in BR-based biochar catalysts for activating PMS were systematically investigated. From Figure 3A, in the NBC750 system, TBA, EtOH, and DMSO inhibited BPA removal by 33.8%, 44.4%, and 36.8%, respectively, indicating that free and surface-bound •OH and $•SO_4^-$ were the dominant ROSs, while $CHCl_3$ caused only 5.3% inhibition, suggesting a minor role for $•O_2^-$. In the NBC750-5th system, TBA, EtOH, DMSO, and $CHCl_3$ inhibited BPA removal by 17.1%, 22.0%, and 6.5%, 8.4%, respectively, reflecting reduced the production of •OH and $•SO_4^-$. Therefore, •OH and $•SO_4^-$ were identified as the primary ROSs in BPA degradation by PMS catalyzed with BR-based biochar, and their decreased production with repeated catalyst use led to lower BPA removal efficiency.

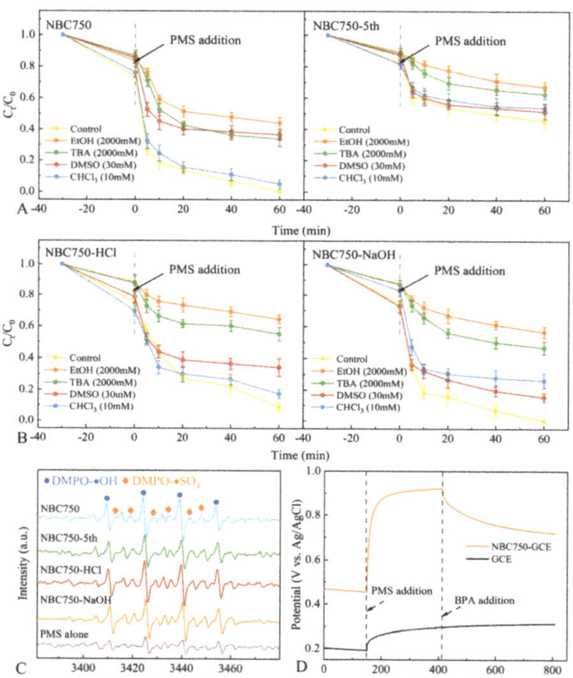

Figure 3. (**A**) Effects of different radical scavengers on removal of BPA in NBC750/PMS/BPA and NBC750-5th/PMS/BPA systems. (**B**) Effects of different radical scavengers on removal of BPA in NBC750-HCl/PMS/BPA and NBC750-NaOH/PMS/BPA systems. (**C**) EPR spectra of DMPO adducts (DMPO-•OH, DMPO-$•SO_4^-$). (**D**) Open-circuit potential curves of the GCE and NBC750-GCE in different systems.

In NBC750-HCl and NBC750-NaOH systems (Figure 3B), TBA and EtOH inhibited BPA degradation by 46.3% and 56.6% for NBC750-HCl, and 46.5% and 56.7% for NBC750-NaOH, while DMSO reduced BPA degradation by 25.5% and 25.8%, indicating an increased production of •OH and $•SO_4^-$ after HCl or NaOH regeneration. Moreover, $CHCl_3$ inhibited degradation by 9.2% in the NBC750-HCl system and 15.4% in the NBC750-NaOH system, showing that NaOH regeneration favored $•O_2^-$ production. Overall, HCl and

NaOH regeneration enhanced •OH and •SO_4^- production, which were crucial factors in improving BPA degradation efficiency.

The EPR technology confirmed the presence of •OH and •SO_4^- by using DMPO as a spin-trapping agent [33]. From Figure 3C, the weaker signals of NBC750-5th confirmed that the production of •OH and •SO_4^- decreased with repeated use of the catalyst. In contrast, the enhanced signals of NBC750-HCl and NBC750-NaOH further demonstrated restored catalytic activity through HCl or NaOH regeneration.

Additionally, it has been reported that the BPA degradation did not depend exclusively on free radical reactions [34]. Consequently, electrochemical methods were utilized to examine the non-radical mechanism [29]. As presented in Figure 3D, there was no reduction in open circuit potential when BPA was introduced to the bare GCE electrode, which inferred that direct electron transfer from BPA to PMS was not feasible without a catalyst to mediate the process. The open circuit potential of the NBC750-coated glassy carbon electrode (NBC750-GCE) rose immediately after PMS addition, followed by a significant drop with BPA addition, revealing electron transfer with BPA as an electron donor [35]. However, the quenching experiments demonstrated a potent inhibition of BPA degradation by EtOH and DMSO. Therefore, non-radical pathways might also play a subsidiary role in BPA degradation.

3.4. Characteristic Changes in BR-Based Biochar Catalysts in PMS System

3.4.1. Surface Morphology and Porosity Analysis

The elemental composition of BR (Figure S3) indicated that BR was rich in organic and inorganic materials, which helped form biochar with abundant functional groups [36]. The morphology of the BR-based biochar catalysts was observed using SEM (Figure 4A–F) [37], revealing the surface adorned with irregular nanoparticles. Additionally, in Figures 4G and S4, the EDS images and corresponding EDS spectrum showed a uniform distribution of N and O in NBC750. At the same time, minimal surface texture differences across samples confirmed the stability of the BR-based biochar catalysts during use and regeneration.

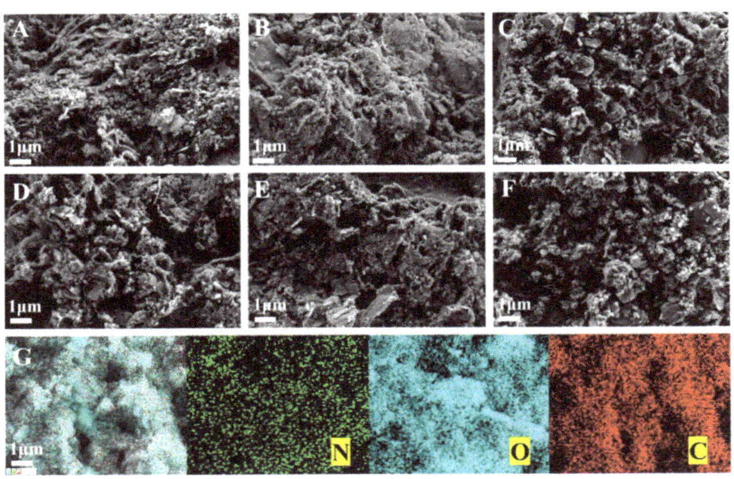

Figure 4. (**A–F**) SEM images of (**A**) BC750; (**B**) NBC750; (**C**) NBC750-5th; (**D**) NBC750-HCl; (**E**) NBC750-NaOH; (**F**) NBC750-500 °C; (**G**) EDS image of NBC750 and corresponding N, O, and C elemental mapping images.

The SSA and porous structure of BR-based biochar catalysts were characterized using N_2 adsorption–desorption isotherms [38]. As shown in Figure 5A, all samples exhibited type-IV isotherms with significant hysteresis loops, indicating a wide distribution of micropores and mesopores [39]. Moreover, the pore size of the catalysts was mainly within 0–10 nm (Figure 5B), supporting the above conclusions. From Table S1, the SSA of NBC750-5th decreased to 28.70 m^2/g, and the pore size also increased to 18.81 nm compared to NBC750 (64.56 m^2/g, 8.13 nm), denoting that the changes in SSA and pore structure might contribute to the deactivation of catalysts. For regenerated catalysts, NBC750-NaOH achieved the highest SSA (149.76 m^2/g) and smallest pore size (7.50 nm), followed by NBC750-HCl (127.45 m^2/g, 7.86 nm) and NBC750-500 °C (94.21 m^2/g, 7.92 nm), denoting that HCl, NaOH, and thermal regeneration all significantly adjusted the SSA and pore size of catalysts.

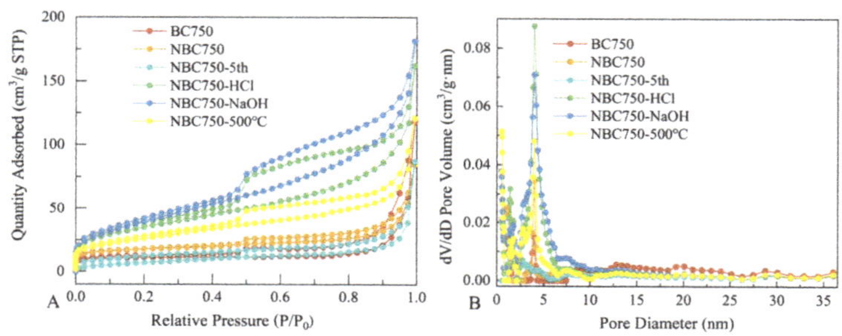

Figure 5. (**A**) N_2 adsorption–desorption isotherms and (**B**) pore size distribution of different BR-based biochar catalysts and regenerated catalysts.

3.4.2. Molecular Structure Analysis

XRD was used to check the crystallinity of the BR-based biochar catalysts [29]. As depicted in Figure 6A, the XRD patterns of all samples were similar, indicating that the reuses and regeneration of catalysts had minimal influence on the crystallinity.

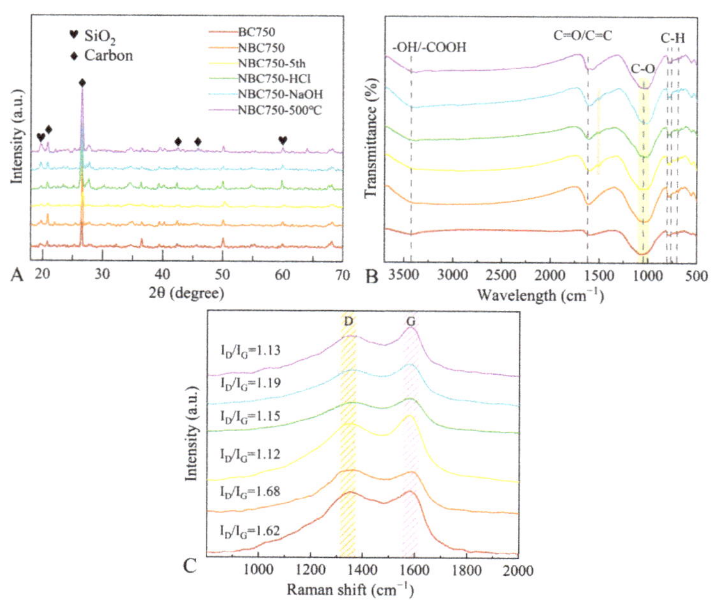

Figure 6. (**A**) XRD patterns, (**B**) FT-IR spectra, and (**C**) Raman spectra of different BR-based biochar catalysts and regenerated catalysts.

The FT-IR was employed to analyze the surface functional groups of the BR-based biochar catalysts [40]. From Figure 6B, all samples showed consistent functional groups, including -OH/-COOH, C=O/C=C, C-O, and C-H bonds [41,42]. Thermal regeneration removed a new peak at 1506 cm^{-1} in NBC750-5th, caused by adsorbed organics. Considering the regeneration experimental results, thermal treatment eliminated these adsorbed organics, but probably only released the adsorption sites with little effect on the catalytic performance of catalysts.

Raman spectroscopy was used to analyze the structural characteristics of the BR-based biochar catalysts. The D band (1350 cm^{-1}) indicated defects, and the G band (1580 cm^{-1}) reflected graphitization [43]. Based on previous studies, the I_D/I_G ratio was calculated by fitting the D and G peaks and determining the ratio of the integrated areas of these two peaks [44]. In Figure 6C, the I_D/I_G value of NBC750-5th decreased to 1.12 with increased catalyst reuse times, symbolizing that the defect sites were destroyed. Moreover, none of the regeneration methods could restore these sites.

3.4.3. Chemical Components on the BR-Based Biochar Catalyst Surface

The XPS spectra of catalysts were further analyzed to determine the key species involved in the deactivation and regeneration mechanisms of the BR-based biochar catalysts (Figure 7A and Table S2). The C 1 s XPS spectrum (Figure S5) displayed four peaks centered around 284.8, 285.8, 286.6, and 293.6 eV, corresponding to C-C, C-O, O-C=O, and C=C [45]. As shown in Figure 7B, binding energies of 398.6, 399.8, 400.8, and 402.5 eV in N 1 s XPS spectrum indicated pyridinic N, pyrrolic N, graphitic N, and NOx [23]. The total N content of NBC750-5th decreased from 4.04 at% to 2.84 at% with the increase in catalyst reuse times (Table S2), suggesting that the N loss impacted the catalytic activity of BR-based biochar. However, the N content of NBC750-HCl, NBC750-NaOH, and NBC750-500 °C was measured at 2.76 at%, 2.52 at%, and 2.66 at%, respectively, indicating that none of the three regeneration methods were able to restore N content.

Equally, the O 1 s XPS spectrum (Figure 7C) peaked at 531.3, 532.3, and 533.4 eV, matching -COOH, C=O, and C-OH, respectively [22]. Due to multiple reuses of catalysts, the O content of NBC750-5th decreased from 31.98 at% to 25.59 at%, with -COOH, C=O, and C-OH dropping from 8.92 at% to 5.06 at%, 15.69 at% to 14.49 at%, and 7.37 at% to 6.04 at%, respectively, reflecting the consumption of oxygen-containing functional groups. It has been demonstrated that -COOH and C-OH could promote the generation of •OH and $•SO_4^-$ [46], while the C=O could serve as the active sites for $•O_2^-$ generation [22]. Therefore, the loss of active sites was the primary cause of catalyst deactivation. After regeneration, O content decreased to 23.14 at% in NBC750-HCl and increased to 28.74 at% in NBC750-NaOH. Specifically, the content of -COOH increased to 6.18 at% in NBC750-HCl, and the percentage of C-OH increased to 7.08 at% after NaOH regeneration. However, thermal regeneration could not restore the oxygen-containing functional groups (-COOH or C-OH), which limited the recovery of the catalytic performance of BR-based biochar. Generally, the (O + N)/C and O/C ratios of BR-based catalysts indicated the hydrophilicity of biochar [47]. The (O + N)/C and O/C ratios of NBC750-5th decreased from 0.79 and 0.64 to 0.44 and 0.39, respectively, but improved to 0.50 and 0.46 after NaOH regeneration, unlike HCl or thermal regeneration (Table S2). This suggested that enhanced hydrophilicity may contribute to the higher catalytic effect of NBC750-NaOH.

The linear sweep voltammetry (LSV) result (Figure 7D) showed reduced electron transfer capacity in NBC750-5th, fully restored by NaOH, partially by HCl, and unaffected by thermal regeneration. Since functional groups such as -COOH, C-OH, and C=O on BR-based biochar could facilitate the electron transfer process [48], this result further confirmed that these oxygen-containing functional groups of biochar could be recovered by HCl or NaOH regeneration.

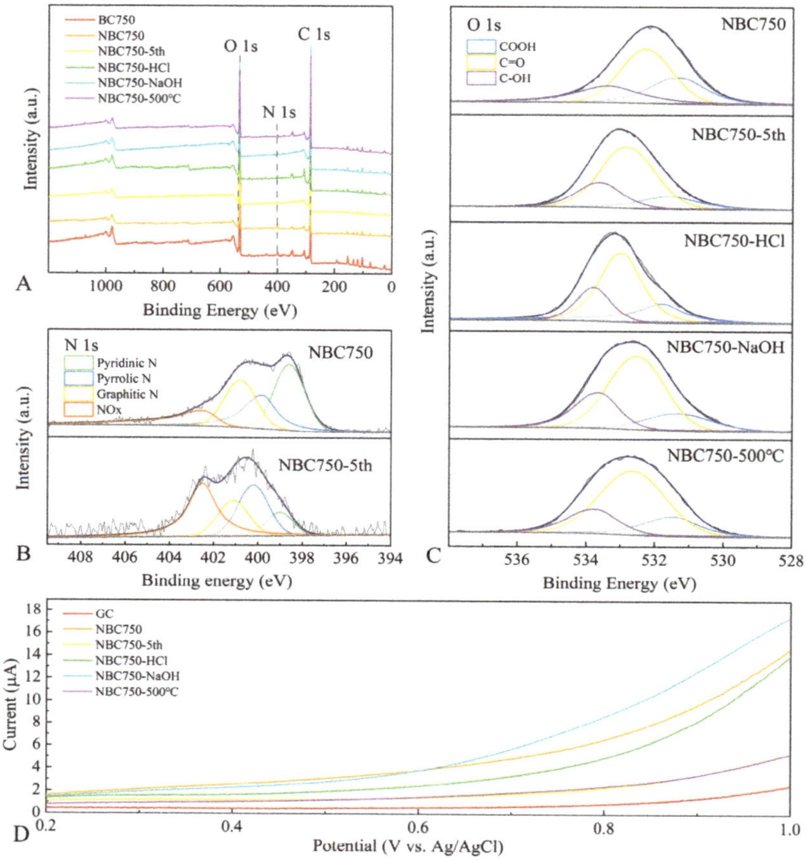

Figure 7. (**A**) Wide XPS spectrum of different BR-based biochar catalysts and regenerated catalysts. High-resolution XPS spectra for catalysts (**B**) N 1 s and (**C**) O 1 s. (**D**) Linear sweep voltammograms under different conditions.

3.4.4. Mechanism of Catalytic and Regeneration of the BR-Based Biochar Catalysts

In this study, the BPA removal rate gradually decreased with increased reuse times in stability tests for BR-based biochar catalysts. Subsequently, HCl, NaOH, and thermal regeneration were used to recover the catalytic activity of catalysts. Compared with HCl and NaOH regeneration, thermal regeneration was unsatisfactory in restoring the catalytic activity of the catalysts.

Based on the experimental results, BPA degradation mainly relied on •OH and $\bullet SO_4^-$, and the decrease in BPA removal was closely related to the decline in ROS production, which could effectively be restored by the HCl or NaOH regeneration. XPS analysis revealed that the depletion of -COOH and C-OH, key active sites for •OH and $\bullet SO_4^-$ production [38], was the main cause of inactivation, with HCl regenerating -COOH and NaOH restoring C-OH, while thermal regeneration had no effect on any oxygen-containing functional groups. In conclusion, the gradual depletion of -COOH and C-OH reduced ROS production during the repeated uses of biochar catalysts, leading to the deactivation of the

BR-based catalysts. Through HCl generation, -COOH functional groups could be restored, and the NaOH method effectively regenerated C-OH. However, thermal regeneration did not have any beneficial effects on the above either, thus failing to recover the catalytic performance of BR-based biochar.

Moreover, BPA strongly interacted with the defect sites on the BR-based biochar catalysts during PMS activation, enabling PMS to extract electrons and directly oxidize BPA [21]. Unfortunately, none of the three regeneration methods restored these sites. Although thermal regeneration partially restored adsorption by increasing SSA, it did not recover catalytic active sites.

Cost-effectiveness is crucial for ensuring the sustainability and practical application of SR-AOPs. The low cost and easy availability of BR make BR-based biochar catalysts cost-effective, reducing raw material expenses. Additionally, the BR-based biochar catalysts are regenerable, which helps lower operational costs. Both HCl and NaOH regeneration methods are economically viable, as the regenerated liquids could be recycled. Thermal regeneration was excluded due to its poor regeneration performance. Future studies could focus on optimizing the biochar production process to reduce its environmental footprint, as well as assessing its long-term sustainability and potential for large-scale applications in engineering.

4. Conclusions

In this study, BR-based biochar was prepared to activate PMS for BPA degradation. It was found that the activation of PMS was associated with oxygen-containing functional groups, N content, and defect sites of the BR-based biochar catalysts, and with the increase in catalyst reuse times, all the above were greatly affected or inhibited. Through the recovery of oxygen-containing functional groups and SSA, HCl or NaOH regeneration can effectively or even completely restore the catalytic activity, thus greatly improving the removal rate of BPA. However, thermal regeneration had little impact on the recovery of the catalytic activity. This research provided an efficient approach to improving the stability and recyclability of BR-based biochar catalysts, and provided new insights into the advancement of high-performance water treatment technologies for the removal of emerging contaminants.

Supplementary Materials: The following supporting information can be downloaded at: https://www.mdpi.com/article/10.3390/w17050744/s1, Text S1: Chemicals and materials; Text S2: Measurement of PMS concentration; Text S3: Determination of ROSs; Text S4: Reaction rate constant measurement equation; Text S5: Measurement of LSV and open-circuit potential; Figure S1: TOC removal rate of NBC750; Figure S2: Effect of catalyst dosage, PMS dosage, and initial pH on removal of BPA; Figure S3: The elemental composition of BR; Figure S4: The EDS spectrum of NBC750; Figure S5: C 1 s XPS spectrum; Table S1: Surface physical structure parameters of catalysts; Table S2: Atom contents (at%) of BR-derived biochar.

Author Contributions: Conceptualization, X.L. and M.L.; methodology, X.Y. and M.L.; investigation, Y.P.; resources, H.W., P.W. and N.D.; data curation, Y.P.; writing—original draft preparation, Y.P.; writing—review and editing, X.L.; supervision, X.L.; funding acquisition, X.Y. All authors have read and agreed to the published version of the manuscript.

Funding: This work is financially supported by Shanghai Rising-Star Program (No. 23QB1404700).

Data Availability Statement: The data used in this study are available on request.

Conflicts of Interest: Xue Yang, Wang and Nina Duan were employed by Shanghai Municipal Engineering Design Institute (Group) Co., Ltd. Haijuan Wei was employed by Shanghai Chengtou Wastewater Treatment Co., Ltd. The research was conducted in the absence of any commercial or financial relationships that could be construed as a potential conflict of interest. All of the authors have declared that there are no conflicts of interests.

References

1. Zhao, L.; Sun, Z.F.; Pan, X.W.; Tan, J.Y.; Yang, S.S.; Wu, J.T.; Chen, C.; Yuan, Y.; Ren, N.Q. Sewage sludge derived biochar for environmental improvement: Advances, challenges, and solutions. *Water Res. X* **2023**, *18*, 12. [CrossRef]
2. Chen, L.Y.; Qin, Y.J.; Chen, B.Q.; Wu, C.L.; Zheng, S.H.; Chen, R.L.; Yang, S.H.; Yang, L.; Liu, Z.J. Enhancing degradation and biogas production during anaerobic digestion of food waste using alkali pretreatment. *Environ. Res.* **2020**, *188*, 9. [CrossRef]
3. Ho, S.H.; Chen, Y.D.; Yang, Z.K.; Nagarajan, D.; Chang, J.S.; Ren, N.Q. High-efficiency removal of lead from wastewater by biochar derived from anaerobic digestion sludge. *Bioresour. Technol.* **2017**, *246*, 142–149. [CrossRef] [PubMed]
4. Kummu, M.; Guillaume, J.H.A.; De Moel, H.; Eisner, S.; Flörke, M.; Porkka, M.; Siebert, S.; Veldkamp, T.I.E.; Ward, P.J. The world's road to water scarcity: Shortage and stress in the 20th century and pathways towards sustainability. *Sci. Rep.* **2016**, *6*, 38495. [CrossRef]
5. Xiao, H.P.; Li, K.; Zhang, D.Q.; Tang, Z.H.; Niu, X.J.; Yi, L.Z.; Lin, Z.; Fu, M.L. Environmental, energy, and economic impact assessment of sludge management alternatives based on incineration. *J. Environ. Manag.* **2022**, *321*, 115848. [CrossRef]
6. Zeng, L.; Chen, Q.; Liang, N.; Ji, P.X.; Lu, M.; Wu, M.; Oleszczuk, P.; Pan, B.; Xing, B.S. The promoted degradation of biochar-adsorbed 2,4-dichlorophenol in the presence of Fe(III). *J. Hazard. Mater.* **2023**, *458*, 131774. [CrossRef] [PubMed]
7. Han, H.K.; Chen, M.F.; Sun, C.T.; Han, Y.Y.; Xu, L.L.; Zhao, Y.M. Synergistic enhancement in hydrodynamic cavitation combined with peroxymonosulfate fenton-like process for bpa degradation: New insights into the role of cavitation bubbles in regulation reaction pathway. *Water Res.* **2025**, *268*, 13. [CrossRef] [PubMed]
8. Deblonde, T.; Cossu-Leguille, C.; Hartemann, P. Emerging pollutants in wastewater: A review of the literature. *Int. J. Hyg. Environ. Health* **2011**, *214*, 442–448. [CrossRef]
9. Guo, Y.J.; Ma, C.Y.; Gao, Z.Y.; Wu, M.Z.; Shen, C.C.; Xu, Z.H. Insights into mechanism of peroxymonosufate activation by Mo single-atom catalysts: Singlet oxygen evolution and role of Mo-N coordination. *J. Environ. Manag.* **2024**, *358*, 120846. [CrossRef]
10. Jiang, S.F.; Ling, L.L.; Chen, W.J.; Liu, W.J.; Li, D.C.; Jiang, H. High efficient removal of bisphenol A in a peroxymonosulfate/iron functionalized biochar system: Mechanistic elucidation and quantification of the contributors. *Chem. Eng. J.* **2019**, *359*, 572–583. [CrossRef]
11. Hou, S.Y.; Hu, H.L.; Fu, Q.; Xiao, T.Y.; Xie, J.Q.; Chan, S.H.; He, M.J.; Miao, B.; Zhang, L. *Undaria pinnatifida* (wakame)-derived Fe, N co-doped graphene-like hierarchical porous carbon as highly efficient catalyst for activation of peroxymonosulfate (PMS) toward degradation of tetracycline (TC). *Sep. Purif. Technol.* **2024**, *333*, 13. [CrossRef]
12. Zhang, H.; Zhao, J.; Liu, F.; Yin, J.L.; Li, W.; Dai, X.Q.; Zhou, P.; Liu, Y.; Lai, B. Immobilizing cobalt-iron bimetal on Al_2O_3 by electroless plating-calcination for peroxymonosulfate activation: Performance, mechanistic and practicality. *Chem. Eng. J.* **2024**, *488*, 11. [CrossRef]
13. Yang, L.; Jiao, Y.; Xu, X.M.; Pan, Y.L.; Su, C.; Duan, X.G.; Sun, H.Q.; Liu, S.M.; Wang, S.B.; Shao, Z.P. Superstructures with Atomic-Level Arranged Perovskite and Oxide Layers for Advanced Oxidation with an Enhanced Non-Free Radical Pathway. *ACS Sustain. Chem. Eng.* **2022**, *10*, 1899–1909. [CrossRef]
14. Huang, S.M.; Wang, T.; Chen, K.; Mei, M.; Liu, J.X.; Li, J.P. Engineered biochar derived from food waste digestate for activation of peroxymonosulfate to remove organic pollutants. *Waste Manag.* **2020**, *107*, 211–218. [CrossRef]
15. Kohantorabi, M.; Moussavi, G.; Giannakis, S. A review of the innovations in metal- and carbon-based catalysts explored for heterogeneous peroxymonosulfate (PMS) activation, with focus on radical vs. non-radical degradation pathways of organic contaminants. *Chem. Eng. J.* **2021**, *411*, 26. [CrossRef]
16. He, C.; Wang, J.B.; Wang, C.R.; Zhang, C.H.; Hou, P.; Xu, X.Y. Catalytic ozonation of bio-treated coking wastewater in continuous pilot- and full-scale system: Efficiency, catalyst deactivation and in-situ regeneration. *Water Res.* **2020**, *183*, 116090. [CrossRef]
17. Kong, X.T.; Garg, S.; Chen, G.F.; Waite, D. Investigation of the deactivation and regeneration of an $Fe_2O_3/Al_2O_3{\cdot}SiO_2$ catalyst used in catalytic ozonation of coal chemical industry wastewater. *J. Hazard. Mater.* **2023**, *451*, 131194. [CrossRef]
18. Yang, S.J.; Qiu, X.J.; Jin, P.K.; Dzakpasu, M.; Wang, X.C.C.; Zhang, Q.H.; Zhang, L.; Yang, L.; Ding, D.H.; Wang, W.D.; et al. MOF-templated synthesis of $CoFe_2O_4$ nanocrystals and its coupling with peroxymonosulfate for degradation of bisphenol A. *Chem. Eng. J.* **2018**, *353*, 329–339. [CrossRef]
19. Hou, J.F.; Xu, L.X.; Han, Y.X.; Tang, Y.Q.; Wan, H.Q.; Xu, Z.Y.; Zheng, S.R. Deactivation and regeneration of carbon nanotubes and nitrogen-doped carbon nanotubes in catalytic peroxymonosulfate activation for phenol degradation: Variation of surface functionalities. *RSC Adv.* **2019**, *9*, 974–983. [CrossRef] [PubMed]
20. Hu, W.R.; Tong, W.H.; Li, Y.L.; Xie, Y.; Chen, Y.D.; Wen, Z.Q.; Feng, S.F.; Wang, X.Q.; Li, P.Y.; Wang, Y.B.; et al. Hydrothermal route-enabled synthesis of sludge-derived carbon with oxygen functional groups for bisphenol A degradation through activation of peroxymonosulfate. *J. Hazard. Mater.* **2020**, *388*, 121801. [CrossRef]

21. Cheng, C.; Li, J.P.; Wen, Y.Z.; Wang, J.L.; Jin, C.Y.; Sun, C.L.; Wang, H.L.; Wei, H.Z.; Yang, X.J. Deactivation mechanism of Fe/Al_2O_3 catalyst during the ozonation of reverse osmosis concentrates (ROCs): Effect of silicate. *Chem. Eng. J. Adv.* **2020**, *1*, 8. [CrossRef]
22. Wang, G.L.; Chen, S.; Quan, X.; Yu, H.T.; Zhang, Y.B. Enhanced activation of peroxymonosulfate by nitrogen doped porous carbon for effective removal of organic pollutants. *Carbon* **2017**, *115*, 730–739. [CrossRef]
23. Mian, M.M.; Liu, G.J. Activation of peroxymonosulfate by chemically modified sludge biochar for the removal of organic pollutants: Understanding the role of active sites and mechanism. *Chem. Eng. J.* **2020**, *392*, 123681. [CrossRef]
24. Chen, M.; Yang, C.X.; Yu, M.H.; Han, M.Y.; Meng, Z.H.; Zhao, T.; Niu, J.R.; Mu, S.T.; Zhang, J.; Ma, J.J.; et al. N-doped carbon achieved by pyrolyzing urea and active carbon for the capacitive deionization coupled PMS oxidation system. *Desalination* **2024**, *587*, 12. [CrossRef]
25. Edwin, N.N.; Garcia, T.; Solsona, B.; Taylor, S.H. The influence of cerium to urea preparation ratio of nanocrystalline ceria catalysts for the total oxidation of naphthalene. *Catal. Today* **2008**, *137*, 373–378. [CrossRef]
26. Chen, Z.G.; Lei, C.; Yao, L.L.; Mo, Y.; Li, J.X.; Qu, H.W.; Zhou, Z.; Luo, W. Synergistic pyrolysis of rice and chili straw under N_2/CO_2 atmosphere: Nutritional elements (N/P/K) migration and transformation from straw to pyrolysis products. *Energy* **2025**, *316*, 13. [CrossRef]
27. Xiao, T.; Zhou, P.K.; Liu, Y.; Zhang, K.K.; Liu, F.Y.; Guo, G.; Ni, F.Q.; Deng, Y. Impact of pyrolysis temperature on heavy metals environmental risk in biochar derived from co-pyrolysis of *Alternanthera philoxeroides* and sludge. *J. Environ. Chem. Eng.* **2024**, *12*, 9. [CrossRef]
28. Zhao, C.H.; Shao, B.B.; Yan, M.; Liu, Z.F.; Liang, Q.H.; He, Q.Y.; Wu, T.; Liu, Y.; Pan, Y.; Huang, J.; et al. Activation of peroxymonosulfate by biochar-based catalysts and applications in the degradation of organic contaminants: A review. *Chem. Eng. J.* **2021**, *416*, 128829. [CrossRef]
29. Wang, J.; Duan, X.G.; Gao, J.; Shen, Y.; Feng, X.H.; Yu, Z.J.; Tan, X.Y.; Liu, S.M.; Wang, S.B. Roles of structure defect, oxygen groups and heteroatom doping on carbon in nonradical oxidation of water contaminants. *Water Res.* **2020**, *185*, 116244. [CrossRef] [PubMed]
30. Wang, L.L.; Lan, X.; Peng, W.Y.; Wang, Z.H. Uncertainty and misinterpretation over identification, quantification and transformation of reactive species generated in catalytic oxidation processes: A review. *J. Hazard. Mater.* **2021**, *408*, 124436. [CrossRef]
31. Tang, W.Q.; Liu, Y.Y.; You, Q.L.; Yang, X.F.; Liao, G.Y.; Wang, D.S.; Yan, Y.; Shang, Q.G. Constructing a 3D ordered macroporous cobalt monoatomic catalyst for efficient SMX degradation via PMS activation. *J. Environ. Chem. Eng.* **2024**, *12*, 12. [CrossRef]
32. Oh, W.D.; Lim, T.T. Design and application of heterogeneous catalysts as peroxydisulfate activator for organics removal: An overview. *Chem. Eng. J.* **2019**, *358*, 110–133. [CrossRef]
33. Zhu, K.M.; Wang, X.S.; Geng, M.Z.; Chen, D.; Lin, H.; Zhang, H. Catalytic oxidation of clofibric acid by peroxydisulfate activated with wood-based biochar: Effect of biochar pyrolysis temperature, performance and mechanism. *Chem. Eng. J.* **2019**, *374*, 1253–1263. [CrossRef]
34. Ren, W.; Xiong, L.L.; Yuan, X.H.; Yu, Z.W.; Zhang, H.; Duan, X.G.; Wang, S.B. Activation of Peroxydisulfate on Carbon Nanotubes: Electron-Transfer Mechanism. *Environ. Sci. Technol.* **2019**, *53*, 14595–14603. [CrossRef] [PubMed]
35. Chen, H.; Li, X.; Li, W.Z.; Feng, J.; Zhao, Y.; Zhang, H.X.; Ren, Y.M. Nitrogen-doped biochar/MnO_2 as an efficient PMS activator for synergistic BPA degradation via non-free radical pathways in the water. *J. Environ. Chem. Eng.* **2024**, *12*, 112446. [CrossRef]
36. Ahmed, M.J.; Hameed, B.H. Insight into the co-pyrolysis of different blended feedstocks to biochar for the adsorption of organic and inorganic pollutants: A review. *J. Clean. Prod.* **2020**, *265*, 17. [CrossRef]
37. Cai, S.; Zhang, Q.; Wang, Z.Q.; Hua, S.; Ding, D.H.; Cai, T.M.; Zhang, R.H. Pyrrolic N-rich biochar without exogenous nitrogen doping as a functional material for bisphenol A removal: Performance and mechanism. *Appl. Catal. B Environ.* **2021**, *291*, 120093. [CrossRef]
38. Fan, H.L.; Wang, J.X.; Wu, P.P.; Zheng, L.; Xiang, J.F.; Liu, H.L.; Han, B.X.; Jiang, L. Hydrophobic ionic liquid tuning hydrophobic carbon to superamphiphilicity for reducing diffusion resistance in liquid-liquid catalysis systems. *Chem* **2021**, *7*, 1852–1869. [CrossRef]
39. Sing, K.S.W.; Everett, D.H.; Haul, R.A.W.; Moscou, L.; Pierotti, R.A.; Rouquerol, J.; Siemieniewska, T. Reporting Physisorption Data for Gas Solid Systems with Special Reference To the Determination of Surface-Area and Porosity (Recommendations 1984). *Pure Appl. Chem.* **1985**, *57*, 603–619. [CrossRef]
40. Song, W.; Yu, Z.H.; Li, H.; Ji, Y.Q.; Cao, L.L.; Ren, L.Y.; Li, X.G.; Li, Y.F.; Xu, X.; Yan, L.G. Insights into the factors influencing the oxidation of antibiotic pollutants in nitrogen-doped biochar/PMS system: The roles of physicochemical properties and reaction pathways. *Chem. Eng. J.* **2024**, *498*, 155601. [CrossRef]
41. Silva, T.L.; Ronix, A.; Pezoti, O.; Souza, L.S.; Leandro, P.K.T.; Bedin, K.C.; Beltrame, K.K.; Cazetta, A.L.; Almeida, V.C. Mesoporous activated carbon from industrial laundry sewage sludge: Adsorption studies of reactive dye Remazol Brilliant Blue R. *Chem. Eng. J.* **2016**, *303*, 467–476. [CrossRef]

42. Liu, W.J.; Nie, C.Y.; Li, W.L.; Ao, Z.M.; Wang, S.B.; An, T.C. Oily sludge derived carbons as peroxymonosulfate activators for removing aqueous organic pollutants: Performances and the key role of carbonyl groups in electron-transfer mechanism. *J. Hazard. Mater.* **2021**, *414*, 125552. [CrossRef]
43. Wang, J.; Chen, Z.M.; Chen, B.L. Adsorption of Polycyclic Aromatic Hydrocarbons by Graphene and Graphene Oxide Nanosheets. *Environ. Sci. Technol.* **2014**, *48*, 4817–4825. [CrossRef]
44. Shi, M.M.; Bao, D.; Li, S.J.; Wulan, B.R.; Yan, J.M.; Jiang, Q. Anchoring PdCu Amorphous Nanocluster on Graphene for Electrochemical Reduction of N_2 to NH_3 under Ambient Conditions in Aqueous Solution. *Adv. Energy Mater.* **2018**, *8*, 6. [CrossRef]
45. Wu, C.X.; Li, L.F.; Zhou, H.; Ai, J.; Zhang, H.T.; Tao, J.L.; Wang, D.S.; Zhang, W.J. Effects of chemical modification on physicochemical properties and adsorption behavior of sludge-based activated carbon. *J. Environ. Sci.* **2021**, *100*, 340–352. [CrossRef]
46. Chen, X.; Oh, W.D.; Hu, Z.T.; Sun, Y.M.; Webster, R.D.; Li, S.Z.; Lim, T.T. Enhancing sulfacetamide degradation by peroxymonosulfate activation with N-doped graphene produced through delicately-controlled nitrogen functionalization via tweaking thermal annealing processes. *Appl. Catal. B Environ.* **2018**, *225*, 243–257. [CrossRef]
47. Shaheen, S.M.; Niazi, N.K.; Hassan, N.E.E.; Bibi, I.; Wang, H.L.; Tsang, D.C.W.; Ok, Y.S.; Bolan, N.; Rinklebe, J. Wood-based biochar for the removal of potentially toxic elements in water and wastewater: A critical review. *Int. Mater. Rev.* **2019**, *64*, 216–247. [CrossRef]
48. Li, M.; Li, D.Y.; Guan, Z.Y.; Xu, Q.Q.; Shi, Y.T.; Xia, D.S. Carboxy-functionalized sludge-derived biochar for efficiently activating peroxymonosulfate to degrade bisphenol A. *Sep. Purif. Technol.* **2022**, *297*, 121525. [CrossRef]

processes

MDPI

Article

Numerical Study on Force Mechanisms and Eddy Current Effect of Rayleigh–Taylor Instability in the Intrinsic Model of Activated Carbon

Li Ye *, Qidong Zhang, Yizeng Wang, Zhipeng Ye and Haoran He

School of Energy and Power Engineering, University of Shanghai for Science and Technology, Shanghai 200093, China; 222210140@st.usst.edu.cn (Q.Z.); 222210129@st.usst.edu.cn (Y.W.); 223370216@st.usst.edu.cn (Z.Y.); 223370263@st.usst.edu.cn (H.H.)
* Correspondence: yeli@usst.edu.cn

Abstract: In industrial operations, fluid transport within activated carbon—as one kind of porous media—often involves nonlinear instabilities, notably Rayleigh–Taylor instability (RTI). Herein it has been investigated for the flows in its intrinsic model to analyze the generation of adverse factors in practical applications. By studying the mechanism of force influence, it can explore how porous medium parameters impact RTI dynamics. The results show that horizontal forces dominate RTI's initial evolution, with porosity influencing peak flow velocity timing delays (0% to 22.39% delay). The varying pore scales, delays peak velocity timing (7.63% to 1.46% delay) during initial stages, converging at later stages. Notably, the vortex line rupture at $\varepsilon = 0.999$ triggers Taylor vortex formation, enhancing velocity by 9.8816 times and promoting bidirectional mass transfer. This study enhances understanding of mass transfer efficiency of catalyst in industrial applications, enabling tailored porous media designs for optimized heat and mass transfer.

Keywords: Rayleigh–Taylor instability; lattice Boltzmann method; heat and mass transfer; fluid mechanics; activated carbon; numerical simulation

1. Introduction

The study of fluid dynamics in porous media has garnered significant attention in recent years due to its widespread applications across various fields, including petroleum engineering, cooling systems, geothermal energy extraction, and so on [1–3]. Activated carbon is a substance with strong adsorption capacity, which can adsorb various harmful gases and particulate matter. Adding activated carbon to fluidized bed boilers is commonly used to reduce pollutant emissions, minimize ash accumulation inside the boiler, improve operational stability, and extend boiler lifespan. Its performance is fundamentally dependent on mass transfer efficiency, which is closely related to the internal flow characteristics. Rayleigh–Taylor instability (RTI), a typical fluid interface instability phenomenon, occurs between the interface of two fluids with different densities. It controls the mixing and transport of the entrainable fluid at interfaces with different densities, significantly affecting the mass transfer process. Therefore, studying the evolution of Rayleigh–Taylor instability within activated carbon can provide new perspectives and methods for optimizing mass transfer behavior.

RTI is characterized by the interpenetration of heavier and lighter fluids influenced by gravity and interfacial tension, and it has been extensively studied in non-porous environments [4]. The complex interplay between the structural characteristics of porous media, such as porosity and permeability, poses significant challenges for both theoretical and experimental investigations of RTI. Recent advancements in computational fluid dynamics, particularly the lattice Boltzmann method (LBM), have emerged as powerful tools for simulating fluid flow and transport processes in porous media [4–7]. Shijin D. [8] simulated the RTI phenomenon in the case of the slip boundary of the generalized N–S equation (Navier–Stokes equation). To accurately capture the complex interactions between the porosity, permeability, and fluid dynamics of the medium, the study needs to appropriately select the time step, grid size, and suitable physical models (such as boundary conditions), while a proper verification method should be used to validate the LBM model. For example, the LBM simulations under high Reynolds number face challenges regarding time steps, grid resolution, and model parameter settings. Hong L. [9] improved the phase-field lattice Boltzmann method and conducted high-resolution direct numerical simulations of multi-mode immiscible Rayleigh–Taylor instability (RTI) at a low Atwood number (At = 0.1). This work extended the range of Reynolds numbers for which LBM can accurately capture RTI flow and provided recommendations on the grid resolution and boundary conditions in simulation. Besides, Rui L. [10] simulated the concentration evolution of hydroxides during the charging process of a zinc-nickel single-liquid flow battery cathode using the LBM and velocity-concentration dual-distribution model. It has revealed the dynamic processes of ions and protons during charging and extended the model's applicability. The method of using has simulated the flow of a specific substance within a selected region, which is suitable for processes within fixed solid-formed porous media where reactions occur. However, this study provides accurate results in a specific region, and the overall flow pattern can only be inferred in a general manner. For the research topic of activated carbon water purification herein, it is necessary to predict the internal flow behavior within a large-scale porous media to obtain the flow results. And the primary focus is on adsorption and other related effects, without considering changes in the concentration of reaction products.

Therefore, we consider using the volume averaging method to study the averaged range separately, namely the REV (representative elementary volume). The resulting model does not rely on a specific microscopic particle distribution but instead describes the behavior of the overall medium. This approach can be applied to various types of porous media, including but not limited to soil, rocks, battery electrodes, and filtration materials. And while appropriately selecting the time step, grid size, and suitable physical models (such as boundary conditions), the dual-distribution model provides a good control approach. And the further improvements to the model are also required to ensure the accuracy of the LBM simulation on RTI in activated carbon.

In addition to designing the accuracy and rationality of simulations, many studies have also selected different parameters to predict or evaluate instability phenomena. Several studies have sought to elucidate how the properties of porous media influence fluid instabilities by discussing finger position, flow regimes, or flow rate, including the studies focused on buoyancy-driven flows and fingering instabilities [11–13] or building methods [14–16]. However, in many cases, the evolution of forces determines the changing trends of flow within porous media. A comprehensive analysis of the force mechanisms driving and modulating RTI in porous media, particularly in activated carbon applications, remains insufficient. To address this gap, this study aims to simulate the evolution of RTI within activated carbon at the mesoscopic scale using the LBM and to predict the resulting internal flow dynamics. The theoretical foundation is provided by the generalized Navier–Stokes (N–S) equations, which incorporate additional terms to account for the influence of the porous matrix on fluid behavior. Considering the diverse porosity and pore size (key factors influencing permeability) characteristics of activated carbon, this study systematically investigates the effects of these two properties on RTI flow behavior, especially in flow and force. By elucidating how porous media properties govern the development and evolution of RTI, this research provides a deeper understanding of the underlying mechanisms and offers valuable insights for optimizing fluid dynamics in porous media. The findings not only advance the fundamental understanding of RTI but also have practical implications for enhancing mass transfer in industrial applications, such as environmental remediation and energy storage. Furthermore, a deeper understanding of fluid mixing and transport within activated carbon can provide valuable information for designing remediation strategies for contaminated sites in environmental engineering.

In summary, this study represents a significant advancement in elucidating the complex interplay between fluid dynamics and the properties of activated carbon in the context of RTI. By revealing the impact of instability on fluid interfaces and flow structures, this study enables a more accurate description of concentration distribution and mass transfer rates, thereby improving the predictive accuracy of mass transfer models. Furthermore, investigating the evolution of RTI facilitates the optimization of activated carbon's pore structure design, such as by controlling pore size and distribution to suppress or harness instability, ultimately enhancing mass transfer efficiency. By leveraging the capabilities of the LBM, this research elucidates the complex interplay between fluid dynamics and the properties of activated carbon, offering new insights into the force mechanisms governing RTI in activated carbon and other porous media. These findings not only advance fundamental understanding but also provide theoretical guidance and technical support for practical applications in environmental management, energy, and chemical engineering, thereby bridging the gap between fundamental research and real-world engineering challenges.

2. Model Description

2.1. Flow Model

The convection diffusion of fluid in porous media can be described by the generalized Navier–Stokes equation coupled with the convection diffusion equation. The fluid flow is also calculated in the model Nithiarasu [17] proposed, containing a linear drag term (Darcy), a viscous term (Brinkman), and a nonlinear drag term (Forchheimer), finally in this research, which is expressed in the LBGK (lattice Bhatnagar–Gross–Krook) model. The fluid heat transfer is described by a temperature-dependent convection diffusion equation. The specific form of the equation set is as follows:

$$\frac{\partial \rho}{\partial t} + \nabla \cdot (\rho \boldsymbol{u}) = 0 \tag{1}$$

$$\frac{\partial \boldsymbol{u}}{\partial t} + (\boldsymbol{u} \cdot \nabla)\left(\frac{\boldsymbol{u}}{\varepsilon}\right) = -\frac{1}{\rho}\nabla(\varepsilon p) + \nu_e \cdot \nabla^2 \boldsymbol{u} + \boldsymbol{F} \tag{2}$$

$$\sigma\frac{\partial T}{\partial t} + \boldsymbol{u} \cdot \nabla T = \nabla \cdot (\alpha_m \nabla T) \tag{3}$$

where ν_e is effective viscosity, as $\nu_e = \mu_e/\rho$, heat capacity ratio of wall to fluid $\sigma = \varepsilon + (1-\varepsilon)\rho_s c_{ps}/\rho_f c_{pf}$, ρ_f is the density of the fluid, ρ_s is the physical density of the wall, c_{pf} is the specific heat at constant pressure of the fluid, c_{ps} is the specific heat at constant pressure of the wall. α_m is the thermal diffusivity coefficient, as $\alpha_m = k_m/\rho_f c_{pf}$, The effective thermal conductivity $k_m = k_e + k_d$, k_e is the stagnation thermal conductivity, and k_d is the thermal conductivity caused by thermal dispersion. u is the velocity of the fluid and ε is the porosity. F is the total force including medium resistance and external force:

$$\boldsymbol{F} = -\frac{\varepsilon \nu}{K}\mathbf{u} - \frac{\varepsilon F_e}{\sqrt{K}}|\mathbf{u}|\mathbf{u} + \varepsilon \boldsymbol{G} \tag{4}$$

where ν is the fluid viscosity coefficient, and $\boldsymbol{G}$ is the external volume force. In the definition of total force in Formula (4), the first term on the right side of the equation is linear medium resistance (Darcy), and the second term on the right side of the equation is nonlinear medium resistance (Forchheimer). The structural function is also involved $F_e = 1.75/\sqrt{150 \cdot \varepsilon^3}$.

To control the pore size dp, the permeability K of the porous medium in this paper is expressed using the Ergun empirical formula [18]:

$$K = \frac{\varepsilon^3 {D_p}^2}{150(1-\varepsilon)^2} \tag{5}$$

where D_p represents the particle scale, and then to be measured in [mm]. In this model, dp = 0.414 D_p, so the relationships between dp and K is $K \propto \mathrm{dp}^2$.

A double-distribution model has been employed to solve the velocity and temperature fields separately. The evolution equation is:

$$f_i(x + c_i\delta_t, t + \delta_t) - f_i(x,t) = -\frac{1}{\tau}\left(f_i(x,t) - f_i^{(eq)}(x,t)\right) + \delta_t F_i \tag{6}$$

where f_i is the velocity distribution function, and the equilibrium distribution function $f_i^{(eq)}$ is expressed as [19]:

$$f_i^{(eq)} = \omega_i \rho \left[1 + \frac{c_i \mathbf{u}}{c_s^2} + \frac{\mathbf{uu} : (c_i c_i - c_s^2 I)}{2\varepsilon c_s^4}\right] \tag{7}$$

where τ is the relaxation time, ρ is the macroscopic density, and the force term F_i is expressed as:

$$F_i = \omega_i \rho \left(1 - \frac{1}{2\tau}\right)\left[\frac{c_i \boldsymbol{F}}{c_s^2} + \frac{\mathbf{u}\boldsymbol{F} : (c_i c_i - c_s^2 I)}{\varepsilon c_s^4}\right] \tag{8}$$

The D2Q9 (two-dimensional nine-velocity) model is a commonly used discrete velocity model (DVM), typically employed in the LBM for fluid dynamics simulations, in which fluid particles move on a two-dimensional lattice grid and have nine possible velocity directions. The schematic diagram is shown in Figure 1. In the LBGK model with D2Q9 [20], the discrete velocity $c_i = ce_i$, and e_i is a unit direction vector, which takes the following form:

$$e_i = \begin{cases} (0,0) & i = 0 \\ \left(\cos\left[\frac{(i-1)\pi}{2}\right], \sin\left[\frac{(i-1)\pi}{2}\right]\right) & i = 1,2,3,4 \\ \left(\sqrt{2}\cos\left[\frac{(i-1)\pi}{2} + \frac{\pi}{4}\right], \sin\left[\frac{(i-1)\pi}{2} + \frac{\pi}{4}\right]\right) & i = 5,6,7,8 \end{cases} \tag{9}$$

where ${c_s}^2 = c^2/3$, and the value of the weight coefficient ω_i is determined based on the requirements of fluid uniformity and symmetry. In order to ensure the accuracy and stability of the simulation results, ω_i of the D2Q9 model is as follows:

$$\omega_0 = \frac{4}{9}, \omega_{1-4} = \frac{1}{9}, \omega_{5-8} = \frac{1}{36} \tag{10}$$

These weighting values ensure that the D2Q9 model exhibits consistent physical behavior in all directions, maintaining symmetry and uniformity. This guarantees the conservation of mass, momentum, and energy during simulations while also providing stable numerical results and accurately capturing fluid dynamic behaviors.

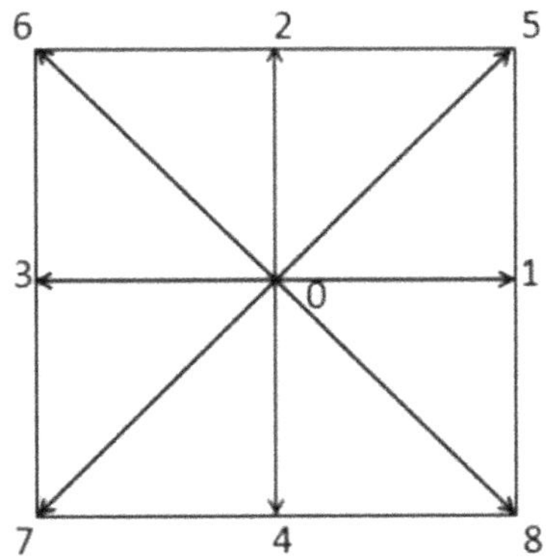

Figure 1. D2Q9 (two-dimensional nine-velocity) model.

The macroscopic density and velocity are calculated using the following equations:

$$\rho = \sum_{i=0}^{b-1} f_i(x,t) \tag{11}$$

$$\rho\mathbf{u} = \sum_{i=0}^{b-1} c_i f_i(x,t) \tag{12}$$

The permeability density is defined as:

$$\rho = \sum_i f_i \tag{13}$$

In the calculation of the velocity distribution, the effective kinematic viscosity coefficient ν_e, which is related to the relaxation time τ, is given by the following equation:

$$\tau = \frac{\nu_e}{c_s^2 \delta t} + 0.5 \tag{14}$$

For solving the temperature-dependent convection–diffusion equation, its evolution equation is:

$$g_i(x + c_i\delta_t, t + \delta_t) - g_i(x,t) = -\frac{1}{\tau'}\left(g_i(x,t) - g_i^{(eq)}(x,t)\right) \tag{15}$$

where g_i is the temperature distribution function, and ω_i is the weight coefficient, and the equilibrium distribution function of temperature is:

$$g_i^{(eq)}(x,t) = \omega_i T\left[1 + \frac{3}{c^2}(e_i \cdot \mathbf{u})\right] \tag{16}$$

In the calculation of the temperature distribution, the diffusion coefficient D is related to the relaxation time τ', is given by the following equation:

$$\tau' = \frac{3D\delta_t}{dx^2} + 0.5 \tag{17}$$

The LBGK model simplifies parameter adjustment and extends the range of model applications flexibly by associating effective viscosity coefficients and diffusivity coefficients with relaxation time while maintaining physical consistency and numerical stability. Furthermore, since the relaxation time essentially reflects the timescale of particle collision and recovery of equilibrium, it is consistent with the microscopic dynamics of viscosity and thermal diffusion, ensuring that the model has a clear physical meaning. Equations (6) and (15) involved in this model can be expanded using the Chapman–Enskog method into Equations (1)–(3), demonstrating the strict recovery of the Navier–Stokes equations and the heat equation, thereby validating the theoretical correctness of the relationship between relaxation time and transport coefficients.

2.2. Model Validation

Guo Z. L. proposed a method to validate the REV-scale lattice Boltzmann model, which has been applied in various studies [21]. Therefore, this paper also selects this method to verify the accuracy of the established model.

The system settings are as follows: The upper and lower interfaces of the porous medium are impermeable plates. The grid size is set to $Nx \times Ny = 64 \times 8$, and the density of the entire field is set to 1.0. The initial flow field model is driven by a constant force, with the boundary velocity set to 0.1. The upper and lower boundaries are treated using the half-way bounce-back scheme. Figure 2 shows the agreement between the analytical solution and the numerical results based on the model proposed herein. The error between the simulation data and the literature data in the figure does not exceed 3%, indicating that the flow model is both effective and accurate, thus demonstrating the model's accuracy preliminarily.

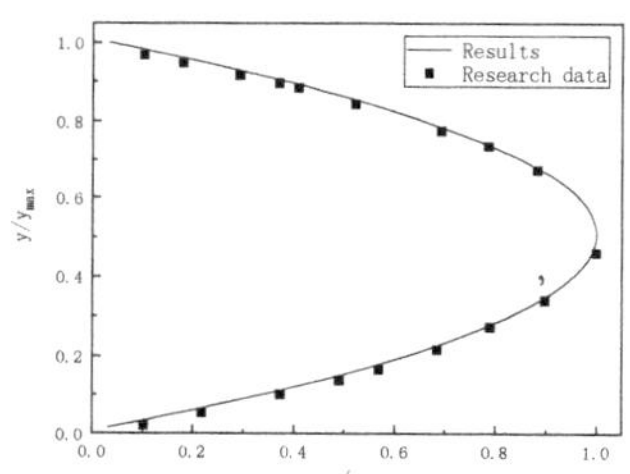

Figure 2. Dimensionless velocity profiles between impermeable plates.

To validate the accuracy of the lattice Boltzmann equation (LBE) model for fluid convection and diffusion in porous media at the REV scale, this study performs numerical simulations of natural convection in a closed square cavity with varying dimensionless parameters. The simulation results are compared with existing literature. The temperatures on the left and right walls of the cavity are set to $T_h = 0.5$ and $T_c = -0.5$, with the initial flow field at rest and $Pr = 1.0$. The grid size used is $Nx \times Ny = 128 \times 128$. For this problem, the average Nusselt number (Nu) on the cold wall is used to quantify the convective heat transfer intensity. The formula for calculating the average Nusselt number on the cold wall is given as follows [17].

$$\overline{\mathrm{Nu}} = \frac{1}{H}\int_0^H -\frac{H}{\Delta T}\frac{\partial T}{\partial y}dy \tag{18}$$

Table 1 presents the average Nusselt number on the cold wall for $K = 10^{-2}$ and 10^{-4} at different ***Ra*** numbers. The results show that the LBE model's numerical predictions at the REV scale align well with those from literature [17], with errors within 3%, indicating the model's accuracy in describing fluid flow and diffusion in porous media.

Table 1. Comparison of the average Nusselt number on the hot wall under different conditions (a: LBE simulation results; b: results from literature [17]).

		ε = 0.4			ε = 0.6			ε = 0.9		
K	Ra	a	b	error%	a	b	error%	a	b	error%
10^{-2}	10^3	1.008	1.010	0.20	1.012	1.015	0.30	1.018	1.023	0.49
	10^4	1.383	1.408	1.78	1.515	1.530	0.98	1.654	1.640	0.85
	10^5	3.021	2.983	1.27	3.461	3.555	2.64	3.905	3.910	0.13
	5×10^5	4.966	4.990	0.28	5.716	5.740	0.42	6.615	6.700	1.27
10^{-4}	10^5	1.050	1.067	1.59	1.074	1.071	0.28	1.076	1.072	0.37
	10^6	2.547	2.550	0.12	2.651	2.725	2.71	2.717	2.740	0.84
	10^7	7.634	7.810	2.25	7.985	8.183	2.42	9.031	9.202	1.86

2.3. Numerical Model of RTI

The correctness of the LBE model was proven in the previous section. In this section, the LBE model is used to solve the coupled system of governing equations for the generalized N–S equations and the convection–diffusion equation to study the RTI phenomenon in porous media. The model allows for the independent analysis of the effects of porosity and permeability on the RTI phenomenon. Since permeability is related to pore size, the model is further optimized to be applicable to the simulation and analysis of RTI in porous media relevant to this study.

2.3.1. Boundary Treatment

The periodic unit in the porous medium has been simulated. Two miscible fluids have different densities at different temperatures; after disturbance at the interface, the RTI occurs. In the simulation, the left and right boundaries of the model are treated using periodic boundary conditions, expressed as:

$$f_i(0, y, z, t + \delta_t) = f_i^+(L, y, z, t) \tag{19}$$

$$f_i(L, y, z, t + \delta_t) = f_i^+(0, y, z, t) \tag{20}$$

The upper and lower boundaries are treated using the half-way bounce-back scheme. Ladd [22] proposed a half-step bounce-back scheme suitable for general moving boundaries. The fixed boundary can be located at the midpoint of the lattice nodes, with second-order accuracy, expressed as:

$$f_{\bar{i}}\left(x_f, t + \delta_t\right) = f_i^+(x_f, t) - 2\omega_i \rho \frac{c_i \cdot \mathbf{u}_w}{c_s^2} \tag{21}$$

where c_i represent the original and reflected velocity directions pointing towards the wall, respectively. After the half-bounce, the velocity direction at that lattice point reverses, and $\mathbf{u}_w = \left(u_{wx}, u_{wy}, u_{wz}\right)$ is the wall velocity, ω_i is the weight coefficient of the equilibrium distribution function.

The collision and streaming steps in the model evolution equation are [23]:

$$\text{Collision}: \ f_i^+(x, t) - f_i(x, t) = -\frac{1}{\tau}\left(f_i(x, t) - f_i^{(eq)}(x, t)\right) \tag{22}$$

$$\text{Streaming}: \ f_i(x + ce_i\delta_t, t + \delta_t) = f_i^+(x, t) \tag{23}$$

The non-equilibrium extrapolation scheme has second-order accuracy, strong applicability, and good numerical stability. It is assumed that O is on the boundary, C is inside the flow field, and P is outside the flow field. According to the extrapolation format, which divides the distribution function on the boundary into equilibrium and non-equilibrium states, the O distribution function of a lattice point on the boundary can be expressed as:

$$f_i(O,t) = f_i^{(eq)}(O,t) + f_i^{(neq)}(O,t) \tag{24}$$

This study is based on the porous medium seepage model, introducing a reasonable range of activated carbon porosity and calculating the appropriate permeability based on the pore size, thereby improving and adapting the model for liquid-liquid applications in activated carbon reactors. The schematic diagram of the distribution area and selected configuration source of activated carbon in the straight tube reactor is shown in Figure 3. The simulated real-world application involves a cylindrical reactor filled with activated carbon. Initially, a portion of cold fluid is present in the upper part of the cylinder, separated by a barrier. Hot fluid is then introduced from the bottom of the cylinder. After the barrier is removed, the RTI phenomenon occurs inside the activated carbon. Through controlling the conditions, the fluid within the activated carbon undergoes full convective mixing, followed by the discharge of part of the fluid, and the process is repeated.

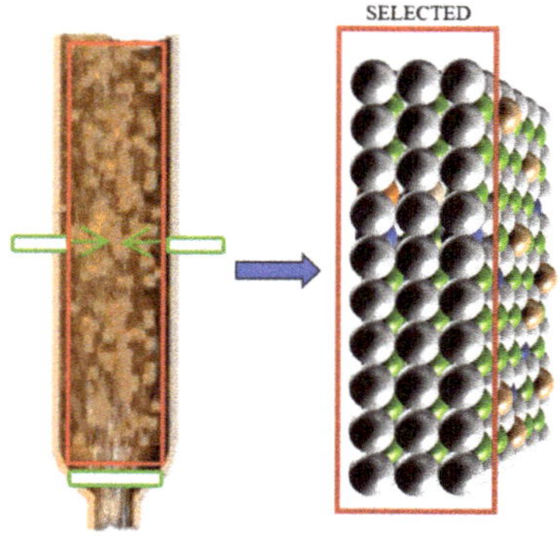

Figure 3. Distribution area and selected configuration source of activated carbon in straight tube reactor. (The green arrow indicating the direction of partition closure, and the green box represents the partition and the bottom valve, while the red box highlights the selected region of activated carbon inside the reactor and its microscopic image.).

It is noteworthy that during the process, two fluids of different temperatures interact within the porous medium without any external force or flow velocity except for gravity, with adiabatic walls at the top and bottom having zero velocity. The advantage of this reactor lies in its reliance on RTI as the driving force, eliminating the need for stirring and additional heating. Moreover, the reactor's size is larger than that of microchannel reactors, making it promising for applications such as active fluid or biomass reactions. The study primarily focuses on the qualitative analysis of the RTI evolution behavior inside activated carbon at this aspect ratio. Using periodic boundary conditions on the left and right boundaries reduces the accuracy loss caused by wall effects and provides a reference for future studies on scaling up the reactor size.

2.3.2. Feature Parameter Processing

Here, assuming that density is constant in both the time derivative and convective terms, flow is driven by buoyancy, so we used the Boussinesq approximation, in which all fluid parameters, except for body forces, can be considered constants. The system can be described by the following set of dimensionless equations:

$$\nabla \cdot \boldsymbol{u} = 0 \tag{25}$$

$$\frac{\partial \boldsymbol{u}}{\partial t} + (\boldsymbol{u} \cdot \nabla)\left(\frac{\boldsymbol{u}}{\varepsilon}\right) = -\frac{1}{\rho}\nabla(\varepsilon p) + \sqrt{\frac{Pr}{Ra}}\nabla^2 \boldsymbol{u} + \boldsymbol{F} \tag{26}$$

$$\frac{\partial T}{\partial t} + \boldsymbol{u} \cdot \nabla T = \frac{1}{\sqrt{PrRa}}\nabla^2 T \tag{27}$$

where u is the fluid velocity, ϵ is the porosity of the porous medium, and F is the sum of the resistance of the porous medium and the gravitational forces caused by the density difference, given by:

$$\boldsymbol{F} = -\frac{\varepsilon\sqrt{Pr/Ra}}{Da}\boldsymbol{u} - \frac{\varepsilon F_\varepsilon}{\sqrt{Da}}|\boldsymbol{u}|\boldsymbol{u} - \varepsilon T\boldsymbol{k} \tag{28}$$

In this equation, Pr is the Prandtl number, which is the ratio of the kinematic viscosity and the temperature diffusion coefficient, representing the relative relationship between momentum diffusion and molecular diffusion; Ra is the Rayleigh number, representing the relative magnitude of the convective rate to the diffusive rate; Da is the Darcy number, a parameter representing the permeability of the porous medium. And the dimensionless characteristic parameters of the generalized N–S equations include the Reynolds number (Re) and viscosity ratio (J). The definitions of the above dimensionless parameters are as follows:

$$Pr = \frac{v}{\alpha},\ Ra = \frac{g\beta\Delta T H^3}{vD}, Da = \frac{K}{L^2}, Re = \frac{LU}{\nu}, J = \frac{\nu_e}{\nu} \tag{29}$$

where ν is the fluid kinematic viscosity, α is the fluid diffusion coefficient, g is the gravitational acceleration, $\beta = \partial\rho/\partial T$ is the temperature expansion coefficient, K is the permeability of the porous media, and L is the characteristic length. The characteristic velocity U and characteristic time T are:

$$U = \sqrt{g\beta\Delta C H},\ T = \sqrt{H/g\beta\Delta C} \tag{30}$$

According to the definition of F, the ratio of nonlinear to linear resistance is:

$$\frac{F_e U/\sqrt{K}}{\nu/K} = \sqrt{Da}Re \tag{31}$$

when Re or Da is small, the nonlinear resistance term can almost be neglected in the calculation of seepage phenomena.

The initial positions of hot and cold fluids affect the subsequent viscous behavior of Rayleigh–Taylor instability (RTI), but the conditions for RTI mainly depend on porosity and permeability. This study focuses on the impact of porosity and pore size on RTI in the flow, force, and mass fields. The main focus of this research is to improve the model for activated carbon applications. The role of initial fluid positions will be explored in future studies. In the straight pipe reactor considered in this study, the partition is placed in the center of the simulation area, with equal proportions of hot and cold fluids (50% each) at the start of RTI. The interface disturbance stream function is given by:

$$y = 2.0 - 0.05\sin\left(2\pi x + \frac{1}{2}\pi\right) \tag{32}$$

Initially, the density of the entire field is set to 1.0, with no external forces or additional flow velocities applied. The model features periodic boundaries on the left and right sides, while the upper and lower boundaries are defined as adiabatic walls with a velocity of 0 m/s. The temperatures of two fluids in the porous medium are denoted as *Tc* and *Th* (*Tc* < *Th*), then to be measured in [°C].

The porous medium has a length of *Lx* and a width of *Ly*, with the horizontal direction represented by *x* and the vertical direction by *y*. Inside the medium, two miscible fluids are present, both treated as incompressible. Fluid 1, which is the hotter fluid, is positioned below fluid 2, the colder fluid. Initially, both fluids are at rest relative to each other. Figure 3 has shown the distribution area and selected configuration source of activated carbon in the straight tube reactor.

2.4. Grid Independence Verification

The grid independence verification is shown in Figure 4. It shows the fluid temperature profiles at the same proportional position with a ratio of 1:4 at time $t = 5$ for different grid numbers: 200 × 800 = 160,000, 256 × 1024 = 262,144, and 300 × 1200 = 360,000. The grid independence verification reveals that the temperature obtained with 262,144 and 360,000 grid numbers are essentially the same.

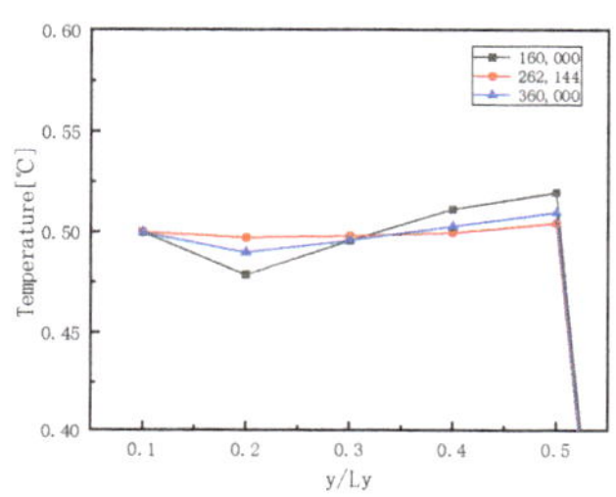

Figure 4. Grid independence verification.

Therefore, 262,144 grid numbers are sufficient to achieve the required accuracy, and this number is selected for data simulation. When the relative position is 0.2 times the length position, the difference ratio between the small grid and the maximum grid is 2.349%, while the difference ratio between the selected grid and the latter is 1.424%. Similarly, at the position of $1/Ly = 0.4$, the difference ratio of the small grid is 1.654%, while the one of the selected grid is only 0.629%. So here the model with the grid of $Nx \times Ny = 256 \times 1024$ has been selected to simulate, which ensures accuracy and reduces computational difficulty.

3. Results

The influence in force of different porosities and pore scales has been investigated herein. The parameters selected for the numerical simulation are as follows: $Pr = 1.0$, $Ra = 10^6$, $\Delta x = \Delta y = 1/256$, with porosities ranging from 0.2 to 0.999 ($\varepsilon = 0.999$ called hollow media, here equal to no media), and pore sizes (dp) ranging from 0.1 to 0.6. The corresponding permeability (*K*) has been calculated based on Equation (5). The temperatures are defined as $T_h = 0.5$ °C, $T_c = -0.5$ °C to ensure a uniform density trend at non-interface locations, facilitating the simulation of the behavior of two miscible fluids, as shown in Figure 5. Figure 5a represents the geometry of the system, and Figure 5b represents the mesh in the simulation.

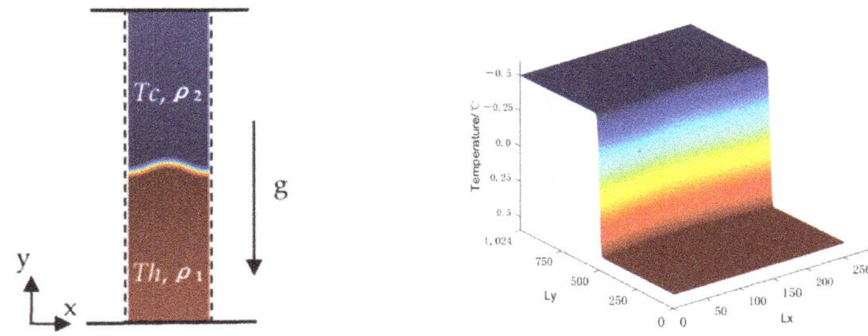

(a) (b)

Figure 5. Geometry of the flow and mesh. (**a**) Geometry of system. (**b**) Geometry of mesh in simulation.

The pore structure of porous media is generally very complex, and it is difficult to obtain its fine structure [24,25]. The study is based on mesoscopic, so active control of parameters can be achieved to incorporate the characteristics of activated carbon. To ensure rationality, a simulation of activated carbon with a porosity of 0.3, 0.6, and 0.999 (equal to no media) is first conducted to observe its internal RT evolution and then compared with the results of hollow media. The results are shown in Figures 6–8.

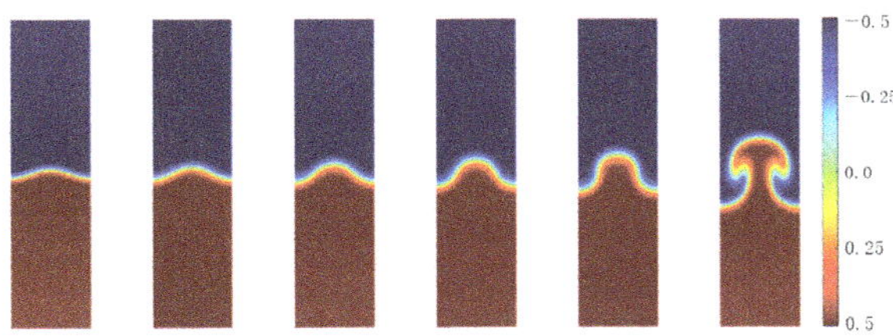

Figure 6. RT time sequence of temperature in $\varepsilon = 0.3$ ($t = 1, 2, 3, 4, 5, 8$).

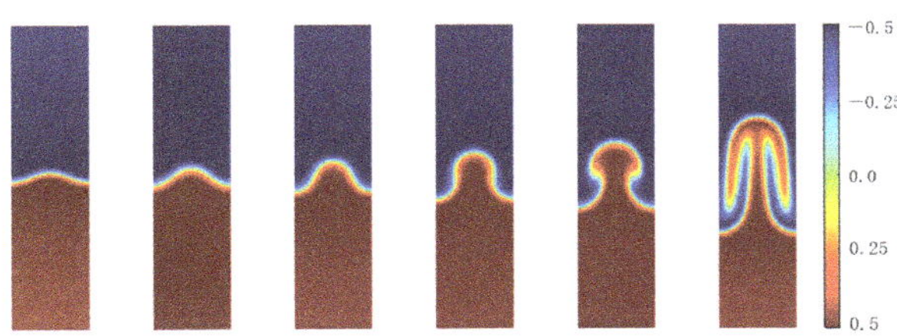

Figure 7. RT time sequence of temperature in $\varepsilon = 0.6$ ($t = 1, 2, 3, 4, 5, 8$).

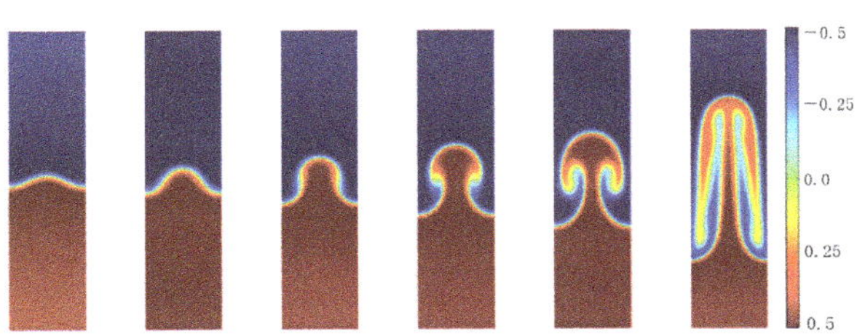

Figure 8. RT time sequence of temperature in hollow media ($t = 1, 2, 3, 4, 5, 8$).

The center temperature of the finger and bubble formed by convective diffusion of two fluids is approximately zero. After the effect of viscosity and adjacent convection, the shape of the fluid bubble is gradually slender until mixing completely. When porosity of porous media is at the initial stage with the media in $\varepsilon = 0.3$ and hollow media, the fluid RTI phenomenon is basically the same. And the change in porosity has little effect on the instability phenomenon. This is because, during the initial phase, fluid motion is primarily driven by density differences and initial perturbations, while the resistance effect of the porous media has not yet significantly influenced the overall behavior. However, as time progresses, the impact of porosity becomes more pronounced. In high-porosity media, the resistance is lower, and the effect of gravity on the fluid is more significant, leading to smoother fluid flow and an earlier onset of instability. In contrast, in low-porosity media, the higher resistance suppresses fluid flow, delaying the onset of instability. Additionally, the faster fluid mixing in high-porosity media further accelerates the development of instability. As a result, RTI under the condition of large porosity appears earlier than that under the condition of small porosity. Due to the downward action of gravity, the heavier liquid gradually forms a fluid similar to a finger, which flows unstably to the area where the light fluid is located, and the unstable phenomenon of the light fluid bulging upward is reflected quickly. In the later convection process, the seepage velocity of the two fluids is also faster under the condition of large porosity. Therefore, porosity primarily affects the onset time of instability and the speed of fluid convection, while its influence on the similarity of the initial stage is minimal. This difference arises from the varying resistance effects of porous media on fluid motion and the response speed of fluid behavior under gravity-driven conditions.

The effect on RTI from porosity is mainly reflected during the time for the overall system to turn from unstable into stable status. As porosity increases, the resistance from the porous media on the flow (working on F) weakens, especially under gravity. The parameter values appear to be more influenced by backflow caused by instability. At $t = 5$, RTI occurs, causing the data trend to differ significantly from others. Due to mutual solubility and mixing of the fluids in the RTI zone under flow action, with the streamlines affected by opposing vortices (seems like one mushroom), leading to complex flow. Without porous media, the flow is unobstructed, and RTI development is completely free. As a result, the flow process deepens at the same time step. In combination, the time required to obtain the same process is about 1~3 times different from that for the development of RTI in porosity of 0.6 than without media. It can be concluded that the change of porosity has an impact on the time of occurrence of the RTI process and the speed of fluid convection after instability.

The numerical results of axial and radial velocity at different grid heights ($0.4Ly$, $0.5Ly$, and $0.6Ly$) have been shown in Figures 9 and 10 about the RTI occurrence and development in $t = 5$ as the media in $\varepsilon = 0.6$ and hollow media (equal to no media). The fluid with higher density is relatively stable above the fluid with lower density in the initial stage. The fluid is considered as natural convection in hollow media. It can be seen that when $t = 5$, the fluid without a porous medium has undergone relatively obvious mass transfer, indicating that RTI has occurred, while at the same time in $\varepsilon = 0.6$, there is no significant change in fluid shape at the interface. Comparing the time steps required to reach the same manifold at the interface of the two conditions, it is found the porosity of 0.6 is nearly 1.8 times longer than that of no porous media. In the convection process, the seepage velocity of the two fluids is faster under the condition of large porosity. It is because the influence of the big porosity leads to the decrease in the resistance force between two fluids under gravity. The instability occurred early, and the fluid convection velocity after instability becomes fast, so the time duration has been shortened in bigger porosity.

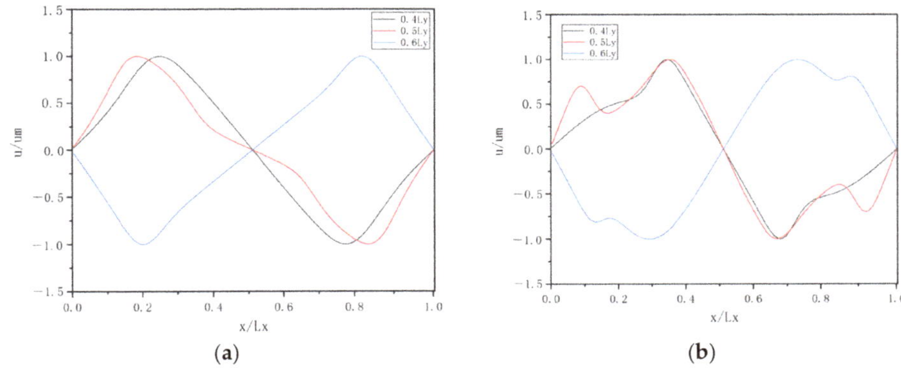

Figure 9. (**a**) u diagram in ε = 0.6. (**b**) u diagram in hollow media.

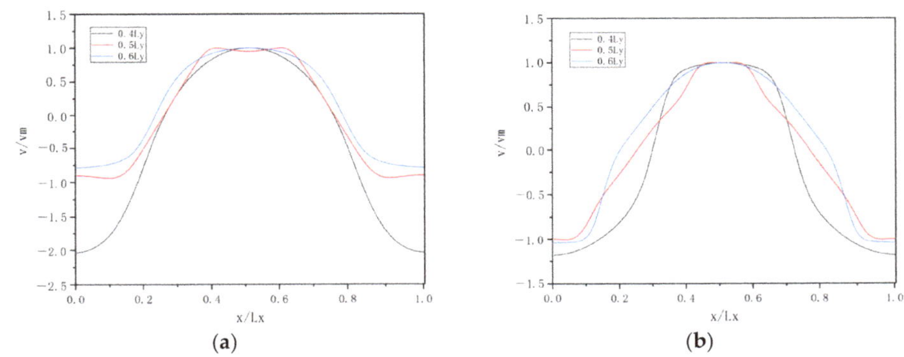

Figure 10. (**a**) v diagram in ε = 0.6. (**b**) v diagram in hollow media.

The presence of porous media significantly affects fluid interactions by reducing friction and resistance between fluids under gravity. While increased porosity may promote mixing, it also slows fluid velocity within the medium, particularly under gravity, due to added resistance, thus slowing RTI progression. However, higher porosity regions, with lower resistance, accelerate mixing by facilitating interactions between counter-rotating vortices and flow lines, enhancing fluid mixing despite slower RTI development. In regions of higher porosity, the increased fluid velocity leads to faster mass transfer and stronger interactions. Higher porosity also destabilizes fluid interfaces, making early fluid instabilities (such as RTI) more likely to occur and intensify. This is due to reduced resistance between fluids, which enables instabilities to develop earlier and more rapidly.

These findings provide valuable insights into fluid behavior in porous media, particularly for geological formations and engineering filtration systems. In geological formations, higher porosity accelerates fluid convection and instability, influencing subsurface fluid migration and distribution. In filtration systems, porosity's effect on fluid flow and mixing is crucial for optimizing filtration efficiency. Larger porosity may reduce resistance and enhance fluid penetration but could also accelerate instability. Understanding porosity's impact helps design more efficient filtration systems, reduce energy consumption, and optimize fluid distribution and mixing.

The evolutionary results are consistent with common sense, indicating that the model and simulation results are reasonable. Due to the wide range of porosity of activated carbon, porosity is the primary topic of discussion in this article to investigate the influence mechanism of force and eddy current effects on RTI. Next, we will discuss the simulation results of media with different porosities.

3.1. Effect of Porosity

The RTI in activated carbon under varying porosity conditions shares similarities with the diffusion of volatile fatty acids (VFAs) into concrete waste. Trancone G. [26] studied the bioleaching of concrete waste using dark fermentation (DF) and moving bed biofilm reactors (MBBR), where VFAs dissolve calcium-based compounds, reducing the concrete's weight and volume. This bioleaching process is analogous to fluid permeation and mixing in activated carbon pores. Akin to high VFA concentrations, high-porosity activated carbon enhances the dissolution of calcium-based compounds, leading to greater weight and volume reduction. Conversely, low-porosity activated carbon results in less effective dissolution. Simulating activated carbon at different porosity levels allows observation of RT instability manifestations, such as interface wave generation, mixing layer thickness, and mixing rates. These simulations not only aid in understanding activated carbon's performance but also provide a theoretical basis for optimizing its structure and functionality. Due to the uneven proportion of solids formed by porous media or the influence of compression processes on activated carbon, its porosity will change. For activated carbon alone, its porosity range before expansion is generally around 0.6 to 0.9, with a very small number of cases having a porosity of 0.2. To obtain more specific simulation results for comparison, this section comprehensively compares and analyzes the results of flow and force under different porosities with a dp of 0.2 and explores the influence mechanism of the internal force field of activated carbon on flow under various conditions.

Figures 11 and 12 present the temperature contours, streamlines, horizontal force component distributions, force lines, and density fields for the convective diffusion at porosities of 0.2 and 0.4 for $t = 10$, $t = 50$, and $t = 100$ (t = temp/1000; the time is measured in [s]). In the figures, red represents high values of the studied quantity, while blue represents low values. The studied quantities in figures a, b, and c correspond to temperature, horizontal force, and density, respectively. Observations indicate that at low porosities, the streamlines and density distributions at the interface are uniform and consistent. Under low-porosity conditions, the primary reason RTI does not develop is the significantly increased resistance of the porous media to fluid motion. This resistance suppresses fluid flow and interfacial disturbances, resulting in uniform and constant streamline and density distributions at the interface. And it is difficult for the fluid to form significant perturbations or interfacial waves, thereby maintaining a relatively balanced state. The lack of sufficient driving forces (such as buoyancy or inertial forces) prevents the breakdown of this equilibrium. A stable linear heat transfer interface and regular vortex structures form, indicating that fluid motion is dominated by conduction rather than convection, further suppressing the development of instability. At the same time, minor disturbances are quickly dissipated under the effect of resistance, preventing their amplification into significant RTI phenomena. Furthermore, the stable and paired distribution of horizontal forces further constrains unstable fluid motion. Therefore, under these conditions, RTI is unlikely to initiate or develop. In contrast, at a porosity of 0.6, changes in various parameters remain insignificant over time. This is due to the increased resistance experienced by fluids in the media at lower porosity, requiring more time or additional perturbations to disrupt a relatively balanced state.

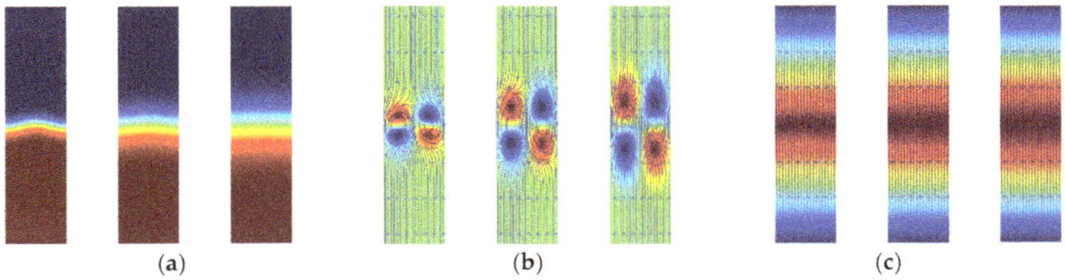

Figure 11. Evolution of RTI at $\varepsilon = 0.2$ for time $t = 10, 50, 100$. ((a) Temperature, (b) streamlines and horizontal force distribution, (c) force lines and density distribution).

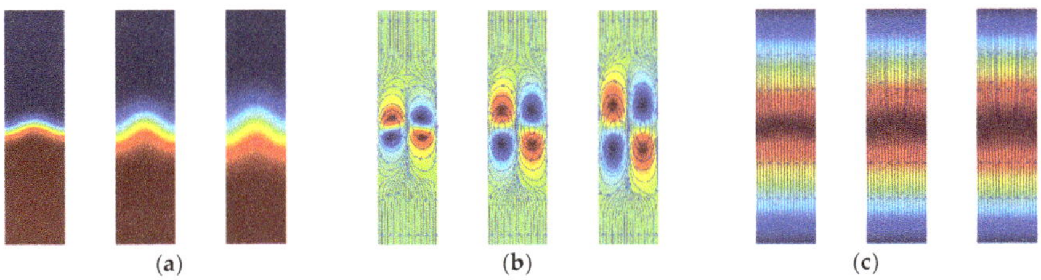

Figure 12. Evolution of RTI at $\varepsilon = 0.4$ for time $t = 10, 50, 100$. ((a) Temperature, (b) streamlines and horizontal force distribution, (c) force lines and density distribution).

Notably, at a porosity of 0.2, a straight heat transfer interface gradually forms. Figure 11 shows regular vortices at the interface, with relatively stable horizontal forces appearing in pairs. This stable state facilitates heat exchange between miscible fluids at the interface, rather than convection following the formation of fingers, indicating that the RTI phenomenon has not yet initiated.

Figures 13–15 depict the results of temperature, streamlines, horizontal force, and density in the media of porosity of 0.6 and 0.8 with the same pore size and in the hollow media at time intervals of $t = 10$, 50, and 100. For the porous media with moderate porosities, the evolution process occurred relatively quickly. As the RTI process progresses, fluid within the bubbles at the interface gradually rises to collide with the upper interface, facilitating the splitting of fluids near the interface and the formation of new vortices. This action pushes more fluid particles outward from the vortex lines, away from the interface. The density distribution indicates a more concentrated particle distribution at the upper and lower interfaces of the cavity.

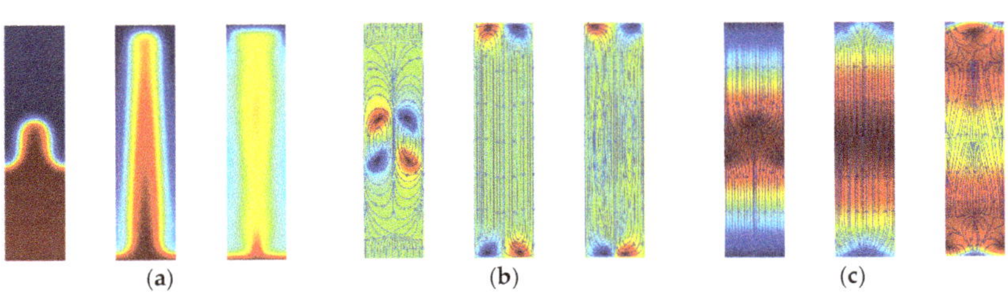

Figure 13. Evolution of RTI at $\varepsilon = 0.6$ for time $t = 10, 50, 100$. ((a) Temperature, (b) streamlines and horizontal force distribution, (c) force lines and density distribution).

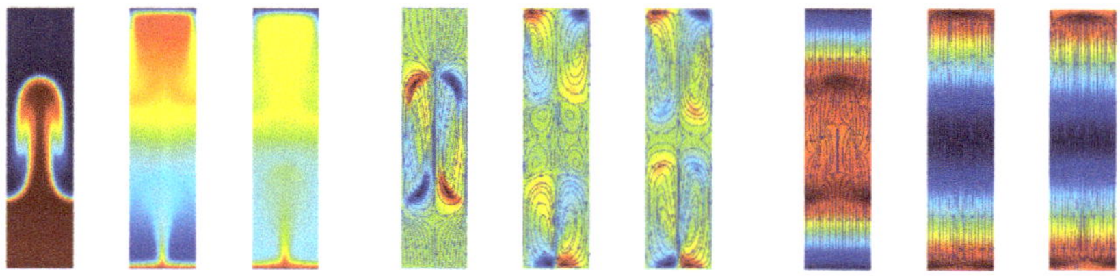

Figure 14. Evolution of RTI at $\varepsilon = 0.8$ for time $t = 10, 50, 100$. ((**a**) Temperature, (**b**) streamlines and horizontal force distribution, (**c**) force lines and density distribution).

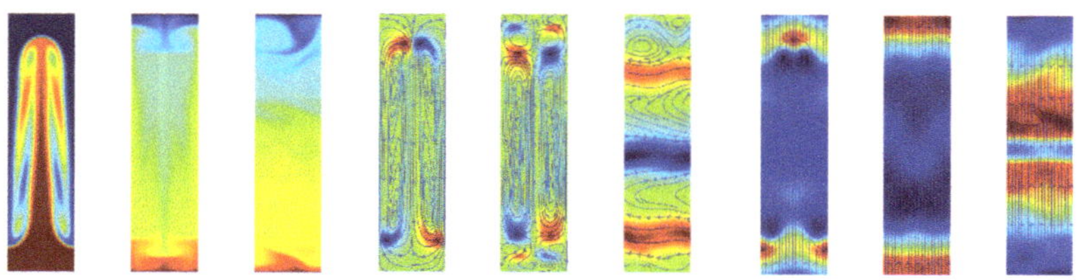

Figure 15. Evolution of RTI at hollow media for time $t = 10, 50, 100$. ((**a**) Temperature, (**b**) streamlines and horizontal force distribution, (**c**) force lines and density distribution).

Notably, at $t = 100$, the hollow media exhibits a state resembling wavelike Taylor vortices, with the horizontal force distribution showing symmetry about the central axis. Adjacent circumferential waves in the horizontal direction exhibit opposite vortex line orientations. This state enhances mass transfer between vortices, while the density contour plot indicates a steep decline in particle density near the original interface compared to the surrounding areas, as particle density is highest in the model.

Through a comparative analysis of flow and force in porous media under different porosities, the study reveals that porosity significantly influences fluid motion and instability development. At low porosity, the high resistance of the porous media suppresses fluid flow, resulting in uniform streamline and density distributions, preventing RTI development. Due to the high resistance, the fluid struggles to generate significant perturbations or interfacial waves, maintaining a balanced state. The formation of a stable linear heat transfer interface and regular vortex structures suggests that heat conduction dominates over convection, further inhibiting instability. At higher porosities (e.g., 0.6 and above), fluid motion becomes more dynamic, allowing RTI to develop and form new vortices, which concentrate particle density at the upper and lower interfaces. Hollow media at high porosity exhibit wave-like Taylor vortices, enhancing mass transfer efficiency. Understanding this phenomenon can improve mass transfer efficiency in industrial applications by facilitating the provision of working fluids that enhance mass transfer and selectively block it, allowing only heat transfer through the continuous exchange of different media. In summary, porosity plays a crucial role in fluid flow and heat exchange, improving mass transfer and optimizing thermal processes in industrial applications.

3.2. Effect of Pore Size

In addition to porosity, permeability is an important factor affecting the flow velocity and linear and nonlinear resistance in a medium. According to the equation introduced in Section 2, it can be seen that the permeability K is related to porosity and particle size (related to pore size). In the design of activated carbon, pore size is a key parameter that affects fluid permeability and mass transfer efficiency. K is proportional to the square of the pore size ($K \propto dp^2$). This means that as the pore size increases, the permeability increases significantly. Pore size has a direct impact on flow resistance from channels and on nonlinear resistance from force. In small pore size, the nonlinear effects (such as turbulence or vortices) of fluid flow are more pronounced, which further increases flow resistance. In contrast, fluid flow in larger pores approaches laminar flow, with weakened nonlinear effects and relatively lower flow resistance. And both porosity and pore size jointly determine the permeability and flow resistance characteristics of a porous media. Under the same permeability, changes in porosity indirectly affect flow resistance by adjusting pore size. Conversely, under the same porosity, changes in pore size directly affect flow resistance by altering permeability. Therefore, even with higher porosity, if the pore size is smaller, the flow resistance of the fluid may still be relatively high.

This section focuses on the simulation analysis of general activated carbon based on a medium with a porosity of 0.6 and compares the force and its influence results under different pore sizes. It considers the impact of six pore sizes on the RTI process in dp = 0.1, 0.2, 0.3, 0.4, 0.5, and 0.6 [mm], corresponding to permeability values of $K = 9 \times 10^{-5}$, 3.6×10^{-4}, 8.1×10^{-4}, 1.44×10^{-3}, 2.25×10^{-3}, 3.24×10^{-3}. Figures 16–18 presented the temperature contours, streamlines, and horizontal force component contours in t = 10, 50, and 100. Figure a shows the temperature cloud map, while Figure b shows the streamlines and horizontal force distribution for dp = 0.1 to 0.6. In the figures, red represents high values of the studied quantity, while blue represents low values. The studied quantities in figures a and b correspond to temperature and horizontal force, respectively. Figure 19 presented the results of force lines and density distribution for dp = 0.1~0.6 at two developing times, where Figure 19a represents the results at t = 10, and Figure 19b represents the results at t = 100. In the Figure 19, red represents high values of density, while blue represents low values.

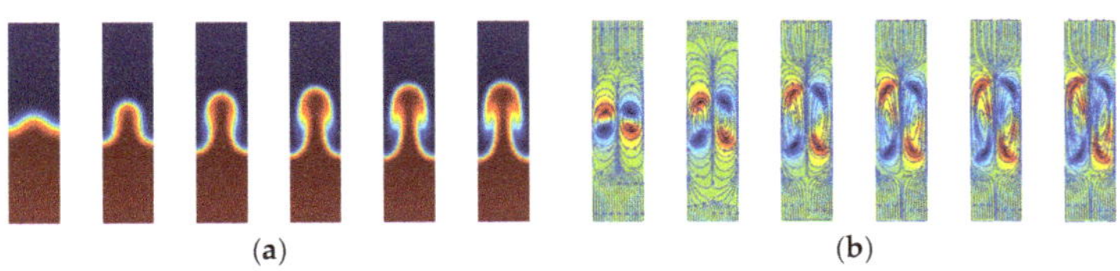

Figure 16. Evolution of RTI at dp = 0.1~0.6 as t = 10. ((**a**) Temperature, (**b**) streamlines and horizontal force distribution).

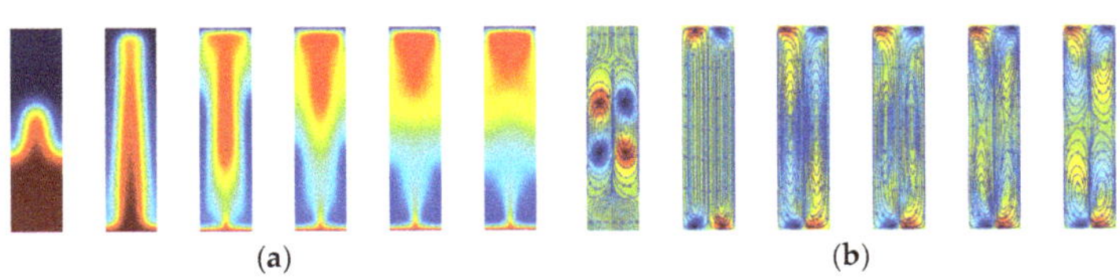

Figure 17. Evolution of RTI at dp = 0.1~0.6 as t = 50. ((**a**) Temperature, (**b**) streamlines and horizontal force distribution).

Observations indicate that for the same porous media, changes in pore sizes have a more significant impact on horizontal forces. At t = 10, when the dp is greater than 0.2, the development progress of RT is similar between the working conditions, and the trend of streamlines is basically the same. The horizontal force also concentrates in the middle finger-shaped fluid formation area to form force vortices, while under the action of force, fluid particles are mainly concentrated in the middle area. With the passage of time, the dp working conditions can be roughly divided into three categories. As dp < 0.3, the RT viscous fingerings developed slowly, thus not touching the top surface, but the miscible displacement has already occurred at the edges of the fingers. When $0.3 \leq dp \leq 0.4$, the temperature cloud map shows obvious differences in development depth, but the flow streamlines are basically a pair of vortices. The peak and valley distribution positions of horizontal forces are the same, but the value of internal forces in the region is greater in the working condition with dp = 0.3. The difference between the two operating conditions with dp > 0.4 is not significant.

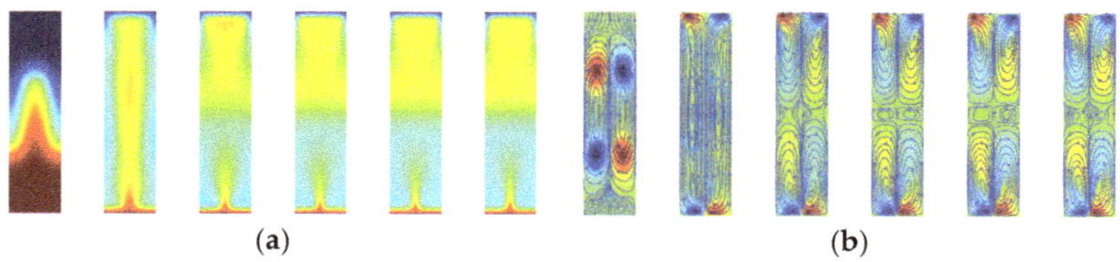

Figure 18. Evolution of RTI at dp = 0.1~0.6 as $t = 100$. ((**a**) Temperature, (**b**) streamlines and horizontal force distribution).

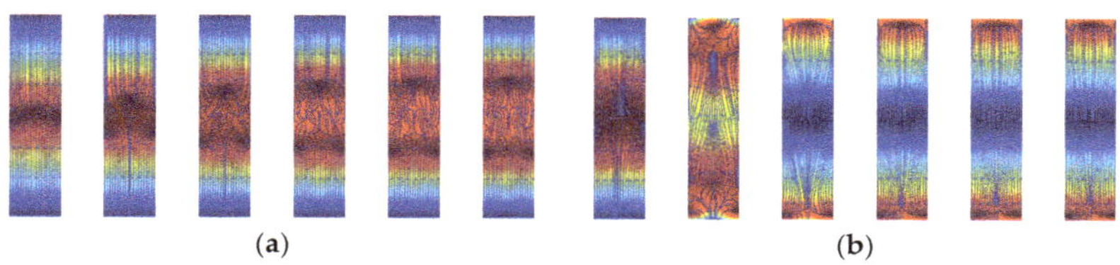

Figure 19. Force lines and density contours at dp = 0.1~0.6 ((**a**) $t = 10$, (**b**) $t = 100$).

At $t = 100$, the three fields of the working condition with dp > 0.2 are similar. Significantly, at larger pore sizes, the flow dynamics are dominated by vortex structures and patterns generated during the flow process. When $dp \geq 0.3$, the degree of variation in operating conditions becomes less pronounced, and with increasing development time, its influence on temperature, streamlines, and density diminishes. Further discussion is provided through quantitative analysis below.

When dp exceeds a specific threshold (dp > 0.4), the influence of further increasing dp on flow resistance and hydrodynamic behavior gradually becomes negligible. According to the Ergun equation, an increase in dp significantly enhances the permeability of the media, thereby reducing flow resistance. However, once dp exceeds the threshold, inertial forces (proportional to the square of the flow velocity) gradually replace viscous forces (linearly related to the flow velocity) as the dominant mechanism governing the flow. The flow state of the fluid approaches that of open-channel flow, and the restrictive effect of the porous media's microstructure on the flow diminishes. As a result, further increases in dp contribute to a saturation effect in resistance variation. The similarity in mass transfer for dp > 0.4 in Figure 19 further validates this threshold effect. It indicates that under large dp conditions, flow field characteristics (such as vortex structures and force distributions) tend to stabilize, with significant reductions in observed differences.

The study is crucial for simulating activated carbon in straight-tube reactors, particularly in optimizing the balance between adsorption efficiency and flow resistance. The study shows that activated carbon particles with $dp \approx 0.3$ achieve the optimal balance between high surface area (increasing adsorption efficiency) and low flow resistance. This finding directly informs the design of activated carbon in straight-tube reactors: at low dp (dp < 0.3), high resistance delays pollutant diffusion but may cause local concentration buildup; at high dp ($dp \geq 0.4$), high permeability facilitates rapid fluid exchange but may reduce adsorption efficiency. Thus, activated carbon particles with dp ≈ 0.3 are ideal for scenarios requiring efficient adsorption and stable flow, such as wastewater treatment or gas purification. In environmental applications, for example, in pollutant migration, low dp media (dp < 0.3) delay pollutant diffusion but may cause concentration buildup, while high dp media (dp ≥ 0.4) are suited for rapid fluid exchange. In engineering applications of chemical reactors, low dp media are suited for precise fluid distribution, though their high pressure drop may increase energy consumption; high dp media are suitable for high flow demands, but mechanical strength and clogging risks must be considered. These findings reveal the nonlinear relationship between dp and flow behavior in porous media, offering a theoretical basis for optimization design.

Additionally, the dp threshold can avoid excessive particle size increases, thus reducing cost or performance redundancy, or require multi-parameter optimization considering porosity and dp synergy. For example, in smart responsive materials, real-time adjustment of equivalent dp via external stimuli (e.g., temperature, electric field) enables on-demand flow control to optimize reactor performance. In practical engineering, activated carbon adsorption design can use the dp ≈ 0.3 conclusion to balance adsorption efficiency and flow resistance; in scenarios like geothermal reservoir enhancement, injecting proppants to increase dp requires strict control of the threshold to avoid stability issues from excessive fracture extension. In conclusion, this study enhances the understanding of fluid dynamics in porous media and provides scientific support for optimizing activated carbon design in straight-tube reactors, with broad environmental and engineering applications.

4. Discussion

In Section 3, we compared the results of the influence of activated carbon porosity and pore size on force, temperature, and flow. In fact, whether it is changes in porosity or pore size, what ultimately changes is the resistance to fluid flow inside activated carbon. It causes the changing force field and the resulting eddy current effect to affect the final flow result, resulting in the process of the RT instability phenomenon. Qualitative analysis can provide intuitive results, but it is difficult to provide specific data on the degree of change. So, in this chapter, we will start from a quantitative analysis perspective to explore the specific changes in RT caused by the effects of porosity and pore size on the changes in horizontal force Fx and flow velocity.

4.1. Analysis on Effect of Porosity

Analyzing data at the same longitude ($x/Lx = 0.25$), Figures 20 and 21 show the results of Fx and $|u/umax|$ values in various positions for five different porosities, where Figure a shows the horizontal force for different porosities ($\varepsilon = 0.2, 0.4, 0.6, 0.8$) and no media ($\varepsilon = 0.999$) at the same longitude ($x/Lx = 0.25$) across the entire axial scale at $t = 10$, while Figure b shows the horizontal force results at $t = 100$. At the same longitude ($x/Lx = 0.25$), despite varying porosities, the force distribution (Fx) and horizontal velocity ($|u/umax|$) both exhibit symmetry relative to the center point. This symmetry originates from the development pattern of RTI in porous media. In low-porosity media, the fluid seepage resistance is relatively high, resulting in slower flow development. In contrast, in high-porosity media, the fluid seepage resistance is minimal, allowing the flow to rapidly convect along the direction of gravity and form symmetric vortex structures.

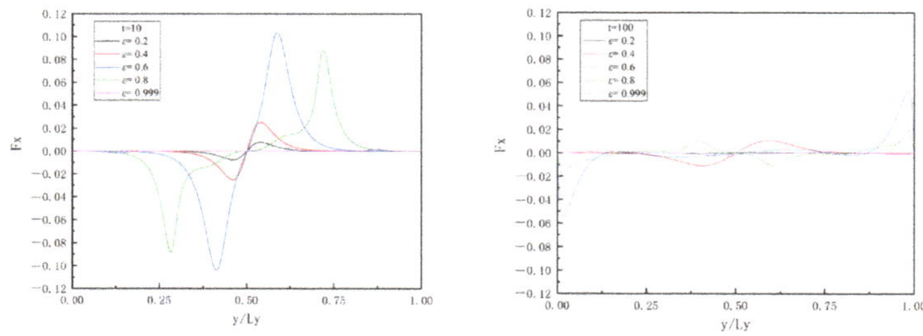

(a) (b)

Figure 20. Plots of y/Ly-Fx for five porosities ((**a**) t = 10, (**b**) t = 100).

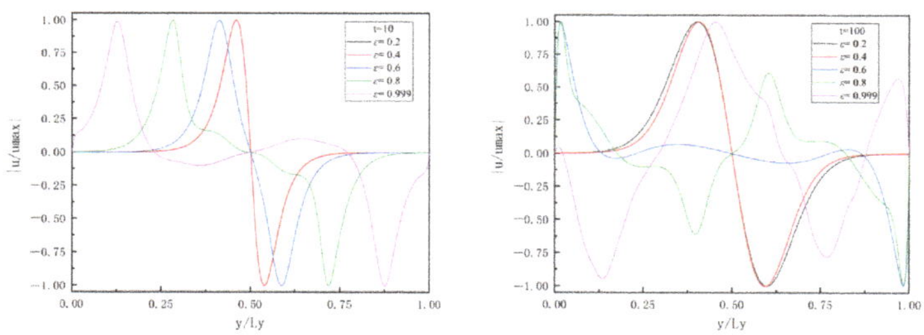

(a) (b)

Figure 21. Plots of y/Ly-|u/umax| for five porosities ((**a**) t = 10, (**b**) t = 100).

At t = 10, the generation of symmetric forces causes a symmetrical distribution of horizontal velocities. This symmetry is indicative of a balanced pressure difference within the system, influenced by the media's porosity. As porosity increases, the delay in extrema velocity suggests a smoother, less resistive flow path, which impacts how fluid is distributed and how fast the flow reaches its peak velocity. The force distributions, although symmetric, become less pronounced as porosity increases, leading to smoother velocity profiles and more evenly distributed flow. Influenced by these forces, the positions where the flow velocity reaches its extrema are delayed by 0%, 8.84%, 22.39%, and 21.82%, respectively, as the porosity increases from 0.2 to 0.999. The simulation results between the porosity of 0.2 and no media are consistent with the resistance effect predicted by the Darcy–Forchheimer model, also verifying the accuracy of the resistance term in the model.

At t = 100, the behavior evolves as the vortex structures in low-porosity media are much more stable and resistant to rupture, whereas in high-porosity media, the flow is less hindered by resistance, leading to the rupture of vortex lines and the development of Taylor vortices, accelerating mass transfer between vortices. The maximum speed can be 9.8816 times that of other operating conditions, accompanied by the generation of oppositely directed transverse wave-shaped mass transfer.

During the two stages of RTI development, medium-porosity media are most affected by horizontal forces, while fluids in low-porosity media face greater seepage resistance, resulting in slower development with similar patterns. Hollow media, experiencing minimal resistance to fluid seepage, develop convection rapidly along the direction of gravity, generating near-zero horizontal forces initially and achieving peak horizontal velocities more quickly. The symmetry of RTI-generated fingers shows that the relative distance of extreme values increases with porosity. Combined with the previous evolution plots, it can be observed that, unlike the single vortex pair in other streamline plots, ε = 0.8 has already generated four pairs of horizontal forces (i.e., two vortex pairs) at this point, while the vortex lines for $\varepsilon = 0.999$ have broken down, resulting in instantaneous horizontal flow velocities far exceeding those in other conditions.

The study of porosity highlights how porosity affects mass transfer and fluid dynamics, offering guidance for designing and optimizing activated carbon-based systems. Performance can be enhanced by focusing on the nonlinear effects of porosity on flow and mass transfer, using porosity gradation, vortex control, and dynamic response integration. Activated carbon with porosity between 0.3 and 0.6 balances high surface area with moderate flow resistance, preventing concentration accumulation or excessive pressure drop in adsorption. Gradient porosity design (such as core–shell structure) optimizes microstructure and regulates vortex behavior to enhance mass transfer efficiency, such as preventing dead zones in gas purification reactors. Although these strategies focus on activated carbon, the porosity–vortex–mass transfer principles apply to other porous systems. In filtration, low-porosity media ($\varepsilon < 0.4$) delay diffusion but risk local concentration accumulation. High-porosity media ($\varepsilon \geq 0.8$) enhance backwashing efficiency through rapid fluid exchange. In enhanced oil recovery, medium-porosity media ($\varepsilon = 0.4$–0.8) balance permeability and strength, optimizing water/gas injection to improve recovery.

These findings show that porosity selection directly impacts mass transfer and flow control, supporting multi-scale optimization of activated carbon systems for improved adsorption efficiency and energy reduction in reactors. They also expand the potential of porous media in environmental and energy engineering, offering key design parameters for system development.

4.2. Analysis on Effect of Pore Size

Using data analysis at the same longitude ($x/Lx = 0.25$), Figures 22 and 23 present the results of Fx and $|u/umax|$ at various positions for different pore sizes in the same porous media, where Figure a shows the horizontal force for each pore size (dp = 0.1–0.6 mm) at the same longitude ($x/Lx = 0.25$) across the entire axial scale at $t = 10$, and Figure b shows the horizontal force results of each situation at $t = 100$. Altering the pore scale does not modify the model structure or the disturbance effects, thus preserving the symmetry of the forces. Influenced by forces, the positions where the flow velocity reaches its extrema are delayed by 7.63%, 3.85%, 2.22%, and 1.46%, respectively, as dp increases from 0.1 to 0.6.

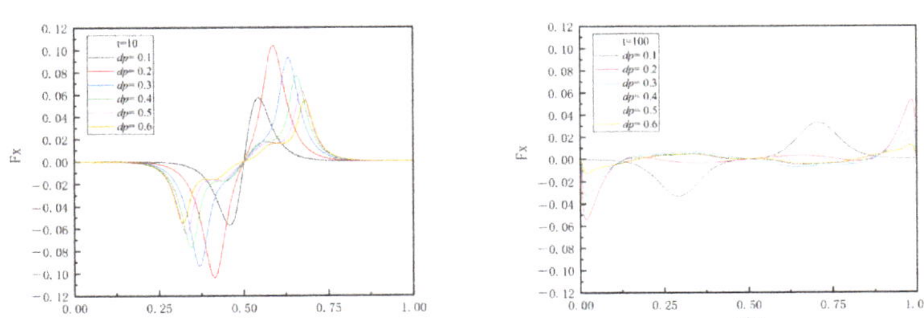

(**a**) (**b**)

Figure 22. Plots of y/Ly-Fx for different pore sizes ((**a**) $t = 10$, (**b**) $t = 100$).

At $t = 100$, combining with the previous evolution figures, it can be observed that unlike the results for dp = 0.1 and dp = 0.2, when dp ≥ 0.3, two pairs of vortices have already formed at this time. Despite the increase in dp, the development is sufficiently advanced that the force field results and flow field outcomes are nearly identical. And the results of dp ≥ 0.3 develop similarly, indicating similar flow patterns in Figure 23b. The conclusions are almost the same as the results above.

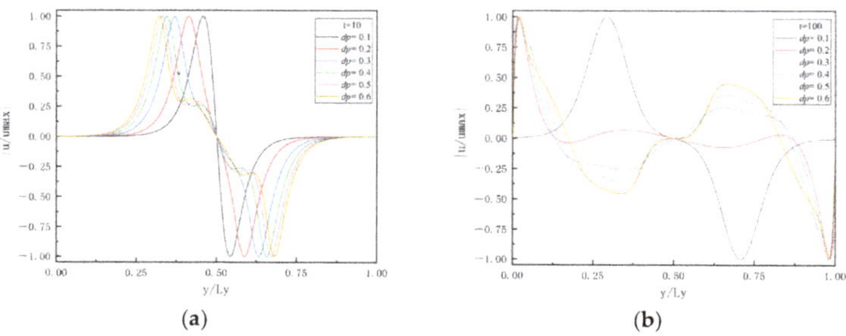

Figure 23. Plots of y/Ly-|u/umax| for different pore sizes ((a) t = 10, (b) t = 100).

Comparing the results at the same latitude under different pore scales, due to the presence of inverse symmetry, only the upper half of the latitude values are analyzed. Figure 24 shows the Fx corresponding to different pore sizes and time instances for the same porous media, using data from y/Ly = 0.1 to 0.5 at the middle of the model. Figure 24a shows the relationships between dp and Fx at the point of x/Lx = 0.25 for different latitudes (y/Ly = 0.1, 0.2, 0.3, 0.4, and 0.5) at t = 10, while Figure 24b shows the results for the same points at t = 100. Under the influence of varying pore sizes, the evolution of RTI morphology from sharp to rounded is closely related to the redistribution of horizontal forces. During the initial stages of RTI development, the bubbles generated by the instability have not yet propagated far from the interface, and the fluid particles are primarily subjected to gravity, with the horizontal component of force approaching 0. From the perspective of dp changes, at low pore scales (dp < 0.3), small pores result in low permeability ($K \propto dp^2$), leading to high fluid seepage resistance and slower RTI development. Under these conditions, the fluid interface is dominated by gravity, forming sharp fingering structures. The horizontal force distribution is concentrated in narrow regions, near the interface, manifesting as a localized high-gradient force field (Figure 24a). The high-gradient force field drives rapid vertical motion of fluid particles, while momentum exchange in the horizontal direction is restricted, resulting in sharp interfacial shapes. For high pore scales (dp $\geq$ 0.3), as dp increases, permeability significantly improves, and fluid resistance decreases. The flow regime transitions from confined laminar flow to more freely developed vortices, with enhanced symmetry in horizontal force distribution (Figure 24b). So, the fluid particles gain more momentum in the horizontal direction, and the formation of vortex structures promotes more uniform interfacial diffusion, leading to a gradual rounding of the RTI shape. Meanwhile, when dp is large enough, the development is sufficiently advanced at the same time, resulting in similar horizontal force outcomes at the same latitude.

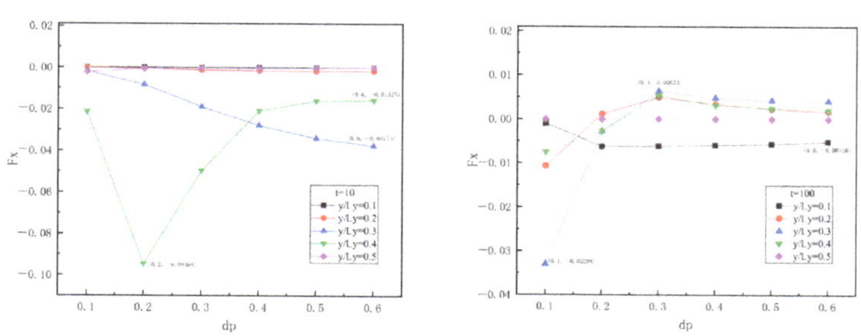

Figure 24. Plots of dp-Fx at different latitudes and pore sizes ((a) t = 10, (b) t = 100).

The particles at the interface experience collisions and compressions due to the density difference between the upper and lower fluids, resulting in a relatively balanced horizontal force. Fluid particles near the interface are more influenced by the traction accompanying the RTI, especially with a slight increase in dp when it is relatively small, causing the RTI shape to gradually become rounded from sharp, and a sudden change in the horizontal force occurs. The mechanism and sudden change of horizontal force with varying dp in the figure are related to the dynamic response of the horizontal force component and its nonlinear relationship with the pore scale. At t = 10, when dp is small (dp = 0.1), the fluid is primarily driven by vertical gravity, and the horizontal force component is nearly zero (Figure 24a). And the RTI has not fully developed, and lateral momentum exchange among fluid particles is weak, resulting in a sharp interface shape. When dp exceeds the threshold (0.3), the nonlinear increase in permeability significantly reduces flow resistance, and the fluid accelerates vortex formation, causing a redistribution of horizontal force due to momentum exchange between vortices. So at dp = 0.4, a sudden change in horizontal force appears near the interface (Figure 24a), corresponding to the transition of the RTI shape from sharp to rounded. This sudden change results from the transition in the flow state—from vertically dominated constrained flow to vortex-driven mixing flow. As dp increases further and permeability approaches open channel conditions, at dp = 0.6, flow resistance becomes negligible. The horizontal force distribution becomes symmetric and uniform, the interface shape is fully rounded, and the mass transfer efficiency between vortices reaches its peak.

The study uncovers the nonlinear control of flow resistance through permeability via dp, linking RTI development to interface shape. Vortex formation and breakdown at higher porosities, like Taylor vortices, enhance mass transfer, offering insights for optimizing porous media mixing. For activated carbon fluid separation, dp $\approx$ 0.3 balances adsorption efficiency and flow resistance, while thermosensitive materials allow dynamic porosity adjustment, improving separation efficiency and adapting to pollutant load changes. These findings also have other practical applications, such as in CO_2 sequestration, where dp $\geq$ 0.3 improves fluid mixing and reduces the risk of finger-like intrusion. In chemical reactors, high-porosity media (dp $\geq$ 0.4) enhance vortex breakdown, promoting reactant mixing and improving reactor efficiency. By regulating permeability and flow resistance, dp profoundly influences RTI dynamics and horizontal force distribution. Low dp favors sharp interfaces and high-gradient force fields, while high dp promotes vortex-driven flow, enhancing mass transfer. These insights are key for designing porous media in applications requiring efficient mixing and flow stability, such as in geological sequestration and fluid separation.

5. Conclusions

This numerical study investigates the force mechanisms and eddy current effects of Rayleigh–Taylor instability (RTI) in the intrinsic model of activated carbon, employing the lattice Boltzmann method (LBM). The results reveal that porosity and pore size (dp) significantly influence fluid dynamics, mass transfer, and instability development in porous media. The results indicate:

(1) Low porosity ($\varepsilon < 0.3$) suppresses RTI due to high fluid resistance, resulting in uniform streamline and density distributions. Heat transfer is dominated by conduction, and stable vortex structures prevent instability initiation. High porosity ($\varepsilon \geq 0.6$) facilitates RTI development, with vortex formation and rupture enhancing mass transfer. In the initial stage of RTI development (t = 10), influenced by these forces, the positions where the flow velocity reaches its extrema are delayed by 0%, 8.84%, 22.39%, and 21.82%, respectively, as the porosity increases from 0.2 to 0.999. After a

certain duration of RTI development (t = 100), the vortex lines at big-pore media have ruptured in force charts. At ε = 0.999, Taylor vortices emerge, increasing flow velocity by up to 9.88 times and promoting bidirectional mass transfer. Medium media are most affected by horizontal forces, while small-pore media exhibited higher resistance, resulting in slower and similar patterns. Hollow media generate near-zero horizontal forces during the initial stages of RTI development. The study of this phenomenon can aid in enhancing mass transfer efficiency in industrial applications and achieving the on-demand provision of working fluids for enhanced mass transfer and heat transfer without mass transfer through continuous exchange of different porous media.

(2) Low dp (dp < 0.3) leads to high resistance and sharp interfacial structures, delaying RTI development. High dp (dp $\geq$ 0.3) reduces resistance, enabling vortex-driven flow and smoother interfaces. Above dp = 0.4, further increases in dp have negligible effects on flow resistance, as inertial forces dominate viscous forces. Influenced by forces, the positions where the flow velocity reaches its extrema are delayed by 7.63%, 3.85%, 2.22%, and 1.46%, respectively, as dp increases from 0.1 to 0.6. The decrease in relative position changes indicates that flow results gradually converge as dp increases. After enough time growing, the development is sufficiently advanced, resulting in nearly identical force field outcomes.

(3) Horizontal forces exhibit symmetry and redistribute with increasing dp, transitioning from sharp to rounded RTI shapes. At dp $\geq$ 0.3, vortex structures enhance momentum exchange, leading to balanced horizontal forces and efficient mass transfer. The threshold effect at dp $\approx$ 0.3 highlights the nonlinear relationship between pore size and flow resistance, providing a basis for optimizing porous media design.

(4) In implications for activated carbon applications, it has been investigated that activated carbon with dp $\approx$ 0.3 balances adsorption efficiency and flow resistance, making it ideal for pollutant adsorption and gas purification. Furthermore, dynamic control of porosity through external stimuli (e.g., temperature, electric fields) enables real-time optimization of flow behavior in smart responsive materials.

In summary, this research elucidates the complex interplay between porosity, pore scale, and fluid dynamics in activated carbon, offering theoretical insights and practical guidelines for optimizing mass transfer and flow control in industrial applications. The findings underscore the importance of porosity and pore size in designing efficient porous media systems, with broad implications for environmental engineering, chemical reactors, and energy storage.

Author Contributions: L.Y. responsible for resources, formal analysis, program compilation and writing original draft; Q.Z. responsible for, resources, formal analysis, writing—review and editing; Y.W. responsible for methodology, project administration; Z.Y. responsible for investigation and resources; H.H. responsible for resources, formal analysis. All authors have read and agreed to the published version of the manuscript.

Funding: This research was funded by National Natural Science Foundation of China (NSFC), with the project leader and important participant being Li Ye. The funding types are the Young Scientists Fund (Project No. 51306122) and the General Program (Project No. 52376068), both of which can be verified through the NSFC official website.

Data Availability Statement: The original contributions presented in this study are included in the article.

Conflicts of Interest: The authors declare no conflicts of interest.

Abbreviations

The following abbreviations are used in this manuscript:

RTI	Rayleigh–Taylor Instability
LBM	Lattice Boltzmann method

References

1. Cao, R. Mechanical energy loss and Rayleigh-Taylor instability in free discharge of vertical sharp-edged orifices. *J. Chem. Eng. Res. Des.* **2023**, *190*, 282–295. [CrossRef]
2. Boyaci, A.; Banerjee, A. Transition to plastic regime for Rayleigh-Taylor instability in soft solids. *J. Phys. Rev. E* **2024**, *109*, 055103. [CrossRef] [PubMed]
3. Wang, J.; Li, H.; Guo, W.; Wang, Z.; Du, T.; Wang, Y.; Abe, A.; Huang, C. Rayleigh–Taylor instability of cylindrical water droplet induced by laser-produced cavitation bubble. *J. Fluid Mech.* **2021**, *919*, A42. [CrossRef]
4. Zhou, Y. Rayleigh-Taylor and Richtmyer-Meshkov instability: Similarity and difference. *J. Phys. Plasmas* **2017**, *24*, 055601.
5. Pan, C.; Hilpert, M.; Miller, C.T. Lattice-Boltzmann simulation of two-phase flow in porous media. *J. Water Resour. Res.* **2006**, *42*, W01412. [CrossRef]
6. Li, J.; Yang, J. Local volume-conservation-improved diffuse interface model for simulation of Rayleigh–Plateau fluid instability. *J. Comput. Phys. Commun.* **2024**, *296*, 109050. [CrossRef]
7. Su, Q. On the Transition of the Rayleigh-Taylor Instability in 2d Water Waves with Point Vortices. *J. Ann. PDE* **2023**, *9*, 2. [CrossRef]
8. Ding, S.; Ji, Z.; Li, Q. Rayleigh–Taylor instability for non-homogeneous incompressible fluids with Navier-slip boundary conditions. *J. Math. Methods Appl. Sci.* **2020**, *43*, 6338–6362. [CrossRef]
9. Hong, L. Direct numerical simulations of multi-mode immiscible Rayleigh-Taylor instability with high Reynolds numbers. *J. Phys. Fluids* **2019**, *31*, 112104.
10. Liu, R.; Yao, S.; Xu, L.; Sun, Z.; Cheng, J. Transient simulation of porous cathodes of zinc-nickel single-flow batteries based on lattice Boltzmann method. *J. Energy Storage* **2020**, *32*, 101937. [CrossRef]
11. Scovazzi, G.; Gerstenberger, A.; Collis, S.S. A discontinuous Galerkin method for gravity-driven viscous fingering instabilities in porous media. *J. Comput. Phys.* **2013**, *233*, 373–399. [CrossRef]
12. Zhang, T.; Juanes, R. Dynamics of viscous fingering in porous media: Pore-scale simulations and upscaling. *J. Adv. Water Resour.* **2015**, *38*, 969–981.
13. Lu, C.; Zhang, M.; Luo, K.; Wu, J.; Yi, H. Rayleigh–Bénard instability in the presence of phase boundary and shear—CORRIGENDUM. *J. Fluid Mech.* **2023**, *975*, E3. [CrossRef]
14. Zhou, D.; Huang, R. Numerical simulation of Rayleigh-Taylor instability in porous media using a phase-field model. *J. Fluid Mech.* **2010**, *651*, 455–490.
15. Ma, Y.; Xiao, X.; Li, W.; Desbrun, M.; Liu, X. Hybrid LBM-FVM solver for two-phase flow simulation. *J. Comput. Phys.* **2024**, *506*, 112920. [CrossRef]
16. Yos, P.S.; Takayuki, A. A filtered cumulant lattice Boltzmann method for violent two-phase flows. *J. Comput. Phys.* **2019**, *390*, 93–120.
17. Nithiarasu, P.; Seetharamu, K.N.; Sundararajan, T. Natural convective heat transfer in a fluid saturated variable porosity medium. *J. Int. J. Heat Mass Transf.* **1997**, *40*, 3955–3967. [CrossRef]
18. Ergun, S.; Orning, A.A. Fluid flow through randomly packed columns and fluidized beds. *Ind. Eng. Chem.* **1949**, *41*, 1179–1184. [CrossRef]
19. Ren, J.J.; Guo, P.; Guo, Z.L.; Wang, Z.H. A lattice Boltzmann model for simulating gas flow in Kerogen pores. *J. Transp. Porous Media* **2015**, *106*, 285–301. [CrossRef]
20. Zhang, T.; Li, X.F.; Wang, X.Z.; Li, J.; Li, W.; Zhao, W.; Yao, T.F. Modelling the water transport behavior in organic-rich nanoporous shale with generalized lattice Boltzmann method. *J. Int. J. Heat Mass Transf.* **2018**, *127*, 123–134. [CrossRef]
21. Fu, D.; Li, Y.; Zhao, J.; Jiang, Y.; Chen, X.; Xu, W. Seepage law of shale gas reservoir based on lattice Boltzmann method. *J. Lithol. Oil Gas Reserv.* **2020**, *32*, 151–160.
22. Ladd, A.J.C. Numerical simulations of particulate suspensions via a discretized Boltzmann equation. 1. Theoretical foundation. *J. Fluid Mech.* **1994**, *271*, 285–309. [CrossRef]
23. Zhao, J.; Fu, D.; Li, Y.; Jiang, Y.; Xu, W.; Chen, X. REV-scale simulation of gas transport in shale matrix with lattice Boltzmann method. *J. Nat. Gas Sci. Eng.* **2018**, *57*, 224–237. [CrossRef]
24. Marina, T.; Natalia, C.; Antonina, C. Different-Scale Simulation of Flows in Porous Media. *J. EPJ Web Conf.* **2021**, *248*, 01016.

25. Alhejaili, W.; Aly, M.A. Artificial intelligence and numerical simulations for Cattaneo–Christov heat and mass fluxes of nano-encapsulated phase change materials in a zigzag porous cavity. *J. Energy Storage* **2024**, *90*, 111750. [CrossRef]
26. Trancone, G.; Policastro, G.; Spasiano, D. Treatment of concrete waste from construction and demolition activities: Application of organic acids from continuous dark fermentation in moving bed biofilm reactors. *J. Chem. Eng. J.* **2025**, *505*, 159536. [CrossRef]

water

MDPI

Article

Membrane Foulant Removal by Ozone-Biocarrier Pretreatment Technology for Industrial Wastewater Reclamation

Ting-Ting Chang [1,2,†], Sheng-Yi Chiu [1,*,†], Chun-Chi Lee [1], Yuan-Liang Tai [1], Guan-You Lin [1], Chun-Hsi Lai [3] and Po-Yu Chen [2]

1 Water Technology Division, Material and Chemical Research Laboratories, Industrial Technology Research Institute, Hsinchu 300044, Taiwan; nina75423@itri.org.tw (T.-T.C.)
2 Department of Materials Science and Engineering, National Tsing Hua University, Hsinchu 300044, Taiwan
3 Department of Environmental Engineering, National Cheng Kung University, Tainan 701401, Taiwan
* Correspondence: sy1012@itri.org.tw
† These authors contributed equally to this work.

Abstract: During wastewater reclamation, organic matter is considered the dominant foulant that shortens the lifetime of ultrafiltration (UF) membranes during operation. Additionally, the mineralization efficiency of organic matter in secondary effluent is typically low due to nonbiodegradable carbon sources. Herein, a combination of ozone and a porous biocarrier reactor was applied as a novel pretreatment system to enhance organic matter removal in the effluent in a lab-scale evaluation and pilot test. The results indicated that 70% of the biopolymer was removed, and the chemical oxygen demand (COD) removal efficiency was 1.8 times higher in this combined process than in the process with a porous biocarrier alone. The UF flux increased by 16% after the combined ozonation and porous biocarrier pretreatment process compared with the process with no pretreatment. Interestingly, the genus *Flavobacterium* (15.59%), containing biopolymer-degrading bacteria, was observed only in the combined ozone plus porous biocarrier process. Moreover, the results show that biopolymers can be removed through the combined ozone and porous biocarrier process due to partial ozone degradation, confirming that this combined process is one of the better pretreatment procedures for organic matter removal and improves the flux of UF during the wastewater reclamation process.

Keywords: porous biocarrier; ozone; membrane fouling; pretreatment; industrial wastewater; reclamation

1. Introduction

With extensive regulatory efforts aiming to mitigate water scarcity problems, secondary effluent from wastewater treatment plants has become an alternative water resource due to its stable water supply and acceptable water quality that meets the effluent discharge standards. Membrane filtration is considered a highly promising process for wastewater reclamation due to its compactness, easy operation, and high pollutant removal rate [1]. Recently, low-pressure membranes, including microfiltration (MF) and UF membranes, have experienced significant growth in drinking water purification and industrial wastewater reclamation. The UF process has been widely used to remove contaminants from industrial wastewater at a relatively low cost. However, despite various advantages of UF technology and substantial progress in module design, the performance of UF membranes in wastewater treatment and reclamation is still hindered by membrane fouling, which could lead to higher operating costs and a shorter membrane lifespan [2].

The secondary effluent from wastewater treatment plants contains complex organic compounds, including natural organic matter (NOM) and soluble microbial products (SMPs), generated during biological treatment processes [3]. Studies in the literature reported that high-molecular-weight organic matter, such as hydrophilic SMPs and protein-like extracellular matter, which belongs to biopolymers, caused a dramatic flux decline in membrane processes [4–6]. Zheng et al. [7] reported that the reversibility of the membrane was inversely related to the biopolymer concentration in the effluent water. These studies showed that biopolymers could be the major foulants that cause flux decline in ultrafiltration. Therefore, determining how to remove these organic substances that may cause serious film fouling at a reasonable cost should be a priority in the development and application of pretreatment technology for wastewater reclamation. In the pretreatment processes, preoxidation of secondary effluent by ozone shows promising potential in preventing or reducing organic fouling during membrane operation [5]. Lee et al. [8] investigated the membrane permeability and fouling mechanism by preozonation of wastewater treatment plant effluent and observed a 50% reduction in the thickness of the fouling layer on the membrane. Barry et al. [9] also reported that preozonation altered the size distribution of effluent organic matter (EfOM) by partial oxidation of high-molecular-weight (MW) organic macromolecules, and a reduction in UF membrane flux decline was observed. Ikehata et al. [10] concluded that partial degradation of organic compounds could greatly enhance their biodegradability, and this approach is preferable for the pretreatment of wastewater containing refractory organic matter prior to biological treatment.

Therefore, in this research, an attempt was made to use the combined preozonation and biological treatment process with a porous biocarrier made of polyurethane (PU) to remove organics and potential fouling materials in wastewater before UF to evaluate the effects on the performance of peroxidation prior to UF. The objectives of this study are to (1) determine the fouling potential and major foulants in secondary effluent via both a lab pilot test and pilot field verification unit; (2) investigate the efficiency of the biological pretreatment process with or without preozonation in the removal of organic foulants prior to UF; and (3) identify major species of microorganisms associated with the degradation of biopolymers in a porous biocarrier reactor.

2. Materials and Methods

2.1. Equipment Setup for Lab Testing System

The secondary effluent was collected from the Linhai Industrial Park wastewater treatment plant (Kaohsiung, Taiwan) for the evaluation of pretreatment. A lab-scale pretreatment unit including an ozonation unit and a porous medium bioreactor was set up for pretreatment process evaluation, and the process flow chart is shown in Figure S1. Ozone was produced by a Sumitomo SGA-01A-PSA4 ozone generator (Tokyo, Japan) with a maximum output of 27 g/h, and the ozone dose was increased from 0.5 to 1.0 mg O_3/mg dissolved organic carbon (DOC) for partial oxidation of the residual organics in the secondary effluent. For the porous biocarrier, a PU foam biocarrier patented by Industrial Technology Research Institute (ITRI) was used as the biological medium, with approximately 80% of the bioreactor volume filled with porous biocarrier. The hydraulic retention time (HRT) through the bioreactor was 4 h, and the COD loading was 0.2 kg/m^3-day. The inoculum of the activated sludge in the biological process was obtained from the wastewater treatment in ITRI campus. The lab-scale pretreatment test was performed for 3 months.

2.2. Equipment Setup for Pilot Testing System

The pilot system was established at the Linhai Industrial Park wastewater treatment plant (Kaohsiung, Taiwan). The secondary effluent was collected and treated by the wastewater reclamation process pilot system. The wastewater reclamation process pilot system was composed of an ozone oxidation reaction tank, a biocarrier process reaction tank, and UF membrane filtration units. A set of hollow fiber UF modules was used for continuous monitoring of the effect of pretreatment processes on the membrane flux performance. In the pilot system, the hollow fiber membrane in the UF module was made of polyvinylidene difluoride (PVDF) with an area of 45 m^2 and capacity of 112 m^3 per day (CMD). The average pore size was between 0.38 and 0.39 μm. The internal structure of the UF membrane was characterized by a Thermo Scientific Helios G4 Plasma FIB (PFIB) DualBeam (focused ion beam scanning electron microscope, or FIB-SEM) (Thermo Scientific, Waltham, MA, USA). A SEM image of the UF membrane is shown in Figure S2.

2.3. Water Quality Analysis

Effluent samples from the pretreatment system were collected once weekly, and DOC, NH_4^+, COD, and biological oxygen demand (BOD) were measured, as shown in Table 1. All water samples were filtered through a 0.22 μm filter before analysis. DOC analysis was conducted using a Sievers 900 TOC analyzer (Suez, Paris, France), and NH_4^+ was measured by Dionex ion chromatography (Thermo Scientific, MA, USA).

Table 1. Water sampling results for secondary effluent before and after pretreatment processes in lab testing.

Process	Secondary Effluent	Porous Biocarrier Reactor Effluent	Ozonation + Porous Biocarrier Reactor Effluent
COD (mg/L)	46.45 ± 17.95	43.85 ± 18.57	41.41 ± 16.39
DOC (mg/L)	21.44 ± 10.90	18.83 ± 7.86	17.67 ± 6.70
BOD_5 (mg/L)	7.54 ± 1.99	n.a. *	n.a. *
NH_4^+-N(mg/L)	32.84 ± 17.53	14.83 ± 9.68	14.45 ± 9.89
NO_3^--N(mg/L)	122.47 ± 32.26	139.38 ± 27.39	138.69 ± 26.91
$\Delta NH_4^+/\Delta NO_3^-$	n.a. *	0.93	0.98

Note: * n.a. represents not analyzed.

2.4. Excitation–Emission Matrix (EEM) Fluorescence Spectroscopy

The EEM scans of each sample were conducted using a fluorescence spectrophotometer (Model: F-7000 FL Spectrophotometer, Hitachi, Tokyo, Japan). Samples were placed in 1 cm quartz cuvettes. EEMs were generated by scanning over excitation wavelengths of 200 to 450 nm at intervals of 5 nm and emission wavelengths of 250 to 550 nm at intervals of 5 nm. The excitation slit width was set at 2.5 nm, and the emission slit width was set at 5 nm. The photomultiplier tube lamp voltage was 700 V, and the scanning speed was set to 30,000 nm/min for all measurements.

2.5. Assimilable Organic Carbon (AOC)

Water samples were filtered through a 0.22 μm membrane (Millipore, Darmstadt, Germany) to remove particulate matter and bacteria. AOC was determined with *Pseudomonas fluorescens* strain P17 (ATCC 49642). The strains were stored at −80 °C in culture medium with 25% glycerol. P17 cells were grown in a medium containing sodium acetate solution (2000 μg/L acetate-C) at 25 °C for 7 days prior to use. Cell concentrations of the stock solution were determined by plate count on R2A agar (25 °C, 3 days). The BacTiter-GloTM microbial Cell Viability Assay (G8231, Promega Corporation, Fitchburg, WI, USA) and Synergy HTX Multi-Mode Reader (Biotek Instruments, Winooski, VT, USA) were used to measure adenosine triphosphate (ATP) luminescence (relative light unit, RLU). The protocol recommends a volume of reagent equal to the volume of cell culture present in each well (100 μL reagent to 100 μL of sample for the 96-well plate) and incubating the mixture at room temperature for 5 min. The internal standard was composed of distilled water with sodium acetate. The water samples were supplemented with K_2HPO_4 (7 mg/L), KH_2PO_4 (3 mg/L), $MgSO_4{\cdot}7H_2O$ (0.1 mg/L), $(NH_4)_2SO_4$ (1 mg/L), NaCl (0.1 mg/L), and $FeSO_4{\cdot}7H_2O$ (1.8 μg/L). Duplicate centrifuge tubes contained indigenous microbes cultured at 10,000 cells/mL density with acetate-C concentrations of 0, 100, 200, 500, 1000, and 2000 μg/L at 25 °C for 3 days. Samples were taken at various intervals and used for ATP analysis.

2.6. Characterization of Molecular Weight Distribution

Characterization of the apparent molecular weight (AMW) distribution of water samples was conducted using high-performance liquid chromatography (HPLC, LC-20ATV, Shimadzu, Kyoto, Japan). Size exclusion chromatography (SEC) was conducted with sequential online detectors consisting of an ultraviolet detector (UVD) (254 nm, SPD-20A, UV–vis detector, Shimadzu, Kyoto, Japan) and organic carbon detector (OCD) (modified Sievers Total Organic Carbon Analyzer 900Turbo, GE Water &Process Technologies, Trevose, PA, USA). The chromatographic column (20 (I.D.) × 200 (L) mm TSK HW-50S, Tosoh, Tokyo, Japan) utilized a weak cation exchange resin based on polymethacrylate. The eluent system consisted of phosphate buffer (2.4 mM NaH_2PO_4 + 1.6 mM Na_2HPO_4; pH 6.8) containing sodium sulfate (25 mM) to achieve an ionic strength of 100 mM. The eluent flow rate was 0.5 mL/min, and the sample volume was 2 mL. PEG standard was used to determine the AMW. Prior to chromatographic separation, samples were made particle-free by being passed through a 0.45 μm filter (Advantec, Tokyo, Japan). Chromatograms were analyzed using a peak-fitting technique to resolve the overlapping peaks and to determine the area under each peak. PeakFit (Version 4.12, Systat Software Inc., San Jose, CA, USA), a commercially available software package, was used for this analysis. The procedure was adapted from the method described by Chow's research [11].

2.7. Lab Testing of Pretreatment Processes for UF Membrane Filtration

The influence of different pretreatment processes (porous biocarrier only, combined ozonation and porous medium biological process) on subsequent thin-film systems was evaluated. The water sample was filtered through a 0.45 μm membrane (Advantec, Tokyo, Japan) before the UF filtration test. Flat-sheet polyethersulfone (PES) UF membranes (NADIR@UP150, Wiesbaden, Germany) have a molecular weight cutoff of 150 kDa. After trimming the UF membranes to a suitable size, they were soaked in ultrapure water for 24 h. Flux tests were then performed on each membrane with ultrapure water, and the membranes with a measured flux difference within 5% were selected. After the screening procedure, the selected UF membrane was soaked in ultrapure water and stored in a refrigerator at 4 °C, and subsequent filtration tests were conducted within 3 days. The UF experiments were conducted in a filtration cell (Amicon 8200, (Millipore, Darmstadt, Germany) at room temperature (24 ± 1 °C). A constant pressure of 14 psi was applied to drive the water sample through the membrane in dead-end filtration mode. The volume of filtrate was measured using an electronic balance and automatically recorded every 30 s. The filtration experiment was stopped after 500 mL of permeate was produced. The initial flux, Jo, is the average permeate flux calculated by filtration of 200 mL ultrapure water. The apparent flux, J, is monitored by present flux. We used J/Jo to represent the decline ratio of UF flux.

2.8. Microbial Analysis

For the porous biocarrier system, the biomass produced in the processes was collected, followed by DNA extraction and analysis by next-generation sequencing (NGS). Raw 16S rRNA sequences of 40,132 gene amplicons were generated by Illumina sequencing analysis. After the low-quality sequences were screened, 40,132 high-quality 16S rRNA gene sequences were grouped into 49 bacterial operational taxonomic units (OTUs). For the combined ozonation and porous biocarrier system, 32,860 raw 16S rRNA gene amplicons were generated by Illumina sequencing analysis. After the low-quality sequences were screened, 32,860 high-quality 16S rRNA gene sequences were grouped into 32 bacterial OTUs. Bowtie2 was used to align sequencing reads against the clusters of the V4 sequence. A 97% similarity standard was applied to V4 sequence clusters. To characterize the dominant bacterial genus, we classified the OTUs taxonomically based on their 16S rRNA-encoding gene sequences for each of the bacterial communities developed under the different pretreatment processes. Biocarrier samples on the top (0.75 m from the bottom of the reactor) and middle (0.45 m from the bottom of the reactor) layers of the biological reactor were taken through sampling ports. The Polymerase Chain Reaction (PCR) CCR primers F515 (5′-GTGCCAGCMGCCGCGGTAA-3′) and R806 (5′-GGACTACHVGGGTWTCTAAT-3′) were designed to amplify the V4 domain of bacterial 16S ribosomal DNA. PCR amplification was performed in a 50 μL reaction volume containing 25 μL 2X Taq Master Mix (Thermo Scientific, MA, USA), 0.2 μM of each forward and reverse primer, and 20 ng DNA template. The reaction conditions consisted of an initial temperature of 95 °C for 5 min; followed by 30 cycles of 95 °C for 30 sec, 54 °C for 1 min, and 72 °C for 1 min; and then a final extension of 72 °C for 5 min. Next, the amplified products were checked by 2% agarose gel electrophoresis and were purified using the AMPure XP PCR Purification Kit (Beckman Coulter, San Jose, CA, USA). Quantification was performed by using a Qubit dsDNA HS Assay Kit (Thermo Scientific, MA, USA) on a Qubit 2.0 Fluorometer (Thermo Scientific, MA, USA) according to the manufacturer's instructions. For V4 library preparation, Illumina adapters were attached to the amplicons using the Illumina TruSeq DNA Sample Preparation v2 Kit (San Diego, CA, USA). Purified libraries were applied for cluster generation and sequencing on the MiSeq system. Sequencing reads from different samples were separated and identified according to specific barcodes at the 5′ end of the sequence (two mismatches allowed). FASTX-Toolkit (https://github.com/agordon/fastx_toolkit (accessed on 22 June 2019)) was employed to process the raw read data files. There were three steps of sequence quality processing: 1. The command was "fastq_quality_filter -Q33 -q 20 -p 70". "-q 20" means that the minimum quality score to keep is 20. "-p 70" indicates that the minimum percentage of bases must have "-q" quality over or equal to 70%. 2. The command was "fastq_quality_trimmer -t 20 -l 100 -Q33". "-t 20" bases with lower quality (<20) were trimmed (checking from the end of the sequence). "-l 100" indicates that the minimum acceptable length of the sequence is 100 after trimming the sequence. 3. Sequences were retained if both forward and reverse sequencing reads passed the first and second steps. To generate taxonomy assignments, the collection of 16S rRNA sequences was retrieved from the SILVA ribosomal RNA sequence database. These sequences were extracted by V4 forward and reverse primers. Then, UCLUST was used to create representative sequence clusters over or equal to 97% similarity. Bowtie2 was used to align sequencing reads against the clusters of the V4 sequence. A 97% similarity standard was applied to V4 sequence clusters. 16S rRNA gene sequences of the microbial community have been deposited in GenBank with accession numbers SRR17332969 and SRR17332970.

3. Results and Discussion

3.1. Lab Evaluation of the Pretreatment Process and UF Membrane Filtration

The basic characteristics of the secondary effluent from the Linhai Industrial Park wastewater treatment plant (Linhai WWTP) are summarized in Table 1. The low BOD_5/COD ratio, which was roughly 0.16, indicated that the residual organic matter in the secondary effluent of the industrial wastewater mainly consisted of recalcitrant organic compounds.

For the removal of COD, NH_4^+-N, and NO_3^--N by the combined ozonation and porous biocarrier pretreatment process, an average COD of 41.41 mg/L, an average NH_4^+-N of 14.45 mg/L, an average NO_3^--N of 138.69 mg/L, and a DOC of 17.67 mg/L were observed. Compared with the porous biocarrier biological process alone, the combined ozonation and porous biocarrier pretreatment process resulted in slightly increased COD and DOC removal. The NH_4^+-N removal was also maintained at levels above 50%. Furthermore, based on total nitrification equilibrium, the mole ratio of ammonium nitrogen converted to nitrate nitrogen should be 1. In the nitrogen conversion performance of the combined ozonation and porous biocarrier process, the ratio of ΔNH_4^+-N to ΔNO_3^--N was 0.98, which is very close to the stoichiometric theoretical value. For investigation of the nitrogen conversion, both pretreatment processes—biocarrier only and ozone combined biocarrier processes—showed that the recalcitrant organic compounds would not have an effect on nitrification, even though the ozone combined biocarrier process could convert the recalcitrant organic compounds into biodegradable organic compounds.

In further investigation of the pretreatment in terms of its effect on the change in flux of the UF membrane, the secondary effluent was filtered through the UF membrane with or without the pretreatment. Figure 1 shows a comparison of the decline in the UF flux of wastewater effluent, with the pretreatment of porous biocarrier and ozone combined porous biocarrier. In this experiment, we used J/Jo to represent the decline ratio of UF flux. The results showed that the lowest flux decline was observed with UF after the combined low-dose ozonation and porous biocarrier pretreatment process compared with the porous biocarrier pretreatment alone or no pretreatment. It was confirmed that the UF flux increased by 16% after the combined ozonation and porous biocarrier pretreatment process compared with the process with no pretreatment. Hamid et al. [12] showed that the decolorization of wastewater with an insignificant change in DOC after ozonation provides evidence of the decomposition of the carbon–carbon double bonds in aromatic structures and the partial degradation of high-MW organic substances with the use of an ozone pretreatment process before membrane filtration. In this study, we demonstrated both a small reduction in DOC and a substantial decrease in the high-MW fraction of organic substances with ozone pretreatment coupled with a porous biocarrier process. Our observation also provides supporting evidence that ozonation partially degraded macromolecules in the secondary effluent, and this transformation resulted in an increase in the permeate flux and alleviated the UF pore blocking, as shown in Figure 1. To further understand the cause of flux improvement, secondary effluent treated with different pretreatment processes was also evaluated with the following organic component characterization.

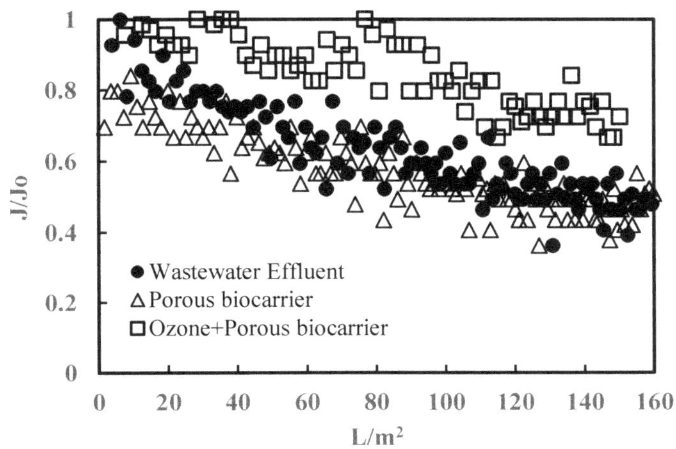

Figure 1. Comparison of decline in flux of wastewater effluent, porous biocarrier, and ozone combined porous biocarrier. The closed circles, open triangles, and open squares indicate the wastewater effluent, porous biocarrier, and ozone combined porous biocarrier, respectively.

3.2. Organic Component Characterization

Based on the results of lab testing of the combined ozonation and porous biocarrier pretreatment process, the ozone dose plays a crucial role in the removal of secondary wastewater EfOM. Therefore, an ozonation test was conducted on the Linhai secondary effluent with different ozone doses based on the O_3/DOC ratio, and EEM analysis was used to evaluate the decomposition of EfOM fractions by ozonation.

As shown in Figure 2, with increasing ozone dose, no more than 10% of DOC could be removed through direct ozonation of Linhai secondary effluent even by increasing the dosage to 75 mgO_3/mg DOC. It showed that the organic compounds in the effluent could not be completely removed but could be undergoing possible partial bond breaking. Organic substances in the wastewater treated with partial ozonation would result in the effectiveness of ozonation on the degradation of a wide variety of compounds with functional groups or moieties, including carbon–carbon double bonds, amines, and activated aromatic rings. The EEM analysis could provide the information related to functional groups breaking down in aromatic proteins, fulvic acid-like compounds, soluble microbial product-like compounds, and humic acid-like compounds. Figure S3 shows the EEM analysis with different doses of ozonation of water samples from secondary effluent before and after porous biocarrier pretreatment and combined ozonation and porous biocarrier pretreatment. The results showed that ozonation at a dose of 0.5 mg O_3/mg DOC resulted in more than 50% removal of DOC compounds, including aromatic protein II (AP), fulvic acid-like (FA), SMP-like, and humic acid-like (HA) fluorescence shown in Table 2. Table S1 shows the excitation and emission wavelength pairs of the main peak of the component compared with the compound descriptions of similar components reported in published studies [13–15]. The EEM fluorescence spectra of the water samples of the secondary effluent and the effluent of the porous biocarrier process were similar. However, the EEM fluorescence intensity decreased significantly after the secondary effluent was treated with the combined ozonation and porous biocarrier process. A fluorescence intensity 30% above reduction in protein-like extracellular matter was observed with the ozone doses of 0.2 and 0.5 mg O_3/mg TOC. Meng et al. [16] studied the reaction mechanism between ozone and extracellular polymeric substances (EPS) extracted from activated sludge and found that the protein-like extracellular matter concentration decreased sharply after treatment with ozone and that the macromolecular contents broke down into small molecules, including carboxylic acids, aldehydes, and ketones, which are more soluble and readily utilized as substrates by activated sludge. Therefore, through proper control of the ozone dose, disruption of the organic macromolecule structure with oxidation of carbon–carbon double bonds was achieved with minimum or partial mineralization of EfOM.

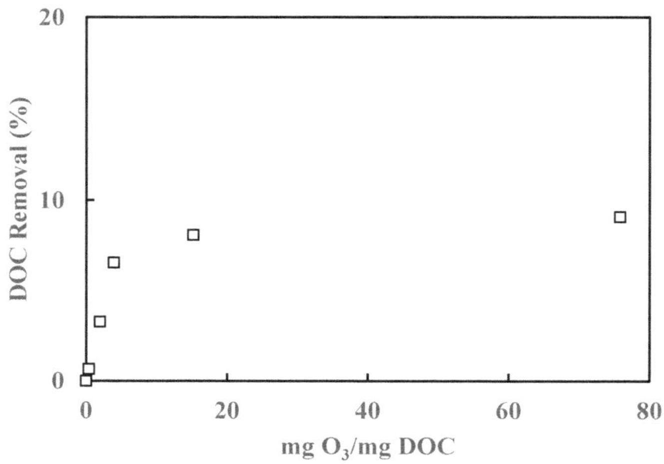

Figure 2. Effect of ozonation on DOC removal from Linhai Industrial Park secondary effluent.

Table 2. The removal of fluorescence: EEM fluorescence results with different O_3/DOC ratios.

	Ozone Dose (mg O_3/mg TOC)		
Fluorescence Intensity Removal (%)	**0.2**	**0.5**	**1**
API	8.3	29	71
APII	28	55	78
FA	30	52	69
SMP	33	67	82
HA	35	63	79

Organic components in different molecular weight fractions in wastewater were also characterized by SEC-OCD spectra. After the secondary effluent was treated with the combined ozonation and porous biocarrier pretreatment process, the intensity of the biopolymer fraction in the SEC-OCD spectra decreased significantly. This shows that the combined pretreatment process removed biopolymers (>10,000 Da) efficiently.

Table 3 summarizes the pollutant removal efficiency results obtained by two different pretreatment processes in lab testing. The removal efficiencies of the combined ozonation and porous biocarrier process for protein-like substances, biopolymers, and AOC were 32%, 70.2%, and 75%, respectively. In addition, less than 300 μg/L AOC was detected in the effluent from the combined ozonation and porous biocarrier pretreatment process. Based on the research of Huber et al. [17], compounds with molecular weights over 10,000 Da were characterized as biopolymers, including amino sugars, polypeptides, and SMP; compounds with molecular weights between 500 and 1200 Da were characterized as humic hydrophobic substances; and compounds with molecular weights lower than 350 Da were denoted as low-molecular-weight (LMW) neutrals, including alcohols, aldehydes, ketones, and amino acids. Hammes et al. [18] have shown that ozonation of drinking water increased the concentration of LMW organic byproducts, and the compounds were more easily utilizable by microorganisms. In our study, the significant reduction in biopolymers on the SEC-OCD footprint of secondary effluent after combined ozonation and porous biocarrier pretreatment provides evidence that ozonation converted larger molecules of EfOMs into LMW compounds that were more easily utilized by microorganisms in porous biocarrier systems, corresponding to a decrease in biopolymers and a slight increase in LMW compounds.

Table 3. Results for SEC-OCD and AOC in secondary effluent from different pretreatment processes in lab testing.

		Secondary Effluent	**Porous Biocarrier Reactor Effluent**	**Ozonation + Porous Biocarrier Reactor Effluent**
Concentration of DOC, in molecular weight	>10,000 Da (μg-C/L)	0.285	0.561	0.085
	500–1200 Da (μg-C/L)	6.74	7.62	7.67
	<350 Da (μg-C/L)	1.90	3.35	2.24
	Protein-like substance removal (%)	n.a. *	3.8	32.0
	Biopolymer removal (%)	n.a. *	−96.9	70.2
	AOC (μg/L)	1196.11 ± 186.93	2512.36 ± 105.97	253.82 ± 31.15

Note: * n.a. represents not analyzed.

3.3. Analysis of Microbial Community Diversity

To gain a deeper understanding of how microbial communities relate to biopolymer removal, we used NGS to investigate the composition of the microbial community associated with each pretreatment process. The results of an analysis of the diversity of the microbial community resulting from the two pretreatments are shown in Figure 3. The major species in the bacterial communities developed in the porous biocarrier system were identified as the nitrite-oxidizing bacteria *Nitrospira* (73.92%), with some ammonia-oxidizing bacteria *Nitrosovibrio* (6.06%) and *Nitrosomonas* (5.82%), whereas the relative abundance of *Geobacter* (3.13%) was low. The combined ozonation and porous biocarrier system supported bacterial communities composed mainly of nitrite-oxidizing *Nitrospira* bacteria (64.21%), with some interesting biopolymer-degrading *Flavobacterium* bacteria (15.59%), whereas the relative abundance of ammonia-oxidizing *Nitrosomonas* (5.74%), *Dokdonella* (5.24%), and *Geobacter* (4.15%) bacteria was low.

In the porous biocarrier process, 44.1% of ammonia was removed through the pretreatment process. In fact, *Nitrospira defluvii* (50.6%) and *Nitrospira* sp. (21.07%), known as nitrite-oxidizing bacteria, were found to be the dominant species in the porous biocarrier system. In addition, several different ammonium-oxidizing bacteria, including *Nitrosovibrio* sp. (6.06%) and *Nitrosomonas nitrosa* (3.25%), were identified in the porous biocarrier system. Uncultured *Geobacter* sp. (3.13%), an anaerobic bacterium that usually plays an important role in the bioremediation of groundwater, was also found in the porous carriers. For the combined ozonation and porous biocarrier process, with more than 50% of the average ammonium removal rate, the dominant biofilm bacteria found in the biocarrier medium included *Nitrospira defluvii* (44.57%), *Nitrospira* sp. (16.35%), and *Nitrospira marina* (3.28%), known as nitrite-oxidizing bacteria.

According to the results of microbial community analysis, several different bacteria in the combined ozonation and porous biocarrier process were identified. *Dokdonella immobilis* (5.24%) has also been isolated from activated sludge in a sequencing batch reactor used for the treatment of triphenylmethane dye effluent by [19]. Interestingly, *Flavobacterium* sp. (15.59%), which are known as biopolymer-degrading bacteria, were found only in the combined ozonation and porous biocarrier process. In earlier studies, [20] found that the classes *Cytophagia-Flavobacteria* seemed to play an important role in the degradation of biopolymers in marine and freshwater environments, and according to a whole-genome sequencing study, the biopolymer-degrading bacterium *Flavobacterium johnsoniae* has been increasingly used as a model microorganism to identify genes involved in biopolymer utilization processes [21]. A recent study by Sack et al. [22] showed the high affinity of *Flavobacterium johnsoniae* strain A3 for 5–6 kDa linear laminarin in oligotrophic conditions and concluded that strain A3 is involved in the bacterial utilization of biopolymers at low concentrations in oligotrophic freshwater environments. Therefore, the microbial community of the class *Flavobacteria* found in the ozone/porous biocarrier process could play a central role in the degradation of biopolymers in secondary effluent. The low-dose ozonation might have transformed the larger-molecule compounds into smaller-molecule biopolymers (ex., laminarin), which could be more easily utilized by *Flavobacterium* sp. This result highlights the promising use of the combined ozonation and porous biocarrier process for the effective reduction of secondary effluent biopolymers to alleviate membrane fouling of UF.

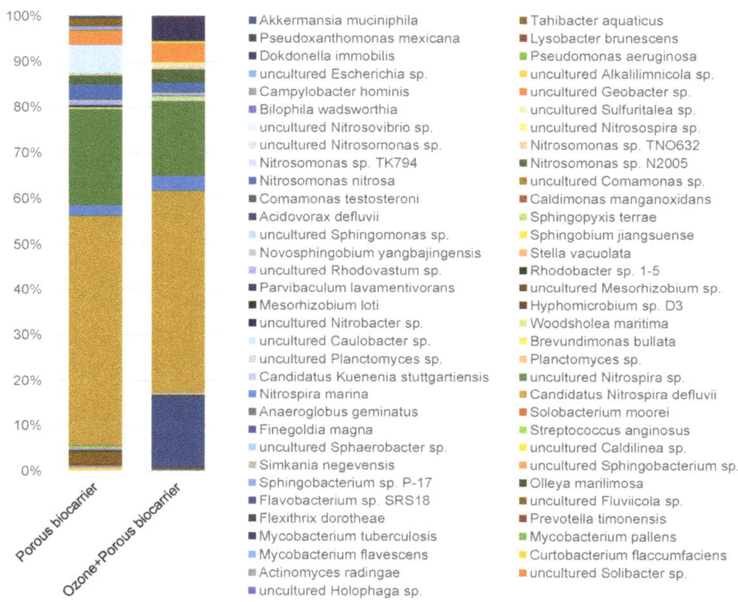

Figure 3. Results of NGS analysis showing the diversity of microbial communities in biomass samples from the biological porous biocarrier only process and the ozone combined with porous biocarrier process.

3.4. Pilot Testing of the Combined Ozonation and Porous Biocarrier Pretreatment Process

In the laboratory test, we observed that the decline in the flux of UF membrane filtration was effectively reduced by the combined low-dose ozonation and porous biocarrier pretreatment process. Subsequently, on-site pilot testing with 200 CMD pretreatment capacity was conducted to study the effectiveness of pretreatment processes on UF membrane filtration. Three sets of pretreatment processes were tested. The first and second sets of tests were separate tests of ozonation and porous biocarrier pretreatment processes individually, and finally, the combined ozonation and porous biocarrier pretreatment process was tested with an ozone dose of 0.08 g O_3/g COD as preozonation. The results of the pilot study are shown in Table 4. When secondary effluent was directly fed into the porous biocarrier reactor without any pretreatment, a 13% DOC removal was observed. On the other hand, the DOC concentration did not change substantially after the secondary effluent was treated with ozone, but the DOC removal was significantly increased to 28% when ozonation was coupled with the porous biocarrier treatment process. The result suggests that preozonation might have changed the size and structure of EfOMs and thus increased the biodegradability of EfOMs by transforming larger organic molecules into smaller molecules, which were more readily decomposed by the following biological treatment process. From the measurement of the BOD_5 (Table 4), it was found that the raw water BOD_5 increased from 8.2 mg/L to 10.8 mg/L after ozone treatment. This provides additional evidence that ozone could convert refractory organic matter into more biodegradable forms, resulting in an increase in EfOM removal by the combined pretreatment process. Notably, the porous biocarrier process demonstrated a significant BOD_5 removal of 50% without ozone pretreatment of wastewater.

Table 4. Results for secondary effluent before and after pretreatment processes in pilot testing.

	Process	Wastewater Effluent	Ozone	Porous Biocarrier	Ozone + Porous Biocarrier
	DOC (mg/L)	9.5	9.7	8.2	6.8
	BOD_5 (mg/L)	8.2	10.8	4.1	4.0
	Biopolymer (mg/L)	0.87	0.89	n.a. *	0.52
Concentration of DOC, in molecular weight	>10,000 Da (μg-C/L)	0.87	0.89	n.a. *	0.52
	500–1200 Da (μg-C/L)	6.49	6.64	n.a. *	5.01
	<350 Da (μg-C/L)	2.20	2.13	n.a. *	1.33

Note: * n.a. represents not analyzed.

In the pilot test, a set of hollow fiber UF modules was used for the study of the effect of pretreatment processes on the membrane flux performance. Figure 4 shows the change in the flux of the UF system operating in constant pressure mode with two types of pretreatment processes. When the wastewater treatment plant secondary effluent was treated only by the porous biocarrier process, a significant decline in the flux was observed after approximately 10 tons of permeate production. However, when porous biocarrier pretreatment was coupled with preozonation, a slight decline in flux was observed only after 26 tons of permeate production, with only 1% decline. The results indeed showed that the flux decline due to membrane fouling was significantly reduced by the combined ozonation and porous biocarrier system. Table 4 also shows that the concentration of biopolymer, which resulted in membrane fouling, as reported by previous researchers [7,15], decreased significantly from 0.87 ppm to 0.52 ppm after the combined ozonation and porous biocarrier pretreatment process. A decrease in DOC concentration from 9.56 ppm to 6.86 ppm was also observed. Wang et al. [23] investigated the impacts of preozonation on membrane fouling for secondary effluent reclamation and concluded that the decomposition of macromolecules such as biopolymers into smaller fragments at an ozone dose of 0.9–2.0 mg O_3/mg DOC increased the free energy of cohesion of EfOM and led to a decrease in attraction and an increase in repulsion between foulants and the hydrophobic PES membrane surface, which might alleviate the decline in flux for PES UF membrane filtration. Our pilot test results demonstrated that the use of ozone in secondary effluent treatment before the porous biocarrier process effectively decomposed fouling-causing biopolymers, delayed the development of UF membrane fouling, and thereby significantly increased permeate production. The reduction in the biopolymer concentration in the secondary effluent verifies that ozone as a pretreatment procedure effectively degraded macromolecule biopolymers into smaller molecules that could more easily be utilized by microorganisms in the subsequent porous biocarrier process and mitigate the fouling potential of UF membranes.

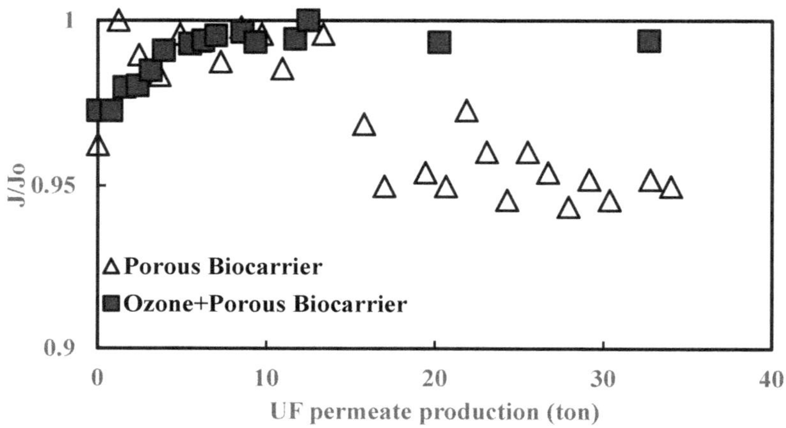

Figure 4. Comparison of decline in flux with porous biocarrier and ozone-combined porous biocarrier. The open triangles and open squares indicate porous biocarriers and ozone-combined porous biocarriers.

In the pilot testing, COD mineralization will relate to biomass increase and CO_2 production as well as DOC removal. However, in this research, we did not conduct the long-term experiment. We tried to use the lab testing evaluation and pilot testing to confirm the effects of the ozone-combined biocarrier process on UF water reclamation. A long-term experiment will be conducted to formulate equations for the biological process and perform a long-term analysis of UF module flux in our future research.

4. Conclusions

The process combining low-dose ozone with advanced porous biocarrier treatment as the pretreatment process for industrial secondary effluent reclamation can effectively slow membrane fouling in the UF process. The results indicated that 70% of the biopolymers were removed, and the COD removal efficiency was 1.8 times higher in this combined process than in the process with a porous biocarrier alone. Compared with the process with no pretreatment, the UF flux increased by 16% after the combined ozonation and porous biocarrier pretreatment process. In addition, species in the genus *Flavobacterium* (15.59%), which are known as biopolymer-degrading bacteria, were observed only in the ozone-combined porous biocarrier process. The results show that biopolymers can be removed through the ozone-combined porous biocarrier process due to partial ozone degradation, confirming that this combined process is one of the better pretreatment procedures for organic matter removal.

Supplementary Materials: The following supporting information can be downloaded at: https://www.mdpi.com/article/10.3390/w17020272/s1, Figure S1: Flow chart of the pretreatment system for secondary effluent from the Linhai industrial park wastewater treatment plant; Figure S2: UF characterization by SEM. The average pore size of top layer and bottom layer of UF is 0.39 μm and 0.38 μm; Figure S3: Typical excitation-emission matrix (EEM) fluorescence spectra. (a) Wastewater effluent; (b) porous biocarrier; (c) ozone+porous biocarrier. Table S1: The spectral characteristics of the component identified in this study and comparisons with previously identified components.

Author Contributions: T.-T.C.: Writing—original draft, data collection and curation, S.-Y.C.: Writing—original draft, data collection and curation, C.-C.L.: Data collection and curation, Y.-L.T.: Conceptualization, Funding acquisition, G.-Y.L.: Conceptualization, Funding acquisition, C.-H.L.: Data collection and curation, P.-Y.C.: Review and editing. All authors have read and agreed to the published version of the manuscript.

Funding: This research was founded by the Ministry of Economic Affairs, A+ Industrial Innovative RD Program and the Ministry of Science and Technology (MOST 109-2221-E-007-058, 110-2221-E-007-054 and 110-2731-M-007-001).

Data Availability Statement: Data is unavailable due to privacy.

Acknowledgments: The authors would like to thank the Ministry of Economic Affairs, A+ Industrial Innovative RD Program and the Ministry of Science and Technology for supporting this research.

Conflicts of Interest: All authors declare that there are no conflicts of interest. The authors report no commercial or proprietary interest in any product or concept discussed in this article.

References

1. Davey, J.; Schäfer, A.I. *Ultrafiltration to Supply Drinking Water in International Development: A Review of Opportunities. Appropriate Technologies for Environmental Protection in the Developing World: Selected Papers from ERTEP 2007, Ghana, Africa, 17–19 July 2007*; Springer: Dordrecht, The Netherlands, 2009; pp. 1–359.
2. Peiris, R.H.; Hallé, C.; Budman, H.; Moresoli, C.; Peldszus, S.; Huck, P.M.; Legge, R.L. Identifying fouling events in a membrane-based drinking water treatment process using principal component analysis of fluorescence excitation-emission matrices. *Water Res.* **2010**, *44*, 185–194. [CrossRef] [PubMed]
3. Shon, H.K.; Vigneswaran, S.; Snyder, S.A. Effluent organic matter (EfOM) in wastewater: Constituents, effects, and treatment. *Crit. Rev. Environ. Sci. Technol.* **2006**, *36*, 327–374. [CrossRef]
4. Zhang, J.; Li, G.; Yuan, X.; Li, P.; Yu, Y.; Yang, W.; Zhao, S. Reduction of Ultrafiltration Membrane Fouling by the Pretreatment Removal of Emerging Pollutants: A Review. *Membrane* **2023**, *13*, 77. [CrossRef]
5. Jeong, S.; Rice, S.A.; Vigneswaran, S. Long-term effect on membrane fouling in a new membrane bioreactor as a pretreatment to seawater desalination. *Bioresour. Technol.* **2014**, *165*, 60–68. [CrossRef] [PubMed]
6. Henderson, R.K.; Subhi, N.; Antony, A.; Khan, S.J.; Murphy, K.R.; Leslie, G.L.; Chen, V.; Stuetz, R.M.; Le-Clech, P. Evaluation of effluent organic matter fouling in ultrafiltration treatment using advanced organic characterization techniques. *J. Membr. Sci.* **2011**, *382*, 50–59. [CrossRef]
7. Zheng, X.; Ernst, M.; Huck, P.M.; Jekel, M. Biopolymer fouling in dead-end ultrafiltration of treated domestic wastewater. *Water Res.* **2010**, *44*, 5212–5221. [CrossRef]
8. Lee, S.; Lee, K.; Wan, W.M.; Choi, Y. Comparison of membrane permeability and a fouling mechanism by pre-ozonation followed by membrane filtration and residual ozone in membrane cells. *Desalination* **2005**, *178*, 287–294. [CrossRef]
9. Barry, M.C.; Hristovski, K.; Westerhoff, P. Membrane fouling by vesicles and prevention through ozonation. *Environ. Sci. Technol.* **2014**, *48*, 7349–7356. [CrossRef] [PubMed]
10. Ikehata, K.; Jodeiri Naghashkar, N.; Gamal El-Din, M. Degradation of aqueous pharmaceuticals by ozonation and advanced oxidation processes: A review. *Ozone Sci. Eng.* **2006**, *28*, 353–414. [CrossRef]
11. Chow, C.W.K.; Fabris, R.; VanLeeuwen, J.; Wang, D.; Drikas, M. Assessing natural organic matter treatability using high performance size exclusion chromatography. *Environ. Sci. Technol.* **2008**, *42*, 6683–6689. [CrossRef] [PubMed]
12. Hamid, K.A.; Sanciolo, P.; Gary, S.; Duke, M.; Muthukumaran, S. Comparison of the effects of ozone, biological activated carbon (BAC) filtration and combined ozone-BAC pre-treatments on the microfiltration of secondary effluent. *Sep. Purif. Technol.* **2019**, *215*, 308–316. [CrossRef]
13. Hudson, N.; Baker, A.; Ward, D.; Reynolds, D.M.; Brunsdon, C.; Carliell-Marquet, C.; Browning, S. Can fluorescence spectrometry be used as a surrogate for the Biochemical Oxygen Demand (BOD) test in water quality assessment? An example from South West England. *Sci. Total Environ.* **2008**, *391*, 149–158. [CrossRef]
14. Murphy, K.R.; Hambly, A.; Singh, S.; Henderson, R.K.; Baker, A.; Stuetz, R.; Khan, S.J. Organic matter fluorescence in municipal water recycling schemes: Toward a unified PARAFAC model. *Environ. Sci. Technol.* **2011**, *45*, 2909–2916. [CrossRef] [PubMed]
15. Yu, H.; Qu, F.; Chang, H.; Shao, S.; Zou, X.; Li, G.; Liang, H. Understanding ultrafiltration membrane fouling by soluble microbial product and effluent organic matter using fluorescence excitation-emission matrix coupled with parallel factor analysis. *Int. Biodeterior. Biodegrad.* **2015**, *102*, 56–63. [CrossRef]
16. Meng, L.; Xi, J.; Yeung, M. Degradation of extracellular polymeric substances (EPS) extracted from activated sludge by low-concentration ozonation. *Chemosphere* **2016**, *147*, 248–255. [CrossRef] [PubMed]
17. Huber, S.A.; Balz, A.; Abert, M.; Pronk, W. Characterization of aquatic humic and non-humic matter with size-exclusion chromatography—Organic carbon detection—Organic nitrogen detection (LC-OCD-OND). *Water Res.* **2011**, *45*, 879–885. [CrossRef]
18. Hammes, F.; Salhi, E.; Koster, O.; Kaiser, H.P.; Egli, T.; Gunten, U.G. Mechanistic and kinetic evaluation of organic disinfection by-product and assimilable organic carbon (AOC) formation during the ozonation of drinking water. *Water Res.* **2006**, *40*, 2275–2286. [CrossRef]

19. Liu, Y.; Jin, J.H.; Liu, H.C.; Liu, Z.P. *Dokdonella immobilis* sp. nov., isolated from a batch reactor for the treatment of triphenylmethane dye effluent. *Int. J. Syst. Evol. Microbiol.* **2013**, *63*, 1557–1561. [CrossRef] [PubMed]
20. Kirchman, D.L. The ecology of Cytophaga–Flavobacteria in aquatic environments. *FEMS Microbiol. Ecol.* **2002**, *39*, 91–100. [CrossRef]
21. McBride, M.J.; Xie, G.; Martens, E.C.; Lapidus, A.; Henrissat, B.; Rhodes, R.G.; Goltsman, E.; Wang, W.; Xu, J.; Hunnicutt, D.W.; et al. Novel features of the polysaccharide-digesting gliding bacterium Flavobacterium johnsoniae as revealed by genome sequence analysis. *Appl. Environ. Microbiol.* **2009**, *75*, 6864–6875. [CrossRef]
22. Sack, E.L.W.; van der Wielen, P.W.J.J.; van der Kooij, D. *Flavobacterium johnsoniae* as a model organism for characterizing biopolymer utilization in oligotrophic freshwater environments. *Appl. Environ. Microbiol.* **2011**, *77*, 6931–6938. [CrossRef]
23. Wang, H.; Park, M.; Liang, H.; Wu, S.; Lopez, I.J.; Ji, W.; Li, G.; Snyder, S.A. Reducing ultrafiltration membrane fouling during potable water reuse using pre-ozonation. *Water Res.* **2017**, *125*, 42–51. [CrossRef] [PubMed]

sustainability MDPI

Article

Indirect Detection of Degradation-Resistant Compounds on Groundwaters Forward-Facing to Current Global Consumerism and Climate Change

Norzagaray Campos Mariano [1,*], López Rocha Saul [1,*], Muñoz Sevilla Patricia [2], Montiel Montoya Jorge [1], Llanes Cárdenas Omar [1], María de los Ángeles Ladrón de Guevara Torres [3] and Luz Arcelia Serrano García [3]

1 Instituto Politécnico Nacional (IPN), Centro Interdisciplinario de Investigación para el Desarrollo Integral Regional, Unidad Sinaloa (IPN-CIIDIR-Unidad Sinaloa), Blvd. Juan de Dios Bátiz Paredes No. 250, Colonia San Joachín, Guasave 81049, Mexico

2 Instituto Politécnico Nacional (IPN), Centro Interdisciplinario de Investigación para el Desarrollo Integral Regional (CIIDIR-IPN-Unidad Oaxaca), Municipio de Santa Cruz Xoxocotlán, Oaxaca 71230, Mexico; nmunozs@ipn.mx

3 Instituto Politécnico Nacional (IPN), Centro Interdisciplinario de Investigaciones y Estudios sobre Medio Ambiente y Desarrollo (IPN-CIIEMAD Unidad CDMX) Calle 30 de Junio de 1520 s/n, La Laguna Ticomán, Gustavo A. Madero 07340, Mexico; maladron@ipn.mx (M.d.l.Á.L.d.G.T.); lugarcias@ipn.mx (L.A.S.G.)

* Correspondence: mnorzagarayc@ipn.mx (N.C.M.); saul_rocha94@hotmail.com (L.R.S.); Tel.: +52-6871217072 (N.C.M.)

Abstract: This study addresses the environmental challenges posed by consumerism, evaluating the impact of Degradation-Resistant Organic Compounds (*DROCs*), such as fats and oils, on surface environments, the subsurface, groundwater, and aquifers. Climate variability has intensified the dispersion of these compounds, particularly in open landfills and poorly managed sites, making it urgent to identify affected areas to plan remediation efforts and mitigate their effects on ecosystems and human health. The objective was to analyze the dispersion of *DROCs* in an abandoned landfill in northwestern Mexico and develop strategies to characterize the decomposition stages of these compounds. In 2021 and 2022, a two-dimensional geoelectric tomography (*GT*) was conducted using a SARIS-Scintrex resistivimeter, recording variations in the apparent resistivity of subsoil. Using resistivity measures, the progressive and regressive numerical interpolation of Newton for finite differences, combined with the Lévy-type diffusion operator, classified the invasion areas into three principal ranges: high (recent invasions), intermediate (transition phase), and low (mature invasions).. These ranges indicated how pollutants migrate from the surface to the saturated zones of the aquifer. To validate the presence of fats and oils, a 24-m-deep well was drilled, revealing a positive correlation (R^2 = 0.863) between the areas covered by the tomograms and the detected contaminants. The results emphasize the need for improved waste management and the careful selection of disposal sites to reduce environmental degradation. The methodology proved effective and rapid, facilitating remediation planning and highlighting the importance of sustainable practices in final disposal to mitigate the impacts of *DROCs*, reduce greenhouse gas emissions, and protect public and environmental health.

Keywords: organic compounds; apparent resistivity; waste; groundwater; remediation strategies

1. Introduction

The economic growth worldwide has positively correlated with an exponential increase in population, urbanization, technological advancement, industrialization, and economic development; however, it also has a direct association with increased consumerism. Globally, the demographic increase in recent decades has exacerbated current consumerism and the rapid rise in the generation of solid and liquid waste (*SLW*), raising concerns about its potential impact on the atmosphere and contributions to climate change [1,2].

The generation of this waste significantly contributes to greenhouse gas (*GHG*) emissions, exacerbating climate change. The worldwide average production of *SLW* by a person is approximately 0.74 kg per day, though this figure varies widely between countries, ranging from 0.11 to 4.54 kg per capita per day [3]. The generation of *SLW* for 2025, is expected to reach 2.2 billion tons, not considering more than 1/3 of the remaining uncollected garbage [4]. By 2050, the global production of *SLW* is projected to reach 3.40 billion tons, with growth rates of 19% in industrialized nations and 40% in developing countries [5]. The major amounts of this waste stems from daily activities in residential, commercial, and institutional sectors, driven largely by consumerism and intensified by climate change, rather than from exceptional events. Integrally, the problem poses a significant worldwide challenge to the sustainability of superficial waters, groundwater, air, and public health. Despite regulatory frameworks and infrastructure investments, consumerism continues with environmental degradation and social inequities, largely due to inadequate liquid and solid waste management (*LSWM*) [6]. In Mexico, despite the regulations outlined in the "*Official Daily of Federation*" in 2023 [7], waste management practices are often implemented indiscriminately, further exacerbating the issue. This problem is intensified by the uncontrolled expansion of many Open–Air Garbage Dumps (*OAGD*s), frequently used for final waste disposal. These sites generate emissions of *GHGs*, contributing to environmental degradation and posing significant public health risks [8]. Since the latter half of the 20th century, the industrialization of Mexico has intensified and increased the demand for raw materials to support rising consumption patterns among a growing population [9]. This has resulted in an uncontrolled expansion of *OAGDs*, which contribute not only to environmental degradation through leachate production but also generate *GHG* emissions to the atmosphere [10]. The issue has extended beyond rural areas, exacerbating environmental changes and contributing to climate change. Therefore, it could be said that in Mexico, rapid urbanization, industrialization, and resource consumption have intensified these problems, highlighting the urgent need to address *LSWM* more effectively and sustainably [11]. Like other regions, Mexico now faces significant environmental, social, and economic challenges linked to its growing population and consumerism, which are closely associated with inadequate *LSWM* practices [4].

Mexican consumerism is evident in the proliferation of various sites by final deposit of *LSW*: Sanitary Landfills (*SLs*), Controlled Fills (*CFs*), and the uncontrolled areas occupied by *OAGD*s, where *LSW* does not receive any treatment [3]. Although sanitary *SLs* are globally recognized as effective for final disposal *LSWM*, they have become inadequate for handling the large volumes of waste generated over decades due to demographic growth. This issue, exacerbated by consumerism and poor *LSWM* practices, has led to an increase in *GHG* emissions globally, contributing to climate change [5,12,13]. *DROC* within porous media of subsoil undergo anaerobic decomposition, producing fats and oils (*DROCDROCs*). On *OAGD*s, leachates from *LSW*, when unmanaged, become significant sources of *GHG* emissions, including methane (CH_4) and hexane (C_6H_{14}), among other gasses. Consequently, inadequate *LSWM* at *OAGD*s has made them major contributors to global *GHG* emissions, exacerbating climate change [5,12–14]. *DROC* contaminate groundwater and air, destroy habitats, and obstruct the groundwater flow (ϕ) when their concentration and density are high. This *OAGD* behavior in the best manner possible highlights the need for *LSW* management practices to be sustainable [5,15].

To address these challenges, engineered landfilling techniques, advanced infrastructure, and recycling systems have emerged as promising solutions to minimize environmental harm and optimize natural resource use [16,17]. Proper site selection for landfills is crucial to prevent leachates from contaminating groundwater; however, not all sites have the ideal hydrological and granulometric conditions for landfill construction [18,19].

In Mexico, *GHG* emissions in 2000 were approximately 600×10^6 tons of carbon dioxide equivalent (tCO_2e) [20] (p. 185). For 2024, taking the emission levels produced in the 2000s as a baseline, emissions are projected to increase by a factor of 33.75, reaching approximately $54{,}000 \times 10^6$ tCO_2e. This significant rise is largely due to diverse factors, such as increased mobility, industrial activity, consumerism, and the recent economic recovery following the COVID-19 pandemic, which was categorized as a 'public health emergency of international concern' from 2020 to 2023 [13]. In 2020, during the pandemic, the *GHG* emissions in Mexico were 804×10^6 tCO_2e, revealing that, regardless of the presence of world problematic health, the emissions reflect the economic development and population growth. However, if it is supposed that the primary source of *GHG* emissions in Mexico will continue to be fossil fuel combustion, and persistently inadequate *LSWM* practices will remain, consumerism, and other effects in the environment are likely to lead to unremedied *GHG* emissions that could rise by 62.5% by 2050, reaching approximately $87{,}750 \times 10^6$ tCO_2e. This figure reflects economic development and population growth.

High concentrations of leachates and the *DROC's* decomposition in *OAGDs* could spread to aquifers in Mexico. Preventing environmental degradation by controlling and separating the leachates through effective *LSMW* practices is essential. In parallel, the detection of *DROCs*, as well as the monitoring of subsoil and groundwater systems, is crucial for site remediation. Understanding the distribution of *DROC* transformation stages on subsoil is critical because some compounds may adhere to aquifer walls and others interact with subsoil properties and the hydraulic parameters from the aquifer.

Factors such as the depth, porous media, the mass transport in porous media (*TIMP*), and groundwater ϕ direction, affecting the concentration of *DROC* are mentioned in [17,21,22]. Understanding and addressing areas with DROC (Degradation-Resistant Organic Compound) anomalies in the subsoil, particularly where these alterations pose potential risks to ecosystems, is a matter of significant social interest. However, the recovery of groundwater, which provides a collective well–being to society, is an activity that society must tackle in collaboration with governmental institutions [23]. Effective remediation requires a multidisciplinary approach, including public awareness, policy reforms, infrastructure improvements, and technological innovations [23]. The geoelectrical and geo–radar methods have been used to characterize landfills [24,25]. Geoelectrical methods have been successfully applied to detect groundwater contamination within surface–subsoil (*S–S*) systems [25]. Studies in various countries have detected and detailed contaminants [26–29]. Notably, high lead contamination in buried waste was identified in Egypt [30]. The impact of environmental conditions on *S–S* systems was studied in Italy [31]. However, the effects of *DROC* in *S–S* structures in full have not yet been studied, mainly due to the nonlinearity in the mass diffusion equations as they travel through anisotropic *TIPM*. Therefore, the ability to describe the stability, equicontinuity, and convergence of the methods used to define each *DROC* stage is essential. The stability, equicontinuity, and convergence, are due to the leachate compounds being denser than water, determining their adherence to aquifer walls or movement according to *TIMP* and ϕ. Some contaminants remain above the water table, while others migrate through the saturated zone, increasing their decomposition and adherence. The *DROC* highlight issues related to the infiltration and accumulation of leachates from poorly managed landfills [23]. This study provides key insights into the spatial and temporal distribution of *DROCs* within coastal *OAGDs*, specifically focusing on an old dump site in Guasave Municipality that has served as the final disposal site for *LSW* for over 60 years. Data from two geoelectrical dipole–dipole surveys conducted in 2022 and 2023 were analyzed, using ρa values to examine the dispersion patterns of *DROCs* in this abandoned landfill in northwestern Mexico. The primary goal was to develop strategies to characterize the stages of decomposition of these compounds within the subsoil.

2. Context of the Study Area

2.1. The OAGD and Area Study

In the northwest of Mexico, an *OAGD* situated approximately 8.5 km from Guasave City and 8.15 km from Sinaloa River, better known as "The Agricultural Heart of Mexico," was selected for this study (Figure 1A). The selection of this site was based on its potential to produce *DROCs* and its ecosystem transitioning from mountainous to coastal, enabling the study of the *DROC's* transformation in the movement anisotropically *TIPM* originating in the Sierra Madre Occidental (*SMO*) traveling to the coast through the free and unconfined aquifer. In the late 1990s, one *SL* was built adjacent to this old *OAGD*, reducing its dependence.

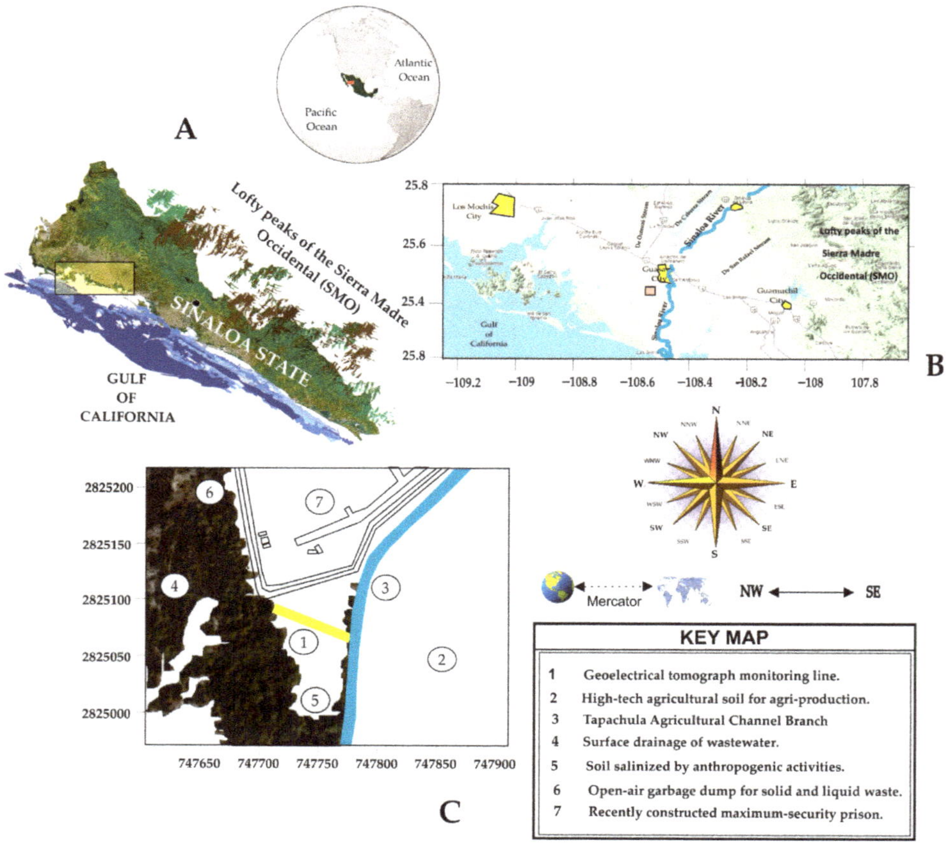

Figure 1. In the ransition environment between mountain and coast, immersed in an anisotropic movement of the *TIMP* in the aquifer of the Sinaloa River Basin, located in the northwest of Mexico, the following is illustrated: (**A**) The state of Sinaloa, with the Sierra Madre Occidental to the north; (**B**) The macro-localization of the old *OAGD* selected for the study of *DROC* facies (red box), and the main cities of the northwest of the state of Sinaloa; (**C**) The micro-localization of the *OAGD* and the description of the main surface and surrounding features..

Indistinctively, the *OAGD* has continued operating without any on–site remediation. The prolonged lack of remediation efforts has likely led to the accumulation of various contaminants mixed with leachates in the subsoil. The Sinaloa River Basin—where this *OAGD* is located—is vital to Mexico's economy due to the significant agricultural activities in which an important number of food products are produced (Figure 1A,B). The total population in 2020 of Guasave Municipally was 289,370 people, with 50.6% being women and 49.4% men. In 2024, there has been an estimated population increase to approximately 327,012 people [32].

2.2. Historical Geology and Tectonic Activity of Region

The historical geology of the region has been significantly influenced by tectonic activity that occurred approximately 60 million years ago, marking the transition from the Late Cretaceous to the Paleogene period. This era is characterized by the rapid movement of the Pacific Plate colliding with the Farallon Plate, which subsequently subducted beneath the North American Plate. This tectonic interaction generated tensile stresses that resulted in folding and ultimately gave rise to the majestic peaks of the Sierra Madre Occidental (*SMO*) [33–35].

Hydric and eolic erosion of igneous rocks from the *SMO* has transported sediments to the surrounding areas. These unconsolidated sediments that form the subsoil and surface soil today are essential for agriculture in the region. The weathering processes enhance nutrient availability and water retention in the soils of the area where this old *OAGD* is situated, significantly impacting agricultural production [36–38].

2.3. Groundwater Studies in the Study Area

The *OAGD*s selected are located over the free or unconfined Sinaloa River aquifer, where the groundwaters are essential for resource demand in the western agricultural part of Mexico, particularly in the state of Sinaloa (Figure 1A,C). Groundwater in this region plays a critical role in agricultural, industrial, and municipal uses, especially considering the semi-arid climate. However, various reports have highlighted that the groundwater reserves of the Sinaloa River aquifer have significantly declined over the past few decades. This reduction has been primarily driven by the increasing demand for agricultural irrigation, exacerbated by recurring droughts [39]. Despite ongoing management efforts, the aquifer continues to face significant challenges, including over-exploitation and decreasing recharge rates. Recharge is mainly through rainfall, which is unevenly distributed throughout the year, as well as from surface water sources, including the Sinaloa River.

However, the excessive extraction of groundwater for irrigation has led to a declining water table and contamination risks. The annual average groundwater availability in the Sinaloa River aquifer (key 2502) is reported to be approximately 48.8 Mm^3/year, according to the most recent CONAGUA update [40,41]. These data are crucial to understanding the capacity for sustainable extraction of groundwater, but, given the intensification of agriculture and rising water demand, the aquifer is facing a substantial reduction in its reserves, threatening the long-term sustainability of these water resources [40]. Additionally, the vulnerability of the aquifer to contamination is a growing concern. Groundwater quality has deteriorated, particularly in areas with intensive agricultural activity. This underscores the need for improved groundwater management strategies to ensure the sustainable use of this vital resource [39]. The coastal plain of the Sinaloa River Basin, where the *OAGD* site selected for this research is located, is particularly susceptible to contamination, primarily due to the shallow depth of the groundwater table, which ranges from 0.1 to 9.2 m. This proximity to the surface increases the risk of contamination. Furthermore, the area's proximity to the Sonora Desert, combined with a high evapotranspiration rate of 834.32 mm/year, which exceeds the historical annual average precipitation of 577.9 mm, heightens the ongoing risk of desertification and degradation of the land due to inadequate water quantity and quality for both economic and human consumption needs [42]. The situation of the aquifer is particularly critical because of the intensive and highly mechanized agricultural practices in the region. Since the economic downturn caused by the *COVID-19* pandemic in 2020, the state of Sinaloa has experienced an exponential decrease in its Gross Domestic Product (*GDP*), with a decline from 4.5% in 2021 to just 0.6% in 2023. This economic contraction has also affected the region's water resources, both surface and groundwater. One of the strategies employed to mitigate these effects on the aquifer and prevent excessive water extraction is the implementation of restrictions (or water use bans), aimed at preserving the long-term sustainability of groundwater resources. This is particularly relevant in regions with intensive agricultural and industrial demands driven by consumerism, similar to the area studied.

2.4. Weathering Conditions for the Study Area

The climatic conditions in the coastal region of northwestern Sinaloa are semi-arid, feature extremely high temperatures in summer, with average maximums between 35 °C and 40 °C, and on exceptional days, reaching up to 45 °C, and in winter, minimum temperatures vary between 12 °C and 18 °C, although in specific areas, they can drop to around 10 °C [43]. The annual precipitation ranges from 231.1 to 760.0 mm, with croplands replacing the original vegetation and the Sinaloa River serving as the primary water source for agriculture and human consumption [44,45]. The total of annual rainfall is primarily concentrated between July and September when monsoons and cyclones contribute 70% to 80% of the total precipitation [46]. Relative humidity fluctuates significantly, reaching between 60% and 80% during the rainy season and dropping to 30–40% in the driest months [47]. Typical wind speeds in the region average between 10 and 20 km/h, but they can exceed 60 km/h during tropical storms or hurricanes (CENAPRED, 2022). In the semi-arid regions of Mexico, the high evaporation rates, with annual averages exceeding 2000 mm, exacerbate moisture loss in the soil and water bodies [46]. These conditions in the north of Mexico significantly influence the dispersion and degradation of pollutants in landfill areas, as high evaporation and seasonal rains affect the transport and infiltration of compounds into the subsoil and aquifers, increasing the risks of groundwater contamination on the unconfined free aquifers of this region [48].

3. Material and Methods

3.1. Data Collection

Two geoelectrical tomographies (*GTs*) were performed using a multielectrode dipole–dipole array to measure ρa values in the subsoil surrounding the selected OAGD. The measurements around the OAGD were acquired over two consecutive years: one set was taken in October 2022 ($\rho a2022$), and the other in August 2023 ($\rho a2023$). The *GT* method involved injecting a direct current (*DC*) of 1 ampere (*A*) into the stratigraphic sedimentary layers of the subsoil. This *DC* was supplied by a SARIS–Scintrex resistivity meter, using two pairs of electrodes: one fixed (*A–B*) and one mobile (*M–N*). The separation between the fixed and mobile electrodes was maintained at 0.5 m, consistent with the Wenner-type geoelectric method. The resistivity meter used had the following specifications: 1*A–DC* output, 100 W power, 500 volts (V) for voltage injection, 11 MΩ internal impedance, and a frequency range from 5 to 6 Hz during each *DC* injection.

The initial and final points of lines in each *GTs* lines were accurately positioned using a portable 13–channel Hammer/GPS device, ensuring precise *UTM* coordinates. The *GTs* lines started in the *SW* direction near the agricultural Tapachula Branch Channel and finished in the *NE* direction (*SW–NE*) just before the Hidalgo Drain, running nearly parallel to the penitentiary building located to the right of the *OAGD* (Figure 1C).

Despite the extensive data collection, some transcription errors occurred during the fieldwork. These errors were meticulously corrected during data processing, and the data were electronically processed and securely stored. To visualize the spatial distribution of annual ρa values and enhance the finite-difference methodology, the ρa values from each *GT* survey were organized in x/z coordinates, where x represented length, and z represented the vertical or depth. Subsequently, geoelectrical pseudo-sections were generated. The annual geoelectrical pseudo-sections were obtained using software that included parameters such as electrode separation *A–B* and *M–N* (for calculating the geometric constant array), the number of required steps, and the distances of each measurement from the initial coordinates. Kriging interpolation with *SURFER 10.0* software was used to visualize the spatial distribution of each annual ρa value [49].

3.2. Distribution and Identification of DROCs in the OAGD

The probability of finding products associated with each phase of *DROC* in the subsoil of the old *OAGD* is attributed to the prolonged accumulation of decomposing organic matter throughout the landfill's lifespan. Leachates containing organic matter descend from the surface through the geological layers of the subsoil; in the absence of oxygen, some of these compounds may degrade through bacterial activity and undergo anaerobic digestion, generating biogas, such as methane (CH_4, which retains 90% of the energy available through direct oxidation), and hexane (C_6H_{14}), among others [50]. Over time, this biogas may have reacted to form new compounds distributed across different geological layers following the *IR* phase. Biogas is primarily generated by decomposition in the saturated zone and, to a lesser extent, in the unsaturated zone of the aquifer. It may also produce oxidation-reduction (*REDOX*) reactions that create other compounds, increasing the complexity of the isotropic dynamics involved in the subsoil transport of *DROCs*. Thus, the *DROCs* present in the subsoil, whether originating directly from LSW or resulting from dissolution in biogas, have enabled field studies to focus on detecting *DROCs* using ρa from product mixtures in the subsoil, leading to phases of *DROCs* immersed in the anisotropy of *TIMP*.

The measurements allowed us to determine the electrical properties of subsoil materials a foundational framework for interpreting the distribution of *DROCs* across different geological layers of the subsoil was provided, in both saturated and unsaturated zones. The anomalies analysis reveals the determination of the temporal variation of ϱa values enabled, indicating the movement or retention of *DROCs* in various geological layers. Additionally, the anisotropy of the porous media was evaluated to determine its influence on the subsurface's ability to retain or remove these contaminants. Zones with high and low *DROC* concentrations were also differentiated, allowing for the identification of anomalies corresponding to areas where contaminants might be migrating toward the geological basement in the free-type aquifer. Additionally, the anisotropy of the porous media was evaluated to determine its influence on the subsurface's ability to retain or remove these contaminants. Zones with high and low *DROC* concentrations were also differentiated, allowing for the identification of areas where contaminants might be migrating toward saturated, or not saturated zona, traveling over the water table or to the geological basement into the bottom from free-type aquifers.

3.3. Methodology TIMP

For this purpose, ρa resistivity measurements were taken with the *1–A–DC* method, as outlined in methodologies from previous studies [51–53]. These measurements were specifically used to capture the spatial distribution of ρa in the studied areas, providing key insights into the underlying geological and environmental conditions. A key element in modeling the anomalous transport process is the incorporation of Lévy-type coefficients, which allow the capture of irregular trajectories and random jumps in contaminant movement. Unlike classical diffusion models, which assume smooth and predictable transport, the Lévy model is particularly suited for heterogeneous and anisotropic media, where the contaminant movement is often erratic and influenced by multiple factors. This approach enables a more accurate representation of the transport dynamics in complex subsoil environments, as demonstrated by the analysis of the spatial and temporal variability of ρa in the study area [54].The coefficients also helped deepen our understanding of the complexity of ϕ in the subsoil, as the movement of *DROCs* did not follow a straightforward isotropic diffusion pattern. The justification for using nonlinear diffusion models, such as Lévy–type coefficients, to adequately represent the behavior of anomalies observed in the annual ρa measurements was based on the anisotropy of the subsoil, the natural variations in ϕ, as well as the influence of environmental factors and the potential changing conditions in the porous media of subsoil.

In Scheme 1, the methodology employed in this research is outlined, covering the entire process from data collection to analysis and modeling with a focus on *DROCs*. This scheme visually represents the core steps, such as the measurement of ρa and the application of nonlinear diffusion models, like Lévy-type coefficients. The drawing scheme aids in visualizing the transformation of data into results that feed into the model, highlighting how subsurface anisotropy and environmental factors influence the behavior of contaminants in the subsoil. By using this scheme, the methodological steps are integrated and clearly represented, allowing for a more comprehensive understanding of the results and the dynamics of *DROCs* in the subsurface.

The following section will develop the equations that respond to the previously described observations, linking ρa with the behavior of *DROCs* in the subsoil under the influence of *TIMP* anisotropy and Lévy–type diffusion coefficients. To avoid inadequate formulations that present issues due to the lack of stability, equicontinuity, compactness, and convergence—factors that may arise when considering minimal assumptions or excluding attributes related to the annual variability of ρa—it was necessary to adopt approaches beyond classical diffusion models.

To represent the journey of the *OAGD* through the subsurface geological layers via *TIMP* anisotropy and to formulate the transformation of ρa for each stage of *OAGD*, it was established that the ф in the areas surrounding the *OAGD* occurs as the sum of three types of ф: (1) an *OAGD* with ф produced by molecular diffusion, (2) another *OAGD* with ф produced by turbulent diffusion, and (3) an *OAGD* with ф produced by shear ф dispersion. Of these, the ф*s* with the greatest magnitude in transport through the porous medium are molecular diffusion and turbulent diffusion, which can be modeled using functional operators suited to the anisotropic behavior of the porous system [55–57]. In this context, the traditional Laplace operators (∇^2), which do not account for abrupt changes in the *S-S* system, proved insufficient to represent turbulent ф.

It was essential to consider that the movement of the *TIMP* through porous media is not smooth but includes jumps and discontinuities, necessitating the use of a generalized Laplace operator [58].

This generalized in operator approach extended the classic Laplacian operator through the Lévy-type-diffusion operator, capable of simulating these abrupt jumps and discontinuities in the transport processes through porous media. Incorporating this diffusion model allowed for better handling of the complex spatiotemporal dynamics of *DROC* concentration levels, whose behavior shows both linear (continuous) and nonlinear (discontinuous) characteristics depending on stochastic conditions [59–61].

This approach enabled precise simulation of transport phenomena in complex porous media, accounting for both predictable, continuous changes (ф) and abrupt, unpredictable shifts caused by nonlinear dynamics. To reduce uncertainty in predicting the convergence of both movements within the porous medium, a monotonic finite-difference scheme was used to analyze the spatiotemporal variability of ρa. This scheme was enhanced by combining general nonlinear Lévy-type diffusion equations with constant equations, which exhibit little to no spatiotemporal variability. The combination of nonlinear and linear elements was represented in the following equation [61,62]:

$$\rho a(x,\ t) = \partial_t u - \mathcal{L}^{\sigma,\,\mu}\ [\varphi(u)],\ in\ R^N = RN\ x\ (0,\ T\ ) \quad (1)$$

where $\rho a\ (x, t)$ was the annual solution obtained in Ωm from $\rho a2022$ and $\rho a2023$, and ∂tu the nonlinear equation associated with the behavior from *TIMP* and $\rho a\ (x, t)$, both reflecting the changes in the complex system of the *OAGD*; $L\sigma,\ \mu$ was the numerical operator used to simulate Lévy-type diffusion in the porous media of subsoil, both linear and non-linear movements of the *TIMP* and, defining the spatiotemporal dynamics of anisotropic transport in the areas adjacent to the *OAGD*, and $\varphi(u)$ the numerical operator fort represent the linear movement utilizing non-decreasing and completely continuous equations.

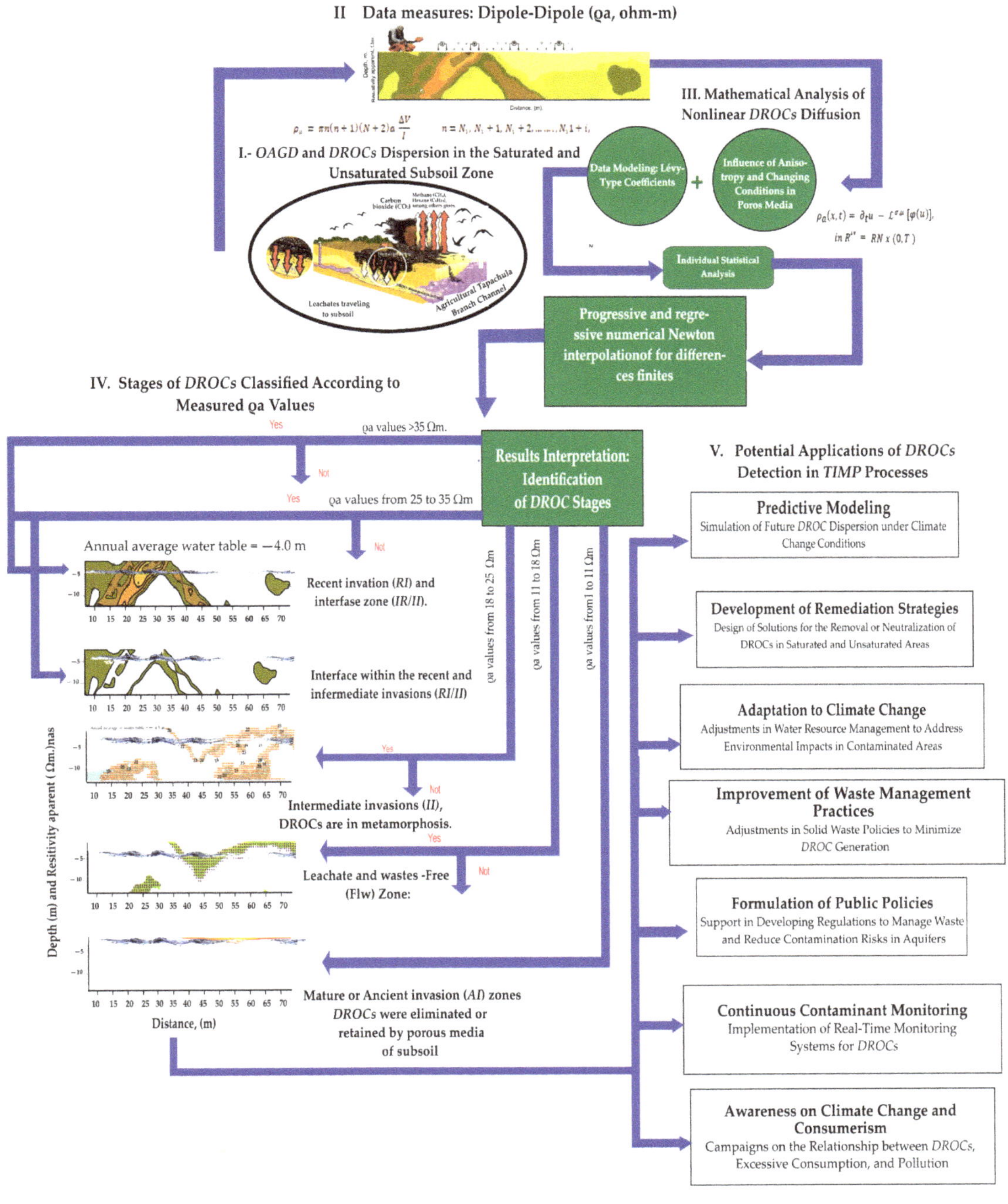

Scheme 1. Methodology for the collection, analysis, and modeling of data related to the behavior of *DROCs* in the subsurface. This scheme illustrates the sequential steps, from apparent resistivity (ρa) measurements to the application of nonlinear diffusion models using Lévy-type coefficients, considering the influence of subsurface anisotropy and environmental factors on contaminant dynamics.

The first term in Equation (1) is generally known as the porous media equation and is commonly associated with predetermined non-negative scalar magnitudes in space and time; in this case, for $\rho a = \rho a\ (x, t)$, it was s represented as follows [62]:

$$\partial_t \rho a = \partial_t u = \Delta(u^m),\ \text{with}\ m > 1\ \text{and}\ t \in \mathrm{R} \tag{2}$$

where $\partial tu = \Delta(u^m)$ occurs in dimension d equal to or greater than the unit ($\partial tu = \Delta(u^m) \in d \geq 1$), m was as a constant greater than the unitary value ($m > 1$) and, $\Delta = \Delta x$ was the Laplace operator interacting with the spatial variables. Given that $\rho a = u$ and $\Delta = \Delta x$ the representation from Laplace operator expression ($\nabla.\nabla = \nabla^2$), then, in generalized terms Equation (2) can be represented as follows [62]:

$$\partial_t \rho a\ (x,\ t) = \nabla \cdot (D\ (\rho a(x,\ t)) \cdot \nabla \rho a(x,\ t)) \tag{3}$$

where $\partial_t \rho a(x,\ t\ )$ was a term that refers to the temporal derivative by $\rho a(x,\ t)$ measuring the changes in mass transporter over time at a given spatial point between the compatible distance of both dipole–dipole *GTs* measured (from 25 to 73 m).

The representation of mass transport using Equations (1) and (3) plays a critical role in modeling the temporal dynamics of *DROCs* transport and dispersion within an anisotropic TIMP in the subsurface. The behavior of the diffusion coefficient, $D\ (\rho a\ (x,\ t))$, is dependent on the values of $\rho a\ (x,\ t)$ obtained at different distances and depths within the subsoil. These variations are important, as they describe how mass or solute propagates through the porous media. Specifically, the diffusion coefficient $D\ (\rho a\ (x,\ t))$ can change based on the concentration of *DROCs* in the subsoil of *OAGDs*. The spatial gradient, $\nabla \rho a\ (x, t)$, represents how the concentration of ρa varies in relation to its position in the subsurface and is a key term for modeling the anisotropic transport of contaminants in porous media.

The term $\nabla\ ^2$, the Laplacian operator, expresses the pointwise product of ∇ (gradient) and $\nabla \rho a\ (x,\ t)$, and it is crucial for describing the net flow of ρa value through any given volume. This approach provides a means to understand how the concentration of *DROCs* is influenced by both local conditions and external forces, which affect the dispersion and movement of these substances within the subsoil.

Equation (3) describes changes in resistivity present with the diffuse in space and describes the evolution form of ρa based on its spatial and temporal gradients, enabling the characterization of contaminant transport in complex porous media, such as those near *OAGDs* sites [63].

In the second term $\mathcal{L}^{\sigma,\ \mu}\ [\varphi(u)]$ of Equation (1), the $\mathcal{L}^{\sigma,\ \mu}$ operator was nonlinear diffusion, and its combined dynamics was obtained, summing the terms of the following equation:

$$\mathcal{L}^{\sigma,\ \mu} = \mathcal{L}^{\sigma}\ + \mathcal{L}^{\ \mu} \tag{4}$$

where L^{σ} represented the simulated portion of the probabilistic model that described random movements within the subsoil in the porous media using the parameters ϕ and *TIMP*. On the other hand, L^{μ} was the local probabilistic representation of anomalies or noise, which masked the information about real movement that was not considered in the acquired value of $\rho a(x,\ t)$.

In this study, the diffusion operator modeled processes involving jumps, corresponding to a generalization of Laplace operators, commonly associated with Levy processes—probabilistic models that describe random movements with a discontinuous increment to obtain the variations ρa within geological layers due to interconnection parameters like density, weight, texture, porosity, and aquifer hydraulic properties, affecting *DROC* degradation.

This presentation through Equation (1), was a methodology that enabled the examination of traction with movements that are not strictly smooth, where the anisotropy of *TIMP* could also exhibit abrupt changes. It also allowed us to map these factors for low qualitative assessment of *DROC* coverage and degradation through ρa distribution. This mapping helps discuss results to define the *DROC* phases in the *S-S* system and link them to *GHG* emissions (CO_2 and CH_4). Identifying these areas is essential for estimating *GHG* emissions and addressing their environmental impacts, which climate change is accelerating. Substances such as C_6H_{14}, fatty acids, soaps, fats, waxes, hydrocarbons, and oils play a significant role in *LSWM* strategies to mitigate these effects.

3.4. Dynamic Nature of DROC and Their Association with ρa Values

The literature indicates that changes in the typology of ρa anomalies at specific sites are often correlated with anisotropy in the porous media of the subsoil, especially in contaminated *DROC* sites. According to Shevnin et al. (2005) [64] and Atekwana and Atekwana (2010) [65], high ρa values typically characterize "aged" (*IA*) or "mature" *DROC* invasions. These highly conductive anomalies are associated with *DROC* entering porous media, where they undergo metamorphosis or transformation. Analyzing the spatial distribution of ρa values over time helps identify areas traversed by *DROC* in saturated and unsaturated zones from the aquifer. Three types of displacements influenced by *TIPM* anisotropy and three configurations of final deposition are observed. The following scenarios describe these possible final *DROC* distributions:

- Controlled Displacement from the Surface: Parameters such as transmissibility, storage capacity, effective porosity, and hydraulic conductivity of the unconfined aquifer determine the *DROCs'* movement to deeper zones, characterized by heavier masses.
- Continued Travel of Substances: *DROC* may have different arrival times at the water table. This could involve two types of movements over the water table or continued travel within the aquifer, contributing to the spread of contamination.
- Arrival of the Heaviest *DROC* at the Aquifer Bottom: The heaviest *DROC* may settle at the bottom of the aquifer.
- To analyze the associations between $\rho a2022$ and $\rho a2023$ and their respective *DROC* phases, we followed the principles established in Equations (1)–(4), which account for both linear and nonlinear representations simulating the anisotropy of *TIMP*. In this context, we considered the measurements of ρa, incorporating both previously reported limits and ranges found in the literature, as well as new ranges proposed for the first time in this study:
- *RI* Phase: Characterized by high ρa values (>35 Ωm), following the criteria of [24].
- *RI/II* Phase: For the *DROC* interface within the old/intermediate zone the range of 25 to 35 Ωm was proposed.
- *II* Phase: Associated with the zone between *RI/AI*, where *DROC* metamorphosis occurs, with a range of 18–25 Ωm.
- Leachate-Free (*Flw*) Zone: Defined by a range of 11 to 18 Ωm, as reported by [29].
- *AI* Phase: Indicative of mature invasion zones, with a range of 1–11 Ωm, based on the criteria of Shevnin et al. (2005) [64] and Atekwana and Atekwana (2010) [65].

3.5. Correlation Analysis of ρa2022 and ρa2023 with DROC Transformation Phases

To analyze the annual ρa measurements, a detailed statistical assessment was conducted for each dataset. For both $\rho a2022$ and $\rho a2023$, key statistical parameters were evaluated, including minimum (ρa_{min}) and maximum (ρa_{max}) values, mean ($\overline{\rho a}$), median (ρam), mode (ρaMo), and total sum ($\Sigma\rho a$) [66]. Dispersion parameters included absolute range (Rabs), standard deviation (σ_0), variance (σ^2), determination coefficient (R^2), non-determination or alienation coefficient (k = $1 - R^2$), coefficient of variation (Cv), standard error of $\overline{\rho a}$ ($SE\overline{\rho a}$), and correlation coefficient (R). Symmetry was assessed using skewness ($Swew\_\rho a$) and kurtosis ($Kurt\_\rho a$), as well as positional statistical parameters *Q25th* and *Q75th*. Given that only two annual samples were available, estimating the 95% confidence intervals *(95% CI)* was necessary to represent the full range of *DROC* distribution across the site [66].

Differences between $\overline{\rho a}$, σ_0, and σ^2 across years were attributed to the behavior of ρa in specific zones within porous media, reflecting transformations and interactions with

TIPM. These variations between ρa_{2022} and ρa_{2023} represent the cumulative impact of *Flw* zones and *DROC* invasion/transformation phases (*RI*, *II*, *AI*, *Flw* zones), which can vary in magnitude and shape over time under *TIPM* influence. Statistical software packages, STATISTIC 7.0 [67] and PASS 2.7b [68], were used to compute these parameters. The coefficient of alienation $(1 - R^2)$ at each stage (%) characterized the transformed mass and annual coverage (m^2) affected by *TIPM*. Linear regression analyses were performed to validate the association between ρa values and *DROC* invasions, focusing on vertical variations in *DROC* within materials drilled to depths of 22 m and 58 m outside the garbage dump during 2023 monitoring.

4. Results

4.1. Annual Dynamic Processes of DROC Stages

In the *NW–SE* direction, similar to the regional groundwater ф in the area, the annual measurements from ρa highlighted dynamic processes in the subsoil zones adjacent to the *OAGD*. These zones provided valuable information on the spatial variation in *DROC* stages in 2022 and 2023, illustrated in of Figure 2A,B. The 16 m length disparity between the lines can be attributed to the uncontrolled expansion of the garbage dump one year after the initial measurements, complicating the full analysis of *DROC* dynamic processes in the initially proposed exploration area of 846.6 m^2.

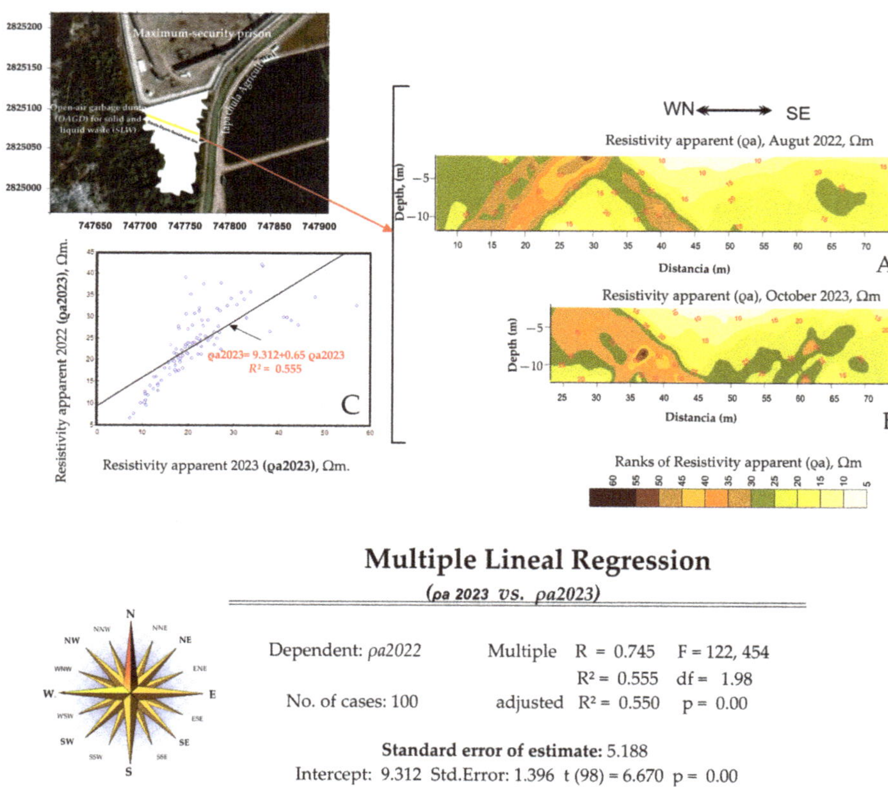

Figure 2. (**A**) Geoelectrical psudoresistivity section (dipole–dipole) in Ωm, performed in October 2022 (longuitude = 78 m); (**B**) Samesection type for August 2023 (longuitude = 54 m), and (**C**) Multiple Lineal Regression: (ρa2022 vs. ρa2023). The exploration depth was 12.5 m in the *OAGD* from Guasave Municipality, Sinaloa, Mexico.

To address this, the focus shifted to a smaller, more consistent area of 647.40 m^2, where both annual datasets provided comparable information. This refined focus allowed for a more accurate analysis of the spatial variability and dynamic processes of *DROCs*, improving the detection and examination of transformation processes and phase interfaces.

The regions identified with significant changes under varying environmental conditions provided valuable insights into *DROC* behavior. This analysis is crucial for understanding the impact of uncontrolled waste expansion on subsoil properties and developing effective *LSWM* strategies. The normality test confirmed that both *ρa2022* and *ρa*2023 datasets followed a normal distribution, with Shapiro–Wilk values of 0.91 and 0.99, respectively. Statistical parameters derived from the PASS 2.7 program are summarized in Table 1. The minimum and maximum *ρa* values were 7.31 and 6.54 Ωm (*ρa*min) and 56.88 and 42.18 Ωm (*ρamax*) for 2022 and 2023, respectively. The mean *ρa* ($\bar{\rho a}$) was 20.5 Ωm in 2022 and 23.43 Ωm in 2023, with mode values (*ρaMo*) of 21.93 and 23.65 Ωm.

Table 1. (**A**) Normality tests, and (**B**) statistical evaluation and central, tendency, and dispersion parameters statistical for analysis *ρa2022* and *ρa2023*.

	A			B		
	ρa2022	*ρa2023*		*ρa2022*		*ρa2023*.
Normal test	Ωm	Ωm	***Parameter statistic***	Ωm		Ωm
Shapiro–Wilk W	0.91	0.99	ρa_{min}	7.31		6.54
p (normal)	0	0.92	ρa_{max}	56.88		42.18
Jarque–Bera JB	49.36	0.62	$\Sigma\rho a$	2193.34		2365.66
p (normal)	0	0.73	$\bar{\rho a}$	21.93		23.66
p (Monte Carlo)	0	0.7	$SE\varrho a$	0.88		0.77
Chi^2	8.48	0.24	σ^2	77.87		59.95
p (normal)	0	0.62	σ_0	8.82		7.74
Anderson–Darling A	2.28	0.12	ρa_m	20.51		23.43
p (normal)	0	0.99	ρa_{Mo}	21.93		23.65
			R_{abs}	49.67		35.64
			R^2 [*ρa*2022 vs. *ρa*2022]		0.555	
			R^2*_adjusted*		0.551	
			$1 - R^2$		0.49	
			Cv	40.23		32.73
			R [*ρa*2022 vs. *ρa*2022]		0.745	
			Q25th	16.86		18.17
			Q75th	25.6		29.17
			Swew_ρa	1.3		0.08
			Kurt_ρa	2.48		−0.31
			Mg	20.37		22.24
			Q25th	16.86		18.17
			Q75th	25.6		29.17
			95%IC	20.20–23.66		22.13–25.17

Keys for individual annual statistical parameters (punctual, dispersion, symmetry, and confidence interval) evaluated: minimum (ρa_{min}) and maximum (ρa_{max}) values, total sum (Σ*ρa*) mean ($\bar{\rho a}$), standard error of ($SE\bar{\rho a}$), standard deviation (σ_0), variance (σ^2), median (*ρam*), mode (*ρaMo*), absolute range (*Rabs*), standard deviation (σ_0), variance (σ^2), determination coefficient (R^2), non-determination or alienation *K* coefficient ($1 - R^2$), coefficient of variation (*Cv*), correlation coefficient (*R*), symmetry skewness (*Swew_ρa*), symmetry kurtosis (*Kurt_ρa*), positional percentile 25% parameters (*Q25th*), positional percentile 75% (*Q75th*) and 95% confidence intervals (*95%CI*).

The sum of *ρa* (Σρa) was 2193.34 Ωm for 2022 and 2365.66 Ωm for 2023. The dispersion parameters, absolute range (*Rabs*) in Ωm for 2022 was 49.67 and in 2023 35.64, σ_0 respectively, in the same years were 8.82 Ωm and 7.74 Ωm, while the variance (σ^2) was 77.87 Ωm^2 in 2022 and 59.95 Ωm^2 in 2023. The coefficient of variation (*CV*) values were 40.23 in 2022 and 32.73 in 2023, with determination coefficients (R^2) of 0.88 and 0.77, respectively. The skewness (*Swew_ρa*) ranged from 1.30 to 0.082, and kurtosis (*Kurt_ρa*) ranged from 2.48 to −0.31. The geometric mean (*Mg*) varied from 20.37 to 22.24 Ωm. The

25th percentile (*Q25th*) ranged from 16.86 to 18.17 Ωm, and the 75th percentile (*Q75th*) ranged from 25.60 to 29.17 Ωm.

The results suggest that the *DROC* at the *RI/II* interface experienced minimal changes in characteristics between 2022 and 2023. The consistency of ρa values over time indicates a stable condition in the subsoil's electrical and potentially chemical properties. The refined focus on a smaller, consistent area allowed for a more accurate analysis of *DROCs'* dynamic processes and highlighted the need for ongoing monitoring and remediation in *OAGD* environments. This stability is essential for understanding the long-term impact of waste decomposition on subsoil properties and serving to inform more effective *LSWM* strategies.

4.2. Assessment of Subsoil Dynamics Using 95% Confidence Interval

According to Bermejo (2005) [66] criteria, the *95% CI* range for $\rho a2022$ was from 20.20 to 23.66 Ωm and for $\rho a2023$ it was from 22.13 to 25.17 Ωm. Notably, both *95% CIs* have true $\overline{\rho a}$ values with relatively narrow ranges (3.46 for 2022 and 3.04 for 2023). The differences in ranges were slightly higher in 2022 (0.42). Overall, the $\overline{\rho a}$ values for $\rho a2022$ and $\rho a2023$ exhibit very similar behavior; however, in 2022, the *95% CI* of the ρa distribution was less compact than in 2023 (21.93 Ωm < 23.66 Ωm). Consequently, σ_0 and σ^2 showed significantly higher values (greater dispersion) in 2022 (8.82 Ωm > 7.74 Ωm). Despite the minimal variations observed in the two sets of *ICI95%* results, notable disparities were detected in the behavior of σ^2 values, with higher values noted in $\rho a2022$ (77.87 Ωm^2 > 59.95 Ωm^2). The general variation in measurements deviated from $\overline{\rho a}$ by approximately 8.82 Ωm in 2022 and 7.74 Ωm in 2023. The 2023 distribution presented a higher grade of homogeneity, characterized by lower σ^2 values (59.95 Ωm^2). These findings suggest a dynamic in the subsoil related to *DROC* degradation, where its aggregates persist within the porous media of the subsoil since the inception of operations at the *OAGD* in the area. To associate σ_0 and σ^2 of annual variation in ρa as an indicator of increases or decreases on the entire *DROC* was considered not incorrect because the ϕ intensity and the *TIMP* are not constant and have heterogeneity and anisotropy mediated in subsoil by *TIMP*.

4.3. Evaluation of Linear Relationships and Model Fit for Subsoil DROC Variability

In the area where annual measurements coincided, Figure 2C depicts the Coefficient of Correlation R (1, −1) and coefficients of determination R^2 (0, 1) for [ρa2022 vs. ρa2023], illustrating a linear association. The correlation coefficient $R = -0.745$ indicates an inverse relationship: as ρa2022 values increased, ρa2023 values decreased. The graphical representation shows dispersion values starting with minimal density in the lower left and increasing toward greater dispersion at the upper right.

This correlation pattern revealed two distinct regions with varying adjustment levels. The lower left zone showed a good fit, with minimal and unbiased differences between observed values and model predictions. In contrast, the upper right zone exhibited undesirable residual patterns, indicating bias in the results.

The data graphically in Figure 2C show a clear decreasing linear trend. However, the correlation coefficient of $R = 0.745$ indicates that while 74.5% of the total information follows a linear trend, the remaining 25.5% does not align perfectly. The misalignment in the dataset over time is attributed to changes in the subsoil of *DROC*, leading to new massif conditions influenced by *TIMP* and the potential of ϕ. The precise determination of how to validate the adjusted model and understand how much the annual measure information could describe the annual changes indicative of the transformation process of one phase to other phases of *DROC* was carried out through the magnitudes acquired between the R^2 and R^2_ *adjusted* coefficients.

The information related to the annual changes in *DROC*, represented through the factor K, could be obtained using the R^2 factor. However, on this occasion, it was not obtained in that manner. The best option, due to its precision, was the R^2*_adjusted*. Therefore, the results for factor K were presented using the expression $K = 1 - R^2\_adjusted$. The results of the R^2 and R^2*_adjusted* coefficients appear in Figure 2, this graphical representation of the model, showing the data distribution adjusted to a linear trend. Note the approximate similarity in the magnitudes of both coefficients represented by a minimal difference of 0.05 units, defined by the values $R^2 = 0.55 \approx R^2\_adjusted = 0.50$.

The proximity in the magnitudes of coefficients R^2 and $adjustmen\_R^2$ allowed establishing the model's fit certainty with the annual data used, regarding its linear trend behavior through the equation $\rho a2023 = 9.312 \times 0.65\ \rho a2022$. Considering this similarity between coefficients indicating a high probability that the model will have good predictive performance in the annual distribution of the areas occupied by the transformed *DROC*, and the measurements provide a well-structured model, the adjusted R^2 value was chosen to define the sites that occupy the new zones with *DROC* in the subsoil porous system. Thus, both measurements significantly contributed to the model's fit, and observing no penalty for the variables, as both are explanatory in the model, suggested that the model variables were relevant and necessary. It was indicated how much of the information shared its σ^2 and also the remaining information K $(1 - R^2\_adjusted = 0.05)$ of 50%, which represents the information within the porous system that occupies the space with the annual changes in the different *DROC* phases distributed in the subsoil. Regarding time, the amount of leachate generated in the *OAGD* since the moment that primary arrives at the porous media of subsoil is dispersing and transforming to constitute the new zones with changes in the concentration of *DROC*, and this similarity between coefficients allowed the establishment of a model that does not represent noise in the adjustment relevant and robust information over these new zones represent the phases of *DROC* transformed. Concerning $R^2 = 0.55$, the shared σ^2 between $\rho a2022$ and $\varrho a2023$ was practically the same result as $R^2\_adjustet$.

4.4. DROC Transformation Dynamics and Invaded Areas: Linear Nonlinear Analysis

According to the model results, perceptually 50% have not shared information on measures corresponding to *DROC* transformed and one year after on subsoil occupying the areas invaded according to the next process:

(1) The transformation of *DROC* begins with the arrival of leachates from the surface, originating from the *OAGD*. These total invasions characterize the *IR* zone, which can encompass both the saturated and unsaturated zones of the aquifer and serve as a continuous source feeding the subsoil system of *DROC* that will be transformed through the dynamics of the porous medium (Figure 3A,B).

(2) An interface *IR/II* indicating changes in the concentration of leachates arriving before the transformation process and tending to reach the next intermediate phase *I* and, in the case shown in Figure 3A,B, is presented in the same graph.

(3) The intermediate phase *II* usually accumulates the most amount of mass/time due to the retard and prolonged process required to reach the conditions to reach the next phase (Figure 3C,D).

The *Flw* zone is usually found in the aquifer's unsaturated zone close to the surface. Here, there is a risk that it could break through the impermeable layer of its geological structure, which might turn it into a leachate zone (Figure 3E,F).

The final *AI* stage of the process determines the status of *DROCs*: either it is retained within the porous medium or completely purged from the groundwater system. Independently of all these transformation processes of *DROC*, there is a zone characterized by its granulometric conditions, predominantly represented by the presence of clayey bodies. This gives it high non-permeability against the invasion of leachate-laden water according to ϕ, allowing it to remain with unpolluted zones (Figure 3G,H); the presence of leachate (*Fwl*) as of August 2022; (F) area covered by *Fwl* as of October 2023; (G) zones of mature or ancient invasions (*AI*) as of August 2022; and (H) area covered by *AI* as of October 2023. The ρa values are shown at the bottom of each diagram for the respective *DROCs* stage. Annual variations in Ωm of apparent resistivities (ρa) within anisotropic *TIMP* travel in the subsoil surrounding *OAGDs* were measured along a *NW–SE* direction intersects the preferential direction of groundwater flow. Measurement reflects the spatiotemporal association of changes in concentration for each stage of *DROCs* on subsoil.

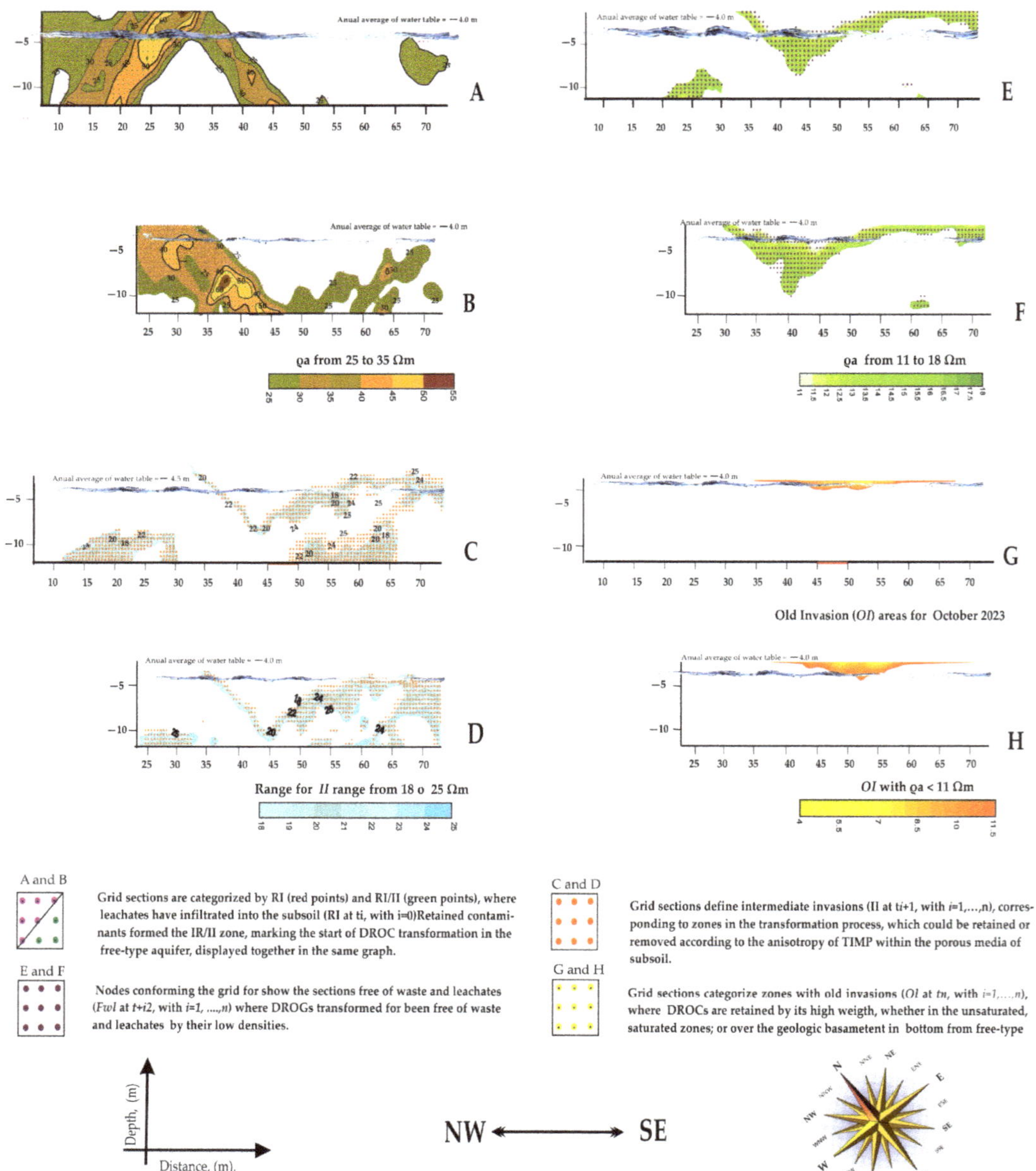

Figure 3. Annual variations spatial in Ωm of *ρa* as indicators of changes mass and it transport within anisotropic *TIMP* zones, where the preferential *NW—SE* direction of groundwater flow intersects each *DROCs* stage. The illustrations show: (**A**) Recent invasions (*RI*) and their interface (*RI/II*) as of August 2023; (**B**) area covered by *RI* and *RI/II* as of October 2023; (**C**) Intermediate Invasions (*II*) zones as of August 2022; (**D**) area covered by *II* zones as of October 2023; (**E**) zones without the presence of leachate (*Fwl*) as of August 2022; (**F**) area covered by *Fwl* as of October 2023; (**G**) zones of mature or ancient invasions (*AI*) as of August 2022; and (**H**) area covered by *AI* as of October 2023. The ρa values are shown at the bottom of each panel for the respective DROCs stage.

4.5. Analysis ρa in Association with RI Stage

The *IR* began promptly upon the entry of leachates into the porous media, initiating their progression from the *RI* stage toward their final state—*AI* or complete degradation—where they are either dispersed or retained within the unsaturated or saturated zones of the aquifer (Figure 3A,B).

For the *RI* phase, measurements of ρa in 2022 showed a range of ρa_{min} to ρa_{max} between 25.17 and 56.87 Ωm, with a total $\Sigma\rho a$ = 893.02 Ωm, an average $\overline{\rho a}$ = 33.07 Ωm, a σ = 8.1 Ωm, and σ^2 = 66.73 Ωm^2. These values indicate substantial heterogeneity and variability in *DROC* distribution across the subsoil, here evidenced by the relatively high dispersion in resistivity values. In 2023, however, ρa values for *RI* under *TIMP* dynamic anisotropy exhibited a narrower range, with ρa_{min} to ρa_{max} x falling between 25.01 Ωm and 42.18 Ωm, and $\Sigma\rho a$ = 1271.295. The average $\overline{\rho a}$ slightly declined to 31.0 Ωm, and σ_0 decreased to 4.49 Ωm, suggesting that resistivity measurements were more homogeneous than in 2022. Additionally, the reduced variance (σ^2 = 20.20 Ωm^2) reflects less variability in ρa values. These observations suggest a more stable presence of *DROCs* in 2023, as increased homogeneity (lower σ_0) and reduced variability (lower σ^2) likely indicate a more consistent and confined distribution of *DROC* concentrations within the system under *TIMP* anisotropy's influence.

This change in ρa distribution can be attributed to interactions between *DROC* and *TIPM* anisotropy, leading to a more stable spread of contaminants. In 2023, in the *RI* phase was recorded for $\Delta\Sigma\rho a$ = 378.27 Ωm, average $\Delta\overline{\rho a}$ = 2.06 Ωm, $\Delta\sigma_0$ = 3.67, and $\Delta\sigma^2$ = f 46.52 Ωm^2. This represents a shift from 2022, where dispersion was higher, suggesting that *DROC* dispersion within the *RI* phase decreased as the transformation process advanced.

The decrease in resistivity in 2023 indicates a stabilization in the *IR* transformation process of *DROC* within the subsurface. Finally, the changes in $\Delta\Sigma\rho a$, σ_0, and σ^2 between 2022 and 2023 underscore the evolving subsurface conditions, particularly in the initial phase of *DROC* transformation. The higher dispersion observed in 2022 may be associated with the initial arrival and transformation of leachates within the porous media of subsoil a, whereas 2023 revealed a more stable and uniform distribution. These findings are essential for understanding leachate dynamics in the *OAGD* environment and contribute to developing effective environmental management strategies.

4.6. Annual Variation ρa on Subsoil for Interface RI/II

Similar to the IR stage, the findings reveal a stable profile of *DROCs* in the *RI/II* interface over the study period. Specifically, ρa values at this boundary showed a narrow variation between 2022 and 2023, with ranges of 25.17–33.63 Ωm and 25.01–34.62 Ωm, respectively.

This consistency suggests that the interface's physical and chemical characteristics did not undergo significant change, indicating minimal *DROC* degradation or transformation within the observed timeframe. This persistence in ρa values suggests that environmental factors, such as electrical resistivity, remained relatively constant, possibly due to the subsoil's composition and structure, which may create a stable environment less conducive to rapid degradation of *DROCs*. The similar behavior observed sin *DROC* distribution and resistivity values over time highlights the resilience of these compounds within the OAGD subsoil environment. This stability implies that *DROCs*, being degradation-resistant, are not quickly transformed or broken down, leading to long-term retention within the soil matrix. Such results point to a low interaction rate between *DROCs* and the surrounding soil minerals or microorganisms, which could be due to limited biological activity, reduced moisture, or other environmental constraints affecting *DROC* breakdown.

Furthermore, the minimal shifts in resistivity values in *RI/II* suggest that *DROCs'* degradation or transformation processes might be slower in these types of aquifer and soil environments, particularly under conditions resembling those observed in the *OAGD* area. This indicates the necessity of developing long-term and tailored remediation strategies to address the slow degradation rates in similar settings. The findings underscore the importance of ongoing monitoring, as the consistent *DROC* presence can contribute to cumulative environmental impacts and may pose challenges in maintaining groundwater quality in the long term in the region.

4.7. Characterization II Invasions and Degradation

Stage *II* of *DROCs* developed within the subsoil environment and porous media conditions, representing the transition from the *RI* phase to the *II* phase. In 2022, ρa_{min} and ρa_{max} ranged from 18 to 24.89 Ωm, and in 2023, the range was slightly adjusted to 18.01–24.87 Ωm. The ρa values in 2022 included $\Sigma\rho a$ = 799.83 Ωm, with $\overline{\rho a}$ =21.04 Ωm, σ_0 = 1.81, and σ^2 = 3.27 Ωm^2. In 2023, similar parameters were observed, with $\Sigma\rho a$ = 768.68 Ωm, an increased average $\overline{\rho a}$ = f 21.96 Ωm, σ_0 = f 1.88, and σ^2 = 3.56 Ωm^2. The central tendency and dispersion statistical parameters a high similarity between the two years, with only slight increases in the average resistivity and minor variations in dispersion, suggesting a consistent profile of *DROCs*.

Notably, the data for the two years overlapped by 96% (R^2 = 0.96), showing a small annual variation of just 4% (approximately 24.5 m^2). This stability implies that *DROC* properties in this stage were likely transitioning gradually towards the AI phase.

The Finite element analysis, supported by Newton interpolation, confirms that Stage *II* showed no significant area evolution differences. However, a reduction in area of 18.72 m^2 per year was recorded (from 237.12 m^2 to 218.4 m^2), indicating that mass transfer processes might be prompting the transition of *DROCs* to the *AI* phase. The similarity in central tendency and dispersion measures across both years suggests minimal changes in ρa properties, indicating a stable presence of *DROCs* within the subsoil system.

The results of annual differences in statistical parameters between 2022 and 2023 were as follows: $\Delta\rho a_{min}$ = 6.54 Ωm, $\Delta\rho a_{max}$ = 10.61 Ωm, $\Delta\Sigma\rho a$ = 54.46 Ωm, $\Delta\overline{\rho a}$ = 9.07 Ωm, $\Delta\sigma_0$ = 1.63, and $\Delta\sigma^2$ = 2.65 Ωm^2.

These differences highlight a significant decrease in ρa values and variability from 2022 to 2023, indicating changes in the subsoil conditions or composition, potentially due to environmental factors or ongoing *OAGD* processes. Furthermore, the reduction in area coverage for the *II* phase, monitored at a rate of 12 m^2/year (from 49.92 to 37.44 m^2), underscores the resilience of the subsoil in these areas. This minimal decrease in area coverage can be linked to natural *Flw* processes, with 112.32 m^2 transitioning into *Flw* zones in 2023.

However, the observed decrease in $\Delta\Sigma\rho a$ in *II* stage, as shown in Table 1, implies that some *DROCs* may have transitioned to the *AI* phase or possibly may have migrated into *Flw* with a gradual transformation influenced by *TIPM* anisotropy and *S-S* interactions. This dynamic process suggests that *DROCs* are being redistributed or transformed within the subsoil environment, likely progressing to more advanced stages in their degradation profile. Such findings are critical for understanding *DROC* behavior and environmental persistence, offering insights into the subsoil's capacity to retain or transform *DROCs* over time.

4.8. Characterization AI and Reduction ρa and Final Degradation

The *DROC* in the *AI* phase, showed a notable shift in both the distribution and concentration between 2022 and 2023. In 2022, the ρa_{min} and $\rho a_{\mathbf{max}}$ ranged from 7.30 to 10.79 Ωm, with a $\Sigma\rho a$ = 78.07 Ωm, $\overline{\rho a}$ = 9.75 Ωm. The standard deviations (σ_0 = 1.35 Ωm, σ^2 = 1.82 Ωm^2); suggest a moderate variability, and indicate that the *DROC* distribution remained relatively stable across the study area. In 2023, there was a slight decrease in the range of ρa_{min} and $\rho a_{\mathbf{max}}$ to 6.54–10.61 Ωm, with $\Sigma\rho a$ = 54.46 Ωm and $\overline{\rho a}$ = 9.07 Ωm decreased signaling a shift towards lower resistivity values. Additionally, the increase in standard deviations (σ_0 = 1.63, σ^2 = 2.65 Ωm^2) indicates greater variability in the resistivity values, suggesting that while the resistivity distribution became more spread out, the overall quantity of *DROC* present in the area had diminished. The significant decrease in $\Sigma\rho a$, along with the increased dispersion measures, implies that *DROC's* resistivity values became more dispersed, but the total presence of *DROC* decreased. This suggests that *DROC* in the *AI* phase may have undergone depletion or transformation due to processes such as degradation, dispersion, or other environmental factors, possibly accelerated by the *TIPM* anisotropy. This influence likely affected the rate and pattern of *DROC* spread and concentration. The statistical changes observed indicate a marked alteration in the characteristics and distribution of *DROC* in *AI* phase, with a reduction in the overall amount but an increase in variability. This variability may be linked to interactions influenced by external environmental factors, such as the ongoing *OAGD* processes and natural *Flw* processes. The annual differences in statistical parameters between 2022 and 2023 were as follows: $\Delta\rho a_{min}$ = 6.54 Ωm, $\Delta\rho a_{max}$ = 10.61 Ωm, $\Delta\Sigma\rho a$ = 54.46 Ωm, $\Delta\overline{\rho a}$ = 9.07 Ωm, $\Delta\sigma_0$ = 1.63, and $\Delta\sigma^2$ = 2.65 Ωm^2. These differences in the anterior statistical parameters highlight a significant decrease in ρa values and variability, indicating changes in subsoil conditions likely influenced by environmental factors or ongoing *OAGD* processes. Furthermore, the reduction in area coverage for the AI phase, observed at a rate of 12 m^2/year (from 49.92 to 37.44 m^2), underscores the resilience of the subsoil in these regions.

4.9. Evolution of Flw Zones

The *Flw* zone is showing in Figure 3E,F, illustrating for 2022 and 2023. The *Flw* zone exhibited $\rho a2022$ values ranging from 11.17 to 17.87 Ωm, with $\Sigma\rho a$ = 379.56 Ωm, average $\overline{\rho a}$ = 15.1 Ωm, σ_0 = 2.12 Ωm, and σ^2 = 4.52 Ωm^2. In 2023, the range was from 11.88 to 17.98 Ωm, with ΣX = 271.21 Ωm, average $\overline{\rho a}$ = 15.06 Ωm, σ_0 = 2.08 Ωm, and σ^2 = 4.33 Ωm^2. The slight differences in $\Delta\overline{\rho a}$ = 0.11 Ωm, $\Delta\sigma_0$ = 0.046 Ωm, and $\Delta\sigma^2$ = 0.19 Ωm^2 between the two years suggest a stable trend in the *Flw* zones, hence, the ρa characteristics of these zones remained largely unchanged.

The consistency in trends of statistical parameters suggests that ρa characteristics are influenced by underlying geological features and leachate dynamics. These stable ρa values and minimal annual changes in the *Flw* zones highlight the effectiveness of porous media in the subsoil surrounding the *OAGD* in preventing pollution processes. The observed variance in ρa, in a small portion of the area, underscores the need for continuous monitoring to manage new leachate infiltration effectively. The spatial correlation and linear regression analysis further validate the consistent trends in ρa, providing valuable insights into the subsoil dynamics and the behavior of leachate in the *OAGD* environment.

In the *OAGD* site, the annual data, showed a 90% overlap (R^2 = 0.90), indicating that only 10% (61.25 m^2) of the total covered area exhibited increases, which were attributed to the limited arrival of new leachates to the aquifer, supported by the continuous dynamics of *TIPM* anisotropy

These findings emphasize the importance of understanding geological features and their interaction with leachate dynamics to ensure effective *OAGD* management. The consistent ρa measurements indicate a stable subsoil environment, which is crucial for maintaining the integrity of the *Flw* zones. Additionally, the slight increase in ρa in certain areas highlights the dynamic nature of the *OAGD* environment and the potential for new leachate infiltration. Continuous monitoring and analysis of ρa trends are essential for the early detection and management of changes, ensuring the long-term stability and environmental safety of *OAGD* sites. The study's results contribute to a better understanding of the interactions between leachate and geological features, offering a foundation for developing more effective strategies for *OAGD* management and pollution prevention.

4.10. Causes of Degradation in DROC Stages

Delving deeper into the dynamics underlying *DROC* degradation, this study invokes the mass transfer principle to unravel the intricate interplay between environmental dynamics and material migration within *OAGD* ecosystems. This analytical approach facilitates the inference that the *OAGD*'s yearly dynamics, particularly within the zone where annual measurements were shared, act as catalysts for the migration of *DROC* from the surface, crossing geological layers in the subsoil at different depths ($\rho a2022$ = 168.48 m^2

and $\rho a2023$ = 255.84 m^2). This nuanced understanding of *DROC* degradation in the subsoil underscores the complexity of *OAGD* ecosystems, where environmental factors intricately interact with geological materials and groundwater properties to shape subsoil dynamics over time.

Such insights have profound implications for *WMLS* in *OAGS* and propose remediation strategies, advocating for a holistic approach that integrates computational modeling with empirical observations to inform effective decision-making processes [69]. The observed changes are particularly driven by heightened contamination risks during specific periods, due to decreases in annual precipitation average (P_{2022} = 855.82 and P_{2023} = 733.6 mm, with $\Delta p = -122.2$ mm) notably from July to October when the pluvial annual magnitudes averages in Sinaloa state are the highest (189.47 mm in 2022 and 174.8 mm 2023) [70]. Local rainfall interacts with contaminants, potentially displacing leachates and facilitating their migration through soil pores. Variations in precipitation intensity further influence this process, affecting the movement of pollutants and contributing to biomass accumulation across different phases of *DROC*.

On the other hand, over time, variations in density ϕ will alter the structure of *DROC*, impacting each phase of water infiltration dynamics and influencing the movement and retention of these compounds in the subsurface layers. Quantifying these changes in the subsoil involves comparative analyses of initial versus altered water infiltration rates, reflecting shifts in physical properties as new areas emerge [69]. Controlling variations in soil bulk density necessitates adaptive management strategies based on subsoil utilization and vegetation types, aiming to mitigate environmental impacts effectively.

This approach allows scientists and policymakers to anticipate environmental challenges, formulate appropriate remediation strategies, and design better waste management practices to reduce the risks of large-scale *DROC* invasion into porous media. Additionally, these measures contribute to lowering greenhouse gas emissions, addressing both environmental pollution and climate change.

4.11. Direct and Indirect Analysis DROCs Stages Within OAGD

The results of the different phases of *DROC*, detected indirectly from the surface in the subsoil, are shown in Figure 2A,B. These results were validated directly and indirectly in the surrounding subsoil areas of the *OAGD* in zones where the annual ρa measurements shared information. The direct confirmation was conducted with the results of stratigraphic variability in a borehole drilled to a depth of 22 m from the surface (Figure 4A). Part A of this figure presents the stratigraphic column, where it can be observed that most of the sediments are composed of different types of unconsolidated sedimentary materials. Laboratory granulometric analyses through sieving determined the granulometry in the drilled borehole, which consisted of particles of various sizes: clay (<0.002 mm), silt (0.002–0.06 mm), sand (0.06–2 mm), gravel (2–6 mm), and pebbles (60–250 mm). This detailed characterization of the sedimentary material is crucial for understanding leachate dynamics, as each sediment has different permeability and porosity properties that affect the transport and dispersion of contaminants.

The results were compared with the ρa values obtained indirectly from the surface using the Dipole–Dipole geoelectric geophysical tool (Figure 3B). The Dipole–Dipole technique is a non-invasive method that provides a ρa profile of the subsoil, offering an indirect but detailed view of the geological structure and the distribution of sedimentary materials. The coincidence between the ρa values and the stratigraphic information obtained through drilling provides robust validation of the geophysical data.

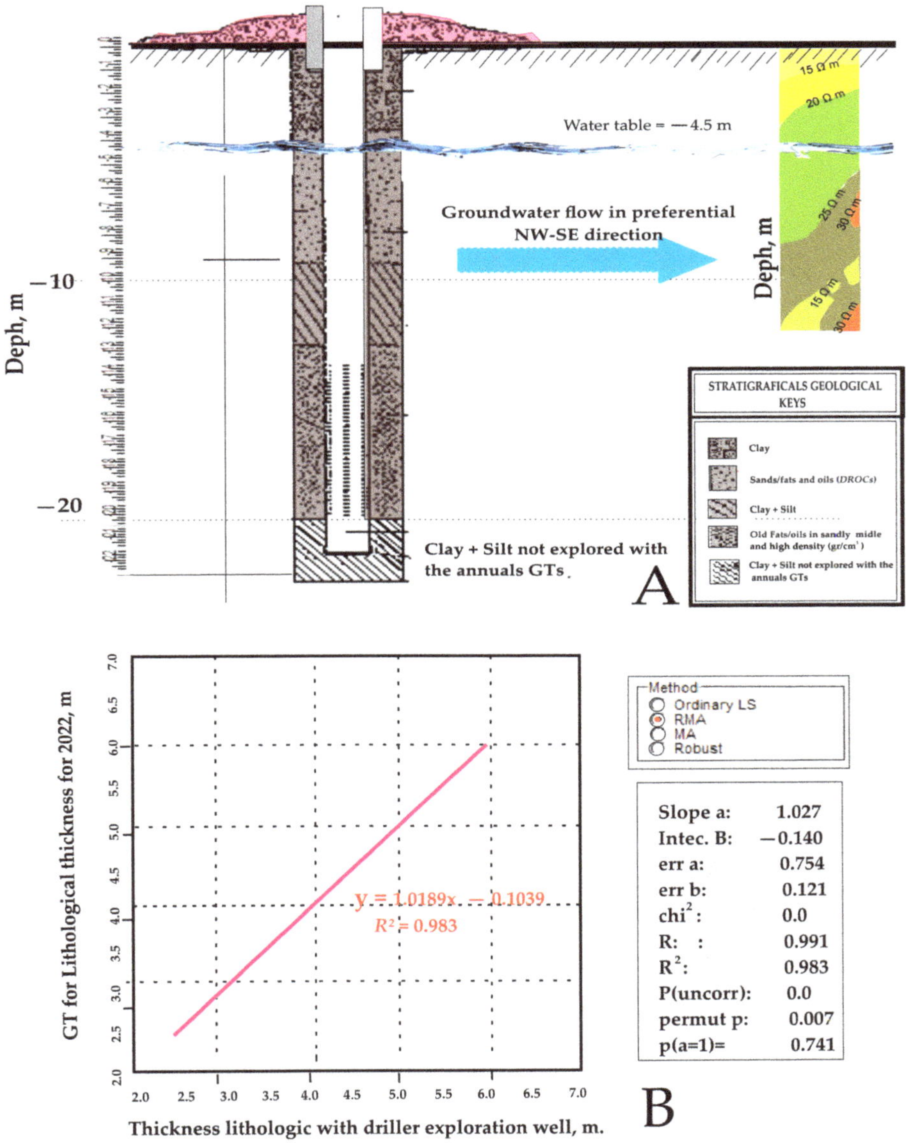

Figure 4. (**A**) Exploratory well drilled to 22 m of depth illustrating one sample from Sinaloa aquifer River for its hydrogeological units on subsoil with presence of fats and oils and, (**B**) linear correlation (R^2 = 0.98) between the hydrogeological thicknesses of the associated with phases of *DROC*.

4.12. Validation and Lineal Correlation of Annual ρa

The correlation between the type of material crossed by the drilling and the thicknesses detected with the Dipole–Dipole technique confirms the indirect detection.

The depth of these strata corresponds to the points in the borehole vertical where water samples were taken for laboratory analysis of *DROC* concentration. The stratigraphic granulometry allowed for the correlation of material type with the probable dispersion velocities of leachates from the source of the *OAGD*.

This information is crucial for understanding how leachates move through the subsoil and how different sedimentary layers can act as barriers or facilitators for contaminant transport. The similarity between the strata, validated by both direct drilling techniques and indirect Dipole–Dipole geoelectric detection, confirms that the methodology for *DROC* detection is effective despite hydrological heterogeneity in aquifer properties and a variable storage coefficient. The ability to correlate directly and indirectly the *DROC* presence validates only the effectiveness of the techniques employed also providing a more comprehensive understanding of the subsoil hydrological system.

4.13. Effectiveness of Methodology and DROCDROC Analysis

Standard gravimetric analysis confirmed the presence of *DROC* in each stratum crossed by the borehole. The procedural blanks used in the laboratory showed a *DROC* residue mass of less than 5 mg, indicating the absence of concentrations greater than 5 mgL^{-1} and ruling out significant contamination during the analysis process. This detailed detection of *DROC* in the stratigraphic samples provides valuable information on the vertical distribution of these compounds in the subsoil, contributing to the understanding of leachate dynamics in the *OAGD* environment.

The *DROC* confirmation through gravimetric analysis was a critical step in validating the findings of the indirect detection techniques and direct drilling. The analysis of gravimetric results allows for precise quantification of contaminants in soil and water samples, ensuring that the interpretations for ρa and stratigraphic data results are accurate and reliable.

4.14. Validation of Detection Methods for DROC in OAGD and LSWM

Delving deeper into the underlying dynamics, the study invoked the mass transfer principle to unravel the interplay between environmental dynamics and material migration within the *OAGD*. This analytical approach facilitated the inference of the *OAGD*'s yearly dynamics, particularly within the *RI* stage, which in subsoil acted as catalysts for the migration of from the surface to varying depths within the subsoil layers. Consequently, this phenomenon resulted in an annual augmentation of invasion surfaces, starting from a modest coverage area of 81.12 m^2 in 2021, which progressively expanded to 168.48 m^2 in 2022. It ultimately culminated at a substantial 255.84 m^2 in 2023.

This nuanced understanding underscores the complex and dynamic nature of *OAGD* ecosystems, wherein environmental factors intertwine with the underlying material properties to dictate the temporal evolution of subsoil dynamics. Such insights hold profound implications for the management and remediation strategies employed within land *OAGD* environments, advocating for a holistic approach that integrates computational modeling with empirical observations to inform effective decision-making processes.

The three analysis methods employed in this study—direct borehole drilling, Dipole–Dipole geoelectric detection, and laboratory gravimetric analysis—, confirmed the validity and effectiveness of the indirect methodology used.

Each method offers a complementary perspective, reinforcing the reliability of the results and emphasizing the importance of a multidisciplinary approach to evaluating leachate contamination in the subsoil of the anywhere *OAGD*, and this integration not only enhances the accuracy of detecting and quantifying *DROC* but also establishes a robust foundation for the correct development of *LSWM* and mitigation strategies by pollutants. By identifying and quantifying *DROCs* across different geological stratigraphy, *LSWM* can design more precise and targeted interventions, reducing environmental impact and improving the sustainability of *LSW* disposal practices.

4.15. Causes of DROC Degradation Phases

The hypothetical scheme presented in Figure 5 illustrates the behavior of *DROCs* through their various phases in the subsurface under static conditions. This framework provides a conceptual tool to predict the future dispersion and accumulation of *DROCs* within porous media. The diagram not only visualizes the causes of *DROCs* degradation at each stage but also addresses their retention in the subsoil of old *OAGD* studied.

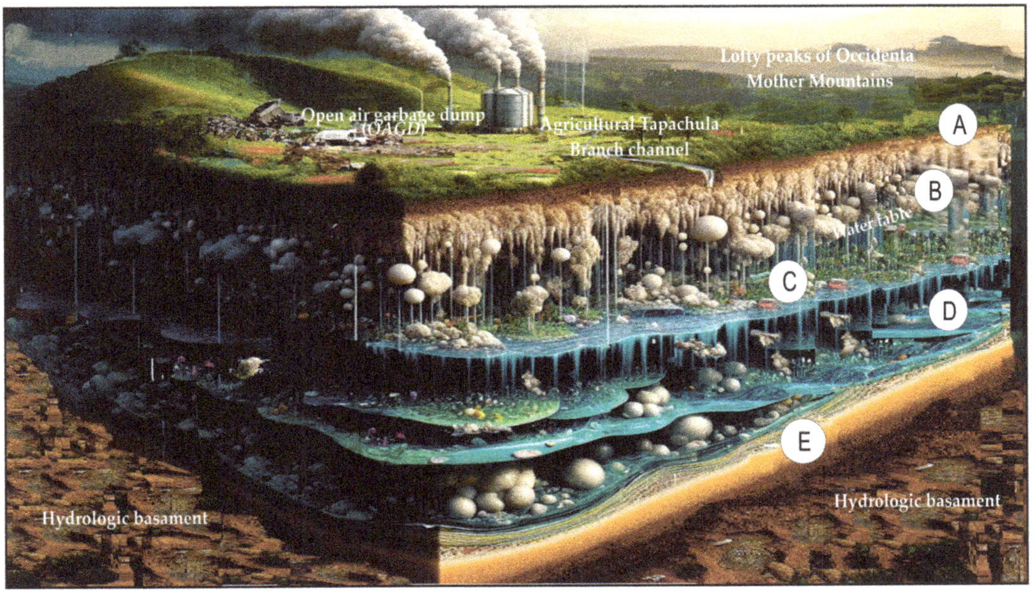

ρa VARIATION FOR COMPOUNDS RESISTANT IN SUBSOIL AS FATS AND OILS (*DROCs*), Ω.m

A Old or Mature Invasion (*OI*) of *DROGs* into subsoil.

B Recent Invasions *(RI)* of *DROGs* with values by ρa > 35.

C Interfase *of DROCs* invasion between *RI* and *II* (*RI/II*) with ρa range from 25 to 35 Ωm.

D Intermediate Invasions (*II*) of DROCs in zones with ρa range from 18 to 25 Ωm.

E Zones without *DROGs* presence and any leachates (*Fwi*) with ρa range from 11 to 18 Ωm.

Figure 5. Scheme hypothetically of porous media in subsoil and the persistent organic compounds as fats and oil, illustrating the degradation stages over time and their movement based on weights and densities simulating the occurrence of dispersion in the old *OAGD* studied. Are observe the gas compound's source and contribution of greenhouse gasses to the atmosphere. The solutes *DROC* are dispersed over the distance and decreased according to the transport *TIMP* in groundwater (Own creation based on geometric attributes generated in SURFER 10.0, vectorized, refined, and projected with two diffusion points in CorelDRAW 18.0. Additional front-facing attributes were edited in Paint 3D, with the final image exported in the .JPG format, also through CorelDRAW 18.0).

Furthermore, it anticipates potential scenarios in which a complete invasion of these substances could occur if no preventive measures are implemented. Based on the classification of *DROC* stages, the dynamic behavior of *S-S* system is illustrated in Figure 5. It shows how each stage changes with respect to depth and the porosity of the subsoil's porous media. Through this model, we aim to understand how *DROCs* interact with the physical properties of the soil, particularly the effects of local topography, variations in subsurface permeability, and the ability of geological layers to either retain or allow the passage of these compounds.

Additionally, the diagram demonstrates how the characteristics of porous media can either facilitate or impede the movement of *DROCs*, ultimately impacting groundwater quality and ecosystem health. The model also emphasizes the significance of waste decomposition and the degradation of *DROCs* in the *OAGDs*, where gas emissions could contribute to global climate change. These gasses, released into the atmosphere, accelerate global warming and disrupt local and regional climate patterns. Consequently, the scheme of Figure 5 not only aids in understanding the current behavior of *DROCs* in the subsurface but also serves as a valuable tool for evaluating the long-term consequences of inadequate waste management.

The model predicts how these processes could worsen future climate variability across different seasons. The scheme also serves as a valuable tool for understanding both the immediate and long-term behavior of *DROCs* in the subsoil e, offering insights into how inadequate waste management practices—often amplified by consumer-driven production—can lead to significant environmental repercussions. The model ultimately aids in assessing potential future impacts, guiding the development of policies and practices that could mitigate environmental degradation and support sustainable resource management.

The scheme is not only an instrumental to mapping the current state of *DROC* behaviors but also serves as a predictive framework for assessing the compounded environmental risks associated with poor waste disposal practices. The insights derived from this model underscore the role of consumerism in perpetuating *DROCs'* environmental impact and emphasize the need for interventions that prioritize environmental preservation and public health in affected regions.

These changes were driven by the heightened risk of contamination during certain periods of the year, particularly from September to October, due to increased precipitation. Local rainfall interacts with contaminants, potentially causing the displacement of leachates and their migration through the pores of the *S-S*. Moreover, this phenomenon can be influenced by variations in the magnitude of the velocity of ϕ. Changes in ϕ, corresponding to the intensity of rainfall, create conditions conducive to the movement of pollutants through the *S-S*, facilitating biomass accumulation across various *DROC* phases. This process also contributes to the restoration of porosity, thereby altering water infiltration dynamics. During the various stages of this process, the changes can be quantified in the *S-S* by conducting a comparative analysis between the initial water infiltration rate and the altered infiltration rate resulting from changes in the physical properties of the *S-S* [69]. In any scenario, the emergence of new areas will lead to variations in soil bulk density within the *S-S*, necessitating the implementation of new management strategies for *S-S* utilization and vegetation types.

However, during *DROC* invasions, the initial porous conditions were altered, leading over time to collapses in the *S-S* system or the cementation across all pore dimensions. This demonstrates that *TIMP* anisotropy can induce variations in water infiltration and alter the sedimentary structure of the *S-S* system in the presence of *DROC* phases.

Figure 4B illustrates the linear regression among the five stratigraphical zones of the well and those observed in ρa_{2023} (with the thickness of the last two beds of the well kept constant). Notably, the last two beds, located between 12.5 and 22 m, exceed the maximum depth explored by both *GTs*, which is 12 m. Despite this, a linear regression value of 0.98 (R^2 = 0.98) was achieved by comparing the sedimentary thicknesses and lithological variations in the well, extending up to 12 m deep. This suggests an acceptable lithological correlation for the saturated and unsaturated zones (Figure 4B). The exploratory well drilled to a depth of 22 m, provided valuable insights into hydrogeological characteristics in deeper formations that could not be reached by the *GTs*. Depths of 12.5 m and 19.5 m were particularly significant, revealing sedimentary facies containing fats and oils. Their high molecular weight and greater density suggested transport from the lateral zone of the garbage dump, where the highest concentration of waste and the primary source of *DROC* were located.

Within the aquifer at greater depths (sometimes associated with aquitards and confined aquifers separated by clays and composed of sandy, lagoon sediments), typical geoelectrical ρa values ranged between 25 and 35 Ωm. Noteworthy, the area has the heaviest rainfall from July to September [71]. Given that the measurements were taken one month after the rainy season had ended, it is possible that the rain contributed to the year-to-year changes in ρa values as a function of *TIMP*.

This phenomenon could elucidate the local behavior of *DROC* dispersion in the *S-S*, as illustrated in Figure 2A,B, where data were collected in October, just one month after the rainy season [72]. Rainfall has a pronounced influence on the dynamics of *S-S* transport, particularly in August, when rainfall typically occurs at its highest. Figure 4 depicts the annual changes in the areas covered by the three phases of *DROC* and in a similar manner in the *Flw* stage due to rainfall.

5. Discussion

The challenges associated with Landfill Solid Waste Management (*LSWM*) primarily arise in areas where suitable land for waste disposal is limited, leading to the use of *OAGDs*. Although subsoil and groundwater systems possess a natural capacity for biological self-recovery—a key factor in defining resilience levels for sustainability—relying solely on single annual linear observations of the entire landfill area, in this case is inadequate. The measurements, numerical processes, and their interpretations, based on the physical laws governing these phenomena, are essential for accurately characterizing subsoil systems. In this study, the numerically defined processes, combined with current linear projections of *DROC* transformation, highlight a high risk of contamination, emphasizing the need for comprehensive preventive measures. This underscores the need for proactive measures and continuous monitoring to ensure the long-term sustainability of the landfill. The results of this study highlight changes in *DROC* at the *OAGD* site, reflecting their dispersion or fixation to the subsoil. The *DROC* degradation could occur under two scenarios: within initial non-static surface degradation phases or in areas initially classified as *Flw* zones but later affected by the saturated zone of the unconfined-free aquifer. It is important to note that dispersion analysis does not necessarily indicate increases or decreases in the overall properties of stability, equicontinuity, and convergence; because the leachate compounds are denser than water, their adherence to aquifer walls or movement is influenced by *TIMP* and ϕ. After that, some contaminants cross the unsaturated zone, reach groundwater, and remain above the water table, while others migrate through the unsaturated zone and travel to the saturated zone, which increases its decomposition and adherence. The issue of *DROC* highlights the infiltration and accumulation of leachates from poorly managed landfills. Moreover, the dispersion analysis, which does not encompass every phase transformation in specific areas, represents the transformation along the measurement lines and indicates the need for comprehensive measurements to understand the behavior and impact on the total subsoil in sites surrounding the *OAGD*.

OAGD deposits pose numerous social, economic, and environmental challenges, and their mitigation cannot be fully addressed by simply transitioning to disposal techniques such as *SLs* and *CFs*. This study emphasizes the critical need for these transitions to prevent leachate production and reduce *GHG* emissions. Such a shift is essential to avoid the proliferation of poorly selected disposal sites and to ensure groundwater sustainability, especially in regions with unsuitable geological conditions and high subsoil permeability.

A key advantage of *SLs* or *CFs* techniques is their involvement in extracting recyclable materials from *LSW*, helping to reduce *GHG* emissions (CO_2 and CH_4). It is irrefutable that effective *SLWM* practices ensure that leachate is properly treated and does not migrate through the subsoil, thus preventing contamination of groundwater and geological layers. Engineer *SL* techniques—including compacted liner bottoms, groundwater monitoring systems, leachate collection, treatment systems, and gas collection systems—are designed to ensure the safe disposal of *LSW* while minimizing environmental impacts.

Geophysical and statistical tools used in this study revealed average behaviors of *DROC* formation and changes. The lack of year-to-year correlation and the differences between statistical measures of ρa in 2022 and 2023 may be attributed to *DROC* dynamics and the presence of *Flw* zones. The strong positive correlation between ρa2023 vs. ρa2023 values in specific areas suggests a consistent pattern across different phases and conditions. This synchronicity indicates that changes in ρa in one zone tend to reflect in others, pointing to interconnected landfill environments where changes propagate due to diverse factors like leachate ϕ pathways, soil composition, and microbial activity. Understanding these interrelations is crucial for accurate forecasting and effective landfill management.

Characterizing contamination based on spill behavior is complex, but identifying spill stages is crucial for remediation. This study presents a means to classify contamination based on ρa values, identifying potential sources and distinguishing between past and recent spills. The landfill site studied was ideal for this model because it gave the facility to define the travel of *DROC* within the porous system. The dispersal of *DROC* from the landfill site accumulated over many years and interacted with groundwater and local stratigraphy, allowing a quantitative determination of *DROC* coverage areas based on incursion and dispersion influenced by the anisotropy of *TIMP*. This analysis of the geometry associated with *DROC* coverage phases illustrated dispersion over time, employing geophysical data to obtain the tomography relatively straightforward.

The extensive history of solid waste storage at the landfill for over 40 years facilitated the continuous invasion of *DROC* into porous media from this *OAGD*, where the wide coverage of the *RI/II* interface at the current time indicates significant waste deposition and mass transformation in the subsoil of sites surrounding the *OAGD*, posing risks to subsoil health and human safety. *DROC* invasion into stratigraphy follows different scenarios: *RI*, characterized by high ρa values not associated with secondary products of landfill cessation, suggests ongoing waste ϕ and contamination transfer. Phase *II*, marked by intermediate ρa values, represents physical–chemical reactions and ion decomposition processes (Figure 5). Phase *AI*, identified by lower ρa values, indicated the mass transformation that alters the properties of *DROCDROCs*, leading to high concentrations that interact with groundwater and the formation of maximum degradation zones.

Monitoring and identifying *SLW* is crucial because the contaminants released during decomposition, pose environmental and health risks. This condition permits us to say that *LSWM* must adhere to strict environmental standards and international treaties signed by Mexico, which require remediation operations based on geographic, geological, and environmental factors. Despite financial constraints not being a barrier, Mexico struggles with fully implementing these regulations; however, some areas have begun to be addressed. Given Mexico's annual growth of 1.4% over the past five years, a targeted review of solid waste management policies is urgent. Federal support for research centers has identified various forms of contamination; however, enforcement remains inadequate. This study contributes to research efforts initiated in the 1990s to characterize coastal aquifers in the Guasave Valley, which is experiencing subsidence from agricultural overexploitation and the recompilation of *LSW* proceeded by this intense activity.

In summary, phases *RI*, *RI/II*, and *II* were identified as having the greatest potential for mass transformation leading to increased *GHG* emissions. This transformation, as indicated by ρa values from 2022 to 2023, suggests a reduction in the total mass within AI zones due to enhanced degradation facilitated by the porous nature of the subsurface, and changes with the precipitation deficit. The observed decrease in ρa values points to significant changes in subsurface properties resulting from the invasion of the AI zone, which impacts leachate mobility, dispersion, and retention. These findings highlight the importance of ongoing monitoring and effective management to track areas affected by *DROC* and to understand the extent of degradation phases in the subsurface.

Moreover, studies in the literature have defined the behavior of resistivity in the subsurface to identify ρa values in an integrated manner, establishing correlations between laboratory studies, drilling, and direct field measurements [73,74].

However, despite existing studies, this correlation analysis defining the total resistivity value in the subsurface has not been an exclusive representation of the type of material or its evolution over time. It is essential to observe the resilience capacity of an ecosystem before performing restoration after the contamination load from leachates, principally those *OAGD* located in areas with a low probability of stopping infiltration from the surface into various subsurface layers. The measurements are crucial for successfully mitigating the environmental impacts through targeted remediation efforts. Effective remediation requires a thorough understanding of contaminants in the dispersion and transformation stage patterns, including the hydrodynamics and hydrogeochemicals properties of subsoil and their interactions with ϕ. Although the federal and state governments have initiated efforts to protect and rehabilitate contaminated lands, many sites in the Mexican territory that require subsoil remediation, particularly in agricultural valleys like Guasave Valley—better known as *"The Agricultural Heart of Mexico"*—remain affected. These efforts often lack the necessary financial support to achieve the objectives outlined in the National Program for the Remediation of Contaminated Sites [14].

Preventing and controlling contamination and preserving aquifer quality requires treatment methodologies that effectively eliminate *DROC* and other chemical substances adhering to pore walls within the porous system, and overall expanding the scope of regulations outlined in the Official Mexican Standards and including additional factors of interest will help comply with international treaties and ensure that Mexican products meet permissible limits for each chemical compound. For example, water quality standards require that *DROC* concentrations do not exceed 25 mgL^{-1} [75]. Establishing essential strategies for proper waste management and site selection is crucial. This includes locating landfills in areas that do not risk contaminating aquifers with studies of hydrogeochemical, hydraulic, geophysical techniques, and others, collaborating to identify useful resources for policy negotiation promoting research to discover and inform other not common remediation techniques and new management practices.

Future research should emphasize extended monitoring, the refinement of the problem as a complex system for the prediction of models, and the incorporation of additional factors such as temperature, moisture content, subterranean ϕ direction, transmissivity, and hydraulic conductivity of the aquifer, among others. In summary, the phases *RI*, *RI/II*, and *II* were identified as having the highest potential for increased greenhouse gas emissions. The *DROC* transformation, indicated by ρa values from 2022 to 2023, suggests a reduction in total *DROC* within the *AI* zones due to enhanced degradation phases through the porous medium. The decline in ρa values points to significant changes in subsoil properties caused by the *AI* zone invasion, affecting leachate dispersion and retention.

6. Conclusions

Groundwater contamination by *DROCs* is a pressing issue, particularly under the intensifying effects of climate change. *DROCs* originate from leachates in open dumps and solid waste sites, and their persistence significantly threatens human health, local biodiversity, and essential water resources. Addressing these risks requires a critical shift from *OAGD* to engineered landfills, with a focus on recycling enhancement and sustainable waste management to mitigate adverse impacts on the subsoil, groundwater, and *GHG* emissions.

This study applies advanced geophysical methods to track and analyze *DROC* contamination. By measuring ρa variations in affected zones, it provides insights into *DROC* degradation stages and their dispersal patterns. Furthermore, correlating these ρa changes with *GHG* emissions allows researchers to pinpoint critical emission periods, creating opportunities for early intervention and mitigation. Such an approach is particularly relevant amid climate change, as each reduction in *GHG* emissions directly supports global climate goals.

Continuous monitoring is crucial to fully capture contamination extent and changes over time. Integrating science-based research with practical applications will advance efforts to rebuild ecosystem resilience and protect vital groundwater sources. This study emphasizes the necessity of transitioning to engineered waste disposal solutions, including landfills with systems for leachate collection and gas emission control. Such measures are especially urgent in regions with high subsoil permeability, where groundwater sustainability depends on preventing leachate spread from improperly managed disposal sites.

This research also highlights the importance of combining geophysical techniques with traditional investigative methods, such as drilling and laboratory analysis, to gain a comprehensive understanding of contamination dynamics. Notably, variations in ρa and contamination patterns across different phases and sites underscore the need for more sophisticated predictive models and extended monitoring. Future studies should aim to improve these models by incorporating environmental factors like temperature and moisture, which affect contaminant behavior and enhance assessment accuracy.

Mexico, in particular, requires stronger regulatory frameworks and improved waste management to tackle the risks posed by *DROCs* effectively. Increased collaboration among researchers, policymakers, and industries will be essential to devise and implement strategies that mitigate environmental contamination in open dumps. The findings of this study suggest that annual observations are insufficient for an accurate understanding of contaminant dispersal. Continuous monitoring, especially in areas surrounding high-permeability coastal dumpsites, is recommended to address and manage environmental risks effectively.

The transition to controlled, engineered waste management systems with safeguards like leachate capture and emission control is pivotal. These engineered approaches prevent leachate generation and reduce GHG emissions, offering a path forward for regions with high subsurface permeability and poor existing waste infrastructure. Implementing such systems on a national scale will be vital for Mexico to meet international environmental standards and protect its aquifers from contamination, especially in agricultural zones contributing significantly to the national economy.

The study concludes by underscoring that remediation of contaminated areas demands a comprehensive approach. Effective contamination management must account for the complex interactions of contaminants within porous media. This includes understanding hydrogeochemical properties, contaminant hydrodynamics, and potential subsurface spread. While federal and state agencies have initiated efforts to rehabilitate polluted lands, areas such as the agriculturally critical Guasave Valley, known as "The Agricultural Heart of Mexico," remain at risk due to insufficient resources and funding. Greater alignment with international waste management standards will be essential to ensure that contaminated sites are remediated effectively and in compliance with global safety thresholds.

Moreover, establishing efficient treatment protocols to remove *DROCs* from aquifers, as well as expanding the scope of environmental regulations to encompass emerging contaminants, will help Mexico meet international treaty requirements. Policies should include guidelines for safe landfill site selection and encourage collaboration in resource identification to support policy formation. Transitioning waste management to authorized and well-regulated enterprises will ensure sustainable practices align with current standards and ultimately foster a more resilient environmental management framework.

This study emphasizes the limitations of traditional environmental engineering perspectives and advocates for transformative changes in landfill management. Adopting comprehensive and cost-effective waste strategies suited to the degradation stages of *DROCs* is imperative. The presented methodology serves as a valuable tool for understanding pollutant transport in both saturated and unsaturated zones, aiding compliance with regulations, and establishing improved standards for waste remediation and management practices [75].

Author Contributions: Conceptualization, N.C.M.; writing and preparation of original draft, L.R.S.; English revision, discussions, and conclusion, M.S.P.; methodology, M.M.J.; data preparation, L.C.O.; writing—review and editing, M.d.l.Á.L.d.G.T. and N.C.M.; introduction and project administration, L.A.S.G.; formal analysis of literature, process, and analysis statistical/mathematical dataset. All authors have read and agreed to the published version of the manuscript.

Funding: The research was funded by the Secretaría de Investigación y Posgrado del IPN through Funding in the annual call for research projects individuals and multidisciplinaries 0220924, 20231294, and 20241198 projects.

Institutional Review Board Statement: Not applicable.

Informed Consent Statement: Not applicable.

Data Availability Statement: The dataset used to obtain the results in this study may be available upon request directly from the corresponding author's email. The data are not included in the document itself; however, the statistical methods and the software used for the analysis are thoroughly described in the manuscript.

Acknowledgments: The authors express their gratitude to the Instituto Politécnico Nacional (IPN) for supporting the Energy Network project through the CALL FOR 'EXTRAORDINARY SUPPORT TO CARRY OUT RESEARCH ACTIVITIES 2024, Category 5, aimed at fostering research, technological development, and innovation projects from the Coordination of Research and Postgraduate Networks of the IPN. We also extend our gratitude to the Secretaría de Investigación y Posgrado (SIP) of the IPN for the financial support provided through individual projects 20220924, 20231294, and 20241198. We deeply appreciate the encouragement received from MDPI's Sustainability journal throughout all stages of the research process, including fieldwork, office work, and resource and material management. Additionally, we sincerely thank the editor and anonymous reviewers for their invaluable feedback. Special appreciation is extended to the CIIDIR department's research initiatives and other affiliated research centers, whose critical comments and insights were instrumental in developing the arguments presented in this work. Finally, we acknowledge our colleagues from the Departments of Environmental Media, Natural Resources, Aquaculture, and Biotechnology at CIIDIR-IPN Sinaloa Unit for their invaluable support.

Conflicts of Interest: To mitigate potential conflicts of interest, several measures have been implemented to ensure transparent and comprehensive disclosure throughout all stages of the research. To prevent any possible conflicts of interest, arising from funding sources, all necessary permissions were obtained for the applied methods in the areas surrounding the *OAGD*. Complete transparency was ensured by disclosing the funding sources. A rigorous approach was considered for data collection and analysis. The methodology employed was based on tools that can be independently verified, ensuring that external influences do not affect the research outcomes. We actively sought collaboration with experts in the field to provide diverse perspectives not tied to any particular interests, focusing solely on objective assessments of the developments and analyses conducted.

References

1. Gupta, N.; Gupta, R. Solid Waste Management and Sustainable Cities in India: The Case of Chandigarh. *Environ. Urban.* **2015**, *27*, 573–588. [CrossRef]
2. Rana, R.; Ganguly, R.; Gupta, A.K. An Assessment of Solid Waste Management System in Chandigarh City, India. *Electron. J. Geotech. Eng.* **2015**, *20*. Available online: https://www.researchgate.net/publication/276275442_An_Assessment_of_Solid_Waste_Management_System_in_Chandigarh_City_India (accessed on 15 January 2024).
3. Ciudades y Transporte Sustentable Residuos y Economía Circular—Ciudades y Transporte Sustentable. Available online: https://ciudadesytransporte.mx/residuos-y-economia-circular/ (accessed on 7 November 2024).
4. Hoornweg, D.; Bhada-Tata, P. *What a Waste a Global Review of Solid Waste Management*; Urban development series; World Bank: Washington, DC, USA, 2012; Volume 15, Available online: https://www.researchgate.net/publication/306201760_What_a_waste_a_global_review_of_solid_waste_management (accessed on 20 May 2024).
5. Kaza, S.; Yao, L.C.; Bhada-Tata, P.; Van Woerden, F. *What a Waste 2.0: A Global Snapshot of Solid Waste Management to 2050*, 1st ed.; The World Bank: Washington, DC, USA, 2018; Volume 1, ISBN 978-1-4648-1347-4. Available online: https://wfp.tind.io/record/59279?ln=es (accessed on 10 July 2024).
6. Hemidat, S.; Achouri, O.; El Fels, L.; Elagroudy, S.; Hafidi, M.; Chaouki, B.; Ahmed, M.; Hodgkinson, I.; Guo, J. Solid Waste Management in the Context of a Circular Economy in the MENA Region. *Sustainability* **2022**, *14*, 480. [CrossRef]

7. Diario Oficial de la Federación. *Ley General Para La Prevención y Gestión Integral de Los Residuos*; Secretaría de Medio Ambiente y Recursos Naturales: Mexico City, México, 2023; pp. 10–37. Available online: https://www.diputados.gob.mx/LeyesBiblio/pdf/LGPGIR.pdf (accessed on 25 May 2023).
8. García-Mondragón, D.; Cervantes-Zepeda, I.; Gómez-Demetrio, W.; Gallego-Alarcón, I.; García-Pulido, D.; González-Blanco, G. Gestión de Los Residuos Sólidos En México: Análisis Cualitativo de Los Diagnósticos Básicos. *Inter Discip.* **2023**, *11*, 215–242. [CrossRef]
9. Hart, J.M. Nineteenth Century Urban Labor Precursors of the Mexican Revolution: The Development of an Ideology. *Americas* **2015**, *30*, 297–318. [CrossRef]
10. SEMARNAT Mexico's State of the Environmental Report. Environmental Statistical Compendium. In *Key and Environmental Performance Indicators; Edicion 2012*; SEMARNAT: Mexico City, Mexico, 2012.
11. Martínez, J.E.D.; Caballero, M.J.L. Densificación Urbana Como Solución Sostenible a Ciudades Difusas. *DELOS* **2022**, *12*, 2–12. Available online: https://ojs.revistadelos.com/ojs/index.php/delos/article/view/605 (accessed on 14 May 2023).
12. SEMARNAT Norma Oficial Mexicana NOM-083-SEMARNAT-2003. Especificaciones de Protección Ambiental Para La Selección Del Sitio, Diseño, Construcción, Operación, Monitoreo, Clausura y Obras Complementarias de Un Sitio de Disposición Final de Residuos Sólidos Urbanos y de Manejo Especial. CDMX. 2021, pp. 1–15. Available online: https://www.dof.gob.mx/nota_detalle.php?codigo=658648&fecha=20/10/2004&print=true (accessed on 14 May 2023).
13. BBVA Research Global | The COVID-19 Impact on Consumption in Real Time and High Definition. Available online: https://www.bbvaresearch.com/wp-content/uploads/2020/12/Weekly_Big_Data_17thDec.pdf (accessed on 12 December 2023).
14. SEMARNAT. *Guía Técnica Para Orientar En La Elaboración de Estudios de Caracterización de Sitios Contaminados*; SEMARNAT: Mexico City, México, 2010; pp. 1–9.
15. The World Bank. What a Waste: A Global Review of Solid Waste Management. Available online: https://hdl.handle.net/10986/17388 (accessed on 25 April 2024).
16. Ferronato, N.; Gorritty Portillo, M.A.; Guisbert Lizarazu, E.G.; Torretta, V.; Bezzi, M.; Ragazzi, M. The Municipal Solid Waste Management of La Paz (Bolivia): Challenges and Opportunities for a Sustainable Development. *Waste Manag. Res. J. A Sustain. Circ. Econ.* **2018**, *36*, 288–299. [CrossRef]
17. Ghafourian, K.; Mohamed, Z.; Ismail, S.; Malakute, R.; Abolghasemi, M. Current Status of the Research on Construction and Demolition Waste Management. *Indian. J. Sci. Technol.* **2016**, *9*. [CrossRef]
18. Josimovi, B.; Mari, I. Methodology for the Regional Landfill Site Selection. In *Sustainable Development—Authoritative and Leading-Edge Content for Environmental Management*; Intechopen: London, UK, 2012. [CrossRef]
19. Arjwech, R.; Somchat, K.; Pondthai, P.; Everett, M.; Schulmeister, M.; Saengchomphu, S. Assessment of Geological, Hydrogeological and Geotechnical Characteristics of a Proposed Waste Disposal Site: A Case Study in Khon Kaen, Thailand. *Geosciences* **2020**, *10*, 109. [CrossRef]
20. Castrejón, D.; Zavala, A.M.; Flores, J.A.; Flores, M.P.; Barrón, D. Analysis of the Contribution of CCS to Achieve the Objectives of Mexico to Reduce GHG Emissions. *Int. J. Greenh. Gas Control.* **2018**, *71*, 184–193. [CrossRef]
21. Mariano, N.-C.; Omar, L.-C.; Patricia, M.-S.; Rosario, R.-G.; de Guevara-Torres María, L.; José, A.-C. Aquifer/Water Sources Drip Factor as Alternative to Measure Climatic Effects on Disponibility of Water Agricultural Irrigation in Semi-Arid/Arid Zones. *IOP Conf. Ser. Earth Environ. Sci.* **2018**, *171*, 012031. [CrossRef]
22. Sellers, K. *Fundamentals of Hazardous Waste Site Remediation*; Routledge: London, UK, 2018. [CrossRef]
23. Sepúlveda, T.V.; Trejo, J.V. Tecnologías de Remediación Para Suelos Contaminados. In *Jiménez Editores*; Jiménez Editores: Mexico City, Mexico, 2002; Volume 65, pp. 1–64, ISBN 968-817-557-9. Available online: https://dama.umh.es/permalink/34CVA_UMH/1c61uvt/alma991000877399706331 (accessed on 25 March 2023).
24. Robledo, F.E.; Martinelli, H.P. Caracterización de Zonas Contaminadas Por Derrames Recientes de Hidrocarburos Mediante Métodos Geofísicos. *Scielo Geoacta* **2011**, *36*, 129–150. Available online: http://sedici.unlp.edu.ar/handle/10915/147552 (accessed on 12 September 2023).
25. Aizebeokhai, A.P. 2D and 3D Geoelectrical Resistivity Imaging: Theory and Field Design. *Sci. Res. Essays* **2010**, *5*, 3592–3605. Available online: https://www.researchgate.net/publication/228563980_2D_and_3D_geoelectrical_resistivity_imaging_Theory_and_field_design (accessed on 25 March 2023).
26. Mota, R.; Monteiro Santos, F.A.; Mateus, A.; Marques, F.O.; Gonçalves, M.A.; Figueiras, J.; Amaral, H. Granite Fracturing and Incipient Pollution beneath a Recent Landfill Facility as Detected by Geoelectrical Surveys. *J. Appl. Geophys* **2004**, *57*, 11–22. [CrossRef]
27. Pomposiello, C.; Dapeña, C.; Boujon, P.; Favetto, A. Tomografías Eléctricas En El Basurero Municipal Ciudad de Gualeguychú, Provincia de Entre Ríos: Evidencias de Contaminación. *Rev. De La Asoc. Geol. Argent.* **2009**, *64*, 1851–8249. Available online: https://www.researchgate.net/publication/262497476_Tomografias_electricas_en_el_basurero_municipal_Ciudadde_Gualeguychu_provincia_de_Entre_Rios_evidencias_de_contaminacion (accessed on 12 October 2023).
28. Belmonte, S.I.; Jiménez, M.E.; Bortolotti, A. Estudio de Una Pluma Contaminante Producida Por Lixiviados, Utilizando Métodos Geoeléctricos. *Geos* **2008**, *28*, 142. Available online: https://www.ugm.org.mx/publicaciones/geos/pdf/geos08-2/sesiones_regulares/GGA.pdf (accessed on 12 August 2023).

29. Suárez, J.; Zúñiga, H. Detection of Waste's Layers with Geoelectric Methods: Case of Relleno Sanitario Río Azul. *Rev. Geol. Am. Central* **2013**, 129–139. Available online: http://www.scielo.sa.cr/scielo.php?script=sci_arttext&pid=S0256-70242013000200011&lng=en&nrm=iso (accessed on 12 October 2023).
30. Khalil, M.H. Magnetic, Geo-Electric, and Groundwater and Soil Quality Analysis over a Landfill from a Lead Smelter, Cairo, Egypt. *J. Appl. Geophys* **2012**, *86*, 146–159. [CrossRef]
31. Pedretti, D.; Masetti, M.; Beretta, G.P.; Vitiello, M. A Revised Conceptual Model to Reproduce the Distribution of Chlorinated Solvents in the Rho Aquifer (Italy). *Groundw. Monit. Remediat.* **2013**, *33*, 69–77. [CrossRef]
32. DMP Guasave: Economía, Empleo, Equidad, Calidad de Vida, Educación, Salud y Seguridad Pública | Data México. Available online: https://www.economia.gob.mx/datamexico/es/profile/geo/guasave (accessed on 2 October 2024).
33. Ferrari, L.; López-Martínez, M.; Orozco-Esquivel, T.; Bryan, S.E.; Duque-Trujillo, J.; Lonsdale, P.; Solari, L. Late Oligocene to Middle Miocene Rifting and Synextensional Magmatism in the Southwestern Sierra Madre Occidental, Mexico: The Beginning of the Gulf of California Rift. *Geosphere* **2013**, *9*. [CrossRef]
34. Bryan, S.E.; Orozco-Esquivel, T.; Ferrari, L.; López-Martínez, M. Pulling Apart the Mid to Late Cenozoic Magmatic Record of the Gulf of California: Is There a Comondú Arc? *Geol. Soc. Lond. Spec. Publ.* **2014**, *385*, 389–407. [CrossRef]
35. Duque-Trujillo, J.; Ferrari, L.; Orozco-Esquivel, T.; López-Martínez, M.; Lonsdale, P.; Bryan, S.E.; Kluesner, J.; Piñero-Lajas, D.; Solari, L. Timing of Rifting in the Southern Gulf of California and Its Conjugate Margins: Insights from the Plutonic Record. *Geol. Soc. Am. Bull.* **2015**, *127*, 702–736. [CrossRef]
36. Cotler, H.; Corona, J.A.; Galeana-Pizaña, J.M. Erosión de Suelos y Carencia Alimentaria En México: Una Primera Aproximación. *Investig. Geográficas* **2020**, *103*, 59976. [CrossRef]
37. Lal, R. Soils and Sustainable Agriculture. A Review. *Agron. Sustain. Dev.* **2008**, *28*, 57–64. [CrossRef]
38. Herrero, M.; Thornton, P.K.; Power, B.; Bogard, J.R.; Remans, R.; Fritz, S.; Gerber, J.S.; Nelson, G.; See, L.; Waha, K.; et al. Farming and the Geography of Nutrient Production for Human Use: A Transdisciplinary Analysis. *Lancet Planet. Health* **2017**, *1*, e33–e42. [CrossRef] [PubMed]
39. Córdoba-Rodríguez, J.; Muñoz-Saavedra, P. *Groundwater Quality and Contamination Risks in the Sinaloa River Aquifer*; Environmental Monitoring and Assessment: Mexico City, Mexico, 2020; Volume 19.
40. Martínez-Hernández, M.; Ramírez, P.; Torres, L. Hydrogeological Properties and Vulnerability to Contamination in the Sinaloa River Aquifer. *J. Hydrol.* **2021**, *594*, 123–135.
41. Comisión Nacional del Agua (CONAGUA). Actualización de la Disponibilidad Media Anual de Agua En El Acuífero Río Sinaloa (2502), Estado de Sinaloa. Available online: https://sigagis.conagua.gob.mx/gas1/Edos_Acuiferos_18/sinaloa/DR_2502.pdf (accessed on 7 November 2024).
42. Llanes-Cárdenas, O.; Norzagaray-Campos, M.; Maya-Delgado, Y.; Muñoz-Sevilla, P.; Beltrán-Morales, F.; Murillo-Amador, B.; Troyo-Diéguez, E. Efectos Hidroambientales de La Extracción de Agua Del Acuífero Del Río Sinaloa. *Univ. Y Cienc.* **2011**, *27*, 239–249. Available online: http://www.scielo.org.mx/scielo.php?script=sci_arttext&pid=S0186-29792011000300001&lng=es&nrm=iso (accessed on 12 November 2024).
43. Rodríguez Miranda, I.A. *Estrategias Pasivas En El Diseño de la Envolvente PARA el Control de la Temperatura Interior En zonas de Clima Cálido Húmedo: Contribuciones a Los Sistemas de Fachada de Edificios Residenciales En Sinaloa*; Universitat Politècnica de Catalunya: Mexico City, Mexico, 2023. Available online: http://hdl.handle.net/2117/399102 (accessed on 12 November 2024).
44. Fernández, A.; Pérez-Campos, X. Lithosphere Thickness in the Gulf of California Region. *Tectonophysics* **2017**, *719–720*, 17–26. [CrossRef]
45. Michaud, F.; Royer, J.Y.; Bourgois, J.; Dyment, J.; Calmus, T.; Bandy, W.; Sosson, M.; Mortera-Gutiérrez, C.; Sichler, B.; Rebolledo-Viera, M.; et al. Oceanic-Ridge Subduction vs. Slab Break off: Plate Tectonic Evolution along the Baja California Sur Continental Margin since 15 Ma. *Geology* **2006**, *34*, 13. [CrossRef]
46. Comisión Nacional del Agua (CONAGUA). Disponibilidad de Agua y Fluctuaciones En El Nivel Freático En El Noroeste de México. Available online: https://www.gob.mx/conagua (accessed on 3 November 2024).
47. Sosa, R.F.S.S.C.; Constantino, T.R. *Sequía En México*; Alejandro, L.M., Nayeli, A.P., Eds.; Libros UAM Del Agua: Mexico City, Mexico, 2023; Volume 1, ISBN 978-607-28-2875-9. Available online: https://dcsh.xoc.uam.mx/repdig/index.php/libros-dcsh/dcsh/item/570-sequia-en-mexico (accessed on 3 November 2024).
48. Muñiz Monzón, D.G.; Figueroa Vázquez, V.S.; Tapia Frías, K.V.; Vallejo Ortiz, A.F.; Anda López, L.; Angel Hernández, A. Impacto Ambiental En Ganadería Bovina de Leche Mediante Filtración de Agua Contaminada En La Región Semiárida Del Norte de México. *Jóvenes En La Cienc.* **2022**, *14*, 1–7. [CrossRef]
49. Emery, X. Conditioning Simulations of Gaussian Random Fields by Ordinary Kriging. *Math. Geol.* **2007**, *39*, 607–623. [CrossRef]
50. Varnero, M.T. *Manual del Biogas*; Food and Agriculture Organization (FAO): Rome, Italy, 2011; Volume 1, ISBN 978-95-306892-0. Available online: https://www.fao.org/4/as400s/as400s.pdf (accessed on 15 October 2024).
51. Corriols, M.; Dahlin, T. Geophysical Characterization of the León-Chinandega Aquifer, Nicaragua. *Hydrogeol. J.* **2008**, *16*, 349–362. [CrossRef]
52. Áridos De La Mesa, V.; Caso, C.; Dulcita, L.; De Ramos, V.; Potosí, L.; Héctor López-Loera, M. Boletín de La Sociedad Geológica Mexicana Volumen 66. *Boletín De La Soc. Geológica Mex.* **2014**, *66*, 165–181. Available online: http://www.jstor.org/stable/24921255 (accessed on 20 May 2024).

53. Guillermo, B.; Salvatierra, Ñ.; Osman, B. Evaluación Del Subsuelo Mediante La Prospección Geofísica—Geoeléctrica En: El Ámbito de Estudio Del Cerco Perimétrico de La Facultad de Ingeniería de Minas Civil Ambiental–UNH. 2020. Available online: http://repositorio.unh.edu.pe/handle/UNH/3471 (accessed on 11 February 2023).
54. Zhou, G.L.; Hou, Z.T. Stochastic Generalized Porous Media Equations with Lévy Jump. *Acta Math. Sin. Engl. Ser.* **2011**, *27*, 1671–1696. [CrossRef]
55. Vázquez, J.L. The Mathematical Theories of Diffusion: Nonlinear and Fractional Diffusion. In *Nonlocal and Nonlinear Diffusions and Interactions: New Methods and Directions*; Springer: Berlin/Heidelberg, Germany, 2017; Volume 2186, pp. 205–278. [CrossRef]
56. Ignat, L.I.; Stan, D. Asymptotic Behavior of Solutions to Fractional Diffusion–Convection Equations. *J. Lond. Math. Soc.* **2018**, *97*, 258–281. [CrossRef]
57. Chen, L.; Holzinger, A.; Jüngel, A.; Zamponi, N. Analysis and mean-field derivation of a porous-medium equation with fractional diffusion. *Commun. Partial. Differ. Equ.* **2017**, *47*, 2217–2269. [CrossRef]
58. Salomoni, V.A.; De Marchi, N. A Fractional Approach to Fluid Flow and Solute Transport within Deformable Saturated Porous Media. *Int. J. Comput. Mater. Sci. Eng.* **2022**, *11*, 2250003. [CrossRef]
59. Benson, D.A.; Meerschaert, M.M. A Simple and Efficient Random Walk Solution of Multi-Rate Mobile/Immobile Mass Transport Equations. *Adv. Water Resour.* **2009**, *32*, 532–539. [CrossRef]
60. Wu, W.; Zhai, J. Stochastic Generalized Porous Media Equations Driven by Lévy Noise with Increasing Lipschitz Nonlinearities. *J. Evol. Equ.* **2021**, *21*, 4845–4871. [CrossRef]
61. Metcalfe, G.; Lester, D.; Trefry, M. A Primer on the Dynamical Systems Approach to Transport in Porous Media. *Transp. Porous Media* **2023**, *146*, 55–84. [CrossRef]
62. Del Teso, F.; Endal, J.; Jakobsen, E.R. Robust Numerical Methods for Nonlocal (and Local) Equations of Porous Medium Type. Part I: Theory. *SIAM J Numer Anal* **2019**, *57*, 2266–2299. [CrossRef]
63. Essaid, H.I.; Bekins, B.A.; Cozzarelli, I.M. Organic Contaminant Transport and Fate in the Subsurface: Evolution of Knowledge and Understanding. *Water Resour. Res.* **2015**, *51*, 4861–4902. [CrossRef]
64. Shevnin, V.; Delgado-Rodríguez, O.; Fernández-Linares, L.; Zegarra Martínez, H.; Mousatov, A.; Ryjov, A. Geoelectrical Characterization of an Oil-Contaminated Site in Tabasco, Mexico. *Geofis. Int.* **2005**, *44*, 251–263. [CrossRef]
65. Atekwana, E.A.; Atekwana, E.A. Geophysical Signatures of Microbial Activity at Hydrocarbon Contaminated Sites: A Review. *Surv. Geophys.* **2010**, *31*, 247–283. [CrossRef]
66. Bermejo Fraile, B. Análisis de Los Datos En Un Proyecto de Investigación. *Matronas Profesión* **2005**, *6*, 30–36.
67. StatSoft, Inc. *STATISTICA (Data Analysis Software System), Version 7.0*; StatSoft, Inc.: Tulsa, OK, USA, 2005; pp. 1–6. Available online: https://www.statsoft.de/en/home/ (accessed on 10 October 2023).
68. Hammer, Ø.; Harper, D.A.T.; Ryan, P.D. Past: Paleontological Statistics Software Package for Education and Data Analysis. *Palaeontol. Electron.* **2001**, *4*, 1–9. Available online: https://palaeo-electronica.org/2001_1/past/past.pdf (accessed on 22 April 2023).
69. Bi, Y.; Zou, H.; Zhu, C. Dynamic Monitoring of Soil Bulk Density and Infiltration Rate during Coal Mining in Sandy Land with Different Vegetation. *Int. J. Coal Sci. Technol.* **2014**, *1*, 198–206. [CrossRef]
70. SMN-conagua Resúmenes Mensuales de Temperaturas y Lluvia. Available online: https://smn.conagua.gob.mx/es/climatologia/temperaturas-y-lluvias/resumenes-mensuales-de-temperaturas-y-lluvias (accessed on 8 October 2024).
71. Campos, M.N.; Sevilla, P.M.; Montoya, J.M.; Cárdenas, O.L.; Torres, M.L.d.G.; García, L.A.S. Rainfall Potential and Consequences on Structural Soil Degradation of the Most Important Agricultural Region of Mexico. *Atmosphere* **2024**, *15*, 581. [CrossRef]
72. Cuevas-Reyes, V.; Meza, A.L.; Gutiérrez, O.G.G.; Bravo, M.B.; Rosales-Nieto, C.A. Tipología de Productor y Efectos Indirectos Del Cambio Climático En La Ganadería Bovina En Sinaloa. *Rev. Mex. Cienc. Pecu.* **2024**, *15*, 555–569. [CrossRef]
73. Reynolds, J. *An Introduction to Applied and Environmental Geophysics*; John Wiley & Sons: Hoboken, NJ, USA, 2011; Volume 2. Available online: https://www.wiley.com/en-us/An+Introduction+to+Applied+and+Environmental+Geophysics,+2nd+Edition-p-9780470975442 (accessed on 28 April 2023).
74. Khoshouei, M.; Bagherpour, R. Measurement, Prediction, and Modeling of the Drilling Specific Energy by Soft Rock Properties during the Drilling Operation. *Measurement* **2023**, *222*, 113679. [CrossRef]
75. SEMARNAT. NOM-001-SEMARNAT-2021. Que Establece Los Límites Permisibles de Contaminantes En Las Descargas de Aguas Residuales En Cuerpos Receptores Propiedad de La Nación. Secretaría de Medio Ambiente y Recursos Naturales (SEMARNAT): Mexico City, México, 2022; Volume 8. Available online: https://www.dof.gob.mx/nota_detalle.php?codigo=5645374&fecha=11/03/2022#gsc.tab=0 (accessed on 13 January 2023).

water

MDPI

Article

Natural Vanadium–Titanium Magnetite Activated Peroxydisulfate and Peroxymonosulfate for Acid Orange II Degradation: Different Activation Mechanisms and Influencing Factors

Zheng Zhang [1], Libin Zhao [2,3], Jingyuan Tian [1], Shaojie Ren [2,3,*] and Wei Zhang [2,3]

1 Henan Academy of Geology, No. 16 Jinshui East Road, Zhengzhou 450016, China
2 School of Ecology and Environment, Zhengzhou University, 100 Kexue Avenue, Zhengzhou 450001, China
3 Henan International Joint Laboratory of Water Cycle Simulation and Environmental Protection, Zhengzhou 450001, China
* Correspondence: rsj@zzu.edu.cn

Abstract: Persulfate-based advanced oxidation processes have emerged as a promising approach for the degradation of organic pollutants in aqueous environments due to their ability to generate sulfate radicals ($SO_4^{-\cdot}$) within catalytic systems. In this study, peroxydisulfate (PDS) and peroxymonosulfate (PMS) were investigated with the natural vanadium–titanium magnetite (VTM) as the activator for the degradation of acid orange II. The degradation efficiency increased with higher dosages of VTM or persulfate (both PDS and PMS) at lower concentrations (below 10 mM). However, excessive PMS (higher than 10 mM) in the PMS/VTM system led to the self-consumption of free radicals, significantly inhibiting the degradation of acid orange II. The VTM-activated PDS or PMS maintained an effective degradation of acid orange II in a wide pH range (3~11), suggesting remarkable pH stability. The $SO_4^{-\cdot}$ was the main active species in the PDS/VTM system, while hydroxyl radical (·OH) also contributed significantly to the PMS/VTM system. In addition, PMS exhibited better thermal stability during VTM activation. Coexisting ions in an aqueous environment such as bicarbonate (HCO_3^-), carbonate (CO_3^{2-}), and hydrogen phosphate (HPO_4^{2-}) had obvious effects on persulfate activation. Our study systematically investigated the different activation processes and influencing factors associated with PDS and PMS when the natural VTM was used as a catalyst, thereby providing new insights into the persulfate-mediated degradation of organic pollutants in aqueous environments.

Keywords: vanadium–titanium magnetite; persulfate; sulfate radical; advanced oxidation process; coexisting ions

1. Introduction

The advancement of industrialization has led to a significant increase in organic contaminants within aquatic ecosystems, posing a considerable risk to human health [1,2]. Conventional biological treatment methods, such as activated sludge systems and constructed wetlands, have proven inadequate for the effective removal of these contaminants [3]. Advanced oxidation processes (AOPs) employ highly reactive radicals to degrade the organic pollutants [4–6]. Commonly, AOPs, including the Fenton reaction [7], photocatalysis [8,9], and electrocatalysis [10,11], primarily rely on hydroxyl radicals (·OH) for their efficacy. Owing to the high reactivity and minimal self-scavenging properties [12], alternative AOPs utilizing sulfate radical ($SO_4^{-\cdot}$) have drawn significant attention in pollutants mitigation [13,14].

In sulfate radical-based AOPs, sulfate radicals exhibit a significantly higher redox potential (E0 = 2.5–3.1 V), which is more desirable than hydroxyl radicals (E0 = 1.8–2.7 V) toward electron-rich recalcitrant contaminants [15,16]. Persulfate, including peroxydisulfate

(PDS) and peroxymonosulfate (PMS), could serve as a precursor for the generation of sulfate radicals [17,18]. The peroxide bond could be cleaved by metal ions [19], thermo [20], and ultraviolet radiation [21]. Previous research has shown that heating is an effective method for activating PDS or PMS [22]. Transition metal ions and metal oxides have been shown to effectively activate PDS and PMS to degrade organic pollutants [23,24]. Compared with transition metals, the Fe element is noted for its environmental friendliness, non-toxicity, and cost-effectiveness, making it a common choice for persulfate activation [21,25]. Natural vanadium–titanium magnetite, which is doped with various metal elements such as iron, manganese, and titanium [26–28], not only possesses conventional adsorption properties but also functions as an activator of persulfate to generate sulfate radical [29,30]. Acid orange II, which contains azo functional groups and a naphthalene structure, is the largest class of synthetic dyes used in the textile industry. It has caused serious ecological and environmental problems due to the toxicity, non-biodegradability, and potential carcinogenicity [31]. Previous research indicated that the acid orange II could be degraded by AOPs effectively [32,33]. Thus, this is an effective strategy for the degradation of acid orange II through VTM-activated persulfate.

Gauging the true potential of persulfate oxidation technology for organic pollutant treatment in aqueous environments requires a careful evaluation of complex factors such as activation mechanisms and water matrix effects of persulfate. In persulfate activation systems, the unique reactivity of PMS/PDS and the involvement of dominant radicals are significantly different under various aqueous environments [34]. Previous research has indicated that when copper oxide is used as an activator, the dominant reactive species in the PDS system depends on pH conditions; specifically, sulfate radicals predominate in acidic conditions, while hydroxyl radicals are favored in alkaline conditions [35]. Furthermore, elevated temperatures may promote the occurrence of side reactions such as sulfate radical or hydroxyl radical recombination, reducing the removal efficiency of organic pollutants [36]. Therefore, a systematic investigation into the activation mechanisms of various persulfate systems is necessary to explore the potential effects of complex aqueous condition factors.

In this study, natural VTM was employed to activate different persulfates (PDS and PMS). A comprehensive analysis of the microstructure and elemental composition of VTM was conducted to elucidate the potential activation mechanisms for persulfate. The distinct activation mechanisms of PMS and PDS by VTM were explored through the comparative analysis of the degradation characteristics of organic dye acid orange II. The primary degradation mechanisms of acid orange II were assessed under multiple aqueous environmental conditions. Meanwhile, the influence of coexisting ions and free radicals was explored to analyze the potential differences between PDS and PMS activation. Our study provided a valuable exploration of the persulfate activation and highlighted the potential application of persulfate-based advanced oxidation to degrade organic pollutants in complex aqueous environments.

2. Materials and Methods

2.1. Materials

Acid orange II was purchased from Yousuo Chemical Reagent Co., Ltd., Linyi, China. Sodium persulfate ($Na_2S_2O_8$) was obtained from Kemiou Chemical Reagent Co., Ltd., Tianjin, China. Potassium peroxymonosulfate ($KHSO_5 \cdot 0.5KHSO_4 \cdot 0.5K_2SO_4$) was supplied by Aladdin Biochemical Technology Co., Ltd., Shanghai, China.

2.2. Characterizations of VTM

The samples of natural VTM were collected from the Pan-Xi region, in Sichuan Province, China. The samples were ground and passed through a 100-mesh screen. The morphology and surface structure of VTM were observed by field emission scanning electron microscopy (SEM, Zeiss, Germany). The distribution of main elements (Fe and Ti) was measured by an energy dispersive X-ray spectrometer (EDS, Zeiss, Germany) with the atomic-resolution chemical mapping method. The crystalline structure and chemical components of VTM were investigated by X-ray diffraction technique (XRD, Empyrean, PANalytical, NLD, Worcestershire, UK). The chemical compositions of VTM were also analyzed by X-ray photoelectron spectroscopy (XPS, Escalab 250Xi, Thermo Scientific, Waltham, MA, USA). The porous properties, surface area, and pore size distribution of VTM were analyzed by Brunauer–Emmett–Teller (BET, ASAP3020, Micromeritics, Norcross, GA, USA).

2.3. Acid Orange II Decolorization Analysis

The acid orange II (0.1 mM and 200 mL) decolorization test was conducted at a temperature of 23 ± 2 °C with 300 r/min shaking. The single variable method was used to explore the effects of different oxidants and environmental factors on the degradation of acid orange II. First, the concentration of oxidants (PDS or PMS) was 10 mM with the various VTM concentrations (0–1.6 g). Acid orange II was detected based on the previous study; after filtration with a 0.45 μm membrane, the concentration of acid orange II was tested at 484 nm by using a spectrophotometer (UV759CRT, Shanghai Youke Instrument CO., LTD, Shanghai, China) [37]. The degradation efficiency of acid orange II can be calculated by Equation (1).

$$Decolorization\ efficiency\ (\%) = \frac{C_0 - C}{C_0} \times 100\% \quad (1)$$

where C_0 is the initial concentration of the acidic orange II solution, and C is the concentration of the acidic orange II solution in the collected sample.

Subsequently, the decolorization of acid orange II was determined at different oxidant concentrations. The dosages of PDS or PMS were 0, 1, 2, 5, 10, 20, 30, and 40 mM, respectively. In order to explore the effect of different pH on the degradation of acid orange II, experimental conditions (VTM 0.4 g, PDS 10 mM, and PMS 10 mM) were used. NaOH and H_2SO_4 solutions were used to adjust the pH of the reaction system.

2.4. Exploration of Degradation Mechanism

2.4.1. Free Radical Quenching Experiment

Anhydrous ethanol (EtOH) and tert-butyl alcohol (TBA) were used as free radical quenchers to explore the activation mechanism of PDS/PMS by VTM at different conditions. Different concentrations of EtOH (0 to 3.0 mmol/L) and TBA (0 to 2.0 mmol/L) were added to the oxidative degradation system. Then, the degradation rate of acid orange II was measured.

2.4.2. Coexisting Ion Effects on the Activation of PDS/PMS by VTM

For investigating the effect of coexisting ions in natural water on the degradation of acid orange II by VTM-activated PDS/PMS, several ions commonly found in natural water, including Cl^-, NH_4^+, HCO_3^-, CO_3^{2-}, Mg^{2+}, Al^{3+}, and HPO_4^{2-}, were selected to explore their interference with the catalytic degradation system. Acid orange II degradation experiments with different concentration ions (0, 1, 10, and 100 mM) were conducted under the specified reaction conditions of VTM (0.4 g), PDS or PMS (2 mM). The concentration of acid orange II in the collected samples was detected to analyze the change in activation efficiency of PDS or PMS by VTM.

3. Results and Discussion

3.1. Characterization of VTM

3.1.1. Microstructure and Chemical Analysis

The morphology and chemical composition of the VTM were analyzed by the SEM and EDS analyses (Figure 1). The VTM exhibited a smooth surface, potentially causing poor adsorption capacity. XRD diffraction measurements were conducted to examine the nature of the crystallinity of the VTM (Figure 2a). The primary component of the VTM was identified as magnetite according to the standard JCPDS 019-0629 (2θ = 24.138°, 33.153°,

35.612°, 39.277°, 43.519°, 49.480°, 57.429°, and 62.451°), which can be attributed to the natural magnetite with spinel structure. In addition, due to the low concentrations of vanadium (V) and titanium (Ti), these elements were not detectable in the XRD analysis.

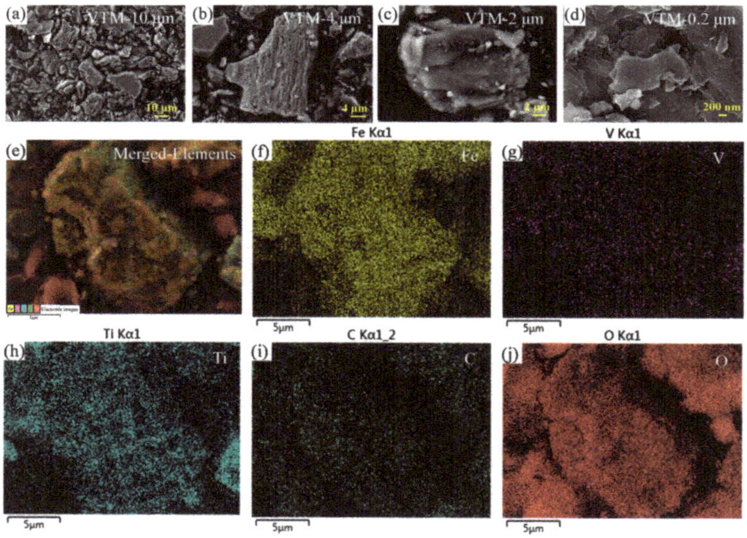

Figure 1. SEM and EDS mapping images of the VTM. (**a–d**) Surface morphology of VTM at different resolutions. EDS results of different elements on the VTM surface: (**e**) merged image, (**f**) Fe, (**g**) V, (**h**) Ti, (**i**) C, and (**j**) O elements.

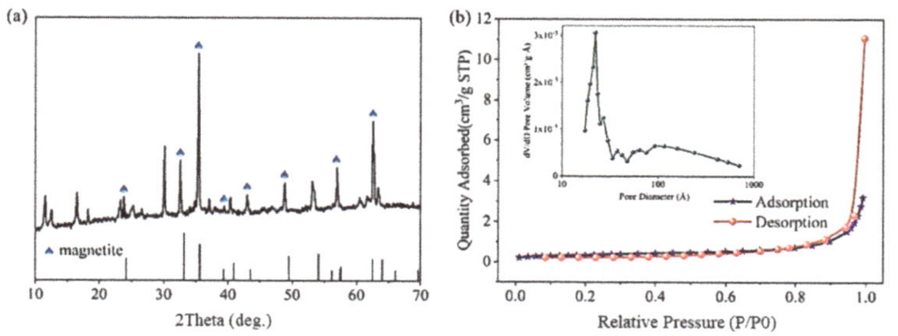

Figure 2. (**a**) XRD patterns **of** the VTM. (**b**) Nitrogen adsorption−desorption isotherms; inset image was the pore size distribution of the VTM.

The specific surface area and pore size of the VTM were measured through nitrogen adsorption–desorption isotherms. As illustrated in Figure 2b, the isotherm did not display a pronounced upward convexity at low relative pressures. The presence of a hysteresis loop at elevated relative pressures suggested that the catalyst possessed irregular mesopores. The analysis revealed a narrow pore size of 30.6 angstroms and a broader aperture range of 17.4 to 695.4 angstroms, which can be attributed to the irregular size and shape of the pores within the aggregated VTM particles [38,39]. In addition, the porosity of VTM was 4.33 g/cm^3, with a bulk density of 1.29 g/cm^3, resulting in a porosity of 70.11%.

3.1.2. XPS Analysis of VTM

The elemental valence bond composition of VTM was investigated by the XPS analysis (Figure 3). XPS is a non-destructive surface analysis technique with a detection depth of around 5–10 nm, revealing the relative content of elements in the samples. The binding energy was calibrated with a C1s peak of 284.8 eV [40], followed by peak fitting for analyzing the functional groups on the natural VTM surface. The peaks at 709.02 eV and 722.12 eV can be attributed to the presence of Fe^{2+}. The peaks at 710.21 eV and 723.31 eV were observed at VTM indicating the presence of Fe^{3+}. In addition, the typical Fe 2p3/2 and Fe 2p1/2 peaks located at 707.71 and 720.10 eV may correspond to Fe-S [41,42], since the S element was detected in the original natural-VTM before the reaction (Table S1). The VTM was identified as the ferrite compound, with a decrease in the Fe^{2+} content and an increase in the Fe^{3+} content observed following the degradation process of acid orange II. The XPS peak spectra of VTM before and after the reaction were slightly shifted compared to the standard peaks, which may be related to the complexity of the natural VTM. From the full-range scan of XPS, the decrease in the elements was observed after the reaction, such as Mg (from 11.48 atom% to 5.95 atom%) and Al (from 9.64 atom% to 7.73 atom%). Meanwhile, the ratios of the main metal elements of VTM were significant, such as Fe (from 1.00 atom% to 6.99 atom%) and Ti (from 0.13 atom% to 2.36 atom%). The changes in the composition of VTM due to the release of surface elements are also clearly reflected in the XPS results. We further made inferences about the O1s on the VTM surface based on the main elemental composition. The results indicate that the peaks at 529.58 eV could be attributed to the Ti-O, Fe-O bond [43] or lattice oxygen peak [44], the peak at nearly 531 $\pm$0.1 eV could be regarded as an Al-O bond, and the peak at 531.8 eV could be assumed as a Si-O bond due to a slight peak shift. The peaks at 531.58 eV corresponded to the C-O bond within the transition metal layer. In the C1s spectrum, the peaks at 285.4 and 288.84 eV were attributed to C–O and C=O, respectively [45]. The relative amount of C=O bond was decreased owing to the oxidation effect of reactive generated by VTM-activated PDS/PMS after the reaction. The significant changes in Fe and O elements also indicated that both of them are involved in the removal of acid orange II by the VTM-activated persulfate reaction.

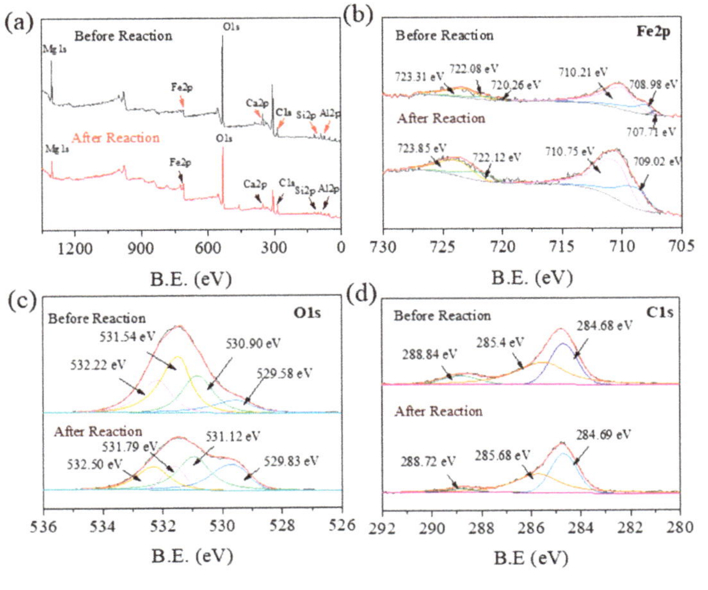

Figure 3. XPS spectra of the VTM before and after acid orange II degradation tests: (**a**) full-range scan, (**b**) Fe 2p spectra, (**c**) O 1s spectra, and (**d**) C 1s spectra.

3.2. Degradation of Acid Orange II Under Different Conditions

3.2.1. Effects of VTM Dosage on Acid Orange II Degradation

The concentration of VTM plays an important role in the activation of PDS and PMS. As shown in Figure 4a,b, there were no significant alterations in the concentration of acid orange II when PDS or PMS was utilized independently, without the presence of VTM. The degradation efficiency of acid orange II in the PDS/VTM system reached 45.5% after 60 min with 0.1 g/L VTM added. However, when the VTM dosage was 0.2 g/L, the degradation efficiency of acid orange II increased to 75.4%. Further increasing the VTM concentration to 0.5 g/L resulted in nearly complete degradation of acid orange II after 60 min. Similarly, in the PMS/VTM system, the degradation efficiency of acid orange II also increased with the VTM concentration. However, its degradation efficiency was much lower than that observed in the PDS/VTM system. When the VTM concentration was 0.1 g/L, the degradation efficiency of acid orange II was only 11%. When the VTM dosage was 0.5 g/L, there was only a 45% degradation of acid orange II. Fe(II) is the primary agent responsible for activating PDS and PMS to generate $SO_4^{-\cdot}$. As the source of Fe(II), the amount of activator VTM can significantly affect the degradation efficiency. Elevated VTM concentrations can enhance the activation of PDS and PMS, thereby generating a greater number of reactive species (Reactions (2) and (3)) [46]. However, in the PDS/VTM system, the decreased degradation efficiency of acid orange II was observed when the VTM dosage increased from 2.0 g/L to 8.0 g/L. The addition of excessive VTM led to the unfavorable recombination of active species $SO_4^{-\cdot}$, as reaction (4) inhibited the degradation of acid orange II. In addition, the $SO_4^{-\cdot}$ may react with extra Fe(II) (Reaction (5)) [47], causing the additional consumption of active substances.

$$S_2O_8{}^{2-} + Fe(II) \rightarrow Fe(III) + SO_4{}^{-\cdot} + SO_4{}^{2-} \tag{2}$$

$$HSO_5{}^{-} + Fe(II) \rightarrow Fe(III) + SO_4{}^{-\cdot} + OH^{-} \tag{3}$$

$$SO_4{}^{-\cdot} + SO_4{}^{-\cdot} \rightarrow 2SO_4{}^{2-} \tag{4}$$

$$SO_4{}^{-\cdot} + Fe(II) \rightarrow SO_4{}^{2-} + Fe(III) \tag{5}$$

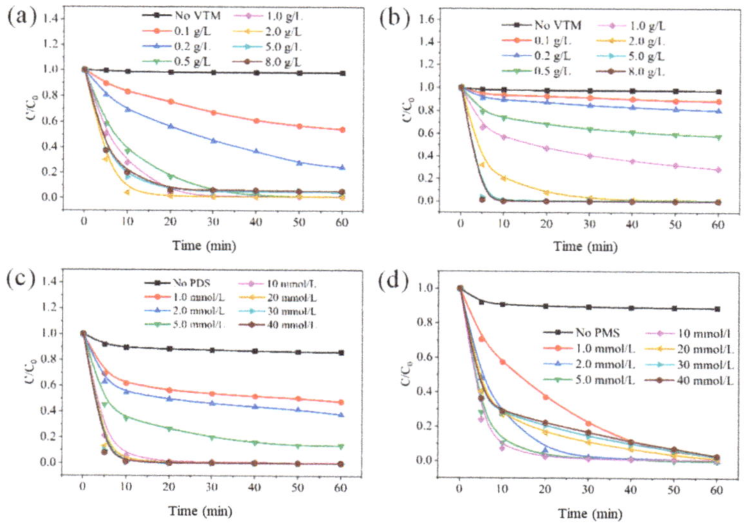

Figure 4. Effect of VTM dosage on the acid orange II degradation in (**a**) VTM/PDS system and (**b**) VTM/PMS system. Effect of (**c**) PDS and (**d**) PMS concentrations on the acid orange II degradation efficiency.

However, owing to the lower activation efficiency of VTM on PMS, no corresponding decrease in the degradation rate of acid orange II was observed in the PMS/VTM system with excessive VTM usage. Considering both the efficiency and cost implications of the activator, a VTM dosage of 2.0 g/L was selected for subsequent experiments.

3.2.2. Effect of PDS or PMS Dosage

As shown in Figure 4c,d, minimal degradation of acid orange II occurred when only VTM was introduced into the reaction system. This rapid removal of acid orange II within ten minutes may be due to the adsorption effect of VTM. The degradation efficiency of acid orange II exhibited an increase in response to higher concentrations of PDS or PMS. When the concentration of PDS increased from 0 to 1.0 mM, the degradation efficiency of acid orange II increased significantly from 13.8% to 52.1%. Further increases in PDS concentration to 10 mM or more resulted in the complete degradation of acid orange II within 30 min (Figure 4c). Compared to PDS, when the concentration of PMS was 1.0 mM, the degraded efficiency reached 97.0% after 60 min, indicating that VTM provided a higher activation efficiency for PMS (Figure 4d). However, when the PMS concentration exceeded 10 mM, a decrease in the degradation efficiency of acid orange II was observed, attributed to the adverse consumption of active species (Reaction (6)) [48]. Based on these findings, subsequent experiments exploring degradation mechanisms and the influence of environmental factors utilized 2.0 mM concentrations of either PDS or PMS.

$$SO_4^{\cdot-} + HSO_5^- \rightarrow SO_5^{\cdot-} + SO_4^{2-} + H^+ \tag{6}$$

3.2.3. Effect of pH, Temperature, and Coexisting Ions on the Reaction System

The activation of PDS/PMS by VTM was influenced by the pH of the reaction system. As illustrated in Figure 5, within the VTM/PDS reaction system, the degradation profiles of acid orange II were nearly identical across a range of initial pH values from 3 to 11, suggesting that the pH had minimal impact on the activation efficiency of VTM for PDS. The $SO_4^{-\cdot}$ generated in the reaction process can react with the H_2O or OH^-, causing the formation of H^+ and decreasing the pH of the reaction system (Reactions (7) to (10)). Finally, despite variations in initial pH, the pH of the reaction system converged to a similar value within the range of pH 3 to 11 (inset image, Figure 5a). It is worth noting that in the VTM/PMS system, when the initial pH of the solution was 1.0, the degradation efficiency of the acid orange II was lower (~82%) than that at other pH levels (Figure 5b). This indicates that the pH of the reaction system has a negligible effect on VTM-activated PDS/PMS, suggesting that the activation process is applicable across a broad pH range (11).

$$SO_4^{-\cdot} + H_2O \rightarrow H^+ + SO_4^{2-} + {}^{\cdot}OH \tag{7}$$

$$SO_4^{-\cdot} + OH^- \rightarrow SO_4^{2-} + {}^{\cdot}OH \tag{8}$$

$$SO_4^{-\cdot} + {}^{\cdot}OH \rightarrow HSO_4^- + 1/2O_2 \tag{9}$$

$$S_2O_8^{2-} + H_2O \rightarrow 2HSO_4^- + 1/2O_2 \tag{10}$$

$$HSO_4^- \rightarrow SO_4^{2-} + H^+ \tag{11}$$

Temperature is an important parameter influencing PDS/PMS activation. As shown in Figure 5c,d, different temperatures (25 to 55 °C) had little effect on the degradation efficiency of acid orange II in the VTM/PDS reaction system which may be related to the faster active reaction in this system. In addition, when there was no PDS/PMS present, acid orange II did not degrade at 55 °C for 60 min, suggesting that the acid orange II was its relative stability and lack of spontaneous decomposition under elevated temperature conditions. In the PDS system, acid orange II exhibited an obvious degradation process (83.7%) even without VTM, indicating that this temperature was sufficient to activate PDS and generate active radicals. Conversely, in the PMS reaction system acid orange II was not substantially degraded (15%) at the same temperature (55 °C) without VTM. PDS and

PMS had different molecular structures. The dissociation energy of the O-O in PMS is 377 kJ/mol, while 92 kJ/mol in PDS [49,50]. Thus, PMS is more difficult to be thermally activated compared to PDS.

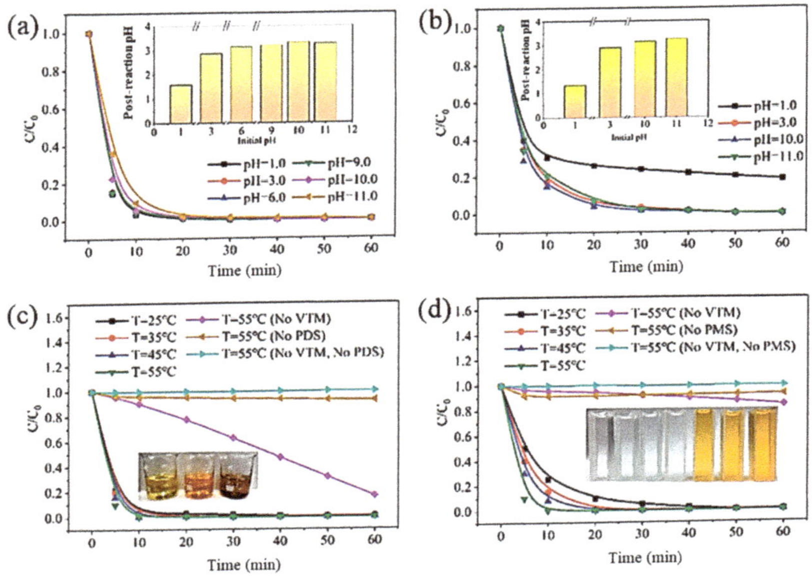

Figure 5. Effect of different initial pH on the degradation efficiency of acid orange II in the (**a**) VTM/PDS system and (**b**) VTM/PMS system. The inset image is the final pH after a 60 min reaction. Effect of different temperatures on the degradation efficiency of acid orange II in the (**c**) VTM/PDS system and (**d**) VTM/PMS system. The inset image is the color change in the acid orange II after a 60 min reaction.

Furthermore, the presence of coexisting ions in the aqueous solution may affect the activation efficiency of PDS/PMS with VTM [51]. A comprehensive investigation was conducted on various ions, including Cl^-, NH_4^+, HCO_3^-, CO_3^{2-}, Mg^{2+}, Al^{3+}, and HPO_4^{2-}. It was observed that as the concentration of these ions increased, their inhibitory effect on the degradation of acid orange II also intensified. In the PDS/VTM system, it was found that the HCO_3^-, CO_3^{2-}, and HPO_4^{2-} exhibited a significant influence on the acid orange II degradation (Figure S1). These ions could consume the active free radicals generated in the system, thereby reducing the occurrence of oxidation. Moreover, when the CO_3^{2-} and HPO_4^{2-} concentrations were higher than 10 mM, the degradation rate of acid orange presented a slight increase, which can be attributed to the change in pH of the reaction system during a large amount of ions addition [52]. For the PMS/VTM system, it was found that HCO_3^- and HPO_4^{2-} presented a significant influence on the acid orange II degradation (Figure S2). Consistent with the findings in the PDS/VTM system, the degradation efficiency of acid orange II decreased with rising ionic concentrations (NH_4^+, Mg^{2+}, and Al^{3+}). However, the high concentration (100 mM) of HCO_3^{2-} and HPO_4^{2-} increased the reaction rate compared to a lower ionic strength of 10 mM.

3.3. Analysis of Degradation Mechanisms by the PDS/PMS with VTM

The primary mechanism underlying the degradation of acid orange II involves the activation of PDS/PMS by VTM, which leads to the release of oxidative free radicals. Here, EtOH and TBA were applied to explore the influence of active free radicals in the reaction system (Figure 6). The reaction rates of EtOH and TBA with $SO_4^{-\cdot}$ and

$\cdot OH$ were markedly different. Previous research indicated that the rate constant of the reaction between EtOH and $\cdot OH$ is 1.2~2.8 × 10^9 $M^{-1}s^{-1}$, and the rate constant of reaction with SO_4^- is 1.6~7.7 × 10^7 $M^{-1}s^{-1}$ [53], while the rate constant of the reaction between TBA and $\cdot OH$ is 3.8~7.6 × 10^8 $M^{-1}s^{-1}$, and the rate constant of reaction with SO_4^- is 4~9.1 × 10^5 $M^{-1}s^{-1}$ [54]. Thus, EtOH is served to quench both $SO_4^{-\cdot}$ and $\cdot OH$, while TBA is specifically employed to quench $\cdot OH$.

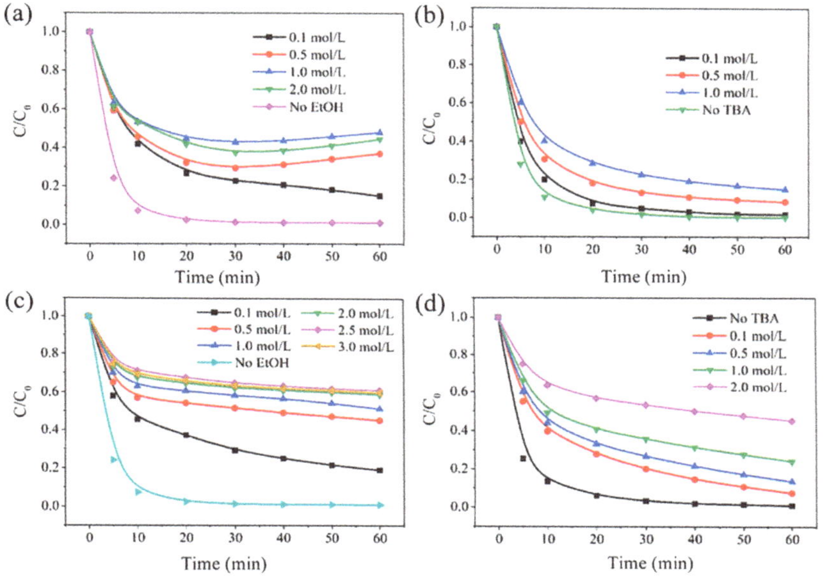

Figure 6. (**a**) Effect of EtOH on acid orange II degradation in the VTM/PDS reaction system. (**b**) Effect of TBA on acid orange II degradation in the VTM/PDS reaction system. (**c**) The influence of EtOH on acid orange II degradation in the VTM/PMS reaction system. (**d**) The influence of TBA on acid orange II degradation in the VTM/PMS reaction system.

In the VTM/PDS system, acid orange II was completely degraded within 60 min without the addition of EtOH or TBA. The introduction of EtOH reduced the degradation efficiency, dropping to 85% with a concentration of 0.1 mM (Figure 6a). TBA exhibited a minimal impact on the degradation of acid orange in the VTM/PDS system, further indicating that $SO_4^{-\cdot}$ was the main active species in the system (Figure 6b). In addition, both EtOH and TBA had similar inhibition effects in the VTM/PMS system, with the degradation rate decreasing to 81% upon the addition of 0.1 mol/L EtOH and to 89% with 0.1 mol/L TBA (Figure 6c). Meanwhile, the degradation rate decreased to 89% when 0.1 mol/L TBA was added (Figure 6d). This suggests that $\cdot OH$ plays a critical role in the VTM/PMS reaction system.

To further analyze the degradation of acid orange II in the various reaction systems, the UV–visible spectrum was used during the degradation process (Figure 7). The absorption spectrum of acid orange in aqueous solution mainly consists of four bands: hydrazone at 484 nm, azo at 430 nm, benzene ring at 230 nm, and naphthalene ring at 310 nm [55]. As the reaction progressed, the disappearance of the visible band absorption peak confirmed the destruction of the azo bond. In the VTM/PMS system, a slight increase in absorption peaks at 310 nm and 228 nm was observed, indicating the degradation of the azo structure and the formation of naphthalene and benzene rings.

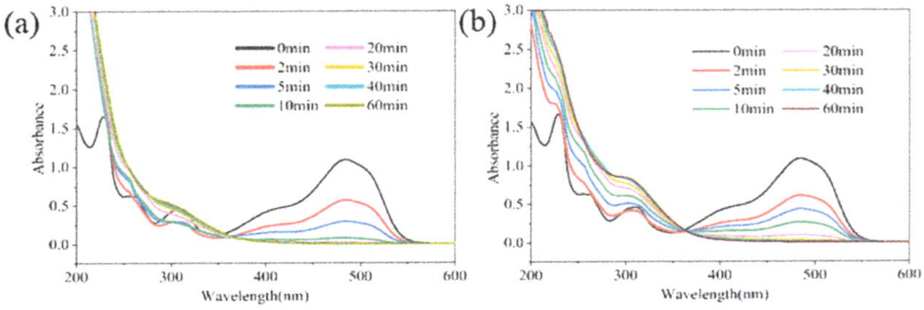

Figure 7. UV-vis analysis of acid orange II in the VTM/PDS reaction system (**a**) and VTM/PMS reaction system (**b**).

4. Conclusions

This study comprehensively investigates differences in the activation performance of VTM on PDS and PMS by determining the degradation of acid orange II. Various persulfates exhibited distinct activation mechanisms for the removal of organic contaminations. Initially, the structure and chemical properties of natural VTM were analyzed, indicating the mesoporous characteristics and multi-metal composition that can facilitate the activation of persulfate catalysts. Subsequently, the degradation of acid orange II was determined, revealing that the degradation efficiency of acid orange II increased with the VTM dosage. When the VTM dosage was 0.2 g/L, the degradation efficiency of acid orange II increased to 75.4%. It is worth noting that excessive VTM inhibited the degradation of acid orange II in persulfate systems, especially in the PDS/VTM system. Extraneous oxidants also affect the degradation performance, such as excess PMS (10 mM), which leads to the self-consumption of free radicals and reduces the reaction efficiency of the system. Temperature and pH were also identified as significant factors affecting persulfate activation. It was found that heating effectively activated PDS but had no impact on PMS at 25 to 55 °C, further proving the greater thermal stability of PMS compared to PDS. The free radical quenching experiments revealed that $SO_4^{-\cdot}$ was the primary active species in the PDS/VTM system, while ·OH predominated in the PMS/VTM system. The 0.1 mM EtOH reduced 85% of the degradation efficiency, further indicating that $SO_4^{-\cdot}$ was the main active species in the system. Moreover, the azo bond of acid orange II was the main target of the active substance in the reaction system. The environmental ions, including HCO_3^-, CO_3^{2-}, and HPO_4^{2-}, significantly influenced the degradation of acid orange by changing the pH of the reaction system and consuming the active radicals. This research enhances the understanding of the distinct mechanisms and influencing factors associated with the activation of PDS and PMS by VTM, promoting the optimal application of persulfate in aqueous environments.

Supplementary Materials: The following supporting information can be downloaded at: https://www.mdpi.com/article/10.3390/w16213109/s1, Table S1: The relative contents of main elements in the natural VTM before and after reaction. Figure S1: Effects of different ions on the degradation of acid orange II by VTM-activated PDS; Figure S2: Effects of different ions on the degradation of acid orange II by VTM-activated PMS.

Author Contributions: Z.Z.: Investigation, Validation, Formal analysis, Writing—original draft, Writing—review and editing. L.Z.: Writing—review and editing. J.T.: Writing—review and editing. S.R.: Conceptualization, Writing—review and editing. W.Z.: Writing—review and editing, Supervision, Funding acquisition. All authors have read and agreed to the published version of the manuscript.

Funding: The authors acknowledge the financial support from the following sources: the Henan Province International Science and Technology Cooperation Project (242102521018); the Training Plan of Young Backbone Teachers in Colleges and Universities of Henan Province under Grant (2021GGJS005); the Chunhui plan of the Ministry of Education of China cooperation project (HZKY20220278); the

Research Fund of Key Laboratory of Water Management and Water Security for Yellow River Basin, Ministry of Water Resources (under construction, 2023-SYSJJ-01); the Youth Talent Lifting Project of Henan Province (2023HYTP035); the Open Research Fund of Henan Key Laboratory of Water Resources Conservation and Intensive Utilization in the Yellow River Basin (grant NO. HAKF202101); the Open Project Funding of Henan Key Laboratory of Ecological Environment Protection and Restoration of Yellow River Basin (LYBEPR202301); the Open Project Funding of Key Laboratory of Intelligent Health Perception and Ecological Restoration of Rivers and Lakes, Ministry of Education, Hubei University of Technology (HGKFYB15); and the Open fund from Henan Key Laboratory of Water Pollution Control and Rehabilitation Technology (CJSP2022001).

Data Availability Statement: The original data are available by request from the authors.

Conflicts of Interest: The authors declare that they have no known competing financial interests or personal relationships that could have appeared to influence the work reported in this paper.

References

1. Kanzari, F.; Syakti, A.D.; Asia, L.; Malleret, L.; Piram, A.; Mille, G.; Doumenq, P. Distributions and sources of persistent organic pollutants (aliphatic hydrocarbons, PAHs, PCBs and pesticides) in surface sediments of an industrialized urban river (Huveaune), France. *Sci. Total Environ.* **2014**, *478*, 141–151. [CrossRef]
2. Dsikowitzky, L.; Schwarzbauer, J. Industrial organic contaminants: Identification, toxicity and fate in the environment. *Environ. Chem. Lett.* **2014**, *12*, 371–386. [CrossRef]
3. Garfí, M.; Flores, L.; Ferrer, I. Life Cycle Assessment of wastewater treatment systems for small communities: Activated sludge, constructed wetlands and high rate algal ponds. *J. Clean. Prod.* **2017**, *161*, 211–219. [CrossRef]
4. Miklos, D.B.; Remy, C.; Jekel, M.; Linden, K.G.; Drewes, J.E.; Hübner, U. Evaluation of advanced oxidation processes for water and wastewater treatment—A critical review. *Water Res.* **2018**, *139*, 118–131. [CrossRef]
5. Gahrouei, A.E.; Vakili, S.; Zandifar, A.; Pourebrahimi, S. From wastewater to clean water: Recent advances on the removal of metronidazole, ciprofloxacin, and sulfamethoxazole antibiotics from water through adsorption and advanced oxidation processes (AOPs). *Environ. Res.* **2024**, *252*, 119029. [CrossRef]
6. Shang, Y.; Xu, X.; Gao, B.; Wang, S.; Duan, X. Single-atom catalysis in advanced oxidation processes for environmental remediation. *Chem. Soc. Rev.* **2021**, *50*, 5281–5322. [CrossRef]
7. Aveiro, L.R.; Da Silva, A.G.M.; Candido, E.G.; Antonin, V.S.; Parreira, L.S.; Papai, R.; Gaubeur, I.; Silva, F.L.; Lanza, M.R.V.; Camargo, P.H.C.; et al. Application and stability of cathodes with manganese dioxide nanoflowers supported on Vulcan by Fenton systems for the degradation of RB5 azo dye. *Chemosphere* **2018**, *208*, 131–138. [CrossRef]
8. Nosaka, Y.; Nosaka, A.Y. Generation and Detection of Reactive Oxygen Species in Photocatalysis. *Chem. Rev.* **2017**, *117*, 11302–11336. [CrossRef] [PubMed]
9. Moradi, A.; Kazemeini, M.; Hosseinpour, V.; Pourebrahimi, S. Efficient degradation of naproxen in wastewater using Ag-deposited ZnO nanoparticles anchored on a house-of-cards-like MFI-type zeolite: Preparation and physicochemical evaluations of the photocatalyst. *J. Water Process Eng.* **2024**, *60*, 105155. [CrossRef]
10. Aran-Ais, R.M.; Yu, Y.; Hovden, R.; Solla-Gullon, J.; Herrero, E.; Feliu, J.M.; Abruna, H.D. Identical Location Transmission Electron Microscopy Imaging of Site-Selective Pt Nanocatalysts: Electrochemical Activation and Surface Disordering. *J. Am. Chem. Soc.* **2015**, *137*, 14992–14998. [CrossRef]
11. Oturan, N.; van Hullebusch, E.D.; Zhang, H.; Mazeas, L.; Budzinski, H.; Le Menach, K.; Oturan, M.A. Occurrence and Removal of Organic Micropollutants in Landfill Leachates Treated by Electrochemical Advanced Oxidation Processes. *Environ. Sci. Technol. Lett.* **2015**, *49*, 12187–12196. [CrossRef] [PubMed]
12. Rastogi, A.; Al-Abed, S.R.; Dionysiou, D.D. Sulfate radical-based ferrous–peroxymonosulfate oxidative system for PCBs degradation in aqueous and sediment systems. *Appl. Catal. B Environ.* **2009**, *85*, 171–179. [CrossRef]
13. Giannakis, S.; Lin, K.-Y.A.; Ghanbari, F. A review of the recent advances on the treatment of industrial wastewaters by Sulfate Radical-based Advanced Oxidation Processes (SR-AOPs). *Chem. Eng. J.* **2021**, *406*, 127083. [CrossRef]
14. Ushani, U.; Lu, X.; Wang, J.; Zhang, Z.; Dai, J.; Tan, Y.; Wang, S.; Li, W.; Niu, C.; Cai, T.; et al. Sulfate radicals-based advanced oxidation technology in various environmental remediation: A state-of-the-art review. *Chem. Eng. J.* **2020**, *402*, 126232. [CrossRef]
15. Oh, W.-D.; Dong, Z.; Lim, T.-T. Generation of sulfate radical through heterogeneous catalysis for organic contaminants removal: Current development, challenges and prospects. *Appl. Catal. B Environ.* **2016**, *194*, 169–201. [CrossRef]
16. Devi, P.; Das, U.; Dalai, A.K. In-situ chemical oxidation: Principle and applications of peroxide and persulfate treatments in wastewater systems. *Sci. Total Environ.* **2016**, *571*, 643–657. [CrossRef]
17. Anipsitakis, G.P.; Dionysiou, D.D. Degradation of Organic Contaminants in Water with Sulfate Radicals Generated by the Conjunction of Peroxymonosulfate with Cobalt. *Environ. Sci. Technol.* **2003**, *37*, 4790–4797. [CrossRef] [PubMed]
18. Song, H.; Yan, L.; Wang, Y.; Jiang, J.; Ma, J.; Li, C.; Wang, G.; Gu, J.; Liu, P. Electrochemically activated PMS and PDS: Radical oxidation versus nonradical oxidation. *Chem. Eng. J.* **2020**, *391*, 123560. [CrossRef]
19. Anipsitakis, G.P.; Dionysiou, D.D. Radical Generation by the Interaction of Transition Metals with Common Oxidants. *Environ. Sci. Technol.* **2004**, *38*, 3705–3712. [CrossRef]

20. Ji, Y.; Fan, Y.; Liu, K.; Kong, D.; Lu, J. Thermo activated persulfate oxidation of antibiotic sulfamethoxazole and structurally related compounds. *Water Res.* **2015**, *87*, 1–9. [CrossRef]
21. An, D.; Westerhoff, P.; Zheng, M.; Wu, M.; Yang, Y.; Chiu, C.-A. UV-activated persulfate oxidation and regeneration of NOM-Saturated granular activated carbon. *Water Res.* **2015**, *73*, 304–310. [CrossRef] [PubMed]
22. Anipsitakis, G.P.; Tufano, T.P.; Dionysiou, D.D. Chemical and microbial decontamination of pool water using activated potassium peroxymonosulfate. *Water Res.* **2008**, *42*, 2899–2910. [CrossRef]
23. Wang, J.; Wang, S. Activation of persulfate (PS) and peroxymonosulfate (PMS) and application for the degradation of emerging contaminants. *Chem. Eng. J.* **2018**, *334*, 1502–1517. [CrossRef]
24. Li, Y.; Dong, H.; Li, L.; Tang, L.; Tian, R.; Li, R.; Chen, J.; Xie, Q.; Jin, Z.; Xiao, J.; et al. Recent advances in waste water treatment through transition metal sulfides-based advanced oxidation processes. *Water Res.* **2021**, *192*, 116850. [CrossRef] [PubMed]
25. Shang, W.; Dong, Z.; Li, M.; Song, X.; Zhang, M.; Jiang, C.; Feiyun, S. Degradation of diatrizoate in water by Fe(II)-activated persulfate oxidation. *Chem. Eng. J.* **2019**, *361*, 1333–1344. [CrossRef]
26. Liang, X.; Zhong, Y.; Zhu, S.; Zhu, J.; Yuan, P.; He, H.; Zhang, J. The decolorization of Acid Orange II in non-homogeneous Fenton reaction catalyzed by natural vanadium–titanium magnetite. *J. Hazard. Mater.* **2010**, *181*, 112–120. [CrossRef]
27. Gan, C.-d.; Yang, J.-y.; Du, X.-y.; Li, J.-l.; Tang, Q.-x.; Nikitin, A. Vanadium mobilization and redistribution during mineral transformation of vanadium-titanium magnetite tailings with different weathering degrees. *Sci. Total Environ.* **2023**, *894*, 165068. [CrossRef]
28. Zhang, W.; Tang, G.; Yan, J.; Zhao, L.; Zhou, X.; Wang, H.; Feng, Y.; Guo, Y.; Wu, J.; Chen, W.; et al. The decolorization of methyl orange by persulfate activated with natural vanadium-titanium magnetite. *Appl. Surf. Sci.* **2020**, *509*, 144886. [CrossRef]
29. Chen, X.; Yang, B.; Oleszczuk, P.; Gao, Y.; Yuan, X.; Ling, W.; Waigi, M.G. Vanadium oxide activates persulfate for degradation of polycyclic aromatic hydrocarbons in aqueous system. *Chem. Eng. J.* **2019**, *364*, 79–88. [CrossRef]
30. Fang, G.; Wu, W.; Liu, C.; Dionysiou, D.D.; Deng, Y.; Zhou, D. Activation of persulfate with vanadium species for PCBs degradation: A mechanistic study. *Appl. Catal. B Environ.* **2017**, *202*, 1–11. [CrossRef]
31. Su, Y.; Wu, Z.; Wu, Y.; Yu, J.; Sun, L.; Lin, C. Acid Orange II degradation through a heterogeneous Fenton-like reaction using Fe–TiO2 nanotube arrays as a photocatalyst. *J. Mater. Chem. A* **2015**, *3*, 8537–8544. [CrossRef]
32. Yu, X.; Sun, J.; Li, G.; Huang, Y.; Li, Y.; Xia, D.; Jiang, F. Integration of •SO4$^-$-based AOP mediated by reusable iron particles and a sulfidogenic process to degrade and detoxify Orange II. *Water Res.* **2020**, *174*, 115622. [CrossRef]
33. Feng, S.; Xiao, B.; Wu, M.; Wang, Y.; Chen, R.; Liu, H. Copper phosphide: A dual-catalysis-center catalyst for the efficient activation of peroxydisulfate and degradation of Orange II. *Sep. Purif. Technol.* **2020**, *248*, 117004. [CrossRef]
34. Lee, J.; von Gunten, U.; Kim, J.-H. Persulfate-Based Advanced Oxidation: Critical Assessment of Opportunities and Roadblocks. *Environ. Sci. Technol.* **2020**, *54*, 3064–3081. [CrossRef]
35. Liang, H.-y.; Zhang, Y.-q.; Huang, S.-b.; Hussain, I. Oxidative degradation of p-chloroaniline by copper oxidate activated persulfate. *Chem. Eng. J.* **2013**, *218*, 384–391. [CrossRef]
36. Johnson, R.L.; Tratnyek, P.G.; Johnson, R.O.B. Persulfate Persistence under Thermal Activation Conditions. *Environ. Sci. Technol.* **2008**, *42*, 9350–9356. [CrossRef]
37. Tian, S.H.; Tu, Y.T.; Chen, D.S.; Chen, X.; Xiong, Y. Degradation of Acid Orange II at neutral pH using Fe2(MoO4)3 as a heterogeneous Fenton-like catalyst. *Chem. Eng. J.* **2011**, *169*, 31–37. [CrossRef]
38. Lv, H.; Ma, L.; Zeng, P.; Ke, D.; Peng, T. Synthesis of floriated $ZnFe_2O_4$ with porous nanorod structures and its photocatalytic hydrogen production under visible light. *J. Mater. Chem.* **2010**, *20*, 3665–3672. [CrossRef]
39. Su, M.; He, C.; Sharma, V.K.; Abou Asi, M.; Xia, D.; Li, X.-z.; Deng, H.; Xiong, Y. Mesoporous zinc ferrite: Synthesis, characterization, and photocatalytic activity with H_2O_2/visible light. *J. Hazard. Mater.* **2012**, *211–212*, 95–103. [CrossRef]
40. Zubir, N.A.; Yacou, C.; Motuzas, J.; Zhang, X.; Diniz da Costa, J.C. Structural and functional investigation of graphene oxide–Fe3O4 nanocomposites for the heterogeneous Fenton-like reaction. *Sci. Rep.* **2014**, *4*, 4594. [CrossRef]
41. Li, Y.; Li, S.; Hu, J.; Zhang, Y.; Du, Y.; Han, X.; Liu, X.; Xu, P. Hollow FeCo-FeCoP@C nanocubes embedded in nitrogen-doped carbon nanocages for efficient overall water splitting. *J. Energy Chem.* **2021**, *53*, 1–8. [CrossRef]
42. Zhang, Y.; Duan, Z.; Jin, Y.; Han, H.; Xu, C. Chemical Bond Bridging across Two Domains: Generation of Fe(II) and In Situ Formation of FeSx on Zerovalent Iron. *Environ. Sci. Technol.* **2023**, *57*, 11336–11344. [CrossRef]
43. Lai, L.; Zhou, H.; Lai, B. Heterogeneous degradation of bisphenol A by peroxymonosulfate activated with vanadium-titanium magnetite: Performance, transformation pathways and mechanism. *Chem. Eng. J.* **2018**, *349*, 633–645. [CrossRef]
44. Pang, Y.; Zhou, X.; Vovk, E.I.; Guan, C.; Li, S.; van Bavel, A.P.; Yang, Y. Understanding lanthanum oxide surface structure by DFT simulation of oxygen 1s calibrated binding energy in XPS after in situ treatment. *Appl. Surf. Sci.* **2021**, *548*, 149214. [CrossRef]
45. Yi, X.; Gao, R.; Liu, X.; Chen, Y.; Zeng, B.; Xu, Y.; Yuan, C.; Dai, L. Phenylenediamine-Guanidine Phosphate Carbon Dots for Flame-Retardant and UV-Resistant Poly(vinyl alcohol) Composite Films. *ACS Appl. Nano Mater.* **2024**, *7*, 13260–13270. [CrossRef]
46. Jiang, M.; Lu, J.; Ji, Y.; Kong, D. Bicarbonate-activated persulfate oxidation of acetaminophen. *Water Res.* **2017**, *116*, 324–331. [CrossRef]
47. Bennedsen, L.R.; Muff, J.; Søgaard, E.G. Influence of chloride and carbonates on the reactivity of activated persulfate. *Chemosphere* **2012**, *86*, 1092–1097. [CrossRef]
48. Cai, C.; Zhang, H.; Zhong, X.; Hou, L. Ultrasound enhanced heterogeneous activation of peroxymonosulfate by a bimetallic Fe–Co/SBA-15 catalyst for the degradation of Orange II in water. *J. Hazard. Mater.* **2015**, *283*, 70–79. [CrossRef]

49. Yang, S.; Wang, P.; Yang, X.; Shan, L.; Zhang, W.; Shao, X.; Niu, R. Degradation efficiencies of azo dye Acid Orange 7 by the interaction of heat, UV and anions with common oxidants: Persulfate, peroxymonosulfate and hydrogen peroxide. *J. Hazard. Mater.* **2010**, *179*, 552–558. [CrossRef]
50. Wacławek, S.; Lutze, H.V.; Grübel, K.; Padil, V.V.T.; Černík, M.; Dionysiou, D.D. Chemistry of persulfates in water and wastewater treatment: A review. *Chem. Eng. J.* **2017**, *330*, 44–62. [CrossRef]
51. Ghauch, A.; Tuqan, A.M.; Kibbi, N. Ibuprofen removal by heated persulfate in aqueous solution: A kinetics study. *Chem. Eng. J.* **2012**, *197*, 483–492. [CrossRef]
52. Ma, J.; Yang, Y.; Jiang, X.; Xie, Z.; Li, X.; Chen, C.; Chen, H. Impacts of inorganic anions and natural organic matter on thermally activated persulfate oxidation of BTEX in water. *Chemosphere* **2018**, *190*, 296–306. [CrossRef]
53. Liu, F.; Yi, P.; Wang, X.; Gao, H.; Zhang, H. Degradation of Acid Orange 7 by an ultrasound/ZnO-GAC/persulfate process. *Sep. Purif. Technol.* **2018**, *194*, 181–187. [CrossRef]
54. Xu, Y.; Ai, J.; Zhang, H. The mechanism of degradation of bisphenol A using the magnetically separable $CuFe_2O_4$/peroxymonosulfate heterogeneous oxidation process. *J. Hazard. Mater.* **2016**, *309*, 87–96. [CrossRef]
55. Yang, S.; Yang, X.; Shao, X.; Niu, R.; Wang, L. Activated carbon catalyzed persulfate oxidation of Azo dye acid orange 7 at ambient temperature. *J. Hazard. Mater.* **2011**, *186*, 659–666. [CrossRef]

water

MDPI

Review

Bibliometric Analysis of Research Status, Hotspots, and Prospects of UV/PS for Environmental Pollutant Removal

Hanying Wang [1,2,3,4], Zheng Ji [1,2,3,4,*], Ruiqian Liu [1,4], Hongfeng Hu [1,4] and Wanting Sun [1,2,3,4]

1 School of Geography and Tourism, Shaanxi Normal University, Xi'an 710119, China; 15829171400@163.com (H.W.); 0711lrq@snnu.edu.cn (R.L.); 1687357153@snnu.edu.cn (H.H.); wtsun1004@163.com (W.S.)

2 International Joint Research Centre of Shaanxi Province for Pollutants Exposure and Eco-Environmental Health, Xi'an 710119, China

3 National Experimental Teaching Demonstration Center of Geography, Shaanxi Normal University, Xi'an 710119, China

4 Provincial Experimental Teaching Demonstration Center of Environmental Science Major, Shaanxi Normal University, Xi'an 710119, China

* Correspondence: jizheng@snnu.edu.cn

Abstract: There are various methods for environmental organic pollutant degradation and removal, among which ultraviolet/persulfate has drawn significant attention due to its excellent oxidizing properties, high maneuverability, and less formation of by-products. To comprehensively assess the development of ultraviolet/persulfate, a bibliometric analysis was conducted based on relevant literature indexed by Web of Science from 2002 to 2024. The findings revealed a growing number of publications, with China, Iran, and the United States being the top three countries with the highest total number of publications. Robust regional collaborations were evident. Additionally, Chinese and American scholars presented more significant activity in this field, and their publications exhibited higher quality. Furthermore, the current research hotspots, as indicated by keywords, primarily focused on the degradation mechanism of ultraviolet/persulfate and the reaction kinetic model simulation. Bibliometric results underscored that ultraviolet/persulfate, as an effective and environmentally friendly disinfection technology, possessed substantial potential for controlling diverse environmental pollutants, such as antibiotics, dyes, natural organic substances, heavy metals, microorganisms, and so on. Future research might concentrate on developing new catalytic composites and optimization of the photoactivation system. The practical application still needs to be investigated due to the complexity of the water matrix. The revival of microorganisms and the variation of toxicity should be considered further.

Keywords: UV/PS; bibliometric analysis; water pollution control; CiteSpace

1. Introduction

With the rapid development of the global economy and the continuous advancement of industrialization, severe environmental burden has become a thorny problem worldwide [1,2]. Traditional environmental contaminants, emerging contaminants (ECs), and trace organic pollutants (TOPs) [3], such as bacteria, dyes, antibiotics, personal care products (PPCPs), perfluoroalkyl acids (PFAAs), polychlorinated biphenyls (PCBs) and so on, have significant adverse effects on the environment and human health [4]. Traditional biochemical treatments are insufficient to remove ECs and TOPs, resulting in the secondary effluent failure to meet current water quality standards, and further advanced treatments are of great necessity. To make up for the limitations of traditional water and wastewater treatment, advanced oxidation processes (AOPs) have been widely employed due to better performance for pollutant removal and higher selectivity, stability, and safety [5]. AOPs could effectively oxidize organic pollutants and inactivate viruses, bacteria, and other microorganisms [6]. Moreover, AOPs might generate less disinfection by-products during the reactions. Thus, AOPs might be a good choice for not only meeting more stringent water quality standards but also protecting human health and reducing environmental burden.

Activation of persulfate (PS) is one of the most widely used advanced oxidation technologies. The activated persulfate could generate sulfate radical ($SO_4^{-}\cdot$) and hydroxyl radical ($\cdot OH$), which possess much stronger oxidizability than unactivated persulfate and can realize the oxidation and even mineralization of organic pollutants at room temperature. There are two types of persulfates: peroxymonosulfate (PMS) and peroxydisulfate (PDS). PMS possesses an asymmetric structure and is easily activated by electron transfer. PDS is not easily activated due to its molecular symmetry, but the activated PDS would produce more $SO_4^{-}\cdot$ resulting in a higher oxidation ability than that of PMS in general. Both PMS and PDS are cheap, stable, and easily available. Therefore, persulfate, as a highly effective oxidant, has excellent advantages in environmental pollution control.

Most traditional activation methods for PS require additional chemicals, which might bring potential adverse effects and production. In addition, with the introduction of Standards for drinking water quality (GB 5749-2022), Water reuse guidelines-Benefits evaluation of reclaimed water use (GB/T 42247-2022) in China, and the promulgation of Environmental management systems by the United Nations (ISO 14002-2:2023), the public have become more concerned about water safety. UV-activated persulfate (UV/PS) technology has received extensive attention recently because of less chemical addition. Much research has demonstrated the significant degradation effect of UV/PS on both traditional and emerging pollutants, and TOPs [7], and this technology presents strong application potential for the removal of toxic and harmful organic pollutants in the environment.

Bibliometrics is a measurement method that helps to analyze the published literature quantitatively and systemically. It is capable of evaluating the development of research status and predicting the future trend in a particular field [8] based on various characteristics of academic publications, such as the number of the literature and their citations, the number of authors (individuals or collectively as a group), and so on [9]. This paper focused on the application of UV/PS technology. Therefore, bibliometrics was applied in the study to summarize the development and research interests of UV/PS technology based on Web of Science (WOS) data from January 2002 to April 2024. CiteSpace was used to visually analyze and present the cooperative network, research topics and trends, aiming to deeply explore and analyze the research development in the field of UV/PS and provide helpful information for researchers in this field.

2. Materials and Methods

2.1. Data Source

In order to improve the representativeness and credibility of the collected literature data, all the data were obtained from the WOS core collection. The WOS data retrieval conditions were: [Topic = (UV AND persulfate) OR (UV AND PDS) OR (UV AND PS) AND (AOPs AND UV)]. The literature type was "Research Article" and they were written in English. The screening conditions for research directions were set as "Chemistry, Engineering, Environmental Sciences Ecology, Water Resources, Science Technology Other Topics, Public Environmental Occupational Health, Biotechnology Applied Microbiology" and so on based on search conditions in WOS. The publication period was from January 2002 to April 2024. A total of 2898 papers were retrieved under this search condition, and 835 valid papers were obtained by manually excluding the irrelevant literature. All documents were downloaded in TXT format files, and the number of data files was prefixed with "download_"as the file name.

2.2. Statistical Analysis and Visualization

CiteSpace was applied for visual analysis, with the time slicing set to 2002–2024 in 1-year increments. Four folders, "input", "output", "data", and "project", were established, respectively. Node types selected included institution, region, author, keywords, and reference. Data analysis was conducted using Microsoft Excel 2019, and related graphics were produced by Origin 2021 and R (version 4.3.3) with the ggplot2 toolkit. The flow chart of this bibliometric analysis is shown in Figure 1.

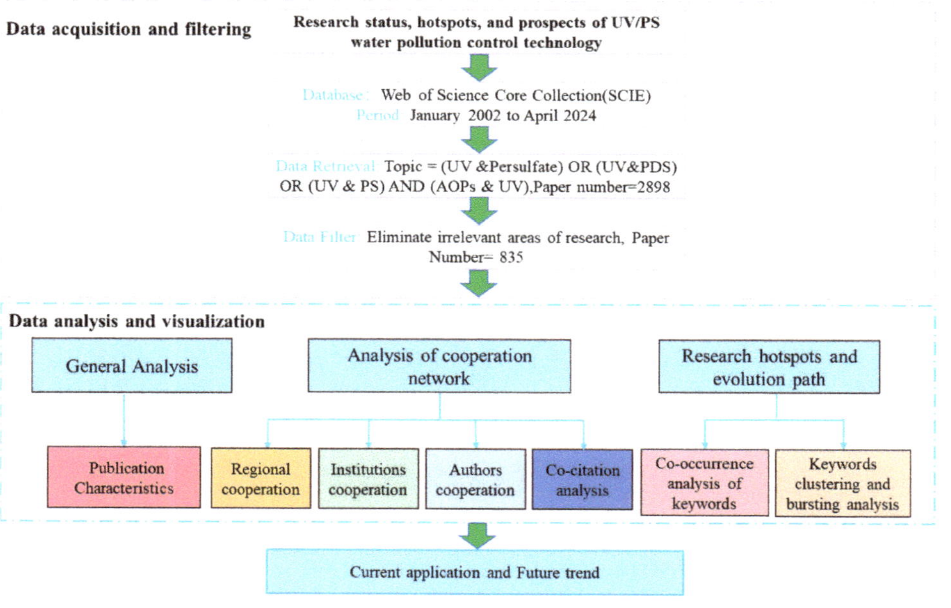

Figure 1. Flow chart of bibliometric analysis.

3. Results

3.1. Publication Characteristics

The annual number of published papers is an important criterion for evaluating the development of related research. In order to understand the development history, current situation, and future trends of UV/PS, the number of research articles issued by the WOS database was assessed year by year, as shown in Figure 2. It could be seen that the number of papers published from 2002 to 2024 is increasing yearly. In general, the development history focusing on UV/PS could be divided into two stages. The first phase was a slow development phase, from 2002 to 2012 with 61 publications. Then, a rapid development stage appeared, and the number of publications has increased rapidly with a total of 774 articles published during that time. The rapid increase in annual publications after 2012 might be closely related to the update of international policy. Guidelines for Drinking-water Quality (4th edition) issued by WHO in July 2011 provided an authoritative basis for more restrictive international water safety standards [10]. The number of publications has increased year by year since then and reached the highest in 2022, reflecting the policies' guidance for the rapid progress of science and technology. Researchers showed much more enthusiasm, and their research confidence has become more and more sufficient.

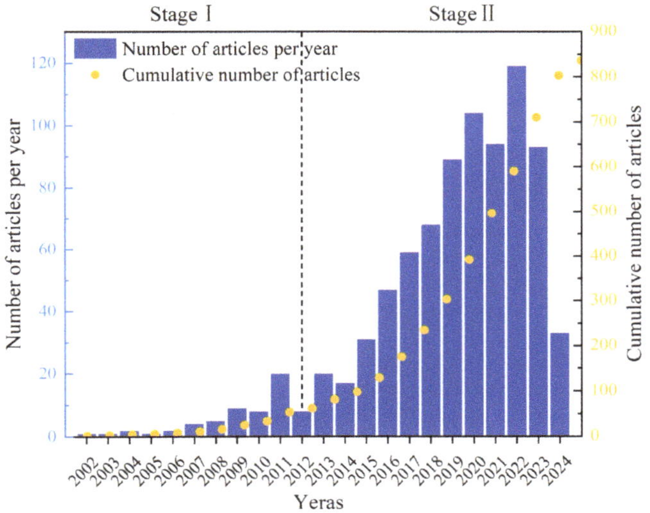

Figure 2. Number of publications on UV/PS per year from 2002 to 2024.

Figure 3 shows a statistical chart of the top ten subject areas or research directions according to the numbers of publications on UV/PS applications based on the WOS search conditions. The number of articles published in the subject area is indicated. It could be seen that UV/PS technology has been applied more frequently in the fields of Environmental Sciences Ecology (100 articles), Engineering (82 articles), Water Resources (66 articles), and Chemistry (62 articles). It could be seen that the application of UV/PS is wide-ranging. Unexpectedly, to facilitate the application of UV/PS, a lot of researchers also estimated the economic cost or cost-effectiveness of UV/PS (referring to Business Economics in Figure 3, 31 articles).

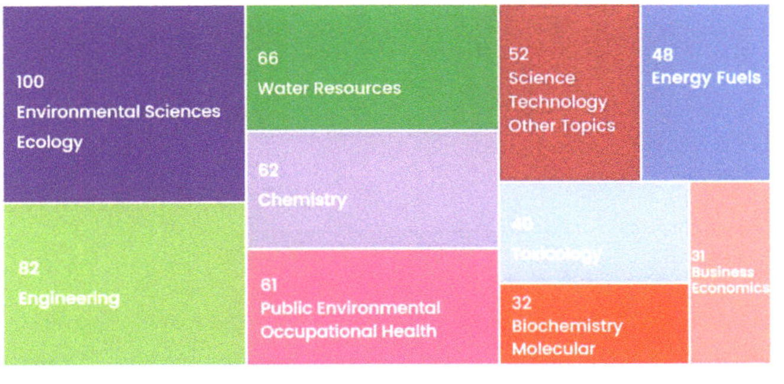

Figure 3. Number of publications on UV/PS in the top ten subject areas.

UV/PS could be widely applied for pollutant removal in environmental governance. During the COVID-19 pandemic and post-epidemic period, potential risks have been presented to the ecological environment and human health resulting from the frequent and excessive use of disinfectants. With the introduction of the Standards for Driving Water Quality (GB 5749-2022) and Standard Test Method for Drinking Water (GB/T 5750-2023) in China in recent years, the use of disinfectants and the supervision of disinfection by-products are becoming more and more strict [11]. It might predict that green disinfection technology based on UV/PS would draw extensive attention in the next few years.

3.2. Analysis of Cooperation Network

Through the analysis of the cooperation network, the communication trajectory of scientific research activities could be obtained, including the communication within or between different countries, regions, institutions and scientific research groups, and so on.

3.2.1. Research Contributions from Different Countries and Regions

Through the statistical analysis of research articles published by different countries and regions, the scientific investment and academic influence of the country or region in certain research fields could be obtained. The results are shown in Figure 4. There are 57 nodes and 174 connections in the regional cooperation network diagram, indicating that a total of 57 countries or regions have published articles in the field of UV/PS. The statistics of the number of publications are shown in Table 1. The top five are China (408 articles), Iran (97 articles), the United States (87 articles), Spain (53 articles), and Chinese Taiwan (39 articles), among which China is the leading country in this field. From the perspective of the number of citations, articles published by China were cited the most, indicating that Chinese research achievements have been highly recognized and Chinese researchers are in a leading position in this field. The number of articles published by the United States were cited the second most. The research on UV/PS in the United States began earlier, and the related research achievements also have a great influence. The cooperation network diagram shows that there is certain cooperation between countries, while the nodes represented by China, Spain, the United States, and Iran are in the purple outer ring, with an intermediary centrality of 0.37, 0.27, 0.23, 0.16, and 0.10, respectively. Whether from the number of publications or intermediary centrality, these four countries had frequent and close international cooperation in this field with other regions, and they all made great efforts on UV/PS.

Table 1. Statistics on the number of national and regional publications and citations in the field of UV/PS.

Regions	Number of Publications	Starting Year	Citation Number
China	408	2006	20,704
Iran	97	2005	3256
USA	87	2004	10,729
Spain	53	2005	1931
Taiwan, China	39	2009	1416
India	37	2006	1644
South Korea	27	2009	991
Turkey	23	2013	917
France	22	2009	1300
Pakistan	17	2013	1490

3.2.2. Research Contributions from Different Institutions

Analysis of contributions from different research institutions could reveal the achievements and cooperation relationships of institutions in certain fields [12]. Figure 5 shows the collaboration map of institutions. According to statistics, 317 research institutions published literature focusing on UV/PS from 2002 to 2024, with a network density of 0.0078. Among all institutions, Tongji University (46 articles), Harbin Institute of Technology (40 articles), University of Cincinnati (24 articles), and Tsinghua University (24 articles) are the top five institutions with the highest number of research articles published in this field. However, the network density value is relatively low, indicating that the cooperation between domestic and foreign institutions is more dispersed, and the cooperation seems to be more regional. The cooperation network is mainly centered on the Harbin Institute of Technology, Tongji University, the University of Cincinnati, Alberts Medical University of Iran, and the Technical University of Madrid. A close cooperative relationship exists between universities and institutions in the geographically close area. The intermediary centrality of the number of papers published by the Technical University of Madrid and Harbin Institute of Technology are the top two, indicating that these two institutions conducted relatively close cooperation and communication with other institutions. The researchers from these two institutions are the main force in this field and provide academic contributions of major significance to the field. To promote technological development, institutional cooperation, especially for academic exchanges, should be encouraged and conducted.

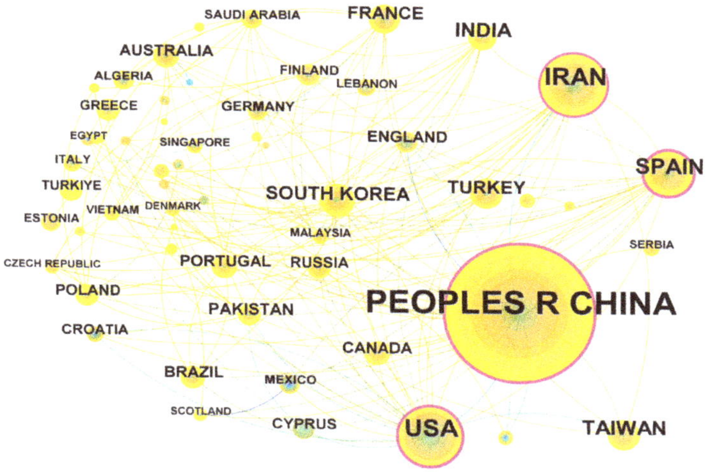

Figure 4. Network map of national and regional cooperation focusing on UV/PS.

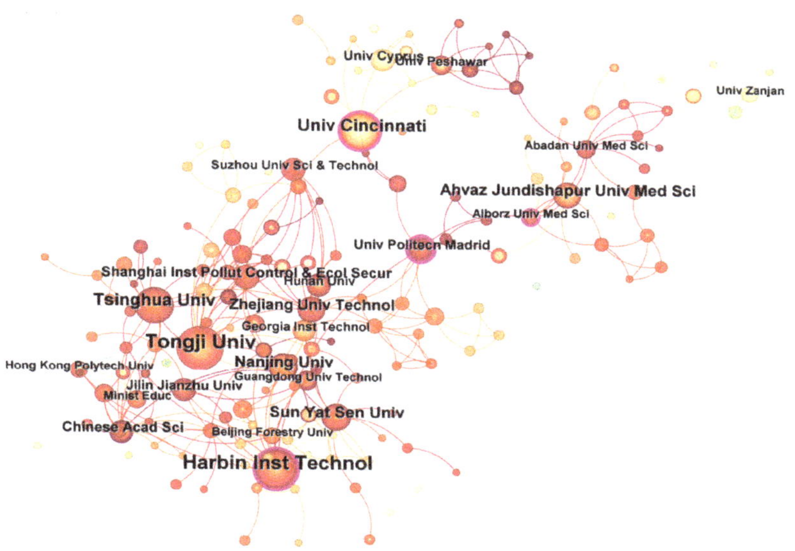

Figure 5. Number of publications and cooperation relationships regarding UV/PS from different institutions.

3.2.3. Research Contributions from Different Authors

Figure 6 shows the cooperative relationship, the number, and the quality of published papers based on the authors. A total of 540 authors are involved in this field, among which Professor Ma Jun from the Harbin Institute of Technology in China and Professor Dionysiou Dionysios from the University of Cincinnati in the United States might be the most prominent contributors in the field. Three authors have published more than 20 articles: Professor Dionysiou Dionysios D from the University of Cincinnati (22 articles), Professor Ma Jun from the Harbin Institute of Technology (22 articles), and Professor Gao Naiyun from Tongji University (21 articles). Independent research groups are led by these three authors, among which Prof. Dionysiou Dionysios D published articles that were cited the most, with 4035 times. The total citations of articles supervised by Professor Ma Jun and Professor Gao Naiyun were 3187 and 1515 times, respectively. As for a single article, the article focusing on the effect of pH value on the degradation of refractory organic matter in UV/PMS systems [12] conducted by Professor Ma Jun's group is the most cited article and has been cited as many as 979 times, suggesting good guidance and inspiration for later scholars. The research on UV/PS drew wide attention and made great improvement in the last 10 years, and 7 of the top 10 authors are Chinese scholars, indicating the fruitful research achievements and high research enthusiasm from Chinese scholars in the field of UV/PS. It also showed huge research progress and worldwide academic influence made by Chinese scholars. However, there seems to be a relatively fixed cooperative relationship between the institutions, groups, and the scientific researchers. It might be seen that the articles are usually organized by certain groups with frequent cooperation. Therefore, strengthening communication between researchers from different regions and institutions would help promote in-depth research in UV/PS and accelerate its application in various fields besides environmental governance.

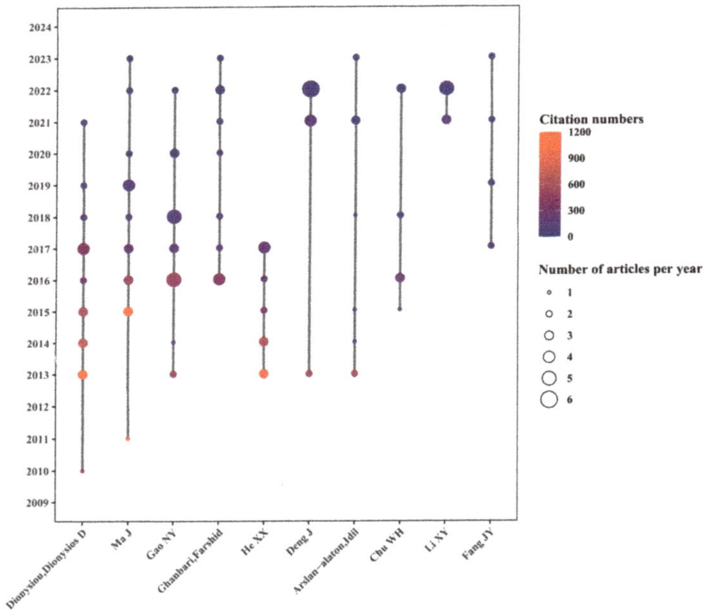

Figure 6. Timeline of authors focusing on UV/PS.

3.2.4. Co-Citation Analysis

Co-citation refers to the fact that two or more papers are cited simultaneously by one or more papers, and the cited papers constitute a co-citation relationship. It would reflect the knowledge structure of certain research fields and is an important index to measure the academic influence of the cited papers. The relationships of those most co-cited papers in the field of UV/PS are plotted in Figure 7 [12–47]. It could be seen that the author with the highest number of co-citations is Wang JL (2018) with a total of 65 citations, followed by Waclawek S (2017) with 74 citations, and Xie PC (2015) with 72 citations in the third place. The specific information of the highly co-cited papers is summarized in Table 2. The research team led by Prof. Wang JL of Tsinghua University studied the catalytic activation of persulfate by metals [46] or visible light [47] to degrade traditional organic pollutants and ECs [48]. Their research made great efforts on the modification of the advanced oxidation technology. The research, supervised by Prof. Xie PC, focused on the degradation of emerging pollutants and disinfection by-products by UV-activated chlorine [49] and UV-activated persulfate [50]. Their findings provide a new insight for future investigations. The articles of these highly cited authors mainly focused on the removal effect of different types of pollutants, the modification of the research methods, and the exploration of the mechanism, which constitutes the basic framework of UV/PS technology.

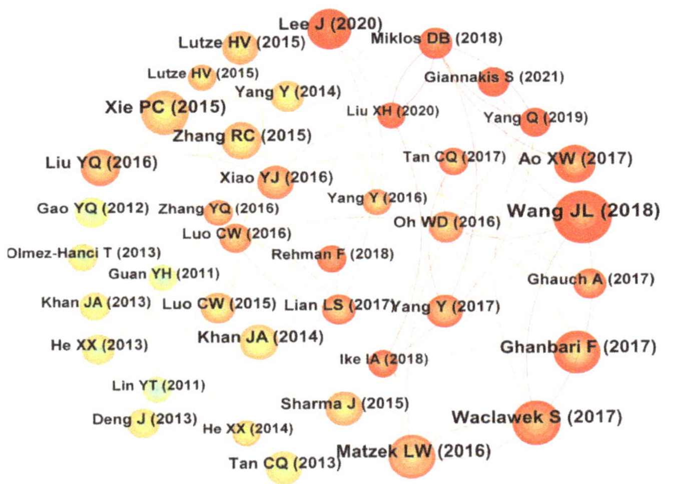

Figure 7. Co-citation graph focusing on UV/PS. The relationships of those most co-cited papers in the field of UV/PS are plotted in Figure 7 [12–47].

Table 2. Highly cited papers focusing on UV/PS.

Citation	Centrality	First Author	Year of Publication	Source Journal	Title of the Articles	References
108	0.03	Wang JL	2018	CHEM ENG J	Activation of persulfate (PS) and peroxymonosulfate (PMS) and application for the degradation of emerging contaminants	[48]
74	0.06	Waclawek S	2017	CHEM ENG J	Chemistry of persulfates in water and wastewater treatment: A review	[51]
72	0.03	Xie PC	2015	WATER RES	Removal of 2-MIB and geosmin using UV/persulfate: Contributions of hydroxyl and sulfate radicals	[49]
64	0.02	Matzek LW	2016	CHEMOSPHERE	Activated persulfate for organic chemical degradation: A review	[52]
64	0.02	Ghanbari F	2017	CHEM ENG J	Application of peroxymonosulfate and its activation methods for degradation of environmental organic pollutants: Review	[53]

3.3. Research Hotspots and Evolution Path

Keyword co-occurrence, clustering, and burst terms analysis could help investigators better identify research hotspots and trends in various fields and give academic support.

3.3.1. Co-Occurrence Analysis of Keywords

Co-occurrence analysis of keywords could help to reveal the research hotspots and development trends in certain fields. The keywords co-occurrence map of the WOS database is shown in Figure 8. The most frequently used keywords are degradation (270 times), removal (216 times), kinetics (215 times), aqueous solution (211 times), and rate constants (204 times). The keywords could be divided into the following three categories according to their research purposes.

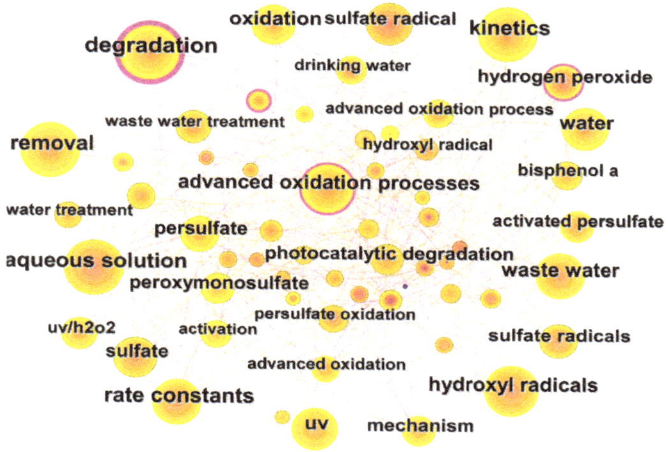

Figure 8. Keywords collaboration network for UV/PS in Web of Science.

(1) Degradation mechanism: "hydroxyl radicals", "sulfate radicals", "UV", and "Persulfate" are the typical keywords. Studies on UV/PS suggest that pollutants might be mainly degraded in two ways: UV direct photolysis and PS activation. As for the direct photolysis, energy input by UV could directly degrade pollutants by destroying certain chemical bonds. Zhang et al. compared the degradation of sulfamethoxazole (SMX) and trimethoprim by UV, UV/PS, and UV/H_2O_2, respectively. It was found that a higher contribution rate and a better degradation effect by UV alone was obtained [16]. The direct photolysis rate of pollutants varies significantly due to their molar absorption coefficient and quantum yield. On the other hand, UV radiation could activate PS to generate $SO_4^{-}\cdot$ and $\cdot OH$, indirectly contributing to the degradation of pollutants. The experiments of PDS or PMS alone without UV activation showed that the removal rate of organic pollutants is almost negligible, but UV activation could accelerate 80–100% of pollutant removal [54]. No matter the direct or indirect effect of UV, it mainly depends on the specific structure of the pollutants.

(2) Reaction kinetics: "kinetics" and "rate constants" are the representative keywords. The order and model of kinetic reaction are commonly used to describe the rate of the reactions. The simulation of degradation models is frequently obtained by comparing the accuracy and similarity of calculation of different established formulations. Feng et al. used different models, such as Chick–Watson, Collins–Selleck, Hom, and Biphasic, to simulate the inactivation rate of *E. coli* under UV irradiation. The experimental results show that the Biphasic model could well reflect the kinetic changes of *E.coli* inactivation under the experimental conditions [55].

(3) The applications: the typical keywords are often expressed as "bisphenol a" and other "dyes", "drinking water", "wastewater", and other environmental media containing various organic pollutants and pathogenic microorganisms [51]. At present, activated persulfate advanced oxidation technologies are often used in the degradation and removal of antibiotics [26], pesticides [56], dyes [37], and other traditional and emerging pollutants and TOPs in water, and it also has huge potential in the soil organic pollution remediation [57].

The current research regarding UV/PS mainly focuses on the exploration of the mechanism, the free radical and non-free radical reaction pathways, the factors that influence the degradation process, and the kinetic simulation expression. Moreover, the research interests have gradually shifted from the degradation of certain single pollutants to the removal of co-existing pollutants which could reflect the true environmental pollution.

3.3.2. Keywords Clustering and Bursting Analysis

The cluster analysis of keywords (as shown in Figure 9) and bursting keywords information (as shown in Figure 10) in WOS are summarized. The keywords could be divided into four categories as follows: UV/PS system, mechanism, pollutants, and kinetics.

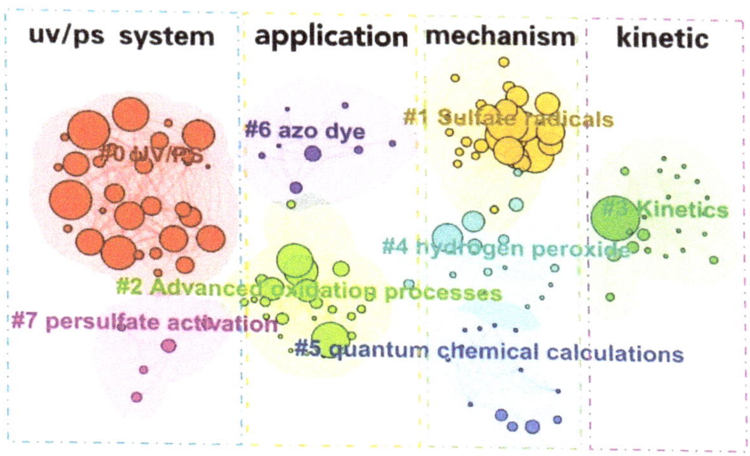

Figure 9. Clustering analysis of keywords in the UV/PS field.

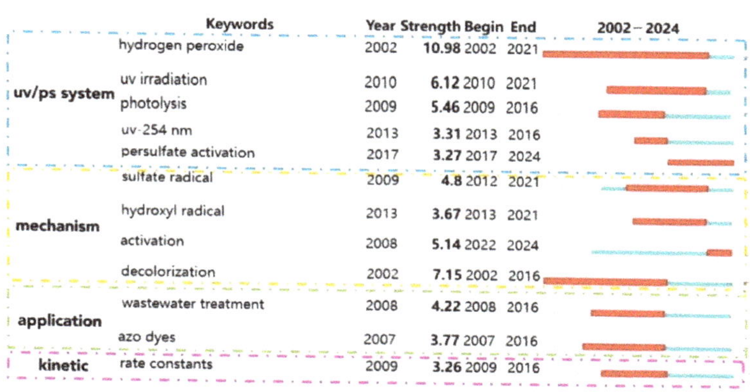

Figure 10. Keyword bursting map of UV/PS technology in WOS.

(1) Optimization of UV/PS

Clusters 0# UV/PS and 7# persulfate activation are the main bodies for these kinds of research. The burst keyword "UV/PS system" also indicates the system optimization. Due to the strong oxidation property, persulfate has shown great advantages in efficiently degrading organic pollutants in water [58]. Traditional activation methods include heating, ultrasonic, transition metal, and ultraviolet irradiation [59]. Luo et al. [39] investigated the removal of 2,4,6-trichloroanisole using persulfate activated by UV254 and found that $SO_4^{-}\cdot$ played a more dominant role than ·OH in the degradation process. Significant progresses have been made in photoactivated PS technology based on UV, visible light, or sunlight irradiation [35,60]. Currently, optimization of the UV/PS system draws a lot of research interest. There are two aspects for the UV/PS system optimization. One is the choice of different UV wavelengths. Due to the high free radical yield, the processes based on the KrCl excimer lamp (UV222-AOPs) have attracted wide attention in water and wastewater treatment. Some studies have applied UV222/PS to degrade antibiotics such as florfenicol and ciprofloxacin [61], pesticides, endocrine-disrupting compounds [62], etc. A good removal efficiency was achieved. Verma et al. found that for the degradation of anatoxin-a, UV260 could activate PMS more effectively than UV270, resulting in up to 98.6% degradation rate under the condition of [PMS]/[anatoxin-a] = 100:1. The process would be inhibited by the presence of carbonate and bicarbonate ions; meanwhile, the presence of humic acids (HA) with a certain concentration range would enhance the degradation of anatoxin-a through the photosensitization effect [63]. That means the choice of different wavelengths would directly impact the effect of pollutant removal. Vacuum ultraviolet (VUV) is also used to activate persulfate, alone or combined with other activation methods. VUV/PS was applied to remove the total odor substances (2-methylisoborneol and geosmin) in water. At a PS concentration of 0.5 mmol/L, the VUV/PS process increased the removal of 2-MIB and GSM by 76% and 74%, respectively, compared to VUV alone or PS alone. The degradation of 2-MIB and GSM would be enhanced with the increase in the PS concentration and VUV intensity within a certain range, but inhibited by HCO_3^- and humic acid [64]. UV/VUV/PS was suggested to be superior to degrade dye methylene blue (MB) regarding organic pollutant removal efficiency and water treatment cost. It was indicated that MB degradation could be fitted well with pseudo-first-order kinetics in all the PS, UV254, VUV/UV254, UV254/PS, and VUV/UV254/PS processes, and reach the highest reaction rate at 327.6 m^2 $einstein^{-1}$ in the UV/VUV/PS system, while the degradation would be significantly affected by pH and the introduction of HA, HCO_3^-, Cl^-, etc., would inhibit MB degradation [65]. Thus, the selection and combination of different UV wavelengths would be of great potential to improve the efficiency of UV/PS.

(2) Application and development of UV/PS

The keywords classified as clusters 6# azo dye and 2# AOPs are mainly focusing on the application of UV/PS, as well as the bursting keywords "wastewater treatment" and "azo dye." Dyes, antibiotics, PPCPs, microorganisms, and other organic pollutants in the water body have a great impact on the water environment. Therefore, AOPs (2# advanced oxidation processes) are widely used in water pollution control. Tan et al. [21] used UV, UV/H_2O_2, and UV/PS to degrade antipyrine in water. UV/PS showed the highest degradation efficiency and is regarded as a priority technology for controlling emerging pollutants such as PPCP. Clustering analysis suggested that researchers not only focused on the removal effect of traditional environmental pollutants by UV/PS, but also extended their interests to the degradation of disinfection products, by-products, and some other ECs and TOPs, indicating the potential of UV/PS as a green disinfectant.

(3) Degradation mechanism of UV/PS

Clusters 1# sulfate radicals, 4# hydrogen peroxide, and 5# quantum chemical calculations, represented the research on the reactive species in the UV/PS process. Khan et al. [15] used UV/PS to degrade atrazine and found that the reactive free radicals produced in AOP were more effective in removing the side alkyl groups of atrazine. The chemical properties of PMS and PDS are shown in Table 3. As mentioned above, under

UV irradiation, the O-O bond in $S_2O_8^{2-}$ (mainly for PDS) is broken to generate two $SO_4^{-}\cdot$, while the O-O bond in HSO_5^- (mainly for PMS) would generate one $SO_4^{-}\cdot$ and one $\cdot OH$ (Equations (1) and (2)) [39]. Due to the oxidability of $SO_4^{-}\cdot$ being stronger than that of $\cdot OH$, PDS presents a higher oxidability. However, when acting with water solution, H_2O could utilize UV light to generate electron transfer resulting in the activation of PDS or PMS (Equations (3)–(5)) [66]. Moreover, the generated $SO_4^{-}\cdot$ could also react with H_2O and OH^- to form $\cdot OH$ (Equations (6) and (7)) [39].

Table 3. Basic features of PMS and PDS.

Oxidizing Agent	Chemical Formula	Molecular Mass	Solubility/(g/L)	Redox Potential/V	-O-O- Key Length/Å	Bond Energy/(kJ/mol)	Activation Method
PMS	HSO_5	114.07	>250	1.82	1.453	140~213.3	transition metal, heating, UV, ultrasonic, and carbon-based materials
PDS	$H_2S_2O_8$	194.13	>520	2.01	1.497	140	

Along with the degradation caused by free radicals, the non-radical pathway in persulfate advanced oxidation processes has recently gained attention due to its advantages of high selectivity and resistance to interference. The researchers investigated the degradation effect of ciprofloxacin by UV/PDS. It was found that $O^{2-}\cdot$ could be converted into singlet oxygen ($^1O_2$) by reacting with H_2O or H^+, and $^1O_2$ was identified as the dominant active species during the degradation process [67]. When PMS activated by the Bimetallic Ni/Fe Atom Cluster Catalysts was used to remove phenol, it was also found that the non-free radical pathway was dominant in the degradation, indicating $^1O_2$ and electron transfer promoted the efficient degradation of phenol in wastewater [68]. It could be seen that non-free radicals also play an important role in the UV/PS oxidization process. However, further studies are still needed due to the diverse and complex mechanisms. Therefore, research focusing on the reaction mechanism would still be one of the hotspots.

$$S_2O_8^{2-} \xrightarrow{UV} 2SO_4^{\cdot-} \tag{1}$$

$$HSO_5^{-} \xrightarrow{UV} SO_4^{\cdot-} + \cdot OH \tag{2}$$

$$H_2O \xrightarrow{UV} H\cdot + \cdot OH \tag{3}$$

$$S_2O_8^{2-} + H\cdot \rightarrow SO_4^{\cdot-} + SO_4^{2-} + H^+ \tag{4}$$

$$HSO_5^{-} + H^+ \rightarrow SO_4^{\cdot-} + H_2O \tag{5}$$

$$SO_4^{\cdot-} + H_2O \rightarrow SO_4^{2-} + \cdot OH + H^+ \tag{6}$$

$$SO_4^{\cdot-} + OH^- \rightarrow SO_4^{2-} + \cdot OH \tag{7}$$

(4) Kinetics simulation of UV/PS

The cluster3# kinetics and the burst keyword "rate constants" represent the research concentrated on the reaction kinetics, which would be directly helpful to understanding the degradation efficiency. Lu et al. found that UV/PS could significantly promote the degradation of sulfadimethoxine, and the degradation rate constant is four times and twenty times higher than that of the single UV and single PS, respectively, and the degradation rate constant increases with the increase in UV intensity and PS dosage, while the degradation decline was observed with the existence of some background inorganic anions [69].

4. Discussion

4.1. Current Application

There is a huge potential for UV/PS application in environment remediation. UV/PS has been mainly applied for organic and microbial pollutants removal, which are shown in Figure 11. Table 4 lists some typical cases of various pollutants removal by UV/PS technology to further understand the effect and mechanism of UV/PS technology. Here, three typical types of pollutants were introduced.

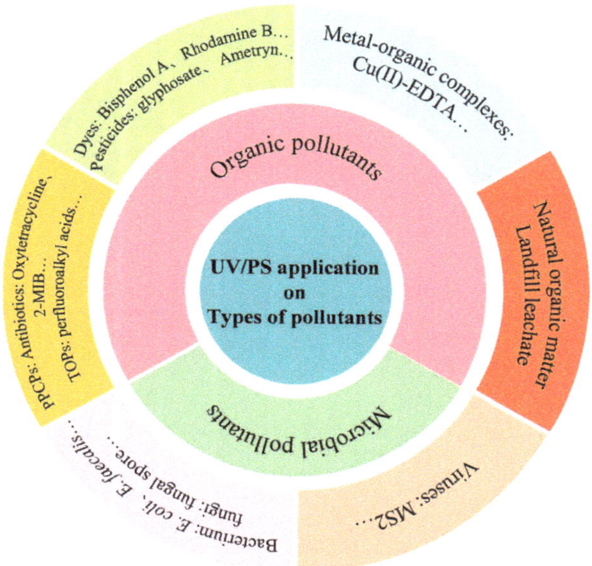

Figure 11. Pollutant categories in UV/PS application.

(1) Natural organic matter

Natural organic matter (NOM) are important components in a water environment. The fluctuation of their composition and concentration would result in various interactions in biogeochemical cycles [70]. The UV/PS technology has shown significant effectiveness in NOM removal. Alayande et al. [71] used UV/PMS to mineralize the transparent exopolymer particles (TEPs), and a removal rate exceeding 44% only within 180 min was achieved. Under UV irradiation, $SO_4^{-}\cdot$ and $\cdot OH$ generated by UV/PMS would attack the unsaturated bonds and aromatic structures in TEPs. Tian et al. [72] investigated the removal efficiency of NOM by UV, PDS, and UV/PS, with removal rates of 35%, 15%, and 91%, respectively. The radicals produced by UV/PS would primarily target the unsaturated NOM fraction, followed by the refractory fraction, resulting in the destruction of unsaturated bonds and aromatic structures. It would suggest that the main mechanism of NOM removal by UV/PS involves the generation of free radicals to break down the unsaturated bonds and aromatic structures. The degradation efficiency of UV/PS technology is not high for certain natural organic compounds due to the complex environmental pollution. For example, the existence of bromide ions would inhibit the degradation of some organic micropollutants by the UV/PS, and the degradation rate might be reduced by almost 89.5%. This indicates that the degradation efficiency of UV/PS technology would be affected by some other components co-existing in the reaction system [73]. Thus, there might be a relatively wide range for the degradation efficiency of NOM by UV/PS, and the specific degradation efficiency depends on a variety of factors, including the reaction conditions, the characteristics of NOM, other components in the reaction system, and so on.

(2) Antibiotics

In recent years, antibiotics have been widely used in medical treatment, animal husbandry, and aquaculture. The consequent antimicrobial resistance and antibiotic resistance genes pose serious threats to the environment and public health [74]. UV/PS exhibits excellent ability to effectively degrade insoluble pollutants such as antibiotics in water [28]. Ao et al. [75] used medium-pressure ultraviolet (MPUV), PMS, and MPUV/PMS to degrade tetracycline (TTC) in water. The results showed that the removal rates of TTC were 14%, 17%, and 82% for MPUV, PMS, and MPUV/PMS, respectively. Hydroxylation, demethylation, and decarboxylation effects were observed in the process. Furthermore, ·OH attacking the double bond in TTC played a leading role in the degradation; meanwhile, MPUV was indicated to directly destroy the ketone/enol bond. In addition, Ghauch et al. [26] studied the chloramphenicol (CAP) degradation by UV/PS in water. Both $SO_4^{-}\cdot$ and ·OH are crucial for CAP degradation. The UV/PS system could significantly improve the degradation and mineralization efficiency compared with the single UV system. Another investigation showed that the degradation of the antiepileptic drug oxcarbazepine (OXC) by UV/PS was more efficient than UV or PS alone [76]. The removal rate of antibiotics might be in the range of 70–100%. The specific degradation efficiency is also influenced by a variety of factors, including the type and concentration of antibiotic, the pH of the reaction system, and the type of catalyst used, and so on. With the global consensus on antibiotic pollution and the implementation of the WHO AWaRe (Access, Watch, Reserve) antibiotic book (2023), the WHO Guidance on wastewater and solid waste management for manufacturing of antibiotics (2024), and WHO bacterial priority pathogens list (2024), UV/PS would still play a huge role in antibiotic removal.

(3) Microorganisms

Microorganisms could not be completely removed by traditional water and wastewater treatment processes, thus resulting in a huge potential health risk to human beings. It is of great necessity to inactivate microorganisms, especially pathogens in water by efficient and safe water treatment technologies. A phage MS2 inactivation experiment showed that a 4.39 log inactivation rate could be achieved for only 4 min by UV/PS treatment, which was 1.44 log higher than that of the single UV treatment [77]. Guerra-Rodriguez et al. [78] studied the inactivation effect of *E. faecalis* in distilled water, salt water, and simulated wastewater by UV/PMS and UV/H_2O_2. The results showed that under the UV/PMS system, the bacteria could be completely inactivated with less dose of chemicals, less energy input, and shorter reaction time. The inactivation mechanisms of microorganisms by UV/PS are suspected as the following three aspects: First, free radicals, such as ·OH and $SO_4^{-}\cdot$, could lead to bacterial inactivation. Free radicals generated in the outside of cells possess strong oxidizing abilities, which could first oxidize lipids and proteins in the cell membrane to cause cell structure damage with intracellular substances leakage, and then those intracellular substances might be attacked resulting in the final bacterial death [79]. The second one might be the direct and indirect destruction of endogenous processes caused by UV radiation, which would result in excessive pressure on cells, so that the reactive oxygen species might penetrate the cell membrane and react with the cytoplasm, leading to oxidative stress much more beyond the tolerance of cells. Thirdly, persulfate might directly oxidize and then penetrate the cell wall and cell membrane, which would promote the reaction between PS and the metals that already existed in the cells [80], and result in cell death. The inactivation efficiency for free-living bacteria by UV/PS could reach more than 90%, but the application of UV/PS to inactivate biofilm is rarely investigated and the repair effect of microorganisms also needs to be further studied.

Table 4. Research on UV/PS application for different kinds of pollutants.

Contaminants			Light Wavelength and Intensity	Concentration of Pollutant	Concentration of PS	Reaction Water Body	Active Species	Removal Rate	Reference
Type I Organic pollutant	antibiotic	chloramphenicol (CAP)	253.6 nm 2.43 mW/cm^2	31 μM	0.25 mM	river water (RW) filtered (FWW) non-filtered (NFWW)	·OH, SO_4^-	NFWW 86.9% FWW 87.3% RW 50%	[26]
		oxytetracycline (OTC)	254 nm 0.1 mW/cm^2	10 μM	1 mM	simulative water	·OH, SO_4^-	97%	[14]
		florfenicol (FLO)	254 nm 398 μW/cm^2	20 μM	2 mM	ultrapure water	·OH, SO_4^-	UV 24.9% PS 0% UV/PS 98.4%	[81]
	dye	orange II	254 nm 0.73 mW/cm^2	20 mg/L	1 mM	ultrapure water	·OH, SO_4^-·	56.3%	[82]
		methylene blue (MB)	185 nm 2.59×10^{-5} einstein m^{-2} s^{-1} 254 nm 5.31×10^{-4} einstein m^{-2} s^{-1}	10 μM	0.5 mM	filtered water surface water secondary wastewater effluent	·OH, SO_4^-·	Compared with UV/PS, the pseudo-first-order kinetic k of VUV/UV/PS increases	[65]
	natural organic matter	HA	254 nm 0.25 mW/cm^2	2 mg/L	0.5 mM	DI water NaCl solution synthetic seawater	·OH, SO_4^-·	DI water100% NaCl solution 78% synthetic seawater 58%	[83]
		UV254/DOC	254 nm 1.53 mW/cm^2	5.41–7.55 mg/L 0.098–0.122 mg/L	0.6 mM/0.6 mM	surface water	·OH, SO_4^-·	UV254 91% DOC 58%	[72]
Type I Inorganic pollutant (metal complexes)		Cu(II)-EDTA	254 nm 0.25 mW/cm^2	0.3 mM	[84]0/[Cu(II)-EDTA]0 = 60 mM	simulative water	·OH, SO_4^-·	about 90%	[85]

Table 4. *Cont.*

Contaminants			Light Wavelength and Intensity	Concentration of Pollutant	Concentration of PS	Reaction Water Body	Active Species	Removal Rate	Reference
Type II Micro-organism	bacteria	macrolides-resistant bacteria (MRB) sulfonamides-resistant bacteria (SRB) tetracyclines-resistant bacteria (TRB) quinolones-resistant bacteria (QRB)	254 nm 400 μW/cm^2	8, 512, 16, 32 mg/L	1 mmol/L	secondary effluent	$\cdot OH, SO_4^{-}\cdot$	MRB 96.6% SRB 94.7% TRB 98.0% QRB 99.9%	[86]
	virus	bacterio-phage (MS2)	253.7 nm 160 μW/cm^2	10^6 PFU/mL	0.3 mmol/L	ultrapure water	$\cdot OH, SO_4^{-}\cdot$	4.39 lg can be removed in 4 min	[77]
	fungal or spore	Penicillium Aspergillus niger Trichoderma harzianum	265 nm, 280 nm, 265 & 280 nm 0.215, 0.214, 0.185 mW/cm^2	2–4 × 10^6 CFU/mL	1 mM	fungal spore suspensions	$\cdot OH, SO_4^{-}\cdot$	The membrane damage and permeability of the three fungi were accelerated, and the level of photoactivation was reduced	[87]

4.2. Future Trend

A large number of studies have shown that there is a great potential for UV/PS in environmental remediation.

(1) From the perspective of the light source system, the ultraviolet light-emitting diode (UV-LED) would be a better alternative light source for UV irradiation due to its low operating cost and excellent sustainability. Studies have shown that the removal rate of humic acid by UV-LED/chlorine is higher than that of traditional UV/chlorine [88]. UV-LED provides some unique functions. For example, UV-LED could emit various radiation wavelengths; meanwhile, periodic high-frequency illumination would occur. Periodic high-frequency illumination could trigger pulse irradiation. Studies have shown that pulse irradiation could help to reach a significantly higher degradation efficiency than that of continuous irradiation [89]. However, the pulsed UV-LED/PS method is affected by the duty cycle, pulse period, and so on [90]. Therefore, pulsed irradiation of UV-LED might be a potential research hotspot. Besides UV activation, visible light activation of PS (Vis/PS) might be more energy-saving and more efficient in removing pollutants. However, due to the insufficient visible light absorption and the electron-hole complexation, Vis/PS usually needs to be mediated by photoresponsive catalytic materials to regulate the optical properties and the activation rate of PS [91,92]. The enhancement of pollutant degradation was suggested due to the synergistic effect between photocatalysis and persulfate activation [60]. Therefore, visible photocatalytic activation should also be further investigated.

(2) The choice of different wavelengths also significantly affects the pollutant's degradation. For example, by comparing the degradation efficiency from the wavelength 255 to 365 nm generated by UV-LED, the highest degradation efficiency of ciprofloxacin was found at 280 nm [93]. VUV has also emerged as a compelling candidate for its high efficiency and environmental friendliness. It was demonstrated that 185 nm VUV irradiation is capable of generating ROS in situ, resulting in a superior performance in the degradation of organic pollutants [94]. Studies have shown that the PS activation rate by 185 nm VUV and the pollutant's removal rate were much higher than those of 254 nm, and it was suggested that 185 nm UV might promote the formation of ·OH. However, the mechanism still needs to be explored [95]. One study compared the removal rate of bioleachate by UV254/PS and UV365/PS, respectively. The results showed that the UV365/PS system was more effective [54]. In addition to what has been mentioned above, the combination of different wavelengths might be another research trend. The VUV/UV/PS process was indicated superior to the UV/PS process regarding organic pollutant removal efficiency and water treatment cost [65]. The exploration of pulse irradiation, the combination of multi-wavelength UV, and the potential synergistic effects for simultaneous removal of various pollutants in water still need to be investigated.

(3) The synthesis and modification of carriers such as bimetallic catalysts [96], magnetic nanocomposites [97,98], and nitrogen-doped carbonaceous materials [99] are also of great importance to optimize persulfate advanced oxidation technology [100]. Lee et al. [101] used UV/PS and UV/PS/Cu^{2+} to assess the degradation effect of non-oxidizing bactericides and fungicides. The results indicated that UV/PS/Cu^{2+} could increase the degradation efficiency by 4.4–22.2%. It would be seen that the combination of metal-based materials and UV/PS technology has a more significant removal effect on certain pollutants. Therefore, the current research and development of UV/PS should focus on the effective activation of persulfate to continuously produce various active species to achieve a better degradation effect of environmental pollutants.

(4) Regarding the oxidant selection, PMS and PDS showed different characteristics in degrading pollutants through UV activation. By comparison of the degradation of imidacloprid by UV/PMS and UV/PDS, different k-values were obtained under the same reaction time, oxidant, and UV dose, with 0.057/min for UV/PDS and 0.041/min for UV/PMS respectively. Meanwhile, the removal rate for UV/PDS was also higher [102]. Similar results were found when comparing the removal rate of pollutants in textile wastewater [103].

However, UV/PMS was sometimes indicated to be more effective than UV/PDS, e.g., the degradation efficiency of SMX by UV/PMS is slightly higher than that of UV/PDS, and the effect of pH might be one potential reason [17]. Therefore, researchers should rationally select the oxidant with careful consideration in the future.

(5) As for complex water bodies application, UV activation of persulfate is affected by various factors, including the concentration of oxidant, pH, inorganic anions, organic substances, actual UV light intensity, and reaction time [104]. In the present studies, natural water bodies are rarely applied, and the experimental condition was frequently set up to simulate the natural water samples which might focus on just a few kinds of pollutants. The subsequent studies should focus on the actual water bodies, such as rivers, lakes, reclaimed water seawater, and so on, to investigate the interference effects of natural water matrix such as humic acid, inorganic anions, as well as some other reactive species generated in the engineering practice. In addition, photoreactivation and dark repair of microorganisms and some resistant types of cells such as viable but nonculturable cells (VBNC) and persisters should also be considered. The risk assessment of biotoxicity and ecotoxicity needs to be comprehensively investigated as well.

5. Conclusions

Based on the bibliometric analysis, UV/PS has been suggested as an effective and environmentally friendly disinfection technology for controlling diverse environmental pollutants. The following conclusions could be drawn:

(1) UV/PS has been attracting the attention of researchers since its invention. Researchers in China, Iran, and the United States made a great contribution to the field of UV/PS, and the research interests are still growing.

(2) The research teams led by Prof. Ma Jun, Prof. Gao Naiyun, and Prof. Dionysiou Dionysiou D showed widespread and profound academic influence for later research. The research article supervised by Prof. Wang JL at Tsinghua University was cited the most, which provides a good insight for the related researchers.

(3) The research in the field of UV/PS mainly focused on the reaction mechanism and kinetics simulation. Moreover, UV/PS, indicated as an effective technology, could be applied for the degradation of traditional pollutants, ECs, and TOPs in various scenarios.

(4) The improvement of the light source system, the combination of multi-wavelengths, the development of reusable and efficient catalysts, and the selection of suitable oxidants might become the hotspots for future research on UV/PS.

(5) To promote the application of UV/PS, the influencing factors (inorganic anions, etc.) in the practical water matrix and some other active species (active chlorine species, etc.), especially generated during the practical engineering application should be focused on. The revival of certain microorganisms, the induce of VBNC and persister cells might be further investigated. Moreover, the potential toxicity risk needs to be assessed.

Author Contributions: Methodology, Z.J.; modeling, H.W.; formal analysis, H.W.; data analysis, H.W. and W.S.; data curation, H.W., R.L. and H.H.; writing—original draft preparation, H.W. and Z.J.; writing—review and editing, Z.J.; visualization, H.W.; supervision, Z.J.; funding acquisition, Z.J. All authors have read and agreed to the published version of the manuscript.

Funding: The research was funded by the National Natural Science Foundation of China (52000127), the Science and Technology Project of Xi'an (22GXFW0018) and (2017071CG/RC034(SXSF002)), the China Scholarship Council (201906875037), and the Natural Science Foundation of Shaanxi Province (2017JQ5074).

Data Availability Statement: The data presented in this study are available on request from the corresponding author.

Conflicts of Interest: The authors declare no conflicts of interest.

Abbreviations

AOPs: advanced oxidation process; AWaRe, Access, Watch, Reserve; CAP, chloramphenicol; ECs, emerging contaminants; GSM, geosmin; ·OH, hydroxyl radical; HA, humic acids; 2-MB, 2-methylisoborneol; MB, methylene blue; MPUV, medium pressure ultraviolet; NOM, natural organic matter; OXC, oxcarbazepine; PPCPs, personal care products; PFAAs, perfluoroalkyl acids; PCBs, polychlorinated biphenyls; PS, persulfate; PMS, peroxymonosulfate; PDS, peroxydisulfate; SMX, sulfamethoxazole; $SO_4^{-}\cdot$, sulfate radical; $^1O_2$, singlet oxygen; TEPs, transparent exopolymer particles; TTC, tetracycline; TOPs, trace organic pollutants; UV/PS, UV-activated persulfate; UV-LED, ultraviolet light-emitting diode; Vis/PS, visible light/PS; VBNC, viable but nonculturable cells; VUV, vacuum ultraviolet; WOS, Web of Science.

References

1. Zhao, F.; Sun, Y.; Zhang, J. Does industrial agglomeration and environmental pollution have a spatial spillover effect?: Taking panel data of resource-based cities in China as an example. *Environ. Sci. Pollut. Res.* **2023**, *30*, 76829–76841. [CrossRef] [PubMed]
2. Zhao, Y.G.; Chen, L.H.; Ye, M.L.; Su, W.S.; Lei, C.; Jin, X.J.; Lu, Y.S. U(VI) removal on polymer adsorbents: Recent development and future challenges. In *Critical Reviews in Environmental Science and Technology*; Taylor & Francis: Abingdon, UK, 2024.
3. Gong, H.; Hu, J.; Rui, X.; Wang, Y.; Zhu, N. Drivers of change behind the spatial distribution and fate of typical trace organic pollutants in fresh waste leachate across China. *Water Res.* **2024**, *263*, 122170. [CrossRef] [PubMed]
4. Morin-Crini, N.; Lichtfouse, E.; Liu, G.; Balaram, V.; Ribeiro, A.R.L.; Lu, Z.; Stock, F.; Carmona, E.; Teixeira, M.R.; Picos-Corrales, L.A.; et al. Worldwide cases of water pollution by emerging contaminants: A review. *Environ. Chem. Lett.* **2022**, *20*, 2311–2338. [CrossRef]
5. Abou Khalil, Z.; Baalbaki, A.; Bejjani, A.; Ghauch, A. MIL88-A as a mediator for the degradation of sulfamethoxazole in PS systems: Implication of solar irradiation for process improvement. *Environ. Sci. Adv.* **2022**, *1*, 797–813. [CrossRef]
6. Meng, X.; Li, F.; Yi, L.; Dieketseng, M.Y.; Wang, X.; Zhou, L.; Zheng, G. Free radicals removing extracellular polymeric substances to enhance the degradation of intracellular antibiotic resistance genes in multi-resistant Pseudomonas Putida by UV/H_2O_2 and UV/peroxydisulfate disinfection processes. *J. Hazard. Mater.* **2022**, *430*, 128502. [CrossRef]
7. Zhang, Y.X.; Xiang, J.L.; Wang, J.J.; Du, H.S.; Wang, T.T.; Huo, Z.Y.; Wang, W.L.; Liu, M.; Du, Y. Ultraviolet-based synergistic processes for wastewater disinfection: A review. *J. Hazard. Mater.* **2023**, *453*, 131393. [CrossRef] [PubMed]
8. Wei, J.; Liang, G.; Alex, J.; Zhang, T.; Ma, C. Research Progress of Energy Utilization of Agricultural Waste in China: Bibliometric Analysis by Citespace. *Sustainability* **2020**, *12*, 812. [CrossRef]
9. Ali, G.; Ahmed, S.; Mubushar, M.; Gulzar, A. Global Research on Ethical Leadership: A Bibliometric Investigation and Knowledge Mapping. *J. Chin. Hum. Resour. Manag.* **2023**, *14*, 51–65. [CrossRef]
10. Li, Z.; Song, L. Interpretation of the Guidelines for drinking-water quality (fourth edition) issued by World Health Organization. *Water Wastewater Eng.* **2012**, *38*, 9–13.
11. Pandian, A.M.K.; Rajamehala, M.; Singh, M.V.P.; Sarojini, G.; Rajamohan, N. Potential risks and approaches to reduce the toxicity of disinfection by-product—A review. *Sci. Total Environ.* **2022**, *822*, 153323. [CrossRef]
12. Guan, Y.H.; Ma, J.; Li, X.C.; Fang, J.Y.; Chen, L.W. Influence of pH on the Formation of Sulfate and Hydroxyl Radicals in the UV/Peroxymonosulfate System. *Environ. Sci. Technol.* **2011**, *45*, 9308–9314. [CrossRef] [PubMed]
13. Lee, J.; Gunten, U.v.; Kim, J.H. Persulfate-Based Advanced Oxidation: Critical Assessment of Opportunities and Roadblocks. *Environ. Sci. Technol.* **2020**, *54*, 3064–3081. [CrossRef] [PubMed]
14. Liu, Y.Q.; He, X.X.; Fu, Y.S.; Dionysiou, D.D. Kinetics and mechanism investigation on the destruction of oxytetracycline by UV-254 nm activation of persulfate. *J. Hazard. Mater.* **2016**, *305*, 229–239. [CrossRef] [PubMed]
15. Khan, J.A.; He, X.; Shah, N.S.; Khan, H.M.; Hapeshi, E.; Fatta-Kassinos, D.; Dionysiou, D.D. Kinetic and mechanism investigation on the photochemical degradation of atrazine with activated H_2O_2, $S_2O_8^{2-}$ and HSO_5^{-}. *Chem. Eng. J.* **2014**, *252*, 393–403. [CrossRef]
16. Zhang, R.C.; Sun, P.Z.; Boyer, T.H.; Zhao, L.; Huang, C.H. Degradation of Pharmaceuticals and Metabolite in Synthetic Human Urine by UV, UV/H_2O_2, and UV/PDS. *Environ. Sci. Technol.* **2015**, *49*, 3056–3066. [CrossRef]
17. Ao, X.W.; Liu, W.J. Degradation of sulfamethoxazole by medium pressure UV and oxidants: Peroxymonosulfate, persulfate, and hydrogen peroxide. *Chem. Eng. J.* **2017**, *313*, 629–637. [CrossRef]
18. Lutze, H.V.; Bircher, S.; Rapp, I.; Kerlin, N.; Bakkour, R.; Geisler, M.; Von Sonntag, C.; Schmidt, T.C. Degradation of Chlorotriazine Pesticides by Sulfate Radicals and the Influence of Organic Matter. *Environ. Sci. Technol.* **2015**, *49*, 1673–1680. [CrossRef]
19. Sharma, J.; Mishra, I.M.; Dionysiou, D.D.; Kumar, V. Oxidative removal of Bisphenol A by UV-C/peroxymonosulfate (PMS): Kinetics, influence of co-existing chemicals and degradation pathway. *Chem. Eng. J.* **2015**, *276*, 193–204. [CrossRef]
20. Xiao, Y.J.; Zhang, L.F.; Zhang, W.; Lim, K.Y.; Webster, R.D.; Lim, T.T. Comparative evaluation of iodoacids removal by UV/persulfate and UV/H_2O_2 processes. *Water Res.* **2016**, *102*, 629–639. [CrossRef]
21. Tan, C.; Gao, N.Y.; Deng, Y.; Zhang, Y.J.; Sui, M.H.; Deng, J.; Zhou, S.Q. Degradation of antipyrine by UV, UV/H_2O_2 and UV/PS. *J. Hazard. Mater.* **2013**, *260*, 1008–1016. [CrossRef]

22. Gao, Y.Q.; Gao, N.Y.; Deng, Y.; Yang, Y.Q.; Ma, Y. Ultraviolet (UV) light-activated persulfate oxidation of sulfamethazine in water. *Chem. Eng. J.* **2012**, *195*, 248–253. [CrossRef]
23. Luo, C.W.; Ma, J.; Jiang, J.; Liu, Y.Z.; Song, Y.; Yang, Y.; Guan, Y.H.; Wu, D.J. Simulation and comparative study on the oxidation kinetics of atrazine by UV/H_2O_2, UV/HSO5 and UV/S2O82-. *Water Res.* **2015**, *80*, 99–108. [CrossRef] [PubMed]
24. Yang, Y.; Lu, X.L.; Jiang, J.; Ma, J.; Liu, G.Q.; Cao, Y.; Liu, W.L.; Li, J.; Pang, S.Y.; Kong, X.J.; et al. Degradation of sulfamethoxazole by UV, UV/H_2O_2 and UV/persulfate (PDS): Formation of oxidation products and effect of bicarbonate. *Water Res.* **2017**, *118*, 196–207. [CrossRef]
25. Lian, L.S.; Yao, B.L.; Hou, S.D.; Fang, J.Y.; Yan, S.W.; Song, W.H. Kinetic Study of Hydroxyl and Sulfate Radical-Mediated Oxidation of Pharmaceuticals in Wastewater Effluents. *Environ. Sci. Technol.* **2017**, *51*, 2954–2962. [CrossRef]
26. Ghauch, A.; Baalbaki, A.; Amasha, M.; El Asmar, R.; Tantawi, O. Contribution of persulfate in UV-254 nm activated systems for complete degradation of chloramphenicol antibiotic in water. *Chem. Eng. J.* **2017**, *317*, 1012–1025. [CrossRef]
27. He, X.X.; de la Cruz, A.A.; Dionysiou, D.D. Destruction of cyanobacterial toxin cylindrospermopsin by hydroxyl radicals and sulfate radicals using UV-254 nm activation of hydrogen peroxide, persulfate and peroxymonosulfate. *J. Photochem. Photobiol. A-Chem.* **2013**, *251*, 160–166. [CrossRef]
28. Miklos, D.B.; Remy, C.; Jekel, M.; Linden, K.G.; Drewes, J.E.; Huebner, U. Evaluation of advanced oxidation processes for water and wastewater treatment-A critical review. *Water Res.* **2018**, *139*, 118–131. [CrossRef]
29. Oh, W.D.; Dong, Z.; Lim, T.T. Generation of sulfate radical through heterogeneous catalysis for organic contaminants removal: Current development, challenges and prospects. *Appl. Catal. B-Environ.* **2016**, *194*, 169–201. [CrossRef]
30. Giannakis, S.; Lin, K.Y.A.; Ghanbari, F. A review of the recent advances on the treatment of industrial wastewaters by Sulfate Radical-based Advanced Oxidation Processes (SR-AOPs). *Chem. Eng. J.* **2021**, *406*, 127083. [CrossRef]
31. Deng, J.; Shao, Y.S.; Gao, N.Y.; Xia, S.J.; Tan, C.Q.; Zhou, S.Q.; Hu, X.H. Degradation of the antiepileptic drug carbamazepine upon different UV-based advanced oxidation processes in water. *Chem. Eng. J.* **2013**, *222*, 150–158. [CrossRef]
32. Khan, J.A.; He, X.X.; Khan, H.M.; Shah, N.S.; Dionysiou, D.D. Oxidative degradation of atrazine in aqueous solution by UV/H_2O_2/Fe^{2+}, UV/$S_2O_8^{2-}$/Fe^{2+} and UV/HSO_5^-/Fe^{2+} processes: A comparative study. *Chem. Eng. J.* **2013**, *218*, 376–383. [CrossRef]
33. Zhang, Y.Q.; Zhang, J.F.; Xiao, Y.J.; Chang, V.W.C.; Lim, T.T. Kinetic and mechanistic investigation of azathioprine degradation in water by UV, UV/H_2O_2 and UV/persulfate. *Chem. Eng. J.* **2016**, *302*, 526–534. [CrossRef]
34. Ike, I.A.; Linden, K.G.; Orbell, J.D.; Duke, M. Critical review of the science and sustainability of persulphate advanced oxidation processes. *Chem. Eng. J.* **2018**, *338*, 651–669. [CrossRef]
35. Yang, Q.; Ma, Y.H.; Chen, F.; Yao, F.B.; Sun, J.; Wang, S.N.; Yi, K.X.; Hou, L.H.; Li, X.M.; Wang, D.B. Recent advances in photo-activated sulfate radical-advanced oxidation process (SR-AOP) for refractory organic pollutants removal in water. *Chem. Eng. J.* **2019**, *378*, 122149. [CrossRef]
36. Lin, Y.T.; Liang, C.J.; Chen, J.H. Feasibility study of ultraviolet activated persulfate oxidation of phenol. *Chemosphere* **2011**, *82*, 1168–1172. [CrossRef]
37. Rehman, F.; Sayed, M.; Khan, J.A.; Shah, N.S.; Khan, H.M.; Dionysiou, D.D. Oxidative removal of brilliant green by UV/S2O82-, UV/HSO5- and UV/H_2O_2 processes in aqueous media: A comparative study. *J. Hazard. Mater.* **2018**, *357*, 506–514. [CrossRef]
38. Yang, Y.; Pignatello, J.J.; Ma, J.; Mitch, W.A. Comparison of Halide Impacts on the Efficiency of Contaminant Degradation by Sulfate and Hydroxyl Radical-Based Advanced Oxidation Processes (AOPs). *Environ. Sci. Technol.* **2014**, *48*, 2344–2351. [CrossRef]
39. Luo, C.W.; Jiang, J.; Ma, J.; Pang, S.Y.; Liu, Y.Z.; Song, Y.; Guan, C.T.; Li, J.; Jin, Y.X.; Wu, D.J. Oxidation of the odorous compound 2,4,6-trichloroanisole by UV activated persulfate: Kinetics, products, and pathways. *Water Res.* **2016**, *96*, 12–21. [CrossRef]
40. He, X.X.; de la Cruz, A.A.; O'Shea, K.E.; Dionysiou, D.D. Kinetics and mechanisms of cylindrospermopsin destruction by sulfate radical-based advanced oxidation processes. *Water Res.* **2014**, *63*, 168–178. [CrossRef]
41. Lutze, H.V.; Kerlin, N.; Schmidt, T.C. Sulfate radical-based water treatment in presence of chloride: Formation of chlorate, inter-conversion of sulfate radicals into hydroxyl radicals and influence of bicarbonate. *Water Res.* **2015**, *72*, 349–360. [CrossRef]
42. Olmez-Hanci, T.; Arslan-Alaton, I. Comparison of sulfate and hydroxyl radical based advanced oxidation of phenol. *Chem. Eng. J.* **2013**, *224*, 10–16. [CrossRef]
43. Liu, X.H.; Liu, Y.; Lu, S.Y.; Wang, Z.; Wang, Y.Q.; Zhang, G.D.; Guo, X.C.; Guo, W.; Zhang, T.T.; Xi, B.D. Degradation difference of ofloxacin and levofloxacin by UV/H_2O_2 and UV/PS (persulfate): Efficiency, factors and mechanism. *Chem. Eng. J.* **2020**, *385*, 13. [CrossRef]
44. Tan, C.Q.; Fu, D.F.; Gao, N.Y.; Qin, Q.D.; Xu, Y.; Xiang, H.M. Kinetic degradation of chloramphenicol in water by UV/persulfate system. *J. Photochem. Photobiol. A-Chem.* **2017**, *332*, 406–412. [CrossRef]
45. Yang, Y.; Pignatello, J.J.; Ma, J.; Mitch, W.A. Effect of matrix components on UV/H_2O_2 and UV/$S_2O_8^{2-}$ advanced oxidation processes for trace organic degradation in reverse osmosis brines from municipal wastewater reuse facilities. *Water Res.* **2016**, *89*, 192–200. [CrossRef]
46. Wang, S.Z.; Liu, Y.; Wang, J.L. Iron and sulfur co-doped graphite carbon nitride (FeOy/S-g-C_3N_4) for activating peroxymonosulfate to enhance sulfamethoxazole degradation. *Chem. Eng. J.* **2020**, *382*, 122836. [CrossRef]
47. Liu, Y.; Zhang, L.; Liu, T.C.; Wang, J.L. Photocatalytic selective oxidation of ammonium to dinitrogen by FeOx-MgO activated persulfate under solar-light irradiation. *Chem. Eng. J.* **2023**, *454*, 140499. [CrossRef]

48. Wang, J.; Wang, S. Activation of persulfate (PS) and peroxymonosulfate (PMS) and application for the degradation of emerging contaminants. *Chem. Eng. J.* **2018**, *334*, 1502–1517. [CrossRef]
49. Xie, P.; Ma, J.; Liu, W.; Zou, J.; Yue, S.; Li, X.; Wiesner, M.R.; Fang, J. Removal of 2-MIB and geosmin using UV/persulfate: Contributions of hydroxyl and sulfate radicals. *Water Res.* **2015**, *69*, 223–233. [CrossRef]
50. Wan, Y.; Xie, P.; Wang, Z.; Ding, J.; Wang, J.; Wang, S.; Wiesner, M.R. Comparative study on the pretreatment of algae-laden water by UV/persulfate, UV/chlorine, and UV/H_2O_2: Variation of characteristics and alleviation of ultrafiltration membrane fouling. *Water Res.* **2019**, *158*, 213–226. [CrossRef]
51. Waclawek, S.; Lutze, H.V.; Grubel, K.; Padil, V.V.T.; Cernik, M.; Dionysiou, D.D. Chemistry of persulfates in water and wastewater treatment: A review. *Chem. Eng. J.* **2017**, *330*, 44–62. [CrossRef]
52. Matzek, L.W.; Carter, K.E. Activated persulfate for organic chemical degradation: A review. *Chemosphere* **2016**, *151*, 178–188. [CrossRef] [PubMed]
53. Ghanbari, F.; Moradi, M. Application of peroxymonosulfate and its activation methods for degradation of environmental organic pollutants: Review. *Chem. Eng. J.* **2017**, *310*, 41–62. [CrossRef]
54. Yang, J.; Zhu, M.; Dionysiou, D.D. What is the role of light in persulfate-based advanced oxidation for water treatment? *Water Res.* **2021**, *189*, 116627. [CrossRef] [PubMed]
55. Feng, H.N.; Xu, D.Y.; Dong, Y.L.; Zhnag, L.; Wang, P.F.; Ren, Z.J. Kinetic Model and Influencing Factors of UV Disinfection. *Water Purif. Technol.* **2020**, *39*, 83–89.
56. Derbalah, A.; Sakugawa, H. Sulfate Radical-Based Advanced Oxidation Technology to Remove Pesticides From Water A Review of the Most Recent Technologies. *Int. J. Environ. Res.* **2024**, *18*, 11. [CrossRef]
57. Zhou, M.Z.; Cang, L. Progress of Chemical Oxidation Mechanism of Peroxymonosulfate and Its Application in Remediation of Organic Contaminated Soil. *Soil* **2022**, *54*, 653–666.
58. Liu, Z.; Ren, X.; Duan, X.; Sarmah, A.K.; Zhao, X. Remediation of environmentally persistent organic pollutants (POPs) by persulfates oxidation system (PS): A review. *Sci. Total Environ.* **2023**, *863*, 160818. [CrossRef]
59. Nidheesh, P.V.; Divyapriya, G.; Titchou, F.E.; Hamdani, M. Treatment of textile wastewater by sulfate radical based advanced oxidation processes. *Sep. Purif. Technol.* **2022**, *293*, 121115. [CrossRef]
60. Heidarpour, H.; Padervand, M.; Soltanieh, M.; Vossoughi, M. Enhanced decolorization of rhodamine B solution through simultaneous photocatalysis and persulfate activation over Fe/C3N4 photocatalyst. *Chem. Eng. Res. Des.* **2020**, *153*, 709–720. [CrossRef]
61. Zhao, M.; Liu, Y.Y.; Feng, M.B.; Yu, X.; Wang, L. Degradation mechanisms of antibiotics in UV222/H_2O_2 and UV222/ persulfate systems: Dual roles of inorganic anions. *Chem. Eng. J.* **2024**, *489*, 151371. [CrossRef]
62. Yin, R.; Zhang, Y.; Wang, Y.; Zhao, J.; Shang, C. Far-UVC Photolysis of Peroxydisulfate for Micropollutant Degradation in Water. *Environ. Sci. Technol.* **2024**, *58*, 6030–6038. [CrossRef] [PubMed]
63. Verma, S.; Nakamura, S.; Sillanpaa, M. Application of UV-C LED activated PMS for the degradation of anatoxin-a. *Chem. Eng. J.* **2016**, *284*, 122–129. [CrossRef]
64. Sun, X.; Shi, L.-X.; Zhang, Y.; Yang, L.; Tang, X. Removal of Odorants in Drinking Water Using VUV/Persulfate. *Huan Jing Ke Xue = Huanjing Kexue* **2018**, *39*, 2195–2201. [PubMed]
65. Wen, D.; Li, W.T.; Lv, J.R.; Qiang, Z.M.; Li, M.K. Methylene blue degradation by the VUV/UV/persulfate process: Effect of pH on the roles of photolysis and oxidation. *J. Hazard. Mater.* **2020**, *391*, 121855. [CrossRef] [PubMed]
66. Lei, X.; Lei, Y.; Zhang, X.R.; Yang, X. Treating disinfection byproducts with UV or solar irradiation and in UV advanced oxidation processes: A review. *J. Hazard. Mater.* **2021**, *408*, 17. [CrossRef]
67. Ma, Y.; Wang, Z.; Yang, W.; Chen, C.; Li, J.; He, R.; Liu, S. Insights into the radical and nonradical oxidation degradation of ciprofloxacin in peroxodisulfate activation by ultraviolet light. *J. Water Process Eng.* **2022**, *49*, 103184. [CrossRef]
68. Liu, J.; Song, Z.; Liu, B.; Zhang, X.; Huang, Y.; Han, D. Bimetallic Ni/Fe Atom Cluster Catalysts Enhance Non-free Radical Degradation of Organic Pollutant Phenol. *Catal. Lett.* **2024**, *154*, 2182–2196. [CrossRef]
69. Lu, J.; Ding, P.; Lyu, Z.; Xu, G.; Peng, M.; Du, E.; Zheng, L. Degradation kinetics and mechanism of SDM by UV/PS process in water. *Environ. Pollut. Control* **2022**, *44*, 183.
70. Sillanpaa, M.; Ncibi, M.C.; Matilainen, A. Advanced oxidation processes for the removal of natural organic matter from drinking water sources: A comprehensive review. *J. Environ. Manag.* **2018**, *208*, 56–76. [CrossRef]
71. Alayande, A.B.; Yun, E.T.; Silva, F.; Hong, S. Mechanistic insights into the potential applicability of a sulfate-based advanced oxidation process for the control of transparent exopolymer particles in membrane-based desalination. *Desalination* **2022**, *522*, 9. [CrossRef]
72. Tian, J.Y.; Wu, C.W.; Yu, H.R.; Gao, S.S.; Li, G.B.; Cui, F.Y.; Qu, F.S. Applying ultraviolet/persulfate (UV/PS) pre-oxidation for controlling ultrafiltration membrane fouling by natural organic matter (NOM) in surface water. *Water Res.* **2018**, *132*, 190–199. [CrossRef] [PubMed]
73. Zhang, Q. Effect of Bromine Ion on Kinetics of Degradation of Organic Micropollutants in Ultraviolet/Persulfate Oxidation System. Master Thesis, Kunming University of Science and Technology, Kunming, China, 2023.
74. Zhang, Y.; Zhao, Y.G.; Maqbool, F.; Hu, Y. Removal of antibiotics pollutants in wastewater by UV-based advanced oxidation processes: Influence of water matrix components, processes optimization and application: A review. *J. Water Process Eng.* **2022**, 45. [CrossRef]

75. Ao, X.W.; Sun, W.J.; Li, S.M.; Yang, C.; Li, C.; Lu, Z.D. Degradation of tetracycline by medium pressure UV-activated peroxymonosulfate process: Influencing factors, degradation pathways, and toxicity evaluation. *Chem. Eng. J.* **2019**, *361*, 1053–1062. [CrossRef]
76. Bu, L.J.; Zhou, S.Q.; Shi, Z.; Deng, L.; Li, G.C.; Yi, Q.H.; Gao, N.Y. Degradation of oxcarbazepine by UV-activated persulfate oxidation: Kinetics, mechanisms, and pathways. *Environ. Sci. Pollut. Res.* **2016**, *23*, 2848–2855. [CrossRef] [PubMed]
77. Zhang, C.M.; Yang, H.M.; Wang, Z. Characteristics and Mechanisms of Bacteriophage MS2 Inactivation in Water by UV Activated Sodium Persulfate. *Huan Jing Ke Xue Huanjing Kexue* **2021**, *42*, 4807–4814.
78. Guerra-Rodriguez, S.; Rodriguez, E.; Moreno-Andres, J.; Rodriguez-Chueca, J. Effect of the water matrix and reactor configuration on Enterococcus sp. inactivation by UV-A activated PMS or H_2O_2. *J. Water Process Eng.* **2022**, *47*, 102740. [CrossRef]
79. Xiao, R.; Liu, K.; Bai, L.; Minakata, D.; Seo, Y.; Goktas, R.K.; Dionysiou, D.D.; Tang, C.J.; Wei, Z.; Spinney, R. Inactivation of pathogenic microorganisms by sulfate radical: Present and future. *Chem. Eng. J.* **2019**, *371*, 222–232. [CrossRef]
80. Berruti, I.; Oller, I.; Inmaculada Polo-Lopez, M. Direct oxidation of peroxymonosulfate under natural solar radiation: Accelerating the simultaneous removal of organic contaminants and pathogens from water. *Chemosphere* **2021**, *279*, 130555. [CrossRef]
81. Serna-Galvis, E.A.; Ferraro, F.; Silva-Agredo, J.; Torres-Palma, R.A. Degradation of highly consumed fluoroquinolones, penicillins and cephalosporins in distilled water and simulated hospital wastewater by UV254 and UV254/persulfate processes. *Water Res.* **2017**, *122*, 128–138. [CrossRef]
82. Chen, Y.Q.; Tong, Y.; Liu, Z.Z.; Huang, L.Z.; Yuan, J.P.; Xue, Y.W.; Fang, Z. Enhanced degradation of Orange II using a novel UV/persulfate/sulfite system. *Environ. Chem. Lett.* **2019**, *17*, 1435–1439. [CrossRef]
83. Alayande, A.B.; Hong, S. Ultraviolet light-activated peroxymonosulfate (UV/PMS) system for humic acid mineralization: Effects of ionic matrix and feasible application in seawater reverse osmosis desalination*. *Environ. Pollut.* **2022**, *307*, 119513. [CrossRef] [PubMed]
84. Anipsitakis, G.P.; Dionysiou, D.D. Transition metal/UV-based advanced oxidation technologies for water decontamination. *Appl. Catal. B-Environ.* **2004**, *54*, 155–163. [CrossRef]
85. Xu, Z.; Shan, C.; Xie, B.H.; Liu, Y.; Pan, B.C. Decomplexation of Cu(II)-EDTA by UV/persulfate and UV/H_2O_2: Efficiency and mechanism. *Appl. Catal. B-Environ.* **2017**, *200*, 439–447. [CrossRef]
86. Zhou, C.S.; Wu, J.W.; Dong, L.L.; Liu, B.F.; Xing, D.F.; Yang, S.S.; Wu, X.K.; Wang, Q.; Fan, J.N.; Feng, L.P.; et al. Removal of antibiotic resistant bacteria and antibiotic resistance genes in wastewater effluent by UV-activated persulfate. *J. Hazard. Mater.* **2020**, *388*, 122070. [CrossRef]
87. Wan, Q.Q.; Cao, R.H.; Wen, G.; Xu, X.Q.; Xia, Y.C.; Wu, G.H.; Li, Y.F.; Wang, J.Y.; Xu, H.N.; Lin, Y.Z.; et al. Efficacy of UV-LED based advanced disinfection processes in the inactivation of waterborne fungal spores: Kinetics, photoreactivation, mechanism and energy requirements. *Sci. Total Environ.* **2022**, *803*, 150107. [CrossRef]
88. Gao, Z.C.; Lin, Y.L.; Xu, B.; Xia, Y.C.; Hu, C.Y.; Zhang, T.Y.; Cao, T.C.; Pan, Y.; Gao, N.Y. A comparison of dissolved organic matter transformation in low pressure ultraviolet (LPUV) and ultraviolet light-emitting diode (UV-LED)/chlorine processes. *Sci. Total Environ.* **2020**, *702*, 134942. [CrossRef]
89. Eskandarian, M.R.; Ganjkhanloo, M.; Rasoulifard, M.H.; Hosseini, S.A. Energy-efficient removal of acid red 14 by UV-LED/persulfate advanced oxidation process: Pulsed irradiation, duty cycle, reaction kinetics, and energy consumption. *J. Taiwan Inst. Chem. Eng.* **2021**, *127*, 129–139. [CrossRef]
90. Hajimiri, I.; Rasoulifard, M.H.; Dorraji, M.S.S.; Eskandarian, M.R. Study of controlled pulsed illumination (CPI) efficiency in the homogeneous UV-LED/S2O82- process for the removal of organic dye from contaminated water. *Results Eng.* **2024**, *21*, 101792. [CrossRef]
91. Liu, Y.; Guo, H.; Zhang, Y.; Cheng, X.; Zhou, P.; Zhang, G.; Wang, J.; Tang, P.; Ke, T.; Li, W. Heterogeneous activation of persulfate for Rhodamine B degradation with 3D flower sphere-like BiOI/Fe_3O_4 microspheres under visible light irradiation. *Sep. Purif. Technol.* **2018**, *192*, 88–98. [CrossRef]
92. Yang, L.; Bai, X.; Shi, J.; Du, X.; Xu, L.; Jin, P. Quasi-full-visible-light absorption by D35-TiO_2g-C_3N_4 for synergistic persulfate activation towards efficient photodegradation of micropollutants. *Appl. Catal. B-Environ.* **2019**, *256*, 117759. [CrossRef]
93. Ou, H.S.; Ye, J.S.; Ma, S.L.; Wei, C.H.; Gao, N.Y.; He, J.Z. Degradation of ciprofloxacin by UV and UV/H_2O_2 via multiple-wavelength ultraviolet light-emitting diodes: Effectiveness, intermediates and antibacterial activity. *Chem. Eng. J.* **2016**, *289*, 391–401. [CrossRef]
94. Chen, S.; Sheng, X.; Zhao, Z.; Cui, F. Chemical-free vacuum ultraviolet irradiation as ultrafiltration membrane pretreatment technique: Performance, mechanisms and DBPs formation. *J. Environ. Manag.* **2024**, *351*, 119785. [CrossRef] [PubMed]
95. An, D.; Westerhoff, P.; Zheng, M.X.; Wu, M.Y.; Yang, Y.; Chiu, C.A. UV-activated persulfate oxidation and regeneration of NOM-Saturated granular activated carbon. *Water Res.* **2015**, *73*, 304–310. [CrossRef] [PubMed]
96. Xu, H.; Zhang, J.; Zhang, G.; Ye, Q.; Zhou, G.; Yang, F. Kinetics research on bimetallic CoMnO_x/UV catalyzed persulfate for the degradation of AO7. *Ind. Water Treat.* **2018**, *38*, 42–45.
97. Eskandarian, M.R.; Rasoulifard, M.H.; Fazli, M.; Ghalamchi, L.; Choi, H. Synergistic decomposition of imidacloprid by TiO2-Fe_3O_4 nanocomposite conjugated with persulfate in a photovoltaic-powered UV-LED photoreactor. *Korean J. Chem. Eng.* **2019**, *36*, 965–974. [CrossRef]
98. Song, T.; Gao, Y.; Hu, R.; Li, G.; Yu, X. Degradation of Methyl Orange in Aqueous Solution via Magnetic TiO_2/Fe_3O_4 Conjugated with Persulfate. *Water Air Soil Pollut.* **2023**, *234*, 508. [CrossRef]

99. Zhong, Q.F.; Lin, Q.T.; He, W.J.; Fu, H.Y.; Huang, Z.F.; Wang, Y.P.; Wu, L.B. Study on the nonradical pathways of nitrogen-doped biochar activating persulfate for tetracycline degradation. *Sep. Purif. Technol.* **2021**, *276*, 119354. [CrossRef]
100. Li, Y.T.; Jin, X.L. *Research Progress of Bismuth-Based Bimetallic Oxides in Advanced Oxidation of Persulfate*; Journal of Tianjin Polytechnic University: Tianjin, China, 2023; pp. 483–495.
101. Lee, M.Y.; Wang, W.L.; Du, Y.; Hu, H.Y.; Huang, N.; Xu, Z.B.; Wu, Q.Y.; Ye, B. Enhancement effect among a UV, persulfate, and copper (UV/PS/Cu^{2+}) system on the degradation of nonoxidizing biocide: The kinetics, radical species, and degradation pathway. *Chem. Eng. J.* **2020**, *382*, 122312. [CrossRef]
102. Wang, Q.; Rao, P.; Li, G.; Dong, L.; Zhang, X.; Shao, Y.; Gao, N.; Chu, W.; Xu, B.; An, N.; et al. Degradation of imidacloprid by UV-activated persulfate and peroxymonosulfate processes: Kinetics, impact of key factors and degradation pathway. *Ecotoxicol. Environ. Saf.* **2020**, *187*, 109779. [CrossRef]
103. Cifci, D.I.; Sensoy, O.; Gunes, E.; Gunes, Y. Treatment of a textile wastewater using sulfate and hydroxyl radicals based oxidation under the UV light: A Comparision between UV/H_2O_2, UV/PS and UV/PMS. *Environ. Eng. Manag. J.* **2023**, *22*, 17–26. [CrossRef]
104. Calaixo, M.R.C.; Ribeirinho-Soares, S.; Madeira, L.M.; Nunes, O.C.; Rodrigues, C.S.D. Catalyst-free persulfate activation by UV/visible radiation for secondary urban wastewater disinfection. *J. Environ. Manag.* **2023**, *348*, 12. [CrossRef] [PubMed]

water

MDPI

Article

Preparation of Bimetallic CoFe@CSC-700 Carbonated Microspheres and Activated Peroxymonosulfate for Degradation of Levofloxacin

Tongke Hu [1], Yazhen Chen [1], Xiaolan Guo [1], Yongjun Peng [1] and Jianhua Cheng [1,2,*]

[1] Ministry of Education Key Laboratory of Pollution Control and Ecological Remediation for Industrial Agglomeration Area, School of Environment and Energy, South China University of Technology, Guangzhou 510006, China; hutongke611@163.com (T.H.); yzchen0114@163.com (Y.C.); lesleylh1@163.com (X.G.); jerry199916@163.com (Y.P.)

[2] South China Institute of Collaborative Innovation, Dongguan 523808, China

* Correspondence: jhcheng@scut.edu.cn

Abstract: The exploration of efficient, low-leaching, and recyclable transition-metal-based catalysts is of great importance for the removal of pollutants from peroxymonosulfate (PMS) in water purification processes. In this study, a bimetallic CoFe@CSC-700 composite was prepared by an alkaline gel pyrolysis reduction method using chitosan as a forming agent and applied to activate PMS to degrade levofloxacin (LEV). The leaching concentration of both cobalt and iron ions in the CoFe@CSC-700 catalyst was reduced by about 8-fold compared to the monometallic composite pellet catalyst. In addition, the removal efficiency of the CoFe@CSC-700 catalyst can still reach 90% after five cycles, showing good recyclability, recoverability and stability. Both free radical pathways ($SO_4\cdot^-$, $\cdot OH$, and $\cdot O_2^-$) and non-free radical pathways ($^1O_2$) were detected in the oxidation reaction, with free radical pathways as the main contributor. The possible degradation pathways of LEV were proposed by LC-MS tests. Overall, this study provides new insights into the construction of efficient and stable PMS catalysts for wastewater treatment.

Keywords: cobalt–iron bimetallic; chitosan carbonized microspheres; peroxymonosulfate; levofloxacin

1. Introduction

Water resource pollution has become a serious challenge globally with rapid economic development and the improvement of people's quality of life. In particular, the abuse and improper discharge of drugs have attracted extensive attention from numerous researchers and the public [1]. Levofloxacin (LEV), a third-generation fluoroquinolone, is a representative pollutant that is used on a large scale in aquaculture and agriculture, and the residues of this drug enter the environment through the food chain and have been detected in various river waters [2]. Its ecotoxicity in the aquatic environment is significant, and when organisms are exposed to LEV-containing water bodies, their behavioral patterns, growth rates, and reproduction may be adversely affected, thus posing significant potential risks to ecosystem stability and human health over time. Further, the presence of LEV in aqueous environments can also induce rapid genetic mutation in bacteria, leading to the development of drug-resistant bacteria [3]. However, LEV shows extreme resistance to traditional biological treatments and cannot be completely metabolized. Therefore, there is an urgent need to explore effective remediation technologies for removing LEV from aqueous environments.

Advanced oxidation processes (AOPs) based on sulfate radicals are an emerging technology that can effectively treat antibiotics in wastewater [4]. In recent years, the sulfate radical ($SO_4\cdot^-$) has garnered recognition as a technology with significant potential due to its high redox potential (2.5~3.1 V), long half-life, and wide pH use range [5,

6]. Typically, $SO_4{\cdot}^-$ is generated from peroxymonosulfate (PMS) and peroxydisulfate (PDS). Considering the asymmetric molecular structure of PMS, it is easier to activate and produce sulfate radical ($SO_4{\cdot}^-$) than PDS [7]. There are many activation methods for PMS, including thermal activation [8], alkali activation [9], electrical activation [10], ultraviolet activation [11], transition metal activation [12] and ultrasonic activation [13]. The transition metal activation method is more applicable than other activation methods.

So far, Co-based metal oxides and their composites have attracted much attention among the many transition metals owing to their excellent catalytic properties [14]. However, the relatively high toxicity and potential carcinogenicity of Co leaching limit the practical application of Co in environmental remediation. Fe-based catalysts are considered as excellent alternatives for activating PMS because they are relatively environmentally friendly, widely available, and low-cost, but their catalytic performance is relatively low [15]. Previous studies have shown that when Co and Fe are combined, there is a synergistic catalytic effect, and Co-Fe bonding can reduce or even prevent Co^{2+} leakage. The study of Yang et al. [16] showed that the strong interaction between Co and Fe inhibited the leaching of Co^{2+}. The construction of bimetallic structures also improves the catalytic activity due to the synergistic effect between different metal species [17]. For example, Zhou et al. [18] found that the degradation efficiency of 2,4-dichlorophenol (2,4-DCP) in the $FeCo_2O_4$/PMS system was higher than that in the monometallic oxide Fe_3O_4/PMS system and Co_3O_4/PMS system. Although many efficient catalysts available for the removal of organics have been reported, the unfavorable usability of powdered catalysts and metal leaching are still issues to be solved. Chitosan (CS) is a green biopolymer that has gradually attracted great interest due to its good molding properties and high biocompatibility [19]. Since CS chains are rich in -NH_2 and -OH groups with high affinity for transition metals, many transition metals can be anchored to CS chains [20,21] and form magnetic chitosan microspheres in alkaline solutions, thus reducing metal loss and facilitating separation.

In this study, bimetallic CoFe@CSC-700 composites were designed and prepared by an alkaline gel pyrolytic reduction method to address the unfavorable recoverability of powder catalysts and metal leaching. Using LEV as the target pollutant, the catalytic activity of the CoFe@CSC-700/PMS system was systematically investigated, and the influencing factors (solution pH, catalyst dosing, PMS concentration) and the effects of different water quality components were explored. In addition, the recycling stability and recoverability of CoFe@CSC-700 were evaluated, and the reactive oxygen species (ROS) in the reaction system and reaction mechanism were described. Finally, the intermediates of LEV degradation were also elucidated by LC-MS, and the degradation pathways of the reaction were considered. Theoretical support and a practical application basis were provided to promote the activation of PMS by bimetallic composites for the purification of water pollutants.

2. Materials and Methods

2.1. Materials

Levofloxacin (LEV, ≥98.0%) was purchased from Sigma-Aldrich Co., Ltd. (Shanghai, China). Cobalt chloride hexahydrate ($CoCl_2{\cdot}6H_2O$, ≥99.0%), ferric chloride hexahydrate ($FeCl_3{\cdot}6H_2O$, ≥99.0%), peroxymonosulfate ($KHSO_5{\cdot}0.5KHSO_4{\cdot}0.5K_2SO_4$, PMS), methanol (MeOH, HPLC, ≥99.9%), tert butyl alcohol (TBA), L-histidine (L-His), and p-Benzoquinone (p-BQ) were purchased from Shanghai Macklin Biochemical Technology Co., Ltd. (Shanghai, China). Chitosan (CS, 95.0% deacetylation), sulfuric acid (H_2SO_4, 95~98%), acetic acid (CH_3COOH, 99%), sodium acetate (CH_3COONa, ≥99%), sodium hydroxide (NaOH, ≥98%), sodium chloride (NaCl), sodium nitrate ($NaNO_3$), and sodium sulfate (Na_2SO_4) were supplied by Aladdin Biochemical Technology Co., Ltd. (Shanghai, China). Humic acid (≥90%) used as natural organic matter (NOM) was provided by Sinopharm Chemical Reagent Beijing Co., Ltd. (Beijing, China). All reagents were not purified and were used directly. Ultrapure water (18.2 MΩ·cm) from the Milli-Q system was used to prepare solutions in all experiments. The river water in this study was taken from the Pearl River.

2.2. Characterization

The phases of the products were characterized by X-ray powder diffraction (XRD, Bruck D8 Advance, Hersbruck, Germany). Scanning electron microscopy (SEM) images were acquired (FE-SEM, ZEISS ULTRA 55, Oberkochen, Germany). The microstructure and crystal structure of the prepared materials were studied by high-resolution transmission electron microscopy (HRTEM, FEI Tecnai TF-20, Hillsboro, OR, USA) under 200 kV voltage. Raman spectra were measured with a Raman microscope (Raman, Renishaw Invia RAMA, Wottonunder-Edge, UK), and the excitation wavelength was 532 nm. The elemental and surface valence states of the synthesized materials were measured by X-ray photoelectron spectroscopy (XPS, Thermo Scientific Nexsa, Waltham, MA, USA). An Inductively Coupled Plasma Mass Spectrometer was used to measure the concentration of iron and cobalt ions (ICP-MS, Agilent 7700, Santa Clara, CA, USA). An electron paramagnetic resonance spectrometer (EPR, Bruker A300, Rheinstetten, Germany) was used to identify the type of reactive oxygen species. The intermediates during the LEV degradation process were identified by liquid chromatography–mass spectrometry (LC/MS, Agilent 1290 Infinity/6460, Santa Clara, CA, USA).

2.3. Preparation of CoFe@CSC-X Carbonized Microspheres

Firstly, 2 mL of acetic acid and 2 g of CS were added to 100 mL of deionized water and stirred to dissolve them completely. Then, an aqueous metal solution with a Fe concentration of 0.5 mol/L and a Co/Fe ratio of 1:1 was prepared from $CoCl_2{\cdot}6H_2O$ and $FeCl_3{\cdot}6H_2O$, and then this solution was added to the above CS solution at a volume ratio of 1:10 and stirred for 30 min; the mixture was recorded as solution A. NaOH and CH_3COONa were dissolved in deionized water with stirring at a ratio of 1.25 M/0.1 M to obtain another solution, which was recorded as solution B. Solution A was injected into 80 °C solution B via syringe, forming reddish-brown gel microspheres. After 2 h of hardening, the microspheres were washed to neutrality with deionized water and oven-dried at 80 °C for 12 h. The resulting microsphere material was recorded as CoFe@CS. CoFe@CS microspheres were then calcined at different calcination temperatures (600 °C, 700 °C, and 800 °C) under a nitrogen atmosphere for 2 h to prepare CoFe@CSC-X, where X denotes the temperature. Similar to the preparation of CoFe@CS microspheres, CS microspheres were successfully prepared without incorporating the metal solution. At the same time, Co@CS microspheres were prepared by replacing $FeCl_3{\cdot}6H_2O$ with $CoSO_4{\cdot}7H_2O$, and Fe@CS microspheres were prepared by replacing $CoCl_2{\cdot}6H_2O$ with $FeSO_4{\cdot}7H_2O$. Under the same conditions as those used for the preparation of CoFe@CSC-X, the products of the calcination of the above three microspheres were named CSC microspheres, Co@CSC microspheres, and Fe@CSC microspheres, respectively.

2.4. Evaluation of the Catalytic Performance

In all batch experiments, 250 mL conical flasks containing 100 mL of LEV solution (10 mg/L) at pH = 6.8 (initial pH) were placed on a shaker (180 rpm, 298 K) to evaluate the catalytic performance. The initial pH of the reaction solution was unadjusted unless otherwise specified. When discussing the effect of initial pH on degradation, the initial pH of the solution could be adjusted to the desired pH with 1 M H_2SO_4 or NaOH solution. A certain amount of catalyst and PMS were simultaneously introduced into the conical flask to initiate the catalytic reaction. At regular time intervals, 1 mL of sample was taken at each time during the reaction, and the solution was filtered through a 0.22 μm filter membrane. Next, 0.5 mL of methanol was immediately injected into sampling bottles to quench the free radical reaction and redundant oxidant. The used catalysts were collected with a magnet and washed alternately with deionized water and ethanol to remove residual organics for repeatability and stability tests. All the experiments were repeated three times, and the average value was taken as the data results.

2.5. Analytical Method

In all experiments, the concentration of LEV was quantified by high-performance liquid chromatography (HPLC) using the standard working curve method. The mobile phase consisted of 20% chromatographic grade acetonitrile and 80% ultrapure water containing 0.3% formic acid in a 10 μL injection volume at a flow rate of 1 mL/min with a detection wavelength of 290 nm.

3. Results

3.1. Catalyst Characterization

The SEM results showed the surface morphology of the catalysts. As shown in Figure 1a, the CoFe@CSC-700 crystals were shaped as microspheres with a diameter of about 1 mm, and the surface of the uncalcined CoFe@CS (Figure 1d) was folded, which was presumed to be the wrapped chitosan molecular chain. After CoFe@CS was calcined at 700 °C, the wrapped chitosan volatilized at high temperature, more voids appeared on the surface of the spheres, and some of the internal cobalt–iron complexes were revealed, as shown in Figure 1b,c. Meanwhile, the microstructure of CoFe@CSC-700 was characterized by TEM. As shown in Figure 1e,f, the sample has an obvious "core–shell" structure. In Figure 1f–i, the typical lattice stripes of graphitic carbon shells are obvious, indicating the presence of graphitized carbon shells. According to Figure 1g,h, the lattice spacing of the 'core' can be calculated, and the results match the (110), (200) and (211) planes of the CoFe alloy. In summary, it can be seen that the CoFe alloy particles were encapsulated within the graphitic carbon shell that was embedded in the graphitic carbon matrix, which reduced the clustering of CoFe alloy particles and enhanced electron mobility [22,23]. Meanwhile, the graphitic carbon matrix can serve as a protective layer for the CoFe alloy particles, protecting them from oxidation and corrosion, which improves the stability of the composites [24]. In addition, the combined EDX spectra showed the elemental distributions, as shown in Figure 1i, for C, N, O, Fe, and Co. Obviously, the range of C and N elements did not differ much, indicating the partial doping of the carbon matrix with N. Meanwhile, the same distribution of Co and Fe elements indicated the successful synthesis of CoFe alloy.

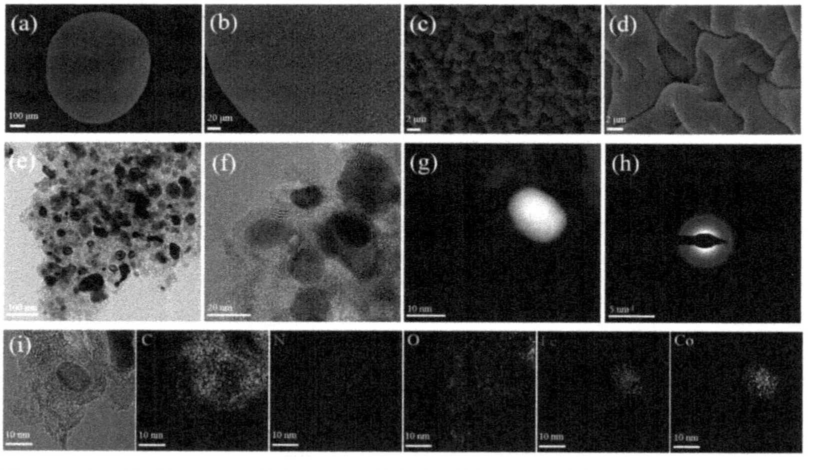

Figure 1. (**a–c**) SEM images of CoFe@CSC-700; (**d**) SEM image of CoFe@CS; (**e**,**f**) CoFe@CSC-700 TEM images; (**g**) HRTEM images; (**h**) SAED images; (**i**) EDX images and corresponding surface scan images of elements of C, N, O, Fe, and Co.

XRD analyses were performed to clarify the crystal structure and composition of the catalysts. As shown in Figure 2a, characteristic peaks of CoFe@CSC-600, CoFe@CSC-700, and CoFe@CSC-800 at 44.8°, 65.3°, and 82.7° were typical of the (110), (200), and (211) lattice planes assigned to CoFe alloys (PDF# 49-1567). It is possible that CoFe@CS was calcined under a N_2 atmosphere under higher temperature conditions when the metal center was reduced to a lower oxidation state, resulting in CoFe alloys [25]. The CoFe@CSC-700 characteristic peaks had the highest intensities, whereas the diffraction peak intensities of CoFe alloys were changed by changing the calcination temperature, suggesting that the degree of crystallinity was affected by the temperature to a certain extent. In addition, an extra characteristic peak can be observed in the XRD pattern of CoFe@CSC-700, which is known from the literature to be a broad peak at 26.3° attributable to the (002) diffraction plane of graphitized carbon [24,26], which corresponds to the previous TEM results.

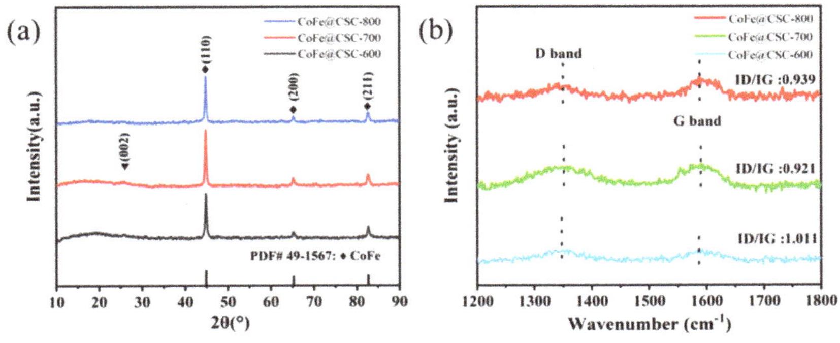

Figure 2. (**a**) XRD patterns of CoFe@CSC-600, CoFe@CSC-700, and CoFe@CSC-800; (**b**) Raman spectra of CoFe@CSC-600, CoFe@CSC-700, and CoFe@CSC-800.

To explore the degree of graphitization at various calcination temperatures, Raman spectroscopy was conducted. The Raman results exhibited two distinctive peaks positioned approximately at 1350 cm^{-1} and 1580 cm^{-1}, which were attributed to the D and G bands, respectively. The D band indicated the contribution of sp^3 defects in the carbon, i.e., the presence of defects or disorder [27], whereas the G band was attributed to the stretching of the C-C bonds within the graphite structure [28], which was related to the graphitized carbon of the material [24]. Therefore, the degree of graphitization of the material can be expressed by the I_D/I_G intensity ratio. The smaller this ratio was, the higher the degree of graphitization was. As shown in Figure 2b, the I_D/I_G ratio values for CoFe@CSC-600, CoFe@CSC-700, and CoFe@CSC-800 were 1.011, 0.921, and 0.939, respectively. With the increase in calcination temperature, the I_D/I_G ratio decreased and then increased, and it was lowest at 700 °C, indicating that the degree of graphitization would be damaged if the temperature was too high [26]. This indicated that the Raman experimental results also corresponded to the experimental results of XRD.

In order to investigate the elemental composition and valence distribution of the CoFe@CSC-700 surface, the catalyst was analyzed by XPS. Firstly, the CoFe@CSC-700 consists of C, O, N, Fe, and Co elements, as seen in the survey spectrum (Figure 3a), which was compatible with the EDX results. Figure 3b shows the high-resolution C1s energy spectra obtained from the test, which corresponded to the C-C characteristic peak at 284.8 eV binding energy, the C-N bond characteristic peak at 285.7 eV binding energy, the C=O bond characteristic peak at 288.1 eV, and the O-C=O bond characteristic peak at 290.0 eV, as shown in the past literature [29], whereas the π-π vibration of carbon corresponds to the π-π vibration at 291.6 eV. The incorporation of C-N bonds indicated that nitrogen had been successfully doped into the carbon molecular structure, constituting a crucial facet of the redox reaction process [30]. The concurrent presence of nitrogen-doped graphitic carbon alongside CoFe alloys can enhance the electron mobility between the metallic components and PMS [31]. The high-resolution energy level spectrum of N 1s for CoFe@CSC-700 (Figure 3c), can be categorized into three separate peaks. The binding energies of these peaks were 398.6 eV, 400.7 eV, and 402.5 eV, which were attributed to pyridine N, pyrrole N, and graphite N. As shown in Figure 3d, the O 1s XPS high-resolution energy spectrum can be fitted to three distinct peaks, the peaks of M-O (lattice oxygen), M-O-H (adsorbed oxygen), and impurity water, which are located at 530.4 eV, 532.3 eV, and 534.7 eV, respectively [32]. Figure 3e demonstrates the electronic state characteristics of Fe 2p, which specifically consists of two spin orbitals, Fe $2p_{1/2}$ and Fe $2p_{3/2}$. The peak at 707.4 eV corresponds to the characteristic peak of Fe^0, 711.0 eV and 724.2 eV correspond to the characteristic peaks of Fe^{II}, 713.8 eV and 726.3 eV are the characteristic peaks of Fe^{III}, and the peak at 719.7 eV corresponds to the satellite peaks of Fe. The high-resolution spectra of Co 2p (Figure 3f) consist mainly of the peaks at Co^{II} at 782.7 eV and 797.1 eV and Co^{III} at 780.3 eV and 795.4 eV, as well as two satellite peaks at 787.0 eV and 803.4 eV. There are five valence states of $Fe^0/\equiv Fe^{II}/\equiv Fe^{III}$ and $\equiv Co^{II}/\equiv Co^{III}$ on the surface of the two elements, and the presence of multiple valence states is favorable for electron transfer according to the redox potentials of Fe cobalt. According to the redox potential of cobalt–iron, the existence of multiple valence states is favorable for electron transfer, and a redox cycle of $\equiv Fe^{II}/\equiv Fe^{III}$ and $\equiv Co^{II}/\equiv Co^{III}$ can exist to accelerate the activation of PMS.

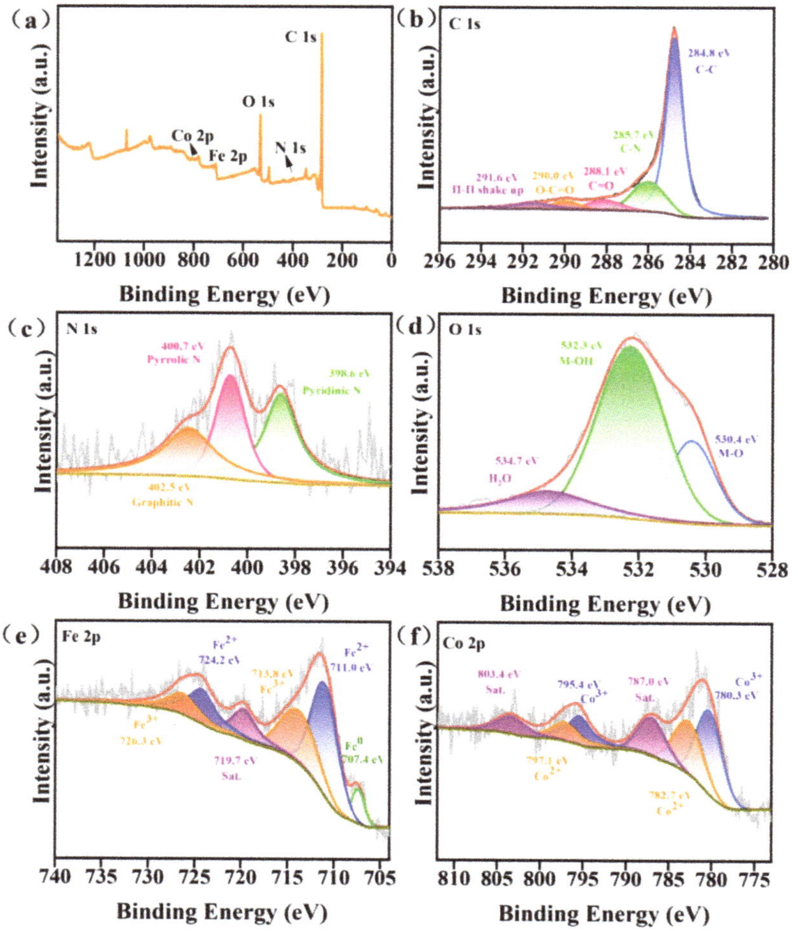

Figure 3. (**a**) Wide XPS spectrum of CoFe@CSC-700; (**b**) C 1s spectra; (**c**) N 1s spectra; (**d**) O 1s spectra; (**e**) Fe 2p spectra; (**f**) Co 2p spectra.

3.2. Catalytic Performance

In order to investigate the practical effects of the five different catalysts prepared (CSC-700, Co@CSC-700, Fe@CSC-700, CoFe@CSC-700, and CoFe@CS) applied for the degradation of LEVs by activated PMS and to select the best-performing samples for the subsequent experiments, firstly, adsorption experiments were carried out without the addition of PMS, and the results are shown in Figure 4. When CSC-700, Co@CSC-700, Fe@CSC-700, CoFe@CSC-700, and CoFe@CS were used alone, none of them adsorbed more than 10% of LEV within 30 min, so the adsorption capacity of these five materials was negligible during the whole degradation process. The effect of the five different catalysts prepared on the removal of LEV within the PMS system was mainly a degradation effect rather than an adsorption effect.

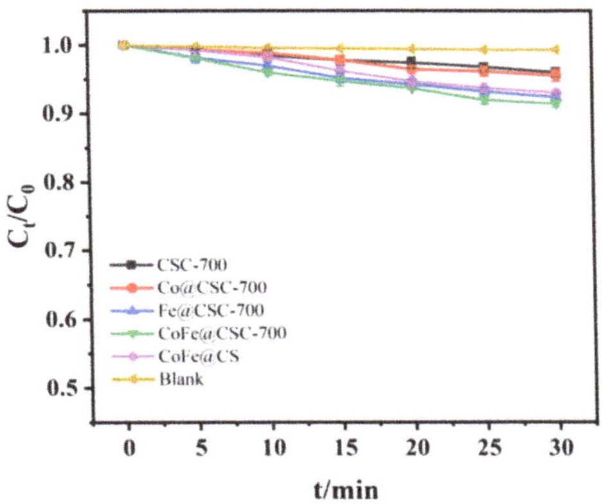

Figure 4. Adsorption effect of different catalysts on LEV.

Based on the results of the above adsorption experiments, the degradation performance of PMS activated by the above five different chitosan microspheres was compared without considering the effect of adsorption effect on LEV degradation. An investigation of the degradation performance without catalyst addition and with pure PMS addition was also carried out. As shown in Figure 5, when there was only pure PMS without a catalyst in the system, PMS alone degraded only 2% of LEV within 30 min, which indicated that PMS alone basically could not generate free radicals and its degradation effect was negligible. When both CSC-700 and PMS were present, there was no significant increase in LEV removal, indicating that CCM was ineffective against PMS. When Co@CSC-700, Fe@CSC-700, CoFe@CSC-700, and CoFe@CS were used as catalysts, the LEV removal rates were 99.8%, 29.3%, 99.9%, and 44.1%, respectively. It can be seen that the combination of Co@CSC-700 and CoFe@CSC-700 with PMS significantly improved the LEV degradation efficiency, reaching 99.9% removal within 30 min. These results unequivocally demonstrated that the carbonated Co@CSC-700 and CoFe@CSC-700 presented a greater number of active sites than the uncarbonated CoFe@CS, whereas Fe@CSC-700 was not as effective as the uncarbonated CoFe@CS, probably due to the presence of cobalt, which had stronger catalytic activity than iron [33]. Compared to the monometallic catalyst Fe@CSC-700, CoFe@CSC-700 clearly had superior catalytic performance; while Co@CSC-700 was certainly also catalytic, its reaction rate was smaller than that of CoFe@CSC-700, confirming the cooperative catalytic impact of Co and Fe. Meanwhile, considering the environmental goodness of the materials, inductively coupled plasma mass spectrometry (ICP-MS) was utilized for determining the concentration of cobalt ions and iron ions leached out from the solution at the end of the reaction for the three systems Co@CSC-700/PMS, Fe@CSC-700/PMS, and CoFe@CSC-700/PMS. The results showed that the leaching of cobalt ions from Co@CSC-700/PMS was up to 3.33 mg/L, and that of iron ions from Fe@CSC-700/PMS was up to 0.85 mg/L. The leaching of cobalt ions and iron ions from CoFe@CSC-700 decreased to 0.43 mg/L and 0.10 mg/L, and these values were reduced by about 8 times; this result was lower than the requirements of the national standard (GB3838-2002) [34]. Additionally, this finding reaffirmed the cooperative catalytic impact of Co and Fe, and the Co/Fe coupling could inhibit ion leaching [16]. Considering the performance, environmental soundness, and reusability, CoFe@CSC-700 was used in this study to carry out the following experiments related to the influence of environmental factors and so on.

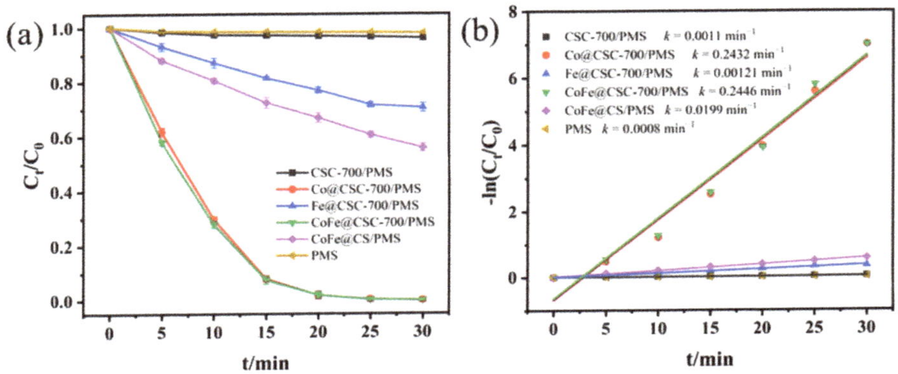

Figure 5. (**a**) Comparison of the effects of LEV degradation in different conditions; (**b**) reaction constants of LEV degradation.

It is well known that the state of presence of contaminants and the surface charge of the catalyst are affected by the solution pH, which in turn affects the PMS system reaction process. In our study, the point of zero charge (pH_{PZC}) was determined by measuring the zeta potential of the catalyst surface, which represents the pH of the solution when the net surface charge of the material is zero [35]. As shown in Figure S1, the pH_{PZC} of CoFe@CSC-700 was 6.28, the catalyst surface is positively charged when the pH of the solution < 6.28 and negatively charged at pH > 6.28. The dissociation constants of the LEV molecule in water are pKa1 = 6.02 and pKa2 = 8.15, respectively. It usually exists in water in the cationic form (pH < 6.02), the amphoteric form (6.02 < pH < 8.15), and the anionic form (pH > 8.5). Figure 6 shows the degradation of LEV by the CoFe@CSC-700/PMS system at different pH conditions. The degradation efficiency of LEV can reach more than 99% when the pH ranges from 3.0 to 6.8, and overall, the degradation rate was faster when it was acidic. Although the surface charge distributions of the catalyst and the pollutant were essentially similar in the acidic pH range and there was electrostatic hindrance, the release of cobalt–iron ions increased with decreasing pH, thus accelerating the catalytic reaction. Therefore, the decrease in LEV degradation rate in systems with pH ranging from 3.0 to 6.8 may be caused by a decrease in cobalt iron dissolution. When the pH was further increased (6.8~11.0), the degradation of LEV decreased continuously. In particular, the degradation effect on LEV decreased to 56% at pH 11.0. This was due to the fact that both catalyst and pollutant surfaces were negatively charged, and the electrostatic effect may have hindered the reaction process between CoFe@CSC-700 and LEV. In summary, considering the removal effect and convenience, the initial pH of 6.8 was chosen for the experiments in this study.

As shown in Figure 7, the effects of different catalyst dosages (30~130 mg/L) on the catalytic degradation of LEV in the CoFe@CSC-700/PMS system were investigated. The degradation efficiency of LEV increased from 42% to more than 99% when the catalyst dosage was increased from 30 mg/L to 130 mg/L. When the catalyst dosage was increased from 100 to 130 mg/L, the overall LEV removal rate was basically the same, but the degradation rate was slightly slower. On one hand, a higher catalyst dosage can provide more active sites for activating the PMS reaction for LEV removal [36,37]. On the other hand, at a certain concentration of PMS, excessive catalyst dosage may compete for the reaction with PMS and inhibit the rate of formation of reactive oxygen components. The appropriate amount of catalyst was necessary to obtain the excellent performance of the reaction system, so the CoFe@CSC-700 dosage of 100 mg/L (0.1 g/L) was selected in the subsequent experiments considering the effect and economic cost.

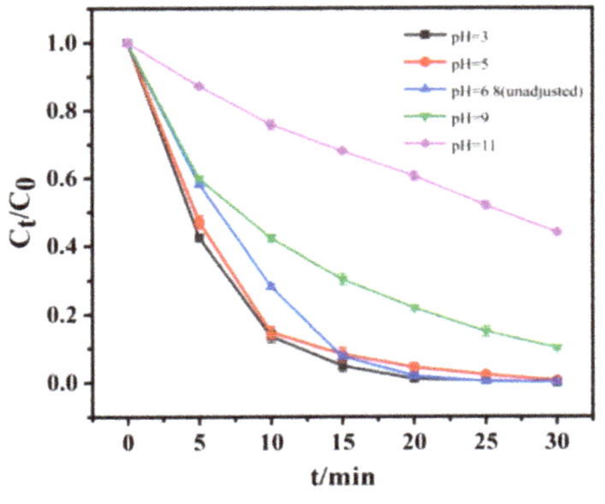

Figure 6. PMS system with CoFe@CSC-700 at different initial pH values to solve LEV rendering.

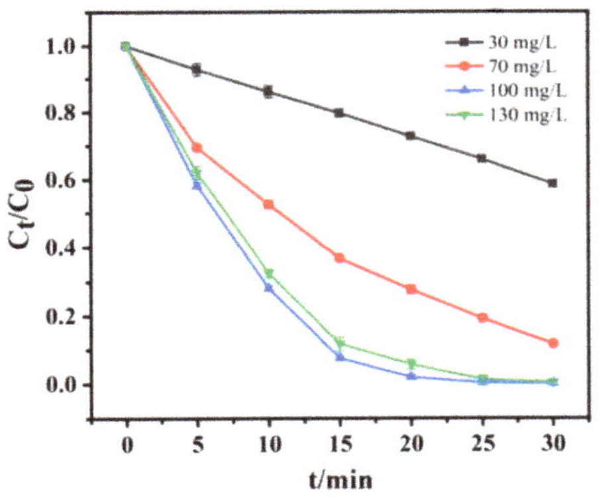

Figure 7. PMS system with CoFe@CSC-700 at different initial pH values to solve LEV rendering.

The concentration of PMS plays a significant role in the efficiency of LEV removal. Figure 8 shows the degradation curves of LEV in the reaction system when different concentrations of PMS were added. The LEV degradation efficiency of the CoFe@CSC-700/PMS system continued to increase with the increase in PMS concentration from 25 mg/L to 100 mg/L and reached a peak at a PMS concentration of 100 mg/L, which was attributed to the fact that a larger concentration of PMS could provide more free radicals for the reaction. However, when the PMS concentration was further escalated, the LEV removal rate remained nearly consistent. This result was due to the fact that at a fixed catalyst dosage, an increase in PMS concentration led to self-scavenging between free radicals [38,39]. Therefore, a concentration of PMS of 100 mg/L (0.1 g/L) was chosen for the reaction in the subsequent experiments.

In the complex environment of natural waters, a diverse range of anions and naturally organic compounds are present, which may have an impact on the chemical reaction process through competitive interactions with the target pollutants or by inducing the generation of novel free radicals. In this study, we selected three ubiquitous inorganic anions (Cl^-, NO_3^-,

SO_4^{2-}) and typical humic acid (HA) to investigate the effect of co-existing components in the system on the removal of LEV by the CoFe@CSC-700/PMS system. As shown in Figure 9, the high concentration of Cl^- had a large inhibitory effect on the catalytic degradation by PMS, and the 10 mM Cl^- concentration caused the degradation efficiency of LEV to decrease from 99.9% to 84.1% within 30 min. This change in degradation efficiency could be caused by two reasons. On one hand, Cl^- consumed highly active ·OH and $SO_4·^-$ in the CoFe@CSC-700/PMS system and generated low-activity Cl· (Equation (1)) and $ClOH·^-$ (Equation (2)). On the other hand, the reaction of Cl^- and HSO_5^- generated HOCl (Equation (3)) and Cl_2 (Equation (4)), leading to a decrease in the production of ·OH and $SO_4·^-$ in the CoFe@CSC-700/PMS system [40]. In addition, it can also be seen from the figure that the addition of NO_3^- and SO_4^{2-} slightly inhibited LEV removal at a concentration of 10 mM, with LEV removal reduced by 5.1% and 6.2%, respectively. The reduced degradation efficiency indicated that they competed with the target pollutants (Equations (5)–(7)). However, they reacted slowly with ·OH and $SO_4·^-$, so they had little effect on the presence of ·OH and $SO_4·^-$ in the system [41,42]. In addition to the anions mentioned above, natural organic matter (NOM) is also present in natural water. HA is one of the common forms of NOM, and its effect on the CoFe@CSC-700/PMS system is also significant. The inhibitory effect of HA on LEV degradation decreased to 86.2% with the addition of HA (Figure 9), which was attributed to the fact that HA, with its rich electronic structure, competes with LEV for ·OH and $SO_4·^-$, thus inhibiting LEV degradation [43,44]. In summary, the order of inhibition of LEV degradation by different water quality backgrounds was $Cl^- > HA > SO_4^{2-} > NO_3^-$.

$$Cl^- + SO_4^{\cdot -} \rightarrow Cl^{\cdot} + SO_4^{2-} \quad (1)$$

$$Cl^- + \cdot OH \rightarrow ClOH^{\cdot -} \quad (2)$$

$$Cl^- + HSO_5^- \leftrightarrow SO_4^{2-} + HOCl \quad (3)$$

$$2Cl^- + HSO_5^- + H^+ \rightarrow SO_4^{2-} + Cl_2 + H_2O \quad (4)$$

$$NO_3^- + SO_4^{\cdot -} \rightarrow NO_3^{\cdot} + SO_4^{2-} \quad (5)$$

$$NO_3^- + \cdot OH \rightarrow OH^- + NO_3^{\cdot} \quad (6)$$

$$SO_4^{2-} + \cdot OH \rightarrow SO_4^{\cdot -} + OH^- \quad (7)$$

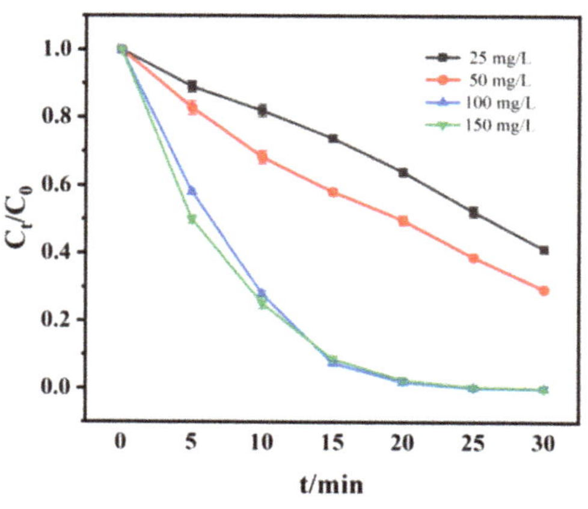

Figure 8. PMS system with CoFe@CSC-700 at different concentrations of PMS to solve LEV rendering.

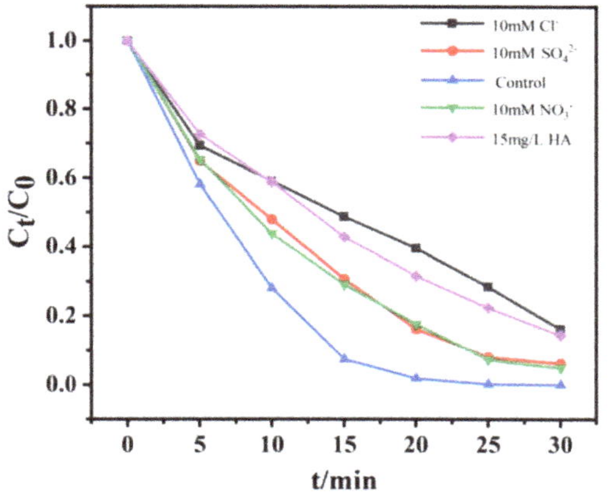

Figure 9. Effect of Cl^-, NO_3^-, SO_4^{2-}, and HA on LEV degradation in the CoFe@CSC-700/PMS system.

3.3. Evaluation of the Stability and Performance of CoFe@CSC-700

The reusability and stability of the catalyst are the key investigation index; in order to explore this performance, CoFe@CSC-700 cyclic reaction experiments were carried out. After each cyclic reaction, the catalyst was washed with water and ethanol and vacuum-dried for reuse. From Figure 10a, it can be seen that the degradation rate of CoFe@CSC-700 for LEV can still be maintained above 90% after five degradation cycles, which indicates that CoFe@CSC-700 was stable in nature. From Section 3.2, it can be seen that the leaching value of cobalt and iron ions in CoFe@CSC-700 was lower than the national standard, but the leaching of metal ions led to the loss of the active site, and thus there was a slight decrease in the catalytic efficiency after five cycles of the experiment. Also, the decrease in catalytic efficiency during cycling was related to the coverage of active sites on the surface of the CoFe@CSC-700 catalyst by degradation intermediates [45]. In addition to the consideration of catalyst efficiency, the more difficult recovery of catalyst materials in water is the key issue that leads to the inability to reuse in the current practical application of the PMS process, which will increase the cost of use. From Figure 10b, it can be seen that CoFe@CSC-700 had a saturation magnetization of 96.12 emu/g. Therefore, the catalyst and the water can be separated very quickly under the application of a magnetic field, which can save costs in practical applications. In addition, as shown in Figure 10c, the XRD results before and after the reaction indicated that the crystal structure and elemental composition of the CoFe@CSC-700 catalyst remained stable, and the slight change in the peak intensity of the XRD could be caused by the loss of metal ions. This further indicated that CoFe@CSC-700 has good reusability and high water stability in the PMS reaction system and has strong potential for environmental applications. Table S1 listed the performance comparison of CoFe@CSC-700 with other previously reported catalysts for LEV degradation [46–52]. As can be seen from Table S1, CoFe@CSC-700 was able to achieve a higher LEV removal rate in a shorter period at a lower catalyst dosage and PMS concentration than the catalysts prepared in other studies, suggesting that the catalyst has superior performance and greater potential application prospects for treating LEV pollution in water.

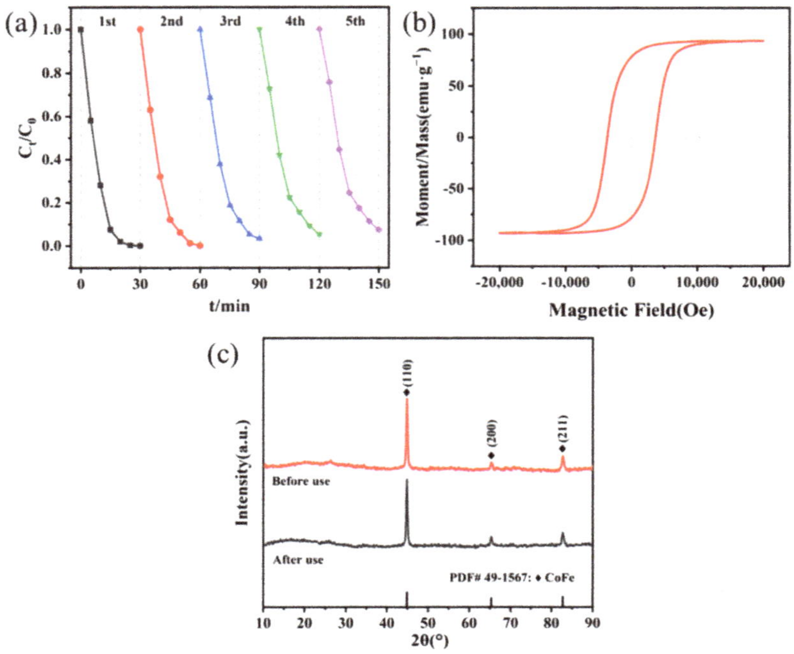

Figure 10. (**a**) The degradation effect of CoFe@CSC-700/PMS system on LEV in 5 cycles; (**b**) the hysteresis loop of CoFe@CSC-700; (**c**) XRD comparison before and after use of CoFe@CSC-700/PMS.

3.4. Identification of ROS and Proposed Catalytic Mechanism

The degradation mechanism of the LEV in the CoFe@CSC-700/PMS system was investigated by free radical quenching experiments, which were able to reveal the types of ROS that removed LEV and the extent of their contribution. In these experiments, sufficient amounts of the corresponding specific quenchers were added at the beginning of the reaction to determine the effect of the main ROS on LEV degradation, while the other experimental conditions were kept constant. MeOH was chosen as the quencher for $SO_4\cdot^-$ and $\cdot OH$ because of the high reaction rates ($K_{SO4\cdot-}$ = (1.6 ~ 7.7) × 10^7 $M^{-1}S^{-1}$, $K_{\cdot OH}$ = (1.2 ~ 2.8) × 10^9 $M^{-1}S^{-1}$). Also, TBA, p−BQ, and L-His were also employed as quenchers corresponding to $\cdot OH$, $\cdot O_2^-$, and $^1O_2$, respectively, with corresponding reaction rates of (3.8 ~ 7.6) × 10^8 $M^{-1}S^{-1}$, (0.9 ~ 1.0) × 10^9 $M^{-1}S^{-1}$, and 5 × 10^7 $M^{-1}S^{-1}$, respectively [53,54]. As shown in Figure 11d, the removal of LEV reached 99.9% within 30 min of the reaction without the addition of a quencher, while the addition of TBA and MeOH resulted in LEV removals of 55.8% and 16.3%, respectively. The results showed that both $SO_4\cdot^-$ and $\cdot OH$ played significant roles in the LEV degradation reaction with comparable contribution importance. After the addition of p-BQ, the degradation rate of LEV decreased to 57.2%, indicating that $\cdot O_2^-$ is another important active species. Compared with the above quenchers, the degradation rate of LEV decreased to 82.8% after the addition of L-His, which had the least inhibitory effect on the catalytic activity in the CoFe@CSC-700/PMS system, but the contribution of $^1O_2$ to the degradation of LEV could not be neglected. In summary, the results showed that $SO_4^{\cdot-}$, $\cdot OH$, $\cdot O_2^-$, and $^1O_2$ were all involved in the degradation process of LEV in the CoFe@CSC-700/PMS system, in which $SO_4^{\cdot-}$, $\cdot OH$, and $\cdot O_2^-$ played the dominant roles, and $^1O_2$ also contributed to LEV degradation.

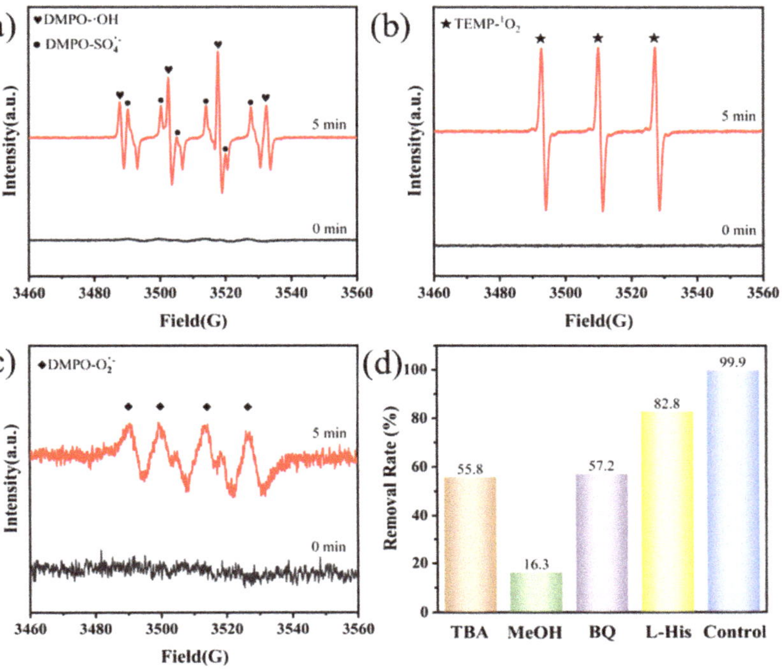

Figure 11. (**a**) EPR spectra of $SO_4\cdot^-$ and $\cdot OH$ captured by DMPO; (**b**) EPR spectra captured by DMPO $\cdot O_2^-$; (**c**) EPR spectra of $^1O_2$ captured by TEMP; (**d**) quenching experiments.

To further identify the active components in the reaction system, the ROS generated during the catalytic degradation process were determined by EPR spectroscopic detection [55]. In this process, DMPO and TEMP were selected as the trapping agents that are capable of efficiently trapping and labeling radicals such as $SO_4^{\cdot-}$, $\cdot OH$, $\cdot O_2^-$, and $^1O_2$. As shown in Figure 11a, when DMPO was used as the spin trapping agent, no significant changes in the $SO_4^{\cdot-}$ radicals were observed at 0 min with only the addition of catalyst and no PMS present. After 5 min of adding PMS, significant characteristic signal peaks for $SO_4\cdot^-$ and $\cdot OH$ were observed where the intensity ratio of DMPO-$\cdot$OH was about 1:2:2:1, which can be attributed to the hyperfine splitting of DMPOX [56]. Similarly, no corresponding characteristic peaks were observed at 0 min, but after 5 min of PMS addition, a broad quadruple peak of DMPO-$\cdot O_2^-$ was observed, as shown in Figure 11c. TEMP was employed as the spin trapping agent, and according to the results in Figure 11b, a distinct triplet signal peak with an intensity ratio of 1:1:1 was observed at 5 min in the catalyst/PMS/TEMP system, which can be attributed to the characteristic peak of $^1O_2$. The analysis of the EPR results showed the presence of $SO_4^{\cdot-}$, $\cdot OH$, $\cdot O_2^-$, and $^1O_2$, which corresponded to the quenching experimental results. In conclusion, CoFe@CSC-700 can rapidly activate PMS and promote the decomposition of PMS molecules and the release of $^1O_2$.

Combined with analyses such as reaction XPS and EPR, the mechanism of LEV degradation by CoFe@CSC-700/PMS system can be roughly deduced. The redox potentials of $Fe^0/{\equiv}Fe^{II}$ and ${\equiv}Fe^{II}/{\equiv}Fe^{III}$ are 0.44 V and 0.77 V, respectively, whereas that of ${\equiv}Co^{II}/{\equiv}Co^{III}$ is 1.81 V. Therefore, Fe^0 can provide electrons to ${\equiv}Co^{III}$ and ${\equiv}Fe^{III}$ [57]. From Equation 8, Fe^0 can activate PMS to generate$\cdot$OH, while Fe^0 can reduce ${\equiv}Fe^{III}/{\equiv}Co^{III}$ to ${\equiv}Fe^{II}$ and ${\equiv}Co^{II}$ as shown in Equations (9) and (10) [56]. Furthermore, since the redox potentials of $HSO_5^-/SO_5\cdot^-$ and $HSO_5^-/SO_4\cdot^-$ are 1.1 V and 2.5~3.1 V, respectively, and the redox potential of ${\equiv}Co^{II}/{\equiv}Co^{III}$ (1.81 V) is in between, thermodynamically, the activation of PMS becomes possible [58,59]. PMS acts as an oxidizing agent and a reducing agent at the same time, participating in the $\equiv Co^{II}/\equiv Co^{III}$ cycle [60]. Since $\equiv Co^{II}/\equiv Co^{III}$ has a low redox potential relative to $\equiv Fe^{II}/\equiv Fe^{III}$, $\equiv Fe^{II}$ can also transfer electrons to $\equiv Co^{III}$, allowing the reduction of $\equiv Co^{III}$ to $\equiv Co^{II}$. Also, $\equiv Fe^{II}$ can activate PMS to produce $SO_4^{\cdot-}$, but there is an undesirable reaction between $\equiv Fe^{III}$ and PMS, since compared to $HSO_5^-/SO_5^{\cdot-}$, the redox potentials $\equiv Fe^{II}/\equiv Fe^{III}$ are both negative compared to $SO_5^{\cdot-}$, and the reaction is not thermodynamically feasible. Therefore, the redox reactions and cycles in $Fe^0/\equiv Fe^{II}/\equiv Fe^{III}$ and $\equiv Co^{II}/\equiv Co^{III}$ activate PMS, which continuously generates ROS. In addition, O_2 in water traps electrons to generate $\cdot O_2^-$, and some of the $SO_5^{\cdot-}$ can further react with H_2O to generate $^1O_2$. Thus, in the presence of these highly reactive $SO_4^{\cdot-}$, $\cdot OH$, and $\cdot O_2^-$ radicals and $^1O_2$, the degradation of LEV by CoFe@CSC-700-catalyst-activated PMS was achieved. Based on the above analysis results, the mechanism of LEV degradation by the CoFe@CSC-700/PMS system is revealed, and the specific reaction mechanism process is shown in the following equations:

$$Fe^0 + HSO_5^- \rightarrow \equiv Fe^{II} + \cdot OH + SO_4^{2-} \tag{8}$$

$$Fe^0 + \equiv Fe^{III} \rightarrow \equiv Fe^{II} \tag{9}$$

$$Fe^0 + \equiv Co^{III} \rightarrow \equiv Fe^{II} + \equiv Co^{II} \tag{10}$$

$$\equiv Fe^{II} + HSO_5^- \rightarrow \equiv Fe^{III} + OH^- + SO_4^{\bullet-} \tag{11}$$

$$\equiv Co^{II} + HSO_5^- \rightarrow \equiv Co^{III} + OH^- + SO_4^{\bullet-} \tag{12}$$

$$\equiv Fe^{II} + \equiv Co^{III} \rightarrow \equiv Fe^{III} + \equiv Co^{II} \tag{13}$$

$$\equiv Co^{III} + HSO_5^- \rightarrow \equiv Co^{II} + H^+ + SO_5^{\bullet-} \tag{14}$$

$$O_2 + e^- \rightarrow \cdot O_2^- \tag{15}$$

$$2SO_5^{\bullet-} + H_2O \rightarrow 1.5\,^1O_2 + 2HSO_4^- \tag{16}$$

3.5. LEV Degradation Products and Pathways

To explore the potential degradation pathways of LEV in the CoFe@CSC-700/PMS system, LC-MS was utilized to identify the intermediates produced during LEV degradation in the system. According to the mass spectrum information obtained by the test and related reports, the spectrum of the intermediate product of LEV partial degradation (Figure 12) and the possible LEV degradation pathway (Figure 13) could be obtained. On this basis, four possible pathways were proposed. The first pathway was the disconnection of the piperazine ring to generate N-methylpiperazine with m/z = 101 [61]. The second pathway was the destruction of the N-methyl group on the piperazine ring of LEV and oxidation of the N-methyl group to the carboxyl group by ROS to generate the m/z = 392 degradation product, which was subsequently dehydrated to generate N-formyl LEV (m/z = 376), and then decarboxylation and the disconnection of piperazine took place to generate m/z = 202. Then, demethylation finally produced degradation products with m/z = 188 [62]. The third pathway involved the removal of the N-methyl group from the piperazine ring of LEV generating a degradation product with m/z = 348, followed by the N oxidation of the piperazine ring to a hydroxylamine generating a degradation product with m/z = 364, and subsequently, the removal of hydroxyl generated a degradation product with m/z = 346, which was subsequently decarboxylated to generate a degradation product with m/z = 288 [63]. The fourth pathway involved the oxidative ring opening of the LEV piperazine ring to generate a bis-N-formyl LEV (m/z = 392), followed by the removal of the two formyl groups to generate a degradation product with m/z = 336, followed by the removal of the N-methyl group to generate a degradation product with m/z = 322, followed by the removal of methylamino group to generate a degradation product with m/z = 279, followed by the removal of the amino group and the removal of fluorine to generate a degradation product with m/z = 262 [64]. All these intermediates can be further decomposed into other smaller molecules of organic and inorganic compounds and eventually mineralized into water and carbon dioxide.

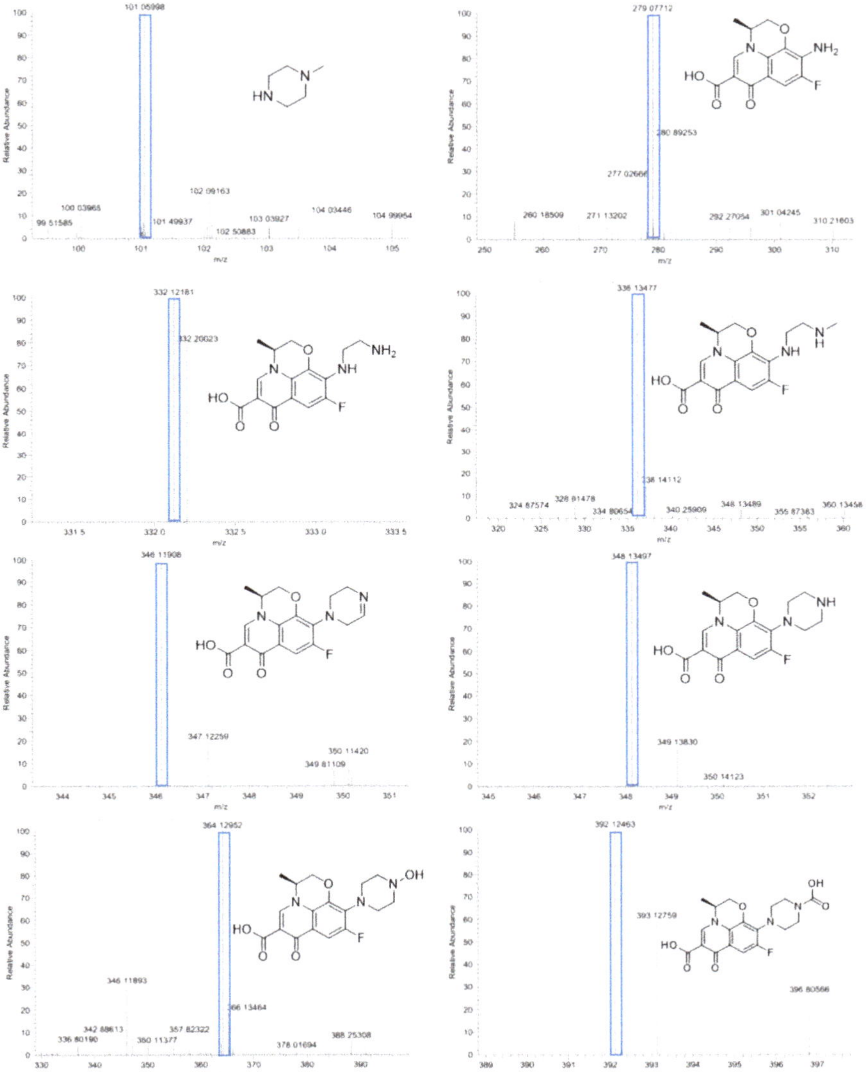

Figure 12. Spectrum of partial intermediate products for activated persulfate degradation of LEV.

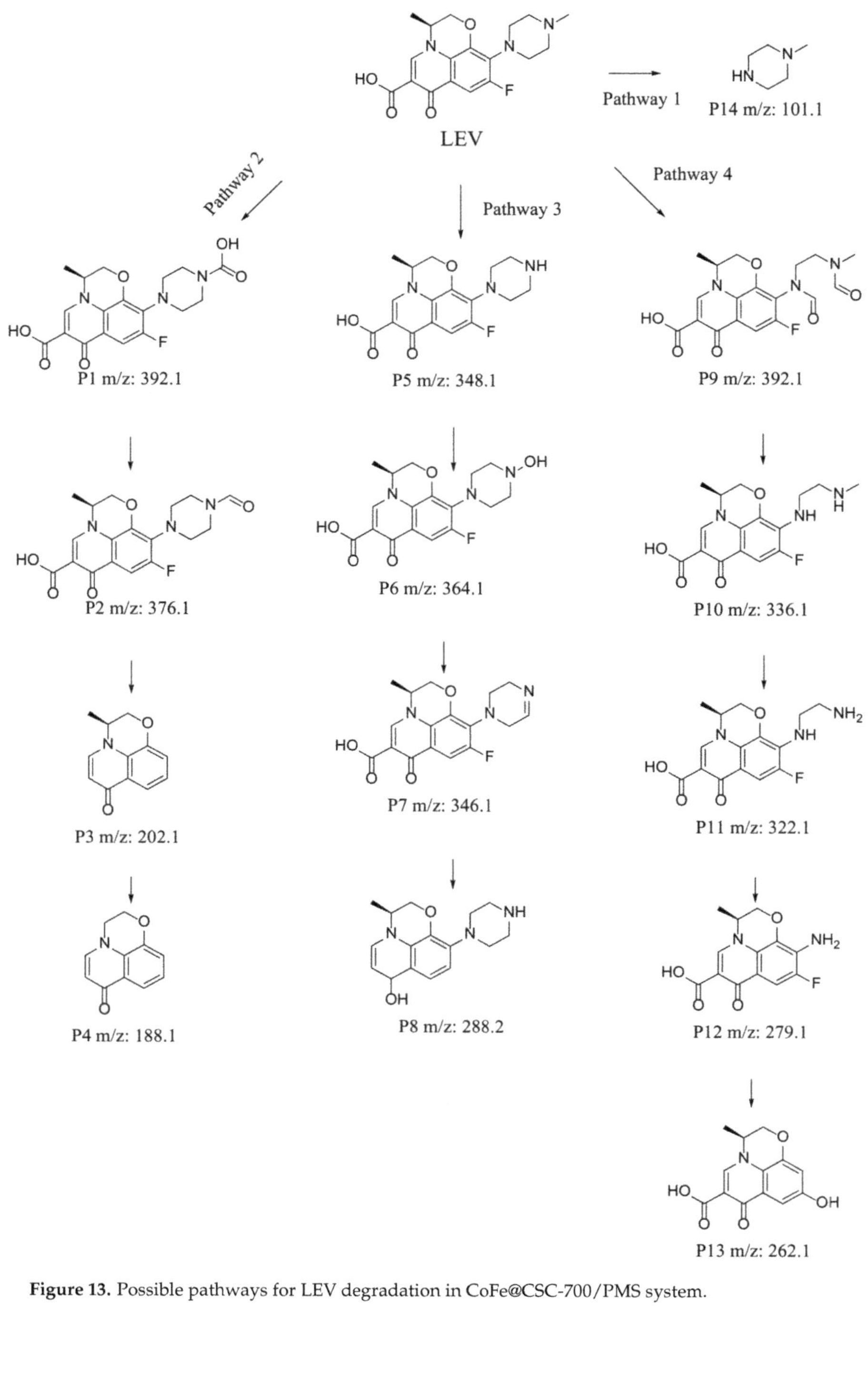

Figure 13. Possible pathways for LEV degradation in CoFe@CSC-700/PMS system.

3.6. Toxicity Analysis

In order to reveal whether LEV and its degradation products in this study pose a threat to human health or not, the developmental toxicity and mutagenicity of the LEV and intermediates were predicted by quantitative structure–activity relationship (QSAR) using Toxicity Estimation Software Tool (T.E.S.T). As shown in Figure 14, LEV was considered developmentally toxic (1.03) and mutagenicity positive (0.65). The toxicity of the intermediates changed as the degradation reaction progressed. As shown in Figure 14a, the developmental toxicity of the intermediates of P7 and P13 was slightly higher than that of LEV, while the developmental toxicity of all the other intermediates was reduced, with P14 being non-toxic (0.49), which implies a reduction in toxicity. In addition, except for P13, P9, and P2, the mutagenicity results showed reduced toxicity for other intermediates, and the P8 and P14 intermediates showed negative mutagenicity (Figure 14b). Therefore, the overall biotoxicity of the intermediates in the PMS-catalyzed system was reduced compared to LEV, indicating that the catalytic degradation of LEV by the CoFe@CSC-700/PMS system is an environmentally friendly process.

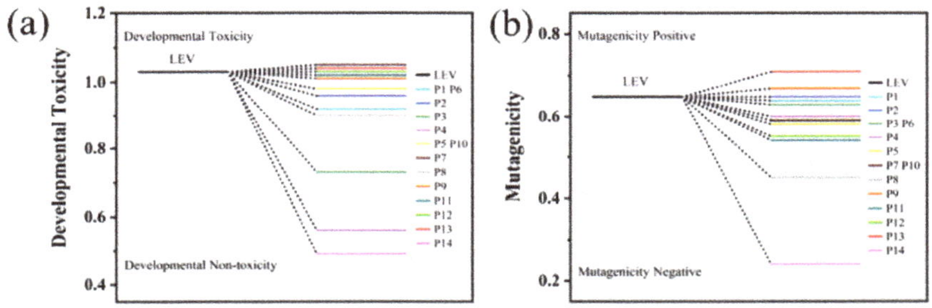

Figure 14. QSAR analysis based on (**a**) developmental toxicity and (**b**) mutagenicity.

4. Conclusions

In summary, the CoFe@CSC-700 microspheres prepared in this study were mainly composed of CoFe alloy and graphitized carbon, which were microspheres of about 1 mm with an apparent "core–shell" structure, and the calcined chitosan acted as a protective coating for the CoFe alloy. The CoFe@CSC-700 had a high degree of graphitization (I_D/I_G = 0.921) resulting in excellent properties. Under the conditions of pH = 6.8, PMS = 0.1 g/L, and CoFe@CSC-700 = 0.1 g/L, the degradation of LEV was nearly complete within 30 min, and the metal leaching was in accordance with national standards. CoFe@CSC-700 showed excellent activity after five cycles, and more than 90% degradation efficiency was maintained. During the catalytic reaction, the relavant redox reactions of $Fe^0/\equiv Fe^{II}/\equiv Fe^{III}$ and $\equiv Co^{II}/\equiv Co^{II}$ activated PMS. The quenching experiment and EPR test results indicated the involvement of $SO_4\cdot^-$, $\cdot OH$, $\cdot O_2^-$, and $^1O_2$ in the degradation process of LEV, with $SO_4\cdot^-$, $\cdot OH$, and $\cdot O_2^-$ playing a leading role and $^1O_2$ also contributing to the degradation of LEV. Four possible degradation pathways were postulated by LC-MS analysis. CoFe@CSC-700 exhibited outstanding catalytic activity, as well as remarkable reusability and recyclability. This study provides a potential method for preparing bimetallic catalysts that are easy to separate and recover for application in AOPs.

Supplementary Materials: The following supporting information can be downloaded at https://www.mdpi.com/article/10.3390/w16131818/s1: Figure S1: Zeta potential of CoFe@CSC-700; Table S1: Comparisons of CoFe@CSC-700 with other catalysts for the degradation of LEV.

Author Contributions: T.H.: conceptualization, experiment setup, sampling, investigation, writing—original draft and editing. Y.C., X.G. and Y.P.: investigation, formal analysis, data curation. J.C.: conceptualization, methodology, writing—review and editing, funding acquisition. All authors have read and agreed to the published version of the manuscript.

Funding: This work was financed by the National Natural Science Foundation of China (No. 21976060), the Dongguan Social Science and Technology Development Project (20221800905482).

Data Availability Statement: The original contributions presented in the study are included in the article/Supplementary Materials, further inquiries can be directed to the corresponding author.

Conflicts of Interest: The authors declare no conflicts of interest.

References

1. Niu, J.; Chen, Y.; Li, X.; Lin, J.; Cheng, J.; Hu, Y. A "sandwich layer" of N-doped carbon nanotubes coated on the surface of oxidized iron-foam is used to drive peroxymonosulfate activation. *Sep. Purif. Technol.* **2023**, *311*, 123273. [CrossRef]
2. Aldred, K.J.; Kerns, R.J.; Osheroff, N. Mechanism of quinolone action and resistance. *Biochemistry* **2014**, *53*, 1565–1574. [CrossRef] [PubMed]
3. Yang, Y.; Li, H.; Hou, Y.L.; Wang, C.; Zhang, K.Y.; Man, Z.H.; Shang, J.W.; Cheng, X.W. Green synthesis of Co-Ce ZIF derivatives for enhanced O_3/PMS degradation oflevofloxacin. *J. Environ. Chem. Eng.* **2024**, *12*, 112158. [CrossRef]
4. Guerra-Rodríguez, S.; Rodríguez, E.; Singh, D.N.; Rodríguez-Chueca, J. Assessment of Sulfate Radical-Based Advanced Oxidation Processes for Water and Wastewater Treatment: A Review. *Water* **2018**, *10*, 1828. [CrossRef]
5. Wang, J.; Wang, Q.; Gao, P.; Sun, D.; Jin, L.; Ma, L.; Yang, L.; Zhao, J. Enhanced PMS Activation by Highly Dispersed Mn-Ce Bimetallic Oxide on Carbon Nanotubes for Degradation of Phenol. *Water* **2023**, *15*, 2243. [CrossRef]
6. Wang, C.; Zhou, G.; Xu, Y.; Yu, P.; Sun, Y. Nitrogen Doped Cobalt Anchored on the Used Resin-Based Catalyst to Activate Peroxymonosulfate for the Removal of Ibuprofen. *Water* **2022**, *14*, 3754. [CrossRef]
7. Mo, Y.; Xu, W.; Zhang, X.; Zhou, S. Enhanced Degradation of Rhodamine B through Peroxymonosulfate Activated by a Metal Oxide/Carbon Nitride Composite. *Water* **2022**, *14*, 2054. [CrossRef]
8. Huang, K.C.; Zhao, Z.; Hoag, G.; Dahmani, A.; Block, P. Degradation of volatile organic compounds with thermally activated persulfate oxidation. *Chemosphere* **2005**, *61*, 551–560. [CrossRef] [PubMed]
9. Qi, C.D.; Liu, X.T.; Ma, J.; Lin, C.Y.; Li, X.W.; Zhang, H.J. Activation of peroxymonosulfate by base: Implications for the degradation of organic pollutants. *Chemosphere* **2016**, *151*, 280–288. [CrossRef]
10. Kohantorabi, M.; Moussavi, G.; Giannakis, S. A review of the innovations in metal- and carbon-based catalysts explored for heterogeneous peroxymonosulfate (PMS) activation, with focus on radical vs. non-radical degradation pathways of organic contaminants. *Chem. Eng. J.* **2021**, *411*, 127957. [CrossRef]
11. He, X.; Mezyk, S.P.; Michael, I.; Fatta-Kassinos, D.; Dionysiou, D.D. Degradation kinetics and mechanism of beta-lactam antibiotics by the activation of H_2O_2 and $Na_2S_2O_8$ under UV-254 nm irradiation. *J. Hazard. Mater.* **2014**, *279*, 375–383. [CrossRef] [PubMed]
12. Fan, Y.; Ji, Y.; Zheng, G.; Lu, J.; Kong, D.; Yin, X.; Zhou, Q.S. Degradation of atrazine in heterogeneous Co_3O_4 activated peroxymonosulfate oxidation process: Kinetics, mechanisms, and reaction pathways. *Chem. Eng. J.* **2017**, *330*, 831–839. [CrossRef]
13. Hao, F.; Guo, W.; Wang, A.; Leng, Y.; Li, H. Intensifcation of sonochemical degradation of ammonium perffuorooctanoate by persulfate oxidant. *Ultrason. Sonochem.* **2014**, *21*, 554–558. [CrossRef] [PubMed]
14. Chen, C.; Liu, L.; Li, Y.X.; Li, W.; Zhou, L.X.; Lan, Y.Q.; Li, Y. Insight into heterogeneous catalytic degradation of sulfamethazine by peroxymonosulfate activated with $CuCo_2O_4$ derived from bimetallic oxalate. *Chem. Eng. J.* **2020**, *384*, 123257. [CrossRef]
15. Huang, J.Z.; Zhang, H.C. Mn-based catalysts for sulfate radical-based advanced oxidation processes: A Review. *Environ. Int.* **2019**, *133*, 105141. [CrossRef] [PubMed]
16. Yang, Q.J.; Choi, H.; Al-Abed, S.R.; Dionysiou, D.D. Iron–cobalt mixed oxide nanocatalysts: Heterogeneous peroxymonosulfate activation, cobalt leaching, and ferromagnetic properties for environmental applications. *Appl. Catal. B Environ.* **2009**, *88*, 462–469. [CrossRef]
17. Zhou, X.Q.; Luo, C.G.; Luo, M.G.; Wang, Q.L.; Wang, J.; Liao, Z.W.; Chen, Z.L.; Chen, Z.Q. Understanding the synergetic effect from foreign metals in bimetallic oxides for PMS activation: A common strategy to increase the stoichiometric efficiency of oxidants. *Chem. Eng. J.* **2020**, *381*, 122587. [CrossRef]
18. Zhou, R.; Zhao, J.; Shen, N.F.; Ma, T.G.; Su, Y. Efficient degradation of 2,4-dichlorophenol in aqueous solution by peroxymonosulfate activated with magnetic spinel $FeCo_2O_4$ nanoparticles. *Chemosphere* **2018**, *197*, 670–679. [CrossRef] [PubMed]
19. Pujol, A.A.; Leo'n, I.; Ca'rdenas, J.; Sepúlveda-Guzmán, S.; Manríquez, J.; Sirés, I.; Bustos, E. Degradation of phenols by heterogeneous electro-Fenton with a Fe_3O_4- chitosan composite and a boron-doped diamond anode. *Electrochim. Acta* **2020**, *337*, 135784. [CrossRef]
20. Baran, T. Pd NPs@Fe_3O_4/chitosan/pumice hybrid beads: A highly active, magnetically retrievable, and reusable nanocatalyst for cyanation of aryl halides. *Carbohyd. Polym.* **2020**, *237*, 116105. [CrossRef]
21. Chen, X.; Xu, X.J.; Zheng, X.C.; Guan, X.X.; Liu, P.U. Chitosan supported palladium nanoparticles: The novel catalysts for hydrogen generation from hydrolysis of ammonia borane. *Mater. Res. Bull.* **2018**, *103*, 89–95. [CrossRef]
22. Shang, Z.X.; Chen, Z.L.; Zhang, Z.B.; Yu, J.; Tan, S.Z. CoFe nanoalloy particles encapsulated in nitrogen-doped carbon layers as bifunctional oxygen catalyst derived from a Prussian blue analogue. *J. Alloy. Compd.* **2018**, *740*, 743–753. [CrossRef]
23. Ge, B.C.; Li, K.X.; Fu, Z.; Pu, L.T.; Zhang, X. The addition of ortho-hexagon nano spinel Co_3O_4 to improve the performance of activated carbon air cathode microbial fuel cell. *Bioresour. Technol.* **2015**, *95*, 180–187. [CrossRef] [PubMed]

24. Wang, Y.; Wang, H.; Ye, J.; Shi, L.; Feng, X. Magnetic CoFe alloy@C nanocomposites derived from ZnCo-MOF for electromagnetic wave absorption. *Chem. Eng. J.* **2020**, *383*, 123096. [CrossRef]
25. Zakaria, M.B.; Chikyow, T. Recent advances in Prussian blue and Prussian blue analogues: Synthesis and thermal treatments. *Coord. Chem. Rev.* **2017**, *352*, 328–345. [CrossRef]
26. Li, M.Q.; Luo, R.; Wang, C.H.; Zhang, M.; Zhang, W.X.; Klu, P.K.; Yan, Y.B.; Qi, J.W.; Sun, X.Y.; Wang, L.J.; et al. Iron-tannic modified cotton derived Fe^0/graphitized carbon with enhanced catalytic activity for bisphenol A degradation. *Chem. Eng. J.* **2019**, *372*, 774–784. [CrossRef]
27. Shang, Y.N.; Chen, C.; Zhang, P.; Yue, Q.Y.; Li, Y.W.; Gao, B.Y.; Xu, X. Removal of sulfamethoxazole from water via activation of persulfate by Fe_3C@NCNTs including mechanism of radical and nonradical process. *Chem. Eng. J.* **2019**, *375*, 122004. [CrossRef]
28. Dresselhaus, M.S.; Jorio, A.; Hofmann, M.; Dresselhaus, G.; Saito, R. Perspectives on carbon nanotubes and graphene Raman spectroscopy. *Nano. Lett.* **2010**, *10*, 751–758. [CrossRef] [PubMed]
29. Xu, D.Y.; Liu, B.B.; Liu, G.Y.; Su, K.D.; Li, J.H. N-doped bamboo-like CNTs combined with CoFe-$CoFe_2O_4$ as a highly efficient electrocatalyst towards oxygen evolution. *Int. J. Hydrogen Energy* **2020**, *45*, 6629–6635. [CrossRef]
30. Zhang, L.J.; Su, Z.X.; Jiang, F.L.; Yang, L.L.; Qian, J.J.; Zhou, Y.Y.; Li, W.M.; Hong, M.C. Highly graphitized nitrogen-doped porous carbon nanopolyhedra derived from ZIF-8 nanocrystals as efficient electrocatalysts for oxygen reduction reactions. *Nanoscale* **2014**, *6*, 6590–6602. [CrossRef]
31. Duan, X.G.; Ao, Z.M.; Li, D.G.; Sun, H.Q.; Suvorova, A.; Saunders, M.; Wang, G.X.; Wan, S.B. Surface-tailored nanodiamonds as excellent metal-free catalysts for organic oxidation. *Carbon* **2016**, *103*, 404–411. [CrossRef]
32. Huo, L.M.; Wang, T.; Xuan, K.; Li, L.; Pu, Y.F.; Li, C.X.; Qiao, C.Z.; Yang, H.; Bai, Y. Synthesis of dimethyl carbonate from CO_2 and methanol over Zr-based catalysts with different chemical environments. *Catalysts* **2021**, *11*, 710. [CrossRef]
33. Zhou, X.C.; Chen, S.Q.; Zhou, M.J.; Li, M.; Lan, S.; Feng, T. Highly efficient cobalt-based amorphous catalyst for peroxymonosulfate activation toward wastewater remediation. *Rare Met.* **2023**, *42*, 1160–1174. [CrossRef]
34. Zhao, L.; Li, W.; Lin, L.; Guo, W.; Zhao, W.; Tang, X.; Gong, D.; Li, Q.; Xu, P. Field investigation on river hydrochemical characteristics and larval and juvenile fish in the source region of the Yangtze River. *Water* **2019**, *11*, 1342. [CrossRef]
35. Mashentseva, A.A.; Aimanova, N.A.; Parmanbek, N.; Temirgaziyev, B.S.; Barsbay, M.; Zdorovets, M.V. Serratula coronata L. mediated synthesis of ZnO nanoparticles and their application for the removal of alizarin Yellow R by photocatalytic degradation and ddsorption. *Nanomaterials* **2022**, *12*, 3293. [CrossRef] [PubMed]
36. Deng, J.; Feng, S.F.; Ma, X.; Tan, C.; Wang, H.; Zhou, S.; Zhang, T.; Li, J. Heterogeneous degradation of Orange II with peroxymonosulfate activated by ordered mesoporous $MnFe_2O_4$. *Sep. Purif. Technol.* **2016**, *167*, 181–189. [CrossRef]
37. Zeng, Q.Y.; Tan, J.; Gao, B.B.; Cai, T.; Zhang, Q.Y.; Liu, Y.L.; Chang, S.; Zhao, S.F.; Wu, S.Q. Embedding Co in perovskite MoO_3 for superior catalytic oxidation of refractory organic pollutants with peroxymonosulfate. *Chemosphere* **2023**, *314*, 137726. [CrossRef]
38. Zhao, W.; Duan, Z.Y.; Zheng, Z.; Li, B. Cobalt bismuth oxide with cobalt(II/III) as a new stable peroxymonosulfate activator for effective degradation, mineralization, and detoxification of diclofenac in water. *J. Clean. Prod.* **2023**, *365*, 132781. [CrossRef]
39. Zhu, S.J.; Wang, Z.W.; Ye, C.; Deng, J.; Ma, X.Y.; Xu, Y.P.; Wang, L.; Tang, Z.; Luo, H.; Li, X.Y. Magnetic Co/Fe nanocomposites derived from ferric sludge as an efficient peroxymonosulfate catalyst for ciprofloxacin degradation. *Chem. Eng. J.* **2022**, *432*, 134180. [CrossRef]
40. Yuan, R.X.; Jiang, M.L.; Gao, S.M.; Wang, Z.H.; Wang, H.Y.; Boczkaj, G.; Liu, Z.J.; Ma, J.; Li, Z.J. 3D mesoporous α-$Co(OH)_2$ nanosheets electrodeposited on nickel foam: A new generation of macroscopic cobalt-based hybrid for peroxymonosulfate activation. *Chem. Eng. J.* **2020**, *80*, 122447. [CrossRef]
41. Yang, Y.Q.; Ji, W.Q.; Li, X.Y.; Lin, H.D.; Chen, H.J.; Bi, F.K.; Zheng, Z.H.; Xu, J.C.; Zhang, X.D. Insights into the mechanism of enhanced peroxymonosulfate degraded tetracycline using metal organic framework derived carbonyl modified carbon-coated Fe^0. *J. Hazard. Mater.* **2022**, *424*, 127640. [CrossRef] [PubMed]
42. Cheng, M.; Ma, R.; Chai, G.; Chen, Y.; Bai, L.; Wang, D.; Qian, J.; Chen, G.H. Nitrogen-doped carbonized polyaniline(N-CPANI) for peroxydisulfate(PDS) activation towards efficient degradation of doxycycline(DOX) via the non-radical pathway dominated by electron transfer. *Chem. Eng. J.* **2023**, *453*, 139810. [CrossRef]
43. Wang, A.Q.; Zheng, Z.K.; Wang, H.; Chen, Y.W.; Luo, C.H.; Liang, D.J.; Hu, B.; Qiu, R.L.; Yan, K. 3D hierarchical H_2-reduced Mn-doped CeO_2 microflowers assembled from nanotubes as a high-performance Fenton-like photocatalyst for tetracycline antibiotics degradation. *Appl. Catal. B-Environ.* **2020**, *277*, 119171. [CrossRef]
44. Ye, F.; Su, Y.M.; Li, R.P.; Sun, W.; Pu, M.J.; Yang, C.; Yang, W.C.; Huang, H.M.; Zhang, Q.C.; Wong, J.W.C. Activation of persulfate on fluorinated carbon: Role of semi-ionic C-F in inducing mechanism transition from radical to electron-transfer nonradical pathway. *Appl. Catal. B-Environ.* **2023**, *337*, 122992. [CrossRef]
45. Zhang, Z.; Dai, Y.Z. Co_3O_4/C-PC composite derived from pomelo peel-loaded ZIF-67 for activating peroxymonosulfate(PMS) to degrade ciprofloxacin. *J. Water Process. Eng.* **2022**, *49*, 103043. [CrossRef]
46. Zhong, Y.W.; Shih, K.M.; Diao, Z.H.; Gang, S.; Su, M.H.; Hou, L.A.; Chen, D.Y.; Kong, L.J. Peroxymonosulfate activation through LED-induced $ZnFe_2O_4$ for levofloxacin degradation. *Chem. Eng. J.* **2021**, *417*, 129225–129235. [CrossRef]
47. Pi, Y.Q.; Gao, H.Q.; Cao, Y.D.; Cao, R.L.; Wang, Y.B.; Sun, J.H. Cobalt ferrite supported on carbon nitride matrix prepared using wastebattery materials as a peroxymonosulfate activator for the degradation of levofloxacin hydrochloride. *Chem. Eng. J.* **2020**, *379*, 122377–122385. [CrossRef]

48. Liu, L.L.; Zhan, R.; Zhang, M.; Li, J.N.; Wang, Z.P.; Mi, H.S.; Zhang, Y.X. Insights into the performance, mechanism, and ecotoxicity of levofloxacin degradation in $CoFe_2O_4$ catalytic peroxymonosulfate process. *J. Environ. Chem. Eng.* **2022**, *10*, 107435–107445. [CrossRef]
49. Zhao, J.; Xiao, P.F.; Han, S.; Zulhumar, M.; Wu, D.D. Preparation of magnetic copper ferrite nanoparticle as peroxymonosulfate activatingcatalyst for effective degradation of levofloxacin. *Water Sci. Technol.* **2022**, *85*, 645–663. [CrossRef]
50. Zhao, Q.Z.; Wu, Y.Z.; Zhang, X.X.; Zhou, L.; Lu, S.; Zhang, J.L.; Liu, Y.D.; Lei, J.Y. Enhanced PMS activation by Mn_2O_3-loaded h-BN for levofloxacin removal: Unveiling the dominant influence of non-free radical pathway and N-Mn-mediated promotion of stable, long-lived Mn(IV) species. *Appl. Surf. Sci.* **2024**, *657*, 159716. [CrossRef]
51. Li, M.K.; Huang, F.L.; Hu, L.; Sun, W.; Li, E.P.; Xiong, D.L.; Zhong, H.; He, Z.G. Efficient activation of peroxymonosulfate by a novel catalyst prepared directly from electrolytic manganese slag for degradation of recalcitrant organic pollutes. *Chem. Eng. J.* **2020**, *401*, 126085. [CrossRef]
52. He, Y.X.; Qian, J.; Wang, P.F.; Wu, J.; Lu, B.H.; Tang, S.J.; Gao, P. Acceleration of levofloxacin degradation by combination of multiple free radicals via MoS_2 anchored in manganese ferrite doped perovskite activated PMS under visible light. *Chem. Eng. J.* **2022**, *431*, 133933. [CrossRef]
53. Yuan, S.H.; Liao, P.; Alshawabkeh, A.N. Electrolytic manipulation of persulfate reactivity by iron electrodes for trichloroethylene degradation in groundwater. *Environ. Sci. Technol.* **2014**, *48*, 656–663. [CrossRef]
54. Chen, L.W.; Zuo, X.; Zhou, L.; Huang, Y.; Yang, S.J.; Cai, T.M.; Ding, D.H. Efficient heterogeneous activation of peroxymonosulfate by facilely prepared Co/Fe bimetallic oxides: Kinetics and mechanism. *Chem. Eng. J.* **2018**, *345*, 364–374. [CrossRef]
55. Zhu, J.L.; Wang, J.; Shan, C.; Zhang, J.; Lv, L.; Pan, B.C. Durable activation of peroxymonosulfate mediated by Co-doped mesoporous $FePO_4$ via charge redistribution for atrazine degradation. *Chem. Eng. J.* **2019**, *375*, 122009. [CrossRef]
56. Liu, Y.; Guo, H.G.; Zhang, Y.L.; Tang, W.H.; Cheng, X.; Li, W. Heterogeneous activation of peroxymonosulfate by sillenite $Bi_{25}FeO_{40}$: Singlet oxygen generation and degradation for aquatic levofloxacin. *Chem. Eng. J.* **2018**, *343*, 128–137. [CrossRef]
57. Sheng, J.L.; Guo, S.K.; Yuan, C.; Nie, X.R.; Cui, P.L.; Jiang, H.M. Degradation bensulfuron-methyl by magnetic CoFe alloy@N-doped graphitized carbon derived from $CoFe_2O4$ activated by peroxymonosulfate. *Chem. Eng. J.* **2023**, *466*, 143158. [CrossRef]
58. Su, S.N.; Guo, W.L.; Leng, Y.Q.; Yi, C.L.; Ma, Z.M. Heterogeneous activation of Oxone by $Co_xFe_{3x}O_4$ nanocatalysts for degradation of rhodamine B. *J. Hazard. Mater.* **2013**, *244–245*, 736–742. [CrossRef]
59. Ma, W.J.; Wang, N.; Du, Y.C.; Tong, T.Z.; Zhang, L.J.; Andrew Lin, K.Y.; Han, X.J. One-step synthesis of novel Fe_3C@nitrogen-doped carbon nanotubes/graphene nanosheets for catalytic degradation of Bisphenol A in the presence of peroxymonosulfate. *Chem. Eng. J.* **2019**, *356*, 1022–1031. [CrossRef]
60. Li, X.N.; Rykov, A.I.; Zhang, B.; Zhang, Y.J.; Wang, J.H. Graphene encapsulated FexCoy nanocages derived from metal-organic frameworks as efficient activators for peroxymonosulfate. *Catal. Sci. Technol.* **2016**, *6*, 7486–7494. [CrossRef]
61. Yan, J.W.; Lin, G.; Chai, S.Y.; Guo, C.; Zhang, W.; Wan, H.Y. ZIF-67 loaded lotus leaf-derived biochar for efficient peroxymonosulfate activation for sustained levofloxacin degradation. *Chem. Eng. J.* **2023**, *458*, 141456. [CrossRef]
62. Zhang, G.S.; Wang, Y.; Chen, M.; Xu, J.X.; Wang, L. ZIF-67-derived carbon@Co_3S_4/$CoSO_4$/MnO polyhedron to activate peroxymonosulfate for degrading levofloxacin: Synergistic effect and mechanism. *Chem. Eng. J.* **2023**, *451*, 138976. [CrossRef]
63. Xue, X.J.; Liao, W.D.; Liu, D.L.; Zhang, X.D.; Huang, Y.M. MgO/Co_3O_4 composite activated peroxymonosulfate for levofloxacin degradation: Role of surface hydroxyl and oxygen vacancies. *Sep. Purif. Technol.* **2023**, *306*, 122560. [CrossRef]
64. Deng, Q.C.; Zhang, X.D.; Chang, L.; Chai, H.X.; Huang, Y.M. The MOF/LDH derived heterostructured Co_3O_4/$MnCo_2O_4$ composite for enhanced degradation of levofloxacin by peroxymonosulfate activation. *Sep. Purif. Technol.* **2022**, *294*, 121182. [CrossRef]

catalysts

MDPI

Review

Performance of Traditional and Emerging Water-Treatment Technologies in the Removal of Tetracycline Antibiotics

Fanjin Wang [†], Ziyi Wang [†], Yue Zhao and Jian Zhang *

School of Chemistry Engineering, Northeast Electric Power University, Jilin City 132012, China; 2022304040205@neepu.edu.cn (F.W.); 2020304040135@neepu.edu.cn (Z.W.); zhaoyue@ciac.ac.cn (Y.Z.)

* Correspondence: zhangjian82916@163.com

† These authors contributed equally to this work.

Abstract: Tetracycline antibiotics are widely used in human medical treatment, control of animal disease, and agricultural feed because of their broad spectrum of action, high efficiency, and low cost. The excessive use of antibiotics and arbitrary discharge of antibiotic wastewater have become increasingly serious problems, and the current sewage-treatment process is not ideal for treating water contaminated with tetracycline antibiotics, leading to increasingly prominent antibiotic pollution in water and the imminent need for its removal. In order to understand the necessity of removing tetracycline antibiotics from the water environment, this paper first expounds on their source, harms, and pollution status in oceans and in surface water, groundwater, wastewater, and drinking water. It next introduces the research status of conventional treatment methods such as adsorption methods, biological methods, and physical and chemical methods, then introduces new treatment methods such as advanced oxidation methods and comprehensive treatment technology in sewage plants. The degradation effects, mechanisms of action, and challenges of these methods were summarized. The advantages and disadvantages of each treatment technology are compared. Finally, potential future processing technologies are discussed.

Keywords: tetracycline antibiotics; environment; wastewater; pollution; treatment technology

1. Introduction

Antibiotics are chemical substances that can interfere with the growth of other living cells and inhibit the growth of or kill some pathogenic microorganisms. They are widely used in human and animal medical treatment and in aquaculture [1]. With the continuous development of low-income countries, the demand for antibiotics is increasing. From 2000 to 2015, the global human consumption of antibiotics increased by 65% and the rate of antibiotic consumption increased by 39% [2]. It is reported that tetracycline antibiotics are the second most widely used antibiotics in the world [3]. Tetracycline antibiotics at high concentrations can inhibit the synthesis of microbial proteins and kill a variety of atypical organisms, such as Gram-positive bacteria, Gram-negative bacteria, protozoan parasites, mycoplasma, chlamydia, Rickettsia, etc. [4]. Aureomycin, oxytetracycline, and tetracycline, as well as tetracycline's semi-synthetic derivatives, such as metheomycin, doxycycline, and minocycline, are common tetracycline antibiotics that are mainly used to treat acne, trachoma, and other diseases [5,6]. However, only a small amount of antibiotics can be metabolized or absorbed by organisms, and most antibiotics are discharged into the water environment through various routes [7]. This discharge has seriously polluted the environment and destroyed ecological diversity.

Tetracycline antibiotics, although they are a new environmental pollutant, are ubiquitous in oceans and in surface water, groundwater, wastewater, and drinking water because of their persistence. It is necessary to understand the hazards associated with tetracycline antibiotics in the water environment and the methods used for their removal. This paper expounds on the sources, hazards, and pollution status of tetracycline antibiotics, introduces treatment/removal technologies and their mechanisms of action, and compares and analyzes the advantages and disadvantages of each treatment technology so as to provide a reference for the future treatment of tetracycline antibiotics, which may involve combining technologies or developing new technologies.

2. Harm of Tetracycline Antibiotic Residues in Water

Tetracycline antibiotic residues in water mainly come from two types of sources: domestic and industrial. Tetracycline-like antibiotics are difficult for organisms (humans and animals) to completely digest and absorb, and about 50–80% of antibiotics are converted into more toxic metabolites [3]. These metabolites are mainly excreted through feces and urine, which inevitably leads to antibiotics existing in urban and rural domestic sewage. Pharmaceutical wastewater contains more antibiotics than does urban domestic wastewater, showing that existing sewage treatment plants are not able to effectively remove them [8]. The wastewater produced in the pharmaceutical manufacturing process and tetracycline antibiotics that are not used by organisms enter sewage treatment plants and landfills as raw medicines and incomplete metabolites, thus polluting the water and soil and destroying the ecological balance. To understand the hazards posed by tetracycline antibiotic residues in the water environment, ECOSAR software (Ecosar.v.2.2-application) was used to predict the acute and chronic toxicity of tetracycline antibiotics and their metabolites and degradation products in the water environment, as shown in Table 1, where P1–P4 are typical photolysis products [9] and P5–P11 are typical degradation products [10].

Table 1. ECOSAR software predicts the acute and chronic toxicity of some tetracycline antibiotics and their metabolites and degradation products in the water environment (extremely toxic chemicals are shown in red, poisonous in yellow, harmful in black, and non-poisonous in green.).

Name	Formula	Acute Toxicity (mg/L)			Chronic Toxicity (mg/L)		
		Fish (96 h-LC_{50})	Daphnia (48 h-LC_{50})	Green Alga (96 h-EC_{50})	Fish (ChV)	Daphnia (ChV)	Green Alga (ChV)
Aureomycin	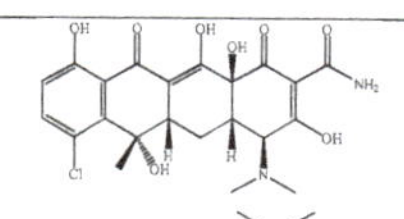	4.58×10^3	590	5.22×10^3	345	40.4	408
Oxytetracycline	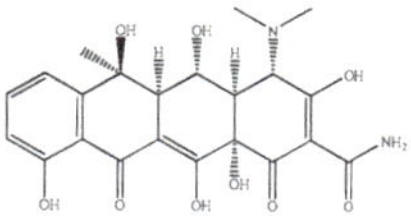	1.39×10^5	6.70×10^3	1.47×10^5	8.41×10^3	314	6.80×10^3
Doxycycline	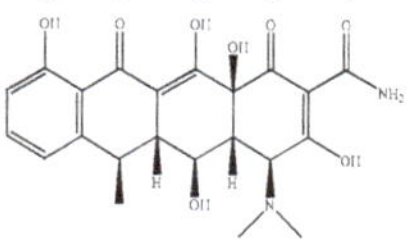	1.25×10^4	1.18×10^3	1.39×10^4	877	71.9	921
Minocycline	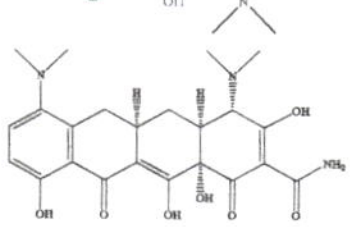	2.86×10^3	416	3.30×10^3	222	29.9	275
Tetracycline	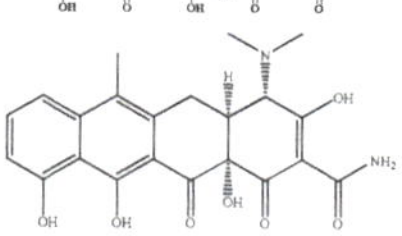	1.18×10^4	1.13×10^3	1.31×10^4	831	69.5	879
P1	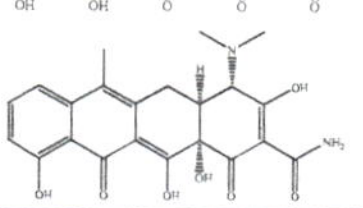	151	49.9	187	14.1	4.91	23.9

Table 1. *Cont.*

Name	Formula	Acute Toxicity (mg/L)			Chronic Toxicity (mg/L)		
		Fish (96 h-LC_{50})	Daphnia (48 h-LC_{50})	Green Alga (96 h-EC_{50})	Fish (ChV)	Daphnia (ChV)	Green Alga (ChV)
P2	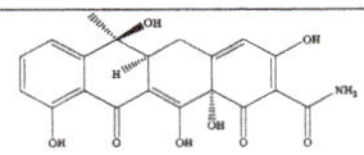	3.07×10^3	421	3.52×10^3	234	29.5	284
P3	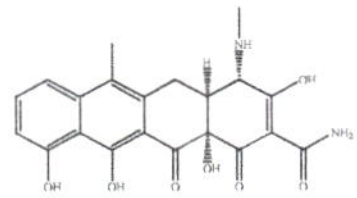	36.2	2.22	5.34	0.301	0.477	0.861
P4	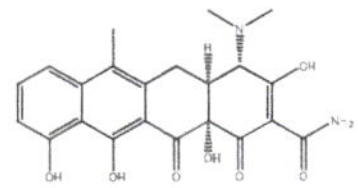	149	2.01×10^3	25.9	101	817	2.83
P5	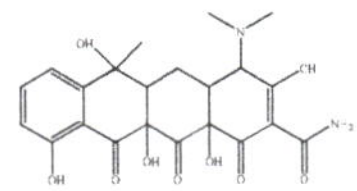	587	134	703	50.4	11.5	74.3
P6	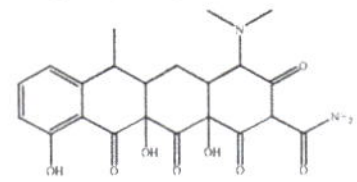	28.6	15.4	36.8	2.98	1.83	6.10
P7	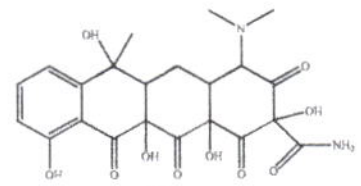	39.4	19.7	50.3	4.03	2.27	8.03
P8	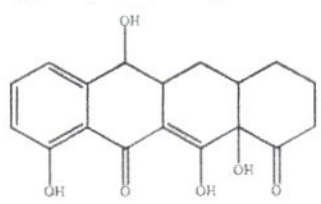	274	70.8	331	24.1	6.36	37.3
P9	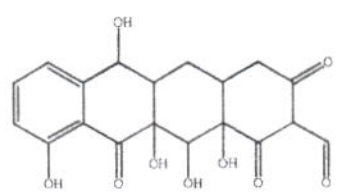	7.89×10^3	821	8.84×10^3	566	51.8	617
P10	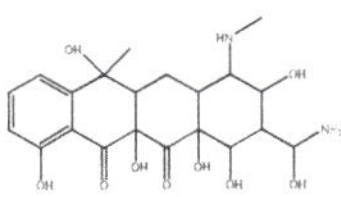	5.14×10^4	3.27×10^3	5.53×10^4	3.30×10^3	170	2.97×10^3
P11	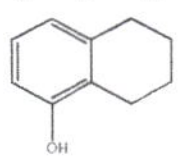	1.95	1.64	2.60	0.224	0.233	0.547
		Extremely toxic	Poisonous		Harmful	Non-poisonous	

Tetracycline antibiotics have stable properties and can accumulate in feces, soil, the water environment, and the atmosphere, where they can persist for a long time. The concentration level of tetracycline antibiotics in the water environment is mostly measured in $\mu g \cdot L^{-1}$, a level that inhibits the growth of bacteria and the survival of organisms. These chemicals accumulate in the human body through the food chain and food web, threatening human health. Tetracycline antibiotics have been detected in most of the foods we eat [11]. In the long run, this exposure will slow down the metabolism of the human body; prevent the synthesis of proteins in lymphocytes; inhibit the immune system [12]; cause risk of superinfection [13]; have teratogenic [14], carcinogenic, and mutagenic effects [15]; and lead to to joint diseases, nephropathy, central nervous system defects, endocrine disorders, and other diseases [16].

3. Pollution Status of Tetracycline Antibiotics in Water

The use of antibiotics in China accounts for about half of the antibiotic use in the world [17]. Tetracycline antibiotics are widely used because of their broad spectrum of action, high efficiency, and low cost. Because of the abuse of antibiotics by human beings, antibiotics are discharged into the environment in the form of medicines and metabolites. Tetracycline antibiotics are ubiquitous in the water environment because of their strong hydrophilicity and low steam pressure, and they are easily soluble in water. Figure 1 shows the sources of tetracycline antibiotics in the water environment.

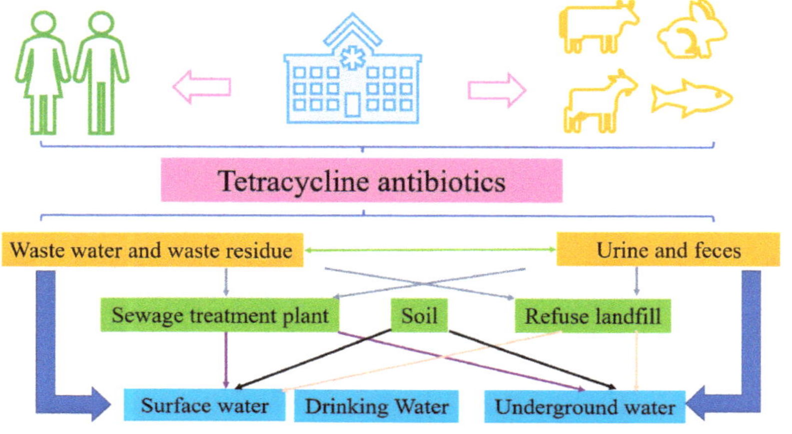

Figure 1. Sources of tetracycline antibiotics in the water environment.

In recent years, with increasing awareness of environmental protection, more and more attention has been paid to the problem of antibiotic residues in the water environment. A large number of studies at home and abroad have reported antibiotic residues in the water environment (oceans, surface water, groundwater, wastewater, drinking water). Table 2 lists the reports on the mass concentrations of tetracycline antibiotics in the water environment.

Table 2. Mass concentrations of tetracycline antibiotics in water environments.

Antibiotic Source	Region	Mass Concentration [1)]	Reference
Ocean	Bohai Gulf	16 ng·L^{-1} TC, 93 ng·L^{-1} OTC	[18]
	Iran Persian Gulf	4.0~71 ng·L^{-1} TET	[19]
	Costa Rica	74~73,722 ng·L^{-1} DOX	[20]
Surface water	Jiangxi Jinjiang	ND~86.1 ng·L^{-1} OTC, ND~5.92 ng·L^{-1} DOC, ND~5.92 ng·L^{-1} TC	[21]
	Nanjing	ND~160 ng·L^{-1} DOX	[22]
	Tianjin	0~9.74 ng·L^{-1} TET, 0~34.5 ng·L^{-1} OTC 0~3.74 ng·L^{-1} DOX CTC	[23]
	Huangpu River	15.07~113.89 ng·L^{-1} TC	[24]
Underground water	Jiangxi Jinjiang	ND~4.18 ng·L^{-1} CTC, ND~2.65 ng·L^{-1} OTC, ND~1.56 ng·L^{-1} DOC	[21]
	Harbin	0.35~3.91 ng·L^{-1} DOX	[25]
waste water	Jiangxi Jinjiang	ND~58.6 ng·L^{-1} CTC, 54.0~2.72 × 10^3 ng·L^{-1} OTC, 10.9~49.3 ng·L^{-1} DOC, ND~27.4 ng·L^{-1} TC	[21]
Drinking Water	Huaihe river basin (wet season)	Mengxian 3.38 ng·L^{-1} TCs, Yingdong 4.73 ng·L^{-1} TCs, Yongqiao 5.99 ng·L^{-1} TCs, Lingbi 8.60 ng·L^{-1} TCs, Shouxian 2.02 ng·L^{-1} TCs, Hexian 3.28 ng·L^{-1} TCs	[26]

Note: [1)] ND stands for not detected.

4. Research Status of Treatment Technology Used for the Removal of Tetracycline Antibiotics from Water Environments

Tetracycline antibiotics migrate and transform in the environment and eventually affect the atmosphere, water, and soil. With people paying more and more attention to the ecological environment, research on tetracycline antibiotics is growing at home and abroad. The epimers and dehydrated bodies of tetracycline antibiotics may appear at pH 3.0–6.5, and the epimerization may occur at pH 6.5–9.0 [27]. To remove pollutants more effectively, it is very important to understand both the methods currently used for removal and their mechanisms of action.

In the water environment, the treatment technologies used to remove tetracycline antibiotics include adsorption; biological, physical and chemical methods; advanced oxidation; and the comprehensive treatment technologies used in sewage plants.

4.1. Adsorption Method

Adsorption removes tetracycline antibiotics by transferring them to the adsorbent by π-interactions [28], electrostatic interactions [29], and hydrogen bonding [30]. The adsorbent surface forms a zone with low potential energy and high molecular density near the surface [31,32]. There are four steps in the adsorption of tetracycline antibiotics: transport through the stagnant membrane near the adsorbent, diffusion across the membrane, diffusion through the pores, and adsorption [33,34]. The mechanism by which adsorption removes tetracycline antibiotics is shown in Figure 2.

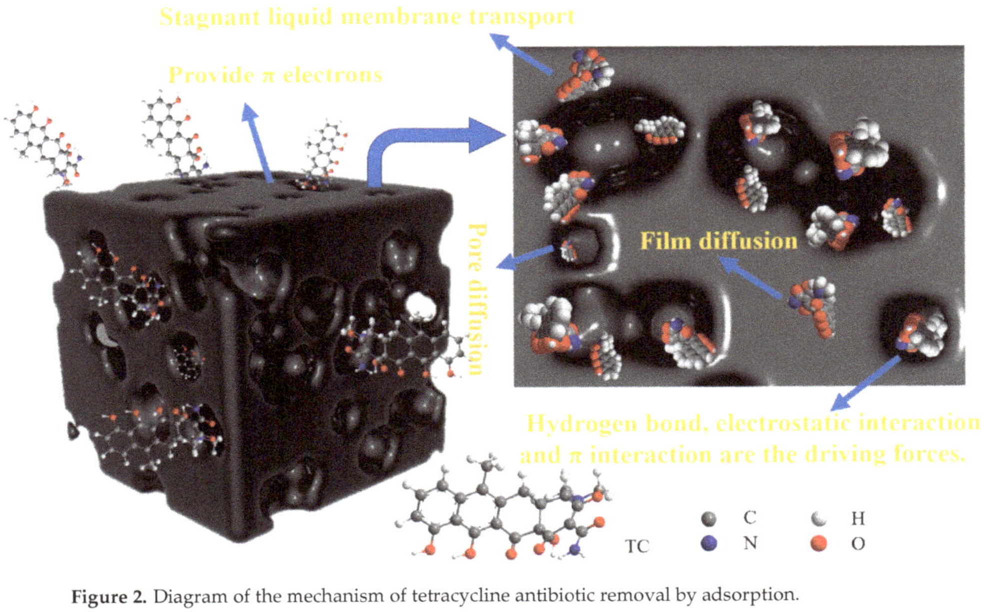

Figure 2. Diagram of the mechanism of tetracycline antibiotic removal by adsorption.

Nanoparticles produced by microorganisms in sewage treatment plants are natural adsorption materials that can be combined with tetracycline antibiotics in the water environment. Yu et al. [35] used asymmetric flow field-flow separation and multi-angle static light scattering to separate nanoparticles with different particle sizes produced by microorganisms from sewage treatment plant wastewater. These nanoparticles have a strong capacity for adsorption of tetracycline. If the adsorbate of nanoparticles and antibiotics is not properly treated, the adsorbate will further migrate and transform in the water environment, posing a great threat. Carbon materials [36], biochar [37], mineral materials [38], metal skeleton materials [39], nano-materials [40], and other new materials can be used as adsorption materials. Qiao et al. [41] developed Mn (II)-coated mesoporous silica nanoparticles to effectively remove tetracycline from an aqueous solution. It was found that Mn–O complexation was the dominant mechanism of tetracycline absorption. Electrostatic attraction and cation–π interactions also contributed to the absorption of tetracycline. Liang et al. [42] used poly-dopamine-polystyrene nanofibers to adsorb tetracycline antibiotics in water. The adsorption of this material reached equilibrium within 5 min, and the removal rate was greater than 85%. This method does not require pretreatment of water samples, but the adsorption process is most efficient at high concentrations and the residual concentration in water environment is low, so it has certain limitations. In evaluating adsorption, we should also consider whether the adsorption material can be easily separated from the water body to avoid secondary pollution. Wang et al. [43] used magnetic ion exchange resin (MIEX) to remove tetracycline from water. With this approach, the separation speed of tetracycline is fast, the reusability is high, and the capacity and rate of adsorption are high.

Adsorption collects pollutants, but it only changes their form and cannot remove them completely. Some adsorbents with small particle sizes are difficult to separate from water bodies, and it is easy for subsequent treatment to cause secondary pollution. At present, there is little research on the toxicity of adsorption materials. Therefore, the development of low-toxicity adsorption materials with high specific surface area and high adsorption capacity represents the future of tetracycline antibiotic removal by adsorption.

4.2. Biological Method

The biological method is the mainstream technology used to remove tetracycline antibiotics and mainly includes the microbial method and the plant method.

In the microbial method, tetracycline antibiotics are decomposed into small molecular substances via the action of microorganisms [44]. The activated-sludge method is a representative example of a method involving the microbial degradation of tetracycline antibiotics. It uses microorganisms, mainly bacterial micelles, to adsorb and degrade antibiotics, and this method is characterized by high adsorption and low degradation [45]. Liu et al. [46] used a method that involved adding exogenous surfactant and in-situ biosurfactant to cultivate microorganisms that could be used to remove tetracycline antibiotics. The method converts antibiotics instead of absorbing them and thus significantly reduces the toxicity of antibiotics. M.de Cazes et al. [47] successfully removed tetracycline antibiotics in an aqueous solution by immobilizing laccase at room temperature, with results showing good reactivity and stability. The use of a membrane bioreactor improves the reaction rate of activated sludge, reduces the amount of excess sludge, and results in good treatment of tetracycline antibiotic wastewater. However, membrane pollution and high energy consumption have always been the aspects of membrane bioreactors that need to be improved. Microbial degradation is the most widely used method for the removal of antibiotics, but the molecular mechanism of microbial decomposition and transformation of tetracycline antibiotics under different nutritional conditions is still unclear [48–50]. Shao et al. [51] proposed one route for the biotransformation of tetracycline (removal of methyl, carbonyl, and amine groups). Qi [52] proposed that the degradation paths of oxytetracycline are hydrolysis and biodegradation, respectively. Through dehydration, demethylation, decarbonylation, esterification, deamination, enol ketone isomerization, and enzymatic activity, oxytetracycline is degraded; the degradation process is shown in Figure 3.

The plant method uses a combination of plants and microorganisms to achieve pollution control and has the advantages of low treatment cost and environmental friendliness. Gujarathi et al. [53] used sunflower cultures to treat tetracycline and oxytetracycline, and the treatment effect decreased as the antibiotic concentration increased. Chen et al. [54] cultivated Dapiao and Eichhornia crassipes by hydroponics, with a good effect on the removal rate of tetracycline hydrochloride. Plant degradation involves high environmental requirements, and at present, there are few studies on how antibiotics enter plants and the mode of transportation, so it is still necessary to further research the plant-based degradation of tetracycline antibiotics.

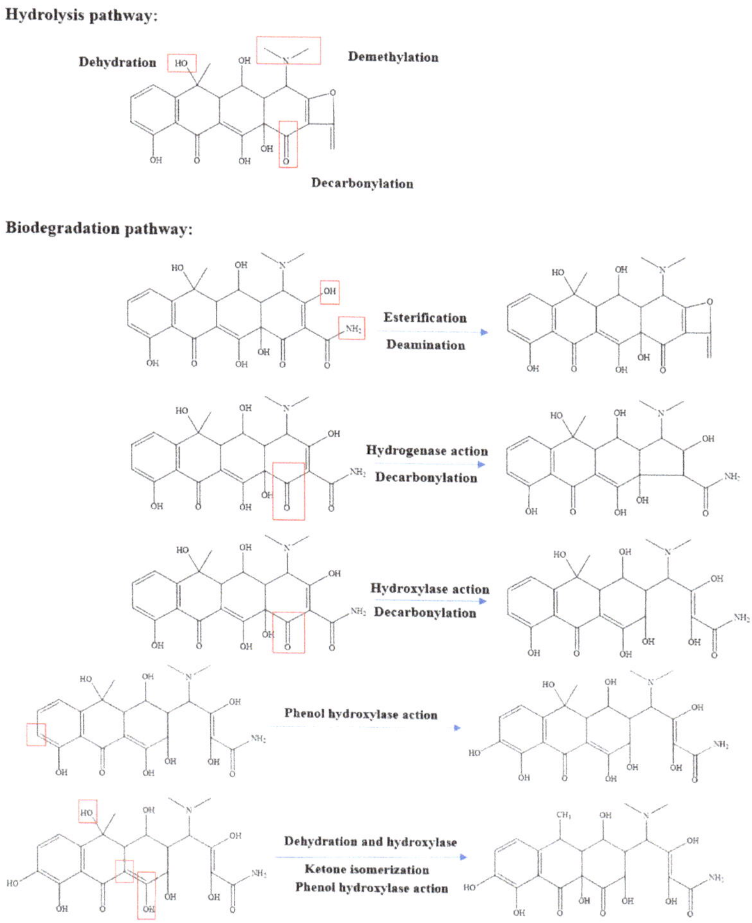

Figure 3. Mechanisms of biological treatment of tetracycline antibiotics [52].

4.3. Physical and Chemical Method

Water contaminated with tetracycline antibiotics is difficult to treat directly by biological methods because of the antibiotics' high biological toxicity. Physical and chemical methods are usually used as pretreatment, with advanced treatment methods mainly including coagulation, sedimentation, air flotation, reverse osmosis, ultrafiltration, and membrane filtration. Wojnárovits et al. [55] found that radiation therapy can make these non-biodegradable antibiotic-containing matrices biodegradable, reducing their toxicity and eliminating their antimicrobial activity, and found that the reduction of the technical parameters (COD, TOC) demonstrates the applicability of radiation technology for the degradation of recalcitrant compounds even in highly complex aqueous matrices. Zhang et al. [56] used reverse osmosis and ultrafiltration to treat tetracycline production-waste liquid, with remarkable effectiveness. Gholam Hossein Safari et al. [57] showed that RSM was a suitable method by which to optimize the operating conditions for maximum TC degradation. Under optimum conditions, the TC degradation, COD and TOC removal were 95.01%, 72.8%, and 59.7%, respectively. The $US/S_2O_8^{2-}$ process was found to be a feasible technology for TC degradation in aqueous solution. Saitoh et al. [58] used the coagulation–flotation method to quickly remove tetracycline antibiotics from water. In the presence of Al(III) ions, tetracycline in water was collected in the coagulum as hydrophobic ion pairs ($m/z = 736$) between Al(III) chelate and dodecyl sulfate ions. Tetracycline was completely removed from the water within 5 min. Liang et al. [59] used integrated membrane-filtration technology, including ultrafiltration and two-stage reverse osmosis, to treat pig wastewater and found that these methods could remove bacteria and 99.79% of antibiotic-resistance genes. This method is effective for the removal of antibiotics, but both water quality and membrane properties will affect the effectiveness. Membrane fouling and the existence of high concentration retentate after treatment are the main problems in the application of this technology.

The physicochemical method is simple and easy to implement and can be used as both pretreatment and advanced treatment. In the future, the physicochemical method will be used to develop a treatment method for water contaminated with tetracycline antibiotics with reduced operation complexity and operation costs, high development efficiency, and environmentally friendly qualities.

4.4. Advanced Oxidation Processes

The advanced oxidation method is being widely studied by researchers because of its advanced technology [60]. There are two main common features of advanced oxidation methods. First, the hydroxyl radicals generated by the reaction will effectively decompose the toxic organic pollutants that are difficult to degrade until they are completely converted into harmless inorganic substances, such as CO_2, N_2, SO_2^{4-}, PO_3^{4-}, O_2, H_2O, etc., without secondary pollution, which is difficult to achieve by other oxidation methods. Second, the reaction time is short, the reaction speed is fast, and the process is controllable and non-selective and can thus degrade all kinds of organic pollutants. According to the different reaction conditions used, it can be divided into ozone oxidation, Fenton oxidation, photocatalysis, electrocatalysis, and so on.

4.4.1. Ozone Oxidation Method

The ozonation method produces strong oxidizing free radicals (such as ·OH, etc.) through indirect oxidation reactions between ozone molecules and antibiotics or via decomposition reactions in water that can be used to mineralize them [61,62]. O_3 has a strong oxidizing ability and can form carboxylic acids with tetracycline antibiotics, thus improving the biodegradability of wastewater [63]. The mechanism by which ozone oxidation can be used to degrade tetracycline antibiotics is shown in Figure 4.

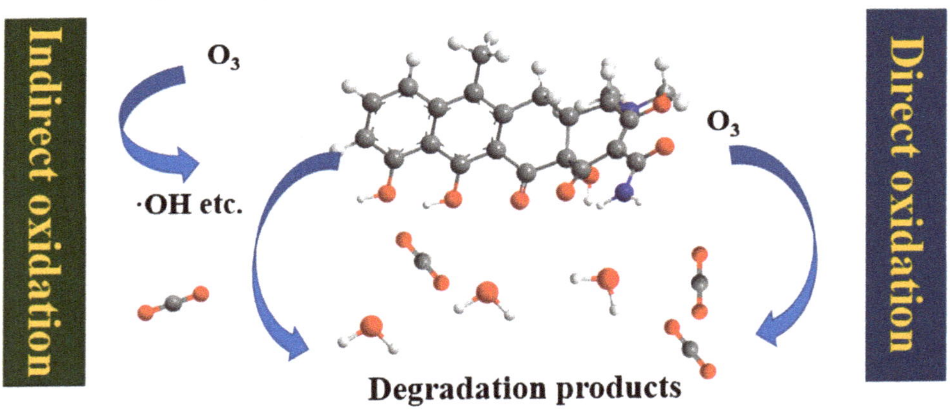

Figure 4. Mechanism of ozone oxidation degradation of tetracycline antibiotics.

Ozone is highly oxidizing and has a good effect on the degradation of organic matter. The treatment effect of ozone oxidation alone is not impressive, and catalytic ozone oxidation is often used. Luu et al. [64] used O_3, H_2O_2, and ultraviolet (UV) to degrade tetracycline. After 5 days of treatment, the biodegradability of wastewater was improved. Li et al. [65] used the combined process of denitrification biofilter + ozone + biological aeration filtration to produce reclaimed water. The combination process shows good removal performance of conventional pollutants, which meets the local discharge standards and the purpose of water reuse. The ozone method has the characteristics of quick reaction, simple process, and environmental friendliness. Developing catalysts with good catalytic effects and low cost represents the future of research on ozone oxidation.

4.4.2. Fenton Oxidation Method

Tetracycline antibiotics have strong biological toxicity, and the biodegradability of wastewater is low. Free radicals and other active oxygen substances produced by the Fenton oxidation process are effective in treatment. Hydrogen peroxide and Fe^{2+} generate OH at pH 2–5 and gradually produce other reactive oxygen species and reaction intermediates, thus contributing to the degradation of tetracycline antibiotics [66]. The reaction mechanism is as follows [67]:

$$Fe^{2+} + H_2O_2 \rightarrow Fe^{3+} + OH^- + \cdot OH \quad (1)$$

$$Fe^{3+} + H_2O_2 \rightarrow Fe^{2+} + H^+ + HO_2 \cdot \quad (2)$$

$$Fe^{2+} + \cdot OH \rightarrow Fe^{3+} + OH^- \quad (3)$$

$$Fe^{3+} + HO_2 \cdot \rightarrow Fe^{2+} + H^+ + O_2 \quad (4)$$

$$H_2O_2 + \cdot OH \rightarrow H_2O + HO_2 \cdot \quad (5)$$

$$Fe^{2+} + HO_2 \cdot \rightarrow Fe^{3+} + HO^{2-} \quad (6)$$

$$RH + \cdot OH \rightarrow R \cdot + H_2O \quad (7)$$

$$R \cdot + Fe^{3+} \rightarrow \cdot R^+ + Fe^{2+} \quad (8)$$

$$R \cdot + H_2O_2 \rightarrow OH + \cdot OH \quad (9)$$

Fenton oxidation can be used as a pretreatment process and advanced treatment process. Bai [68] used the Fenton oxidation method to treat antibiotic wastewater in the pharmaceutical industry and achieved a good removal effect. Wu et al. [66] treated SBR effluent by the Fenton oxidation method, and the removal rate of COD reached 77.8%. Shao et al. [69] introduced nitrogen vacancies and potassium atoms into g-C_3N_4, and these molecules were then combined with FeOCl to fabricate the Z-type heterojunction catalyst FeOCl/NvCN, which was used to remove 95.74% of tetracycline (TC) within 60 min. The Fenton oxidation process can significantly improve the biodegradability of tetracycline antibiotic wastewater in the practice of systematic treatment. The Fenton oxidation process can significantly improve the biodegradability of tetracycline antibiotic wastewater in the practice of systematic treatment. However, the amount of Fe^{2+} and H_2O_2 used in the treatment process is large, and a large amount of acid and alkali are used in the pH-adjustment process, which undoubtedly increases the cost of wastewater treatment. Making catalysts with high effectiveness and low cost, exploring the best amounts to use, improving the methods, and treating large-scale wastewater are the future directions of research involving Fenton oxidation.

4.4.3. Photocatalytic Method

In the photocatalytic method, under specific light conditions, the catalyst generates electron–hole pairs, which react with H_2O and other molecules to generate ·OH, which oxidizes and decomposes tetracycline antibiotics and mineralizes them [70,71]. The ·OH acts on the double bond, keto group, amino group, and keto-enol group of tetracycline antibiotics, and the mechanism is shown in Figure 5.

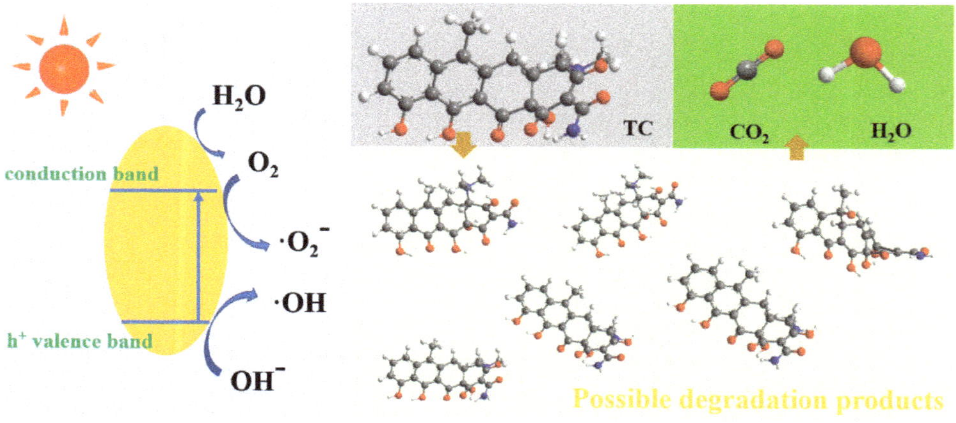

Figure 5. Mechanism of photocatalytic treatment of tetracycline antibiotics [9].

The recovery of the catalyst and the study of its stability have always been the focus of research on the photocatalytic treatment of tetracycline antibiotics. C. Reyes et al. [72] used a TiO_2 aqueous suspension to irradiate a tetracycline solution with three different light sources (an ultraviolet lamp, a sunbathing device and a UV-A lamp). The results showed that tetracycline was clearly mineralized under the action of the ultraviolet lamp and the solarium. Das et al. [73] synthesized semiconducting CdS nanorods using a hydrothermal procedure with ethylenediamine as a ligand and found higher catalytic efficiency in the presence of blue-light as compared to white-light irradiation and demonstrated that the photodegradation process did not require any supplemental oxygen source. Nagamine et al. [74] successfully synthesized cadmium sulfide semiconductor nanoparticles by a simple coprecipitation method, and these nanoparticles were used to degrade tetracycline. After exposure for 1 h, the photocatalytic efficiency reached nearly 80% for TC degradation. Zhang et al. [75] prepared TiC-SOH/g-CN (TiCSOHCN) composite photocatalysts by acid etching and sonochemistry, and the removal of TC reached 75.42% within 2 h. Ma et al. [76] prepared La/Co@TiO_2 nanospheres by co-impregnation and increased the oxygen vacancies on the catalyst surface by photocatalytic activation of molecular oxygen to accelerate the photogenerated electron-hole separation and charge transfer at the interface. The photocatalytic degradation experiments using 3 wt% La/Co@TiO_2 nanospheres showed that the efficiency of tetracycline (TC) degradation was 100% under visible light. Tang et al. [77] designed a novel p-n heterostructure visible-light photocatalyst by anchoring p-type Bi_2O_3 on n-type Ti^{3+}-TiO_2 porous material by a simple photodeposition method with subsequent calcination. Complete removal of tetracycline from various aqueous matrices under visible light was realized. Li et al. [78] successfully synthesized a novel direct Z-type SnS_2@$ZnIn_2S_4$@kaolinite heterostructured photocatalyst by a hydrothermal method. Under visible-light irradiation, in an experiment carried out in parallel with SnS_2 and $ZnIn_2S_4$, the as-obtained heterostructure displayed greatly improved photocatalytic degradation of TCH, with an apparent reaction rate of 0.0231 min^{-1}, which is 20.81 times and 2.31 times higher than those of SnS_2 and $ZnIn_2S_4$, respectively. Zong et al. [79] achieved efficient degradation of tetracycline by combining TiO_2 fragmented shells (TFNs) catalysts with convergent pulsed ultrasound in a customized ultrasonic reactor. TFNs were rapidly removed from solutions with low concentrations of tetracycline in as little as 6 min at low power input (21 W). Photocatalysis has the advantages of simple operation and mild reaction conditions, but the utilization rate of light energy is low, the cost is high, and the chromaticity of wastewater will also affect the efficiency of treatment of water contaminated with tetracycline antibiotics by light.

4.4.4. Electrocatalytic Method

In the electrocatalytic method, under the action of an electric field, ·OH radicals attack the double bonds, phenolic groups, and amine groups of tetracycline antibiotics and mineralize them [80–82]. There are four ways to treat tetracycline antibiotics by electrocatalysis, namely, direct oxidation, direct reduction, indirect oxidation, and indirect reduction. The mechanism is shown in Figure 6.

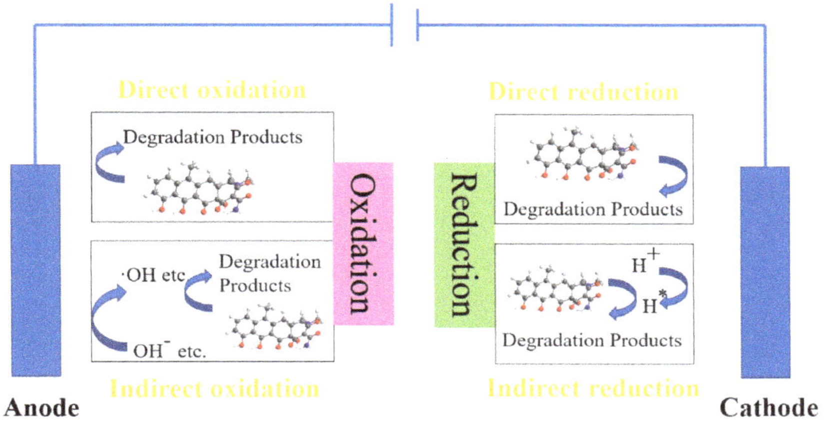

Figure 6. Mechanism of electrocatalytic treatment of tetracycline antibiotics (Which H* show Atom H) [83].

Electrocatalysis is a method with a wide application range and simple operation. Ni et al. [81] used FeNi/NF system to treat aquaculture wastewater discharged into seawater, and after 2% NaCl was added, 99.8% tetracycline could be treated in 30 min. The choice of the electrode is closely related to the effectiveness of antibiotic degradation. Oturan et al. [84] found that the process with a BDD anode showed excellent oxidation/mineralization ability. Li et al. [85] used electrocatalysis to pretreat antibiotic wastewater with salt content, and the treatment effect was good. Sun et al. [86] prepared an efficient Bi-Sn-Sb/γ-Al_2O_3 particle electrode for the degradation of tetracycline. Lu et al. [87] utilized a Pd/Ru/MXene/NF electrode for the electrocatalytic treatment of tetracycline hydrochloride wastewater, achieving a degradation rate exceeding 91% within 60 min. Shao et al. [88] loaded agarose (AG) onto ITO using the dipping method and deposited Pd onto the AG/ITO electrode through electrodeposition to prepare a Pd/AG/ITO composite electrode. The TC degradation efficiency of the AG/ITO composite electrode reached 85.21% within 120 min. It was found that the electrode has excellent electrocatalytic activity. Although the electrocatalysis method has advantages, there is little research on the treatment of pharmaceutical intermediates produced by antibiotics over different time frames, and ogoing effort is thus still needed from researchers.

4.5. Comprehensive Treatment Technology of Sewage Plant

Traditional sewage treatment plants cannot completely remove antibiotics from sewage by biodegradation or adsorption, which leads to environmental pollution and a high risk associated with the water [89]. Most sewage treatment plants are not designed to remove antibiotics themselves. Because of the limitations of individual water treatment methods, they cannot completely treat water contaminated with antibiotics. Usually, such water is treated by a combination of various methods [90,91]. Xu et al. [92] used UVC/persulfate processes to treat antibiotic wastewater, and TC was removed by hydroxylation, demethylation, decarbonylation, photochemistry, and dehydration.

The SWOT method was used to analyze the adsorption method, the biological method, the physical and chemical methods, the advanced oxidation method, and the comprehensive treatment technology used in sewage plants, as shown in Table 3.

Table 3. SWOT analysis of tetracycline antibiotics treatment technology in water environment.

Technology	Treatment Efficiency	Cost	Advantage	Disadvantage
Adsorption method	High	Low	Multifunctional, low energy consumption, easy operation and no by-products	Easily causes secondary pollution
Biological method	Low	Low	Low cost and mature technology	Unsafe
Materialization method	Low	High	Simple operation	General effect
Advanced oxidation processes	High	High	High degree of mineralization of pollutants and stable effects	Pharmaceutical intermediate
Comprehensive treatment technology used in sewage plants	High	Low	Mature technology, simple operation, etc.	Sludge treatment is difficult

5. Conclusions and Future Perspectives

Tetracycline antibiotics exist in various water environments (oceans, surface water, groundwater, wastewater, drinking water) due to the difficultly of their degradation and thus pose a threat to human health and the ecological environment. At present, the sources and distribution of tetracycline antibiotics in water resources are still unclear, so it is necessary to further monitor tetracycline antibiotics in water environments to determine the extent of this pollution and design treatment approaches.

At present, research on the degradation of tetracycline antibiotics in water environment has not been widely carried out at home and abroad. Adsorption can only concentrate pollutants, not remove them, and thus may cause secondary pollution in the process of adsorbent regeneration. The biological method is widely used and has low treatment costs and mature technology, but it is still necessary to consider the generation of medicines-resistant superbacteria in the treatment process; Antibiotics cannot be completely degraded/removed by a single method, so combined treatment methods are widely used. The physicochemical method, advanced oxidation technology and the comprehensive treatment technology used in sewage plants all have their own advantages. The physicochemical method can be used as a pretreatment technology or as advanced treatment technology. Advanced oxidation technology is efficient and easy to operate, but the research on the toxicity of pharmaceutical intermediates and their complex environmental behaviors should also be strengthened. In the comprehensive treatment of sewage, several technologies are often used to remove pollutants, with good effectiveness. In summary, in the context of aiming to improve existing technologies, future developments will involve combining technologies or developing new technologies to remove tetracycline antibiotics from water environments, thus protecting human health and helping to maintain a good ecological environment.

Funding: This research received no external funding.

Data Availability Statement: Not applicable.

Conflicts of Interest: The authors declare no conflict of interest.

References

1. Shao, S.; Hu, Y.; Cheng, J.; Chen, Y. Research Progress on Distribution, Migration, Transformation of Antibiotics and Antibiotic Resistance Genes (ARGs) in Aquatic Environment. *Crit. Rev. Biotechnol.* **2018**, *38*, 1195–1208. [CrossRef] [PubMed]
2. Klein, E.Y.; Boeckel, T.V.; Martinez, E.M.; Pant, S.; Gandra, S.; Levin, S.; Goossens, H.; Laxminarayan, R. Global Increase and Geographic Convergence in Antibiotic Consumption between 2000 and 2015. *Proc. Natl. Acad. Sci. USA* **2018**, *115*, E3463–E3470. [CrossRef] [PubMed]
3. Jandira, L.; Yasmin, V.; Nicoly, W.; Siara, S.; Guilherme, L.D.; Elvis, C. A Review of the Occurrence, Disposal, Determination, Toxicity and Remediation Technologies of the Tetracycline Antibiotic. *Process Saf. Environ. Prot.* **2022**, *160*, 25–40.
4. Haiyin, Z.; Qixing, Z. Research Progress on Treatment Technology of Tetracycline Antibiotics Pollution in the Environment. *J. Environ. Eng. Technol.* **2021**, *11*, 571–581.
5. Chopra, I.; Roberts, M.C. Tetracycline Antibiotics: Mode of Action, Applications, Molecular Biology, and Epidemiology of Bacterial Resistance. *Microbiol. Mol. Biol. Rev.* **2001**, *65*, 232–260. [CrossRef] [PubMed]

6. Chopra, I. New Developments in Tetracycline Antibiotics: Glycylcyclines and Tetracycline Efflux Pump Inhibitors. *Med. Resist. Updates* **2002**, *5*, 119–125. [CrossRef] [PubMed]
7. Bilal, M.; Mehmood, S.; Rasheed, T.; Iqbal, H.N. Antibiotics Traces in the Aquatic Environment: Persistence and Adverse Environmental Impact. *Curr. Opin. Environ. Sci. Health* **2020**, *13*, 68–74. [CrossRef]
8. Xinyan, G.; Zheng, Y.; Yi, Z.; Weili, X.; Deyang, K.; Zhengjun, S.; Na, W. Behavior of Antibiotic Resistance Genes under Extremely High-Level Antibiotic Selection Pressures in Pharmaceutical Wastewater Treatment Plants. *Sci. Total Environ.* **2018**, *612*, 119–128.
9. Krakkó, D.; Heieren, B.T.; Illés, Á.; Kvamme, K.; Dóbé, S.; Záray, G. (V) UV Degradation of the Antibiotic Tetracycline: Kinetics, Transformation Products and Pathway. *Process Saf. Environ. Prot.* **2022**, *163*, 395–404. [CrossRef]
10. Zhang, H.; Bian, J.; Yang, C.; Hu, Z.; Liu, F.; Zhang, C. Removal of Tetracycline from Livestock Wastewater by Positive Single Pulse Current Electrocoagulation: Mechanism, Toxicity Assessment and Cost Evaluation. *Sci. Total Environ.* **2022**, *810*, 151955. [CrossRef]
11. Li, Y.X.; Zhang, X.L.; Li, W.; Lu, X.F.; Liu, B.; Wang, J. The Residues and Environmental Risks of Multiple Veterinary Antibiotics in Animal Faeces. *Environ. Monit. Assess.* **2013**, *185*, 2211–2220. [CrossRef] [PubMed]
12. Lindberg, R.H.; Björklund, K.; Rendahl, P.; Johansson, M.I.; Tysklind, M.; Andersson, B.V. Environmental Risk Assessment of Antibiotics in the Swedish Environment with Emphasis on Sewage Treatment Plants. *Water Res.* **2007**, *41*, 613–619. [CrossRef] [PubMed]
13. Wang, J.; Chu, L.; Wojnárovits, L.; Takács, E. Occurrence and Fate of Antibiotics, Antibiotic Resistant Genes (ARGs) and Antibiotic Resistant Bacteria (ARB) in Municipal Wastewater Treatment Plant: An Overview. *Sci. Total Environ.* **2020**, *744*, 140997. [CrossRef] [PubMed]
14. Beyene, T. Veterinary Medicines Residues in Food-Animal Products: Its Risk Factors and Potential Effects on Public Health. *J. Vet. Sci. Technol.* **2016**, *7*, 285.
15. Klimek, L.; Aderhold, C.; Sperl, A.; Trautmann, A. Allergic Reactions to Antibiotics—Two Sides of the Same Coin: Clearly Diagnose or Reliably Rule Out. *Allergo J. Int.* **2017**, *26*, 212–218. [CrossRef]
16. Tadesse, T. Public Health Impacts of Antibiotic Residues in Foods of Animal Origin: A Review. *Public Policy Adm. Res.* **2017**, *7*, 6–11.
17. Shao, Y.; Wang, Y.; Yuan, Y.; Xie, Y. A systematic review on antibiotics misuse in livestock and aquaculture and regulation implications in China. *Sci. Total Environ.* **2021**, *798*, 149205. [CrossRef] [PubMed]
18. Zou, S.; Xu, W.; Zhang, R.; Tang, J.; Chen, Y.; Zhang, G. Occurrence and Distribution of Antibiotics in Coastal Water of the Bohai Bay, China: Impacts of River Discharge and Aquaculture Activities. *Environ. Pollut.* **2011**, *159*, 2913–2920. [CrossRef] [PubMed]
19. Raheleh, K.; Fatemeh, P.; Mohammad, S.; Sahebi, S.; Tahmasebi, R.; Ahmadi, M.; Sorial, G.; Asgari, G.; Ramavandi, B. Occurrence, Distribution, and Potential Sources of Antibiotics Pollution in the Water-Sediment of the Northern Coastline of the Persian Gulf, Iran. *Sci. Total Environ.* **2018**, *627*, 703–712.
20. Spongberg, A.L.; Witter, J.D.; Acuña, J.; Vargas, J.; Murillo, M.; Umaña, G.; Gómez, E.; Perez, G. Reconnaissance of Selected PPCP Compounds in Costa Rican Surface Waters. *Water Res.* **2011**, *45*, 6709–6717. [CrossRef]
21. Li, J.L.; Wang, M.; Hu, F.W.; Dong, Y.H.; Sun, Z.X.; Wang, Y. Antibiotic Pollution Characteristics and Ecological Risk Assessment in Jinjiang River Basin, Jiangxi Province. *Environ. Sci.* **2022**, *43*, 4064–4073.
22. Li, H.; Chen, Y.; Feng, M.J.; Wang, B.; Bu, Y.Q.; Zhang, S.H. Pollution characteristics and risk assessment of antibiotics in Nanjing drinking water sources. *Acta Sci. Circumstantiae* **2020**, *40*, 1269–1277.
23. Yin, C.N.; Xie, P.; Jiao, M.; Guo, B.Y.; Yu, C.H. Distribution characteristics of 54 kinds of PPCPs in surface water in Tianjin. *Environ. Chem.* **2021**, *40*, 2820–2831.
24. Lei, J.; Hu, X.; Yin, D.; Zhang, H.; Yu, Z. Occurrence, Distribution and Seasonal Variation of Antibiotics in the Huangpu River, Shanghai, China. *Chemosphere* **2011**, *82*, 822–828.
25. Ma, J.S.; Wang, Z.; Zhang, Z.Y.; Liu, Q.; Li, L.J. Distribution Characteristics of 29 Antibiotics in Groundwater in Harbin. *Rock Miner. Anal.* **2021**, *40*, 944–953.
26. Zhou, P.P. Investigation of Pollution and Health Risk Assessment on Tetracyclines in Drinking Water and Animal Derived Food of Anhui. Master's Thesis, Anhui Medical University, Hefei, China, 2019.
27. Turan, B.; Sarigol, G.; Demircivi, P. Adsorption of Tetracycline Antibiotics Using Metal and Clay Embedded Cross-Linked Chitosan. *Mater. Chem. Phys.* **2022**, *279*, 125781. [CrossRef]
28. Ai, Y.; Liu, Y.; Huo, Y.; Zhao, C.; Sun, L.; Han, B.; Cao, X.; Wang, X. Insights into the Adsorption Mechanism and Dynamic Behavior of Tetracycline Antibiotics on Reduced Graphene Oxide (RGO) and Graphene Oxide (GO) Materials. *Environ. Sci. Nano* **2019**, *6*, 3336–3348. [CrossRef]
29. Shen, Q.; Wang, Z.; Yu, Q.; Cheng, Y.; Liu, Z.; Zhang, T.; Zhou, S. Removal of Tetracycline from an Aqueous Solution Using Manganese Dioxide Modified Biochar Derived from Chinese Herbal Medicines Residues. *Environ. Res. Lett.* **2020**, *183*, 109195. [CrossRef] [PubMed]
30. Jang, H.M.; Yoo, S.; Choi, Y.-K.; Park, S.; Kan, E. Adsorption Isotherm, Kinetic Modeling and Mechanism of Tetracycline on Pinus Taeda-Derived Activated Biochar. *Bioresour. Technol.* **2018**, *259*, 24–31. [CrossRef]
31. Van Der Spoel, D.; Lindahl, E.; Hess, B.; Groenhof, G.; Mark, A.; Berendsen, H.C. Gromacs: Fast, Flexible, and Free. *J. Comput. Chem.* **2005**, *26*, 1701–1718. [CrossRef]
32. Rostamian, R.; Amiri, N.; Behnejad, H. How Does Graphene Nanosheet Affect the Pharmaceutical Adsorption? A Comprehensive Insight from Molecular Dynamics Simulation, Quantum Mechanics and Experimental Study. *J. Mol. Liq.* **2018**, *269*, 29–37. [CrossRef]

33. Ninwiwek, N.; Hongsawat, P.; Punyapalakul, P.; Prarat, P. Removal of the Antibiotic Sulfamethoxazole from Environmental Water by Mesoporous Silica-Magnetic Graphene Oxide Nanocomposite Technology: Adsorption Characteristics, Coadsorption and Uptake Mechanism. *Colloids Surf. A Physicochem. Eng. Asp.* **2019**, *580*, 123716. [CrossRef]
34. Prarat, P.; Hongsawat, P.; Punyapalakul, P. Amino-Functionalized Mesoporous Silica-Magnetic Graphene Oxide Nanocomposites as Water-Dispersible Adsorbents for the Removal of the Oxytetracycline Antibiotic from Aqueous Solutions: Adsorption Performance, Effects of Coexisting Ions, and Natural Organic Matter. *Environ. Sci. Pollut. Res.* **2020**, *27*, 6560–6576.
35. Yu, K.; Sun, C.; Zhang, B.; Hassan, M.; He, Y. Size-Dependent Adsorption of Antibiotics onto Nanoparticles in a Field-Scale Wastewater Treatment Plant. *Environ. Pollut.* **2019**, *248*, 1079–1087. [CrossRef] [PubMed]
36. Ahmed, M.B.; Zhou, J.L.; Ngo, H.H.; Guo, W. Adsorptive Removal of Antibiotics from Water and Wastewater: Progress and Challenges. *Sci. Total Environ.* **2015**, *532*, 112–126. [CrossRef] [PubMed]
37. Qiu, B.; Shao, Q.; Shi, J.; Yang, C.; Chu, H. Application of Biochar for the Adsorption of Organic Pollutants from Wastewater: Modification Strategies, Mechanisms and Challenges. *Sep. Purif. Technol.* **2022**, *300*, 121925. [CrossRef]
38. Hu, Y.; Chen, C.; Yang, L.; Cui, J.; Hao, Q.; Sun, D. Handy Purifier Based on Bacterial Cellulose and Ca-Montmorillonite Composites for Efficient Removal of Dyes and Antibiotics. *Carbohydr. Polym.* **2019**, *222*, 115017. [CrossRef]
39. Yu, L.-L.; Cao, W.; Wu, S.-C.; Yang, C.; Cheng, J.-H. Removal of Tetracycline from Aqueous Solution by MOF/Graphite Oxide Pellets: Preparation, Characteristic, Adsorption Performance and Mechanism. *Ecotoxicol. Environ. Saf.* **2018**, *164*, 289–296. [CrossRef]
40. Huang, B.; Liu, Y.; Li, B.; Liu, S.; Zeng, G.; Zeng, Z.; Wang, X.; Ning, Q.; Zheng, B.; Yang, C. Effect of Cu (II) Ions on the Enhancement of Tetracycline Adsorption by Fe_3O_4@ SiO_2-Chitosan/Graphene Oxide Nanocomposite. *Carbohydr. Polym.* **2017**, *157*, 576–585. [CrossRef]
41. Qiao, H.; Wang, X.; Liao, P.; Zhang, C.; Liu, C. Enhanced Sequestration of Tetracycline by Mn(II) Encapsulated Mesoporous Silica Nanoparticles: Synergistic Sorption and Mechanism. *Chemosphere* **2021**, *284*, 131334. [CrossRef]
42. Liang, S.; Zhang, H.; Dai, H.; Wan, X.; Zhu, F.; Xu, Q.; Ji, W. Efficient, rapid and simple adsorption method by polydopamine polystyrene nanofibers mat for removal of multi-class antibiotic residues in environmental water. *Chemosphere* **2022**, *288*, 132616. [CrossRef] [PubMed]
43. Wang, T.; Pan, X.; Ben, W.; Wang, J.; Hou, P.; Qiang, Z. Adsorptive Removal of Antibiotics from Water Using Magneticion Exchange Resin. *J. Environ. Sci.* **2017**, *52*, 111–117. [CrossRef] [PubMed]
44. Shao, S.; Wu, X. Microbial Degradation of Tetracycline in the Aquatic Environment: A Review. *Crit. Rev. Biotechnol.* **2020**, *40*, 1010–1018. [CrossRef]
45. Scaria, J.; Anupama, K.; Nidheesh, P. Tetracyclines in the Environment: An Overview on the Occurrence, Fate, Toxicity, Detection, Removal Methods, and Sludge Management. *Sci. Total Environ.* **2021**, *771*, 145291. [CrossRef]
46. Liu, C.-X.; Xu, Q.-M.; Yu, S.-C.; Cheng, J.-S.; Yuan, Y.-J. Bio-Removal of Tetracycline Antibiotics under the Consortium with Probiotics Bacillus Clausii T and Bacillus Amyloliquefaciens Producing Biosurfactants. *Sci. Total Environ.* **2020**, *710*, 136329. [CrossRef]
47. De Cazes, M.; Belleville, M.P.; Petit, E.; Llorca, M.; Rodríguez-Mozaz, S.; De Gunzburg, J. Design and Optimization of an Enzymatic Membrane Reactor for Tetracycline Degradation. *Catal. Today* **2014**, *236*, 146–152. [CrossRef]
48. Zhang, X.; Yajuan, W.; Hong, T.; Zhang, R.; Ma, Z. Screened and Degradation Characteristics of a Four Tetracycline Antibiotics Degrading Bacterium. *Environ. Chem.* **2022**, *41*, 2761–2770.
49. Liao, Q.; Rong, H.; Zhao, M.; Luo, H.; Chu, Z.; Wang, R. Interaction between Tetracycline and Microorganisms During Wastewater Treatment: A Review. *Sci. Total Environ.* **2021**, *757*, 143981. [CrossRef] [PubMed]
50. Leng, Y.; Bao, J.; Xiao, H.; Song, D.; Du, J.; Mohapatra, S.; Werner, D.; Wang, J. Transformation Mechanisms of Tetracycline by Horseradish Peroxidase with/without Redox Mediator Abts for Variable Water Chemistry. *Chemosphere* **2020**, *258*, 127306. [CrossRef] [PubMed]
51. Shao, S.; Hu, Y.; Cheng, J.; Chen, Y. Biodegradation Mechanism of Tetracycline (TEC) by Strain Klebsiella sp. SQY5 as Revealed through Products Analysis and Genomics. *Ecotoxicol. Environ. Saf.* **2019**, *185*, 109676. [CrossRef]
52. Qi, W.-N. Microbial Degradation Mechanism of Tetracycline Antibiotics. Master's Thesis, Guizhou Normal University, Guiyang, China, 2019.
53. Gujarathi, N.P.; Haney, B.J.; Park, H.J.; Wickramasinghe, S.R.; Linden, J.C. Hairy Roots of Helianthus Annuus: A Model System to Study Phytoremediation of Tetracycline and Oxytetracycline. *Biotechnol. Prog.* **2010**, *21*, 775–780. [CrossRef] [PubMed]
54. Chen, X.J.; Li, F.Y.; Hao, Y.B. The Preliminary Exploration of Remediation the Antibiotics Polluted Water by Two Hydrophytes. *Subtrop. Plant Sci.* **2012**, *41*, 1–7.
55. Wojnárovits, L.; Wang, J.-L.; Chu, L.-B.; Tóth, T.; Kovács, K.; Bezsenyi, A.; Szabó, L.; Homlok, R.; Takács, E. Matrix effect in the hydroxyl radical induced degradation of β-lactam and tetracycline type antibiotics. *Radiat. Phys. Chem.* **2022**, *193*, 109980. [CrossRef]
56. Zhang, J.; Giorno, L.; Drioli, E. Study of a Hybrid Process Combining PACs and Membrane Operations for Antibiotic Wastewater Treatment. *Desalination* **2005**, *194*, 101–107. [CrossRef]
57. Safari, G.H.; Nasseri, S.; Mahvi, A.H.; Yaghmaeian, K.; Nabizadeh, R.; Alimohammadi, M. Optimization of sonochemical degradation of tetracycline in aqueous solution using sono-activated persulfate process. *J. Environ. Health Sci. Eng.* **2015**, *13*, 76. [CrossRef] [PubMed]

58. Saitoh, T.; Shibata, K.; Fujimori, K.; Ohtani, Y. Rapid Removal of Tetracycline Antibiotics from Water by Coagulation-Flotation of Sodium Dodecyl Sulfate and Poly(Allylamine Hydrochloride) in the Presence of Al(III) Ions. *Sep. Purif. Technol.* **2017**, *187*, 76–83. [CrossRef]
59. Liang, C.; Wei, D.; Zhang, S.; Ren, Q.; Shi, J.; Liu, L. Removal of Antibiotic Resistance Genes from Swine Wastewater by Membrane Filtration Treatment. *Ecotoxicol. Environ. Saf.* **2021**, *210*, 111885. [CrossRef] [PubMed]
60. Xujian, Q.I.; Fanhao, W.E.I.; Jiawei, F.A.N. Research progress on treatment of antibiotics and their resistance genes by advanced oxidation technologies. *Ind. Water Treat.* **2022**, *42*, 55–64.
61. Lopez Penalver, J.J.; Gomez Pacheco, C.V.; Sanchez Polo, M.; Rivera Utrilla, J. Degradation of Tetracyclines in Different Water Matrices by Advanced Oxidation/Reduction Processes Based on Gamma Radiation. *J. Chem. Technol. Biotechnol. Adv.* **2013**, *88*, 1096–1108. [CrossRef]
62. Wang, J.-S.; Yi, X.-H.; Xu, X.; Ji, H.; Alanazi, A.M.; Wang, C.-C.; Zhao, C.; Kaneti, Y.V.; Wang, P.; Liu, W.; et al. Eliminating Tetracycline Antibiotics Matrix Via Photoactivated Sulfate Radical-Based Advanced Oxidation Process over the Immobilized MIL-88A: Batch and Continuous Experiments. *Chem. Eng. J.* **2022**, *431*, 133213. [CrossRef]
63. Zhi, D.; Wang, J.; Zhou, Y.; Luo, Z.; Sun, Y.; Wan, Z.; Luo, L.; Tsang, D.W.; Dionysiou, D. Development of Ozonation and Reactive Electrochemical Membrane Coupled Process: Enhanced Tetracycline Mineralization and Toxicity Reduction. *Chem. Eng. J.* **2020**, *383*, 123149. [CrossRef]
64. Luu, H.T.; Lee, K. Degradation and Changes in Toxicity and Biodegradability of Tetracycline During Ozone/Ultraviolet-Based Advanced Oxidation. *Water Sci. Technol.* **2014**, *70*, 1229–1235. [CrossRef] [PubMed]
65. Li, X.; Shi, H.; Li, K.; Zhang, L. Combined Process of Biofiltration and Ozone Oxidation as an Advanced Treatment Process for Wastewater Reuse. *Front. Environ. Sci. Eng.* **2015**, *9*, 1076–1083. [CrossRef]
66. Wu, X.F.; Yang, K. Experimental study on advanced treatment of antibiotic wastewater by Fenton process. *Straits Sci.* **2017**, *1*, 19–21.
67. Bishop, D.F.; Stern, G.; Fleischman, M.; Marshall, L. Hydrogen Peroxide Catalytic Oxidation of Refractory Organics in Municipal Waste Waters. *Ind. Eng. Chem. Process Des. Dev.* **1968**, *7*, 110–117. [CrossRef]
68. Bai, J.L. Study on the Application of Iron-carbon Micro-electrolysis and Fenton Oxidation in Antibiotic Wastewate Treatment. *Sichuan Nonferrous Met.* **2019**, *2*, 51–54.
69. Shao, C.; Zhang, J.; Wang, Z.; Zhang, L.; Wang, B.; Ren, J.; Zhang, X.; He, W. Photo-Fenton degradation of tetracycline on nitrogen vacancy and potassium-doped Z-scheme FeOCl/NvCN heterojunction with low H_2O_2 consumption: Activity and mechanism. *J. Alloys Compd.* **2023**, *970*, 172532. [CrossRef]
70. Lv, J.; Zhang, B.; Qi, G.H.; Yan, L.; Li, C.; Zhang, Z. Research Progress of Photocatalytic Oxidation Process to Antibiotic Wastewater Treatment with TiO_2. *Shanghai Chem. Ind.* **2014**, *39*, 7–10.
71. Xiao, M.W. A New Technology for Treatment of Antibiotic Wastewater by Photocatalytic Oxidation. Master's Thesis, Guangdong University of Technology, Guangzhou, China, 2005.
72. Reyes, C.; Fernandez, J.; Freer, J.; Mondaca, M.A.; Zaror, C.; Malato, S.; Mansilla, H.D. Degradation and Inactivation of Tetracycline by TiO_2 Photocatalysis. *J. Photochem. Photobiol. A Chem.* **2006**, *184*, 141–146. [CrossRef]
73. Das, S.; Ahn, Y.H. Synthesis and application of CdS nanorods for LED-based photocatalytic degradation of tetracycline antibiotic. *Chemosphere* **2022**, *291*, 132870. [CrossRef]
74. Nagamine, M.; Osial, M.; Jackowska, K.; Krysinski, P.; Widera-Kalinowska, J. Tetracycline Photocatalytic Degradation under CdS Treatment. *J. Mar. Sci. Eng.* **2020**, *8*, 483. [CrossRef]
75. Zhang, J.; Shao, C.; Zhen, L.; Li, Y.; Bai, H.; Zhang, L.; Ren, G.; Wang, X. Treatment of antibiotics in water by SO_3H-modified Ti_3C_2 Mxene photocatalytic collaboration with g-C_3N_4. *J. Mater. Sci. Technol.* **2024**, *194*, 124–137. [CrossRef]
76. Ma, Y.-Y.; Qi, P.; Meng, S.; Zuo, N.; Mominou, N.; Li, S.Z.; Jing, C.Y.; Wang, L. Photocatalytic oxidation degradation of tetracycline over La/Co@TiO_2 nanospheres under visible light. *Environ. Res.* **2022**, *215*, 114297. [CrossRef] [PubMed]
77. Tang, T.; Yin, Z.-L.; Chen, J.-R.; Zhang, S.; Sheng, W.-C.; Wei, W.-X.; Xiao, Y.-G.; Shi, Q.-Y.; Cao, S.-S. Novel p-n heterojunction Bi_2O_3/Ti^{3+}-TiO_2 photocatalyst enables the complete removal of tetracyclines under visible light. *Chem. Eng. J.* **2020**, *417*, 128058. [CrossRef]
78. Li, Y.-C.; Yu, B.; Hu, Z.-Q.; Wang, H. Construction of direct Z-scheme SnS_2@$ZnIn_2S_4$@kaolinite heterostructure photocatalyst for efficient photocatalytic degradation of tetracycline hydrochloride. *Chem. Eng. J.* **2022**, *429*, 132105. [CrossRef]
79. Zong, Z.-Y.; Gilbert, E.; Wong, C.Y.; Usadi, L.; Qin, Y.; Huang, Y.-H.; Raymond, J.; Hankins, N.; Kwan, J. Efficient sonochemical catalytic degradation of tetracycline using TiO_2 fractured nanoshells. *Ultrason. Sonochem.* **2023**, *101*, 106669. [CrossRef] [PubMed]
80. Liu, Z.; Zhu, M.; Wang, Z.; Wang, H.; Deng, C.; Li, K. Effective Degradation of Aqueous Tetracycline Using a Nano-TiO_2/Carbon Electrocatalytic Membrane. *Materials* **2016**, *9*, 364. [CrossRef] [PubMed]
81. Ni, T.; Feng, H.; Tang, J.; Wang, J.; Yu, J.; Yi, Y.; Wu, Y.; Guo, Y.; Tang, L. A Novel Electrocatalytic System with High Reactive Chlorine Species Utilization Capacity to Degrade Tetracycline in Marine Aquaculture Wastewater. *Chemosphere* **2022**, *300*, 134449. [CrossRef] [PubMed]
82. Xie, W.; Shi, Y.; Wang, Y.; Zheng, Y.; Liu, H.; Hu, Q.; Wei, S.-Y.; Gu, H.-B.; Guo, Z.-H. Electrospun Iron/Cobalt Alloy Nanoparticles on Carbon Nanofibers Towards Exhaustive Electrocatalytic Degradation of Tetracycline in Wastewater. *Chem. Eng. J.* **2021**, *405*, 126585. [CrossRef]

83. Wu, S.; Hu, Y.H. A Comprehensive Review on Catalysts for Electrocatalytic and Photoelectrocatalytic Degradation of Antibiotics. *Chem. Eng. J.* **2021**, *409*, 127739. [CrossRef]
84. Oturan, N.; Wu, J.; Zhang, H.; Sharma, V.K.; Oturan, M.A. Electrocatalytic Destruction of the Antibiotic Tetracycline in Aqueous Medium by Electrochemical Advanced Oxidation Processes: Effect of Electrode Materials. *Appl. Catal. B Environ.* **2013**, *140–141*, 92–97. [CrossRef]
85. Li, Z.P.; Xu, G.H.; Cao, H.; Dai, X. On the Anaerobic Treatment Technology of Antibiotic Wastewater by Three Dimensional Electro Catalytic Coupling. *Value Eng.* **2014**, *33*, 305–306.
86. Sun, W.; Sun, Y.; Shah, K.J.; Chiang, P.-C.; Zheng, H. Electrocatalytic Oxidation of Tetracycline by Bi-Sn-Sb/γ-Al_2O_3 Three-Dimensional Particle Electrode. *J. Hazard. Mater.* **2019**, *370*, 24–32. [CrossRef]
87. Lu, M.; Sun, J.; Cui, B.; Zhang, J.; Ren, J.; Li, R. Pd, Ru/2D MXene nanosheets/3D self-supporting nickel foam composite electrode and its electrocatalytic synergistic degradation of antibiotics. *Sep. Purif. Technol.* **2024**, *340*, 126736. [CrossRef]
88. Shao, C.; Zhang, J.; Liu, Y.; Jiang, Y.T.; Jia, Y.P.; Li, G.M.; Sun, Z. Effective degradation of tetracycline by Pd/AG/ITO electrode: Electrode preparation, characterization, kinetics, degradation mechanism and toxicity assessment. *J. Environ. Chem. Eng.* **2023**, *11*, 110344. [CrossRef]
89. Jing, H.; Juan, Z.; Pan, W.; Tsang, Y.F. Occurrence and Fate of Antibiotics in a Wastewater Treatment Plant and Their Biological Effects on Receiving Waters in Guizhou. *Process Saf. Environ. Prot.* **2018**, *113*, 483–490.
90. Chen, J.; Liang, J.; Li, C.; Dai, J.; Mai, W.; Wei, Y. An Enriched Ammonia-Oxidizing Microbiota Enables High Removal Efficiency of Ammonia in Antibiotic Production Wastewater. *Chemosphere* **2023**, *310*, 136854. [CrossRef] [PubMed]
91. Pei, M.; Zhang, B.; He, Y.; Su, J.; Gin, K.; Lev, O.; Shen, G.; Hu, S. State of the Art of Tertiary Treatment Technologies for Controlling Antibiotic Resistance in Wastewater Treatment Plants. *Environ. Int.* **2019**, *131*, 105026. [CrossRef] [PubMed]
92. Xu, M.; Deng, J.; Cai, A.; Ma, X.; Li, J.; Li, Q.; Li, X. Comparison of UVC and UVC/Persulfate Processes for Tetracycline Removal in Water. *Chem. Eng. J.* **2020**, *384*, 123320. [CrossRef]

catalysts

MDPI

Review

An Overview of Environmental Catalysis Mediated by Hydrogen Peroxide

Monica Rigoletto [1], Enzo Laurenti [1,*] and Maria Laura Tummino [2,*]

[1] Department of Chemistry, University of Torino, Via P. Giuria 7, 10125 Torino, Italy; monica.rigoletto@unito.it

[2] Institute of Intelligent Industrial Technologies and Systems for Advanced Manufacturing, National Research Council of Italy (CNR-STIIMA), Corso G. Pella 16, 13900 Biella, Italy

* Correspondence: enzo.laurenti@unito.it (E.L.); marialaura.tummino@cnr.it (M.L.T.)

Abstract: The use of hydrogen peroxide (produced in situ or ex situ) as the main agent in oxidative processes of environmental pollutant removal is widely studied. The degradation of water pollutants, such as dyes, pharmaceuticals, cosmetics, petroleum derivatives, and even pathogens, has been successfully obtained by different techniques. This review gives an overview of the more recent methods developed to apply oxidative processes mediated by H_2O_2 and other reactive oxygen species (ROS) in environmental catalysis, with particular attention to the strategies (Fenton-like and Bio-Fenton, photo- and electro-catalysis) and the materials employed. A wide discussion about the characteristics of the materials specifically studied for hydrogen peroxide activation, as well as about their chemical composition and morphology, was carried out. Moreover, recent interesting methods for the generation and use of hydrogen peroxide by enzymes were also presented and their efficiency and applicability compared with the Fenton and electro-Fenton methods discussed above. The use of Bio-Fenton and bi-enzymatic methods for the in situ generation of ROS seems to be attractive and scalable, although not yet applied in full-scale plants. A critical discussion about the feasibility, criticalities, and perspectives of all the methods considered completes this review.

Keywords: Bio-Fenton; electro-catalysis; Fenton; H_2O_2 detection; hydrogen peroxide; photo-catalysis; water remediation

1. Introduction

The depletion of water resources is a theme of global concern and it is more and more exacerbated by increasing water consumption (especially connected to certain agricultural and industrial activities [1,2]), contamination sources and climate changes that bring about aridity/desertification issues [3,4]. As a consequence, access to clean water is progressively more limited and the need for the reuse of wastewater is mandatory worldwide [3,4]. Wastewater has to be treated to fulfill the quality requirements before the reintroduction in specific systems (e.g., agricultural, industrial, potable, etc.) [5].

Wastewater depuration is typically a multi-step procedure in which most of the chemical and pathogenic contaminants are successfully removed [4]. Nevertheless, recalcitrant and new kinds of not commonly monitored pollutants (Contaminants of Emerging Concern, such as pharmaceuticals and personal care products), detected at trace/sub-trace levels in water bodies and often not removed by traditional depuration processes, represent a further hurdle to the achievement of high-quality water standards [4,6,7]. For this reason, advanced methods should support these processes, among which the most studied ones since the 1980s are the so-called Advanced Oxidation Processes (AOPs) [8]. The methods involving the presence of Reactive Oxygen Species (ROS) with strong oxidizing properties underlie AOPs' definition and are aimed at bringing about the partial/total mineralization of chemical pollutants to carbon dioxide, water and inorganic ions up to disinfection from pathogens [7,9,10]. The main ROS are non-radical species such as H_2O_2 and singlet oxygen ($^1O_2$), and highly reactive free radicals, such as hydroxyl ($^{\bullet}OH$), hydroperoxide ($HOO^{\bullet}$), superoxide ($O_2^{\bullet-}$), carbonate anion ($CO_3^{\bullet-}$), and sulfate ($SO_4^{\bullet-}$) radical, deriving from the activation of precursors (hydrogen peroxide, persulfate/peroxydisulfate, peroxymonosulfate and sodium percarbonate) via oxidation/reduction reactions [7,11–14].

The oxidation processes exploiting H_2O_2 and the derived ROS are the focus of the present review. Many different treatments have been proposed [11,15–18], also in a variety of combinations and with different triggering sources (ultrasounds, UV/vis light, heat and electrochemical energy) [7,15,19–24], as summarized in Figure 1.

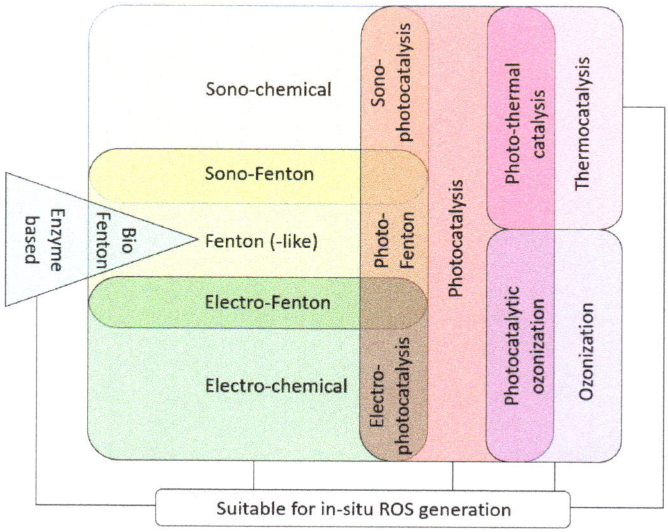

Figure 1. Scenario of AOPs exploiting hydrogen peroxide and derived ROS.

In Figure 1, the methods that are mainly recognized to be capable of inducing an in situ generation of ROS [17,25–27] have been underlined since the use of such methods can overcome the problems related to high costs and excessive consumption of hydrogen peroxide, as well as avoid the hazards associated with its transport, handling, and storage in significant quantities [25,27,28]. A further distinction can be made considering homogeneous and heterogeneous processes. In this latter category, solid catalytic materials are applied for the generation of active species in the aqueous medium through catalytic ozonation, photocatalytic and electrochemical systems, heterogeneous Fenton-like processes, etc. [29–31]. Heterogeneous catalyst-based methods are considered greener to permit both easier recycling of active materials and a decrease in costs and pollution [4]. For this reason, these will be more explore more in depth in this excursus.

From what has been seen so far, the two most common methods to produce hydrogen peroxide in heterogeneous systems are photocatalysis and electrocatalysis (Figure 2). In order to briefly introduce the basilar mechanism of a photocatalytic process (Figure 2A), it is possible to use the well-known semiconductor TiO_2 as an example. The mechanism of the UV/TiO_2 implies that when titania is irradiated with light energy equal to or higher than its band gap, an electron (e^-) can be excited from the valence band to the conduction band, leaving a hole (h^+) in the valence band. If charge separation is maintained, the paired e^--h^+ may migrate to the surface of the photocatalyst. In the aqueous phase, the photoinduced h^+ can oxidize surface hydroxyl groups or surface-bond water molecules to produce hydroxyl radicals and other ROS [32,33]. Regarding electrocatalysts, they have to initiate an Oxygen Reduction Reaction (ORR) to generate hydrogen peroxide via two e^- or four e^- processes (Figure 2B). Therefore, the involved materials have to possess suitable electronic configuration and redox properties, conductivity, low overpotential, fast kinetics and good Faraday efficiency to allow for oxygen reduction [34–37].

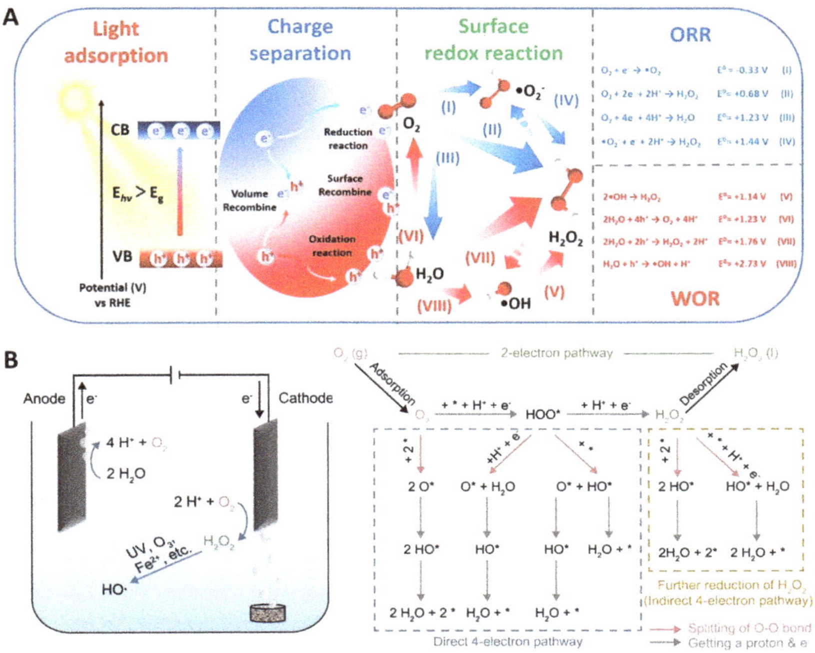

Figure 2. (**A**) Photocatalytic H_2O_2 production with a semiconducting-based process (ORR = Oxygen Reduction Reaction; WOR = Water Oxidation Reaction). (**B**) Electrochemical H_2O_2 generation and related two- and four-electron oxygen reduction reaction pathways. Images reproduced with permission of the authors of [38,39].

The methods based on the use/production of hydrogen peroxide and related ROS are mainly founded on the following reactions (1–19) [8,17,21,22,40]. Equation (1) is related to ozone-related systems, Equations (2)–(6) to the peroxone (O_3/H_2O_2) system, Equations (5)–(11) to photolysis and photocatalytic (hν)/thermocatalytic (ΔT) reactions in the presence of a heterogeneous catalyst ("Cat." refers to a generic semiconductor), and Equations (6) and (12)–(18) to Fenton and Fenton-like processes. Equation (18), in particular, is an example of radical recombination that can restore H_2O_2 [15]. Equation (19) is related to ultrasound (US)-assisted reactions, in which sound waves can lead to cavitation phenomena, involving vapor- and gas-filled microbubbles, and to the generation of high temperature and high pressure. Finally, the formation of organic radicals ($R^\bullet$) during these processes introduces new variables into this chain of reactions [41].

$$3O_3 + H_2O \rightarrow 2\,{}^\bullet OH + 4O_2 \quad (1)$$

$$H_2O_2 \rightarrow HO_2^- + H^+ \quad (2)$$

$$HO_2^- + O_3 \rightarrow {}^\bullet OH + O_2^- + O_2 \quad (3)$$

$$O_3 + H_2O + h\nu \rightarrow H_2O_2 + O_2 \quad (4)$$

$$H_2O_2 + h\nu \rightarrow 2\,{}^\bullet OH \quad (5)$$

$$H_2O_2 + {}^\bullet OH \rightarrow H_2O + HOO^\bullet \quad (6)$$

$$\text{Cat.} + h\nu/\Delta T \rightarrow e^-{}_{\text{(conduction band)}} + \text{hole}^+{}_{\text{(valence band)}} \quad (7)$$

$$\text{hole}^+{}_{\text{(valence band)}} + OH^-{}_{\text{(surface)}} \rightarrow {}^\bullet OH \quad (8)$$

$$\text{hole}^+{}_{\text{(valence band)}} + H_2O_{\text{(adsorbed)}} \rightarrow {}^\bullet OH + H^+ \quad (9)$$

$$e^-{}_{\text{(conduction band)}} + O_{2\,\text{(adsorbed)}} \rightarrow O_2{}^{\bullet-} \quad (10)$$

$$H^+ + O_2{}^{\bullet-} \rightarrow HOO^\bullet \quad (11)$$

$$H_2O + h\nu \rightarrow {}^\bullet OH + H^\bullet \text{ when } \lambda(h\nu) < 242 \text{ nm} \quad (12)$$

$$Fe^{2+} + H_2O_2 \rightarrow Fe^{3+} + {}^\bullet OH + OH^- \quad (13)$$

$$Fe^{3+} + H_2O_2 \rightarrow Fe^{2+} + HOO^\bullet + H^+ \quad (14)$$

$${}^\bullet OH + Fe^{2+} \rightarrow Fe^{3+} + OH^- \quad (15)$$

$$Fe^{3+} + HOO^\bullet \rightarrow Fe^{2+} + O_2H^+ \quad (16)$$

$$Fe^{2+} + HOO^\bullet + H^+ \rightarrow Fe^{3+} + H_2O_2 \quad (17)$$

$$2HOO^\bullet \rightarrow H_2O_2 + O_2 \quad (18)$$

$$H_2O + US \rightarrow {}^\bullet OH + H^\bullet \quad (19)$$

It is worth briefly discussing the Fenton processes and the distinction between Fenton and Fenton-like ones. In the classical homogeneous Fenton process, low pH values (around 3.0) are necessary to avoid the precipitation of iron oxyhydroxides from Fe(II), resulting in higher costs [42–44]. Furthermore, when complex real matrices have to be depolluted, there is the risk of precipitation of the iron catalyst, leading to the formation of sludge and undesired by-products [7]. To overcome these limitations, Fenton-like processes are designed to use oxidants other than hydrogen peroxide and/or transition metals other than Fe^{2+} (e.g., Fe^{3+}, Cu^{2+}/Cu^+, iron-based minerals, nano zero-valent iron, etc.), and/or employ heterogeneous catalysts and/or external energy sources to create similar reactions [29,44–47]. All the aspects concerning the chemical development of AOP methods involving H_2O_2 will be addressed in Section 2.1.

As anticipated, another important point that must be considered concerns the production of H_2O_2 to be used in these reactions. In recent years, biological systems for the in situ generation of H_2O_2 have been carefully taken into account since they allow for the production of hydrogen peroxide under mild conditions, avoiding the use of critical solvents and reducing or avoiding the formation of toxic by-products.

One of the most exploited enzymes for this purpose is Glucose Oxidase (GOx), which catalyzes the oxidation of β-*d*-glucose to *d*-glucono-δ-lactone (further hydrolyzed into gluconic acid) and hydrogen peroxide in the presence of molecular dioxygen, as summarized in Equation (20).

$$\beta\text{-}d\text{-glucose} + O_2 + H_2O \rightarrow \text{gluconic acid} + H_2O_2 \quad (20)$$

This catalytic mechanism is also widely exploited in Bio-Fenton processes that actually use H_2O_2 enzymatically produced as a reagent for the Fenton reaction [48]. Unlike the classic Fenton reaction, the Bio-Fenton process also occurs at pH values near neutrality owing to the synergy with all the GOx reaction products [49]. A wide discussion about the environmental application of the Bio-Fenton technique will be made in Section 2.2 of this review, together with a description of innovative bi-enzymatic systems coupling GOx and other enzymes.

Finally, a section about H_2O_2 detection techniques and another on the critical assessment of the whole H_2O_2-mediated systems will be presented to enrich the overview of the practical implications of these depuration treatments. The global structure of this review will cover the topics indicated in Figure 3.

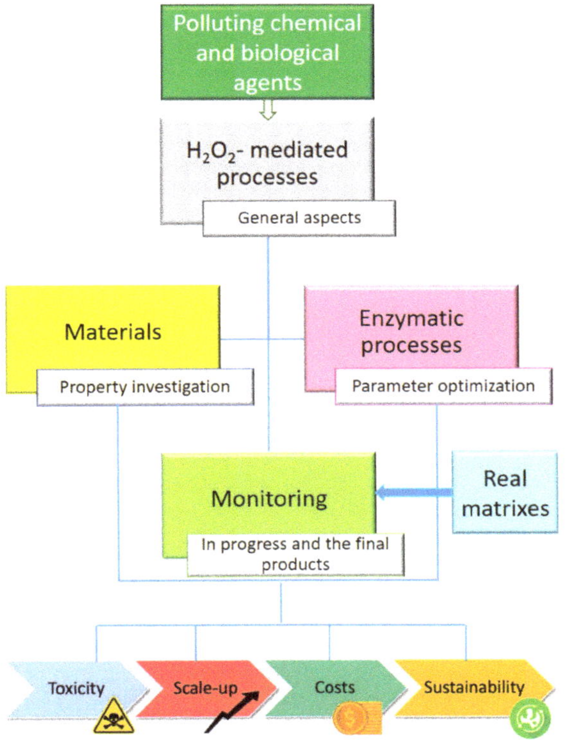

Figure 3. Workflow at the basis of this review, indicating the core topics.

2. Recent Developments in AOP Systems Involving H_2O_2

This section will examine different material- and enzyme-based systems involving hydrogen peroxide and its derived ROS for decontaminating wastewater from several types of contaminating agents, ranging from synthetic dyes (often found downstream textile industries) to pharmaceuticals and other harmful chemicals derived from different production activities to biological pathogens.

2.1. Advanced Materials

By searching on the Scopus database for studies with the query "H_2O_2 AND generation AND wastewater" and further limiting the search to the keywords "Hydrogen peroxide", 407 documents were found. Of these research studies, 96 possess "Iron", 77 "Iron compounds", 32 "Ferric Ion", 17 "Ferric compounds", 38 "Ferrous Ion", 109 "Fenton", 23 "Photo-Fenton", and 42 "Electro-Fenton" as further keywords. This circumstance makes clear the strong contribution of iron-based chemistry and Fenton-type reactions when environmental decontamination mediated by hydrogen peroxide is attempted. Table 1 shows a selection of recent works reporting iron as the main component or coadjuvant in remediation routes. Indeed, the opportunities of obtaining Fe-containing materials from natural sources (entries 10, 23 and 24 in Table 1) or preparing different kinds of iron-based compounds are uncountable [41,47,50–53]. Moreover, even in Fenton and photo-Fenton processes defined as "heterogeneous", iron can be added as a homogeneous reactant (entries 13, 15 and 21 in Table 1).

Table 1. Iron-based systems for pollutant removal mediated by H_2O_2 (except for two cases, all the works reported were published within the last 5 years) *.

System	Pollutants	Conditions	Performance	Mechanism and Notes	Ref.
1. Fe(III)-functionalized polyacrylonitrile/polypropylene fiber mesh	Bisphenol A (BPA)	Best setup: 75 ppm of BPA, 300 ppm of H_2O_2, pH = 3, 60 °C.	100% BPA abatement in <30 min; effects of temperature and pH. Fast degradation in Rotating Catalytic Reactor.	Fenton process; reusable catalyst but poisoning by the intermediates.	[54]
2. Zn-Carbon Nanotubes (CNTs) in the presence of Fe^{2+}	4-chlorophenol (4-CP)	Conc. 4-CP = 50 mg L^{-1}; Conc. Fe^{2+} 20 mg L^{-1}, pH = 2.0. O_2 was fed in the reaction mixture (400 mL min^{-1}).	Abatement of 4-CP and TOC: 98.8% and 87.4%, respectively (20 min). When 4-CP was spiked in real wastewater, the abatements of 4-CP and TOC were 47.0% and 45.6%. Effects of pH, Zn-CNTs dosage and Fe^{2+} amount.	Fenton: in situ generation of a high concentration of H_2O_2, rapid regeneration of Fe^{2+} from the reduction of Fe^{3+} by Zn and high adsorption ability of Zn-CNT towards pollutants.	[55]
3. Graphitic-C_3N_4 QDs with FeOOH	Tetracycline (TC), p-nitrophenol (PNP), 2,4-dinitrophenol (2,4-DNP)	500 W Xe lamp with a 420 nm cut-off; addition of H_2O_2; optimal pH = 7.	Abatement: TC ca. 90% (2 min), PNP 90% (10 min), 2,4-DNP ca. 90% (5 min). Dependence on catalyst/pollutant ratio, pH, conc. H_2O_2.	Photo-Fenton: main action of $^{\bullet}OH$ with the aid of $O_2^{\bullet-}$ and h^+; photogenerated e^- in CB favored Fe^{3+}/Fe^{2+} cycling.	[56]
4. Ultrathin porous Graphitic-C_3N_4 nanosheets with amorphous FeOOH QDs	Oxytetracycline (OTC)	300 W Xe lamp with a 420 nm cut-off; pH = 7.	Highest efficiency for 20%FeOOH-composite degrading 86.23% of OTC (120 min) and TOC removal of 48.6%.	Photo-Fenton; Graphitic-C_3N_4 in situ produced H_2O_2, improving transport of photogenerated e^--h^+ pairs; FeOOH generated $^{\bullet}OH$.	[57]
5. α-Fe_2O_3/Graphitic-C_3N_4	Rhodamine B (RhB), tetracycline hydrochloride (TC-H)	Simulated solar light with 300 W Xe lamp; neutral pH.	Degradation of RhB 96% (90 min) and TC-H 95% (150 min).	Photo-Fenton: effective separation and transfer of photogenerated charge carriers; H_2O_2 photoproduction on g-C_3N_4; $^{\bullet}OH$ generation from H_2O_2 decomposition on α-Fe_2O_3; $O_2^{\bullet-}$ and h+ played supporting role.	[58]
6. Magnetite/maghemite NPs coated with waste-sourced bio-based substances (BBS)	Phenol (PH)	For Fenton: addition of H_2O_2 5×10^{-4} M; photoactivation by a lamp with max emission at 365 nm; pH = 3.5.	100% PH degradation (5 min). Reusability allowed by the constant Fe release from NPs.	Fenton and photo-Fenton processes, but higher efficiency in Fenton mode. In photo-Fenton $^{\bullet}OH$ are generated. BBS acted preventing catalyst oxidation and Fe precipitation.	[59]

Table 1. *Cont.*

System	Pollutants	Conditions	Performance	Mechanism and Notes	Ref.
7. Glucose-mediated Fe_3O_4 magnetic NPs	Methylene blue (MB), Cr(VI)	For Fenton and photo-Fenton: addition of H_2O_2; photocatalytic reduction of Cr(VI); photoactivation with vis. light (250 W).	MB degradation at pH 9: 93% in 75 min by photo-Fenton and 92% in 120 min by Fenton process. 100% photoreduction of Cr(VI) in 25 min. Influence of H_2O_2 concentration, pH, catalyst/pollutant dosage.	Higher efficiency of both Fenton and photo-Fenton ascribed to the novel synthesis method. $^{\bullet}OH$ was detected as primary ROS. The catalyst was stable and reusable.	[60]
8. CuO-Fe_2O_3 heterojunction	Quinoline Yellow (QY)	Best setup: H_2O_2 and QY concentrations 27.6 mM and 100 mg L^{-1}, respectively; pH = 3, 40 °C.	100% QY removal in ca. 60 min. Dependence on pH, conc. H_2O_2 and dye, catalyst dose, temperature.	Photo-Fenton and Fenton (no strong effect of irradiation); recyclability.	[61]
9. Fe_3O_4/Cu magnetic NPs prepared using *Rosmarinus officinalis* leaves aqueous extract	Methyl Orange (MO), imipenem (IMI), imatinib mesylate (IMA)	300 W Xe lamp with a 420 nm cut-off.	MO, IMI, IMA degradation of 96.6%, 81.8% and 84%, respectively, after 5400 s; TOC decrement.	Photocatalytic process with production of H_2O_2 and $O_2^{\bullet-}$ as main ROS; reusability of the catalyst; beneficial effect of *R. officinalis* leaves extract.	[62]
10. Mexican Natural Zeolite-based Membrane	Reactive Black 5 (RB5)	Best setup: pH= 3, conc. H_2O_2 = 3 g L^{-1}, conc. RB5 = 100 mg L^{-1}, added $FeCl_3$ = 0.013 g L^{-1}, LED lamp emission at 405 nm (2.2 W), permeation flux ≅ 467 cm^3 m^{-2} h^{-1}.	92.3% discoloration in 30 min; progressive TOC decrement.	Photo-Fenton; reusable membranes, but possible Fe leaching.	[63]
11. Triphase MIL-101(Fe)/Graphitic-C_3N_4/hydrophobic carbon cloth	Methyl Orange (MO), methylene blue (MB), rhodamine B (RhB), rhodamine 6G (Rh6G)	300 W Xe lamp; dye solution = 10 ppm, pH = 3. Catalyst (size: 2.5 cm^2) hydrophilic surface immersed in solution, while the hydrophobic part was exposed to air for O_2 supply.	Abatement: MO 99%, MB 99%, RhB 98%, Rh6G 97% (130 min); high reaction rate constant.	Photo-Fenton with photoactivated in situ production of H_2O_2, promoted by triphase design of the Z-scheme heterojunction with a favored pathway for O_2 transfer. Photoinduced e^- and h^+ separation efficiency; $^{\bullet}OH$ and $O_2^{\bullet-}$ were the main ROS.	[64]

Table 1. *Cont.*

System	Pollutants	Conditions	Performance	Mechanism and Notes	Ref.
12. MoS_2-Fe_x composite	Sulfadiazine (SD)	Best catalyst: MoS_2-Fe_{75}; Optimal conditions: H_2O_2 addition, SD = 10 mg L^{-1}, pH = 6.5.	SD degradation 91.1% (90 min); effect of catalyst dosage, pH, H_2O_2 feeding way.	Fenton: MoS_2-Fe_x selectivity for $^1O_2$. "Small amount for multiple times" feeding way of H_2O_2 increased MoS_2-Fe_x stability and SD degradation rate, reducing H_2O_2 decomposition. Formation of Fe sludge was much reduced than nano-iron powder. Long-term effectiveness of the MoS_2-Fe_x/H_2O_2 system.	[65]
13. MoS_2 in the presence of Fe^{3+}	Bisphenol A (BPA), benzoic acid (BA), sulfadiazine (SDZ), rhodamine B (RhB), carbamazepine (CBZ), 4-acetamidophenol (APAP), ciprofloxacin (CIP), tetracycline hydrochloride (TC-H)	Optimal system: 1.0 mM H_2O_2, 0.3 g L^{-1} MoS_2 and 0.15 mM Fe^{3+}, pH = 3.0.	Abatement (60 min): CBZ 65.9%, APAP 79.1%, SDZ 84.1%, BA 86.0%, RhB 90.8%, CIP 92.5%, BPA 93.0%, TC 100%. Effects of pH and concentrations of pollutant, H_2O_2, Fe^{3+}, MoS_2.	Fenton: strong oxidative intermediate Mo^{6+} peroxo-complexes besides $^{\bullet}OH$ radicals. Stable and reusable catalyst.	[66]
14. TiO_2-supported Fe (FeTi-ox)	Acetaminophen (AAP), benzoic acid (BA), carbamazepine (CBZ), phenol (PH)	Conc. H_2O_2 = 10 mM and conc. pollutant = 5−10 μM, pH = 7.	Degradations (2 h): AAP 100%, PH 55%, while BA and CBZ were not removed. Pollutants selectively reacted with $^{\bullet}OH$ (BA and CBZ) or with both $^{\bullet}OH$ and Fe(IV) (AAP and PH).	Fenton: relevant interaction H_2O_2-TiO_2 forms a peroxo−titania complex Fe(III)-Ti-OOH, which reacted further with H_2O to give surface oxidant Fe[IV]-O^{2+} even in the presence of Cl^-, HCO_3^- ions and organic matter. Reusability of FeTi-ox.	[67]
15. P25 (TiO_2) in the presence of US and Fe^{2+}/Fe^{3+}	Bisphenol A (BPA), sulfadiazine (SDZ)	US at 400 kHz alone or in the presence of P25 under vis. light (LED lamp emitting at 400–630 nm).	With US and Fe^{2+}: >90% degradation of BPA and >80% for SDZ in 30 min. In the presence of Fe^{2+}/Fe^{3+}, US, P25 and vis. light, 100% SDZ abatement in 60 min.	Homogeneous sono-Fenton: in situ generation of H_2O_2 and $^{\bullet}OH$. Sono-photo-Fenton: P25 promoted Fe^{2+}/Fe^{3+} cycling by the photoproduced e^-. P25 favored reaction at circumneutral pH and pollutant mineralization.	[68]

Table 1. *Cont.*

System	Pollutants	Conditions	Performance	Mechanism and Notes	Ref.
16. Iron-cobalt oxide nanosheets (CoFe-ONSs)	Tetracycline (TC)	Conc. TC 50 mg L^{-1}, conc. H_2O_2 20mM, neutral pH.	TC removal 83.5% (50 min). Effects of catalyst dosage, conc. H_2O_2, pH, temperature, conc. TC, anions and water sources.	Fenton: $^{\bullet}OH$ were the main ROS. Redox cycles of FeII/FeIII and CoII/CoIII enhanced $^{\bullet}OH$ generation. Negligible Fe ions leaching from catalyst (reusability).	[69]
17. Magnetic ZnO@Fe_3O_4 composite	p-nitrophenol (p-NP)	Conc. p-NP 35 mg L^{-1}, pH = 3 (optimal); 100 W incandescent lamp (400–1700 nm).	100% p-NP removal in 60 min. Effect of temperature.	Photo-Fenton: self-generation of H_2O_2, with primary role of $^{\bullet}OH$; catalyst reusability.	[70]
18. $ZnFe_2O_4/BiVO_4$ heterojunction	Methylene blue (MB).	Best composition: $0.15ZnFe_2O_4/BiVO_4$.MB conc. 40 mg L^{-1}, eventual addition of H_2O_2; photoactivation with 300 W Xe lamp.	MB degradation with photocatalytic method (2 h): 83.7%; with Photo-Fenton (1 h): 98.8%.	Photocatalysis: h^+ had a main role in degrading MB; photo-Fenton: activation of H_2O_2 ($^{\bullet}OH$ production) by photogenerated e^-; photogenerated carriers separation efficiency.	[71]
19. $LaFeO_3$ prepared from citric acid and $LaFeO_3$ synthesized from waste-sourced bio-based substances	4-methylphenol (4-MP) and crystal violet (CV)	Conc. pollutant 10 mg L^{-1}; pH = 8–10; 1500 W Xe lamp with a 340 nm cut-off.	$LaFeO_3$ prepared from waste-sourced bio-based substances removed 100% CV and ca. 40% 4-MP. The citric acid-derived $LaFeO_3$ photodegraded 90% of 4-MP and 30% of CV.	Photocatalysis: different $LaFeO_3$ efficiencies were ascribed to different ζ-potentials. A homogeneous photo-Fenton process can occur when $LaFeO_3$ synthesized from waste-derived substances releases Fe and carboxylate ions.	[72, 73]
20. Cerium, Cobalt, Copper-doped Strontium Ferrate (SCFCC)	*Escherichia Coli*	Best material with 20% Cu doping. Bacteria conc. 1.0–3.0 $\times 10^5$ CFU/mL. SCFCC/inoculum ratio = 1 g/50 mL. Tests in the dark, and after thermo- or UV-activation (max 70 °C heating and UV-A 300 W lamp, respectively).	Max. bacterial removal 55% in the dark, 98% after UV, 40% after thermal activation.	Photo/Thermo-catalysis: formation of H_2O_2 and $^{\bullet}OH$. Activity was also influenced by SCFCC metal ions' redox couples and oxygen vacancies.	[74]

Table 1. *Cont.*

System	Pollutants	Conditions	Performance	Mechanism and Notes	Ref.
21. Peanut shell-derived biochar (PBC) in the presence of Fe^{2+}	Bisphenol A (BPA), dimethyl phthalate (DMP), sulfamethoxazole (SMX)	Degradation tests conducted in a three-electrode cell aerated with O_2 (0.4 L min^{-1}); voltage applied 0.5–1.1 V; pH = 3.0, conc. Fe^{2+} 0.2 mM and conc. pollutants 20 mg L^{-1}.	Removal efficiencies of 98–100% within 15 min; mineralization efficiencies of 83–100%.	Electro-Fenton: PBC hierarchical porous structure and defects caused a high surface area, electrical and ionic conductivity. The presence of OOH/C–O–C and N on the surface ensured a high two-electron ORR selectivity for H_2O_2 production, accelerated Fe^{2+} regeneration, also enabling $^{\bullet}OH$ accumulation.	[75]
22. Ferromagnetic activated carbon from rubber seed hull	Bezaktiv Brilliant Blue (BBB)	Best setup: pH = 3, conc. H_2O_2= 17 mol L^{-1}, conc. BBB = 100 mg L^{-1}.	BBB removal > 75% in all conditions (4 h). Effects of pH, conc. H_2O_2 and pollutant, catalyst dosage.	Fenton; reusable catalyst.	[76]
23. Natural pyrite (FeS_2)	Carbamazepine (CBZ)	Addition of tartaric acid (TA), citric acid (CA), ascorbic acid (AA); simulated sunlight (300 W Xe lamp).	No CBZ degradation with pyrite; CBZ abatement: 70%, 60%, 53% in pyrite/TA, pyrite/CA, pyrite/AA systems, respectively under irradiation (30 min). Effect of catalyst dosage, pH, conc. CBZ.	Photo-Fenton; in situ generated H_2O_2 without extra pH adjustment; organic acids can form complex with Fe in pyrite, promoting Fe(II) dissolution. Upon irradiation, pyrite is excited to generate photo-e^-, able to reduce oxygen to produce H_2O_2 and $^{\bullet}OH$.	[77]
24. Goethite (α-FeOOH)	Bisphenol A (BPA)	H_2O_2 addition (1.0 mM), conc. BPA 0.1 mM; reaction allowed under acidic, neutral and weakly alkaline conditions.	BPA degradation 75.9% after 240 min; pH dependency.	Fenton; $^{\bullet}OH$ production. Good structural stability of catalyst; higher efficiency with H_2O_2 in comparison to persulfate oxidation system due to the limited radical scavenging.	[78]

Table 1. *Cont.*

System	Pollutants	Conditions	Performance	Mechanism and Notes	Ref.
25. Heterogeneous Nanoscale Zero Valent Ion (n-ZVI)	Glyphosate (GLY)	Eventual addition of H_2O_2. Optimal pH 3–4.	Up to 87% of GLY removal (30 min) in Fenton mode. Effects of pH, conc. H_2O_2, n-ZVI dosage. Efficiency also in tap water (100% degradation in 40 min), despite potentially interfering ions.	Adsorption and Fenton (after H_2O_2 addition).	[79]

* Legend: QDs = Quantum Dots; CB = conduction band; e^- = electrons; h^+ = holes; λ cut-off < 420 nm indicates visible light; TOC = Total Organic Carbon; NPs = nanoparticles; ORR = Oxygen Reduction Reaction. For the sake of clarity, when mechanisms are indicated as "Fenton" for brevity, they have to be considered Fenton-like processes.

In the next subsections, the discussion, starting from the data of Table 1, will bring up several subjects, citing material types from 0D to 3D, from inorganic catalysts to organic substances, and from mono-element to binary and mixed systems, referring to the possibility of tuning the compositions with dopants or with the formation of composites/hybrids. The surface physical properties (area, porosity, surface charge), the chemical defects, such as oxygen vacancies, and a hint regarding the role of synthesis have also been taken into account. In Figure 4, the main points of this study are depicted.

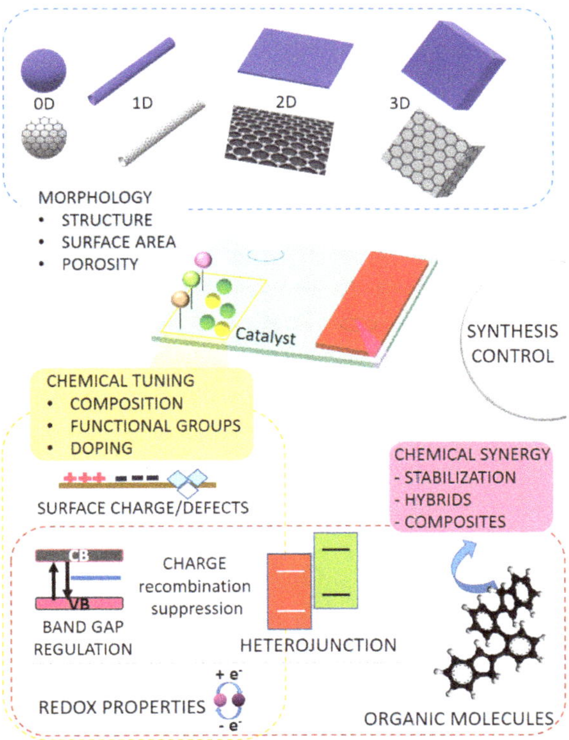

Figure 4. Focus on inorganic catalytic material properties that can be subjected to fine-tuning devoted to enhancing pollutant degradation efficiencies.

2.1.1. Considerations on Structures and Morphologies

In heterogeneous systems, the role of materials is determined not only by their chemical activity in a strict sense, but also by other physical–chemical characteristics, such as the materials' structures and morphologies. As the catalyst surface in the H_2O_2-mediated processes is involved in the adsorption and dissociation of oxygen/water molecules and oxygenated intermediates, a favorable surface charge, and the increase in the specific surface area and, thus, the number of active sites has been pursued as strategies to accelerate the reactions [80,81]. In this regard, Farhadian et al. [82] highlighted the importance of the support within composites/functionalized materials in a review focused on mesoporous silica with high surface area and high pore volume, capable of loading and dispersing catalytically active iron species for heterogeneous Fenton oxidation reactions. Indeed, the large cavities of the silica support can facilitate the mass transfer processes, and the high surface area of these materials increases the number of active sites, allowing for a more intimate contact with contaminants. Strategies for enhancing the catalytic performance of iron/silica composites were indicate, such as the simultaneous loading of different metals and the design of multimodal pore supports, surface functionalization to favor the distribution of active sites, and the use of radiation, ultrasound, and electrolysis.

In the same direction, entries 1–5 and 11 in Table 1 report the presence of 0D (quantum dots), 1D (carbon nanotubes) and 2D materials (polyacrylonitrile fiber mesh, graphitic carbon nitride, carbon cloth), also in combination. Salunkhe et al. [83] dedicated a review to 0D–2D synergistic nanocomposites for photocatalytic applications: the size-dependent bandgap tunability and effective charge carrier production of quantum dots can be exploited together with graphitic-C_3N_4 features, such as low toxicity, chemical and thermal stability, visible light absorption, n-type semiconducting property, and a configurable bandgap. Graphitic-C_3N_4 has been widely used for the photocatalytic degradation of pollutants alone or in co-presence with other compounds, such as MgO, Bi_2O_3, MoO_3/Ag, Au/Ni_2P, etc. [84–91]. Among the attempts to exploit graphitic-C_3N_4, Torres-Pinto et al. prepared these materials as metal-free photocatalysts able to generate hydrogen peroxide and, in turn, $^{\bullet}OH$ in the presence of dissolved oxygen when irradiated by visible light [92]. Adding low dosages of dissolved iron promoted a Fenton process and the C_3N_4 photocatalytic system improved the mineralization of resorcinol, phenol, gallic acid, and benzoic acid at a natural pH by approximately 1.2 times.

Graphene and reduced graphene oxide (rGO), 2D versatile carbon materials, have demonstrated excellent mechanical properties, high charge-carrier mobility, high specific surface area, a large number of functional groups and good electrocatalytic activity, attracting increasing attention for environmental applications as adsorbents, photocatalysts and electrocatalysts [93]. Many composites have been designed, as well, with doping elements, nano zero-valent iron, oxides, mixed oxides, and sulfides in binary and ternary systems [94–100]. An interesting application of these materials is also the possibility of transforming them from 2D to 3D systems, such as aerogels, sponges and foams, which possess enhanced porosity and lightness as inner characteristics [101,102]. For instance, a novel cathode of macroporous graphene aerogel (GA) with a high specific surface area was proposed for the electro-Fenton reaction, efficiently and continuously producing in situ H_2O_2 to degrade antibiotic ciprofloxacin [97]. The good mineralization efficiency was attributed to the relevant number of macropores of GA that acted as reaction traps to accelerate the electro-generation of H_2O_2, subsequently decomposed by Fe^{2+} to form $^{\bullet}OH$. Simultaneously, the strong charge transfer ability of GA was beneficial to the conversion of Fe^{3+}/Fe^{2+}, making this material a promising candidate material for electro-Fenton cathode due to its low cost, high efficiency and corrosion resistance.

For similar reasons connected to morphology, biochars and active carbons have been widely employed in H_2O_2-mediated processes (entries 21 and 22 in Table 1). Indeed, active carbons are produced from carbon-based compounds—often non-renewable (i.e., coal, petroleum residues, peat, lignite and polymers)—by the combination of a pyrolytic process and chemical or physical treatments through activators such as chemicals (acids, alkalis, salts) or gases (i.e., steam, CO_2, N_2) [103,104]. Biochars, instead, were initially defined as carbonized biomass (formed under a low O_2 environment from agricultural residue, animal waste, or refuse of woody plants) and they can also be subjected to further physical–chemical transformations to improve surface features [103,104]. For both active carbons and biochars, the raw material, together with the production technique and operating temperature, have an important influence on product yield and composition [101]. In their review, for instance, Ribeiro et al. [105] discussed the various activities of carbon materials as catalysts on their own and hybrid magnetic carbon nanocomposites in catalytic wet peroxide oxidation reactions for the degradation of organic pollutants. Apart from purely iron-based compounds that are currently being developed [106,107], several other doping/functionalizing moieties have been used on biochar and active carbons, such as Ag_3PO_4 [108], MnO_2, nitrogen [109], boron [110], B/N/graphene [111], N/S/Fe [112], FeAl-layered double hydroxide [113], copper oxides [114–116], CuNi [117], and Sr/Ce [118].

Other efforts to improve the morphological and textural features have been made by acting directly on inorganic active phases, as in the examples of entries 10 and 11 in Table 1, where crystalline porous solids [119] (zeolites and a metal–organic framework, MOF) have been used, taking advantage of to their large surface area, tunable porosity and chemistry [120–122]. In particular, in entry 11, MIL (Materials of Institute Lavoisier) indicates one of the most popular MOFs and MIL-101 (Fe) is the one assembled by 1,4-benzene dicarboxylate ligand and iron(III) cation [123].

2.1.2. Considerations on Chemical Compositions and Stability

Among the inorganic species used as active phases in hydrogen-peroxide-mediated processes, the simplest compositions are zero-valent metal nanoparticles (e.g., Fe^0, Al^0, Zn^0, Cu^0) [124–130], which bear strong chemical reducibility, high efficiency, and large specific surface although the main drawbacks are the tendency towards agglomeration and oxidization [131]. Nanoscale zero-valent iron (n-ZVI) is certainly the most utilized in environmental applications, see entry 25 in Table 1. The typical reaction initiating the whole process is the slow-releasing source of dissolved Fe^{2+} in acidic solutions, starting a Fenton-like pathway [41,79,127]. Al^0 was reported to be able to produce hydrogen peroxide in situ in water at neutral conditions and further decompose H_2O_2 into $^{\bullet}OH$ in an acidic solution [125]. Other reaction types that zero-valent metal nanoparticles can support are, for example, simple photocatalysis or ozone-based processes [124,128].

Other important categories of materials exploited in H_2O_2-mediated processes are metal sulfides, metal oxides and mixed oxides. Metal sulfides have been presented in Table 1 (entries 12 and 13) in combination with iron since they can accelerate the Fe^{3+}/Fe^{2+} cycle through the exposed reducing metal active sites or directly participate as a co-catalyst in the reaction, activating the oxidant to generate active radicals [41]. However, these compounds can be employed as principal active phases. According to the previous findings, there are two general mechanisms commonly proposed for metal sulfide-based AOPs. One is the improvement of the electron transfer efficiency resulting from the reductive S^{2-} on the catalyst surface; the other one is the protons' capture by unsaturated S atoms on the surface of the metal sulfide, resulting in the formation of H_2S and exposition of metal active sites with reducing properties [132]. In general, the behavior of these compounds can be designed by choosing the most adequate metal (Fe, W, Co, Zn, Mo, Pb, Cr, Cu) in the sulfide since this factor has been reported to influence the performances [41,132]. An example of a further fine-tuning effort was conducted by Bai et al. [133], who synthesized MoS_2 photocatalysts with sulfur vacancies that in situ produced H_2O_2 and hydroxyl radicals. The reaction was favored by the transformation of variable metal molybdenum atom, Mo(IV)/Mo(VI), that effectively enhanced the degradation efficiency towards tetracycline and diclofenac.

Among binary oxides, one of the first and most studied is TiO_2, which is very well known as a semiconductor photocatalyst, as explained in the Introduction. The limitations of this system are mainly attributable to the high band gap of TiO_2, which requires UV light to trigger the titania activity, and the rapid recombination of e^--h^+ pairs, with consequent photocatalytic efficiency reduction and radiation energy loss [134]. For this reason, many efforts have been made to overcome these issues through metal and non-metal doping, surface modification, and fabrication of composites with other materials. Iron and/or H_2O_2 assistance is a popular strategy in the literature [135–137]; here, two significant examples of the combination of titania with iron, H_2O_2 and ultrasound to start different Fenton-like processes are reported in Table 1, entries 14 and 15. More detailed information on the modulation titania band gap, surface reactivity, charge transfer, and stability, among other properties, has been extensively reported in ad hoc reviews [138–142].

To break free from titanium oxide in photocatalysis and iron oxides in Fenton-like systems (entries 5–9, 16 and 17 in Table 1), a plethora of other binary oxides and their various combinations have been developed, such as cobalt, copper, zinc, cerium, tin, and tungsten oxides [143–152]. Different but interesting studies also concern metal peroxides able to release H_2O_2, as recently reported for MgO_2 in dye degradation and bacterial decontamination [153], and for CaO_2 in trichloroethylene abatement [154].

Mixed oxides in different crystalline structures, such as perovskite (ABO_3), scheelite (ABO_4) and spinel types (AB_2O_4) [155–157], have been widely employed in environmental depollution (entries 18–20 in Table 1). They are exploited for their ability to degrade pollutants both in the presence and in the absence of hydrogen peroxide through their inherent semiconducting and redox properties, or by forming heterojunction structures. Often, they are employed as photocatalysts and electrocatalysts [158–165]. These materials with at least two metal elements in the basic structure have the advantage of being regulable in many characteristics, starting from the composition, which can bear multiple doping elements, creating several combinations. Moreover, in many studies, the possibility of inducing non-stoichiometry and charge compensation mechanisms is considered another important tool to create defects such as oxygen vacancies inducing high oxygen mobility, unusual element valence states, such as Fe(IV), and active redox couples [18]. For perovskite-type oxides, for instance, activity descriptors based on the occupancy by electrons in the orbitals of the active metal were proposed to explain material catalytic properties [166].

In general, when oxide materials are involved, catalytic activities have also been related to the degree and symmetry of the crystalline structure [80,167].

Regarding the use of magnetic iron-based oxides, they have been presented as a convenient tool not only to perform heterogeneous Fenton-like reactions, but also to allow for a better and easier separation from the water medium [50]. Many works have described these systems; for instance, in Table 1, they are exemplified in entries 6, 7, 9, and 17. Actually, the possibility of recovery, regeneration and reuse of the active phase in decontamination procedures is a fundamental parameter to define their sustainability. This aspect has been stressed in many studies and is highlighted in Table 1 (entries 1, 6–10, 13, 14, 16, 17, 22).

As partially anticipated in Section 2.1.1 discussing 2D carbon materials, such as graphitic carbon nitride, a novel frontier of H_2O_2 generation through photocatalysis is the replacement of traditional inorganic photocatalysts with organic ones, addressing remarkable advantages, such as narrow bandgap, adjustable band edge potentials, ability to control surface configurations, and tunability of structural units to promote efficient charge separation and transfer [168]. Various strategies to improve the activity and stability of organic photocatalysts have been explored, such as the construction of donor–acceptor structures, the design of conjugated structures, the incorporation of heteroatoms, the enhancement of the internal electric field, and the substitution of functional groups. Some examples of these compounds considered by Zhang et al. [168] are resorcinol-formaldehyde resins, poly(3-(4-ethylnylphenyl)ethynyl)pyridine, and Covalent Heptazine Frameworks, among others. Hydrogen peroxide generation was reported to occur by the anthraquinone-mediated oxygen reduction reaction (ORR), the radical-related ORR, the water oxidation reaction (WOR), and the dual ORR and WOR pathways.

Moreover, the role of natural-inspired organic substances as coadjutants in catalytic and Fenton-like reactions for water depollution has been explored both in homogeneous and heterogeneous systems. In particular, humic-like substances (HLSs), given the intrinsic influence of humic acids in natural ecosystems, have been investigated [4]. In a recent review, García-Ballesteros et al. [169] pointed out that HLSs, which can even be isolated from different wastes, are photosensitive and are able to generate ROS such as hydroxyl radicals and singlet oxygen or triplet excited states upon irradiation (Equations (21)–(24)) [4,170]. This circumstance is particularly convenient when HLSs are complexed with iron to promote homogeneous (photo)-Fenton-like processes at mild pH, preventing Fe deactivation [171].

$$HLS + h\nu \rightarrow {}^3HLS^* \quad (21)$$

$$HLS + h\nu \rightarrow HLS^{+\bullet} + e^-_{aq} \quad (22)$$

$${}^3HLS^* + \text{H-Substrate} \rightarrow HLS^\bullet + \text{Substrate}^\bullet \quad (23)$$

$$HLSH^\bullet + O_2 \rightarrow HLS + HOO^\bullet \quad (24)$$

Novel hybrid materials for heterogeneous catalysis have also been synthesized using HLSs to induce photocatalytic properties [172,173]. In some cases, these materials were produced by combining HLSs with iron-based oxides to drive heterogeneous (photo)-Fenton (or mixed heterogeneous/homogeneous) processes (as in entries 6 and 19 in Table 1) [170].

The effects of ligands in heterogeneous Fenton-like systems have also been explored with different chemical substances, like organic acids (entry 23 in Table 1). For instance, EDTA (ethylenediaminetetraacetic acid) was found to be able to enhance the activation rate of H_2O_2 via the decrease in the redox potential of the Fe(III)/Fe(II) pair. Additionally, it improved the utilization efficiency of hydrogen peroxide by preserving the electron-rich Fe(II) under the attack of $O_2^{\bullet-}$ and $HOO^\bullet$ [174]. Also, in the case of semiconductor oxides, the influence of ligands and complexes has been studied to enhance the photocatalytic performances [175–177], and this further underlines the importance of even more careful engineering of materials to make decontamination processes more efficient.

Lastly, the modulation of the physical–chemical properties of catalysts has also been carried out by fine-regulating, during the synthesis, both the operating parameters and the precursors' types, which act to stabilize the final materials as well as impart specific properties, for instance, specific surface charge and exposed functional groups [178–180] (see entries 7, 9 and 19 in Table 1). A simplified roadmap for the development of functional materials for H_2O_2-mediated processes is summarized in Figure 5.

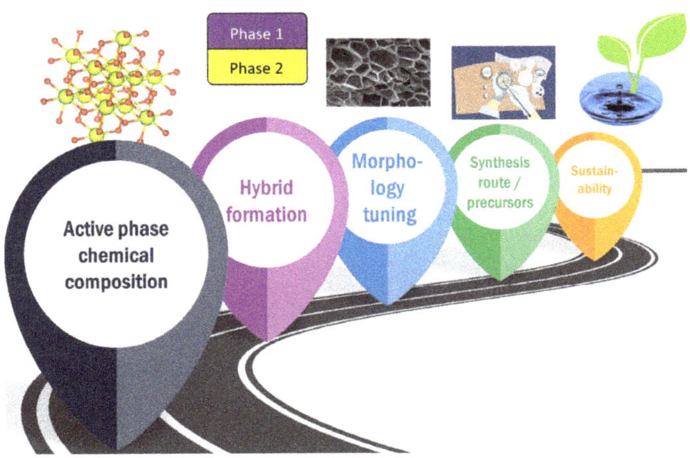

Figure 5. Strategies to obtain environmentally active materials in H_2O_2-based reactions.

2.2. Enzyme-Driven Processes

2.2.1. Glucose-Oxidase-Based Bio-Fenton

The applications of the Bio-Fenton approach in water remediation are increasing since the range of treatable pollutants is quite broad. In Bio-Fenton remediation processes, the hydrogen peroxide necessary to activate the Fenton reaction is mainly provided by an enzymatic reaction catalyzed by Glucose Oxidase (GOx) that occurs in mild pH conditions. GOx is a stable oxidoreductase able that use oxygen as an electron acceptor to catalyze β-*d*-glucose to *d*-glucono-δ-lactone by using the coenzyme FAD (Flavin Adenine Dinucleotide) as an electron carrier [181]. The oxidation of β-*d*-glucose leads to the reduction of one FAD molecule to the hydrogenated form $FADH_2$; successively, *d*-glucono-δ-lactone is non-enzymatically hydrolyzed to gluconic acid and the reduced coenzyme is re-oxidized to FAD by a molecule of dioxygen, producing a molecule of H_2O_2 for each reaction cycle, as in Figure 6.

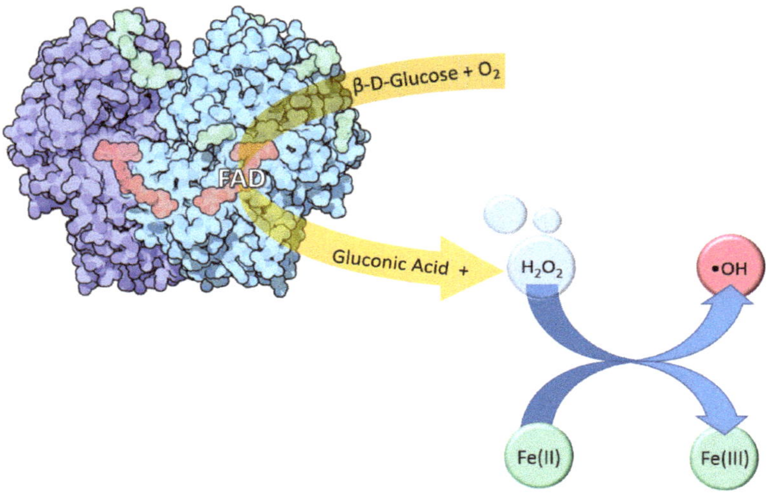

Figure 6. Schematic representation of Bio-Fenton reaction mediated by GO_x (3D structure of GOX by Protein Data Bank, entry: 1GPE) [182].

GOx in vivo is usually coupled with Catalase, which rapidly causes the dismutation of H_2O_2; on the contrary, in environmental applications, GOx is mainly exploited when coupled to a system able to activate Fenton or Fenton-like reactions. Table 2 provides a summary of recent studies concerning several classes of environmental contaminants removed by the Bio-Fenton reaction. To optimize the hydrogen peroxide production, different aspects have to be taken into consideration. The amount of H_2O_2 produced depends on both the concentration of the biocatalyst and glucose. Indeed, according to the Michaelis–Menten model, at low concentrations of glucose, the rate of H_2O_2 production is linearly proportional to its initial amount, but the reaction rate decreases when the glucose concentration increases until reaching a maximum value, which, in turn, depends on the concentration of GOx [48,183].

Huang and co-workers [184] highlighted the complex role of glucose in the Bio-Fenton reaction, reporting that, at a too-high concentration, it could also act as a hydroxyl radical scavenger, inhibiting the oxidation reaction. The optimization of hydrogen peroxide generation could be obtained, therefore, by using specific techniques, which allow for the evaluation of the influence of different variables on the reaction outcomes [185].

Gluconic acid formed from the oxidation of glucose makes the process a chelate-Bio-Fenton, since it is a well-known chelating agent for ferrous and ferrate ions and is also used in non-enzymatic AOP processes to avoid precipitation of iron hydroxides at neutral pH [186]. Ahuja et al. [49] observed a 100% Fe(III) chelation with a ligand–ion ratio of 1:1 at pH 4 and an increasing concentration of chelate complexes with pH, demonstrating the advantage of using Bio-Fenton at circumneutral pH. The same authors [49] also demonstrated that the gluconate-complex could directly react with H_2O_2 or be the source of iron ions.

In order to ensure high efficiency in milder conditions, the Bio-Fenton reaction can be performed with iron citrate instead of iron sulfate [187–190]. However, in general, the introduction of iron as a homogeneous reactant can lead to the formation of iron-containing sludge [191,192]. Therefore, heterogeneous Fe sources such as iron oxides and oxyhydroxides, like Fe_3O_4 (magnetite), γ-Fe_2O_3 (maghemite), α-Fe_2O_3 (hematite), α-FeOOH (goethite) [193], or green rust [194] have also been employed. These solids prevent sludge formation, improve Fenton efficiency in a wider range of pH, and allow for their reuse.

A change in concentration and availability of iron ions with pH is a fundamental factor that influences the efficiency of the Fenton reaction [195]; furthermore, it is well known that pH affects the enzymatic activity. Wang et al. [196] found that the removal of trichloroethene by a Bio-Fenton reaction carried out with GOx immobilized on magnetic particles at the equilibrium state did not depend on the pH, but this parameter affected the reaction rate that decreased by approximately 60% when the pH increased from 3.6 to 9. Liu and co-workers [197] demonstrated that strongly acidic and alkaline environments significantly inhibited the 4-chlorophenol degradation, decreasing from 87.6% at pH 7 to only 55.2% at pH 3 and 35.8% at pH 11. This degradation trend follows the hydrogen peroxide production, which is minimal at pH 3 and 11, while it has a maximum at pH 7. The authors underlined that during the Bio-Fenton reaction, whatever the initial pH value, there was an increase in the acidity of the solution due to the production of gluconic acid. Therefore, the efficiency of the reaction was only maintained if the variation in pH was limited within the activity range (4.5–7).

Another crucial factor is the temperature. An optimal range between 25 and 60 °C for GOx catalytic capability was reported. As for the pH, the temperature mainly influences the kinetics of the reaction more than its final outcome. Huang et al. [184] found a 2.8-fold increase in k_{obs} for trichloroethene degradation when the temperature increased from 15 to 40 °C. A similar trend was also reported by Karimi et al. [198] for the decolorization of malachite green; in this case, a slight reduction in the decolorization kinetic was already observed at 35 °C, probably due to a partial denaturation of the enzyme. On the contrary, Wang et al. [196] showed that enzyme immobilization could improve enzyme thermal stability, reporting for GOx immobilized on magnetic particles an enhancement in activity with increasing the temperature from 15 to 40 °C, with consequent reduction in the time needed to reach the equilibrium from 96 h to 24 h.

The coupled effect of thermal stability and improvement in the enzymatic activity was also reported by Zhao et al., who found that 3,4-dimethylaniline was removed with a higher rate by GOx immobilized on Kaoline (Kaoline@GOx) than a free GOx system [183]. The authors suggested that this effect was due to an enhancement in GOx spatial conformation when the enzyme was fixed on the support.

Recently, some studies have reported an improvement in Bio-Fenton efficiency when appropriate UV-light irradiation was applied. Liu and co-workers [197] observed the evolution of hydrogen peroxide produced by GOx immobilized on Kaoline in the presence of organic green rust as a source of iron. Changing the UV-light intensity, the equilibrium concentration of H_2O_2 first increased with the irradiation (from 0 to 150 $\mu W\ cm^{-2}$) and then decreased with further raising the irradiance up to 400 $\mu W\ cm^{-2}$. Consequently, the degradation percentage of 4-chlorophenol followed the same trend. The authors proposed that the mechanism at the basis of the variation in the enzymatic activity could be related to the photo-excitation of the cofactor FAD into FAD*, which induced a glucose oxidase stronger reactivity. Indeed, the FAD cofactor shows high absorption capacity at 250–300 nm and strong UV light tolerance [199]. Obviously, excessive light intensity causes enzyme inhibition and denaturation.

Very few works investigate the Bio-photo-Fenton strategy for environmental remediation. Ghatge and co-workers [200] studied the degradation of sulfonated polyethylene employing glucose oxidase immobilized on TiO_2. They demonstrate the synergy between Bio-Fenton and light irradiation in the production of reactive species following the concentration of the degradation products. Indeed, compared to the Bio-Fenton degradation using free GOx, 21-fold and 17-fold higher amounts of acetic acid and butanoic acid were, respectively, released by the Bio-photo-Fenton reaction after incubation for 6 h.

Table 2. GOx Bio-Fenton systems for pollutant removal mediated by H_2O_2 (acronym not explicated elsewhere: COD = Chemical Oxygen Demand, U = unit).

Pollutant	Performances	Conditions	Notes	Ref
1.Trichloroethene (TCE)	Removal after 192 h - TCE 5 mg L^{-1}: 76.2% - TCE 50 mg L^{-1}: 94.1%.	200 mg magnetic nanoparticles (MIG); 2.5 mM Glucose.	- Gox immobilized on MIG - Recycled for 4 cycles - Effectiveness in the ranges of T 15–45 °C and of pH 3.6–9.0 - Influence of inorganic ions: Ca^{2+} >Mg^{2+} >Cu^{2+} and $H_2PO_4^-$> Cl^-> SO_4^{2-}.	[196]
2. 4-chlorophenol (4-CP)	Removal after 250 min - GOx@Kaolin/OGR: 77.5% - GOx@Kaolin/OGR/UV: 96.1%.	4-CP: 5 mg L^{-1} UV: 150 μW cm^{-2} T: 25 ± 1 °C OGR: 250 mg L^{-1} GOx@Kaolin: 2.5 U mL^{-1} Glucose: 5 mmol L^{-1}.	- GOx immobilized on Kaolin (GOx@Kaolin) coupled with organic green rust (OGR, source of iron) and enhanced by UV light - Reusability 6 cycles.	[197]
3.Trichloroethylene (TCE)	- After 24 h at pH 7 Ground water: 30% or 48% if doped with additional H_2O_2 (after 3 h).	TCE: 60 mg L^{-1} Glucose: 60 mM GOx: 1 mg mL^{-1} Fe: 25 mg L^{-1}.	- GOx in solution - Organic matter of ground water acts as a radical scavenger.	[201]
4. Polycyclic aromatic hydrocarbons (PAHs): Naphthalene (NAP), Anthracene (ANT), Pyrene (PYR)	- After 48 h, neutral pH, removal%: 95.1%, 75.4%, and 85.2% for NAP, ANT, and PYR - COD reduction%: 28.6%, 13.8%, and 30.8% for NAP, ANT, and PYR.	PAH: 50 mg L^{-1} each GOx: 10 U Fe(III)citrate: 0.1 mM Glucose: 2 mM.	- GOx in solution - Evaluation of PAHs concentration and COD to evaluate the mineralization - Bio-Fenton as pre-treatment to enhance PAHs removal by activated sludge: COD removal from 33% to 72%.	[189]
5. Bisphenol A (BPA)	After 10 days of incubation Removal of 80%.	BPA: 0.1 mM GOx: 10 U Glucose: 32 mM Fe(III)citrate: 0.5 mM pH 5.3.	- GOx in solution.	[188]
6. Chloro-acetanilide herbicides: acetochlor, alachlor, metolachlor, propachlor, butachlor	After 5 days, degradation %: acetochlor: 72.8% alachlor: 73.4% metolachlor: 74.0% propachlor: 47.4% butachlor: 43.8%.	GOx: 10 U pH 5.5 Fe(III)citrate: 0.5 mM Glucose: 32 mM Each herbicide 0.1 mM.	- GOx in solution - Influence of chemical structure of herbicides and particularly of R-O-R′ groups.	[187]

Table 2. *Cont.*

Pollutant	Performances	Conditions	Notes	Ref
7. Trace organic contaminant (TrOCs) (mefenamic acid, ketoprofen, caffeine, carbamazepine, trimethoprim, fenofibrate, diuron, carbendazim, thiabendazole)	After 360 min pH 7 mefenamic acid: 68.54% ketoprofene: 44.7% caffeine: 36.1% carbamazepine: 44.1% trimethoprim: 46.4% fenofibrate: 20.3% diuron: 89.4% carbendazim: 73.1% thiabendazole: 88.9%.	Glucose: 1 M GOx: 100 U mL^{-1} pH 7 50 μg L^{-1} mix of TrOCs T: 30 °C H_2O_2:$FeSO_4$ = 50:1.	- GOx produced from *Aspergillus niger* using *Casuarina equisetifolia* biomass in a pilot-scale - Municipal wastewater as a matrix - Influence of the rate H_2O_2: $FeSO_4$.	[185]
8. Trichloroethylene (TCE)	After 8 h Removal of 78%.	Glucose: 2.5 mM Fe(II): 0.5 mM GOx: 10 U mL^{-1}.	- GOx in solution - Efficiency maintained in the pH range 3–6 and in a T range 15–30 °C.	[184]
9. Sulfonated polyethylene (SPE)	After 6 h with free GOx, concentration degradation products: Acetic acid: 0.22 mM Butanoic acid: 0.01 mM; After 6 h with TiO_2-GOx: degradation product conc.: Acetic acid 4.78 mM Butanoic acid 0.17 mM.	GOx free: Glucose: 32 mM GOx: 1 U mL^{-1} pH 5.5, T: 30 °C SPE: 1 mg mL^{-1}; TiO_2-GOx: As above, except for 10 U of TiO_2-GOx Xe lamp 150 W, 400 nm cut off.	- GOx both free and immobilized on TiO_2 particles - Degradation follows studying the product formation (acetic acid, butanoic acid, isovaleric acid, 1,2-ethanediol monoacetate).	[200]
10. Atrazine (ATZ)	After 360 min 72.8% removal.	Phosphate buffer: 5 mM GOx: 10 μmol min^{-1} Glucose: 3 mM Ferric citrate: 0.5 mM ATZ: 0.1 mM pH 5.8.	- GOx in solution - Toxicological assay and by-products study.	[190]
11. 3, 4-Dimethylaniline (3, 4 DMA)	After 180 min, removal of 86.55%.	3, 4-DMA: 30 mg L^{-1} Green Rush: 1 mM in Fe(II) and 1 mM in Fe(III) Glucose: 5 mM pH 7 Kaolin@GOx: 2.5 U mL^{-1}.	- GOx immobilized on Kaolin (Kaolin@GOx) - Organic green rust as a source of iron.	[183]

2.2.2. Bi-Enzymatic Processes for Water Treatment

Enzymatic Cycles Activated by H_2O_2

Many enzymes require H_2O_2 to carry out their catalytic activity. Among them, peroxidases are the most studied for environmental applications.

Peroxidases are able to bind H_2O_2 to their Fe-heme prosthetic group. In the resting state, the Fe(III) is coordinated by four nitrogen atoms belonging to heme pyrrolic groups and one N atom of the side chain of an amino acid. As shown in Figure 7, the fifth ligand can be the nitrogen of the distal histidine or the sulfur of a cysteine residue. The sixth coordination position is usually free or occupied by a water molecule [202], which is easily substituted by an H_2O_2 molecule. In a typical peroxidase active site, the distal side of the heme is characterized by the presence of one histidine and one arginine residues, which stabilize the coordination with the H_2O_2 molecule and favor the subsequent steps of the catalytic cycle [203,204].

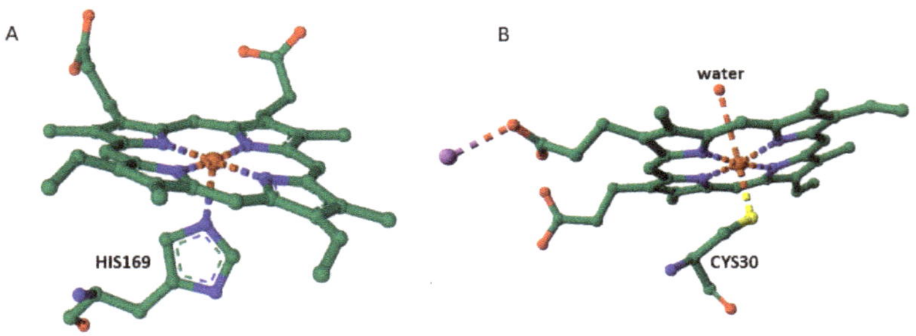

Figure 7. 3D structure of Fe-heme site in (**A**) soybean peroxidase; (**B**) chloroperoxidase. Images built by data from Protein Data Bank. Color legend: orange = iron; blue = nitrogen; red = oxygen; green = carbon, yellow = sulfur; violet = manganese.

In the presence of organic or inorganic substrates, H_2O_2 oxidizes the Fe(III) to a ferryl (Fe(IV)=O) radical, which can be reduced by inducing two steps of one-electron oxidation of a wide variety of compounds with the production of two molecules of organic or inorganic radicals and water (Figure 8A) [205].

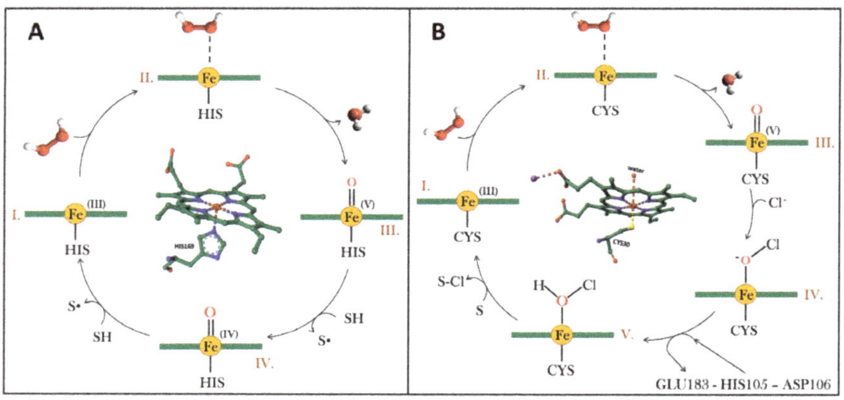

Figure 8. Summary of catalytic cycles. (**A**) Horseradish peroxidase (HIS = histidine, SH = substrate; $S^{•}$ = radical product); (**B**) Chloroperoxidase (CYS = cysteine, GLU = glutamic acid, HIS = histidine, ASP = aspartic acid, S = substrate, S-Cl = chlorinated product).

A quite different mechanism was reported for the heme-thiolate peroxidases, such as chloroperoxidase (CPOs) and aromatic-peroxygenase (APOs).

CPO from *Caldariomyces fumago* fungi was the first discovered halogenating enzyme [206,207]. In addition to substrate halogenation, it also exhibits peroxidase, catalase, and cytochrome P450-like activities [208]. The catalytic cycle leads to halide (chloride, bromide, and iodide, but not fluoride) oxidation, but this is not fully understood.

Wagenknecht and Woggon [209] proposed the mechanism reported in Figure 8B, in which the ferryl cation radical reacts with halides, reducing to Fe(III) and forming a complex capable of catalyzing the chlorination of the substrate. The Fe(V)=O reduction is favored by the three amino acid residues (GLU183, HIS105 and ASP106) that directly take part in its protonation. In the absence of halides, this enzyme could act as a normal peroxidase and follow the scheme in Figure 8A.

APOs are peroxygenases, enzymes able to transfer oxygen from peroxides to organic substrates (aliphatic, aromatic and heterocyclic) [210]. For these reasons, they are very useful for organic functionalization and synthesis reactions.

Dye-decolorizing peroxidases (DyPs) are probably the most recent heme-containing enzyme family discovered. Indeed, they were isolated in 1999 from the basidiomycetous fungus *Bjerkandera adusta* Dec 1 strain [211]. DyPs are so called due to their capability to oxidize synthetic high-redox potential anthraquinone-type dyes [212], but they can also react with other substrates like β-carotene and aromatic sulfides [213]. This different catalytic activity is probably related to some differences in the active site, which contains an aspartate residue in place of the histidine on the distal side [213], and a different arrangement of the secondary structure with the presence of β-sheets. Recent studies on the catalytic cycle suggest an electron transfer from the active site to the enzyme's surface occurs, allowing for the oxidation of large molecules that cannot physically enter the active site and favoring the relatively low specificity of these enzymes. This long electron transfer pathway seems to involve tryptophan and tyrosine residues that can form stable radicals and directly participate in the transfer [214].

Glucose Oxidase Coupling with Peroxidases

There is extensive research about the employment of free enzymes, such as peroxidases, for environmental applications [215], but in the last few decades, articles relating to immobilized enzymes have become increasingly widespread. As reported by Rigoletto et al. [216], different inorganic supports have been developed not only to improve enzymes' stability but also to produce in situ H_2O_2 to initiate the catalytic cycle [217–219].

It is well known that peroxidases could be inactivated by high concentrations of hydrogen peroxide [220–222]; therefore, a controlled in situ H_2O_2 production could be useful to avoid these drawbacks and maintain a stable enzymatic activity.

The possibility of exploiting the enzymatic generation of hydrogen peroxide coupling a peroxidase with GOx in a self-sustaining bi-enzymatic system has become widely studied, mainly for bio-sensing [223,224], but also more recently for environmental application. Although, in some studies, GOx is added to the solution as a homogeneous reagent together with glucose [225], the most widespread trend is to immobilize GOx on a solid support or to co-immobilize it with peroxidase in order to guarantee its reuse.

Co-immobilization of enzymes could enhance the system performances due to a reduction in mass transfer between peroxidase and H_2O_2. Furthermore, the formation of gluconic acid induced by GOx in the peroxidase microenvironment helps to reach a favorable pH value for peroxidase, thereby accelerating the reaction and enhancing the catalytic cycle. The resulting consumption of H_2O_2 prevents its accumulation on GOx, avoiding its inactivation [226,227]. This synergy ensures improved system performances, as reported by Gao and co-workers, who observed a higher Orange G decolorization with GOx and CPO co-immobilized on magnetic graphene oxide (MGO) with respect to those obtained employing GOx and CPO immobilized individually [228]. The co-immobilization on magnetic material also permitted an easy recovery of the support and the reusability of the system. Indeed, the authors reported that GOx-CPO-MGO could be employed for six cycles of water treatment with more than 38% residual activity retained after the last cycle.

Enhanced performances of co-immobilized enzymes have also been reported by Gu and co-workers [229]. The authors employed dopamine and modified cellulose–chitosan composite beads for a covalent loading of Horseradish peroxidase (HRP) and GOx and tested them to degrade acridine in wastewater, obtaining an almost complete removal. However, the removal efficiency was reduced from 99.0% to 61.2% after six reaction cycles; the decrease was tentatively ascribed to the mediator attack on the amino acid groups on the surface of the bi-enzyme system or to the loss of enzymes during application and washing.

Studies on reusability and stability for long storage periods are fundamental to estimating the real applicability of the bi-enzymatic systems in industrial applications. Actually, the literature confirms the possibility of adopting this strategy for the treatment of a large number of contaminants, from dyes [230,231] to other organic contaminants such as ferulic and caffeic acids [232], Bisphenol A [226], and endocrine disruptors [233].

3. Detection Methods for H_2O_2

In this work, different strategies for in situ hydrogen peroxide production have been identified. Important tools for the optimization of these reactions are represented by H_2O_2 detection and quantification methods.

According to most of the studies cited in this review, the commonest strategies employed in the environmental field are colorimetric assays. They are widely used because of their simple setup and easy spectrophotometric detection. They are usually based on chemical reactions involving hydrogen peroxide as a reagent with the formation of colored products. A summary of the literature concerning these methods is shown in Table 3.

Some colorimetric assays date back to the 1980s, but they are still employed, such as the I^{3-} assay, in which the H_2O_2 concentration is determined following the changes in absorption at 352 nm wavelength [57,133,184,234,235], or the reaction between 4-aminoantipyrine and phenol in the presence of peroxidase where the maximum absorption at 505 nm of the red product is monitored [234,236,237]. Another method is the vanadate one, based on the measure of 450 nm [238–240]. Moreover, some colorimetric assays are commercialized as a kit, like the Amplex^{TM} red hydrogen peroxide/peroxidase assay kit [241] (Thermofisher, Waltham, MA, USA) and the PAKTEST [242] (Kiouritsu Chemical Check Lab., Yokohama, Japan). However, not all colorimetric methods are exploitable for monitoring hydrogen peroxide production during AOP treatments. For example, permanganate titration cannot be used since Fe^{2+} ions react with permanganate, affecting H_2O_2 detection [240].

Recently, enzyme-mimetic materials have been developed for luminescent sensing of H_2O_2 in bioassays. They show a very low limit of detection (LOD) and are based on fluorescence [243,244].

Another important class of hydrogen peroxide detection methods is represented by electrochemical sensors in which H_2O_2 can be oxidized or reduced on the electrodes' surface. They show higher sensitivity and selectivity, cost-effectiveness, relatively shorter response time, and miniaturization capabilities. The research in this field is focused on overcoming drawbacks such as overpotentials and low kinetics by modifying the electrode surfaces with different materials, including dyes, redox proteins, metal oxides, and redox polymers, among others [245]. Recently, the interest in nanomaterials has grown and there is no shortage of combinations with the materials listed above that can improve H_2O_2 electrocatalytic sensing.

As seen in the previous paragraphs, a wide range of heme-containing proteins exists in nature that can use H_2O_2 in their catalytic cycles (peroxidases, cytochrome-c, hemoglobin, myoglobin, hemin, etc.), and, for this reason, they are also exploitable in H_2O_2 detection. Different studies suggest that HRP-based electrodes show good affinity with H_2O_2 and noticeable performances [246,247]. In order to ensure an optimized electron transfer, these proteins are often immobilized on highly conductive materials, both organic and inorganic, with a large surface area [248]. Interesting examples of these materials that also find application in the bio-medical sector are gold nanoparticles [249–251], nanometric metal oxides [252,253], metal nanoparticle–MOF [254–257], carbon nanotube–MOF [258] and metal nanoparticle–polymers [259,260].

Despite their efficiency, redox proteins could degrade over time and in harsh environmental conditions. To overcome these drawbacks, the development of inorganic materials with enzyme-mimetic activity is becoming more and more attractive. Most of these inorganic materials are mono- and bi-metals nanoparticles (AuNPs, Ag-NPs, Pt-NPs, Ag-AuNPs, Ni-CoNPs, etc.) or transition metals oxides and, often, to avoid aggregation on the electrode surface, they have been dispersed in polymeric matrices or introduced in MOFs. A broad description of these electrodes can be found in the comprehensive review carried out by Duanghathaipornsuk and co-workers [248].

Among electrochemical sensors with peroxidase-like activity, those based on Prussian Blue (PB) and its Analogues (PBAs) are largely used. They are metals hexacyanoferrate and can catalyze the reduction of H_2O_2 at low potentials. There is a wide variety of PBAs containing one or two transition metals that occupy the outer coordination sphere and that define the selectivity for the detection of different analytes, both organic and inorganic [261]. PBAs employed for H_2O_2 detection usually contain iron coupled with another transition metal like Ni [262,263], Co [264,265] or Mn [264,266].

With the enhancement of nanotechnologies, researchers developed PB nanoparticles [267] to improve pH stability and sensitivity. Zang and co-workers studied the influence of nanoparticle size on electrocatalytic performance, finding that the sensitivity drops exponentially with increasing particle dimensions [268]. These nanozyme-based electrodes could also be affected by PB degradation during the electrochemical H_2O_2 reduction due to the solubilization of ferric hexacyanoferrate induced by the formation of reactive species [269]. Since this could limit the real use of the sensor, stabilization strategies, such as nanoparticles' coverage or dispersion in different polymers, have been designed. For example, Unzuncar et al. [270] developed a two-layer interface based on carboxymethyl cellulose and poly(3,4-ethylene dioxythiophene) with high stability and sensing behavior, not affected by interfering molecules present in tap water. More detailed syntheses of PB and PBA for electrochemical sensing can be found in some recent reviews [261,271].

Finally, many other enzyme-free electrochemical-based sensors for H_2O_2 are based on a wide range of materials, i.e., metal–organic frameworks [272], carbon nanocomposites [273], graphene oxide [274], and MXenes [275].

It must also be mentioned that some methods employing liquid-chromatographic analysis for the detection and quantification of hydrogen peroxide concentration in solution were developed. Some of them are based on the deprotonation of H_2O_2 to HO_2^-, which occurs when the pH of the eluent is higher than the acid-dissociation coefficient of H_2O_2 (pKa = 11.6). The revelation via a UV-detector can be carried out after separation in an ionic chromatography column with a LOD of 0.027 mg L^{-1} [276]. Other studies proposed the use of acidified potassium iodide solution as a mobile phase and the replacement of the reverse phase column with a series of capillary columns. Tantawi et al. [277] found a LOD of 8.29×10^{-4} mM employing this detection strategy. The authors reported that the method also shows high robustness by maintaining a high regression coefficient and excellent sensitivity in real matrices.

Since H_2O_2-mediated processes lead to the formation of ROS, their quantification is also extremely important for monitoring the depuration stages. Although it is not within the scope of this review, we can cite Electron Paramagnetic Resonance (EPR) and scavenging techniques as the main strategies for ROS identification. Specific papers have been devoted to summarizing the state-of-the-art on these topics [278,279].

Table 3. Spectrophotometric methods for H_2O_2 detection and quantification (acronyms not explicated elsewhere: NADPH = reduced Nicotinamide Adenine Dinucleotide Phosphate; TMB = 3,3′,5,5′-tetramethylbenzidine; ABTS = 2,2′-Azino-bis(3-ethylbenzothiazoline-6-sulfonic acid).

System	Wavelength	Conditions	Ref
1. Phenolphthalein method	552 nm	- 0.24 mL of phenolphthalein stock solution: 0.02 g mL^{-1} Phenolphthalein solution containing 10 g of NaOH, 5 g of zinc - $CuSO_4$ solution: 0.48 mL, 0.01 M - 100 mL of water.	[234]
2. Iodide method	352 nm	- 2 mL of solution A (66 g L^{-1} KI; 0.2 g L^{-1} of $(NH_4)_6Mo_7O_{24}$; 2 g L^{-1} of NaOH) - 2 mL of solution B (20 g L^{-1} of potassium hydrogen phthalate) - 6 mL of Peroxide samples.	[57,133,234,235]

Table 3. *Cont.*

System	Wavelength	Conditions	Ref
3. Oxidation of NADPH	340 nm	- Phosphate buffer: 0.04 M, pH 7.75 - EDTA: 4×10^{-4} M - Sodium azide: 4.2×10^{-3} M - Glutathione peroxidase: 8×10^{-8} M - Reduced glutathione: 3×10^{-3} M - NADPH: 5.6×10^{-5} M - Glutathione reductase: 1 U -Peroxide sample.	[234]
4. DMAB/MBTH/HRP	590 nm	- 3-(dimethylamino)benzoic acid (DMAB): 5×10^{-4} M - 3-methyl-2-benzothiazolinonehydrazone (MBTH): 2×10^{-5} M - Acetate buffer: 0.1 M, pH 5.5 -HRP.	[280]
5. 4-aminoantypirine/phenol/HRP	505 nm	- 4 mL of 4-aminoantipyrine/phenol reagent (2.34 g L^{-1} of phenol, 1g L^{-1} of 4-aminoantipyrine, 0.001 M phosphate buffer pH 6.9, 2.5 μM HRP) - 6 mL of peroxide sample.	[234,236,237]
6. Nanoparticles decorated $Ce_2(WO_4)_3$ nanosheets (CWNSs)	652 nm	- TMB: 100 μL 8 mM - CWNSs: 70 μL, 1000 μg mL^{-1} - Phosphate buffer: 400 μL, 50 mM pH 4 - MilliQ water: 330 μL - H_2O_2: 100 μL - LOD: 0.15 μM.	[281]
7. Ammonium metavanadate	450 nm	- Metavanadate 6.2 mM - Sulfuric acid 0.058 M - LOD: 143 μM.	[238–240]
8. Fe_3O_4 magnetic nanoparticles (MNPs) with peroxidase mimetics	545 nm	- ABTS: 24 μL, 60 mM - Fe_3O_4 MNPs: 10 μL, 3.74 mg mL^{-1} - Acetate buffer: pH 4185 μL - H_2O_2: 24 μL - Incubation 45 °C for 10 min and then diluted with 900 μL of water (after MNPs removal) and analyzed.	[282]
9. Peroxidase-mimicking metal—organic framework containing catalytic Cu^{2+} and luminescent Tb^{3+}: PA-Tb-Cu MOF (PA = m-phthalic acid)	545 nm Fluorescence (310 nm exciting wavelength)	- Acetate buffer: 960 μL, 10 mM, pH 5.05 - 20 μL of PA-Tb-Cu - MOF suspension: 11.73 mg mL^{-1} - Ascorbic acid: 10 μL, 200 mM - H_2O_2 sample + water up to a final volume of 1 mL - Measure after 20 min incubation - LOD: 0.2 μM.	[244]
10. CeO_2 nanoparticles doped with Eu^{3+}	590 nm Fluorescence (330 nm exciting wavelength)	- Samples prepared in potassium phosphate buffer (KPi) or Phosphate Buffered Saline (PBS) or 10% Farmigene Stain Buffer (FBS) - 125 g L^{-1} of nanoparticles - Measure after 30 min of incubation - LOD: 150 nM.	[243]

Table 3. *Cont.*

System	Wavelength	Conditions	Ref
11. Cobalt/bicarbonate system	260 nm	- Sodium oxalate: 25 μL,16.34 mM - Cobalt chloride: 25 μL, 67.8 mM - 2 mL of sample Finally, the reaction volume is made up to 2.5 mL with 270 μL of saturated sodium bicarbonate solution.	[185,201,283,284]
12. Titanium oxalate	385 nm	- 10 mL of peroxide sample - 1 mL of 1 M sulfuric acid - 1 mL of 50 g L^{-1} potassium titanium oxalate solution - 13 mL of water - Measure after 5 min of incubation - LOD: 29 μM.	[77,125,285]
13. *N*,*N*-diethyl-p-phenylenediamine (DPD)	551 nm	- DPD reagent (27 mL water, 3 mL phosphate buffer, 6 μL methanol, 50 μL of 10 g L^{-1} of DPD solution prepared in sulfuric acid 0.5 M, 50 μL HRP 1g L^{-1}) - LOD: 0.77 μM.	[64,285,286]
14. p-hydroxyphenyl acetic acid (POHPAA)	406–410 nm Fluorescence (315 nm exciting wavelength)	- POHPAA reagent (POHPAA 270 mg L^{-1}, HRP 30 mg L^{-1}, NaOH 1 M, potassium hydrogen phthalate 8.2 g L^{-1} pH 5.8) - LOD: 0.16 μM.	[285,287,288]

4. Criticalities and Perspectives

The processes based on H_2O_2 reactivity for environmental remediation have been investigated in depth in terms of general lab-scale degrading performances, as revealed from the significant number of research documents reported in this review. However, for applicability in real-scale conditions, other aspects have to be taken into account, whose schematic representation is depicted in Figure 9.

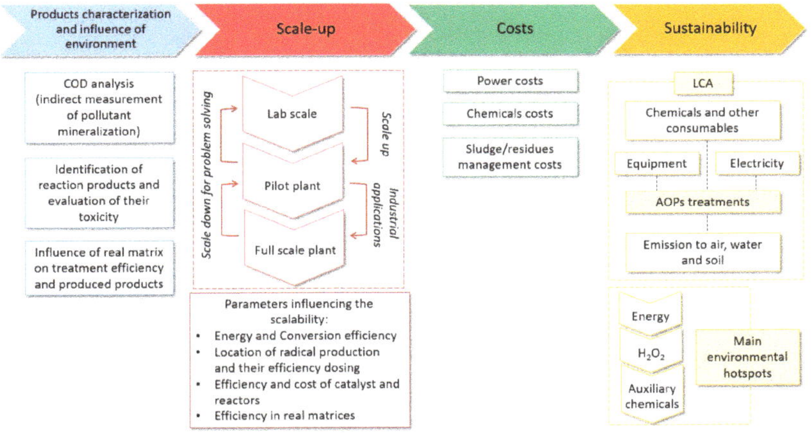

Figure 9. Aspects to be deepened in H_2O_2-mediated processes for real applications.

4.1. *Product Characterization, Toxicity and Influence of Environmental Factors*

The first point concerns the properties' study of the products formed during the depuration procedure and their impact on the environment. The optimum should be to reach the complete mineralization of the pollutants into water and carbon dioxide, but this is not straightforward, not only because the degradation efficiency can be lower than 100%, but also because some treatments, such as those based on enzymes, occur through radical reactions, which do not lead to the mineralization of the contaminants.

Therefore, toxic intermediates and by-products can be produced [289,290]. In this sense, Hofman-Caris et al. [291] found that UV/hydrogen peroxide processes in drinking water treatment may generate mutagenic by-products in particular conditions, mainly influenced by parameters such as nitrate concentration (that can be subjected to photolysis), the presence of natural organic matter, the UV spectrum of the lamps, and the UV dose applied. The role of UV light has also been described in [292]. The comparison among the toxicity levels of H_2O_2/UV, photo-Fenton and heterogeneous photocatalysis (TiO_2/H_2O_2/UV) processes for the treatment of strongly polluted colored wastewater was analyzed as well [293,294]. The use of the system with titania resulted in the most effective method to remove toxins, organic material, and color.

It is, therefore, necessary not only to identify the formed by-products [187,188], but also to evaluate the toxicity of the resulting solution [190] and estimate COD values as an indirect measurement of pollutant mineralization [189,295]. In this regard, Liu et al. compared the COD values after the classic Fenton treatment and Bio-Fenton pre-treatment, followed by active sludge treatment of real wastewater spiked with chlorophenol [197]. The authors found that the resulting COD after the combined treatment was two times lower than that obtained with classic Fenton only, suggesting that the combination of different approaches improves the biodegradability of the selected contaminant.

Regarding biological contaminants, urban wastewater treatment plants are among the main hotspots of antibiotic resistance spread into the environment. It was found that UV/H_2O_2 processes may not be effective in minimizing the antibiotic resistance spread potential into the environment since the death of bacterial cells, which results in DNA release into treated water, may pose a risk for antibiotic resistance transfer to other bacteria present in the receiving water body [296].

Moreover, the control of the hydrogen peroxide quantity used as the oxidant is important since it was found that toxicity increases by increasing H_2O_2 addition, causing not only its undesirable excess in the effluent, but, in the absence of effective separation from the heterogeneous active phase, also catalyst poisoning [38,293].

In general, applying remediation strategies to real conditions means considering the influence of all the environmental factors that are intrinsic to the complexity of real systems, such as the co-presence of ions and organic matter in water matrices that can compete with the target chemical pollutants [201,297,298].

Indeed, wastewater or natural water contains inorganic ions (Ca^{2+}, Mg^{2+}, Cl^-, $H_2PO_4^-$, etc.) and dissolved organic matter that can influence the efficiency of the treatments. For instance, the reduction of trichloroethene removal by Bio-Fenton treatment in the presence of Cu^{2+} was demonstrated [196], as well as the inhibition of the GOx activity due to the reaction with H_2O_2, which resulted in a reduced production of hydroxyl radical [299]. A similar effect was also observed for Mg^{2+}, which is known to hinder the Fenton reaction [300]. Anions, such as $H_2PO_4^-$, Cl^- and SO_4^{2-}, can form complexes with iron ions, reducing their availability for the Fenton reaction [196]. Dissolved organic matter (DOM) in natural water, similarly, can act as a radical scavenger, limiting the number of reactive species available for substrate oxidation and leading to a decrement in pollutant degradation [301]. This kind of competition was also observed by Ravi et al., who reported a reduction in trichloroethylene degradation by the Bio-Fenton reaction carried out with free GOx in groundwater [201].

Also, in the case of pathogen contamination, this control is fundamental. Malvestiti et al. [302] verified the efficiency of O_3, O_3/H_2O_2 and UV/H_2O_2 treatments to disinfect municipal effluents and the influence of carbonate, nitrate and industrial contaminants. The results showed that all AOP treatments were affected by the presence of nitrate and, particularly, carbonate. These ions reduced the inactivation of total coliforms and *Escherichia coli*. Ozone disinfection was the most affected method by scavenging compounds.

Additionally, the selectivity of the decontamination process towards specific pollutants is, in general, another rising theme in environmental research to ensure a higher low-cost/effectiveness ratio and lower secondary pollution [303].

4.2. Processes' Scalability, Cost Analysis and Environmental Impact

Real applications presuppose the scalability of the processes moving from laboratory scale to pilot plants and full-scale plants. At the basis of the scalability process, there are several critical points that have to be considered, such as oxidant conversion efficiency, location of radical production and delivery (i.e., in situ generation or need for a transport step) [304], and the study of kinetic parameters and degradation pathways [305]. All of these have to be initially optimized through the development of proper catalysts with high stability and activity; for instance, in the case of electrodes, they have to be characterized by high activity and corrosion resistance, long working life span and limited synthesis costs [305].

At the same time, the geometry and configuration of the reactors (e.g., batch mode vs. continuous-flow systems), integration with different depuration systems (like biological treatments), and operational costs have to be considered [304,306–308].

Obviously, each strategy shows intrinsic strengths and critical aspects that can limit its development on a large scale. For instance, the high operational costs of lamps, together with the difficult design of photoreactors and lower performances at neutral pH, are challenges for photo-Fenton full-scale applications [45], whereas the difficulty in distributing the formation of cavities in the reactor could be the limiting factor of sono-Fenton scale-up [309]. More complete considerations on AOP scalability have been discussed in [289,304], which also reported lists of pilot plants and full-scale plants of these wastewater remediation approaches.

It is clear that the overall evaluation of the feasibility of depuration procedures must involve the techno-economic aspects together with the environmental impact [101,310].

One widespread factor in evaluating the AOP processes is the consideration of the demand for energy. An example of compared systems was reported by Maniakova et al. [298]; the authors claimed that a solar-driven process catalyzed by N-doped TiO_2 was not yet competitive with the system of Fe^{2+}/H_2O_2/ethylenediamine-*N*,*N*′-disuccinic acid (EDDS, chelating agent) under sunlight in the removal of CECs from secondary treated urban wastewater because higher irradiation times were necessary, which would result in a larger surface area for solar reactors.

From an economic point of view, the criterion of Electric Energy per Order (EEO), which takes into account the electric power, the removal efficiency and the flow rate [311], could be used to compare the cost of different AOP strategies [311,312], but it does not consider other operating costs.

To have a more comprehensive overview of total process expense, several authors proposed to evaluate the sum of the single source of cost: (i) power, (ii) chemicals, and (iii) sludge/waste management (including transportation and disposal) [313,314].

A new criterion to compare AOPs has been proposed, namely the Accumulated Oxygen-equivalent Chemical-oxidation Dose (AOCD), which comprises the accumulation of the oxidant dose normalized to the treatment time. AOCD also takes into account other important parameters such as current density, irradiance, wavelength, active surface area and lamp, distance of the lamp, faradaic and quantum yields, number of electrons exchanged, residence time, etc., depending on the AOP technique considered. The development of tools for AOP comparison is fundamental to help legislators determine and choose the best available techniques for water reuse as defined by the Industrial Emissions European Union Directive 2010/75/EU [315].

Regarding the evaluation of the processes' environmental impacts, the life cycle assessment (LCA) methodology is a valuable and widespread support tool [316]. What emerges from most of the LCA outputs is that the main environmental hotspots of AOP strategies are, in order of relevance, the energy supply, the H_2O_2 employment, and, finally, the auxiliary chemicals.

Since energy is recognized as the main impacting factor, an appropriate choice of energy source is fundamental. For instance, García-Montaño et al. [317] compared different photo-Fenton strategies and noticed that the use of solar irradiation decreased the environmental impacts between 8.3 and 57.3%. Similar results have been reported by Ttofa et al. [318], who found that using solar energy to also power auxiliary equipment, such as pumps, lessened the calculated environmental impacts by up to 83%.

It can, therefore, be stated that investment in renewable energy could be effective in decreasing the environmental repercussions of AOPs.

In studies that report a lower energy imprint, the contribution of chemicals to the environmental impact grows, with a particular influence brought about by the use of hydrogen peroxide and auxiliary chemicals (i.e., NaOH necessary to reach a proper water pH after the treatment), or because of the substances released during the process [316].

From the materials' and chemicals' point of view, Costamagna et al. [319] investigated the photo-Fenton processes through LCA methodology. The photo-Fenton reaction carried out at circumneutral conditions, using the already mentioned soluble bio-organic substances (BBS) as auxiliary agents, was compared with the traditional photo-Fenton run at pH 2.8. The evaluation was carried out both at the laboratory level and at the pilot plant scale. Working in mild conditions reduced the environmental burden associated with the use of chemicals, but the decrement in effectiveness significantly increased the overall impact of the system.

In the case of heterogeneous catalysis, the focus is primarily on the sustainability of the materials, from their synthesis, possibly using greener strategies, to performances and recyclability [101,320,321]. Some studies highlight that the choice of heterogeneous Fenton can significantly reduce the environmental impact of the water treatments owing to the reuse of the catalyst. On the other hand, a higher amount of H_2O_2 seems to be required to degrade the same amount of pollutant, making the research of the least impactful compromise necessary [322].

4.3. Final Considerations

As emerges from the discussion of the recent literature, the panorama of AOP techniques is very diversified and effective for the remediation of different contaminated aqueous matrices. An example of the evolution of depuration systems over time is represented by Fenton-like processes, which can overcome the main issues of classical Fenton reactions, such as the addition of hydrogen peroxide, careful pH regulation in a narrow acid pH range and formation of sludge, although other drawbacks are still required to be solved. The regulation of the amount and type of energy involved, the low on-site production efficiency of hydrogen peroxide, the design of reactors (not only in terms of operation modes, but also chemical stability) and operational costs have to be optimized [16,27,45,323–326]. Issues related to the best compromise among efficiency, energy and chemical demand, operating and monitoring aspects together with the economic convenience must similarly be adjusted in processes based on photo-, electro-catalysis, etc.

It is, therefore, difficult to establish an absolute ranking of all the different techniques since there are a multitude of facets that should be taken into consideration. However, it is possible to summarize the main aspects emerging from the comparative reading of the literature as follows:

- The use of AOPs in mild pH conditions can reduce additional process costs related to the salinity increase, as it is induced in classic Fenton treatment by acidification and further neutralization [315];
- In situ hydrogen peroxide production avoids high costs and hazards associated with its transport, handling, and storage;

- Homogeneous processes such as ozonization have been investigated and already applied in the full-scale treatment of urban wastewater effluents [327];
- Heterogeneous catalysis shows several advantages with respect to homogeneous processes, such as the recyclability of the catalyst, which can contribute both to cost reduction and limitation of the environmental impact. However, this strategy does not seem completely ready for full-scale application due to catalyst preparation costs, effectiveness, stability, and reactor configurations [328];
- The main hotspot of the environmental impact is energy and, consequently, an appropriate choice of energy source can significantly reduce the total impact of the process; thus, investment in renewable energies should be preferred;
- Since wastewaters are characterized by a complex composition, their contamination cannot be solved by employing a single remediation technique. Effective hybrid systems combining AOP techniques and/or biological treatments are reported in the literature [315];
- AOP strategies have been reported among the best available technologies in a recent review about wastewater reuse in European countries [315];
- Limited scale-up and techno-economic analysis are available for Bio-Fenton treatments. Enzyme cost and stability could represent critical points for its real applications; however, some studies suggest that GOx production could be cost-effective and scalable [185]. Moreover, as widely described above, the immobilization of glucose oxidase improves its stability towards oxidation and deactivation due to environmental factors and ensures its reusability, making Bio-Fenton suitable for actual wastewater treatment;
- Although, to the best of the authors' knowledge, there are no reported pilot or full-scale plants that integrate Bio-Fenton in wastewater treatments, its effectiveness as a pretreatment to be coupled with conventional active sludge was highlighted by Wang et al. [189].

Author Contributions: Conceptualization, M.R., E.L. and M.L.T.; writing—original draft preparation, M.R. and M.L.T.; writing—review and editing, M.R., E.L. and M.L.T. All authors have read and agreed to the published version of the manuscript.

Funding: This research received no external funding.

Data Availability Statement: No new data were created or analyzed in this study. Data sharing is not applicable to this article.

Acknowledgments: This work was supported by Project CH4.0 under the MUR program "Dipartimenti di Eccellenza 2023–2027" (CUP: D13C22003520001).

Conflicts of Interest: The authors declare no conflicts of interest.

References

1. Hossain, M.F. Water. In *Sustainable Design and Build;* Butterworth-Heinemann: Oxford, UK, 2019; pp. 301–418. ISBN 9780128167229.
2. Velasco-Muñoz, J.; Aznar-Sánchez, J.; Belmonte-Ureña, L.; Román-Sánchez, I. Sustainable Water Use in Agriculture: A Review of Worldwide Research. *Sustainability* **2018**, *10*, 1084. [CrossRef]
3. Kokkinos, P.; Venieri, D.; Mantzavinos, D. Advanced Oxidation Processes for Water and Wastewater Viral Disinfection. A Systematic Review. *Food Environ. Virol.* **2021**, *13*, 283–302. [CrossRef]
4. Testa, M.L.; Tummino, M.L. Lignocellulose Biomass as a Multifunctional Tool for Sustainable Catalysis and Chemicals: An Overview. *Catalysts* **2021**, *11*, 125. [CrossRef]
5. Abou-Shady, A.; Siddique, M.S.; Yu, W. A Critical Review of Recent Progress in Global Water Reuse during 2019–2021 and Perspectives to Overcome Future Water Crisis. *Environments* **2023**, *10*, 159. [CrossRef]
6. Roy, A.; Sharma, A.; Yadav, S.; Jule, L.T.; Krishnaraj, R. Nanomaterials for Remediation of Environmental Pollutants. *Bioinorg. Chem. Appl.* **2021**, *2021*, 1764647. [CrossRef]
7. Lama, G.; Meijide, J.; Sanromán, A.; Pazos, M. Heterogeneous Advanced Oxidation Processes: Current Approaches for Wastewater Treatment. *Catalysts* **2022**, *12*, 344. [CrossRef]
8. Deng, Y.; Zhao, R. Advanced Oxidation Processes (AOPs) in Wastewater Treatment. *Curr. Pollut. Rep.* **2015**, *1*, 167–176. [CrossRef]

9. Arifin, M.N.; Jusoh, R.; Abdullah, H.; Ainirazali, N.; Setiabudi, H.D. Recent Advances in Advanced Oxidation Processes (AOPs) for the Treatment of Nitro- and Alkyl-Phenolic Compounds. *Environ. Res.* **2023**, *229*, 115936. [CrossRef]
10. Kumari, P.; Kumar, A. ADVANCED OXIDATION PROCESS: A Remediation Technique for Organic and Non-Biodegradable Pollutant. *Results Surf. Interfaces* **2023**, *11*, 100122. [CrossRef]
11. Garcia-Costa, A.L.; Casas, J.A. Intensification Strategies for Thermal H_2O_2-Based Advanced Oxidation Processes: Current Trends and Future Perspectives. *Chem. Eng. J. Adv.* **2022**, *9*, 100228. [CrossRef]
12. Wang, Y.; Lin, Y.; He, S.; Wu, S.; Yang, C. Singlet Oxygen: Properties, Generation, Detection, and Environmental Applications. *J. Hazard. Mater.* **2024**, *461*, 132538. [CrossRef] [PubMed]
13. Ikhlaq, A.; Brown, D.R.; Kasprzyk-Hordern, B. Mechanisms of Catalytic Ozonation: An Investigation into Superoxide Ion Radical and Hydrogen Peroxide Formation during Catalytic Ozonation on Alumina and Zeolites in Water. *Appl. Catal. B Environ.* **2013**, *129*, 437–449. [CrossRef]
14. Li, W.; Zhang, M.; Wang, H.; Lian, J.; Qiang, Z. Removal of Recalcitrant Organics in Reverse Osmosis Concentrate from Coal Chemical Industry by UV/H_2O_2 and UV/PDS: Efficiency and Kinetic Modeling. *Chemosphere* **2022**, *287*, 131999. [CrossRef] [PubMed]
15. Pandis, P.K.; Kalogirou, C.; Kanellou, E.; Vaitsis, C.; Savvidou, M.G.; Sourkouni, G.; Zorpas, A.A.; Argirusis, C. Key Points of Advanced Oxidation Processes (AOPs) for Wastewater, Organic Pollutants and Pharmaceutical Waste Treatment: A Mini Review. *ChemEngineering* **2022**, *6*, 8. [CrossRef]
16. An, J.; Li, N.; Wu, Y.; Wang, S.; Liao, C.; Zhao, Q.; Zhou, L.; Li, T.; Wang, X.; Feng, Y. Revealing Decay Mechanisms of H_2O_2-Based Electrochemical Advanced Oxidation Processes after Long-Term Operation for Phenol Degradation. *Environ. Sci. Technol.* **2020**, *54*, 10916–10925. [CrossRef] [PubMed]
17. Wang, X.; Jing, J.; Zhou, M.; Dewil, R. Recent Advances in H_2O_2-Based Advanced Oxidation Processes for Removal of Antibiotics from Wastewater. *Chin. Chem. Lett.* **2023**, *34*, 107621. [CrossRef]
18. Tummino, M.L. $SrFeO_3$ Peculiarities and Exploitation in Decontamination Processes and Environmentally-Friendly Energy Applications. *Curr. Res. Green Sustain. Chem.* **2022**, *5*, 100339. [CrossRef]
19. Qu, C.; Liang, D.W. Novel Electrochemical Advanced Oxidation Processes with H_2O_2 Generation Cathode for Water Treatment: A Review. *J. Environ. Chem. Eng.* **2022**, *10*, 107896. [CrossRef]
20. Pi, L.; Cai, J.; Xiong, L.; Cui, J.; Hua, H.; Tang, D.; Mao, X. Generation of H2O2 by On-Site Activation of Molecular Dioxygen for Environmental Remediation Applications: A Review. *Chem. Eng. J.* **2020**, *389*, 123420. [CrossRef]
21. Wang, L.; Zhang, J.; Zhang, Y.; Yu, H.; Qu, Y.; Yu, J. Inorganic Metal-Oxide Photocatalyst for H_2O_2 Production. *Small* **2022**, *18*, 2104561. [CrossRef]
22. Mansour, M.S.; Farid, Y.; Nosier, S.A.; Adli, O.; Abdel-Aziz, M.H. Removal of Eosin Yellow Dye from Industrial Wastewater Using UV/H2O2 and Photoelectro-Fenton Techniques. *J. Photochem. Photobiol. A Chem.* **2023**, *436*, 114411. [CrossRef]
23. Wang, J.; Li, C.; Rauf, M.; Luo, H.; Sun, X.; Jiang, Y. Gas Diffusion Electrodes for H2O2 Production and Their Applications for Electrochemical Degradation of Organic Pollutants in Water: A Review. *Sci. Total Environ.* **2021**, *759*, 143459. [CrossRef]
24. Muruganandham, M.; Suri, R.P.S.; Jafari, S.; Sillanpää, M.; Lee, G.J.; Wu, J.J.; Swaminathan, M. Recent Developments in Homogeneous Advanced Oxidation Processes for Water and Wastewater Treatment. *Int. J. Photoenergy* **2014**, *2014*, 821674. [CrossRef]
25. Asghar, A.; Raman, A.A.A.; Daud, W.M.A.W. Advanced Oxidation Processes for In-Situ Production of Hydrogen Peroxide/Hydroxyl Radical for Textile Wastewater Treatment: A Review. *J. Clean. Prod.* **2015**, *87*, 826–838. [CrossRef]
26. Ji, J.; Wang, Z.; Xu, Q.; Zhu, Q.; Xing, M. In Situ H_2O_2 Generation and Corresponding Pollutant Removal Applications: A Review. *Chem.—A Eur. J.* **2023**, *29*, e202203921. [CrossRef]
27. Liu, Y.; Zhao, Y.; Wang, J. Fenton/Fenton-like Processes with in-Situ Production of Hydrogen Peroxide/Hydroxyl Radical for Degradation of Emerging Contaminants: Advances and Prospects. *J. Hazard. Mater.* **2021**, *404*, 124191. [CrossRef]
28. Popat, A.; Nidheesh, P.V.; Anantha Singh, T.S.; Suresh Kumar, M. Mixed Industrial Wastewater Treatment by Combined Electrochemical Advanced Oxidation and Biological Processes. *Chemosphere* **2019**, *237*, 124419. [CrossRef]
29. Rezaei, F.; Vione, D. Effect of PH on Zero Valent Iron Performance in Heterogeneous Fenton and Fenton-like Processes: A Review. *Molecules* **2018**, *23*, 3127. [CrossRef]
30. Kamali, M.; Aminabhavi, T.M.; Costa, M.E.V.; Ul Islam, S.; Appels, L.; Dewil, R. Heterogeneous Advanced Oxidation Processes (HE-AOPs) for the Removal of Pharmaceutically Active Compounds—Pros and Cons. In *Advanced Wastewater Treatment Technologies for the Removal of Pharmaceutically Active Compounds. Green Energy and Technology*; Springer: Cham, Switzerland, 2023; pp. 211–239.
31. Rodríguez-Chueca, J.; Carbajo, J.; García-Muñoz, P. Intensification of Photo-Assisted Advanced Oxidation Processes for Water Treatment: A Critical Review. *Catalysts* **2023**, *13*, 401. [CrossRef]
32. Mazumder, A.; Bhattacharya, S.; Bhattacharjee, C. Role of Nano-Photocatalysis in Heavy Metal Detoxification. In *Nanophotocatalysis and Environmental Applications*; Springer Nature Switzerland: Cham, Switzerland, 2020; pp. 1–33.
33. Tseng, D.H.; Juang, L.C.; Huang, H.H. Effect of Oxygen and Hydrogen Peroxide on the Photocatalytic Degradation of Monochlorobenzene in TiO2 Aqueous Suspension. *Int. J. Photoenergy* **2012**, *2012*, 328526. [CrossRef]
34. Huang, X.; Song, M.; Zhang, J.; Shen, T.; Luo, G.; Wang, D. Recent Advances of Electrocatalyst and Cell Design for Hydrogen Peroxide Production. *Nano-Micro Lett.* **2023**, *15*, 86. [CrossRef]

35. Kim, D.J.; Zhu, Q.; Rigby, K.; Wu, X.; Kim, J.H.; Kim, J.-H. A Protocol for Electrocatalyst Stability Evaluation: H_2O_2 Electrosynthesis for Industrial Wastewater Treatment. *Environ. Sci. Technol.* **2022**, *56*, 1365–1375. [CrossRef]
36. Zhou, W.; Meng, X.; Gao, J.; Alshawabkeh, A.N. Hydrogen Peroxide Generation from O2 Electroreduction for Environmental Remediation: A State-of-the-Art Review. *Chemosphere* **2019**, *225*, 588–607. [CrossRef]
37. Knowles, R.R. Reaching Your Full (Over)Potential: A Novel Approach to Electrocatalytic Oxygen Reduction. *ACS Cent. Sci.* **2015**, *1*, 224–225. [CrossRef]
38. Chen, Z.; Yao, D.; Chu, C.; Mao, S. Photocatalytic H2O2 Production Systems: Design Strategies and Environmental Applications. *Chem. Eng. J.* **2023**, *451*, 138489. [CrossRef]
39. An, J.; Feng, Y.; Zhao, Q.; Wang, X.; Liu, J.; Li, N. Electrosynthesis of H2O2 through a Two-Electron Oxygen Reduction Reaction by Carbon Based Catalysts: From Mechanism, Catalyst Design to Electrode Fabrication. *Environ. Sci. Ecotechnol.* **2022**, *11*, 100170. [CrossRef]
40. Qian, W.; Wu, Z.; Jia, Y.; Hong, Y.; Xu, X.; You, H.; Zheng, Y.; Xia, Y. Thermo-Electrochemical Coupling for Room Temperature Thermocatalysis in Pyroelectric ZnO Nanorods. *Electrochem. Commun.* **2017**, *81*, 124–127. [CrossRef]
41. Luo, H.; Zeng, Y.; He, D.; Pan, X. Application of Iron-Based Materials in Heterogeneous Advanced Oxidation Processes for Wastewater Treatment: A Review. *Chem. Eng. J.* **2021**, *407*, 127191. [CrossRef]
42. Litter, M.I.; Slodowicz, M. An Overview on Heterogeneous Fenton and PhotoFenton Reactions Using Zerovalent Iron Materials. *J. Adv. Oxid. Technol.* **2017**, *20*, 20160164. [CrossRef]
43. Clarizia, L.; Russo, D.; Di Somma, I.; Marotta, R.; Andreozzi, R. Homogeneous Photo-Fenton Processes at near Neutral PH: A Review. *Appl. Catal. B Environ.* **2017**, *209*, 358–371. [CrossRef]
44. Rodríguez-Narváez, O.M.; Pérez, L.S.; Yee, N.G.; Peralta-Hernández, J.M.; Bandala, E.R. Comparison between Fenton and Fenton-like Reactions for l-Proline Degradation. *Int. J. Environ. Sci. Technol.* **2019**, *16*, 1515–1526. [CrossRef]
45. Wang, N.; Zheng, T.; Zhang, G.; Wang, P. A Review on Fenton-like Processes for Organic Wastewater Treatment. *J. Environ. Chem. Eng.* **2016**, *4*, 762–787. [CrossRef]
46. Jiang, C.; Gao, Z.; Qu, H.; Li, J.; Wang, X.; Li, P.; Liu, H. A New Insight into Fenton and Fenton-like Processes for Water Treatment: Part II. Influence of Organic Compounds on Fe(III)/Fe(II) Interconversion and the Course of Reactions. *J. Hazard. Mater.* **2013**, *250–251*, 76–81. [CrossRef]
47. Lai, L.; He, Y.; Zhou, H.; Huang, B.; Yao, G.; Lai, B. Critical Review of Natural Iron-Based Minerals Used as Heterogeneous Catalysts in Peroxide Activation Processes: Characteristics, Applications and Mechanisms. *J. Hazard. Mater.* **2021**, *416*, 125809. [CrossRef]
48. Kahoush, M.; Behary, N.; Cayla, A.; Nierstrasz, V. Bio-Fenton and Bio-Electro-Fenton as Sustainable Methods for Degrading Organic Pollutants in Wastewater. *Process Biochem.* **2018**, *64*, 237–247. [CrossRef]
49. Ahuja, D.K.; Bachas, L.G.; Bhattacharyya, D. Modified Fenton Reaction for Trichlorophenol Dechlorination by Enzymatically Generated H_2O_2 and Gluconic Acid Chelate. *Chemosphere* **2007**, *66*, 2193–2200. [CrossRef]
50. Tummino, M.L.; Nisticò, R.; Franzoso, F.; Bianco Prevot, A.; Calza, P.; Laurenti, E.; Paganini, M.C.; Scalarone, D.; Magnacca, G. The "Lab4treat" Outreach Experience: Preparation of Sustainable Magnetic Nanomaterials for Remediation of Model Wastewater. *Molecules* **2021**, *26*, 3361. [CrossRef]
51. Zhang, W.; Jiang, X.; Ralston, J.; Cao, J.; Jin, X.; Sun, W.; Gao, Z. Efficient Heterogeneous Photodegradation of Eosin Y by Oxidized Pyrite Using the Photo-Fenton Process. *Miner. Eng.* **2023**, *191*, 107972. [CrossRef]
52. Guo, S.; Huang, R.; Yuan, J.; Chen, R.; Chen, F. Efficient Removal of Aromatic Pollutants via Catalytic Wet Peroxide Oxidation over Synthetic Anisotropic Ilmenite/Carbon Nanocomposites. *npj Clean Water* **2023**, *6*, 74. [CrossRef]
53. Thomas, N.; Dionysiou, D.D.; Pillai, S.C. Heterogeneous Fenton Catalysts: A Review of Recent Advances. *J. Hazard. Mater.* **2021**, *404*, 124082. [CrossRef]
54. Wang, J.; Farias, J.; Tiwary, A.; Tangyie, G.C.; Huddersman, K. Advance Oxidation Process (AOP) of Bisphenol A Using a Novel Surface-Functionalised Polyacrylonitrile (PAN) Fibre Catalyst. *Water* **2022**, *14*, 640. [CrossRef]
55. Liu, Y.; Liu, Y.; Yang, Z.; Wang, J. Fenton Degradation of 4-Chlorophenol Using H2O2: In Situ Generated by Zn-CNTs/O2 System. *RSC Adv.* **2017**, *7*, 49985–49994. [CrossRef]
56. Zhao, P.; Jin, B.; Zhang, Q.; Peng, R. Graphitic-C3N4 Quantum Dots Modified FeOOH for Photo-Fenton Degradation of Organic Pollutants. *Appl. Surf. Sci.* **2022**, *586*, 152792. [CrossRef]
57. Shi, W.; Sun, W.; Liu, Y.; Zhang, K.; Sun, H.; Lin, X.; Hong, Y.; Guo, F. A Self-Sufficient Photo-Fenton System with Coupling in-Situ Production H2O2 of Ultrathin Porous g-C3N4 Nanosheets and Amorphous FeOOH Quantum Dots. *J. Hazard. Mater.* **2022**, *436*, 129141. [CrossRef]
58. Wang, Y.; Song, H.; Chen, J.; Chai, S.; Shi, L.; Chen, C.; Wang, Y.; He, C. A Novel Solar Photo-Fenton System with Self-Synthesizing H2O2: Enhanced Photo-Induced Catalytic Performances and Mechanism Insights. *Appl. Surf. Sci.* **2020**, *512*, 145650. [CrossRef]
59. Calza, P.; Di Sarro, J.; Magnacca, G.; Bianco Prevot, A.; Laurenti, E. Use of Low-Cost Magnetic Materials Containing Waste Derivatives for the (Photo)-Fenton Removal of Organic Pollutants. *Materials* **2019**, *12*, 3942. [CrossRef]
60. Chavan, R.; Bhat, N.; Parit, S.; Narasimharao, K.; Devan, R.S.; Patil, R.B.; Karade, V.C.; Pawar, N.V.; Kim, J.H.; Jadhav, J.P.; et al. Development of Magnetically Recyclable Nanocatalyst for Enhanced Fenton and Photo-Fenton Degradation of MB and Cr(VI) Photo-Reduction. *Mater. Chem. Phys.* **2023**, *293*, 126964. [CrossRef]

61. Bousalah, D.; Zazoua, H.; Boudjemaa, A.; Benmounah, A.; Messaoud-Boureghda, M.Z.; Bachari, K. Enhanced Reactivity of the CuO-Fe_2O_3 Intimate Heterojunction for the Oxidation of Quinoline Yellow Dye (E104). *Environ. Sci. Pollut. Res.* **2022**, *29*, 69988–69999. [CrossRef]
62. Mirsadeghi, S.; Zandavar, H.; Rajabi, H.R.; Sajadiasl, F.; Ganjali, M.R.; Pourmortazavi, S.M. Superior Degradation of Organic Pollutants and H_2O_2 Generation Ability on Environmentally-Sound Constructed Fe_3O_4-Cu Nanocomposite. *J. Mater. Res. Technol.* **2021**, *14*, 808–821. [CrossRef]
63. Domenzain-Gonzalez, J.; Castro-Arellano, J.J.; Galicia-Luna, L.A.; Rodriguez-Cruz, M.; Hernandez-Lopez, R.T.; Lartundo-Rojas, L. Photocatalytic Membrane Reactor Based on Mexican Natural Zeolite: RB5 Dye Removal by Photo-Fenton Process. *J. Environ. Chem. Eng.* **2021**, *9*, 105281. [CrossRef]
64. Ju, Y.; Li, H.; Wang, Z.; Liu, H.; Huo, S.; Jiang, S.; Duan, S.; Yao, Y.; Lu, X.; Chen, F. Solar-Driven on-Site H_2O_2 Generation and Tandem Photo-Fenton Reaction on a Triphase Interface for Rapid Organic Pollutant Degradation. *Chem. Eng. J.* **2022**, *430*, 133168. [CrossRef]
65. Chen, Z.; Lian, C.; Huang, K.; Ji, J.; Yan, Q.; Zhang, J.; Xing, M. "Small Amount for Multiple Times" of H_2O_2 Feeding Way in MoS2-Fex Heterogeneous Fenton for Enhancing Sulfadiazine Degradation. *Chin. Chem. Lett.* **2022**, *33*, 1365–1372. [CrossRef]
66. Luo, H.; Cheng, Y.; Zeng, Y.; Luo, K.; Pan, X. Enhanced Decomposition of H2O2 by Molybdenum Disulfide in a Fenton-like Process for Abatement of Organic Micropollutants. *Sci. Total Environ.* **2020**, *732*, 139335. [CrossRef] [PubMed]
67. Kim, H.H.; Lee, H.; Lee, D.; Ko, Y.J.; Woo, H.; Lee, J.; Lee, C.; Pham, A.L.T. Activation of Hydrogen Peroxide by a Titanium Oxide-Supported Iron Catalyst: Evidence for Surface Fe(IV) and Its Selectivity. *Environ. Sci. Technol.* **2020**, *54*, 15424–15432. [CrossRef] [PubMed]
68. Xu, L.; Zhang, X.; Han, J.; Gong, H.; Meng, L.; Mei, X.; Sun, Y.; Qi, L.; Gan, L. Degradation of Emerging Contaminants by Sono-Fenton Process with in Situ Generated H_2O_2 and the Improvement by P25-Mediated Visible Light Irradiation. *J. Hazard. Mater.* **2020**, *391*, 122229. [CrossRef] [PubMed]
69. Nie, M.; Li, Y.; Li, L.; He, J.; Hong, P.; Zhang, K.; Cai, X.; Kong, L.; Liu, J. Ultrathin Iron-Cobalt Oxide Nanosheets with Enhanced H_2O_2 Activation Performance for Efficient Degradation of Tetracycline. *Appl. Surf. Sci.* **2021**, *535*, 147655. [CrossRef]
70. Liu, Q.; Zhou, L.; Liu, L.; Li, J.; Wang, S.; Znad, H.; Liu, S. Magnetic ZnO@Fe_3O_4 Composite for Self-Generated H2O2 toward Photo-Fenton-like Oxidation of Nitrophenol. *Compos. Part B Eng.* **2020**, *200*, 108345. [CrossRef]
71. Cao, Y.; Ren, Y.; Zhang, J.; Xie, T.; Lin, Y. Activation of H_2O_2 by Photo-Generated Electrons for Enhanced Visible Light Driven Methylene Blue Degradation with $ZnFe_2O_4/BiVO_4$ Heterojunction. *Opt. Mater.* **2021**, *121*, 111637. [CrossRef]
72. Deganello, F.; Tummino, M.L.; Calabrese, C.; Testa, M.L.; Avetta, P.; Fabbri, D.; Prevot, A.B.; Montoneri, E.; Magnacca, G. A New, Sustainable $LaFeO_3$ Material Prepared from Biowaste-Sourced Soluble Substances. *New J. Chem.* **2015**, *39*, 877–885. [CrossRef]
73. Bianco Prevot, A.; Fabbri, D.; Bernardini, E.; Deganello, F.; Tummino, M.L.; Magnacca, G. Insights on the Photocatalytic Performances of LaFeO3 Synthesized by Solution Combustion Synthesis. In *Materials Science in Photocatalysis*; Elsevier: Amsterdam, The Netherlands, 2021; pp. 357–370. ISBN 9780128218594.
74. Tummino, M.L.; Vineis, C.; Varesano, A.; Liotta, L.F.; Rigoletto, M.; Laurenti, E.; Deganello, F. $Sr_{0.85}Ce_{0.15}Fe_{0.67}Co_{0.33}$-$XCu_xO_3$ Perovskite Oxides: Effect of B-Site Copper Codoping on the Physicochemical, Catalytic and Antibacterial Properties upon UV or Thermal Activation. *Front. Environ. Eng.* **2023**, *2*, 1249931. [CrossRef]
75. Luo, Z.; Liu, M.; Tang, D.; Xu, Y.; Ran, H.; He, J.; Chen, K.; Sun, J. High H_2O_2 Selectivity and Enhanced Fe^{2+} Regeneration toward an Effective Electro-Fenton Process Based on a Self-Doped Porous Biochar Cathode. *Appl. Catal. B Environ.* **2022**, *315*, 121523. [CrossRef]
76. Mafo, S.G.M.; Tchuifon, D.R.T.; Ngakou, C.S.; Fotsop, C.G.; Kouteu, P.A.N.; Doungmo, G.; Ndifor-Angwafor, G.N.; Anagho, S.G. Study of the Degradation of Bezaktiv Brilliant Blue by the Fenton Process Using a Prepared Ferromagnetic Activated Carbon from Rubber Seed Hull as Heterogeneous Catalyst. *Desalin. Water Treat.* **2023**, *287*, 200–213. [CrossRef]
77. Guo, Q.; Zhu, W.; Yang, D.; Wang, X.; Li, Y.; Gong, C.; Yan, J.; Zhai, J.; Gao, X.; Luo, Y. A Green Solar Photo-Fenton Process for the Degradation of Carbamazepine Using Natural Pyrite and Organic Acid with in-Situ Generated H_2O_2. *Sci. Total Environ.* **2021**, *784*, 147187. [CrossRef] [PubMed]
78. Ding, J.; Shen, L.; Yan, R.; Lu, S.; Zhang, Y.; Zhang, X.; Zhang, H. Heterogeneously Activation of H_2O_2 and Persulfate with Goethite for Bisphenol A Degradation: A Mechanistic Study. *Chemosphere* **2020**, *261*, 127715. [CrossRef] [PubMed]
79. Ahmed, N.; Vione, D.; Rivoira, L.; Castiglioni, M.; Beldean-Galea, M.S.; Bruzzoniti, M.C. Feasibility of a Heterogeneous Nanoscale Zero-Valent Iron Fenton-like Process for the Removal of Glyphosate from Water. *Molecules* **2023**, *28*, 2214. [CrossRef] [PubMed]
80. Deganello, F.; Oko, D.N.; Testa, M.L.; La Parola, V.; Tummino, M.L.; Soares, C.O.; Rivera, J.G.; Orozco, G.; Guay, D.; Tavares, A.C. Perovskite-Type Catalysts Prepared by Nanocasting: Effect of Metal Silicates on the Electrocatalytic Activity toward Oxygen Evolution and Reduction Reactions. *ACS Appl. Energy Mater.* **2018**, *1*, 2565–2575. [CrossRef]
81. Huang, J.; Climent, V.; Groß, A.; Feliu, J.M. Understanding Surface Charge Effects in Electrocatalysis. Part 2: Hydrogen Peroxide Reactions at Platinum. *Chin. J. Catal.* **2022**, *43*, 2837–2849. [CrossRef]
82. Farhadian, N.; Liu, S.; Asadi, A.; Shahlaei, M.; Moradi, S. Enhanced Heterogeneous Fenton Oxidation of Organic Pollutant via Fe-Containing Mesoporous Silica Composites: A Review. *J. Mol. Liq.* **2021**, *321*, 114896. [CrossRef]
83. Salunkhe, T.T.; Gurugubelli, T.R.; Babu, B.; Yoo, K. Recent Innovative Progress of Metal Oxide Quantum-Dot-Integrated g-C3N4 (0D-2D) Synergistic Nanocomposites for Photocatalytic Applications. *Catalysts* **2023**, *13*, 1414. [CrossRef]

84. Liu, W.; Li, Z.; Kang, Q.; Wen, L. Efficient Photocatalytic Degradation of Doxycycline by Coupling α-Bi_2O_3/g-C_3N_4 Composite and H_2O_2 under Visible Light. *Environ. Res.* **2021**, *197*, 110925. [CrossRef]
85. Adhikari, S.; Lee, H.H.; Kim, D.-H. Efficient Visible-Light Induced Electron-Transfer in z-Scheme MoO_3/Ag/C_3N_4 for Excellent Photocatalytic Removal of Antibiotics of Both Ofloxacin and Tetracycline. *Chem. Eng. J.* **2020**, *391*, 123504. [CrossRef]
86. He, Y.Q.; Zhang, F.; Ma, B.; Xu, N.; Binnah Junior, L.; Yao, B.; Yang, Q.; Liu, D.; Ma, Z. Remarkably Enhanced Visible-Light Photocatalytic Hydrogen Evolution and Antibiotic Degradation over g-C_3N_4 Nanosheets Decorated by Using Nickel Phosphide and Gold Nanoparticles as Cocatalysts. *Appl. Surf. Sci.* **2020**, *517*, 146187. [CrossRef]
87. Jourshabani, M.; Asrami, M.R.; Lee, B.-K. An Efficient and Unique Route for the Fabrication of Highly Condensed Oxygen-Doped Carbon Nitride for the Photodegradation of Synchronous Pollutants and H_2O_2 Production under Ambient Conditions. *Appl. Catal. B Environ.* **2022**, *302*, 120839. [CrossRef]
88. Cui, Y.; Ding, Z.; Liu, P.; Antonietti, M.; Fu, X.; Wang, X. Metal-Free Activation of H_2O_2 by g-C_3N_4 under Visible Light Irradiation for the Degradation of Organic Pollutants. *Phys. Chem. Chem. Phys.* **2012**, *14*, 1455–1462. [CrossRef] [PubMed]
89. Chen, Q.-L.; Liu, Y.-L.; Tong, L.-G. Enhanced Visible Light Photocatalytic Activity of G-C3N4 Assisted by Hydrogen Peroxide. *Mater. Res. Express* **2018**, *5*, 046203. [CrossRef]
90. Li, T.; Ge, L.; Peng, X.; Wang, W.; Zhang, W. Enhanced Degradation of Sulfamethoxazole by a Novel Fenton-like System with Significantly Reduced Consumption of H_2O_2 Activated by g-C_3N_4/MgO Composite. *Water Res.* **2021**, *190*, 116777. [CrossRef] [PubMed]
91. Zhou, L.; Lei, J.; Wang, F.; Wang, L.; Hoffmann, M.R.; Liu, Y.; In, S.-I.; Zhang, J. Carbon Nitride Nanotubes with in Situ Grafted Hydroxyl Groups for Highly Efficient Spontaneous H_2O_2 Production. *Appl. Catal. B Environ.* **2021**, *288*, 119993. [CrossRef]
92. Torres-Pinto, A.; Sampaio, M.J.; Teixo, J.; Silva, C.G.; Faria, J.L.; Silva, A.M.T. Photo-Fenton Degradation Assisted by in Situ Generation of Hydrogen Peroxide Using a Carbon Nitride Photocatalyst. *J. Water Process Eng.* **2020**, *37*, 101467. [CrossRef]
93. Yang, Z.; Zhou, S.; Feng, X.; Wang, N.; Ola, O.; Zhu, Y. Recent Progress in Multifunctional Graphene-Based Nanocomposites for Photocatalysis and Electrocatalysis Application. *Nanomaterials* **2023**, *13*, 2028. [CrossRef] [PubMed]
94. Frindy, S.; Sillanpää, M. Synthesis and Application of Novel α-Fe_2O_3/Graphene for Visible-Light Enhanced Photocatalytic Degradation of RhB. *Mater. Des.* **2020**, *188*, 108461. [CrossRef]
95. Shi, W.; Wang, L.; Wang, J.; Sun, H.; Shi, Y.; Guo, F.; Lu, C. Magnetically Retrievable CdS/Reduced Graphene Oxide/$ZnFe_2O_4$ Ternary Nanocomposite for Self-Generated H_2O_2 towards Photo-Fenton Removal of Tetracycline under Visible Light Irradiation. *Sep. Purif. Technol.* **2022**, *292*, 120987. [CrossRef]
96. Tantubay, K.; Das, P.; Baskey (Sen), M. Hydrogen Peroxide–Assisted Photocatalytic Dye Degradation over Reduced Graphene Oxide Integrated $ZnCr_2O_4$ Nanoparticles. *Environ. Sci. Pollut. Res.* **2022**, *29*, 17309–17318. [CrossRef] [PubMed]
97. Wang, Y.; Chen, J.; Gao, J.; Meng, H.; Chai, S.; Jian, Y.; Shi, L.; Wang, Y.; He, C. Selective Electrochemical H2O2 Generation on the Graphene Aerogel for Efficient Electro-Fenton Degradation of Ciprofloxacin. *Sep. Purif. Technol.* **2021**, *272*, 118884. [CrossRef]
98. Abdelfatah, A.M.; El-Maghrabi, N.; Mahmoud, A.E.D.; Fawzy, M. Synergetic Effect of Green Synthesized Reduced Graphene Oxide and Nano-Zero Valent Iron Composite for the Removal of Doxycycline Antibiotic from Water. *Sci. Rep.* **2022**, *12*, 19372. [CrossRef] [PubMed]
99. Su, P.; Zhou, M.; Song, G.; Du, X.; Lu, X. Efficient H_2O_2 Generation and Spontaneous •OH Conversion for In-Situ Phenol Degradation on Nitrogen-Doped Graphene: Pyrolysis Temperature Regulation and Catalyst Regeneration Mechanism. *J. Hazard. Mater.* **2020**, *397*, 122681. [CrossRef] [PubMed]
100. Su, P.; Zhou, M.; Lu, X.; Yang, W.; Ren, G.; Cai, J. Electrochemical Catalytic Mechanism of N-Doped Graphene for Enhanced H_2O_2 Yield and in-Situ Degradation of Organic Pollutant. *Appl. Catal. B Environ.* **2019**, *245*, 583–595. [CrossRef]
101. Tummino, M.L.; Varesano, A.; Copani, G.; Vineis, C. A Glance at Novel Materials, from the Textile World to Environmental Remediation. *J. Polym. Environ.* **2023**, *31*, 2826–2854. [CrossRef]
102. Umar, E.; Ikram, M.; Haider, J.; Nabgan, W.; Imran, M.; Nazir, G. 3D Graphene-Based Material: Overview, Perspective, Advancement, Energy Storage, Biomedical Engineering and Environmental Applications a Bibliometric Analysis. *J. Environ. Chem. Eng.* **2023**, *11*, 110339. [CrossRef]
103. Heidarinejad, Z.; Dehghani, M.H.; Heidari, M.; Javedan, G.; Ali, I.; Sillanpää, M. Methods for Preparation and Activation of Activated Carbon: A Review. *Environ. Chem. Lett.* **2020**, *18*, 393–415. [CrossRef]
104. Tan, X.F.; Liu, S.B.; Liu, Y.G.; Gu, Y.L.; Zeng, G.M.; Hu, X.J.; Wang, X.; Liu, S.H.; Jiang, L.H. Biochar as Potential Sustainable Precursors for Activated Carbon Production: Multiple Applications in Environmental Protection and Energy Storage. *Bioresour. Technol.* **2017**, *227*, 359–372. [CrossRef]
105. Ribeiro, R.S.; Silva, A.M.T.; Figueiredo, J.L.; Faria, J.L.; Gomes, H.T. Catalytic Wet Peroxide Oxidation: A Route towards the Application of Hybrid Magnetic Carbon Nanocomposites for the Degradation of Organic Pollutants. A Review. *Appl. Catal. B Environ.* **2016**, *187*, 428–460. [CrossRef]
106. Zhang, X.; Sun, P.; Wei, K.; Huang, X.; Zhang, X. Enhanced H_2O_2 Activation and Sulfamethoxazole Degradation by Fe-Impregnated Biochar. *Chem. Eng. J.* **2020**, *385*, 123921. [CrossRef]
107. Li, X.; Jia, Y.; Zhang, J.; Qin, Y.; Wu, Y.; Zhou, M.; Sun, J. Efficient Removal of Tetracycline by H_2O_2 Activated with Iron-Doped Biochar: Performance, Mechanism, and Degradation Pathways. *Chin. Chem. Lett.* **2022**, *33*, 2105–2110. [CrossRef]

108. Bankole, O.M.; Adanlawo, O.S.; Ojubola, K.I.; Adeyemi, F.O.; Achadu, O.J.; Ogunniyi, J.A.; Olaseni, S.E.; Ogunlaja, A.S. Facile Synthesis of Solar Active Charcoal Passivated Ag_3PO_4 and Their Two-Channel Mechanisms for H_2O_2 Formation in Aerated Water. *J. Mol. Struct.* **2024**, *1300*, 137264. [CrossRef]
109. Compton, P.; Dehkordi, N.R.; Larese Casanova, P.; Alshawabkeh, A.N. Activated Carbon Modifications for Heterogeneous Fenton-Like Catalysis. *J. Chem. Eng. Catal.* **2022**, *1*, 203. [PubMed]
110. Chen, X.; Wang, L.; Sun, W.; Yang, Z.; Jin, J.; Huang, Y.; Liu, G. Boron Bifunctional Catalysts for Rapid Degradation of Persistent Organic Pollutants in a Metal-Free Electro-Fenton Process: O_2 and H_2O_2 Activation Process. *Environ. Sci. Technol.* **2023**, *57*, 15693–15702. [CrossRef]
111. Wu, Y.; Gao, Z.; Feng, Y.; Cui, Q.; Du, C.; Yu, C.; Liang, L.; Zhao, W.; Feng, J.; Sun, J.; et al. Harnessing Selective and Durable Electrosynthesis of H_2O_2 over Dual-Defective Yolk-Shell Carbon Nanosphere toward on-Site Pollutant Degradation. *Appl. Catal. B Environ.* **2021**, *298*, 120572. [CrossRef]
112. Yang, G.; Mo, S.; Xing, B.; Dong, J.; Song, X.; Liu, X.; Yuan, J. Effective Degradation of Phenol via Catalytic Wet Peroxide Oxidation over N, S, and Fe-Tridoped Activated Carbon. *Environ. Pollut.* **2020**, *258*, 113687. [CrossRef]
113. Fan, X.; Cao, Q.; Meng, F.; Song, B.; Bai, Z.; Zhao, Y.; Chen, D.; Zhou, Y.; Song, M. A Fenton-like System of Biochar Loading Fe–Al Layered Double Hydroxides (FeAl-LDH@BC)/H_2O_2 for Phenol Removal. *Chemosphere* **2021**, *266*, 128992. [CrossRef]
114. Fu, D.; Kurniawan, T.A.; Gui, H.; Li, H.; Feng, S.; Li, Q.; Wang, Y. Role of Cu x O-Anchored Pyrolyzed Hydrochars on H_2O_2 -Activated Degradation of Tetracycline: Effects of Pyrolysis Temperature and PH. *Ind. Eng. Chem. Res.* **2022**, *61*, 8847–8857. [CrossRef]
115. Liu, Z.; Meng, H.; Li, C.; Zhang, H.; Cao, J.; Meng, G. Degradation of Biologically Treated Coking Wastewater over CuOx/PAC, CuOx/GAC, and CuOx/ACF Catalysts under Microwave Irradiation in the Presence of H_2O_2. *J. Environ. Eng.* **2020**, *146*. [CrossRef]
116. Dong, H.; Zou, Y.; Zhang, K.; Sun, Y.; Hui, B.; Yang, D.; Cai, L.; Li, J. Biomimetic Design of Wood Carbon-Based Heterogeneous Catalysts for Enhanced Organic Pollutants Degradation. *Chem. Eng. J.* **2023**, *451*, 138568. [CrossRef]
117. Li, Z.; Liu, F.; Ding, Y.; Wang, F.; You, H.; Jin, C. Preparation and Properties of Cu-Ni Bimetallic Oxide Catalyst Supported on Activated Carbon for Microwave Assisted Catalytic Wet Hydrogen Peroxide Oxidation for Biologically Pretreated Coal Chemical Industry Wastewater Treatment. *Chemosphere* **2019**, *214*, 17–24. [CrossRef] [PubMed]
118. Sharma, G.; Dionysiou, D.D.; Sharma, S.; Kumar, A.; Al-Muhtaseb, A.H.; Naushad, M.; Stadler, F.J. Highly Efficient Sr/Ce/Activated Carbon Bimetallic Nanocomposite for Photoinduced Degradation of Rhodamine B. *Catal. Today* **2019**, *335*, 437–451. [CrossRef]
119. Valtchev, V.; Mintova, S. Zeolites and MOFs? In *Zeolites and Metal-Organic Frameworks*; Amsterdam University Press: Amsterdam, The Netherlands, 2018; pp. 13–24.
120. Rashed, M.N.; Palanisamy, P.N. Introductory Chapter: Adsorption and Ion Exchange Properties of Zeolites for Treatment of Polluted Water. In *Zeolites and Their Applications*; InTechOpen: London, UK, 2018.
121. Zhang, W.; Taheri-Ledari, R.; Saeidirad, M.; Qazi, F.S.; Kashtiaray, A.; Ganjali, F.; Tian, Y.; Maleki, A. Regulation of Porosity in MOFs: A Review on Tunable Scaffolds and Related Effects and Advances in Different Applications. *J. Environ. Chem. Eng.* **2022**, *10*, 108836. [CrossRef]
122. Jalali, S.; Ardjmand, M.; Ramavandi, B.; Nosratinia, F. Elimination of Amoxicillin Using Zeolite Y-Sea Salt as a Good Catalyst for Activation of Hydrogen Peroxide: Investigating Degradation Pathway and the Effect of Wastewater Chemistry. *J. Environ. Manag.* **2022**, *302*, 114045. [CrossRef]
123. Edebali, S. Synthesis and Characterization of MIL-101 (Fe) as Efficient Catalyst for Tetracycline Degradation by Using NaBH4: Artificial Neural Network Modeling. *Appl. Surf. Sci. Adv.* **2023**, *18*, 100496. [CrossRef]
124. Liú, D.; Wang, G.; Liŭ, D.; Lin, J.; He, Y.; Li, X.; Li, Z. Photocatalysis Using Zero-Valent Nano-Copper for Degrading Methyl Orange under Visible Light Irradiation. *Opt. Mater.* **2016**, *53*, 155–159. [CrossRef]
125. Liu, Y.; Guo, J.; Chen, Y.; Tan, N.; Wang, J. High-Efficient Generation of H_2O_2 by Aluminum-Graphite Composite through Selective Oxygen Reduction for Degradation of Organic Contaminants. *Environ. Sci. Technol.* **2020**, *54*, 14085–14095. [CrossRef]
126. Tan, N.; Yang, Z.; Gong, X.B.; Wang, Z.R.; Fu, T.; Liu, Y. In Situ Generation of H_2O_2 Using MWCNT-Al/O_2 System and Possible Application for Glyphosate Degradation. *Sci. Total Environ.* **2019**, *650*, 2567–2576. [CrossRef]
127. Lin, C.-C.; Cheng, Y.-J. Effectiveness of Using Nanoscale Zero-Valent Iron and Hydrogen Peroxide in Degrading Sulfamethazine in Water. *J. Taiwan Inst. Chem. Eng.* **2021**, *118*, 179–186. [CrossRef]
128. Zhang, J.; Wu, Y.; Liu, L.; Lan, Y. Rapid Removal of P-Chloronitrobenzene from Aqueous Solution by a Combination of Ozone with Zero-Valent Zinc. *Sep. Purif. Technol.* **2015**, *151*, 318–323. [CrossRef]
129. Zhong, W.; Peng, Q.; Liu, K.; Tang, X.; Zhang, Y.; Xing, J. Building Cu_0/$CuFe_2O_4$ Framework to Efficiently Degrade Tetracycline and Improve Utilization of H_2O_2 in Fenton-like System. *Chem. Eng. J.* **2023**, *474*, 145522. [CrossRef]
130. Devi, L.G.; Srinivas, M.; ArunaKumari, M.L. Heterogeneous Advanced Photo- Fenton Process Using Peroxymonosulfate and Peroxydisulfate in Presence of Zero Valent Metallic Iron: A Comparative Study with Hydrogen Peroxide Photo-Fenton Process. *J. Water Process Eng.* **2016**, *13*, 117–126. [CrossRef]
131. Li, L.; Hu, J.; Shi, X.; Fan, M.; Luo, J.; Wei, X. Nanoscale Zero-Valent Metals: A Review of Synthesis, Characterization, and Applications to Environmental Remediation. *Environ. Sci. Pollut. Res.* **2016**, *23*, 17880–17900. [CrossRef]

132. Wu, Z.; Xiong, Z.; Lai, B. Metal Sulfide-Based Catalysts in Advanced Oxidation Processes for Water Decontamination. *Environ. Funct. Mater.* **2022**, *1*, 298–315. [CrossRef]
133. Bai, X.; Wang, X.; Jia, T.; Guo, L.; Hao, D.; Zhang, Z.; Wu, L.; Zhang, X.; Yang, H.; Gong, Y.; et al. Efficient Degradation of PPCPs by Mo1−xS2−y with S Vacancy at Phase-Junction: Promoted by Innergenerate-H_2O_2. *Appl. Catal. B Environ.* **2022**, *310*, 121302. [CrossRef]
134. Moma, J.; Baloyi, J. Modified Titanium Dioxide for Photocatalytic Applications. In *Photocatalysts—Applications and Attributes*; IntechOpen: London, UK, 2019.
135. Zulfiqar, M.; Sufian, S.; Rabat, N.E.; Mansor, N. Photocatalytic Degradation and Adsorption of Phenol by Solvent-Controlled TiO2 Nanosheets Assisted with H_2O_2 and $FeCl_3$: Kinetic, Isotherm and Thermodynamic Analysis. *J. Mol. Liq.* **2020**, *308*, 112941. [CrossRef]
136. Popescu, T.; Oktaviani Matei, C.; Culita, D.C.; Maraloiu, V.-A.; Rostas, A.M.; Diamandescu, L.; Iacob, N.; Savopol, T.; Ilas, M.C.; Feder, M.; et al. Facile Synthesis of Low Toxicity Iron Oxide/TiO_2 Nanocomposites with Hyperthermic and Photo-Oxidation Properties. *Sci. Rep.* **2022**, *12*, 6887. [CrossRef]
137. dela Rosa, F.M.; Popović, M.; Papac Zjačić, J.; Radić, G.; Kraljić Roković, M.; Kovačić, M.; Farré, M.J.; Genorio, B.; Lavrenčič Štangar, U.; Kušić, H.; et al. Visible-Light Activation of Persulfate or H_2O_2 by Fe_2O_3/TiO_2 Immobilized on Glass Support for Photocatalytic Removal of Amoxicillin: Mechanism, Transformation Products, and Toxicity Assessment. *Nanomaterials* **2022**, *12*, 4328. [CrossRef]
138. Farhana Jaafar, N.; Farhana Jaafar, N.; Khairuddean, M.; Nordin, N. A Review on Recent Progression of Modifications on Titania Morphology and Its Photocatalytic Performance. *Acta Chim. Slov.* **2020**, *67*, 361–374. [CrossRef]
139. Daghrir, R.; Drogui, P.; Robert, D. Modified TiO_2 for Environmental Photocatalytic Applications: A Review. *Ind. Eng. Chem. Res.* **2013**, *52*, 3581–3599. [CrossRef]
140. Parangi, T.; Mishra, M.K. Titania Nanoparticles as Modified Photocatalysts: A Review on Design and Development. *Comments Inorg. Chem.* **2019**, *39*, 90–126. [CrossRef]
141. Kumar, S.G.; Devi, L.G. Review on Modified TiO_2 Photocatalysis under UV/Visible Light: Selected Results and Related Mechanisms on Interfacial Charge Carrier Transfer Dynamics. *J. Phys. Chem. A* **2011**, *115*, 13211–13241. [CrossRef]
142. Noureen, L.; Wang, Q.; Humayun, M.; Shah, W.A.; Xu, Q.; Wang, X. Recent Advances in Structural Engineering of Photocatalysts for Environmental Remediation. *Environ. Res.* **2023**, *219*, 115084. [CrossRef]
143. Covinich, L.; Felissia, F.; Massa, P.; Fenoglio, R.; Area, M.C. Kinetic Modeling of a Heterogeneous Fenton-Type Oxidative Treatment of Complex Industrial Effluent. *Int. J. Ind. Chem.* **2018**, *9*, 215–229. [CrossRef]
144. Bokhari, T.H.; Ahmad, N.; Jilani, M.I.; Saeed, M.; Usman, M.; Haq, A.U.; Rehman, R.; Iqbal, M.; Nazir, A.; Javed, T. UV/H_2O_2, UV/H_2O_2/SnO_2 and Fe/H_2O_2 Based Advanced Oxidation Processes for the Degradation of Disperse Violet 63 in Aqueous Medium. *Mater. Res. Express* **2020**, *7*, 015531. [CrossRef]
145. Habib, I.Y.; Burhan, J.; Jaladi, F.; Lim, C.M.; Usman, A.; Kumara, N.T.R.N.; Tsang, S.C.E.; Mahadi, A.H. Effect of Cr Doping in CeO_2 Nanostructures on Photocatalysis and H_2O_2 Assisted Methylene Blue Dye Degradation. *Catal. Today* **2021**, *375*, 506–513. [CrossRef]
146. Valles-Pérez, B.Y.; Badillo-Ávila, M.A.; Torres-Delgado, G.; Castanedo-Pérez, R.; Zelaya-Ángel, O. Photocatalytic Activity of ZnO + CuO Thin Films Deposited by Dip Coating: Coupling Effect between Oxides. *J. Sol-Gel Sci. Technol.* **2020**, *93*, 517–526. [CrossRef]
147. Mohammadzadeh, A.; Khoshghadam-Pireyousefan, M.; Shokrianfard-Ravasjan, B.; Azadbeh, M.; Rashedi, H.; Dibazar, M.; Mostafaei, A. Synergetic Photocatalytic Effect of High Purity ZnO Pod Shaped Nanostructures with H_2O_2 on Methylene Blue Dye Degradation. *J. Alloys Compd.* **2020**, *845*, 156333. [CrossRef]
148. Ene, C.D.; Patrinoiu, G.; Munteanu, C.; Ene, R.; Chifiriuc, M.C.; Carp, O. Multifunctional ZnO Materials Prepared by a Versatile Green Carbohydrate-Assisted Combustion Method for Environmental Remediation Applications. *Ceram. Int.* **2019**, *45*, 2295–2302. [CrossRef]
149. Pourmortazavi, S.M.; Rahimi-Nasrabadi, M.; Ahmadi, F.; Ganjali, M.R. $CuCO_3$ and CuO Nanoparticles; Facile Preparation and Evaluation as Photocatalysts. *J. Mater. Sci. Mater. Electron.* **2018**, *29*, 9442–9451. [CrossRef]
150. Kumar, S.G.; Rao, K.S.R.K. Comparison of Modification Strategies towards Enhanced Charge Carrier Separation and Photocatalytic Degradation Activity of Metal Oxide Semiconductors (TiO_2, WO_3 and ZnO). *Appl. Surf. Sci.* **2017**, *391*, 124–148. [CrossRef]
151. Karthikeyan, C.; Arunachalam, P.; Ramachandran, K.; Al-Mayouf, A.M.; Karuppuchamy, S. Recent Advances in Semiconductor Metal Oxides with Enhanced Methods for Solar Photocatalytic Applications. *J. Alloys Compd.* **2020**, *828*, 154281. [CrossRef]
152. Danish, M.S.S.; Estrella, L.L.; Alemaida, I.M.A.; Lisin, A.; Moiseev, N.; Ahmadi, M.; Nazari, M.; Wali, M.; Zaheb, H.; Senjyu, T. Photocatalytic Applications of Metal Oxides for Sustainable Environmental Remediation. *Metals* **2021**, *11*, 80. [CrossRef]
153. El-Shamy, A.G. Synthesis of New Magnesium Peroxide (MgO_2) Nano-Rods for Pollutant Dye Removal and Antibacterial Applications. *Mater. Chem. Phys.* **2020**, *243*, 122640. [CrossRef]
154. Ali, M.; Farooq, U.; Lyu, S.; Sun, Y.; Li, M.; Ahmad, A.; Shan, A.; Abbas, Z. Synthesis of Controlled Release Calcium Peroxide Nanoparticles (CR-NCPs): Characterizations, H_2O_2 Liberate Performances and Pollutant Degradation Efficiency. *Sep. Purif. Technol.* **2020**, *241*, 116729. [CrossRef]

155. Kowalkińska, M.; Głuchowski, P.; Swebocki, T.; Ossowski, T.; Ostrowski, A.; Bednarski, W.; Karczewski, J.; Zielińska-Jurek, A. Scheelite-Type Wide-Bandgap ABO 4 Compounds (A = Ca, Sr, and Ba; B = Mo and W) as Potential Photocatalysts for Water Treatment. *J. Phys. Chem. C* **2021**, *125*, 25497–25513. [CrossRef]
156. Kong, J.; Yang, T.; Rui, Z.; Ji, H. Perovskite-Based Photocatalysts for Organic Contaminants Removal: Current Status and Future Perspectives. *Catal. Today* **2019**, *327*, 47–63. [CrossRef]
157. Chandrasekaran, S.; Bowen, C.; Zhang, P.; Li, Z.; Yuan, Q.; Ren, X.; Deng, L. Spinel Photocatalysts for Environmental Remediation, Hydrogen Generation, CO 2 Reduction and Photoelectrochemical Water Splitting. *J. Mater. Chem. A* **2018**, *6*, 11078–11104. [CrossRef]
158. Wang, J.; Yang, L.; Zhang, L. Constructed 3D Hierarchical Micro-Flowers $CoWO_4$@Bi_2WO_6 Z-Scheme Heterojunction Catalyzer: Two-Channel Photocatalytic H_2O_2 Production and Antibiotics Degradation. *Chem. Eng. J.* **2021**, *420*, 127639. [CrossRef]
159. Das, K.C.; Dhar, S.S. Rapid Catalytic Degradation of Malachite Green by $MgFe_2O_4$ Nanoparticles in Presence of H_2O_2. *J. Alloys Compd.* **2020**, *828*, 154462. [CrossRef]
160. Hong, P.; Li, Y.; He, J.; Saeed, A.; Zhang, K.; Wang, C.; Kong, L.; Liu, J. Rapid Degradation of Aqueous Doxycycline by Surface $CoFe_2O_4/H_2O_2$ System: Behaviors, Mechanisms, Pathways and DFT Calculation. *Appl. Surf. Sci.* **2020**, *526*, 146557. [CrossRef]
161. Wei, Y.; Zhang, Y.; Miao, J.; Geng, W.; Long, M. In-Situ Utilization of Piezo-Generated Hydrogen Peroxide for Efficient p-Chlorophenol Degradation by Fe Loading Bismuth Vanadate. *Appl. Surf. Sci.* **2021**, *543*, 148791. [CrossRef]
162. Demircivi, P.; Simsek, E.B. Visible-Light-Enhanced Photoactivity of Perovskite-Type W-Doped BaTiO3 Photocatalyst for Photodegradation of Tetracycline. *J. Alloys Compd.* **2019**, *774*, 795–802. [CrossRef]
163. Vieten, J.; Bulfin, B.; Roeb, M.; Sattler, C. Citric Acid Auto-Combustion Synthesis of Ti-Containing Perovskites via Aqueous Precursors. *Solid State Ion.* **2018**, *315*, 92–97. [CrossRef]
164. Li, L.; Hu, Z.; Yu, J.C. On-Demand Synthesis of H_2O_2 by Water Oxidation for Sustainable Resource Production and Organic Pollutant Degradation. *Angew. Chem.* **2020**, *132*, 20719–20725. [CrossRef]
165. Li, M.; Han, N.; Zhang, X.; Wang, S.; Jiang, M.; Bokhari, A.; Zhang, W.; Race, M.; Shen, Z.; Chen, R.; et al. Perovskite Oxide for Emerging Photo(Electro)Catalysis in Energy and Environment. *Environ. Res.* **2022**, *205*, 112544. [CrossRef] [PubMed]
166. Nzuzo, Y.; Oseghale, C.O.; Chike-Ekwughe, A.; Maumela, M.; Bingwa, N. Electronic Distribution and Dynamics as Catalytic Descriptors in Heterogeneous Catalysis: A Mini Review. *Catal. Commun.* **2024**, 106901. [CrossRef]
167. Di Paola, A.; Bellardita, M.; Palmisano, L.; Barbieriková, Z.; Brezová, V. Influence of Crystallinity and OH Surface Density on the Photocatalytic Activity of TiO_2 Powders. *J. Photochem. Photobiol. A Chem.* **2014**, *273*, 59–67. [CrossRef]
168. Zhang, Y.; Pan, C.; Li, J.; Zhu, Y. Recent Progress in Nonsacrificial H_2O_2 Generation Using Organic Photocatalysts and In Situ Applications for Environmental Remediation. *Acc. Mater. Res.* **2024**, *5*, 76–88. [CrossRef]
169. García-Ballesteros, S.; García-Negueroles, P.; Amat, A.M.; Arques, A. Humic-Like Substances as Auxiliaries to Enhance Advanced Oxidation Processes. *ACS Omega* **2022**, *7*, 3151–3157. [CrossRef] [PubMed]
170. Bianco Prevot, A.; Testa, M.L.; Laurenti, E.; Tummino, M.L.; Magnacca, G. Soluble Bioorganic Substances from Compost as Photosensitizers for a Sustainable Homogeneous and Heterogeneous Photocatalysis. In *Materials Science in Photocatalysis*; Elsevier: Amsterdam, The Netherlands, 2021; pp. 589–601. ISBN 9780128218594.
171. Vallés, I.; Santos Juanes, L.; Amat, A.M.; Palma, D.; Laurenti, E.; Bianco Prevot, A.; Arques, A. Humic Acids as Complexing Agents to Drive Photo-Fenton at Mild PH in Saline Matrices: Process Performance and Mechanistic Studies. *J. Environ. Chem. Eng.* **2023**, *11*, 111391. [CrossRef]
172. Tummino, M.L.; Testa, M.L.; Malandrino, M.; Gamberini, R.; Bianco Prevot, A.; Magnacca, G.; Laurenti, E. Green Waste-Derived Substances Immobilized on SBA-15 Silica: Surface Properties, Adsorbing and Photosensitizing Activities towards Organic and Inorganic Substrates. *Nanomaterials* **2019**, *9*, 162. [CrossRef] [PubMed]
173. Testa, M.L.; Tummino, M.L.; Agostini, S.; Avetta, P.; Deganello, F.; Montoneri, E.; Magnacca, G.; Prevot, A.B. Synthesis, Characterization and Environmental Application of Silica Grafted Photoactive Substances Isolated from Urban Biowaste. *RSC Adv.* **2015**, *5*, 47920–47927. [CrossRef]
174. Huang, Q.Q.; Liu, H.Z.; Huang, M.; Wang, J.; Yu, H.Q. Ligand-Assisted Heterogeneous Catalytic H_2O_2 Activation for Pollutant Degradation: The Trade-off between Coordination Site Passivation and Adjacent Site Activation. *Appl. Catal. B Environ.* **2023**, *330*, 122592. [CrossRef]
175. Baran, T.; Wojtyła, S.; Minguzzi, A.; Rondinini, S.; Vertova, A. Achieving Efficient H_2O_2 Production by a Visible-Light Absorbing, Highly Stable Photosensitized TiO_2. *Appl. Catal. B Environ.* **2019**, *244*, 303–312. [CrossRef]
176. Zhang, M.; Lin, X.; Yi, Z.; Xu, X.; Yang, J.; Zhu, M. Enhanced Reactive Oxidation Species Generation by Ligand-to-Metal-Charge Transfer between Oxygen Vacancy-Rich ZnO Mesocrystal with Ciprofloxacin Pollutants. *Appl. Catal. B Environ.* **2023**, *321*, 122033. [CrossRef]
177. Fei, B.; Yan, Q.; Wang, J.; Liu, Q.; Long, J.; Li, Y.; Shao, K.; Su, Z.; Sun, W. Green Oxidative Degradation of Methyl Orange with Copper(II) Schiff Base Complexes as Photo-Fenton-Like Catalysts. *Z. Anorg. Allg. Chem.* **2014**, *640*, 2035–2040. [CrossRef]
178. Zhao, J.; Liu, Y.; Fan, M.; Yuan, L.; Zou, X. From Solid-State Metal Alkoxides to Nanostructured Oxides: A Precursor-Directed Synthetic Route to Functional Inorganic Nanomaterials. *Inorg. Chem. Front.* **2015**, *2*, 198–212. [CrossRef]
179. Paris, C.B.; Howe, A.G.; Lewis, R.J.; Hewes, D.; Morgan, D.J.; He, Q.; Edwards, J.K. Impact of the Experimental Parameters on Catalytic Activity When Preparing Polymer Protected Bimetallic Nanoparticle Catalysts on Activated Carbon. *ACS Catal.* **2022**, *12*, 4440–4454. [CrossRef] [PubMed]

180. Köwitsch, I.; Mehring, M. Carbon Nitride Materials: Impact of Synthetic Method on Photocatalysis and Immobilization for Photocatalytic Pollutant Degradation. *J. Mater. Sci.* **2021**, *56*, 18608–18624. [CrossRef]
181. Wang, F.; Chen, X.; Wang, Y.; Li, X.; Wan, M.; Zhang, G.; Leng, F.; Zhang, H. Insights into the Structures, Inhibitors, and Improvement Strategies of Glucose Oxidase. *Int. J. Mol. Sci.* **2022**, *23*, 9841. [CrossRef] [PubMed]
182. Wohlfahrt, G.; Witt, S.; Hendle, J.; Schomburg, D.; Kalisz, H.M.; Hecht, H.-J. 1.8 and 1.9 Å Resolution Structures of the Penicillium Amagasakiense and Aspergillus Niger Glucose Oxidases as a Basis for Modelling Substrate Complexes. *Acta Crystallogr. Sect. D Biol. Crystallogr.* **1999**, *55*, 969–977. [CrossRef] [PubMed]
183. Zhao, Z.; Li, M.; Du, N.; Li, Z.; Zhang, Y.; Zhang, Q. Fenton-like Reaction of Glucose Oxidase-Glucose@Kaolin Coupled with Green Rust: A Framework Triggering FeIV=O in Refractory Pollutants Degradation. *Sep. Purif. Technol.* **2022**, *301*, 122061. [CrossRef]
184. Huang, Y.; Liu, H.; Liu, S.; Li, C.; Yuan, S. Glucose Oxidase Modified Fenton Reactions for In-Situ ROS Generation and Potential Application in Groundwater Remediation. *Chemosphere* **2020**, *253*, 126648. [CrossRef] [PubMed]
185. Vaidyanathan, V.K.; Alanazi, A.K.; Senthil Kumar, P.; Rajendran, D.S.; Chidambaram, A.; Venkataraman, S.; Kumar, V.V.; Rangasamy, G.; Cabana, H.; Abo-Dief, H.M. Cost-Effective, Scalable Production of Glucose Oxidase Using Casuarina Equisetifolia Biomass and Its Application in the Bio-Fenton Oxidation Process for the Removal of Trace Organic Contaminants from Wastewater. *Bioresour. Technol.* **2023**, *377*, 128958. [CrossRef] [PubMed]
186. Krishnan, S.; Martínez-Huitle, C.A.; Nidheesh, P.V. An Overview of Chelate Modified Electro-Fenton Processes. *J. Environ. Chem. Eng.* **2022**, *10*, 107183. [CrossRef]
187. Yang, Y.; Ghatge, S.; Ko, Y.; Yoon, Y.; Ahn, J.-H.; Kim, J.J.; Hur, H.-G. Non-Specific Degradation of Chloroacetanilide Herbicides by Glucose Oxidase Supported Bio-Fenton Reaction. *Chemosphere* **2022**, *292*, 133417. [CrossRef]
188. An, S.; Yoon, Y.; Ahn, J.-H.; Kim, D.; Weon, H.-Y.; Kim, Y.; Hur, H.-G.; Yang, Y. Oxidative Degradation of Bisphenol A by Bio-Fenton Reaction Equipped with Glucose Oxidase and Ferric Citrate: Degradation Kinetics and Pathway. *J. Environ. Chem. Eng.* **2023**, *11*, 109349. [CrossRef]
189. Wang, X.; Song, C.; Liu, X.; Zhang, J.; Zhang, Y.; Shi, X.; Kim, D. Bio-Fenton-Assisted Biological Process for Efficient Mineralization of Polycyclic Aromatic Hydrocarbons from the Environment. *Processes* **2022**, *10*, 1316. [CrossRef]
190. Yoon, Y.; Cho, M. Understanding Atrazine Elimination via Treatment of the Enzyme-Based Fenton Reaction: Kinetics, Mechanism, Reaction Pathway, and Metabolites Toxicity. *Chemosphere* **2024**, *349*, 140982. [CrossRef] [PubMed]
191. Ioffe, M.; Kundu, S.; Perez-Lapid, N.; Radian, A. Heterogeneous Fenton Catalyst Based on Clay Decorated with Nano-Sized Amorphous Iron Oxides Prevents Oxidant Scavenging through Surface Complexation. *Chem. Eng. J.* **2022**, *433*, 134609. [CrossRef]
192. Wang, J.; Tang, J. Fe-Based Fenton-like Catalysts for Water Treatment: Catalytic Mechanisms and Applications. *J. Mol. Liq.* **2021**, *332*, 115755. [CrossRef]
193. Wang, J.; Tang, J. Fe-Based Fenton-like Catalysts for Water Treatment: Preparation, Characterization and Modification. *Chemosphere* **2021**, *276*, 130177. [CrossRef] [PubMed]
194. Bhave, C.; Shejwalkar, S. A Review on the Synthesis and Applications of Green Rust for Environmental Pollutant Remediation. *Int. J. Environ. Sci. Technol.* **2018**, *15*, 1243–1248. [CrossRef]
195. Neyens, E.; Baeyens, J. A Review of Classic Fenton's Peroxidation as an Advanced Oxidation Technique. *J. Hazard. Mater.* **2003**, *98*, 33–50. [CrossRef] [PubMed]
196. Wang, M.; Huang, Y.; Liu, H. Removal of Trichloroethene by Glucose Oxidase Immobilized on Magnetite Nanoparticles. *RSC Adv.* **2023**, *13*, 11853–11864. [CrossRef] [PubMed]
197. Liu, X.; Zhang, L.; Zhang, Q.; Li, M.; Zhao, Z.; Lin, B.; Peng, J.; Shen, H.; He, Q. Fenton-like System of UV/Glucose-Oxidase@Kaolin Coupled with Organic Green Rust: UV-Enhanced Enzyme Activity and the Mechanism of UV Synergistic Degradation of Photosensitive Pollutants. *Environ. Res.* **2024**, *247*, 118257. [CrossRef]
198. Karimi, A.; Aghbolaghy, M.; Khataee, A.; Shoa Bargh, S. Use of Enzymatic Bio-Fenton as a New Approach in Decolorization of Malachite Green. *Sci. World J.* **2012**, *2012*, 691569. [CrossRef]
199. Hakala, M. Photoinhibition of Manganese Enzymes: Insights into the Mechanism of Photosystem II Photoinhibition. *J. Exp. Bot.* **2006**, *57*, 1809–1816. [CrossRef] [PubMed]
200. Ghatge, S.; Yang, Y.; Ko, Y.; Yoon, Y.; Ahn, J.-H.; Kim, J.J.; Hur, H.-G. Degradation of Sulfonated Polyethylene by a Bio-Photo-Fenton Approach Using Glucose Oxidase Immobilized on Titanium Dioxide. *J. Hazard. Mater.* **2022**, *423*, 127067. [CrossRef] [PubMed]
201. Ravi, S.; Lonappan, L.; Touahar, I.; Fonteneau, É.; Vaidyanathan, V.K.; Cabana, H. Evaluation of Bio-Fenton Oxidation Approach for the Remediation of Trichloroethylene from Aqueous Solutions. *J. Environ. Manag.* **2020**, *270*, 110899. [CrossRef] [PubMed]
202. Banci, L. Structural Properties of Peroxidases. *J. Biotechnol.* **1997**, *53*, 253–263. [CrossRef] [PubMed]
203. Sugano, Y.; Sasaki, K.; Shoda, M. CDNA Cloning and Genetic Analysis of a Novel Decolorizing Enzyme, Peroxidase Gene Dyp from Geotrichum Candidum Dec 1. *J. Biosci. Bioeng.* **1999**, *87*, 411–417. [CrossRef] [PubMed]
204. Faraco, V.; Piscitelli, A.; Sannia, G.; Giardina, P. Identification of a New Member of the Dye-Decolorizing Peroxidase Family from Pleurotus Ostreatus. *World J. Microbiol. Biotechnol.* **2007**, *23*, 889–893. [CrossRef]
205. Battistuzzi, G.; Bellei, M.; Bortolotti, C.A.; Sola, M. Redox Properties of Heme Peroxidases. *Arch. Biochem. Biophys.* **2010**, *500*, 21–36. [CrossRef] [PubMed]

206. Nuell, M.J.; Fang, G.H.; Axley, M.J.; Kenigsberg, P.; Hager, L.P. Isolation and Nucleotide Sequence of the Chloroperoxidase Gene from Caldariomyces Fumago. *J. Bacteriol.* **1988**, *170*, 1007–1011. [CrossRef]
207. Conesa, A.; Punt, P.J.; van den Hondel, C.A.M.J.J. Fungal Peroxidases: Molecular Aspects and Applications. *J. Biotechnol.* **2002**, *93*, 143–158. [CrossRef]
208. Al-Maqdi, K.A.; Hisaindee, S.M.; Rauf, M.A.; Ashraf, S.S. Comparative Degradation of a Thiazole Pollutant by an Advanced Oxidation Process and an Enzymatic Approach. *Biomolecules* **2017**, *7*, 64. [CrossRef]
209. Wagenknecht, H.-A.; Woggon, W.-D. Identification of Intermediates in the Catalytic Cycle of Chloroperoxidase. *Chem. Biol.* **1997**, *4*, 367–372. [CrossRef] [PubMed]
210. Piontek, K.; Strittmatter, E.; Ullrich, R.; Gröbe, G.; Pecyna, M.J.; Kluge, M.; Scheibner, K.; Hofrichter, M.; Plattner, D.A. Structural Basis of Substrate Conversion in a New Aromatic Peroxygenase. *J. Biol. Chem.* **2013**, *288*, 34767–34776. [CrossRef] [PubMed]
211. Yoshida, T.; Sugano, Y. Unexpected Diversity of Dye-Decolorizing Peroxidases. *Biochem. Biophys. Rep.* **2023**, *33*, 101401. [CrossRef] [PubMed]
212. Hofrichter, M.; Ullrich, R.; Pecyna, M.J.; Liers, C.; Lundell, T. New and Classic Families of Secreted Fungal Heme Peroxidases. *Appl. Microbiol. Biotechnol.* **2010**, *87*, 871–897. [CrossRef] [PubMed]
213. Catucci, G.; Valetti, F.; Sadeghi, S.J.; Gilardi, G. Biochemical Features of Dye-decolorizing Peroxidases: Current Impact on Lignin Degradation. *Biotechnol. Appl. Biochem.* **2020**, *67*, 751–759. [CrossRef] [PubMed]
214. Uchida, T.; Sasaki, M.; Tanaka, Y.; Ishimori, K. A Dye-Decolorizing Peroxidase from Vibrio Cholerae. *Biochemistry* **2015**, *54*, 6610–6621. [CrossRef] [PubMed]
215. Sellami, K.; Couvert, A.; Nasrallah, N.; Maachi, R.; Abouseoud, M.; Amrane, A. Peroxidase Enzymes as Green Catalysts for Bioremediation and Biotechnological Applications: A Review. *Sci. Total Environ.* **2022**, *806*, 150500. [CrossRef] [PubMed]
216. Rigoletto, M.; Calza, P.; Gaggero, E.; Laurenti, E. Hybrid Materials for the Removal of Emerging Pollutants in Water: Classification, Synthesis, and Properties. *Chem. Eng. J. Adv.* **2022**, *10*, 100252. [CrossRef]
217. Calza, P.; Zacchigna, D.; Laurenti, E. Degradation of Orange Dyes and Carbamazepine by Soybean Peroxidase Immobilized on Silica Monoliths and Titanium Dioxide. *Environ. Sci. Pollut. Res.* **2016**, *23*, 23742–23749. [CrossRef]
218. Iñarritu, I.; Torres, E.; Topete, A.; Campos-Terán, J. Immobilization Effects on the Photocatalytic Activity of CdS Quantum Dots-Horseradish Peroxidase Hybrid Nanomaterials. *J. Colloid Interface Sci.* **2017**, *506*, 36–45. [CrossRef]
219. Wu, J.; He, T.; Ma, X.; Li, C.; Han, J.; Wang, L.; Dong, H.; Zhang, R.; Wang, Y. A Novel Immobilized Horseradish Peroxidase Platform Driven by Visible Light for the Complete Mineralization of Sulfadiazine in Water. *Int. J. Biol. Macromol.* **2023**, *253*, 127239. [CrossRef] [PubMed]
220. Baynton, K.J.; Bewtra, J.K.; Biswas, N.; Taylor, K.E. Inactivation of Horseradish Peroxidase by Phenol and Hydrogen Peroxide: A Kinetic Investigation. *Biochim. Biophys. Acta—Protein Struct. Mol. Enzymol.* **1994**, *1206*, 272–278. [CrossRef] [PubMed]
221. Nicell, J.A.; Wright, H. A Model of Peroxidase Activity with Inhibition by Hydrogen Peroxide. *Enzym. Microb. Technol.* **1997**, *21*, 302–310. [CrossRef]
222. Hiner, A.N.P.; Hernández-Ruiz, J.; Rodríguez-López, J.N.; Arnao, M.B.; Varón, R.; García-Cánovas, F.; Acosta, M. The Inactivation of Horseradish Peroxidase Isoenzyme AZ by Hydrogen Peroxide: An Example of Partial Resistance Due to the Formation of a Stable Enzyme Intermediate. *JBIC J. Biol. Inorg. Chem.* **2001**, *6*, 504–516. [CrossRef] [PubMed]
223. Chansaenpak, K.; Kamkaew, A.; Lisnund, S.; Prachai, P.; Ratwirunkit, P.; Jingpho, T.; Blay, V.; Pinyou, P. Development of a Sensitive Self-Powered Glucose Biosensor Based on an Enzymatic Biofuel Cell. *Biosensors* **2021**, *11*, 16. [CrossRef] [PubMed]
224. Zhang, H.; Chen, Z.; Dai, J.; Zhang, W.; Jiang, Y.; Zhou, A. A Low-Cost Mobile Platform for Whole Blood Glucose Monitoring Using Colorimetric Method. *Microchem. J.* **2021**, *162*, 105814. [CrossRef]
225. Pantić, N.; Prodanović, R.; Đurđić, K.I.; Polović, N.; Spasojević, M.; Prodanović, O. Optimization of Phenol Removal with Horseradish Peroxidase Encapsulated within Tyramine-Alginate Micro-Beads. *Environ. Technol. Innov.* **2021**, *21*, 101211. [CrossRef]
226. Babaei, H.; Ghobadi Nejad, Z.; Yaghmaei, S.; Farhadi, F. Co-Immobilization of Multi-Enzyme Cascade System into the Metal–Organic Frameworks for the Removal of Bisphenol A. *Chem. Eng. J.* **2023**, *461*, 142050. [CrossRef]
227. Yan, Y.; Liu, X.; Jiang, X.; Zhang, W.; Wang, Y.; Wang, Y.; Zhang, Y.; Luo, H.; Yao, B.; Huang, H.; et al. Surface Charge Modifications Modulate Glucose Oxidase PH-Activity Profiles for Efficient Gluconic Acid Production. *J. Clean. Prod.* **2022**, *372*, 133817. [CrossRef]
228. Gao, F.; Guo, Y.; Fan, X.; Hu, M.; Li, S.; Zhai, Q.; Jiang, Y.; Wang, X. Enhancing the Catalytic Performance of Chloroperoxidase by Co-Immobilization with Glucose Oxidase on Magnetic Graphene Oxide. *Biochem. Eng. J.* **2019**, *143*, 101–109. [CrossRef]
229. Gu, Y.; Yuan, L.; Li, M.; Wang, X.; Rao, D.; Bai, X.; Shi, K.; Xu, H.; Hou, S.; Yao, H. Co-Immobilized Bienzyme of Horseradish Peroxidase and Glucose Oxidase on Dopamine-Modified Cellulose–Chitosan Composite Beads as a High-Efficiency Biocatalyst for Degradation of Acridine. *RSC Adv.* **2022**, *12*, 23006–23016. [CrossRef] [PubMed]
230. Liu, X.; Zhang, Q.; Li, M.; Qin, S.; Zhao, Z.; Lin, B.; Ding, Y.; Xiang, Y.; Li, C. Horseradish Peroxidase (HRP) and Glucose Oxidase (GOX) Based Dual-Enzyme System: Sustainable Release of H_2O_2 and Its Effect on the Desirable Ping Pong Bibi Degradation Mechanism. *Environ. Res.* **2023**, *229*, 115979. [CrossRef] [PubMed]
231. Razzaghi, M.; Karimi, A.; Aghdasinia, H.; Joghataei, M.-T. Oxidase-Peroxidase Sequential Polymerization for Removal of a Dye from Contaminated Water by Horseradish Peroxidase (HRP)/Glucose Oxidase (GOx)/Polyurethane Hybrid Catalyst. *Korean J. Chem. Eng.* **2017**, *34*, 2870–2878. [CrossRef]

232. Pitzalis, F.; Monduzzi, M.; Salis, A. A Bienzymatic Biocatalyst Constituted by Glucose Oxidase and Horseradish Peroxidase Immobilized on Ordered Mesoporous Silica. *Microporous Mesoporous Mater.* **2017**, *241*, 145–154. [CrossRef]
233. Taboada-Puig, R.; Junghanns, C.; Demarche, P.; Moreira, M.T.; Feijoo, G.; Lema, J.M.; Agathos, S.N. Combined Cross-Linked Enzyme Aggregates from Versatile Peroxidase and Glucose Oxidase: Production, Partial Characterization and Application for the Elimination of Endocrine Disruptors. *Bioresour. Technol.* **2011**, *102*, 6593–6599. [CrossRef] [PubMed]
234. Frew, J.E.; Jones, P.; Scholes, G. Spectrophotometric Determination of Hydrogen Peroxide and Organic Hydroperoxides at Low Concentrations in Aqueous Solution. *Anal. Chim. Acta* **1983**, *155*, 139–150. [CrossRef]
235. Sekar, R.; DiChristina, T.J. Microbially Driven Fenton Reaction for Degradation of the Widespread Environmental Contaminant 1,4-Dioxane. *Environ. Sci. Technol.* **2014**, *48*, 12858–12867. [CrossRef] [PubMed]
236. Zhou, B.; Wang, J.; Guo, Z.; Tan, H.; Zhu, X. A Simple Colorimetric Method for Determination of Hydrogen Peroxide in Plant Tissues. *Plant Growth Regul.* **2006**, *49*, 113–118. [CrossRef]
237. Fernando, C.D.; Soysa, P. Optimized Enzymatic Colorimetric Assay for Determination of Hydrogen Peroxide (H_2O_2) Scavenging Activity of Plant Extracts. *MethodsX* **2015**, *2*, 283–291. [CrossRef]
238. Schüttler, S.; Jolmes, L.; Jeß, E.; Tschulik, K.; Golda, J. Validation of in Situ Diagnostics for the Detection of OH and H_2O_2 in Liquids Treated by a Humid Atmospheric Pressure Plasma Jet. *Plasma Process. Polym.* **2024**, *21*, e2300079. [CrossRef]
239. Oliveira, M.C.; Nogueira, R.F.P.; Gomes Neto, J.A.; Jardim, W.F.; Rohwedder, J.J.R. Sistema de Injeção Em Fluxo Espectrofotométrico Para Monitorar Peróxido de Hidrogênio Em Processo de Fotodegradação Por Reação Foto-Fenton. *Quim. Nova* **2001**, *24*, 188–190. [CrossRef]
240. Nogueira, R.; Oliveira, M.; Paterlini, W. Simple and Fast Spectrophotometric Determination of H_2O_2 in Photo-Fenton Reactions Using Metavanadate. *Talanta* **2005**, *66*, 86–91. [CrossRef] [PubMed]
241. Thermofisher Scientific AmplexTM Red Hydrogen Peroxide/Peroxidase Assay Kit. Available online: https://www.thermofisher.com/order/catalog/product/A22188?gclid=EAIaIQobChMI1eDRvZrGhAMVx6doCR1c2QCCEAAYASAAEgLFlvD_BwE&ef_id=EAIaIQobChMI1eDRvZrGhAMVx6doCR1c2QCCEAAYASAAEgLFlvD_BwE:G:s&s_kwcid=AL!3652!3!447292198748!!!g!!!10506731179!109642167571&cid=bid (accessed on 13 March 2024).
242. Kyoritsu Chemical-Check Lab Packtest Hydrogen Peroxide. Available online: https://en.kyoritsu-lab.co.jp/products/wak_h2o2 (accessed on 13 March 2024).
243. Pratsinis, A.; Kelesidis, G.A.; Zuercher, S.; Krumeich, F.; Bolisetty, S.; Mezzenga, R.; Leroux, J.-C.; Sotiriou, G.A. Enzyme-Mimetic Antioxidant Luminescent Nanoparticles for Highly Sensitive Hydrogen Peroxide Biosensing. *ACS Nano* **2017**, *11*, 12210–12218. [CrossRef] [PubMed]
244. Qi, Z.; Wang, L.; You, Q.; Chen, Y. PA-Tb-Cu MOF as Luminescent Nanoenzyme for Catalytic Assay of Hydrogen Peroxide. *Biosens. Bioelectron.* **2017**, *96*, 227–232. [CrossRef] [PubMed]
245. Chen, W.; Cai, S.; Ren, Q.-Q.; Wen, W.; Zhao, Y.-D. Recent Advances in Electrochemical Sensing for Hydrogen Peroxide: A Review. *Analyst* **2012**, *137*, 49–58. [CrossRef] [PubMed]
246. Peng, H.-P.; Liang, R.-P.; Qiu, J.-D. Facile Synthesis of Fe_3O_4@Al_2O_3 Core–Shell Nanoparticles and Their Application to the Highly Specific Capture of Heme Proteins for Direct Electrochemistry. *Biosens. Bioelectron.* **2011**, *26*, 3005–3011. [CrossRef] [PubMed]
247. WANG, S.; XIE, F.; LIU, G. Direct Electrochemistry and Electrocatalysis of Heme Proteins on SWCNTs-CTAB Modified Electrodes. *Talanta* **2009**, *77*, 1343–1350. [CrossRef]
248. Duanghathaipornsuk, S.; Farrell, E.J.; Alba-Rubio, A.C.; Zelenay, P.; Kim, D.-S. Detection Technologies for Reactive Oxygen Species: Fluorescence and Electrochemical Methods and Their Applications. *Biosensors* **2021**, *11*, 30. [CrossRef] [PubMed]
249. Chávez, M.; Fernandez-Merino, Á.; del Caño, R.; Sánchez-Obrero, G.; Madueño, R.; Blázquez, M.; Pineda, T. Behind the Optimization of the Sensor Film: Bioconjugation of Triangular Gold Nanoparticles with Hemoproteins for Sensitivity Enhancement of Enzymatic Biosensors. *Biosensors* **2023**, *13*, 467. [CrossRef]
250. Kanwal, A.; Saif, B.; Muhammad, A.; Liu, W.; Liu, J.; Ren, H.; Yang, P.; Lei, Z. Hemoglobin-Promoted Growth of Polyhedral Gold Nanoparticles for the Detection of Glucose, H_2O_2, and Ascorbic Acid. *ACS Appl. Nano Mater.* **2023**, *6*, 4734–4746. [CrossRef]
251. Elewi, A.S.; Al-Shammaree, S.A.W.; AL Sammarraie, A.K.M.A. Hydrogen Peroxide Biosensor Based on Hemoglobin-Modified Gold Nanoparticles–Screen Printed Carbon Electrode. *Sens. Bio-Sens. Res.* **2020**, *28*, 100340. [CrossRef]
252. Salimi, A.; Sharifi, E.; Noorbakhsh, A.; Soltanian, S. Direct Electrochemistry and Electrocatalytic Activity of Catalase Immobilized onto Electrodeposited Nano-Scale Islands of Nickel Oxide. *Biophys. Chem.* **2007**, *125*, 540–548. [CrossRef] [PubMed]
253. Rui, Q.; Komori, K.; Tian, Y.; Liu, H.; Luo, Y.; Sakai, Y. Electrochemical Biosensor for the Detection of H2O2 from Living Cancer Cells Based on ZnO Nanosheets. *Anal. Chim. Acta* **2010**, *670*, 57–62. [CrossRef] [PubMed]
254. Soto, D.; Alzate, M.; Gallego, J.; Orozco, J. Hybrid Nanomaterial/Catalase-Modified Electrode for Hydrogen Peroxide Sensing. *J. Electroanal. Chem.* **2021**, *880*, 114826. [CrossRef]
255. Yagati, A.K.; Ngoc Le, H.T.; Cho, S. Bioelectrocatalysis of Hemoglobin on Electrodeposited Ag Nanoflowers toward H2O2 Detection. *Nanomaterials* **2020**, *10*, 1628. [CrossRef] [PubMed]
256. Wang, H.; Chen, W.; Chen, Q.; Liu, N.; Cheng, H.; Li, T. Metal-Organic Framework (MOF)-Au@Pt Nanoflowers Composite Material for Electrochemical Sensing of H_2O_2 in Living Cells. *J. Electroanal. Chem.* **2021**, *897*, 115603. [CrossRef]
257. Dang, W.; Sun, Y.; Jiao, H.; Xu, L.; Lin, M. AuNPs-NH_2/Cu-MOF Modified Glassy Carbon Electrode as Enzyme-Free Electrochemical Sensor Detecting H_2O_2. *J. Electroanal. Chem.* **2020**, *856*, 113592. [CrossRef]

258. Ji, J.; Ko, S.Y.; Choi, K.M.; Kwon, Y. Hydrogen Peroxide Sensor Using the Biomimetic Structure of Peroxidase Including a Metal Organic Framework. *Appl. Surf. Sci.* **2021**, *554*, 148786. [CrossRef]
259. Tong, P.; Asif, M.; Ajmal, M.; Aziz, A.; Sun, Y. A Multicomponent Polymer-Metal-Enzyme System as Electrochemical Biosensor for H_2O_2 Detection. *Front. Chem.* **2022**, *10*, 874965. [CrossRef]
260. Manickam, P.; Vashist, A.; Madhu, S.; Sadasivam, M.; Sakthivel, A.; Kaushik, A.; Nair, M. Gold Nanocubes Embedded Biocompatible Hybrid Hydrogels for Electrochemical Detection of H_2O_2. *Bioelectrochemistry* **2020**, *131*, 107373. [CrossRef]
261. Matos-Peralta, Y.; Antuch, M. Review—Prussian Blue and Its Analogs as Appealing Materials for Electrochemical Sensing and Biosensing. *J. Electrochem. Soc.* **2020**, *167*, 037510. [CrossRef]
262. Narendra Kumar, A.V.; Harish, S.; Joseph, J.; Phani, K.L. Nix–Fe(1−x)Fe(CN)6 Hybrid Thin Films Electrodeposited on Glassy Carbon: Effect of Tuning of Redox Potentials on the Electrocatalysis of Hydrogen Peroxide. *J. Electroanal. Chem.* **2011**, *659*, 128–133. [CrossRef]
263. Ishizaki, M.; Ohshida, E.; Tanno, H.; Kawamoto, T.; Tanaka, H.; Hara, K.; Kominami, H.; Kurihara, M. H_2O_2-Sensing Abilities of Mixed-Metal (Fe-Ni) Prussian Blue Analogs in a Wide PH Range. *Inorg. Chim. Acta* **2020**, *502*, 119314. [CrossRef]
264. Li, J.; He, L.; Jiang, J.; Xu, Z.; Liu, M.; Liu, X.; Tong, H.; Liu, Z.; Qian, D. Facile Syntheses of Bimetallic Prussian Blue Analogues (KxM[Fe(CN)6]·nH_2O, M=Ni, Co, and Mn) for Electrochemical Determination of Toxic 2-Nitrophenol. *Electrochim. Acta* **2020**, *353*, 136579. [CrossRef]
265. Zhao, H.-C.; Zhang, P.; Li, S.-H.; Luo, H.-X. Cobalt Hexacyanoferrate-Modified Graphene Platform Electrode and Its Electrochemical Sensing toward Hydrogen Peroxide. *Chin. J. Anal. Chem.* **2017**, *45*, 830–836. [CrossRef]
266. Pang, H.; Zhang, Y.; Cheng, T.; Lai, W.-Y.; Huang, W. Uniform Manganese Hexacyanoferrate Hydrate Nanocubes Featuring Superior Performance for Low-Cost Supercapacitors and Nonenzymatic Electrochemical Sensors. *Nanoscale* **2015**, *7*, 16012–16019. [CrossRef]
267. Pakrudheen, I.; Bathinapatla, A.; Murugan, E. Amphiphilic Dendrimer Loaded Prussian Blue Nanoparticle for the Detection of Hydrogen Peroxide. *Indian J. Chem. Technol.* **2022**, *29*, 668–677. [CrossRef]
268. Zhang, M.; Zhang, W.; Engelbrekt, C.; Hou, C.; Zhu, N.; Chi, Q. Size-Dependent and Self-Catalytic Gold@Prussian Blue Nanoparticles for the Electrochemical Detection of Hydrogen Peroxide. *ChemElectroChem* **2020**, *7*, 3818–3823. [CrossRef]
269. Noël, J.; Médard, J.; Combellas, C.; Kanoufi, F. Prussian Blue Degradation during Hydrogen Peroxide Reduction: A Scanning Electrochemical Microscopy Study on the Role of the Hydroxide Ion and Hydroxyl Radical. *ChemElectroChem* **2016**, *3*, 1178–1184. [CrossRef]
270. Uzuncar, S.; Ozdogan, N.; Ak, M. Enzyme-Free Detection of Hydrogen Peroxide with a Hybrid Transducing System Based on Sodium Carboxymethyl Cellulose, Poly(3,4-Ethylenedioxythiophene) and Prussian Blue Nanoparticles. *Anal. Chim. Acta* **2021**, *1172*, 338664. [CrossRef]
271. Ying, S.; Chen, C.; Wang, J.; Lu, C.; Liu, T.; Kong, Y.; Yi, F. Synthesis and Applications of Prussian Blue and Its Analogues as Electrochemical Sensors. *Chempluschem* **2021**, *86*, 1608–1622. [CrossRef] [PubMed]
272. Sohrabi, H.; Maleki, F.; Khaaki, P.; Kadhom, M.; Kudaibergenov, N.; Khataee, A. Electrochemical-Based Sensing Platforms for Detection of Glucose and H_2O_2 by Porous Metal–Organic Frameworks: A Review of Status and Prospects. *Biosensors* **2023**, *13*, 347. [CrossRef]
273. Hu, Y.; Hojamberdiev, M.; Geng, D. Recent Advances in Enzyme-Free Electrochemical Hydrogen Peroxide Sensors Based on Carbon Hybrid Nanocomposites. *J. Mater. Chem. C* **2021**, *9*, 6970–6990. [CrossRef]
274. Shamkhalichenar, H.; Choi, J.-W. Review—Non-Enzymatic Hydrogen Peroxide Electrochemical Sensors Based on Reduced Graphene Oxide. *J. Electrochem. Soc.* **2020**, *167*, 037531. [CrossRef]
275. Wu, X.; Ma, P.; Sun, Y.; Du, F.; Song, D.; Xu, G. Application of MXene in Electrochemical Sensors: A Review. *Electroanalysis* **2021**, *33*, 1827–1851. [CrossRef]
276. Song, M.; Wang, J.; Chen, B.; Wang, L. A Facile, Nonreactive Hydrogen Peroxide (H_2O_2) Detection Method Enabled by Ion Chromatography with UV Detector. *Anal. Chem.* **2017**, *89*, 11537–11544. [CrossRef] [PubMed]
277. Tantawi, O.; Baalbaki, A.; El Asmar, R.; Ghauch, A. A Rapid and Economical Method for the Quantification of Hydrogen Peroxide (H_2O_2) Using a Modified HPLC Apparatus. *Sci. Total Environ.* **2019**, *654*, 107–117. [CrossRef]
278. Chiesa, M.; Giamello, E.; Livraghi, S.; Paganini, M.C.; Polliotto, V.; Salvadori, E. Electron Magnetic Resonance in Heterogeneous Photocatalysis Research. *J. Phys. Condens. Matter* **2019**, *31*, 444001. [CrossRef]
279. Fernández-Castro, P.; Vallejo, M.; San Román, M.F.; Ortiz, I. Insight on the Fundamentals of Advanced Oxidation Processes. Role and Review of the Determination Methods of Reactive Oxygen Species. *J. Chem. Technol. Biotechnol.* **2015**, *90*, 796–820. [CrossRef]
280. Ngo, T.T.; Lenhoff, H.M. A Sensitive and Versatile Chromogenic Assay for Peroxidase and Peroxidase-Coupled Reactions. *Anal. Biochem.* **1980**, *105*, 389–397. [CrossRef]
281. Cao, W.; Ju, P.; Wang, Z.; Zhang, Y.; Zhai, X.; Jiang, F.; Sun, C. Colorimetric Detection of H_2O_2 Based on the Enhanced Peroxidase Mimetic Activity of Nanoparticles Decorated Ce2(WO4)3 Nanosheets. *Spectrochim. Acta—Part A Mol. Biomol. Spectrosc.* **2020**, *239*, 118499. [CrossRef] [PubMed]
282. Wei, H.; Wang, E. Fe_3O_4 Magnetic Nanoparticles as Peroxidase Mimetics and Their Applications in H_2O_2 and Glucose Detection. *Anal. Chem.* **2008**, *80*, 2250–2254. [CrossRef] [PubMed]
283. Masschelein, W.; Denis, M.; Ledent, R. Spectrophotometric determination of residual hydrogen peroxide. *Water Sew. Work.* **1977**, *124*, 69–72.

284. Vaithyanathan, V.K.; Ravi, S.; Leduc, R.; Vaidyanathan, V.K.; Cabana, H. Utilization of Biosolids for Glucose Oxidase Production: A Potential Bio-Fenton Reagent for Advanced Oxidation Process for Removal of Pharmaceutically Active Compounds. *J. Environ. Manag.* **2020**, *271*, 110995. [CrossRef]
285. Kosaka, K.; Yamada, H.; Matsui, S.; Echigo, S.; Shishida, K. Comparison among the Methods for Hydrogen Peroxide Measurements To Evaluate Advanced Oxidation Processes: Application of a Spectrophotometric Method Using Copper(II) Ion and 2,9-Dimethyl-1,10-Phenanthroline. *Environ. Sci. Technol.* **1998**, *32*, 3821–3824. [CrossRef]
286. Bader, H.; Sturzenegger, V.; Hoigné, J. Photometric Method for the Determination of Low Concentrations of Hydrogen Peroxide by the Peroxidase Catalyzed Oxidation of N,N-Diethyl-p-Phenylenediamine (DPD). *Water Res.* **1988**, *22*, 1109–1115. [CrossRef]
287. Shaked, Y.; Armoza-Zvuloni, R. Dynamics of Hydrogen Peroxide in a Coral Reef: Sources and Sinks. *J. Geophys. Res. Biogeosci.* **2013**, *118*, 1793–1801. [CrossRef]
288. Schick, R.; Strasser, I.; Stabel, H.-H. Fluorometric Determination of Low Concentrations of H_2O_2 in Water: Comparison with Two Other Methods and Application to Environmental Samples and Drinking-Water Treatment. *Water Res.* **1997**, *31*, 1371–1378. [CrossRef]
289. Priyadarshini, M.; Das, I.; Ghangrekar, M.M.; Blaney, L. Advanced Oxidation Processes: Performance, Advantages, and Scale-up of Emerging Technologies. *J. Environ. Manag.* **2022**, *316*, 115295. [CrossRef]
290. Carvalho Neves, L.; Beber de Souza, J.; de Souza Vidal, C.M.; Herbert, L.T.; de Souza, K.V.; Geronazzo Martins, K.; Young, B.J. Phytotoxicity Indexes and Removal of Color, COD, Phenols and ISA from Pulp and Paper Mill Wastewater Post-Treated by UV/H_2O_2 and Photo-Fenton. *Ecotoxicol. Environ. Saf.* **2020**, *202*, 110939. [CrossRef]
291. Hofman-Caris, R.C.H.M.; Harmsen, D.J.H.; Puijker, L.; Baken, K.A.; Wols, B.A.; Beerendonk, E.F.; Keltjens, L.L.M. Influence of Process Conditions and Water Quality on the Formation of Mutagenic Byproducts in UV/H_2O_2 Processes. *Water Res.* **2015**, *74*, 191–202. [CrossRef]
292. Rozas, O.; Vidal, C.; Baeza, C.; Jardim, W.F.; Rossner, A.; Mansilla, H.D. Organic Micropollutants (OMPs) in Natural Waters: Oxidation by UV/H_2O_2 Treatment and Toxicity Assessment. *Water Res.* **2016**, *98*, 109–118. [CrossRef]
293. Lopez-Lopez, C.; Purswani, J.; Martín-Pascual, J.; Martínez-Toledo, M.V.; Muñío, M.M.; Poyatos, J.M. Toxic Effect of H_2O_2 in H_2O_2/UV, Photo-Fenton and Heterogeneous Photocatalysis ($TiO_2/H_2O_2/UV$) Systems to Treat Textile Wastewater. *Desalin. Water Treat.* **2015**, *56*, 3044–3053. [CrossRef]
294. Arshad, R.; Bokhari, T.H.; Javed, T.; Bhatti, I.A.; Rasheed, S.; Iqbal, M.; Nazir, A.; Naz, S.; Khan, M.I.; Khosa, M.K.K.; et al. Degradation Product Distribution of Reactive Red-147 Dye Treated by $UV/H_2O_2/TiO_2$ Advanced Oxidation Process. *J. Mater. Res. Technol.* **2020**, *9*, 3168–3178. [CrossRef]
295. Tony, M.A.; Lin, L.-S. Iron Coated-Sand from Acid Mine Drainage Waste for Being a Catalytic Oxidant Towards Municipal Wastewater Remediation. *Int. J. Environ. Res.* **2021**, *15*, 191–201. [CrossRef]
296. Ferro, G.; Guarino, F.; Castiglione, S.; Rizzo, L. Antibiotic Resistance Spread Potential in Urban Wastewater Effluents Disinfected by UV/H_2O_2 Process. *Sci. Total Environ.* **2016**, *560–561*, 29–35. [CrossRef]
297. Rioja, N.; Zorita, S.; Peñas, F.J. Effect of Water Matrix on Photocatalytic Degradation and General Kinetic Modeling. *Appl. Catal. B Environ.* **2016**, *180*, 330–335. [CrossRef]
298. Maniakova, G.; Kowalska, K.; Murgolo, S.; Mascolo, G.; Libralato, G.; Lofrano, G.; Sacco, O.; Guida, M.; Rizzo, L. Comparison between Heterogeneous and Homogeneous Solar Driven Advanced Oxidation Processes for Urban Wastewater Treatment: Pharmaceuticals Removal and Toxicity. *Sep. Purif. Technol.* **2020**, *236*, 116249. [CrossRef]
299. Wu, Y.; Chu, L.; Liu, W.; Jiang, L.; Chen, X.; Wang, Y.; Zhao, Y. The Screening of Metal Ion Inhibitors for Glucose Oxidase Based on the Peroxidase-like Activity of Nano-Fe_3O_4. *RSC Adv.* **2017**, *7*, 47309–47315. [CrossRef]
300. Vitale, A.A.; Bernatene, E.A.; Pomilio, A.B. Inhibition of Fenton Reaction of Glucose by Alcohols and Tetrahydrofuran in Catalytic Concentrations: Calculation of the Stability Constants of ROH/Fe^{2+} Complexes. *Curr. Phys. Chem.* **2022**, *12*, 76–87. [CrossRef]
301. Lindsey, M. Inhibited Hydroxyl Radical Degradation of Aromatic Hydrocarbons in the Presence of Dissolved Fulvic Acid. *Water Res.* **2000**, *34*, 2385–2389. [CrossRef]
302. Malvestiti, J.A.; Dantas, R.F. Disinfection of Secondary Effluents by O_3, O_3/H_2O_2 and UV/H_2O_2: Influence of Carbonate, Nitrate, Industrial Contaminants and Regrowth. *J. Environ. Chem. Eng.* **2018**, *6*, 560–567. [CrossRef]
303. Yang, Z.; Qian, J.; Shan, C.; Li, H.; Yin, Y.; Pan, B. Toward Selective Oxidation of Contaminants in Aqueous Systems. *Environ. Sci. Technol.* **2021**, *55*, 14494–14514. [CrossRef]
304. Mahbub, P.; Duke, M. Scalability of Advanced Oxidation Processes (AOPs) in Industrial Applications: A Review. *J. Environ. Manag.* **2023**, *345*, 118861. [CrossRef] [PubMed]
305. Zhang, M.; Dong, H.; Zhao, L.; Wang, D.; Meng, D. A Review on Fenton Process for Organic Wastewater Treatment Based on Optimization Perspective. *Sci. Total Environ.* **2019**, *670*, 110–121. [CrossRef] [PubMed]
306. Ghafoori, S.; Mehrvar, M.; Chan, P.K. Photoreactor Scale-up for Degradation of Aqueous Poly(Vinyl Alcohol) Using UV/H_2O_2 Process. *Chem. Eng. J.* **2014**, *245*, 133–142. [CrossRef]
307. Bar-Niv, N.; Azaizeh, H.; Kuc, M.E.; Azerrad, S.; Haj-Zaroubi, M.; Menashe, O.; Kurzbaum, E. Advanced Oxidation Process $UV-H_2O_2$ Combined with Biological Treatment for the Removal and Detoxification of Phenol. *J. Water Process Eng.* **2022**, *48*, 102923. [CrossRef]
308. Otálvaro-Marín, H.L.; González-Caicedo, F.; Arce-Sarria, A.; Mueses, M.A.; Crittenden, J.C.; Machuca-Martinez, F. Scaling-up a Heterogeneous H_2O_2/TiO_2/Solar-Radiation System Using the DamkÖhler Number. *Chem. Eng. J.* **2019**, *364*, 244–256. [CrossRef]

309. Pliego, G.; Zazo, J.A.; Garcia-Muñoz, P.; Munoz, M.; Casas, J.A.; Rodriguez, J.J. Trends in the Intensification of the Fenton Process for Wastewater Treatment: An Overview. *Crit. Rev. Environ. Sci. Technol.* **2015**, *45*, 2611–2692. [CrossRef]
310. Bashir, Y.; Raj, R.; Ghangrekar, M.M.; Nema, A.K.; Das, S. Critical Assessment of Advanced Oxidation Processes and Bio-Electrochemical Integrated Systems for Removing Emerging Contaminants from Wastewater. *RSC Sustain.* **2023**, *1*, 1912–1931. [CrossRef]
311. Sgroi, M.; Snyder, S.A.; Roccaro, P. Comparison of AOPs at Pilot Scale: Energy Costs for Micro-Pollutants Oxidation, Disinfection by-Products Formation and Pathogens Inactivation. *Chemosphere* **2021**, *273*, 128527. [CrossRef]
312. Yao, W.; Ur Rehman, S.W.; Wang, H.; Yang, H.; Yu, G.; Wang, Y. Pilot-Scale Evaluation of Micropollutant Abatements by Conventional Ozonation, UV/O3, and an Electro-Peroxone Process. *Water Res.* **2018**, *138*, 106–117. [CrossRef]
313. Mousset, E.; Loh, W.H.; Lim, W.S.; Jarry, L.; Wang, Z.; Lefebvre, O. Cost Comparison of Advanced Oxidation Processes for Wastewater Treatment Using Accumulated Oxygen-Equivalent Criteria. *Water Res.* **2021**, *200*, 117234. [CrossRef]
314. Cañizares, P.; Paz, R.; Sáez, C.; Rodrigo, M.A. Costs of the Electrochemical Oxidation of Wastewaters: A Comparison with Ozonation and Fenton Oxidation Processes. *J. Environ. Manag.* **2009**, *90*, 410–420. [CrossRef] [PubMed]
315. Rizzo, L.; Gernjak, W.; Krzeminski, P.; Malato, S.; McArdell, C.S.; Perez, J.A.S.; Schaar, H.; Fatta-Kassinos, D. Best Available Technologies and Treatment Trains to Address Current Challenges in Urban Wastewater Reuse for Irrigation of Crops in EU Countries. *Sci. Total Environ.* **2020**, *710*, 136312. [CrossRef] [PubMed]
316. Ribeiro, J.P.; Sarinho, L.; Nunes, M.I. Application of Life Cycle Assessment to Fenton Processes in Wastewater Treatment—A Review. *J. Water Process Eng.* **2024**, *57*, 104692. [CrossRef]
317. Garcia Montano, J.; Ruiz, N.; Munoz, I.; Domenech, X.; Garciahortal, J.; Torrades, F.; Peral, J. Environmental Assessment of Different Photo-Fenton Approaches for Commercial Reactive Dye Removal. *J. Hazard. Mater.* **2006**, *138*, 218–225. [CrossRef]
318. Ioannou-Ttofa, L.; Foteinis, S.; Chatzisymeon, E.; Michael-Kordatou, I.; Fatta-Kassinos, D. Life Cycle Assessment of Solar-driven Oxidation as a Polishing Step of Secondary-treated Urban Effluents. *J. Chem. Technol. Biotechnol.* **2017**, *92*, 1315–1327. [CrossRef]
319. Costamagna, M.; Arques, A.; Lo-Iacono-Ferreira, V.G.; Bianco Prevot, A. Environmental Assessment of Solar Photo-Fenton Processes at Mild Condition in the Presence of Waste-Derived Bio-Based Substances. *Nanomaterials* **2022**, *12*, 2781. [CrossRef]
320. Raj, R.; Tripathi, A.; Das, S.; Ghangrekar, M.M. Is Waste-Derived Catalyst Mediated Electro-Fenton a Sustainable Option for Mitigating Emerging Contaminants from Wastewater? *Curr. Opin. Environ. Sci. Health* **2024**, *37*, 100523. [CrossRef]
321. Feijoo, S.; González-Rodríguez, J.; Fernández, L.; Vázquez-Vázquez, C.; Feijoo, G.; Moreira, M.T. Fenton and Photo-Fenton Nanocatalysts Revisited from the Perspective of Life Cycle Assessment. *Catalysts* **2019**, *10*, 23. [CrossRef]
322. Rodríguez, R.; Espada, J.J.; Pariente, M.I.; Melero, J.A.; Martínez, F.; Molina, R. Comparative Life Cycle Assessment (LCA) Study of Heterogeneous and Homogenous Fenton Processes for the Treatment of Pharmaceutical Wastewater. *J. Clean. Prod.* **2016**, *124*, 21–29. [CrossRef]
323. Fan, Y.; Ai, Z.; Zhang, L. Design of an Electro-Fenton System with a Novel Sandwich Film Cathode for Wastewater Treatment. *J. Hazard. Mater.* **2010**, *176*, 678–684. [CrossRef] [PubMed]
324. Ting, W.-P.; Lu, M.-C.; Huang, Y.-H. The Reactor Design and Comparison of Fenton, Electro-Fenton and Photoelectro-Fenton Processes for Mineralization of Benzene Sulfonic Acid (BSA). *J. Hazard. Mater.* **2008**, *156*, 421–427. [CrossRef] [PubMed]
325. Hussain, S.; Aneggi, E.; Goi, D. Catalytic Activity of Metals in Heterogeneous Fenton-like Oxidation of Wastewater Contaminants: A Review. *Environ. Chem. Lett.* **2021**, *19*, 2405–2424. [CrossRef]
326. Salmerón, I.; Plakas, K.V.; Sirés, I.; Oller, I.; Maldonado, M.I.; Karabelas, A.J.; Malato, S. Optimization of Electrocatalytic H_2O_2 Production at Pilot Plant Scale for Solar-Assisted Water Treatment. *Appl. Catal. B Environ.* **2019**, *242*, 327–336. [CrossRef]
327. Bourgin, M.; Beck, B.; Boehler, M.; Borowska, E.; Fleiner, J.; Salhi, E.; Teichler, R.; von Gunten, U.; Siegrist, H.; McArdell, C.S. Evaluation of a Full-Scale Wastewater Treatment Plant Upgraded with Ozonation and Biological Post-Treatments: Abatement of Micropollutants, Formation of Transformation Products and Oxidation by-Products. *Water Res.* **2018**, *129*, 486–498. [CrossRef]
328. Iervolino, G.; Vaiano, V.; Pepe, G.; Campiglia, P.; Palma, V. Degradation of Acid Orange 7 Azo Dye in Aqueous Solution by a Catalytic-Assisted, Non-Thermal Plasma Process. *Catalysts* **2020**, *10*, 888. [CrossRef]

catalysts

MDPI

Article

Insight into the Catalytic Performance of a Zinc-Pillared Curcumin/Bentonite Composite for Enhanced Oxidation of Ibuprofen Residuals into Environmental Products: The Pathway and Toxicity

Sarah I. Othman [1], Marwa H. Shemy [2,3], Haifa E. Alfassam [1], Haifa A. Alqhtani [1], Ahmed A. Allam [4,5], Mostafa R. Abukhadra [3,6,*] and Stefano Bellucci [7,*]

1 Biology Department, College of Science, Princess Nourah bint Abdulrahman University, Riyadh 11671, Saudi Arabia; healfassam@pnu.edu.sa (H.E.A.)
2 Chemistry Department, Faculty of Science, Beni-Suef University, Beni-Suef 65211, Egypt
3 Materials Technologies and their Applications Lab, Geology Department, Faculty of Science, Beni-Suef University, Beni-Suef 65211, Egypt
4 Zoology Department, Faculty of Science, Beni-Suef University, Beni-Suef 65211, Egypt; ahmed.aliahmed@science.bsu.edu.eg
5 Department of Biology, College of Science, Imam Mohammad Ibn Saud Islamic University (IMSIU), Riyadh 11623, Saudi Arabia
6 Geology Department, Faculty of Science, Beni-Suef University, Beni-Suef 65211, Egypt
7 INFN, Laboratori Nazionali di Frascati, E. Fermi 54, 00044 Frascati, Italy
* Correspondence: abukhadra89@science.bsu.edu.eg (M.R.A.); stefano.bellucci@lnf.infn.it (S.B.); Tel.: +20-01288447189 (M.R.A.)

Abstract: Environmental organo bentonite synthesis using curcumin-derived chemicals is used as catalyst support for zinc with a Zn-pillaring structure (Zn@CU/BEN). The obtained composite was assessed as an affordable, highly effective, and multifunctional photocatalyst for enhanced oxidation of ibuprofen (IBU) residuals in water supplies. The Zn@CU/BEN composite (0.4 g/L) displayed significant catalytic activities, resulting in the complete oxidation of IBU (25 mg/L) after 80 min. Then, the complete mineralization based on the full elimination of TOC content was recognized after 160 min, with significant indications about the formed intermediates. The identified intermediates, together with the identification of hydroxyl radicals as the essential oxidizing agent, declared an oxidation pathway of IBU over Zn@CU/BEN that involved three steps: hydroxylation, decarboxylation/demethylation, and ring-opening processes. The toxic properties of raw pollutants as well as the oxidizing product at different durations were assessed based on the cell viability results of kidney (HEK293T) and liver (HepG2) cell lines. The partially oxidized sample in the initial duration displayed a higher toxicity impact than the raw IBU. However, the treated sample after 160 min reflected high biosafety and non-toxic properties (cell viability > 97%). The synergetic impact of bentonite and bentonite organo-modified by curcumin extract reflects enhancements in the adsorption as well as the oxidation performance of pillared zinc as a catalyst.

Keywords: bentonite; curcumin; zinc metal; ibuprofen; oxidation; toxicity

1. Introduction

A broad range of medication residues, together with their metabolic byproducts, have been recorded to have shown up extensively in a variety of natural environments, including sewage water, surface water, groundwater, and drinking water [1,2]. The pollution of vital drinking water supplies with these chemicals correlates with the discharging of waste effluents from hospitals, drug industries, and healthcare institutes in addition to human excrement [1,3]. Ibuprofen (IBU) is a frequently utilized non-steroidal medicine with antipyretic, anti-inflammatory, and analgesic properties. It has been extensively identified as an organic contaminant or pharmaceutical residue in water bodies [4,5]. IBU possesses an extremely high yearly consumption of approximately 15,000 tons per year and has been described as one of the top 50 medications with the greatest environmental hazards [6,7]. The leftovers of IBU encountered in water bodies were categorized as hazardous, non-biodegradable, and chemically resistant pollutants [5,8]. The presence of IBU, including its metabolic byproducts, within the water resources, regardless of minimal levels, has a detrimental effect on the endocrine system's functioning in humans and the reproductive abilities of aquatic organisms [8–10]. The IBU-soluble molecules exhibit bioaccumulation characteristics in fish and other aquatic life and have adverse influences on existing freshwater plants and animals [11]. Furthermore, the IBU molecules undergo frequent conversion into types of secondary compounds such as 4-isobutylacetophenone, which exist widely within the freshwater supplies and are more hazardous than the original compounds [11]. Moreover, long-term exposure to IBU has detrimental impacts on both human liver cells and kidney cells, regardless of whether the contact is brief [12].

Due to its classification as a hazardous contaminant, significant endeavors have been launched by environmental and scientific organizations to eliminate IBU from water [1,8]. In recent times, the implementation of adsorption alongside advanced oxidation as effective methods for removing organic chemicals from water has been highly promoted [13,14]. Advanced oxidation techniques, including photocatalytic, electrochemical, sonochemical, and photo-Fenton decomposition, were recently classified as highly successful approaches for decontaminating organic compounds [15–17]. The advanced methods of oxidation result in extensive degrading and mineralization of organic compounds, leading to the formation of environmentally safe compounds [18,19]. In this regard, a variety of individual and combined nano/microstructures were evaluated for their capacity to photocatalytically oxidize and photo-Fenton oxidize IBU residues within water [3,9]. Selecting a suitable catalyst to provide promising oxidation effects is determined by several factors, spanning both environmental and commercial aspects, including recycling potential, biodegradability, bandgap energy, recovery, thermal stability, recombination speeds, security, production cost, and efficiency of adsorption [20,21]. With these in mind, it was strongly recommended that efforts should go toward developing inventive structures, employing naturally existing precursors, which exhibit improved adsorption and oxidizing abilities [22].

Several inorganic materials, including semi-conductive metals as well as metal oxides, were explored as possible photocatalysts. These materials were assessed either as pure in nature or blends with other components [22–24]. The latest studies focused on investigating multifunctional materials that exhibit enhanced optical and retention properties, with the goal being to to develop enhanced photocatalysts [9,24]. Zinc-based nano- and microstructures were extensively analyzed as potential photocatalysts to trigger the oxidative decomposition of organic compounds [22,25]. Overall, this material exhibits outstanding properties such as excellent durability, non-toxicity, biocompatibility, a substantial surface area, sufficient resistance against photo-corrosion, and noteworthy oxidation capacities. Additionally, it possesses considerable potential for free excitation energy [26–28]. The utilization of Zn or ZnO as a photocatalyst can be improved by incorporating them into composite materials, supplementing their structures with rare metals, and embedding their particulates inside appropriate carriers. This approach has been suggested as a straightforward method that can improve the photocatalytic effectiveness, including regeneration, recombination rate, exciting effectiveness, and adsorption capacity [22,29].

However, studies that have been conducted in this area reveal notable impacts of factors such as crystal structure, morphological characteristics, synthesizing techniques, mechanical strength, crystallite dimensions, hybridization, and surface-modifying procedures on the qualities and applications of zinc-based materials [26,30,31]. Elsewhere, in research evaluating zinc-based hybrids with various types of biopolymers produced using naturally occurring phytochemicals derived from plant-based extracts led to the production of environmentally friendly products with improved technical characteristics [27,32]. Thus, it seems that when employing solutions extracted from edible plants that are enriched in various types of widely recognized phytochemicals, the production of zinc-based nanostructures based on these extracts as a green synthesis method may yield materials with improved physicochemical and environmental characteristics. This approach was highly praised for the development of non-agglomerated nanostructures. The resulting nanoparticles have been identified to have affordable manufacturing costs, require effortless fabrication processes, and boast non-toxicity, a reactive functionalized surface, and safety [27,33]. Additionally, the nanoparticles that were produced were thoroughly coated with films of phytochemicals, including phenolic chemicals, alkaloids, amino acids, and proteins, forming types of metal/phytochemical complexes. These phytochemical coatings enhance the exterior properties of the resulting particles, facilitating the adsorption of organic molecules on their surfaces [27].

Against that background, this work entailed a thorough analysis of the implementation of a bentonite multilayered structure as an efficient substrate or carrier of green-formed Zn metal, curcumin-derived phytochemicals, and complexes of Zn/curcumin-based phytochemicals as an improved environmentally friendly catalytic agent. Curcumin is a naturally occurring pigment with a diphenolic β-diketone structure that is obtained directly from the rhizome of *Curcuma longa* L., frequently referred to as turmeric. It is extensively utilized as a spice and a medicinal ingredient [34]. The extremely symmetrical arrangement of delocalized π electrons within curcumin along with its derived compounds gives rise to remarkable electrical together with optical abilities [34,35]. Recently, it has been established as a photosensitive agent for use in photodynamic treatment, demonstrating antibacterial and anticancer properties. The formulation of a curcumin-based complex containing zinc has already been employed for various environmentally friendly uses, including the elimination of arsenite, fluoranthene, perylene, and chrysene [35]. Curcumin has the ability to reduce the level of organic compounds when paired with metal oxide nanoparticles. Consequently, its combination with zinc oxide might augment its capacity to adsorb and boost its photocatalytic activity [35].

Bentonite (BEN) is a versatile type of clay mineral that is frequently employed in various environmental sectors. It has distinctive characteristics that enable it to be utilized as an adsorbent alongside being a support material for metals and metal-oxide-based catalysts [36–38]. BEN is distinguished by its abundant availability, excessive ion exchange potential, reactive surface, large surface area, non-toxic qualities, biocompatibility, and ability to adsorb and absorb chemicals. These qualities make it a valuable carrier for certain metals, as it assists in enhancing both the physical and chemical characteristics of the catalyst [36,39]. In this study, the incorporation of zinc into the bentonite layers has been accomplished by pillaring the metal together with curcumin, which served as both the reduction and capping reagents. This process resulted in the formation of a zinc-capped curcumin or zinc/curcumin complex as pillared units between the bentonite layers, forming a hybrid structure denoted as Zn@CU/BEN. The incorporation of curcumin-derived phytochemicals proceeded through a conventional organic intercalation procedure, and we utilized its extract to produce a specific type of safe organo-modified bentonite.

The synthetic catalyst was assessed experimentally as a potential and effective photocatalyst for enhanced degradation of IBU residuals from water supplies in the presence of a visible light source. The catalytic activity was monitored based on the degradation percentages in terms of the affected experimental variables, the TOC content and mineralization degree, the formed intermediate compounds and suggested pathway, and the kinetic properties. Moreover, our study involved a detailed toxicological investigation of the treated sample after different oxidation intervals, to follow the toxicity of the intermediate compound and the safety limitations of the applied photocatalytic degradation system of IBU based on the Zn@CU/BEN catalyst.

2. Results and Discussion

2.1. Characterization of the Catalyst

The structural impact of synthesis, including the crystallized phases, was evaluated based on XRD patterns (Figure 1). The employed BEN precursor exhibits the distinctive peaks of montmorillonite (5.78° (001) as well as 6.95° (002)), which are its primary constituents, together with specific impurities like kaolinite and quartz (ref. card no. 000-003-0010) (Figure 1A). The estimated crystallite size and d-spacing values of the montmorillonite component are 12.9 nm and 12.71 Å, respectively. The acid treatment of BEN caused significant modifications with respect to the crystalline structure of montmorillonite. The key peaks exhibited a significant shift towards lower levels (5.05° and 6.6°), accompanied by a notable decrease in their magnitudes (Figure 1B). These reflect the partially destroying effect of the acid on the BEN lattice structure, which is often accompanied by improvements in the reactivity, textural properties, and physicochemical behaviors of the modified product [40,41]. The XRD pattern of Zn@CU/BEN confirmed the different integration and modification steps, including the intercalation of the curcumin-related organic phytochemicals and the Zn-pillaring step (Line c in Figure 1). The intercalation step resulted in a marked exfoliation impact on the bentonite layered units, which was reflected in the notable amorphization or destruction of its structure and the significant intensification of the d-spacing to 26.3 Å (Line c in Figure 1). The intercalated curcumin was confirmed by three corresponding detectable peaks at 12.2°, 14.42°, and 19.5° (Line c in Figure 1). Also, the pillaring of Zn metals with 10.8 nm crystallite size was confirmed by the reduced peaks within the area from 30° to 80° (36.41° (002), 39.0° (100), 43.30° (101), 54.41° (102), and 36.40° (101)) (Figure 1B) (JCPDS 36–1451; JCPDS 00-004-0784).

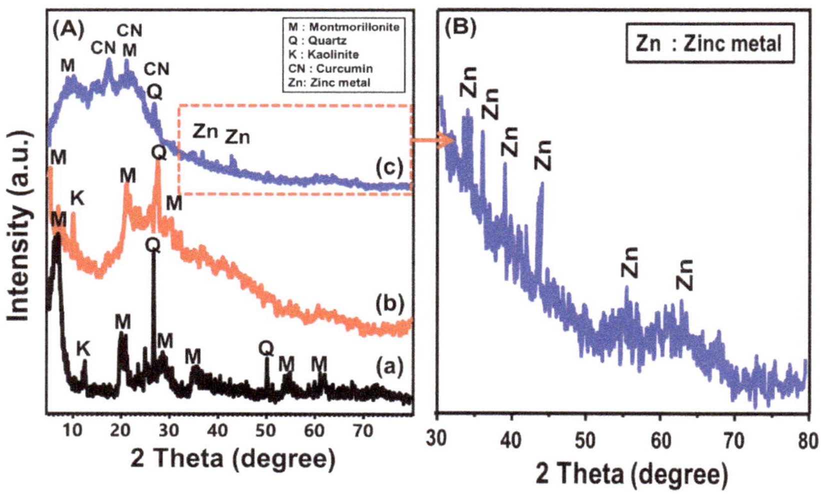

Figure 1. (**A**) XRD patterns of the synthetic structures (raw BEN (a), activated BEN (b), and synthesized Zn@CU/BEN (c)) and (**B**) focus view on the diffraction peaks of zinc metal.

Concerning morphology, the employed bentonite substrate could be distinguished as having prominent aggregation consisting of packed and compressed flake-like particulates (Figure 2A). The distinct cornflake shape of montmorillonite was readily apparent in the high-resolution photos, where the montmorillonite platelets displayed bent and curved shapes (Figure 2B). The HRTEM photo of BEN particulates exhibited an obvious and substantial multilayered interior framework, showcasing the distinctive lattice figures typical of montmorillonite (Figure 2C). Following the addition of curcumin-derived extract, the layered units of BEN expanded and separated notably (Figure 2D). The HRTEM photos also revealed the interplay of the clay sheets with tiny coatings of polymeric organic molecules, which could potentially be associated with the derived chemical from curcumin (Figure 2). The filamentous forms were clearly visible and had a darker gray color relative to the silicate sheets of BEN (Figure 2E). The synthesized Zn@CU/BEN particulates had highly decorative exfoliated layers that were capped with many spherically shaped granules of the synthesized pillared zinc metal (Figure 2F). The development and dispersion of the aforementioned pillared metallic zinc nanoparticles frequently resulted in interconnecting nanopores, which had a substantial beneficial impact on the surface area (Figure 2G,H). The HRTEM scans revealed an array of synthesized zinc nanoparticles distributed inside the BEN layers, confirming their development as pillared nanostructures and surficial-coated sheets (Figure 2I).

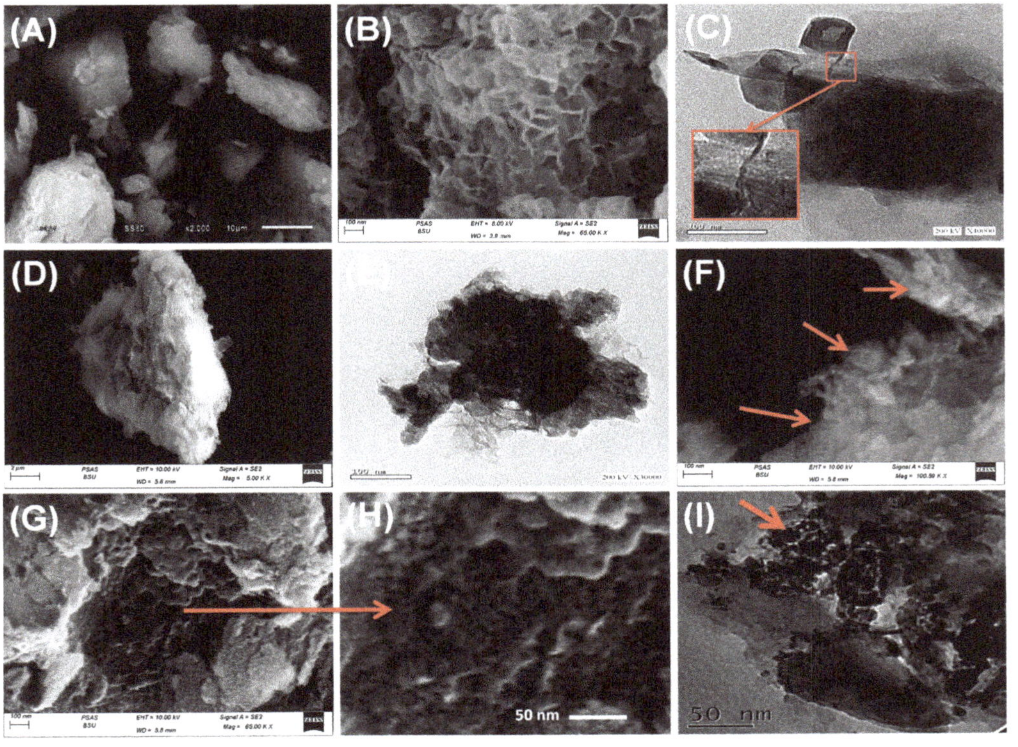

Figure 2. SEM image of raw BEN (**A**,**H**), high-magnification SEM image of BEN surface (**B**), HRTEM images of BEN (**C**), SEM image of CU/BEN (**D**), HRTEM image of CU/BEN (**E**), SEM images of ZN@CU/BEN (**F–H**), and HRTEM image of Zn@CU/BEN (**I**) (the red arrow refer to the Zinc oxide particulates).

The morphological impacts caused substantial modifications in the texture of the hybridized structure, particularly in terms of its porosity and surface area. The measured surface areas of BEN (91.2 m^2/g), A.BEN (98.7 m^2/g), Cu/BEN (106.3 m^2/g), and Zn@CU/BEN (156.7 m^2/g) demonstrated significant enhancement in terms of the different modification phases. The significant increase in surface area resulting from the intercalation procedure with the curcumin-derived extracted solutions may be attributed to the expanding and exfoliating effect caused by these elongated organic molecules between the layered units. Additionally, the pillared nanoparticles of zinc metal that grow within the layers of BEN give the structure a very porous feature that significantly increases the surface area. Furthermore, the documented microscopic characteristics of the capped zinc coating on the exterior of the dispersed BEN sheets are composed of tiny spherical nanoparticles containing large numbers of interstitial nanopores that significantly increase the surface area.

FT-IR spectra were employed to track the chemical functional groups throughout the several transformation operations. The spectrum of the implemented BEN particulates confirmed the presence of the key BEN chemical groups, including coordinated forms of -OH (3400 cm^{-1}) with the structural octahedral ions (AlMg(OH), Al_2OH, and $AlFe^{3+}(OH)$), Si-O (1000 cm^{-1}), interlayer water (1640 cm^{-1}), and Al-O (918 cm^{-1}). The spectral bands ranging from 1000 cm^{-1} to 400 cm^{-1} correspond to the structural bonds of Si-O-Al, Si-O-Mg, Mg-Fe-OH, and Si-O-Si [42,43] (Figure 3A). The spectrum of A.BEN did not show any noticeable appearance of newly formed bands or disappearance of existing bands (Figure 3B). The fundamental impact of the process was demonstrated by the significantly shifted absorption bands, along with the notable enhancement in the magnitude of the distinct bands of the interlayered water and coordinated OH. These changes were a consequence of the hydration influence of the acid washing step (Figure 3B).

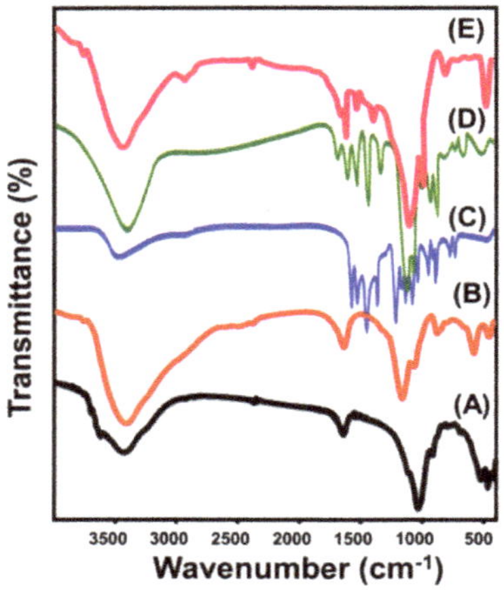

Figure 3. FT–IR spectra of raw BEN (A), activated BEN (B), curcumin (C), CU/BEN (D), and Zn@CU/BEN (E).

The spectrum retrieved for the curcumin powder that was utilized in this study showed the characteristic bands of its principal chemical structure. These include –OH of the phenolic chemicals (3506 cm^{-1}), aromatic C-H (2943–2965 cm^{-1}), C-H of olefin (1432 cm^{-1}), aromatic C-H (1162 and 812 cm^{-1}), cis-CH (716 cm^{-1}), benzoatetrans-CH (963 cm^{-1}), aromatic C-O (1278 cm^{-1}), C=O (1507 cm^{-1}), C-O-C (1023 cm^{-1}), C=C (1626 cm^{-1}), and a benzene ring (1604 cm^{-1}) [37,44]. The CU/BEN spectrum illustrates the effective integration of the inorganic framework of BEN with the organic constituents of the curcumin-derived extract (Figure 3C). The interaction impact is illustrated by the notable displacement of identifying bands representing the crucial chemical groups with respect to the initial materials (Figure 3D). The groups identified in BEN were Si-O (1008 cm^{-1}), Al-O (934 cm^{-1}), Si-O-Si (473.2 cm^{-1}), and Si-O-Al (534 cm^{-1}), while the detectable groups related to curcumin were aromatic C-O (1284 cm^{-1}), benzoatetrans-CH (921 cm^{-1}), C=O (1509 cm^{-1}), olefin C-H (1440 cm^{-1}), and benzene rings (1612 cm^{-1}) (Figure 3D). The established FT-IR spectrum of the synthesized Zn@CU/BEN composite also revealed alterations in the functioning chemical structures (Figure 3E). The organic groups that exist in curcumin-derived phytochemicals, besides the inorganic aluminosilicate-related groups that exist in BEN, were easily detected. The bands revealed significant deviations, indicating the effect of the zinc-pillaring steps (Figure 3E). Furthermore, the weakened bands at 602.7 and 462 cm^{-1} reflected the existence of Zn-O bonds (Figure 3E) [32,33]. This was in good alignment with the -EDX results and validated the detection of Si and Al representing the inorganic BEN host, C represents the organic compounds of the derived extract, and Zn representing the pillared and coated metal.

2.2. Photocatalytic Degradation Results

2.2.1. Effect of Oxidation Parameters

Effect of pH

Our investigation examined the performance of Zn@CU/BEN as a hybrid catalyst in terms of degrading IBU, with particular emphasis on the influence of pH values ranging from pH 3 to 8. The essential variables were consistently kept at certain values during the experiments. These levels were as follows: dosage of 0.25 g/L, volume of 100 mL, duration of 60 min, concentration of 25 mg/L, and temperature of 20 °C. The described oxidizing measurements demonstrated substantial enhancement when raising the pH towards alkaline conditions, particularly up to a pH of 6. The observed elimination percentage increased from 6.9% at pH 2 to 52.6% at pH 6 (Figure 4A). This tendency may be attributed to the insolubility of IBU within acidic environments and its poor retention characteristics during its molecular state [3,45]. Raising the pH level triggers the hydrophilic characteristics of IBU chemicals, leading to enhanced dissolution and retention of their structure [46]. The binding affinities of Zn@CU/BEN for soluble IBU are negatively affected or may be reversed when the pH is set above 6 (Figure 4A). The observed deprotonation of the existing carboxylate groups within IBU structures may be explained by the repelling interactions between these groups and the dominant negative charges throughout Zn@CU/BEN within alkaline environments [47]. Furthermore, the observed impact caused by elevated pH values on the generation of OH groups—which act as key intermediaries to feed the creation of principal hydroxyl radicals, which constitute the most potent oxidizing species—may also contribute to the improved elimination characteristics with regard to pH [48].

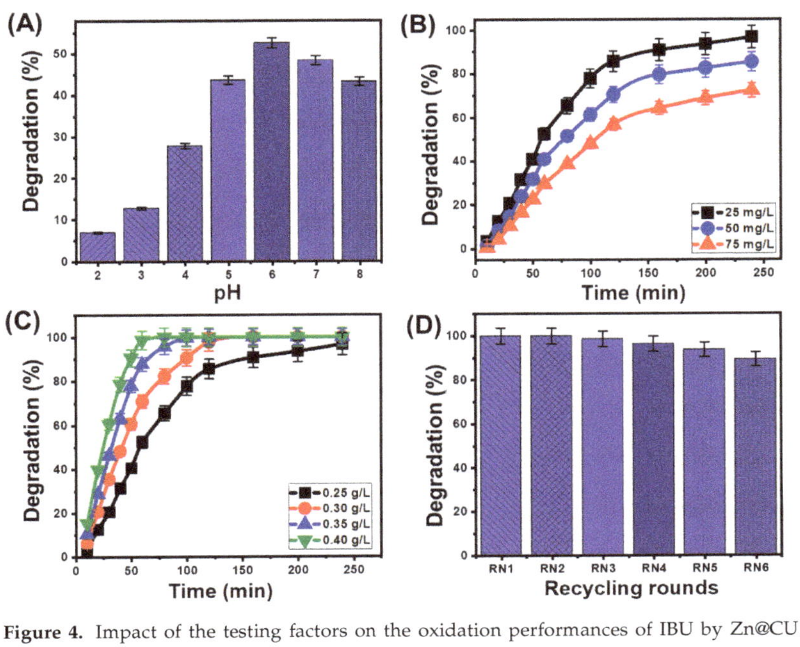

Figure 4. Impact of the testing factors on the oxidation performances of IBU by Zn@CU/BEN including pH (**A**), catalyst dose (**B**), IBU concentration (**C**), and recyclability properties (**D**).

Effects of IBU Concentrations at Different Degradation Durations

The catalytic properties of Zn@CU/BEN were investigated with respect to the period of oxidation (10–240 min) along with the concentrations of IBU (25–75 mg/L). The crucial variables, such as the dosage of 0.25 g/L, overall volume of 100 mL, pH level of 6, and temperature of 20 °C, were unchanged over the duration of the experiments. The Zn@CU/BEN structure exhibited significant oxidative properties throughout the photocatalytic degrading activities of IBU, leading to effective remediation (Figure 4B). The reduction percentages measured for the tested levels of IBU immediately after a period of 240 min were 96.7% (25 mg/L), 85.2% (50 mg/L), and 72.3% (75 mg/L) (Figure 4B). The reduction in the degrading efficiency of Zn@CU/BEN as the content of IBU rises can potentially be credited to the massive formation of thick adsorbed layers of IBU over its surface. The development of extensive coating layers of the adsorbed IBU, in conjunction with the reduced ability of light to penetrate through high IBU levels, ultimately leads to decreased interaction effectiveness between the outer surface of Zn@CU/BEN and incoming light rays [49]. Therefore, the generation of powerful oxidative free radicals is hindered, resulting in an overall decline in the successful completion of the degradation steps. The drop in decomposition rates is particularly notable as the oxidizing time frame is prolonged until reaching a level near the point of equilibrium, around which the degradation system shows a minimal or steady decomposition rate. The observed degradation behavior could potentially be owing to the steady reduction in the quantities of oxidized species generated throughout the IBU disintegration procedures. This decline persists until the complete depletion of these effective species is completed within a particular time frame [48].

Effect of Catalyst Dosage at Different Degradation Durations

The decomposition rates of IBU were evaluated in terms of the implemented dosages of the catalyst within the range of 0.25 g/L to 0.4 g/L of Zn@CU/BEN dosages. This variable is critical for determining the effectiveness of the photocatalytic oxidation reaction at a particular concentration of IBU. The duration of oxidation periods varied between 5 min and 240 min (Figure 4C). The important parameters were maintained at specific values (IBU concentration: 25 mg/L; volume: 100 mL; temperature: 20 °C; pH: 6) throughout testing. Utilizing elevated quantities of Zn@CU/BEN led to a dramatic enhancement in the speed of IBU oxidation, causing a remarkable decrease in the period required to completely degrade the evaluated levels of IBU (Figure 4C). Utilizing Zn@CU/BEN at varying levels (0.3 g/L, 0.35 g/L, and 0.4 g/L) resulted in the complete degradation of 25 mg/L of IBU during particular time periods (160 min, 100 min, and 80 min, respectively) (Figure 4C). The aforementioned behavior has been well-reported in the scientific literature and signifies a notable augmentation in the quantity of operative catalytic sites and the reacting surface area. The previously described investigations revealed that the implementation of Zn@CU/BEN improves the interaction interface involving IBU molecules and its external surface. Furthermore, it results in an increase in the efficiency of the oxidizing agents within the overall system [29,42].

Recyclability of Zn@CU/BEN

Throughout six recycling rounds (RN), the potential of Zn@CU/BEN to be regenerated in numerous cycles of IBU adsorption and degradation was investigated. Following the completion of the tests, the used Zn@CU/BEN portions were filtered and thoroughly rinsed using distilled water for 15 min. Subsequently, they were set aside in a drying furnace at a temperature of 60 °C for a period of 12 h. The potential for recycling of Zn@CU/BEN throughout the oxidation activities of IBU (25 mg/L) was evaluated under identical testing situations corresponding to the oxidizing tests, with a dose of 0.4 g/L and a duration of 100 min at pH 6 (Figure 4D). The findings demonstrated significant sustainability of the Zn@CU/BEN hybrid throughout the oxidization reactions of the IBU medication. The photocatalytic degradation recyclability tests resulted in IBU oxidation levels of 100% (RN1), 100% (RN2), 98.5% (RN3), 96.3% (RN4), 93.7% (RN5), and 89.3% (RN6) (Figure 4D).

This reveals the commercial significance of Zn@CU/BEN, since it could potentially be recycled effectively to eliminate the remaining IBU contaminants in wastewater.

2.2.2. Kinetic and Quantum Yield Studies

The kinetics of the photocatalytic removal reactions of IBU over Zn@CU/BEN were studied using the ideas of first-order (Figure 5A,B) and second-order (Figure 5C,D) models. We adopted regression fitting approaches with the linear formulations in Equations (1) and (2), respectively [22]:

$$\ln \frac{C_o}{C_t} = -kt \tag{1}$$

$$\frac{1}{C_t} = \frac{1}{C_0} + k_2 t \tag{2}$$

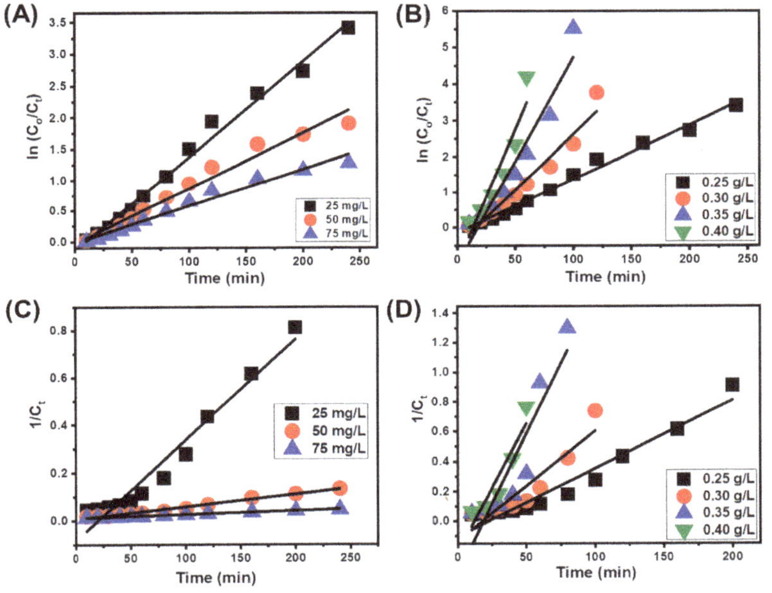

Figure 5. Fitting of the IBU oxidation results with the first-order kinetic model (**A**,**B**) and second-order kinetic model (**C**,**D**).

The results derived from both of the models revealed agreement with the first-order concept for all Zn@CU/BEN dosages (that varied between 0.25 and 0.4 g/L) and IBU concentrations (that ranged between 25 mg/L and 75 mg/L) (Figure 5; Table 1). The outcomes of this investigation suggest that the performances of the reactions are affected by only the dose of Zn@CU/BEN or by the duration of degradation, rather than both of these factors at the same time. The notable situations that exhibited contemporaneous compliance with both kinetic assumptions indicate the probable collaboration of various degrading processes or the participation of various generated oxidative radicals. The kinetic rate constants estimated for the evaluated models exhibited considerable rise and reduction, respectively, as a consequence of the escalation of Zn@CU/BEN dosages and IBU concentrations (Table 1). The aforementioned outcomes were mostly consistent with the results of experimentation.

Table 1. The kinetic models' fitting criteria and quantum yields.

		First Order		Second Order		Quantum Yield
		R^2	K_1	R^2	K_2	(Φ)
Concentration	25 mg/L	0.986	0.015	0.964	0.0042	7.96×10^{-6}
	50 mg/L	0.981	0.0089	0.974	5.3×10^{-4}	4.72×10^{-6}
	75 mg/L	0.982	0.0060	0.980	1.6×10^{-4}	3.18×10^{-6}
Dosage	0.25 g/L	0.98	0.015	0.96	0.0042	7.96×10^{-6}
	0.3 g/L	0.95	0.031	0.86	0.0073	1.64×10^{-6}
	0.35 g/L	0.93	0.057	0.84	0.0187	3.02×10^{-5}
	0.4 g/L	0.89	0.075	0.87	0.0196	3.98×10^{-5}

The quantum yields (Φ) are crucial metrics for assessing the efficiency of the studied photocatalytic fragmentation process employing Zn@CU/BEN. The theoretical concept of quantum yield throughout the oxidation processes signifies the amount of charge carriers sufficient to accomplish successful decomposition of organic structures by ingested light photons [33,48]. Therefore, the computed values acquired can potentially be employed as indications of the effectiveness displayed by the oxidization systems being studied with respect to the parameters being examined. The determination of apparent quantum yields was performed by calculating the ratio of spent carriers of charges to the total number of photons that entered throughout the experimental environment (Equation (3)) [50,51]. The relevant values of Φ could be calculated by applying Equation (4), which relies on the approximated kinetic rate constants determined according to the first-order model [50–52].

$$\Phi = \frac{Number\ of\ LVOX\ molecules}{Number\ of\ absorbed\ photons} \quad (3)$$

$$\Phi = \frac{K_1}{2.303 \times I_{o,\lambda} \times \mathcal{E}\lambda \times \ell} \quad (4)$$

The symbols K_1 (S^{-1}), Ɛλ ($cm^{-1}\ M^{-1}$), Io, λ (Einstein $I^{-1}\ S^{-1}$), and ℓ (cm) denote the kinetic rate constants, molar absorption, intensity of incident light with a specific wavelength λ, and dimension of the used quartz cell, respectively. The corresponding Φ values for the performed tests utilizing different doses of Zn@CU/BEN and IBU levels are displayed in Table 1. The quantum yields during oxidative reactions displayed augmentation with larger doses of Zn@CU/BEN; however, they indicated a decline with rising levels of IBU. The values that were obtained closely aligned with the results of the experiments and the kinetic significance, demonstrating the increasing influence of the Zn@CU/BEN dosages on the photocatalytic oxidative process of IBU (Table 1).

2.2.3. Synergetic Studies

The impact of the various components, such as the BEN substrate, CU/BEN composite, ZnO, and Zn-pillared BEN (Zn@BEN), on the efficacy of the studied Zn@CU/BEN blend as both an adsorbent and photocatalyst was evaluated using the following parameters:

- BEN, CU/BEN, ZnO, Zn@BEN, and Zn@CU/BEN without light source;
- Visible light source without catalyst;
- Visible light source + catalyst (BEN, CU/BEN, ZnO, Zn@BEN, and Zn@CU/BEN).

The experiments were carried out implementing particular settings for the parameters under consideration, including the catalyst dosage (0.4 g/L), entire volume (100 mL), oxidizing time (120 min), temperature (20 °C), and IBU content (25 mg/L) (Figure 6). In the absence of an illumination source, the incorporation of BEN, CU/BEN, ZnO, Zn@BEN, and Zn@CU/BEN resulted in removal levels of 26.5%, 38.3%, 13.8%, 47.5%, and 63.5% through adsorption, respectively (Figure 6A). This highlights the importance of the blending processes involving those with respect to the adsorption properties of IBU, whether through the incorporation of extra-functional chemical groups as reactive binding spots or via increasing the surface area.

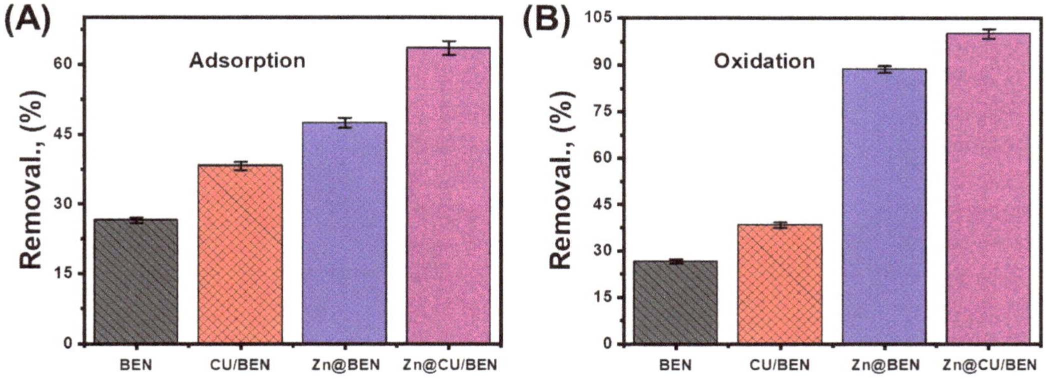

Figure 6. Synergetic effect of the integrated components on the adsorption (**A**) and photocatalytic oxidation (**B**) properties of Zn@CU/BEN.

The IBU compound demonstrated a negligible response to the incident visible light photons, leading to a mere 0.88% reduction in the IBU content. However, employing ZnO, Zn@BEN, and Zn@CU/BEN led to notable catalytic activity and oxidizing impacts on the IBU (Figure 6B), even though the incorporation of BEN and CU/BEN in the context of an illumination source did not demonstrate any significant photocatalytic relevance. The IBU-eliminating percentages obtained with ZnO, Zn@BEN, and Zn@CU/BEN were determined to be 31.7%, 88.6%, and 100%, respectively (Figure 6B). The levels were established after the equilibrium phases of adsorption and desorption. The increase in effectiveness could be attributed to several reasons. Firstly, there was a confirmed increase in the uptake affinity, which is a crucial prerequisite for the effective degradation of pollutants. Furthermore, there was also a rise in the surface area, resulting in an improved interacting interface between the catalytic sites. Furthermore, the consistent dispersion of Zn in the form of pillared nanostructures inside the layers of bentonite promoted the accessibility of the catalytic sites and their interaction with incoming photons. Therefore, it is strongly advised that Zn@CU/BEN should be employed as a cost-effective catalyst when using photocatalytic degradation technology for the remediation of soluble IBU residuals instead of employing the separate constituents BEN, CU/BEN, ZnO, and Zn@BEN.

2.2.4. Mechanism and Pathway

The Affected Oxidizing Species

Determination of the most vital oxidative species that controlled the degradation mechanism of IBU over the Zn@CU/BEN catalyst was performed based on the experimental results of reactive radical entrapping investigations (Figure 7A). Three reagents were employed as entrapping agents to trap the reacting oxidizing species: electron–hole pairs (EDTA-2Na), hydroxyl (Isopropanol (I-P)), and superoxide (1,4-benzoquinone (B-Q)). The tests were conducted with the following adaptations: the Zn@CU/BEN dosage was established at 0.4 g/L, pH at 6, IBU concentration at 25 mg/L, water volume at 100 mL, temperature at 20 °C, duration at 120 min, and volume of trapping chemicals at 1 mmol. The experimental results revealed that the degradation performance or catalytic effectiveness levels of Zn@CU/BEN for IBU dropped to 10.2% (I-P), 90.8% (EDTA-Na), and 65.7% (B-Q) (Figure 7A). Therefore, it seems the oxidative degradation pathway of IBU using Zn@CU/BEN is mainly controlled by the released hydroxyl radicals ($OH^{\bullet}$) generated through photochemical processes, which are followed by the superoxide species.

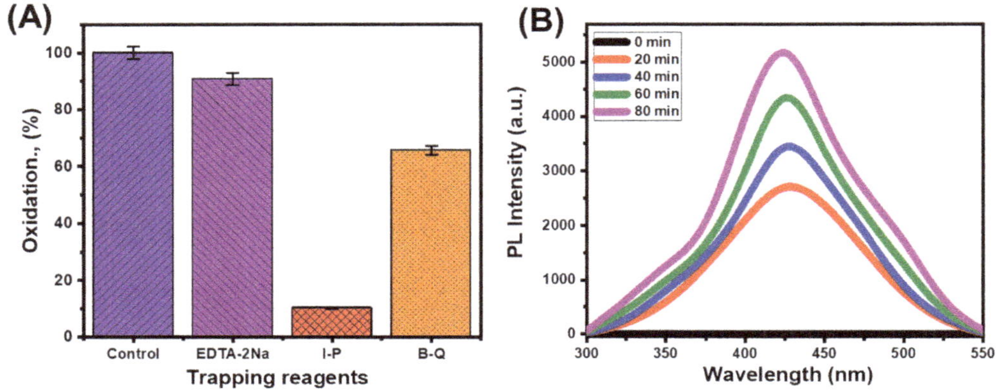

Figure 7. Trapping of the oxidizing radicals during the degradation of IBU by Zn@CU/BEN (**A**) and the change in the measured PL spectra during the oxidation of IBU at different durations (**B**).

Photoluminescence (PL) Detection of the Produced Hydroxyl Radicals

The photoluminescence (PL) method was used to track the presence of $OH^{\bullet}$, employing the probe molecule methodology (Figure 7B). The test involved mixing 20 mL of terephthalic acid (TA) at a concentration of 5×10^{-4} M with a homogenous mixture containing NaOH at a concentration of 2×10^{-3} M and Zn@CU/BEN particulates weighing 0.5 g. The main result of the chemical reactions involving $OH^{\bullet}$ and terephthalic acid was the generation of hydroxyterephthalic acid, which displayed a significant photoluminescent signal. The increase in the intensity of the photoluminescent (PL) signal could potentially be ascribed to the simultaneous increase in the release of $OH^{\bullet}$ provided by the Zn@CU/BEN hybridized form. The photoluminescent (PL) signal was quantified while the oxidation process was underway, by implementing a visible light source positioned at an elevation of about 10 cm distant from the processes. In addition, a 420 nm wavelength cutoff lens was used. The measurement was performed by employing a fluorescence spectrophotometer with a particular wavelength of 425 nm for measurement and an excitation wavelength of 315 nm. The photoluminescence (PL) spectrum of Zn@CU/BEN showed a consistent pattern whenever tested without an illuminating source, suggesting the absence of photocatalytic characteristics. Therefore, it could be stated that no hydroxyl radicals ($OH^{\bullet}$) were produced during the conditions tested. Whenever the hydroxyterephthalic acid was subjected to the light source, it produced a photoluminescence spectrum that was easily observable (Figure 7B). Moreover, the intensity of the spectrum consistently increased when the duration of the test was prolonged (Figure 7B). The increase in the strength of the photoluminescence (PL) spectrum might be ascribed to a higher amount of $OH^{\bullet}$ generated by the Zn@CU/BEN components.

The General Oxidation Mechanism

The starting stage in the photocatalytic oxidation of IBU using Zn@CU/BEN as the applied catalyst is the effective uptake of the soluble IBU molecules over the surface of the composite, especially close contact with the active catalytic sites and the released oxidizing radicals. Simultaneously, the excitation of electrons across the outermost layers of pillared Zn metals results in the establishment of electron–hole pairs owing to the significant absorption of the incident radiation [22]. Some of the excited electrons frequently interact with the oxygen ions throughout the degradation circuit, resulting in the emission of superoxide oxidative radicals ($O_2^{\bullet-}$) [33]. Furthermore, the occurrence of an electron–hole pair (h^+) exerts a disruptive effect on the water molecules, resulting in the emission of hydroxyl oxidative radicals [52]. Afterwards, the oxidative radicals generated, especially the $OH^{\bullet}$ radicals, react immediately with the soluble IBU compounds, causing oxidation and degradation of their organic structures. The continuous interactions result in a sequence of oxidizing steps, leading to the establishment of intermediate compounds. Following a period of complete oxidation, these molecules completely transform into carbon dioxide (CO_2) and water (H_2O), as shown by Equations (5)–(11) and Figure 8 [53,54].

$$\text{Zn@CU/BEN} + h\nu \rightarrow \text{Zn@CU/BEN}\left(e^-_{CB}, h^+_{VB}\right) \quad (5)$$

$$\text{Zn@CU/BEN}\left(e^-\right) + O_2 \rightarrow \text{Zn@CU/BEN} + O_2^{\bullet -} \quad (6)$$

$$(\text{Zn@CU/BEN})e^- + \text{IBU} \rightarrow (\text{Zn@CU/BEN})e^- + \text{IBU}^{+\bullet} \quad (7)$$

$$\text{IBU}^{+\bullet} + O_2^{\bullet -} \rightarrow \text{Oxidized product} \quad (8)$$

$$\text{Zn@CU/BEN}\left(h^+\right) + H_2O \rightarrow \text{Zn@CU/BEN} + H^+ \uparrow + OH^\bullet \quad (9)$$

$$\text{Zn@CU/BEN}\left(h^+\right) + \text{IBU} \rightarrow \text{Oxidized product} \quad (10)$$

$$\text{IBU} + OH^\bullet \rightarrow \text{Oxidized product} \rightarrow CO_2 + H_2O \quad (11)$$

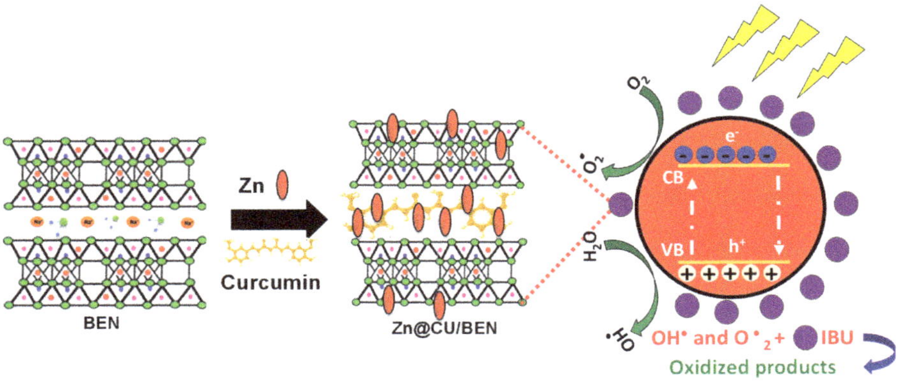

Figure 8. Schematic diagram for the oxidation mechanism of IBU by the Zn@CU/BEN composite.

Mineralization and Oxidation Pathway

TOC Content

The total organic carbon (TOC) concentration in the oxidized IBU-contaminated water samples was monitored to assess the mineralization effectiveness of the degraded IBU compounds and to detect any potential development of intermediate chemicals. The oxidative degradation experiment involved introducing a dose of 0.4 g/L of Zn@CU/BEN. According to the observed curves, there was a significant reduction in the content of total organic carbon (TOC) whenever the oxidation durations increased (Figure 9). Observable disparities existed in the oxidization rates attained for IBU and the TOC elimination percentages established. The observed rates and percentages of TOC elimination in the same situations were significantly lower than the already-established rates and percentages of degradation (Figure 9). The results suggest that the IBU chemicals underwent partial oxidative degradation, which might lead to the formation of secondary intermediate or transitional organic chemicals through modifications or decomposition of the medication's chemical structures. After a period of 160 min, it was established that the photocatalytic treatment, using Zn@CU/BEN as the catalyst, of IBU as a soluble contaminant had led it to become completely degraded, and the TOC content was also removed from the solutions (Figure 9). We can take from this that the Zn@CU/BEN oxidation system of IBU displaced the complete mineralization or decomposition state, producing the final environmentally safe products of CO_2 and H_2O.

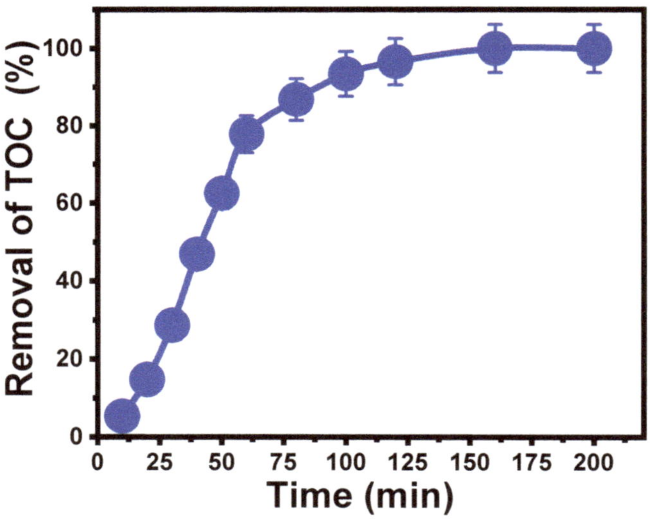

Figure 9. The elimination of the TOC content during the oxidation of IBU by Zn@CU/BEN at different intervals.

Intermediate Compounds and Oxidation Pathway

Several aromatic chemicals were identified in the oxidation of contaminated water samples with IBU, which were treated over the Zn@CU/BEN hybrid material in the presence of visible light (Figure S1, in the Supplementary Materials). These chemicals, which were intermediates or secondary byproducts, included 2-hydroxy-2-(4-isobutyl phenyl) propanoic acid, 1-(4-ethyl phenyl)-2-methyl propan-1-ol, 2-(4-(1-hydroxy-2-methyl propyl) phenyl) propanoic acid, 4-isobutylacetophenone, 2-methyl-butyl benzene, isobutylbenzene, and 4-ethylbenzaldehyde (Figure S1, in the Supplementary Materials). These chemicals or intermediates highlighted the potential oxidation pathways of IBU compounds over the catalytic sites of Zn@CU/BEN. The essential degradation pathway comprises three processes: (1) hydroxylation, (2) decarboxylation/demethylation, and (3) ring-opening phases (Figure 10) [55,56]. The hydroxylation process of the IBU parent chemical structure led to the development of hydroxy-ibuprofen-modified structures, particularly 2-(4-(1-hydroxy-2-methyl propyl) phenyl) propanoic acid and 2-hydroxy-2-(4-isobutyl phenyl) propanoic acid (Figure 10). Subsequently, decarboxylation and demethylation mechanisms occurred, leading to extra oxidation of the intermediates generated. This ultimately resultws in the formation of isobutylbenzene, 1-(4-ethyl phenyl)-2-methyl propan-1-ol, and benzaldehyde, along with various intermediates or secondary chemicals with a small aromatic backbone (Figure 10). The subsequently occurring reaction among the free OH radicals and the ring structures of benzene within the previously generated smaller compounds triggered the formation of different types of carboxylic acids (Figure 10). These carboxylic acids undergo several oxidizing steps, leading to the entire mineralization of IBU into water (H_2O) and carbon dioxide (CO_2) [45].

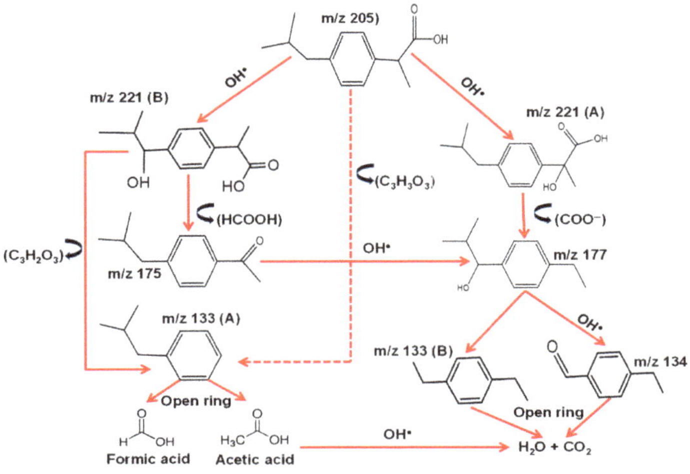

Figure 10. Schematic diagram for the oxidation pathway of IBU by Zn@CU/BEN and the formed intermediates.

2.2.5. Comparison Study

A comparative analysis was conducted to evaluate the efficacy of the synthesized Zn@CU/BEN photocatalyst in eliminating IBU compounds. The analysis considered the oxidizing parameters and the levels of IBU evaluated, and it compared the effectiveness of Zn@CU/BEN with previously studied catalytic structures mentioned in the published literature. The results given demonstrate that the established material has a better level of performance as a photocatalyst for the oxidation of IBU contaminants under visible light. The Zn@CU/BEN catalyst exhibited superior efficacy compared to some synthesized catalysts over brief oxidizing periods, and its application was only necessary in small quantities (Table 2). Consequently, we can surmise that the combination of bentonite and curcumin, along with the incorporation of Zn metal as pillars, allowed us to develop an environmentally friendly blended material with notable enhanced catalytic activities. The resulting material can be implemented effectively for the rapid removal of IBU molecules, which are hazardous medication residues in water.

Table 2. Comparison between the oxidation performance of IBU by Zn@CU/BEN and other investigated catalysts in the literature.

Catalyst	Dosage	Conc.	Light Source	Degradation Performance	Reference
C-N/TiO_2	0.50 g/L	20 mg/L	150 W LED lamp	360 min, ca. 98%	[57]
Bi–TiO_2	2.00 g/L	10 mg/L	250 W UV lamp	360 min, ca. 89%	[58]
Ag/AgCl@MIL-88A(Fe)	0.40 g/L	10 mg/L	500 W Xe lamp	120 min, ca. 81%	[59]
GQD/$AgVO_3$	0.20 g/L	10 mg/L	350 W Xe lamp	120 min, ca. 91%	[60]
g-C_3N_4/MIL-68(In)-NH_2	0.15 g/L	20 mg/L	300 W Xe lamp	120 min, ca. 93%	[45]
g-C_3N_4/ZnFeMMO	0.20 g/L	5 mg/L	500 W Xe lamp	240 min, ca. 92%	[61]
Fe_3O_4@MIL-53(Fe)	0.4 g/L	10 mg/L	500 W Xenon lamp	60 min, ca. 99%	[56]
TiO_2-rGO/SOFs	1.00 g/L	5 mg/L	160 W UV lamp	180 min, ca. 81%	[62]
Biochar-$ZnAl_2O_4$	1 g/L	20 mg/L	125 W UV lamp	120 min, ca. 100%	[63]
MIL-53(Fe)	0.1 g/L	10 mg/L	500 W Xe lamp	60 min, ca. 99%	[9]
Zn@CU/BEN	0.4 g/L	25 mg/L	400 W metal halide lamp	80 min, ca. 100%	This study

2.2.6. Toxicological Properties

Standardizing in vitro proliferation tests were conducted on two human cell lines: HEK293T, derived from the kidney, and HepG2, derived from the liver (Figure 11). The cell lines were subjected to varying doses (50 µM, 150 µM, and 250 µM) of IBU and its oxidized intermediates in water at varied oxidizing durations individually. After 20 min, the intermediate product of IBU's oxidation had a significant inhibiting impact ($p < 0.05$) on the growth of HEK293T cells (Figure 11A). This detrimental effect was observed throughout all the assessed dosages. Remarkably, the oxidized sample after 20 min, at a concentration of 250 µM, resulted in reduction in the cell viability of HEK293T cells by 22.4% (Figure 11A). This demonstrates a greater inhibiting impact in contrast to the original IBU medication. Furthermore, the sample that underwent oxidation for 60 min had a greater toxic effect, resulting in cell viability of 11.8%. This effect was more pronounced in comparison with IBU at a dose of 250 µM. However, the photodegraded sample after 100 min had little detrimental effect on the proliferation of HEK293T cells as compared to the initial treated samples, resulting in 36% cell viability using the highest dosage (250 µM) (Figure 11A). However, the original form of IBU did not exhibit any noteworthy ($p > 0.05$) inhibiting impact on the HEK293T cell line in any of the doses examined. After 160 min, the treated sample showed no inhibitory impact on the examined cell lines, indicating that the completely mineralized material is safe. This was confirmed by the thorough decomposition of IBU into end products. Several studies have demonstrated the toxic impact of IBU on the human kidney [64,65]. However, in this investigation, IBU did not show any cytotoxic effects, most likely because the quantities utilized were low.

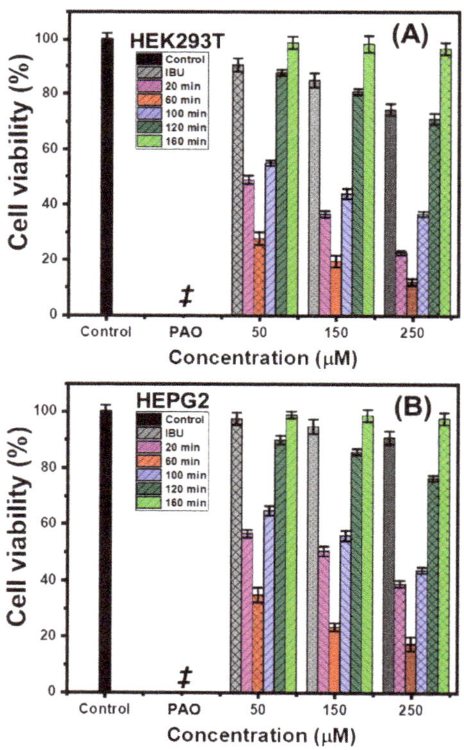

Figure 11. The toxicity effects of the partially and completely oxidized IBU-polluted solutions on HEK293T (**A**) and HEPG2 (**B**) cell lines in the presence of fresh cells as the control and phenylarsine oxide (PAO) as the positive control.

Notably, HEPG2 cells exhibited a comparable effect to the HEK293T cell line (Figure 11B). The growth of HEPG2 cells was inhibited in a dose-dependent way by all the investigated doses collected after 20 min. Remarkably, a dose of 250 μM of the IBU oxidized water sample exhibited a substantially greater inhibition influence on the HEPG2 cell strain in comparison to IBU, resulting in cell viability of 38.8% of the cells (Figure 11B). Moreover, the oxidized sample exhibited a substantial reduction in cell viability at all three doses evaluated after 60 min. Additionally, it had a greater level of inhibition versus IBU at a dosage of 250 μM, resulting in a 17.4% cell viability. The viability of HepG2 cells was not significantly affected by IBU or the completely mineralized IBU-polluted water sample after 160 min, even at the maximum dosage.

In general, the above in vitro experiments established that the photocatalytic degradation byproducts of IBU residuals in the water supply may cause substantial harm to human liver cells and human kidney cells in comparison to the original chemical. Owing to the biological accumulation of such chemicals within human cells and their transfer via food chains, the prolonged consumption of partially oxidized IBU-derived products has the potential to adversely affect liver and kidney functions in humans. Hence, it is necessary to continue the oxidation of IBU using the recommended catalyst for a duration of up to 160 min to be able to accomplish the full mineralization of IBU into a safe end product and prevent the toxic effects of the intermediate compounds.

3. Experimental Work

3.1. Materials and Chemicals

The naturally occurring bentonite sample was collected from a quarry located in the Western Desert of Egypt. The composition of the sample was as follows: 54.82% SiO_2, 2.6% Na_2O, 1.45% TiO_2, 17.56% Al_2O_3, 2.4% CaO, 2.5% MgO, 9.5% Fe_2O_3, and 9.2% LOI (loss on ignition). The hybrid synthesis techniques involved the use of curcumin (>94.99%; its chemical structure is presented in Figure S2, in the Supplementary Materials), sulfuric acid (99%), pure ethanol, zinc nitrate hexahydrate salt (98%), and hydrazine solution (35 wt%), all of which were acquired from Sigma-Aldrich, Cairo, Egypt. For adsorption and oxidation experiments, clinical-grade ibuprofen (98%) was supplied by Sigma-Aldrich, Cairo, Egypt.

3.2. Synthesis of the Catalyst

3.2.1. Acid Activation of Bentonite

To eliminate the existing carbonate and metal impurities, the bentonite fractions were activated using acid. We mixed 10 g of bentonite powdered sample with 100 mL of a dilute (20%) sulfuric acid and mixed them using a magnetic stirrer (650 rpm) over 12 h at 100 °C. Following that, the acid-leached powder was filtered out and then carefully rinsed utilizing distilled water before being dried at room temperature for 48 h (40.3 °C) and stored within specialized containers for subsequent synthesis stages.

3.2.2. Intercalation of Bentonite by Curcumin (CU/BEN)

The bentonite sheets were intercalated using curcumin-extracted organic phytochemicals employing a simple chemical integration process. For this experiment, the curcumin-extracting compounds were obtained by dissolving the curcumin powders (3.7 g) in ethanol (60%) for around 15 min in a sealed vessel at 50 °C and blending the resulting mixture via stirring (1500 rpm). Following this, the resulting mixture was subjected to ultrasound treatment (20 min; 240 W) to increase homogeneity and the yield of the crucial organic phytochemicals. About 50 mL of the resultant extract was homogeneously combined with 4.4 g of BEN as a colloidal solution or suspension in 50 mL of distilled water. To stimulate the diffusion effectiveness of the soluble organic phytochemicals inside the multiple layers of bentonite, the resultant mixture was maintained under rapid stirring (1500 rpm) over 12 h plus ultrasonication over an extra 5 h (240 W). The obtained product after the intercalation (CU/BEN) was subsequently filtrated out of the remaining solution, rinsed carefully for approximately five runs (10 min), and slowly dried at 50 °C for 12 h before being employed in the following stage (Figure 1).

3.2.3. Synthesis of Zn-Pillared CU/BEN Composite (Zn@CU/BEN)

The synthesis methods commenced by producing green extract solution of curcumin, which functions as a crucial green reagent. This extract was implemented as both a reduction agent and a capping agent. Conversely, the BEN/zinc nitrate combination was developed by thoroughly mixing the CU/BEN portions (4.4 g) with 100 mL of an aqueous solution containing zinc nitrate salt (2.2 g). The blending of the two constituents was carried out over a duration of 10 h by employing a magnetic stirrer operating at a speed of 650 rpm, followed by a further 2 h of sonication utilizing a source with a power output of 240 watts. Subsequently, the curcumin-derived extract (100 mL) was combined with the BEN/zinc nitrate blend while stirring vigorously at a constant rate of 1500 rpm till a noticeable reddish precipitate formed. Following that, rapid introduction of 50 mL of hydrazine was performed to verify the effective transformation of previously generated oxides along with hydroxides towards metallic zinc. The process was extended for a further 24 h with intermittent ultrasound treatments (20 min each) to promote adequate mixing among the reactants and maintain uniform dispersion of the pillared metal as well as the immobilized phytochemicals. Subsequently, the Zn-pillared Cu/BEN composite, in the form of solid particulates (Zn@CU/BEN), was successfully separated via filtering with Whitman filter paper. The filtered product underwent five rounds of washing using distilled water, with each successive washing cycle lasting for 10 min. Subsequently, the rinsed product was dried for 12 h at a temperature of 50 °C.

3.3. *Characterization Techniques*

The effects of the modifying and combining operations on the crystal structure of BEN were evaluated by analyzing the X-ray diffraction patterns of the different produced materials, achieved by employing a PANalytical-Empyrean X-ray diffractometer (Malvern Panalytical B.V. Co.; Almelo, The Netherlands). The detection span of the diffractometer extended from 5 to 80 degrees. The effects of the modifying chemical processes and the effective insertion of the active ingredient groups of curcumin-related components and ZnO were monitored using the FT-IR spectra. These spectra were obtained using a Fourier transform infrared spectrometer (FTIR-8400S; Shimadzu Co., Kyoto, Japan) over a frequency spectrum of 400 cm^{-1} to 4000 cm^{-1}. A scanning electron microscope (Zeiss Ultra 55; Carl Zeiss AG Co.; Oberkochen, Baden-Württemberg, Germany) was used to examine the exterior morphology, while a transmission electron microscope (JEM-2100, JEOL Ltd. Co.; Tokyo, Japan) was used to analyze the interior characteristics. A surface area analyzer (SA3100, Beckman Coulter Co.; Brea, CA, USA) was used to track the influence of integration processes on the textural qualities and surface area.

3.4. *Advanced Oxidation of IBU*

The catalytic efficiency of Zn@CU/BEN in oxidizing IBU was assessed after attending to the equilibrium of adsorption as well as desorption inside a Pyrex cylindrical vessel with a diameter equal to 7 cm and an overall length equal to 15 cm. The cell was coupled with a fully embedded visible light bulb positioned 3.5 cm away from the outside surface of the cell. The source of illumination utilized throughout the experiments was a commercially available metal halide bulb with a power output of 400 W and a wavelength of approximately 490 nm. The average incoming illumination intensity was measured to be 18.7 mW/cm^2. Advanced oxidation experiments were conducted using IBU levels of 25, 50, and 75 mg/L, Zn@CU/BEN doses ranging from 0.25 g/L to 0.4 g/L, and oxidizing periods ranging from 10 min to 240 min. The experiments were carried out under constant pH (pH 6), volume (100 mL), and temperature (20 °C) conditions. The overall oxidation performance of IBU was monitored by measuring its concentration within the addressed solutions. The evaluations were performed three times, and the mean for all three findings exhibited a standard deviation of no more than 4.5%.

The residual IBU levels post adsorption and photocatalytic oxidation experiments, including the secondary byproducts resulting from partial oxidation, were quantified by employing liquid chromatography–mass spectrometry (LC-MS). The LC-MS apparatus was outfitted with a mass analyzer, a 3000 HPLC, and a Thermo Fisher Scientific (Waltham, MA, USA) column. The fluid phase associated with the system consisted of a blend of methanol and water, in addition to formic acid (1%). Throughout the analysis, the eluent was dispensed at a velocity of 0.35 mL/min with an injection capacity of 5 μL, and the column itself was kept at 30 °C. The compounds' mass spectra were measured by employing m/z scanning within a 20–400 range, with the nebulizer pressure, temperature, ESI voltage, and nitrogen flow velocity set to 35 psi, 350 °C, 4 kV, and 10 L/min, respectively. The total organic carbon (TOC) contents inside the samples that were oxidized were measured using TOC-VCPH (Shimadzu, Kyoto, Japan) in order to assess the mineralization effectiveness of the oxidation processes.

3.5. *Toxicological Properties*

The toxicological effects of the oxidized intermediates of IBU on the human body were assessed by examining their impacts on the kidney (HEK293T cell lines) and liver (HepG2 cell lines), using cell lines derived from the American Type Cultural Collection (ATCC). The selected cell lines were grown in DMEMs combined with 10% fetal bovine serum in a humidified atmosphere containing 5% CO_2 at 37 °C. Afterwards, the cells were cultivated and seeded in 96-well plates with a density of 4000 cells per well and incubated overnight. Following that, they were subjected to varied concentrations (250 μM, 150 μM, and 50 μM) of IBU and oxidized IBU-contaminated water within a fresh medium before an extra incubation period of 48 h. Phenylarsine oxide (PAO) served as the positive control, while 0.5% dimethyl sulfoxide (DMSO) served as the negative reference. Following a 48 h incubation period, 20 μL of a 3-(4,5-dimethylthiazol-2-yl)-2,5-diphenyltetrazolium bromide (MTT) reagent with a concentration of 5 mg/mL was introduced into each of the wells. The cells thereafter underwent incubation at a temperature of 37 °C for a duration of 2 h. Following this, the fluids were removed, and 100 μL of DMSO (100%) was injected into each well. Consequently, the absorbance was determined at a wavelength of 570 nm using a microplate reader. The data, described as mean ± standard deviation (%), were reported as the ratio of treated samples to control samples (i.e., IBU or its oxidized products/DMSO). The experiments were conducted in at least three repetitions, with each performed at least three times. The observed data were immediately used to compute the cell viability percentage using Equation (12).

$$\text{Cell viability}(\%) = \frac{\text{Mean OD}}{\text{Control OD}} \times 100 \tag{12}$$

4. Conclusions

The Zn@CU/BEN composite was synthesized using simple, environmentally friendly approaches for effective catalytic oxidation of IBU into safe end products under visible light. About 25 mg/L of IBU was completely degraded after 80 min using Zn@CU/BEN at 0.4 g/L and pH 6, reflecting higher activity as compared to several study materials. The synergetic influence of BEN and CU/BEN explains the markedly enhanced properties, achieved by inducing the surface area, adsorption capacity, and interactive interface of the catalytic sites of the pillared Zn. The complete elimination of TOC after 160 min demonstrated complete mineralization and the formation of intermediates. The identified intermediates, in addition to the detection of hydroxyl radicals as the effective oxidizing species, validated the degradation of IBU over Zn@CU/BEN through three progressive processes (hydroxylation, decarboxylation/demethylation, and ring opening). The cell viability of kidney (HEK293T) and liver (HEPG2) cell lines after their treatment with the oxidized samples for 160 min revealed high biosafety, with cell viability greater than 97%.

Supplementary Materials: The following supporting information can be downloaded at: https://www.mdpi.com/article/10.3390/catal14020129/s1, Figure S1. HPLC-MS spectra of the starting IBU drug (A) and its photo-oxidized intermediate products (B and C); Figure S2. The chemical structure of curcumin.

Author Contributions: Conceptualization, S.I.O., M.R.A., S.B., A.A.A. and M.H.S.; methodology, M.H.S., S.I.O., H.E.A. and M.R.A.; software, M.H.S. and H.E.A.; validation, M.R.A., S.B., H.A.A. and S.I.O.; formal analysis, M.H.S., H.E.A. and A.A.A.; writing—review and editing, H.E.A., M.R.A., H.A.A., S.B., M.H.S. and A.A.A.; investigation, S.I.O., S.B., A.A.A. and M.R.A.; resources, M.R.A., M.H.S., H.A.A. and S.I.O.; data curation, M.H.S., H.E.A. and M.R.A.; writing—original draft preparation, M.R.A., H.A.A., S.B., M.H.S., A.A.A., H.E.A. and S.I.O.; visualization, M.R.A., A.A.A. and S.I.O.; supervision, A.A.A. and M.R.A.; project administration, S.I.O.; funding acquisition, S.I.O. All authors have read and agreed to the published version of the manuscript.

Funding: This research was funded by [Deanship of Scientific Research at Princess Nourah bint Abdulrahman University], Research Groups Program Grant no. [(RGP-1440-0002)(4)].

Data Availability Statement: Data are available upon reasonable, by the corresponding authors.

Conflicts of Interest: The authors declare no conflict of interest.

References

1. Ren, Z.; Romar, H.; Varila, T.; Xu, X.; Wang, Z.; Sillanpää, M.; Leiviskä, T. Ibuprofen degradation using a Co-doped carbon matrix derived from peat as a peroxymonosulphate activator. *Environ. Res.* **2020**, *193*, 110564. [CrossRef]
2. Abukhadra, M.R.; Fathallah, W.; El Kashief, F.A.; El-Sherbeeny, A.M.; El-Meligy, M.A.; Awwad, E.M.; Luqman, M. Insight into the antimicrobial and photocatalytic properties of NiO impregnated MCM-48 for effective removal of pathogenic bacteria and toxic levofloxacin residuals. *Microporous Mesoporous Mater.* **2020**, *312*, 110769. [CrossRef]
3. Miranda, M.O.; Cavalcanti, W.E.C.; Barbosa, F.F.; de Sousa, J.A.; da Silva, F.I.; Pergher, S.B.C.; Braga, T.P. Photocatalytic degradation of ibuprofen using titanium oxide: Insights into the mechanism and preferential attack of radicals. *RSC Adv.* **2021**, *11*, 27720–27733. [CrossRef]
4. Lung, I.; Soran, M.L.; Stegarescu, A.; Opris, O.; Gutoiu, S.; Leostean, C.; Lazar, M.D.; Kacso, I.; Silipas, T.D.; Porav, A.S. Evaluation of CNT-COOH/MnO_2/Fe_3O_4 nanocomposite for ibuprofen and paracetamol removal from aqueous solutions. *J. Hazard. Mater* **2021**, *403*, 123528. [CrossRef]
5. Lei, X.; Huang, L.; Liu, K.; Ouyang, L.; Shuai, Q.; Hu, S. Facile one-pot synthesis of hierarchical N-doped porous carbon for efficient ibuprofen removal. *J. Colloid Interface Sci.* **2021**, *604*, 823–831. [CrossRef]
6. Vebber, M.C.; Aguzzoli, C.; Beltrami, L.V.R.; Fetter, G.; Crespo, J.d.S.; Giovanela, M. Self-assembled thin films of PAA/PAH/TiO_2 for the photooxidation of ibuprofen. Part II: Characterization, sensitization, kinetics and reutilization. *Chem. Eng. J.* **2018**, *361*, 1487–1496. [CrossRef]
7. Nawaz, M.; Khan, A.A.; Hussain, A.; Jang, J.; Jung, H.Y.; Lee, D.S. Reduced graphene oxide TiO_2/sodium alginate 3-dimensional structure aerogel for enhanced photocatalytic degradation of ibuprofen and sulfamethoxazole. *Chemosphere* **2020**, *261*, 127702. [CrossRef]
8. Granatto, C.F.; Grosseli, G.M.; Sakamoto, I.K.; Fadini, P.S.; Varesche, M.B.A. Methanogenic potential of diclofenac and ibuprofen in sanitary sewage using metabolic cosubstrates. *Sci. Total Environ.* **2020**, *742*, 140530. [CrossRef]
9. Yang, S.; Wang, K.; Chen, Q.; Wu, Y. Enhanced photocatalytic hydrogen production of S-scheme TiO_2/g-C3N4 heterojunction loaded with single-atom Ni. *J. Mater. Sci. Technol.* **2024**, *175*, 104–114. [CrossRef]
10. Farhadi, N.; Tabatabaie, T.; Ramavandi, B.; Amiri, F. Ibuprofen elimination from water and wastewater using sonication/ultraviolet/hydrogen peroxide/zeolite-titanate photocatalyst system. *Environ. Res.* **2021**, *198*, 111260. [CrossRef]
11. Minella, M.; Bertinetti, S.; Hanna, K.; Minero, C.; Vione, D. Degradation of ibuprofen and phenol with a Fenton-like process triggered by zero-valent iron (ZVI-Fenton). *Environ. Res.* **2019**, *179*, 108750. [CrossRef]
12. Gong, H.; Chu, W.; Huang, Y.; Xu, L.; Chen, M.; Yan, M. Solar photocatalytic degradation of ibuprofen with a magnetic catalyst: Effects of parameters, efficiency in effluent, mechanism and toxicity evolution. *Environ. Pollut.* **2021**, *276*, 116691. [CrossRef]
13. Shanavas, S.; Priyadharsan, A.; Gkanas, E.I.; Acevedo, R.; Anbarasan, P.M. High efficient catalytic degradation of tetracycline and ibuprofen using visible light driven novel Cu/$Bi_2Ti_2O_7$/rGO nanocomposite: Kinetics, intermediates and mechanism. *J. Ind. Eng. Chem.* **2019**, *72*, 512–528.
14. Chen, C.-H.; Lin, Y.-C.; Peng, Y.-P.; Lin, M.-H. Simultaneous hydrogen production and ibuprofen degradation by green synthesized Cu2O/TNTAs photoanode. *Chemosphere* **2021**, *284*, 131360. [CrossRef]
15. Indira, K.; Shanmugam, S.; Hari, A.; Vasantharaj, S.; Sathiyavimal, S.; Brindhadevi, K.; El Askary, A.; Elfasakhany, A.; Pugazhendhi, A. Photocatalytic degradation of congo red dye using nickel–titanium dioxide nanoflakes synthesized by Mukia madrasapatna leaf extract. *Environ. Res.* **2021**, *202*, 111647. [CrossRef]

16. Gaur, J.; Vikrant, K.; Kim, K.-H.; Kumar, S.; Pal, M.; Badru, R.; Masand, S.; Momoh, J. Photocatalytic degradation of Congo red dye using zinc oxide nanoparticles prepared using Carica papaya leaf extract. *Mater. Today Sustain.* **2023**, *22*, 100339. [CrossRef]
17. Xiang, D.; Lu, S.; Ma, Y.; Zhao, L. Synergistic photocatalysis-fenton reaction of flower-shaped CeO_2/Fe_3O_4 magnetic catalyst for decolorization of high concentration congo red dye. *Colloids Surf. A Physicochem. Eng. Asp.* **2022**, *647*, 129021. [CrossRef]
18. Cui, H.; Yu, J.; Zhu, X.; Cui, Y.; Ji, C.; Zhang, C.; Xue, J.; Jia, X.; Qin, S.; Li, R. Advanced treatment of chicken farm flushing wastewater by integrating Fenton oxidation and algal cultivation process for algal growth and nutrients removal. *J. Environ. Manag.* **2021**, *298*, 113543. [CrossRef]
19. Benisha, R.; Amalanathan, M.; Aravind, M.; Mary, M.S.M.; Ahmad, A.; Tabassum, S.; Al-Qahtani, W.H.; Ahmad, I. Catharanthus roseus leaf extract mediated Ag-MgO nanocatalyst for photocatalytic degradation of Congo red dye and their antibacterial activity. *J. Mol. Struct.* **2022**, *1262*, 133005. [CrossRef]
20. Ghanbari, S.; Fatehizadeh, A.; Khiadani, M.; Taheri, E.; Iqbal, H.M.N. Treatment of synthetic dye containing textile raw wastewater effluent using UV/Chlorine/Br photolysis process followed by activated carbon adsorption. *Environ. Sci. Pollut. Res.* **2022**, *29*, 39400–39409. [CrossRef]
21. Abukhadra, M.R.; Saad, I.; Othman, S.I.; Katowah, D.F.; Ajarem, J.S.; Alqarni, S.A.; Allam, A.A.; Al Zoubi, W.; Ko, Y.G. Characterization of Fe0@ Chitosan/Cellulose structure as effective green adsorbent for methyl Parathion, malachite Green, and levofloxacin Removal: Experimental and theoretical studies. *J. Mol. Liq.* **2022**, *368*, 120730. [CrossRef]
22. Yang, X.; Wang, J.; El-Sherbeeny, A.M.; AlHammadi, A.A.; Park, W.-H.; Abukhadra, M.R. Insight into the adsorption and oxidation activity of a ZnO/piezoelectric quartz core-shell for enhanced decontamination of ibuprofen: Steric, energetic, and oxidation studies. *Chem. Eng. J.* **2022**, *431*, 134312. [CrossRef]
23. Yang, Y.; Chen, X.; Pan, Y.; Song, H.; Zhu, B.; Wu, Y. Two-dimensional ZnS (propylamine) photocatalyst for efficient visible light photocatalytic H2 production. *Catal. Today* **2020**, *374*, 4–11. [CrossRef]
24. Abukhadra, M.R.; AlHammadi, A.A.; Khim, J.S.; Ajarem, J.S.; Allam, A.A.; Shaban, M.S. Enhanced adsorption and visible light photocatalytic removal of 5-Fluorouracil residuals using environmental NiO/geopolymer nanocomposite: Steric, energetic, and oxidation studies. *J. Environ. Chem. Eng.* **2022**, *10*, 108569. [CrossRef]
25. Wang, Z.; Srivastava, V.; Ambat, I.; Safaei, Z.; Sillanpää, M. Degradation of Ibuprofen by UV-LED/catalytic advanced oxidation process. *J. Water Process. Eng.* **2019**, *31*, 100808. [CrossRef]
26. Velsankar, K.; Venkatesan, A.; Muthumari, P.; Suganya, S.; Mohandoss, S.; Sudhahar, S. Green inspired synthesis of ZnO nanoparticles and its characterizations with biofilm, antioxidant, anti-inflammatory, and anti-diabetic activities. *J. Mol. Struct.* **2022**, *1255*, 132420. [CrossRef]
27. Ansari, A.; Ali, A.; Khan, N.; Umar, M.S.; Owais, M. Shamsuzzaman Synthesis of steroidal dihydropyrazole derivatives using green ZnO NPs and evaluation of their anticancer and antioxidant activity. *Steroids* **2022**, *188*, 109113. [CrossRef]
28. Sinha, A.; Sahu, S.K.; Biswas, S.; Mandal, M.; Mandal, V.; Ghorai, T.K. Green approach to synthesize MnxZn1-xO nanocomposite with enhanced photocatalytic, fluorescence and antibacterial activity. *Curr. Res. Green Sustain. Chem.* **2021**, *5*, 100244. [CrossRef]
29. Saad, A.M.; Abukhadra, M.R.; Ahmed, S.A.-K.; Elzanaty, A.M.; Mady, A.H.; Betiha, M.A.; Shim, J.-J.; Rabie, A.M. Photocatalytic degradation of malachite green dye using chitosan supported ZnO and Ce–ZnO nano-flowers under visible light. *J. Environ. Manag.* **2020**, *258*, 110043. [CrossRef]
30. Yusof, N.A.A.; Zain, N.M.; Pauzi, N. Synthesis of ZnO nanoparticles with chitosan as stabilizing agent and their antibacterial properties against Gram-positive and Gram-negative bacteria. *Int. J. Biol. Macromol.* **2018**, *124*, 1132–1136. [CrossRef]
31. Rudayni, H.A.; Rabie, A.M.; Aladwani, M.; Alneghery, L.M.; Abu-Taweel, G.M.; Al Zoubi, W.; Allam, A.A.; Abukhadra, M.R.; Bellucci, S. Biological Activities of *Sargassum* Algae Mediated ZnO and Co Doped ZnO Nanoparticles as Enhanced Antioxidant and Anti-Diabetic Agents. *Molecules* **2023**, *28*, 3692. [CrossRef]
32. Meer, B.; Andleeb, A.; Iqbal, J.; Ashraf, H.; Meer, K.; Ali, J.S.; Drouet, S.; Anjum, S.; Mehmood, A.; Khan, T.; et al. Bio-Assisted Synthesis and Characterization of Zinc Oxide Nanoparticles from Lepidium Sativum and Their Potent Antioxidant, Antibacterial, and Anticancer Activities. *Biomolecules* **2022**, *12*, 855. [CrossRef]
33. Abukhadra, M.R.; AlHammadi, A.A.; Khim, J.S.; Ajarem, J.S.; Allam, A.A. Enhanced decontamination of Levofloxacin residuals from water using recycled glass based a green zinc oxide/mesoporous silica nanocomposite; adsorption and advanced oxidation studies. *J. Clean. Prod.* **2022**, *356*, 131836. [CrossRef]
34. de França, B.M.; Oliveira, S.S.; Souza, L.O.; Mello, T.P.; Santos, A.L.; Forero, J.S.B. Synthesis and photophysical properties of metal complexes of curcumin dyes: Solvatochromism, acidochromism, and photoactivity. *Dye Pigment.* **2022**, *198*, 110011. [CrossRef]
35. Arab, C.; El Kurdi, R.; Patra, D. Efficient removal of Congo red using curcumin conjugated zinc oxide nanoparticles as new adsorbent complex. *Chemosphere* **2021**, *276*, 130158. [CrossRef]
36. Abukhadra, M.R.; Adlii, A.; El-Sherbeeny, A.M.; Soliman, A.T.A.; Elgawad, A.E.E.A. Promoting the decontamination of different types of water pollutants (Cd^{2+}, safranin dye, and phosphate) using a novel structure of exfoliated bentonite admixed with cellulose nanofiber. *J. Environ. Manag.* **2020**, *273*, 111130. [CrossRef]
37. Wasim, M.; Shi, F.; Liu, J.; Zhang, H.; Zhu, K.; Tian, Z. Synthesis and characterization of curcumin/MMT-clay-treated bacterial cellulose as an antistatic and ultraviolet-resistive bioscaffold. *J. Polym. Res.* **2022**, *29*, 423. [CrossRef]

38. Arabmofrad, S.; Jafari, S.M.; Lazzara, G.; Ziaiifar, A.M.; Shahiri Tabarestani, H.; Bahlakeh, G.; Cavallaro, G.; Calvino, M.M.; Nasiri Sarvi, M. Preparation and characterization of surface-modified montmorillonite by cationic surfactants for adsorption purposes. *J. Therm. Anal. Calorim.* **2023**, *148*, 13803–13814. [CrossRef]
39. Tong, L.; Liang, T.; Tian, Y.; Zhang, Q.; Pan, Y. Research progress on treatment of mine wastewater by bentonite composite. *Arab. J. Geosci.* **2022**, *15*, 1–23. [CrossRef]
40. Koksal, E.; Afsin, B.; Tabak, A.; Caglar, B. Butylamine-resadiye bentonite composite characterization. *Spectrosc. Lett.* **2020**, *53*, 745–750. [CrossRef]
41. Nagahashi, E.; Ogata, F.; Saenjum, C.; Nakamura, T.; Kawasaki, N. Preparation and Characterization of Acid-Activated Bentonite with Binary Acid Solution and Its Use in Decreasing Electrical Conductivity of Tap Water. *Minerals* **2021**, *11*, 815. [CrossRef]
42. Salam, M.A.; Mokhtar, M.; Albukhari, S.M.; Baamer, D.F.; Palmisano, L.; Jaremko, M.; Abukhadra, M.R. Synthesis and Characterization of Green ZnO@ polynaniline/Bentonite Tripartite Structure (G. Zn@ PN/BE) as Adsorbent for As (V) Ions: Integration, Steric, and Energetic Properties. *Polymers* **2022**, *14*, 2329. [CrossRef]
43. Dardir, F.M.; Mohamed, A.S.; Abukhadra, M.R.; Ahmed, E.A.; Soliman, M.F. Cosmetic and pharmaceutical qualifications of Egyptian bentonite and its suitability as drug carrier for Praziquantel drug. *Eur. J. Pharm. Sci.* **2018**, *115*, 320–329. [CrossRef]
44. Deng, J.; Wang, J.; Hu, H.; Hong, J.; Yang, L.; Zhou, H.; Xu, D. Application of mesoporous calcium silicate nanoparticles as a potential SD carrier to improve the solubility of curcumin. *J. Dispers. Sci. Technol.* **2022**, *44*, 2258–2266. [CrossRef]
45. Cao, W.; Yuan, Y.; Yang, C.; Wu, S.; Cheng, J. In-situ fabrication of g-C3N4/MIL-68 (In)-NH2 heterojunction composites with enhanced visible-light photocatalytic activity for degradation of ibuprofen. *Chem. Eng. Trans.* **2020**, *391*, 123608. [CrossRef]
46. Abukhadra, M.R.; Refay, N.M.; Nadeem, A.; El-Sherbeeny, A.M.; Ibrahim, K.E. Insight into the role of integrated carbohydrate polymers (starch, chitosan, and β-cyclodextrin) with mesoporous silica as carriers for ibuprofen drug; equilibrium and pharmacokinetic properties. *Int. J. Biol.* **2020**, *156*, 537–547. [CrossRef]
47. Martín, J.; Orta, M.d.M.; Medina-Carrasco, S.; Santos, J.L.; Aparicio, I.; Alonso, E. Evaluation of a modified mica and montmorillonite for the adsorption of ibuprofen from aqueous media. *Appl. Clay Sci.* **2019**, *171*, 29–37. [CrossRef]
48. Wang, C.; Sun, R.; Huang, R.; Cao, Y. A novel strategy for enhancing heterogeneous Fenton degradation of dye wastewater using natural pyrite: Kinetics and mechanism. *Chemosphere* **2021**, *272*, 129883. [CrossRef]
49. Abdullah, R.R.; Shabeed, K.M.; Alzubaydi, A.B.; Alsalhy, Q.F. Novel photocatalytic polyether sulphone ultrafiltration (UF) membrane reinforced with oxygen-deficient Tungsten Oxide (WO2.89) for Congo red dye removal. *Chem. Eng. Res. Des.* **2021**, *177*, 526–540. [CrossRef]
50. Huda, A.; Ichwani, R.; Handoko, C.T.; Bustan, M.D.; Yudono, B.; Gulo, F. Comparative photocatalytic performances towards acid yellow 17 (AY17) and direct blue 71 (DB71) degradation using Sn_3O_4 flower-like structure. *J. Phys. Conf. Ser.* **2019**, *1282*, 012097. [CrossRef]
51. Shemy, M.H.; Othman, S.I.; Alfassam, H.E.; Al-Waili, M.A.; Alqhtani, H.A.; Allam, A.A.; Abukhadra, M.R. Synthesis of Green Magnetite/Carbonized Coffee Composite from Natural Pyrite for Effective Decontamination of Congo Red Dye: Steric, Synergetic, Oxidation, and Ecotoxicity Studies. *Catalysts* **2023**, *13*, 264. [CrossRef]
52. Shams-Ghahfarokhi, Z.; Nezamzadeh-Ejhieh, A. As-synthesized ZSM-5 zeolite as a suitable support for increasing the photoactivity of semiconductors in a typical photodegradation process. *Mater. Sci. Semicond. Process.* **2015**, *39*, 265–275. [CrossRef]
53. Adly, E.R.; Shaban, M.S.; El-Sherbeeny, A.M.; Al Zoubi, W.; Abukhadra, M.R. Enhanced Congo Red Adsorption and Photo-Fenton Oxidation over an Iron-Impeded Geopolymer from Ferruginous Kaolinite: Steric, Energetic, Oxidation, and Synergetic Studies. *ACS Omega* **2022**, *7*, 31218–31232. [CrossRef] [PubMed]
54. Sun, Q.; Hu, X.; Zheng, S.; Zhang, J.; Sheng, J. Effect of calcination on structure and photocatalytic property of N-TiO_2/g-C_3N_4@diatomite hybrid photocatalyst for improving reduction of Cr(VI). *Environ. Pollut.* **2019**, *245*, 53–62. [CrossRef]
55. Liu, S.H.; Tang, W.T. Photodecomposition of ibuprofen over g-C_3N_4/Bi_2WO_6/rGO heterostructured composites under visible/solar light. *Sci. Total Environ.* **2020**, *731*, 139172. [CrossRef]
56. Liu, N.; Wang, J.; Wu, J.; Li, Z.; Huang, W.; Zheng, Y.; Lei, J.; Zhang, X.; Tang, L. Magnetic Fe_3O_4@MIL-53(Fe) nanocomposites derived from MIL-53(Fe) for the photocatalytic degradation of ibuprofen under visible light irradiation. *Mater. Res. Bull.* **2020**, *132*, 111000. [CrossRef]
57. Yang, Y.; Ok, Y.S.; Kim, K.-H.; Kwon, E.E.; Tsang, Y.F. Occurrences and removal of pharmaceuticals and personal care products (PPCPs) in drinking water and water/sewage treatment plants: A review. *Sci. Total Environ.* **2017**, *596–597*, 303–320. [CrossRef]
58. Archer, E.; Petrie, B.; Kasprzyk-Hordern, B.; Wolfaardt, G.M. The fate of pharmaceuticals and personal care products (PPCPs), endocrine disrupting contaminants (EDCs), metabolites and illicit drugs in a WWTW and environmental waters. *Chemosphere* **2017**, *174*, 437–446. [CrossRef]
59. Gogoi, A.; Mazumder, P.; Tyagi, V.K.; Tushara Chaminda, G.G.; An, A.K.; Kumar, M. Occurrence and fate of emerging contaminants in water environment: A review. *Groundw. Sustain. Dev.* **2018**, *6*, 169–180. [CrossRef]
60. Liu, J.-L.; Wong, M.-H. Pharmaceuticals and personal care products (PPCPs): A review on environmental contamination in China. *Environ. Int.* **2013**, *59*, 208–224. [CrossRef]
61. Rivera-Utrilla, J.; Sánchez-Polo, M.; Ferro-García, M.Á.; Prados-Joya, G.; Ocampo-Perez, R. Pharmaceuticals as emerging contaminants and their removal from water. A review. *Chemosphere* **2013**, *93*, 1268–1287. [CrossRef] [PubMed]

62. Petrie, B.; Barden, R.; Kasprzyk-Hordern, B. A review on emerging contaminants in wastewaters and the environment: Current knowledge, understudied areas and recommendations for future monitoring. *Water Res.* **2015**, *72*, 3–27. [CrossRef]
63. Siara, S.; Elvis, C.; Harishkumar, R.; Chellam, P.V. $ZnAl_2O_4$ supported on lychee-biochar applied to ibuprofen photodegradation. *Mater. Res. Bull.* **2021**, *145*, 111530. [CrossRef]
64. Balestracci, A.; Ezquer, M.; Elmo, M.E.; Molini, A.; Thorel, C.; Torrents, M.; Toledo, I. Ibuprofen-associated acute kidney injury in dehydrated children with acute gastroenteritis. *Pediatr. Nephrol.* **2015**, *30*, 1873–1878. [CrossRef] [PubMed]
65. Modig, S.; Elmståhl, S. Kidney function and use of nonsteroidal anti-inflammatory drugs among elderly people: A cross-sectional study on potential hazards for an at risk population. *Pharm. Weekbl.* **2018**, *40*, 870–877. [CrossRef] [PubMed]

sustainability

MDPI

Article

Synergism of Artificial Intelligence and Techno-Economic for Sustainable Treatment of Methylene Blue Dye-Containing Wastewater by Photocatalysis

Khumbolake Faith Ngulube [1], Amal Abdelhaleem [1], Manabu Fujii [2] and Mahmoud Nasr [1,3,*]

[1] Environmental Engineering Department, Egypt-Japan University of Science and Technology (E-JUST), Alexandria 21934, Egypt; khumbolake.ngulube@ejust.edu.eg (K.F.N.); amal.elsonbaty@ejust.edu.eg (A.A.)
[2] Civil and Environmental Engineering Department, Tokyo Institute of Technology, Meguro-Ku, Tokyo 152-8552, Japan; fujii.m.ah@m.titech.ac.jp
[3] Sanitary Engineering Department, Faculty of Engineering, Alexandria University, Alexandria 21544, Egypt
* Correspondence: mahmoud.nasr@ejust.edu.eg; Tel.: +20-1006-390-400

Abstract: Recently, removing dyes from wastewater by photocatalysis has been extensively studied by several researchers. However, there exists a research gap in optimizing the photocatalytic process parameters using artificial intelligence to maintain the associated techno-economic feasibility. Hence, this investigation attempts to optimize the photocatalytic degradation of methylene blue (MB) dye using an artificial neural network (ANN) model to minimize the capital and running costs, which is beneficial for industrial applications. A ZnO/MgO photocatalyst was synthesized, showing an energy band gap of 2.96 eV, crystallinity index of 71.92%, pore volume of 0.529 cm^3/g, surface area of 30.536 m^2/g, and multiple surface functional groups. An ANN model, with a 4-8-1 topology, *trainlm* training function, and feed-forward back-propagation algorithm, succeeded in predicting the MB removal efficiency (R^2 = 0.946 and mean squared error = 11.2). The ANN-based optimized condition depicted that over 99% of MB could be removed under C_0 = 16.42 mg/L, pH = 9.95, and catalyst dosage = 905 mg/L within 174 min. This optimum condition corresponded to a treatment cost of USD 8.52/m^3 cheaper than the price estimated from the unoptimized photocatalytic system by ≈7%. The study outputs revealed positive correlations with the sustainable development goals accompanied by pollution reduction, human health protection, and aquatic species conservation.

Keywords: ANN optimization; dye degradation; nanocomposite characterization; photocatalytic performance; treatment cost

1. Introduction

Discharging methylene blue (MB)-laden wastewater into the aquatic environment without proper treatment can impose several negative impacts on the health status of water bodies [1]. For instance, the MB dye can form a colored layer on the water surface [2], preventing light penetration into the marine environment that contains plant and aquatic species [3]. The sunlight absorption and scattering problems could also delay the photosynthetic processes of aquatic plants and reduce the dissolved oxygen availability for marine life [4]. Moreover, exposure of humans to high concentrations of MB can cause serious health problems and dangerous injuries, such as increased heart rate, shock, jaundice, tissue necrosis, vomiting, and cyanosis [5]. As a result of these negative impacts, MB-containing effluents must be efficiently treated before their release into the environmental dimensions.

MB has a complex and recalcitrant molecular structure, allowing it to persist in the environment for long periods and hindering its removal from wastewater by conventional treatment techniques [6]. For instance, the biological degradation of MB requires an extended retention time and certain enzymatic reactions to ensure the microbial breakdown scheme. Moreover, removing MB from wastewater by its adsorption into a certain material creates secondary pollution via disposal of the exhausted adsorbent [7]. Alternatively, various researchers have employed advanced oxidation processes (AOPs) to eliminate MB from aqueous solutions via decomposition and mineralization [3,8,9].

Heterogeneous photocatalysis (semiconductor photocatalysis) is a widely used AOP, having a great potential to degrade multiple pollutants under low energy conditions such as visible light and sunlight irradiation [10,11]. In this process, the photocatalyst material absorbs photon energy equal or greater in magnitude to the energy band gap of the semiconductor material [12]. This absorbed energy excites an electron (e^-) from the valence to the conduction band of the photocatalyst material across the energy band gap [13]. This e^- excitation scheme leaves a hole (h^+) behind in the valence band, where the electron–hole pairs interact with H_2O to form $^{\bullet}OH$ ($H_2O + h^+ \rightarrow {}^{\bullet}OH + H^+$) and O_2 to create $^{\bullet}O_2^-$ ($O_2 + e^- \rightarrow {}^{\bullet}O_2^-$) [14]. These generated highly reactive oxygen species (ROS) participate in a series of REDOX reactions to degrade pollutants (e.g., dyes) into simpler compounds such as CO_2 and H_2O [15].

Zinc oxide (ZnO) and magnesium oxide (MgO) are two semiconductors that have been involved in the photocatalytic degradation of various pollutants. However, these materials suffer from wide band gap energies, as well as rapid e^-—h^+ recombination [16]. These drawbacks could be mitigated by combining the ZnO and MgO pure semiconductor materials to form a ZnO/MgO nanocomposite material, which should be investigated for MB photocatalytic degradation.

Optimizing the photocatalytic process is an essential step to maintain a feasible (reduced cost and time) wastewater treatment scheme [17]. Regression models could be used to predict the treatment system performance and optimize the operational factors. However, by increasing the system complexity and non-linearity (by adding more independent variables), the least squares optimization process could not deliver accurate outputs [3]. Moreover, the regression equations sometimes fail to yield the anticipated preciseness ability and predictability for describing the interactions among the input/output variables [18]. Alternatively, artificial neural network (ANN) has found successful applications in simulating the treatment of various wastewater sources using photocatalytic technology. An ANN is a non-linear prediction model that has a structure consisting of several interconnected processing elements, known as neurons [19]. The elements of this computational (black box) model are also recognized as units or nodes, and they are arranged in layers. Each neuron receives a number of weighted signals from the formerly linked nodes that are further combined and summed up to derive a certain response. This output represents either the last projected finding or an input attribute to the consequent phase, according to the layer ordering in the network topology [20]. A comparison between the product and the actual (target) values is then employed to adjust the network parameters (e.g., weights and biases), enhancing the prediction preciseness. The ANN models could be employed to capture inter-relationships among input and output data pairs that are unclear, nonlinear, or too hard to designate.

While several studies have focused on removing dyes from wastewater by photocatalytic degradation [6,8,21], there is still a research gap in optimizing the associated operational factors to maintain a techno-economically feasible process. Much effort should be exerted on employing artificial intelligence techniques to optimize the photocatalytic degradation process, aiming at minimizing the unit treatment costs. Hence, the current study focuses on creating and characterizing a smart ZnO/MgO nanocomposite material for MB photocatalytic degradation, followed by optimizing the photocatalytic process using ANN. This optimum condition is used to minimize the photocatalytic degradation price expressed by the capital and operational costs. This study's outputs could provide essential steps to find environmental and economic solutions for endorsing photocatalytic wastewater treatment under sustainable industrial-scale implementations.

2. Materials and Methods

2.1. Material Preparation

The ZnO and MgO nanoparticles were synthesized using a co-precipitation method, as previously reported (Figure 1; ref. [22]). The chemicals utilized throughout the study are tabulated in Supplementary Table S1. $ZnSO_4{\cdot}7H_2O$ and $MgCl_2{\cdot}6H_2O$ were used as precursors for ZnO and MgO. The 0.5 M solutions of the precursors were prepared in separate beakers. Under continuous stirring at 500 rpm, a 1 M solution of NaOH was added dropwise to each beaker until a white precipitate started to form. NaOH was added to maintain a pH of 12. Thereafter, stirring proceeded for about 15 min, allowing a complete precipitate formation. The ZnO and MgO precipitates were separated from the solution by filtration and washed multiple times with distilled water. The washed precipitates were then dried overnight at 105 °C. The dry material was crushed into fine powders, followed by calcination in a muffle furnace at 350 °C and 800 °C to achieve the ZnO and MgO nanopowders, respectively.

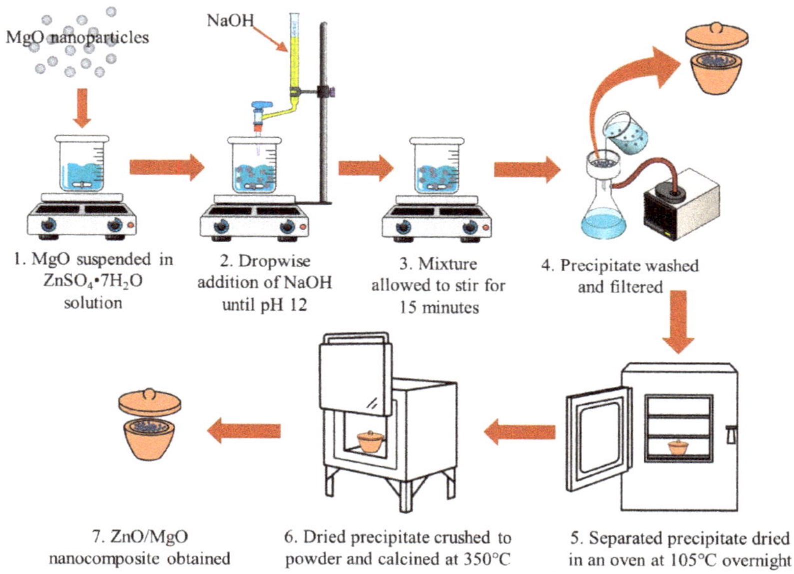

Figure 1. Synthesized technique of ZnO/MgO nanocomposite used for photodegradation of MB dye.

To synthesize the ZnO/MgO nanocomposite, an amount of already synthesized MgO was dispersed into a solution of $ZnSO_4{\cdot}7H_2O$ (100 mL) and mixed continuously for 5 min. Thereafter, a solution of 1 M NaOH was supplemented to the suspension dropwise until the final pH condition reached 12. The formed precipitate was then processed in the same way as ZnO, yielding the ZnO/MgO nanocomposite.

2.2. Experimental Setup and Design

The test was conducted in a rectangular reactor with a 1 m^3 internal volume. The reactor had two metal halide lamps (light intensity 25,000 lx, 400 W), emitting 510 nm UV-vis light. In each experimental run, 100 mL of dye solution was inserted into a 250 mL beaker. A 2 mL sample was withdrawn, transferred through a 0.22 μm nylon syringe filter, and analyzed to obtain the MB concentration. The dye solution's pH was adjusted to the desired level using either H_2SO_4 or NaOH. The required amount of photocatalyst material was then supplemented to the aqueous medium. The beakers were inserted into the reactor and maintained under the dark condition (with mixing) to reach adsorption–desorption equilibrium. Once the equilibrium was observed, another 2 mL sample was withdrawn, and the lamps were turned on and the photocatalytic reaction proceeded for the required period of time. After finishing each experimental run, 2 mL of the sample was again collected and filtered to discard the catalyst material before analysis.

Experimental design and parameter optimization were conducted using Design Expert Trial version 9 software, where a Box–Behnken design (BBD) represented the influence of four independent variables (i.e., catalyst dose, dye level, pH, and duration) on the MB removal efficiency (R%) as the response. Three levels of the input variables (−1, 0, +1) with four factors (Table 1) were used to generate the BBD matrix, giving a total of 29 experimental runs. The assigned ranges were selected based on primary results for MB photocatalytic degradation (see Supplementary Figure S1).

Table 1. Independent variable ranges used in the Box–Behnken design (BBD) experimental design to optimize methylene blue (MB) photocatalytic degradation by the as-prepared ZnO/MgO nanocomposite.

Variable Name	Symbol	Unit	Low	Mid	High
				Coded	
			−1	0	1
				Real Value	
Dye level	A	ppm	10	55	100
Catalyst dose	B	mg/L	100	550	1000
Medium pH	C	-	3	6.5	10
Period	D	min	60	120	180

After finishing the experimental runs, the photocatalysts were subjected to distilled water several times (washing) and then dried at 105 °C overnight to determine the regeneration and recyclability performance [23].

2.3. Analytical Analysis

The MB dye concentrations were detected using a UV spectrophotometer (Jasco V-630 spectrophotometer, Tokyo, Japan) adjusted at a wavelength of 664 nm. The medium pH was monitored using a pH meter (Model Lutron YK-2001 PH). X-ray diffraction (XRD) analysis (Shimadzu XRD-6100, Kyoto, Japan) with a Cu-Kα radiation beam (λ = 0.154 nm) was carried out to obtain the crystallographic data of the synthesized photocatalysts before and after MB photodegradation. Fourier transform infrared spectroscopy (FTIR 8400s, Shimadzu, Kyoto, Japan) was applied to observe the photocatalyst's surface functional groups. The nanocrystalline nature, elemental composition, and surface morphology of the synthesized photocatalysts were given by a transmission electron microscope (TEM) equipped with energy dispersive X-ray spectrometer (EDS) (JEOL, JEM-2100F, Tokyo, Japan) and a scanning electron microscope (SEM) (JEOL, JSM-6360LA, Tokyo, Japan). The Brunauer–Emmett–Teller (BET) N_2 adsorption–desorption method was employed to identify the photocatalyst surface area and associated pore size and volume. The catalyst's optical band gap energy was determined using UV-Vis diffuse reflectance spectrum (UV-Vis DRS, Jasco v-570 spectrometer, Tokyo, Japan) analysis.

2.4. Mathematical and Computational Models

2.4.1. Quadratic Regression Equation

A second-order polynomial model (Equation (1)) was employed to study the correlation between the four variables and the MB R%:

$$y = \beta_o + \sum_{i=1}^{n} \beta_i x_i + \sum_{i=1}^{n-1} \sum_{j=i+1}^{n} \beta_{ij} x_i x_j + \sum_{i=1}^{n} \beta_{ii} {x_i}^2 \quad (1)$$

where y is the response (dye removal efficiency, R%); n represents the number of independent (input) parameters; β_o is the model intercept; β_i, β_{ij}, and β_{ii} represent the first-order (linear) constant, squared, and interaction effects, respectively; and x is the associated independent variables

The model adequacy and significance were assessed via an analysis of variables (ANOVA). Moreover, the impact of each of the four independent variables on the generated model outputs was also assessed using ANOVA under a significance level of 0.05. The adequacy of the generated model (goodness-of-fit) was further validated using the coefficient of determination (R^2) value.

2.4.2. Artificial Neural Network (ANN) Schematic Description

In this investigation, a feed-forward ANN model was applied to estimate and forecast the MB R% using four inputs of photocatalysis experimental factors, i.e., initial MB concentration (C_0), catalyst dosage, pH, and time (Figure 2). Feed-forward was chosen due to its simplicity and widely known network type able to calculate the response directly through single propagation (a forward path) [19]. Furthermore, back-propagation is recognized as a well-organized and promising supervised learning algorithm applied to optimize the network weights. Multiple epoch numbers (iterations) are run to minimize the mean squared error (MSE) between the predicted output and desired vector, further adapting the network architecture. The number of observations (records) used to simulate the ANN model was distributed as 70% (training), 15% (verification), and 15% (testing). MSE was used as the objective function (i.e., the stopping criteria) to describe the magnitude of the gradient of performance for validation. The ANN-related calculations were carried out using MATLAB (R2015a) software.

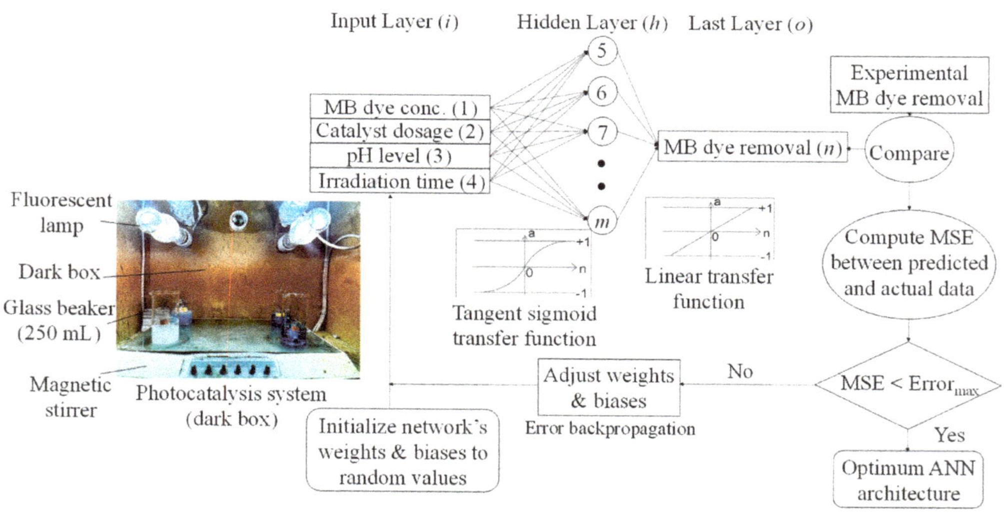

Figure 2. Schematic flowchart of ANN feed−forward back−propagation used to predict methylene blue (MB) removal efficiency under a photocatalytic degradation process.

3. Results and Discussion

3.1. Photocatalyst Characterization

The XRD, FTIR, SEM/EDS, TEM, BET, and UV-vis DRS characterization techniques were applied to the synthesized ZnO and MgO nanopowders, and the results were compared to those obtained from the ZnO/MgO nanocomposite.

3.1.1. XRD Characterization

Figure 3a shows the XRD diffraction profile of pure ZnO, defining peaks at 2θ of 31.6°, 34.3°, 36.1°, 47.5°, 56.5°, 62.7°, 66.2°, 67.8°, and 68.9°. These peaks could correspond to the crystal plane values of hexagonal pure ZnO (JCPDS ref. code 01-079-0207) given by (100), (002), (101), (102), (110), (103), (200), (112), and (201). These XRD peaks complied with those obtained for zinc oxide hollow spheres [24]. The XRD pattern (Figure 3a) also showed characteristic peaks of the cubic structure of pure MgO (JCPDS Ref code 01-074-1225). These peaks were observed at 2θ of 36.8°, 42.8°, 62.2°, 74.5°, and 78.5°, corresponding to the crystal plane values of pure cubic MgO given by (111), (200), (220), (311), and (222), respectively. The diffraction pattern generated complied with those obtained for periclase MgO in an AC-ZnO/MgO nanocomposite [9].

The diffraction pattern of the ZnO/MgO nanocomposite (Figure 3b) showed a combination of the characteristic peaks of hexagonal ZnO and cubic MgO structures. These peaks were denoted by 2θ at 31.8°, 34.1°, 36.1°, 37.8°, 42.6°, 47.5°, 56.4°, 62.1°, and 68.0°. This combination of peaks was indicative of a synergistic effect between ZnO and MgO to obtain the nanocomposite. The presence of extra peaks at 20.2°, 50.7°, and 58.5° could be due to some of the MgO being converted into $Mg(OH)_2$ as a result of moisture adsorption during the synthesis process. These extra peaks corresponded to the (001), (102), and (110) crystal planes of $Mg(OH)_2$ nanosheets [25].

The CIs of nanoparticles were calculated by Equation (2):

$$CI\% = \frac{A_c}{A_t}\,'\,100 \tag{2}$$

where A_c is the area under the crystalline peaks of the XRD pattern and A_t is the total area under the entire XRD pattern of a respective material

The CIs of ZnO and MgO were 67.04% and 60.83%, respectively. The core–shell nanocomposite had a higher CI of 71.92%. The crystalline size (D) of the synthesized catalyst materials was also calculated following the Debye–Scherrer equation (Equation (3)):

$$D = \frac{K\lambda}{\beta cos\theta} \tag{3}$$

where K is the Scherrer constant representing the shape factor (assumed as 0.9), λ is the X-ray beam wavelength (1.541874 Å), β is the full-width half-maximum value of a peak, and θ is the diffraction angle of each peak.

The crystalline size of ZnO was 22.99 nm, complying within the 11.84–24.82 nm range for ZnO synthesized under different pH conditions [26]. The estimated crystalline size of MgO (32.72 nm) obeyed that of MgO synthesized using a 1 M solution of precipitator [27]. Finally, the crystalline size was reduced to 17.24 nm for the synthesized nanocomposite material.

3.1.2. FTIR Characterization

The FTIR peaks observed at 3440.10 cm^{-1} in the three materials (Figure 3c) could correspond to the O–H stretching mode of the hydroxyl functional group [28]. The H–O–H stretching of H_2O could reflect the high affinity of MgO to adsorb water molecules. The peaks detected at 1643.65 cm^{-1} and 1135.76 cm^{-1} were attributed to the O–H stretch vibration [29]. A peak at 1430.05 cm^{-1} was also noticed, indicating the bending vibrations of H_2O adsorbed onto the catalyst material surface [30]. The peaks in the region below 700 cm^{-1} could identify the metal–oxygen bonds [31], e.g., the peaks at 681.21 cm^{-1} and 591.67 cm^{-1} represented the Mg–O bond, and the peak at 424.57 cm^{-1} denoted the Zn–O bond. The presence of these peaks further verified the successful combination of ZnO and MgO.

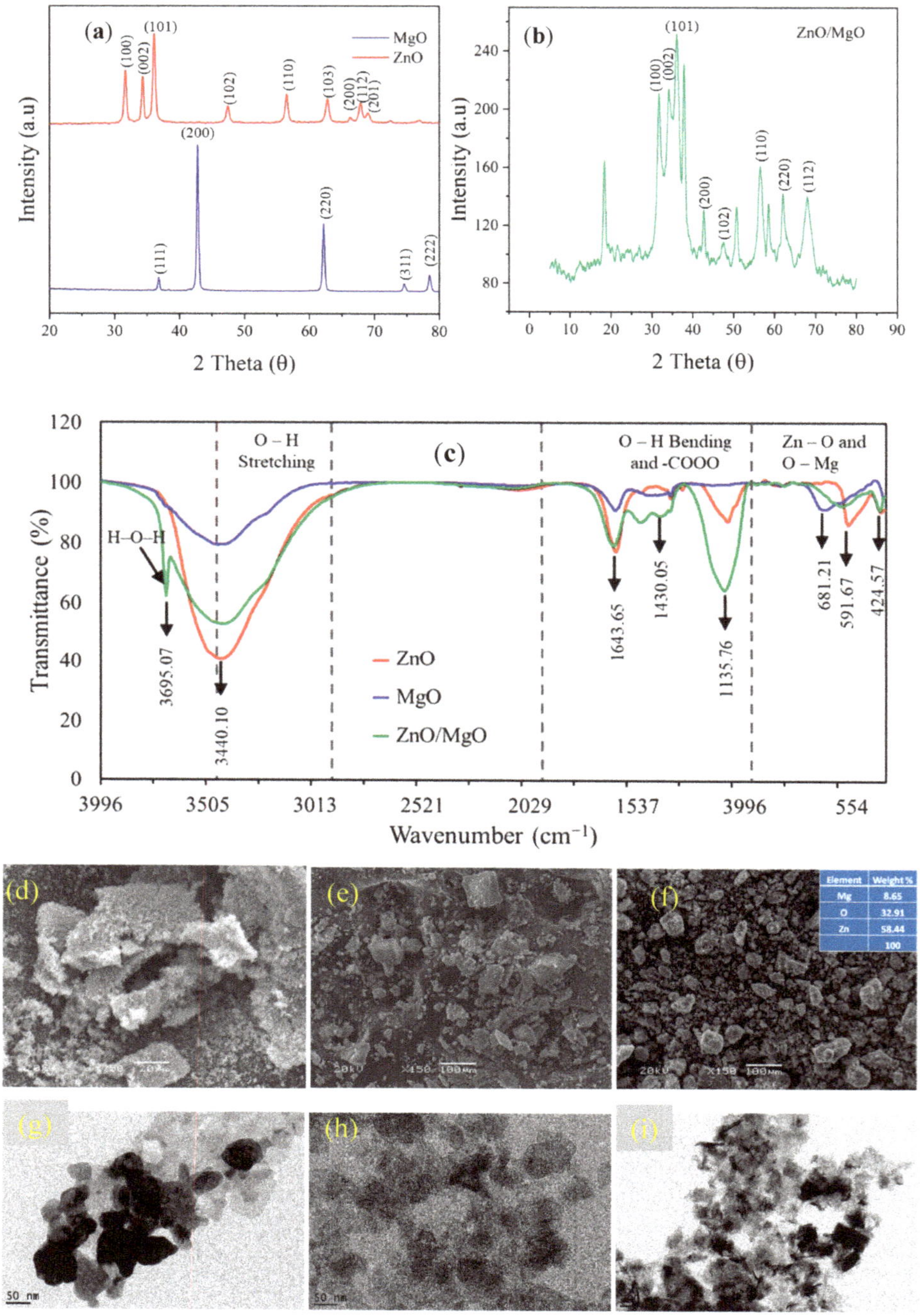

Figure 3. Characterization of synthesized nanocomposites. (**a**) XRD peaks of ZnO and MgO, (**b**) XRD peaks of ZnO/MgO, (**c**) FTIR spectra, (**d**) SEM of ZnO, (**e**) SEM of MgO, (**f**) SEM of ZnO/MgO, (**g**) TEM of ZnO, (**h**) TEM of MgO, and (i) TEM of ZnO/MgO.

3.1.3. SEM/EDS Characterization

The SEM micrograph of pure ZnO (Figure 3d) showed a spongy appearance of agglomerated particles. This result was similar to that obtained for ZnO synthesized via the precipitation technique [32]. The SEM micrograph of pure MgO displayed flat plate-like structures (Figure 3e) of varying sizes. A similar structure was obtained for MgO by employing the co-precipitation method [33]. The SEM micrograph of the ZnO/MgO nanocomposite (Figure 3f) showed a rough external surface that appeared to be a coating of ZnO over MgO. This surface roughness might be due to the gathering and building up of numerous small clusters of ZnO that coated the MgO core nanoparticles. The EDS results showed that Zn had the highest percentage weight followed by Mg and O. This result further verified the successful synthesis of the nanocomposite material, where the ZnO nanoparticles were deposited on MgO (the support material).

3.1.4. TEM Characterization

The TEM image showed that the ZnO particles were irregularly shaped and mildly varied in size (Figure 3g). The agglomeration of ZnO could occur due to the exclusion of capping agents during the particle synthesis process. The TEM image of MgO depicted the existence of roughly circular plate-like structures of a relatively uniform size (Figure 3h). Finally, the TEM image of the ZnO/MgO nanocomposite showed a mixed matrix of the plate-like MgO particles surrounded by the spongy appearance of the ZnO particles (Figure 3i). This result further confirmed that the ZnO/MgO nanocomposite was successfully synthesized, complying with the SEM observations. The particle sizes were measured using Image J free version 1.53t software, revealing 62.96 nm for ZnO, 77.27 nm for MgO, and 114.5 nm for ZnO/MgO.

3.1.5. BET Characterization

The BET analysis exhibited the material N_2 adsorption–desorption isotherms, complying with the existence of mesoporous surfaces (see Supplementary Figure S2a–c). The isotherm shapes of MgO and ZnO fit the type IVb isotherm, whereas the ZnO/MgO nanocomposite shape matched the type IVa isotherm. This finding revealed the presence of a type H3 hysteresis loop, probably depicting the existence of $\geq$4 nm diameter pores. Further, the BET analysis indicated a substantial decrease in the MgO surface area from 80.756 to 30.536 m^2/g after combination with ZnO (Table 2). This finding could be ascribed to the formation of more pores on the ZnO/MgO surface, where interparticle spaces were formed due to the growth of ZnO particles on the MgO surface, as previously reported [34]. This result was further confirmed by the significant increase in the pore diameter from 4.942 nm for MgO to 26.617 nm for ZnO/MgO (Table 2). The pore volumes were also measured; it was observed that the ZnO/MgO nanocomposite possessed the largest pore volume of 0.529 cm^3/g.

Table 2. Summary of BET analysis results for ZnO, MgO, and ZnO/MgO catalyst materials.

Catalyst Material	BET Pore Diameter (nm)	BET Surface Area (m^2/g)	BET Pore Volume (cm^3/g)
ZnO	7.752	11.183	0.021
MgO	4.942	80.756	0.109
ZnO/MgO	26.617	30.536	0.529

3.1.6. UV-Vis DRS Characterization

Figure 4a–c shows the indirect Tauc plots and absorbance curves of the synthesized nanomaterial. The indirect Tauc plots were generated by Equation (4) [35]:

$$(F(R)hv)^n = A(hv - E_g) \tag{4}$$

where F(R) is the Kubelka–Munk function, A is a proportionality constant, hν is the incident photon energy, n = ½, and E_g is the band-gap energy

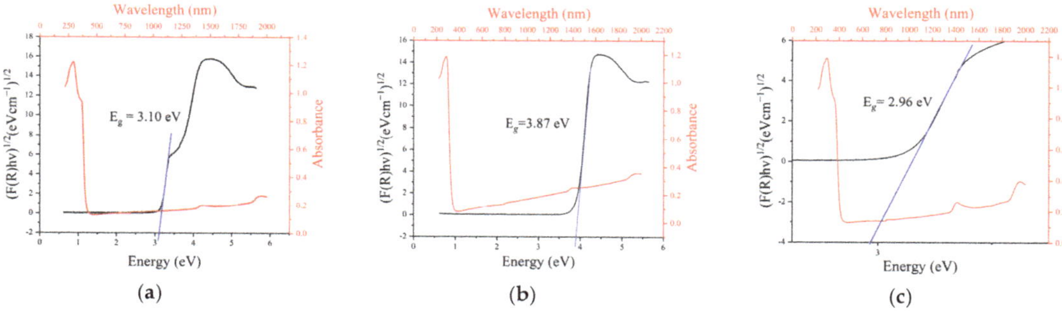

Figure 4. Energy band gaps of (**a**) ZnO, (**b**) MgO, and (**c**) ZnO/MgO nanocomposites.

The linear section of the Tauc plot was extrapolated, cutting the *x*-axis at the energy band gap value. The UV-vis DRS results showed that the optical band gap energies of ZnO, MgO, and ZnO/MgO nanomaterials were 3.10, 3.87, and 2.96 eV, respectively. This implies that the energy band gap was successfully reduced following the combination of ZnO and MgO. This reduced energy band gap of the nanocomposite could facilitate photocatalytic activity under UV light with a wavelength of ≈419 nm, obeying the visible light range.

3.2. Regression Model Application for Predicting MB Removal

The second-order polynomial regression model, governing the R% due to the interactions of the four independent variables, was generated (Equation (5)):

$$\begin{aligned} R\% = 56.342 - \; & 1.084A - 0.02466B + 4.978C - 0.473D - 6.1 \times 10^{-5}AB - 0.1019AC + 4.35 \times 10^{-4}AD \\ & + 3.843 \times 10^{-3}BC + 4.402 \times 10^{-6}BD + 0.0465CD + 9.744 \times 10^{-3}A^2 + 9.362 \times 10^{-6}B^2 \\ & - 0.2177C^2 + 8.92 \times 10^{-4}D^2 \end{aligned} \quad (5)$$

where R% is the MB removal efficiency (%); and the variables A, B, C, and D represent the dye level (ppm), catalyst dosage (mg/L), solution pH, and period (min), respectively.

An ANOVA test was performed to assess the accuracy and adequacy of the model generated (Table 3). The quadratic model exhibited an *F*-value and a *p*-value of 70.33 and <0.0001, respectively. The derived *F*-value was greater than the recognized *F*-value of 2.352 (at 95% significance) [18], signifying the model's appropriateness to forecast the R% observations. Moreover, these statistical results verified that the model was significant, where the *p*-value was below the alpha significance level. The model's prediction accuracy was also demonstrated by the predicted and adjusted R^2 values (0.9195 and 0.9720, respectively). These R^2 values were relatively high and the difference between them was less than 0.2, denoting that the data and the model were appropriately chosen and suitable for estimating the R% values [17]. Moreover, the R^2 value of the model was relatively high, presenting a value of 0.9860. Based on these statistical indicators, the generated quadratic equation could give acceptable predictions of the R% values, i.e., the predicted removal efficiencies were in agreement with the actual measurements (see Supplementary Figure S3). The significance of the variables was also analyzed, giving a *p*-value for each of A, B, C, and D below 0.05.

3.3. Artificial Neural Network (ANN) Application for Predicting MB Removal

3.3.1. Determining the Optimum ANN Architecture

The neurons' number in the hidden layer (N) was altered to examine the ANN prediction performance, evaluating the R^2 values and MSE of each ANN architecture (see Supplementary Table S2). It was noticed that increasing N enhanced the model prediction criterion during and after training. As such, the deficient N caused premature convergence or early termination of the training algorithm, hindering the network learning capability. The N = 8 established the best-hidden neuron arrangement, giving the smallest MSE of 11.20 and the highest R^2 of 0.9464 (see Supplementary Table S2). However, a further increase in N raised the model MSE, resulting in poor network generalization. An excessive N prolonged the calculation period (causing extra fitted functions due to the complex structure), and the training phase became probably accompanied by local minima or over-fitting. Over-fit (or over-learn) represented the network condition that maintained a smaller MSE (e.g., better R^2) for the training procedure, but the test procedure suffered from lower R^2 performances. This optimum N was accompanied by choosing Levenberg–Marquardt (*trainlm*) as the training function. As such, the Levenberg–Marquardt algorithm was highly efficient in simulating complex relationships among input variables within reasonable computation times. Moreover, this learning function is suitable for training moderate-sized feed-forward ANN models and solving nonlinear least squares problems. Hence, a three-layer feed-forward back-propagation ANN, with a "*trainlm*" training function and a "4–8–1" topology to forecast the MB R(%), was selected.

Table 3. Analysis of variables (ANOVA) for prediction of photocatalytic methylene blue (MB) removal by ZnO/MgO nanocomposite using four independent variables, namely initial MB concentration (A), catalyst dosage (B), aqueous solution pH (C), and time (D). The model accuracy is given by R^2 0.9860; adjusted R^2 0.9720; predicted R^2 0.9195; Adeq precision 31.8040.

Source	Sum of Squares	df	Mean Square	*F*-Value	*p*-Value	Significance
Model	17,842.80	14	1274.49	70.33	<0.0001	significant
A	10,467.64	1	10,467.64	577.61	<0.0001	significant
B	147.40	1	147.40	8.13	0.0128	significant
C	2640.01	1	2640.01	145.68	<0.0001	significant
D	208.60	1	208.60	11.51	0.0044	significant
AB	6.11	1	6.11	0.3371	0.5707	insignificant
AC	1030.82	1	1030.82	56.88	<0.0001	significant
AD	5.52	1	5.52	0.3043	0.5899	insignificant
BC	146.57	1	146.57	8.09	0.0130	significant
BD	0.0565	1	0.0565	0.0031	0.9563	insignificant
CD	381.66	1	381.66	21.06	0.0004	significant
A^2	2525.61	1	2525.61	139.36	<0.0001	significant
B^2	23.31	1	23.31	1.29	0.2758	insignificant
C^2	46.14	1	46.14	2.55	0.1329	insignificant
D^2	66.90	1	66.90	3.69	0.0753	significant

3.3.2. Training and Validation Performances

The MSE values were plotted against the epoch numbers (Figure 5a). In this diagram, the simulation procedure was completed by either achieving the highest number of training trials (e.g., training termination after six epochs) or reaching the smallest MSE (e.g., MSE = 10^{-6} as a performance goal), whichever initially ensued [36]. The training phase MSE (performance function) considerably dropped with iteration counts, reflecting the network's capability to modify its weights and biases until reaching the third epoch. After that point, the training curve MSE remained approximately unchanged, depicting the ANN model's ability to learn the relationships between the input and output data pairs. The validation set showed a different behavior, where its MSE curve increased until the first epoch and then remained almost constant. For instance, the best validation profile was monitored at epoch 0 with an MSE of 11.2, after which the ANN model initiated to overfit the data and lose its generalization power (Figure 5b). The MSE of the testing step increased, approximately similar to that of validation, employing that the learning step generated excess parameters. Hence, the zero epoch was used to terminate the training process, ensuring that the network model could precisely forecast R% using unseen input data.

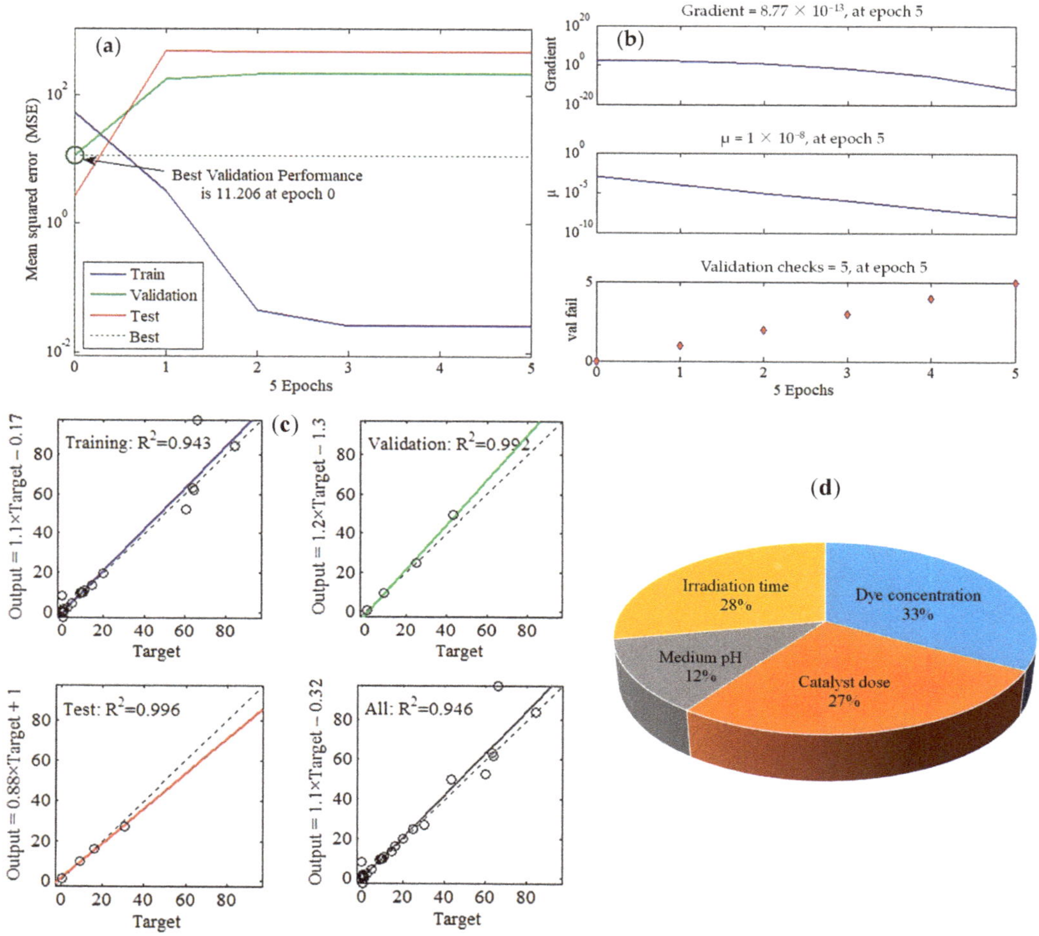

Figure 5. ANN (4-8-1) model employed to predict methylene blue (MB) removal efficiency using four inputs, namely MB concentration, catalyst dosage, pH, and time. (**a**) Training state, (**b**) validation performance, (**c**) regression data, and (**d**) sensitivity analysis.

3.3.3. Optimum Weights and Biases

At epoch = 0, the ANN model was well-learned and the network parameters (i.e., weights and biases) were adequately adapted (see Supplementary Table S3). The values of the linking weights ($W_{8\times4}$ and $W_{1\times8}$) and bias levels ($B_{8\times1}$ and $B_{1\times1}$) were determined based on the *trainlm* optimization algorithm. The layer outputs used to predict R% that matched the target (the best cross-validation performance) were calculated as follows:

Each parameter (independent variables) in the four-dimension input layer ($P_{4\times1}$ vector) was multiplied by its corresponding record in the weight matrix, $W_{8\times4}$, and then cumulated. Then, an eight-dimension bias, $B_{8\times1}$, was included, and a hyperbolic tangent sigmoid "*tansig*" activation function was used to transform the net input into a response signal ($z_{8\times1}$) (Equation (6)):

$$z_{8\times1} = tansig\left(\sum(W_{8\times4} \cdot P_{4\times1}) + B_{8\times1}\right) \quad (6)$$

The products of the eight-dimension middle layer, $z_{8\times1}$, were weighted and integrated into the matrix, $W_{1\times8}$, and, further, a one-dimension bias ($B_{1\times1}$) was included. The last output was introduced to a linear *"purelin"* activation function, yielding the last layer correlation (Equation (7)):

$$y_{1\times1} = tansig\left(\sum(W_{1\times8}\cdot z_{8\times1}) + B_{1\times1}\right) \quad (7)$$

3.3.4. Regression Plot

Figure 5c displays the schemes of the simulated MB (R%) by ANN against their associated experimental data, along with R^2 values of training, validation, and testing. In each regression sub-plot, the solid line denotes the best match of the ANN model. The dashed line represents the model outputs that fit precisely the actual observations (i.e., the theoretical fitting condition with R^2 = 1.000). An overall R^2 of 0.946 designated that the developed ANN (4–8–1) model could explain up to ≈95% of the variability in the MB removal data.

3.3.5. Relative Importance

The obtained weights and biases (see Supplementary Table S3) were involved in computing the relative importance (RI) of the experimental parameters, following Garson's formula [37]. Figure 5d shows that the RI for the dye concentration reached its maximum value at 33%, followed by irradiation time (28%), catalyst dosage (27%), and finally the medium pH (12%). Moreover, the RI values depicted that the investigated independent variables were important for improving the MB photocatalytic degradation performance.

3.4. Photocatalysis Operating Factors Optimization

The created ANN model was considered as the fitness function and combined with a genetic algorithm (GA) to optimize the photocatalytic experimental factors. The assigned GA had 100 iteration numbers, 10 population size, 0.4 crossover probability, and 0.2 mutation probability, and was required to maintain R(%) greater than 90%. This procedure depicted that, at C_0 = 16.42 mg/L and pH = 9.95, the catalyst dosage and time could be 905 mg/L and 174 min, respectively. These values were estimated as 907.35 mg/L and 178 min, respectively, using the quadratic model optimization procedure (see Supplementary Figure S4). Operating the photocatalytic reactor under the ANN-based optimized factors maintained a shorter reaction time and a lower oxidant level compared with the quadratic model-based optimized factors.

According to the verification experiment conducted under the optimum condition, the absolute relative deviation between the ANN model outputs and experimental data was 1.70%, which was smaller than 4.54% for the quadratic model prediction performance. A better predictive accuracy of ANN could be assigned to its universal ability to capture and approximate the non-linearity of the photocatalytic system, whereas the regression model is only limited to second-order polynomial.

3.5. Effect of Independent Variables on MB Degradation

The MB removal efficiencies (R%) were highly influenced by varying the independent variables within the experimental range (Figure 6):

3.5.1. Effect of Initial MB Concentration

Increasing the C_0 within the 10–100 mg/L range was noticed to reduce the R% observations (Figure 6). This decrease in R% could be because of the blocking of active sites available on the nanomaterial surface by a surplus of MB molecules [21]. This pattern was associated with an ineffective photon–catalyst interaction, where a portion of photon energy could be absorbed by the MB molecules rather than by the catalyst phase. This action reduced ROS generation (higher MB:ROS ratios); i.e., there would be an insufficient ROS amount to degrade the MB in the aqueous solution. Another reason for R% reduction at higher C_0 could be the hindering of light penetration through the dye solution and to the photocatalyst surface [12].

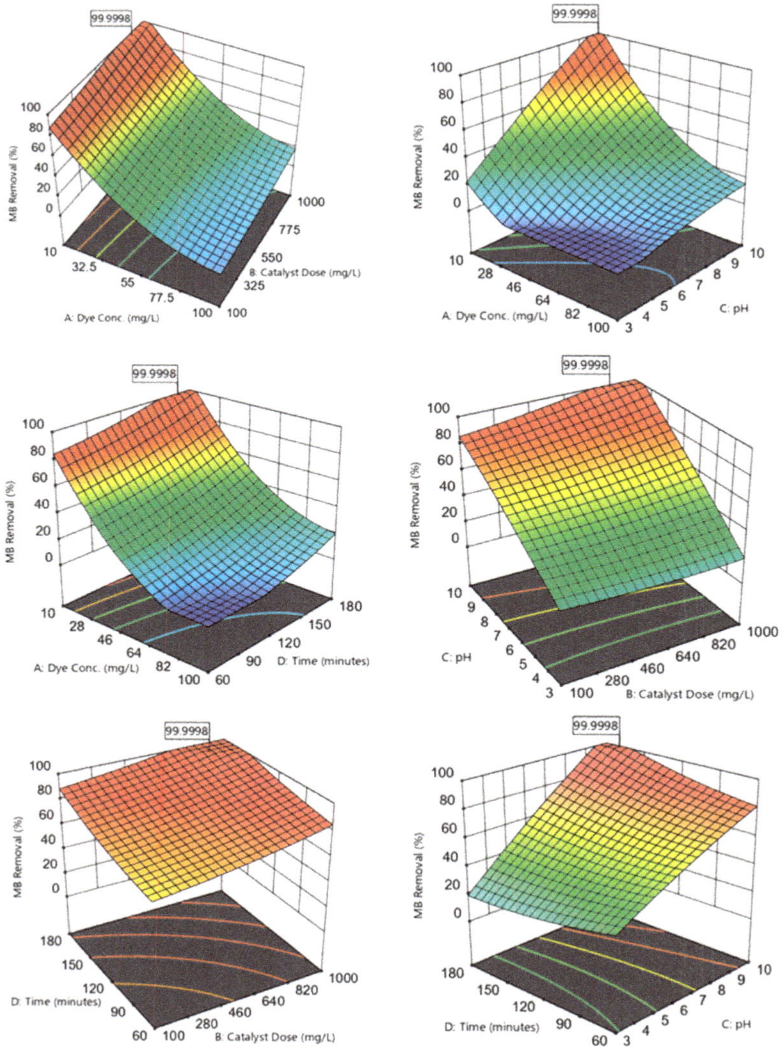

Figure 6. Three-dimensional plots showing the correlations between the MB removal efficiencies and the independent variables; dye concentration, catalyst dose, pH, and photodegradation time.

3.5.2. Effect of Catalyst Dose

The analysis showed that the MB R% was gradually improved by increasing the photocatalyst dosage within the 100–1000 mg/L range (Figure 6). This enhanced performance could be ascribed to an increase in the active site amounts when a high catalyst dosage was introduced to the medium [38]. These active locations could absorb more photons for maintaining the photocatalytic reactions. Increasing the available active sites facilitated the active site–photon interactions, yielding surplus ROS to degrade the MB dye [21].

3.5.3. Effect of pH

The results revealed an improved photocatalytic performance by increasing the solution pH within the 3–10 range (Figure 6). Varying the medium pH directly influenced the photocatalyst surface charge and the associated attraction and repulsion forces between the catalyst and the dye. The catalyst surface could acquire more negative charges as the medium pH became greater than the point of zero charge (pH pzc). Accordingly, as the medium pH became closer to the pH pzc of ≈10.24, the negatively charged sites tended to form strong attraction forces with the cationic MB molecules. When the catalyst was supplemented with the MB solution, it was recognized that the pH would increase to values greater than the initial pH. Increased attraction of the MB onto the photocatalyst surface resulted in better MB–ROS interactions and higher MB degradation efficiencies. Moreover, higher pH values implied the presence of more OH^- present in the solution, resulting in the formation of more •OH radicals to degrade MB [2]. The medium pH could also affect the MB dye dissociation pathway. For instance, the MB dye tended to be ionized and possessed a predominantly cationic species (positive charge) at a solution pH over 3.8 (i.e., the MB dissociation constant, pKa) [39].

3.5.4. Effect of Irradiation Time

The proportional relationship between the irradiation time (60–180 min; Figure 6) and MB R% could be denoted by the prolonged interactions between the dye molecules and the ROS available on the photocatalyst surface [40]. The adsorption of MB onto the material surface could be designated as a function of time until active sites' occupation and saturation, restricting the diffusion mobility of photocatalytic electrons [6].

3.6. Recyclability of the Photocatalyst Material

A recyclability test was performed to investigate the photocatalytic performance of the as-prepared materials for multiple regeneration and reuse cycles (see Supplementary Figure S5). After six successive cycles, the ZnO/MgO nanocomposite exhibited a minor drop in MB removal of about 6.14% compared with 10.56% for ZnO and 13.39% for MgO. It was supposed that ZnO/MgO exhibited higher stability and regeneration capacity than ZnO and MgO individually. However, this photocatalytic efficiency became insufficient in increasing the number of cycles above six. This reduction in performance could be ascribed to the photocatalyst mass loss after repeated use and/or the blocking of active sites by dye traces. This stability was better than B–N–TiO_2/EP photocatalyst material, which showed a reduction in performance >10% for dye degradation after five cycles [41].

3.7. Photocatalytic Degradation Rates and Suggested Degradation Mechanisms

The data in Figure 7a denoted that the MB photocatalytic degradation data fit the pseudo-first order kinetic model [42] with a high R^2 range of 0.957–0.992. The study also depicted that the degradation rate constant was 0.025 min^{-1} for ZnO/MgO. This rate constant was considerably greater than that observed for ZnO (0.003 min^{-1}) and MgO (0.0014 min^{-1}). In particular, this degradation kinetic investigation verified the superiority of the ZnO/MgO nanocomposite for the photocatalytic degradation of MB.

Figure 7b shows the photocatalytic degradation mechanism related to the ZnO/MgO nanocomposite. The conduction and valence band energies of ZnO and MgO were calculated using the equations given by Equations (8) and (9) [43]:

$$E_{CB} = X - E - 0.5E_g \quad (8)$$

$$E_{VB} = E_{CB} + E_g \quad (9)$$

where E_{CB} = conduction band energy; E_{VB} = valence band energy; X = electronegativity value; E = electron energy on the hydrogen scale; E_g = band gap energy.

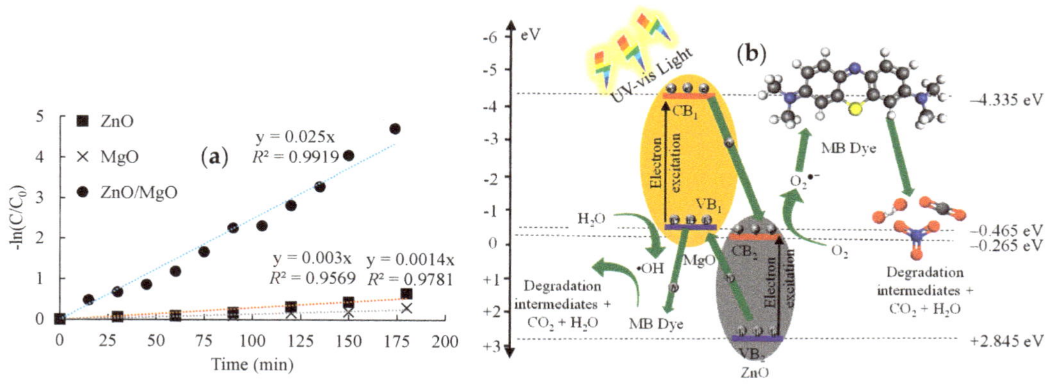

Figure 7. Photocatalytic degradation of MB dye. (**a**) Pseudo−first order fitting and (**b**) suggested degradation mechanisms.

The E_{CB} and E_{VB} of ZnO were determined as −0.265 eV and +2.845 eV, respectively, whereas the E_{CB} and E_{VB} of MgO were calculated as −4.335 eV and −0.465 eV, respectively. The CB and VB values were then used to derive the proposed MB degradation mechanism (Figure 7b). When the ZnO/MgO nanocomposite was exposed to light irradiation, the photocatalytic reaction governed by Equation (10) would occur to generate e^- and h^+:

$$ZnO/MgO + hv \rightarrow e^-(CB) + h^+(VB) \tag{10}$$

where e^-(CB) = electrons in the conduction band; h^+(VB) = the electron vacancy in the valence band.

The VB electrons would absorb photon energy and jump into the CBs. The band energy of the ZnO CB possessed a more positive potential than the VB of MgO. Consequently, the photogenerated e^- in the CB of MgO would have a stronger affinity to migrate and accumulate in the CB of ZnO instead of recombining with the holes in the MgO VB. Simultaneously, the h^+ left in the VB of ZnO would migrate and accumulate in the more negative VB of MgO. This charge transfer tended to reduce the e^-/h^+ recombination, resulting in enhanced photocatalytic activity of the ZnO/MgO nanocomposite.

Following the charge transfer process, the e^- and h^+ accumulated in the ZnO CB and MgO VB, respectively, would be free to interact with molecules in the surrounding MB solution. Electrons would reduce molecular oxygen (O_2) to create superoxide radicals ($O_2^{\bullet-}$), following Equation (11):

$$O_2 + e_{CB} \rightarrow O_2^{\bullet-} \tag{11}$$

This reduction of O_2 could occur because the CB of ZnO possessed a sufficient reduction energy potential (more positive) to facilitate its conversion into $O_2^{\bullet-}$, requiring a reduction energy potential of −0.33 eV [44]. Simultaneously, the h^+ accumulated in the MgO VB would oxidize H_2O molecules present in the MB solution to form hydroxyl radicals (•OH), following Equation (12):

$$h_{VB} + H_2O \rightarrow \bullet OH + H^+ \tag{12}$$

This reaction could occur because the VB of MgO possessed a sufficient oxidation potential (more negative) to facilitate the conversion of H_2O to •OH, which required a reduction potential of about +1.9 eV [45]. The ROS generated ($O_2^{\bullet-}$ and •OH) would easily interact with MB molecules (Equation (13)), breaking it down into less toxic and simpler intermediates and eventually CO_2 and H_2O. Moreover, the photogenerated h^+

and e^- accumulated on MgO VB and ZnO CB, respectively, could directly interact with MB molecules:

$$(h_{VB}, \bullet OH, e_{CB}, O_2^{\bullet -}) + MB \rightarrow Degradation\ intermediates + CO_2 + H_2O \quad (13)$$

3.8. Photocatalytic Degradation of MB Reported in the Literature

The effectiveness of the ZnO/MgO nanomaterial was compared with other ZnO- and MgO-based photocatalyst materials for MB dye removal (Table 4). Other photocatalysts have exhibited greater degradation performances at lower MB levels [24,31,46,47] and/or longer degradation times [48]. Most studies integrated ZnO and MgO into a third material, resulting in ternary photocatalysts [9,48] for improved performance. In our investigation, the photocatalyst material comprised ZnO and MgO, and the operational factors were optimized to obtain high degradation efficiencies for relatively higher dye concentrations. The results obtained in our study showed that the synthesized nanocomposite material could be successfully utilized for MB degradation.

Table 4. Comparison of the performance of the ZnO/MgO nanocomposite with other photocatalysts toward methylene blue (MB) removal.

Photocatalyst	Experimental Condition				Removal Efficiency (%)	Ref.
	MB Conc. (mg/L)	Catalyst Dose (mg/L)	pH	Time (min)		
ZnO/MgO nanocomposite	16.42	907.35	9.95	178	99.99	This study
ZnO/MgO/AC nanoparticle	3.2	1000	7	20	100	Karimi et al. [9]
Novel tri-phase CuO–MgO–ZnO nanocomposite	5	400	7	100	88.5	Munawar et al. [31]
ZnO/MgO nanocomposites	10	40	7	120	89.5	Panchal et al. [47]
Hollow MgO-ZnO microspheres	5	300	7	150	100	Vishwanathan et al. [24]
Ternary MgO/ZnO/In_2O_3	10	500	7	300	98	Xiang et al. [48]
CdO–ZnO–MgO nanocomposite	1	25	7	120	91	Revathi and Karthik [46]

3.9. Economic Feasibility of the ZnO/MgO Nanocomposite Photocatalysis System

A cost-benefit analysis was performed to evaluate the economic feasibility of applying the ZnO/MgO nanocomposite for treating dye-polluted water. The optimum conditions obtained from the optimization of independent variables were utilized to design the photocatalytic system used in the analysis. The photocatalytic system was designed to treat a daily influent volume of 10 m^3. This flow rate is suitable for a compact-sized wastewater treatment facility with simple equipment. The breakdown of the economic analysis comprised estimates of the capital/investment costs and operational/running costs (Table 5). These calculations were derived using a precision of ±20% [49], because cost estimation from the real-world application of photocatalysis could vary among countries based on the fluctuations in the currency exchange rates, assumptions applied, feed-in tariffs, and electricity demand. The preciseness and uncertainty of this economic estimation were also assessed by developing sensitivity analysis, varying the photocatalytic reactor cost, chemical prices, and electrical and water consumption costs between −30% and 30% [50] (see Supplementary Figure S6).

3.9.1. Capital Costs

The one-time expenses incurred to implement the photocatalysis system were represented by the capital costs. About 11.69% of these capital costs accounted for the photocatalytic reactor and tank installations, where the volume (V_c) could be estimated by Equation (14) [20]:

$$V_c = \frac{Q}{D} \times \frac{t_t}{t_w} \tag{14}$$

where Q represents the average volume of wastewater treated annually (m^3/yr), D is the number of working days per year (assuming 300 days per year), t_t is the optimum reaction time (e.g., 90 min), and t_w is the number of plant operation hours per day (e.g., 1440 min/d represents a 24 h daily plant operation time).

The obtained tank volume was then applied to calculate the construction cost by Equation (15) [51]:

$$CC = V_c \times C_P \tag{15}$$

where C_p represents the price per m^3 of a photoreactor unit derived from subcontractor and supplier quotes.

The cost of water collection tanks was also added to the initial investment. Moreover, the prices of mechanical equipment, comprising UV lamps (light source), a stirrer motor, pumps, and a rotary calcination furnace for catalyst synthesis, were also included, as previously reported [52].

Table 5. Cost estimates for the photocatalytic degradation of MB dye-laden wastewater (10 m^3/day) using the ZnO/MgO nanocomposite under unoptimized and ANN-optimized conditions (using a precision of ±20%).

Item	Cost		Justification
	Unoptimized	ANN-Optimized	
Photocatalysis installation (PI)	USD 15,500	USD 14,060	Reactors (USD 1680 × 2 = USD 3360), UV lamps (USD 1150 × 2 = USD 2300), stirrer motors (USD 590 × 2 = USD 1180), pumps (USD 2010 × 2 = USD 4020), and rotary calcination furnace (USD 3200). Optimized reaction time reduced the reactor volume by ~6%
Piping, valves, and wiring	USD 4650	USD 4218	0.3 PI
Site work	USD 1550	USD 1406	0.1 PI
Subtotal (1)	USD 21,700	USD 19,684	1.4 PI = Q
Contractor O&P	USD 3255	USD 2953	0.15 Q
Subtotal (2)	USD 24,955	USD 22,637	1.15 Q = R
Engineering fees	USD 3743	USD 3395	0.15 R
Subtotal (3)	USD 28,698	USD 26,032	1.15 R = S
Contingency fees	USD 5740	USD 5206	0.2 S
Subtotal (4)	USD 34,438	USD 31,239	1.2 S = T
Others	USD 6888	USD 6248	0.2 T
Total capital cost	USD 41,325	USD 37,486	1.2 T
Annual capital cost	USD 8265/yr	USD 7497/yr	USD 37,486/project lifetime (5 yr)
Interest rate consideration	USD 10,056/yr	USD 9122/yr	USD 7497/yr × $(1 + i)^n$
Unit capital cost	USD 3.35/m^3	USD 3.04/m^3	USD 9122/yr ÷ (10 m^3/d × 300 d/yr)
Chemical costs	USD 90/month	USD 75/month	15 kg/month × USD 5/kg
Electrical consumption costs	USD 1157/month	USD 1105/month	8500 kWh/month × USD 0.13/kWh
Water consumption	USD 42/month	USD 40/month	168 m^3/month × USD 0.24/m^3
Labor cost	USD 120/month	USD 120/month	
Land rental costs	USD 42/month	USD 38/month	5% of total capital cost
Repair and maintenance costs	USD 126/month	USD 114/month	15% of total capital cost
Others	USD 168/month	USD 152/month	20% of total capital cost
Total operating cost	USD 1744/month	USD 1644/month	Sum of total operating items
Unit total operating cost	USD 5.81/m^3	USD 5.48/m^3	USD 1644/month ÷ (10 m^3/d × 30 d/month)
Unit total cost	USD 9.17/m^3	USD 8.52/m^3	Unit capital cost + Unit total operating cost

3.9.2. Operational Costs

The daily expenses incurred to run the photocatalysis system were represented by the operational costs. These costs included several items related to consumables, electricity consumption, salaries and wages, overheads, and plant maintenance [53]. The costs of chemicals needed to synthesize and apply the ZnO/MgO nanocomposite were estimated by Equation (16):

$$Cost_{chemicals} = C_i \times P_i \quad (16)$$

where $Cost_{chemicals}$ is the monthly cost of chemicals incurred (USD/month), C_i is the quantity of chemicals supplied (kg/month), and P_i is the chemical price per unit weight (USD/kg)

The electricity utilization costs comprised the power consumption to operate an electronic overhead stirrer (69 W), pumps (55 W), and UV lamps (46 W) [54]. This amount of energy was used to estimate the electrical energy per order (E_{EO}) by Equation (17) [55]:

$$E_{EO} = \frac{38.4 \times P}{V \times k} \quad (17)$$

where P is the power of all electrical equipment (using 170 W), V is the volume of treated water per day (L), and k is the reaction rate constant (ln (C/C_o)).

The consumption per m^3 (E_{EO}) was used to estimate the annual consumption for the plant receiving 10 m^3/day. The annual consumption in KWh and the electricity feed-in tariff (USD 0.13/kWh) were then used to estimate the electricity consumption cost per year.

The costs of water required for photocatalyst synthesis and regeneration were also considered. This estimate was derived from the lab-scale preparation procedures, where ≈1 L of water was required to synthesize and regenerate 1 g of the ZnO/MgO nanocomposite. The water tariff was set at USD 0.24/m^3.

The costs associated with the repair and maintenance of the treatment system were estimated at 1.5% of all capital costs [51]. An employee salary rate of USD 120/month was utilized to determine the labor costs associated with water sampling (three samples/week), cleaning and repair of equipment, and operation of the photocatalysis system. Statutory obligations (taxes) were also considered as 1% of the total capital costs. These operational prices resulted in a total running treatment cost of USD 5.48/m^3.

3.10. Meeting the Sustainability Dimensions from the Photocatalytic Degradation of Dyes

The current study's outputs maintain several benefits that could fulfill the sustainability dimensions. For example, removing dye pollutants (e.g., MB) from the textile industrial effluents before final disposal tends to protect the aquatic environment. This aim has a positive correlation with SDG14 "Life Below Water" by protecting and conserving the aquatic species from marine pollution because dyes negatively affect the fish organs, such as the liver and gills, and reduce sunlight penetration and the related photosynthetic activities [6]. Moreover, this treatment facility could meet the targets of SDG3 "Good Health and Well-being" because textile dyes can enter the food chain through biomagnification and biotransformation [2]. Also, the study demonstrates that ANN is an innovative tool that could optimize the photocatalytic treatment facility and reduce the operational costs, meeting SDG9 "Industry, Innovation, and Infrastructure". Furthermore, using sunlight as an irradiation source (cleaner energy) and reusing the synthesized ZnO/MgO photocatalyst under six consecutive cycles complies with the targets of SDG12 "Responsible Consumption and Production". This SDG could also be improved by conducting essential research on the management and/or disposal of exhausted ZnO/MgO photocatalysts. For Instance, Razzaghian Ghadikolaee et al. [56] found that the photocatalyst material after exhaustion could be managed by using it as an attractive additive to maintain better 3D-printed concrete properties.

Despite the aforementioned benefits provided by photocatalytic dye degradation, some aspects related to the synthesis of the catalyst materials and their disposal after reuse should be addressed for enhancing the process's sustainability. Regarding the manufacturing methods, plant extracts [57] and microorganisms [58] have attracted much attention as reducing agents in the photocatalyst green synthesis process. Applying these eco-friendly and cleaner techniques to synthesize the ZnO/MgO could reduce the chemical costs and minimize exposure to possible material toxicity. Where the recyclability of the spent catalyst material is concerned, the ZnO/MgO nanocomposite should be tested for applications as an additive in the construction industry, following previous research [56]. This benefit would avoid the unrestricted disposal of solid waste to the terrestrial and inland freshwater ecosystems and the related services, meeting the targets of SDG15 "Life on Land". Assuming that the nanocomposite can neither be applied in construction nor in soil amendment, the spent ZnO and MgO can be treated by thorough washing to remove any adsorbed pollutants before disposal into well-designed and properly operated landfills.

4. Conclusions

This study succeeded in providing a promising artificial neural network (ANN) model that could cost-effectively optimize the photocatalytic dye degradation system. Operating the photocatalytic degradation process under the determined optimum conditions enhanced the methylene blue (MB) removal efficiency by $\approx$12%, while the treatment cost was reduced by about 7% compared with the unoptimized condition. The study outcomes also included the successful synthesis of a ZnO/MgO photocatalyst that had a narrower optical band gap and better surface morphology, functional groups, and crystal structure than ZnO and MgO individually. Hence, the proposed artificial intelligence technique can be beneficial for generating more accurate predictions of the photodegradation process in wastewater treatment while maintaining better operational conditions that provide higher economic benefits. Further studies are required to determine the degradation pathway and by-products generated, identify the ROS responsible for the photocatalytic degradation process, use green routes for ZnO/MgO synthesis, and enlarge the data set size for a more accurate ANN training process.

Supplementary Materials: The following supporting information can be downloaded at: https://www.mdpi.com/article/10.3390/su16020529/s1, Figure S1. Initial photocatalysis experiments under light irradiation; Figure S2. N_2 adsorption–desorption isotherms of (a) ZnO, (b) MgO, and (c) ZnO/MgO; Figure S3. Plot of actual versus predicted MB removal efficiencies; Figure S4. Optimization of photocatalysis independent variables using the response surface methodology (RSM) based on the quadratic regression model; Figure S5. Recyclability of the photocatalyst material; Figure S6. Effect of variation from −30% to 30% on the total treatment cost; Table S1. Chemicals and reagents used for experimental procedures. All chemicals were utilized in the experiments without further purification; Table S2. Selection of best number of neurons in the hidden layer (N) and transfer function based on the smallest MSE of different ANN models for predicting MB removal efficiencies. Here, *trainoss*: one step secant; *traincgb*: conjugate gradient with Powell–Beale restart; *trainbfg*: Broyden–Fletcher– Goldfarb–Shanno (BFGS) quasi-Newton; *trainlm*: Levenberg–Marquardt; trainscg: scaled conjugate gradient; *traincgf*: Fletcher–Powell conjugate gradient; *traingd*: gradient descent back-propagation; Table S3. Optimum weights and biases of the feed-forward back-propagation ANN (4–8–1) with *trainlm* training function used to predict the MB R(%). Reference [59] is cited in the supplementary materials.

Author Contributions: Conceptualization, A.A., M.F. and M.N.; methodology, K.F.N.; formal analysis, K.F.N.; writing—original draft preparation, K.F.N.; writing—review and editing, A.A., M.F. and M.N.; visualization, A.A., M.F. and M.N.; supervision, A.A., M.F. and M.N. All authors have read and agreed to the published version of the manuscript.

Funding: This research was funded by funds that were partially supported by TICAD7, Egypt-Japan University of Science and Technology (E-JUST), and Japan International Cooperation Agency (JICA).

Institutional Review Board Statement: Not applicable.

Informed Consent Statement: Not applicable.

Data Availability Statement: The data that support the findings of this study are available within the article (and/or) its Supplementary Materials.

Acknowledgments: The first author is very grateful to TICAD7 for providing financial support in the form of an MSc. scholarship. Also, thanks to JICA-Japan International Cooperation Agency for providing all facilities and equipment to accomplish this study.

Conflicts of Interest: The authors declare no conflicts of interest.

References

1. Wang, B.; Suen, H.; Zhang, B. Hydrodynamic cavitation as a promising route for wastewater treatment—A Review. *Chem. Eng. J.* **2022**, *412*, 128685. [CrossRef]
2. Vasiljevic, Z.Z.; Dojcinovic, M.P.; Vujancevic, J.D.; Jankovic-Castvan, I.; Ognjanovic, M.; Tadic, N.B.; Stojadinovic, S.; Nikolic, M.V.; Brankovic, G.O. Photocatalytic degradation of methylene blue under natural sunlight using iron titanate nanoparticles prepared by a modified sol-gel method: Methylene blue degradation with Fe_2TiO_5. *R. Soc. Open Sci.* **2020**, *7*, 200708. [CrossRef] [PubMed]
3. Yang, L.; Zhao, Z.; Wang, H.; Cai, Z. Synthesis of ZnO/ZnS core/shell microsphere and its photocatalytic activity for methylene blue and eosin dyes degradation. *J. Dispers. Sci. Technol.* **2020**, *41*, 2152–2158. [CrossRef]
4. Moorthy, K.; Rathi, B.G.; Shukla, S.P.; Kumar, K.; Bharti, V.S. Acute toxicity of textile dye Methylene blue on growth and metabolism of selected freshwater microalgae. *Environ. Toxicol. Pharmacol.* **2021**, *82*, 103552. [CrossRef] [PubMed]
5. Khan, I.; Saeed, K.; Zekker, I.; Zhang, B.; Hendi, A.H.; Ahmad, A.; Ahmad, S.; Zada, N.; Ahmad, H.; Shah, L.A.; et al. Review on Methylene Blue: Its Properties, Uses, Toxicity and Photodegradation. *Water* **2022**, *14*, 242. [CrossRef]
6. Alkaykh, S.; Mbarek, A.; Ali-Shattle, E.E. Photocatalytic degradation of methylene blue dye in aqueous solution by $MnTiO_3$ nanoparticles under sunlight irradiation. *Heliyon* **2020**, *6*, e03663. [CrossRef] [PubMed]
7. Karim, M.A.; Aziz, K.H.; Omer, K.M.; Salih, Y.M.; Mustafa, F.; Rahman, K.O.; Mohammad, Y. Degradation of aqueous organic dye pollutants by heterogeneous photo-assisted Fenton-like process using natural mineral activator: Parameter optimization and degradation kinetics. In Proceedings of the 7th International Conference on Water Resource and Environment (WRE 2021), Xi'an, China, 1–4 November 2021; Volume 958.
8. Mendis, A.; Thambiliyagodage, C.; Ekanayake, G.; Liyanaarachchi, H.; Jayanetti, M.; Vigneswaran, S. Fabrication of Naturally Derived Chitosan and Ilmenite Sand-Based TiO_2/Fe_2O_3/Fe-N-Doped Graphitic Carbon Composite for Photocatalytic Degradation of Methylene Blue under Sunlight. *Molecules* **2023**, *28*, 3154. [CrossRef]
9. Karimi, M.A.; Hatefi-Mehrjardi, A.; Askarpour Kabir, A.; Zaydabadi, M. Synthesis, characterization, and application of MgO/ZnO nanocomposite supported on activated carbon for photocatalytic degradation of methylene blue. *Res. Chem. Intermed.* **2014**, *41*, 6157–6168. [CrossRef]
10. Hama Aziz, K.H.; Omer, K.M.; Mahyar, A.; Miessner, H.; Mueller, S.; Moeller, D. Application of Photocatalytic Falling Film Reactor to Elucidate the Degradation Pathways of Pharmaceutical Diclofenac and Ibuprofen in Aqueous Solutions. *Coatings* **2019**, *9*, 465. [CrossRef]
11. Wen, L.; Wang, D.; Xi, J.; Tian, F.; Liu, P.; Bai, Z.W. Heterometal modified Fe_3O_4 hollow nanospheres as efficient catalysts for organic transformations. *J. Catal.* **2022**, *413*, 779–785. [CrossRef]
12. Akpan, U.G.; Hameed, B.H. Parameters affecting the photocatalytic degradation of dyes using TiO_2-based photocatalysts: A review. *J. Hazard. Mater.* **2009**, *170*, 520–529. [CrossRef] [PubMed]
13. Ren, G.; Han, H.; Wang, Y.; Liu, S.; Zhao, J.; Meng, X.; Li, Z. Recent Advances of Photocatalytic Application in Water Treatment: A Review. *Nanomaterials* **2021**, *11*, 1804. [CrossRef] [PubMed]
14. Liyanaarachchi, H.; Thambiliyagodage, C.; Liyanaarachchi, C.; Samarakoon, U. Efficient photocatalysis of Cu doped TiO_2/g-C3N4 for the photodegradation of methylene blue. *Arab. J. Chem.* **2023**, *16*, 104749. [CrossRef]
15. Gracien, E.B.; Jérémie, M.L.; Joseph, L.K.-K.; Omer, M.M.; Antoine, M.K.; Hercule, K.M.; Gerard, M.N. Role of hydroxyl radical scavenger agents in preparing silver nanoparticles under γ-irradiation. *SN Appl. Sci.* **2019**, *1*, 961. [CrossRef]
16. Wu, H.; Gao, C.; Ma, L.; Wang, L. Controllable microemulsion method for the synthesis of $Mg(OH)_2$/PS core–shell structures. *Micro Nano Lett.* **2021**, *16*, 413–418. [CrossRef]
17. Mohamed, O.A.; Masood, S.H.; Bhowmik, J.L. Optimization of fused deposition modeling process parameters for dimensional accuracy using I-optimality criterion. *Meas. J. Int. Meas. Confed.* **2016**, *81*, 174–196. [CrossRef]
18. Amani-Ghadim, A.R.; Aber, S.; Olad, A.; Ashassi-Sorkhabi, H. Optimization of electrocoagulation process for removal of an azo dye using response surface methodology and investigation on the occurrence of destructive side reactions. *Chem. Eng. Process. Process Intensif.* **2013**, *64*, 68–78. [CrossRef]
19. Bakr, M.; Nasr, M.; Ashmawy, M.; Tawfik, A. Predictive performance of auto-aerated immobilized biomass reactor treating anaerobic effluent of cardboard wastewater enriched with bronopol (2-bromo-2-nitropropan-1,3-diol) via artificial neural network. *Environ. Technol. Innov.* **2021**, *21*, 101327. [CrossRef]
20. Alalm, M.G.; Nasr, M. Artificial intelligence, regression model, and cost estimation for removal of chlorothalonil pesticide by activated carbon prepared from casuarina charcoal. *Sustain. Environ. Res.* **2018**, *28*, 101–110. [CrossRef]
21. Sadeghi, M.; Heydari, M.; Javanbakht, V. Photocatalytic and photo-fenton processes by magnetic nanophotocatalysts for efficient dye removal. *J. Mater. Sci. Mater. Electron.* **2021**, *32*, 5065–5081. [CrossRef]
22. Meenakshi, S.; Rajarajan, M.; Rajendran, S.; Kennedy, Z.R.; Brindha, G. Synthesis and characterization of magnesium oxide nanoparticles. *Chem. Mater. Sci.* **2012**, *50*, 10618–10620.

23. Subalakshmi, A.; Kavitha, B.; Srinivasan, N.; Rajarajan, M.; Suganthi, A. An affordable efficient SrWO4 decorated Bi2O3 nanocomposite: Photocatalytic activity for the degradation of methylene blue under visible light irradiation. *Mater. Today Proc.* **2022**, *48*, 409–419. [CrossRef]
24. Vishwanathan, S.; Das, S. Glucose-mediated one-pot hydrothermal synthesis of hollow magnesium oxide-zinc oxide (MgO-ZnO) microspheres with enhanced natural sunlight photocatalytic activity. *Environ. Sci. Pollut. Res.* **2023**, *30*, 8512–8525. [CrossRef] [PubMed]
25. Saoud, K.M.; Saeed, S.; Al-Soubaihi, R.; Bertino, M. Microwave Assisted Preparation of Magnesium Hydroxide Nano-sheets. *Am. J. Nanomater.* **2014**, *2*, 21–25.
26. Mustapha, S.; Ndamitso, M.; Abdulkareem, A.; Tijani, J.; Shuaib, D.; Mohammed, A.; Sumaila, A. Comparative study of crystallite size using Williamson-Hall and Debye-Scherrer plots for ZnO nanoparticles. *Adv. Nat. Sci. Nanosci. Nanotechnol.* **2019**, *10*, 045013. [CrossRef]
27. Yousefi, S.; Ghasemi, B. Precipitator concentration-dependent opto-structural properties of MgO nanoparticles fabricated using natural brine. *SN Appl. Sci.* **2020**, *2*, 852. [CrossRef]
28. Sangeeta, M.; Karthik, K.V.; Ravishankar, R.; Anantharaju, K.S.; Nagabhushana, H.; Jeetendra, K.; Vidya, Y.S.; Renuka, L. Synthesis of ZnO, MgO and ZnO/MgO by Solution Combustion Method: Characterization and Photocatalytic Studies. *Mater. Today Proc.* **2017**, *4*, 11791–11798. [CrossRef]
29. Ansari, A.; Ali, A.; Asif, M.; Shamsuzzaman, A. Microwave-assisted MgO NP catalyzed one-pot multicomponent synthesis of polysubstituted steroidal pyridines. *New J. Chem.* **2018**, *42*, 184–197. [CrossRef]
30. Mensah, K.; Samy, M.; Ezz, H.; Elkady, M.; Shokry, H. Utilization of iron waste from steel industries in persulfate activation for effective degradation of dye solutions. *J. Environ. Manag.* **2022**, *314*, 7. [CrossRef]
31. Munawar, T.; Mukhtar, F.; Yasmeen, S.; Naveed-ur-Rehman, M.; Nadeem, M.S.; Riaz, M.; Mansoor, M.; Iqbal, F. Sunlight-induced photocatalytic degradation of various dyes and bacterial inactivation using CuO–MgO–ZnO nanocomposite. *Environ. Sci. Pollut. Res.* **2021**, *28*, 42243–42260. [CrossRef]
32. Arora, A.K.; Devi, S.; Jaswal, V.S.; Singh, J.; Kinger, M.; Gupta, V.D. Synthesis and characterization of ZnO nanoparticles. *Orient. J. Chem.* **2014**, *30*, 1671–1679. [CrossRef]
33. Ajeesha, T.; Ashwini, A.; George, M.; Manikandan, A.; Mary, J.; Slimani, Y.; Almessiere, M.; Baykal, A. Nickel substituted $MgFe_2O_4$ nanoparticles via co-precipitation method for photocatalytic applications. *Phys. B Condens. Matter* **2021**, *606*, 412660. [CrossRef]
34. Yang, X.-H.; Sun, S.-Y.; Fu, H.-T.; Li, W.-F.; An, X.-Z. Two-Step Fabrication of Porous α-Fe_2O_3@δTiO_2 core-shell Nanostructures with Enhanced Photocatalytic Activity. In Proceedings of the 3rd Annual International Conference on Advanced Material Engineering (AME 2017), Shanghai, China, 14–16 April 2017.
35. Makuła, P.; Pacia, M.; Macyk, W. How To Correctly Determine the Band Gap Energy of Modified Semiconductor Photocatalysts Based on UV–Vis Spectra. *J. Phys. Chem. Lett.* **2018**, *9*, 6814–6817. [CrossRef] [PubMed]
36. Azadi, S.; Karimi-Jashni, A.; Javadpour, S. Modeling and optimization of photocatalytic treatment of landfill leachate using tungsten-doped TiO2 nano-photocatalysts: Application of artificial neural network and genetic algorithm. *Process Saf. Environ. Prot.* **2018**, *117*, 267–277. [CrossRef]
37. Garson, G. Interpreting neural-network connection weights. *AI Expert* **1991**, *6*, 46–51.
38. Kalaycıoğlu, Z.; Uysal, Ö.; Pekcan, Ö.; Erim, F.B. Efficient Photocatalytic Degradation of Methylene Blue Dye from Aqueous Solution with Cerium Oxide Nanoparticles and Graphene Oxide-Doped Polyacrylamide. *ASC Omega J.* **2023**, *8*, 13004–13015. [CrossRef] [PubMed]
39. Sousa, H.R.; Silva, L.S.; Sousa, P.A.A.; Sousa, R.R.M.; Fonseca, M.G.; Osajima, J.A.; Silva-Filho, E.C. Evaluation of methylene blue removal by plasma activated palygorskites. *J. Mater. Res. Technol.* **2019**, *8*, 5432–5442. [CrossRef]
40. Kumar, A.; Pandey, G. The photocatalytic degradation of Methyl Green in presence of Visible light with photoactive $Ni_{0.10}$:$La_{0.05}$:TiO_2 nanocomposites. *IOSR J. Appl. Chem.* **2017**, *10*, 31–44.
41. Xue, H.; Jiang, Y.; Yuan, K.; Yang, T.; Hou, J.; Cao, C.; Feng, K.; Wang, X. Floating photocatalyst of B-N-TiO_2/expanded perlite: A sol-gel synthesis with optimized mesoporous and high photocatalytic activity. *Sci. Rep.* **2016**, *6*, 29902. [CrossRef]
42. Lagergren, S. Zur Theorie der sogenannten Adsorption gelöster Stoffe. *K. Sven. Vetenskapsakad. Handl.* **1898**, *24*, 1–39.
43. Habibi, M.; Habibi-Yangjeh, A.; Pouran, S.R.; Chand, H.; Krishnan, V.; Xu, X.; Wang, C. Visible-light-triggered persulfate activation by CuCo2S4 modified ZnO photocatalyst for degradation of tetracycline hydrochloride. *Colloids Surf. A Physicochem. Eng. Asp.* **2022**, *642*, 128640. [CrossRef]
44. Zhu, F.; Ma, J.; Ji, Q.; Cheng, H.; Komarneni, S. Visible-light-driven activation of sodium persulfate for accelerating orange II degradation using $ZnMn_2O_4$ photocatalyst. *Chemosphere* **2021**, *278*, 130404. [CrossRef] [PubMed]
45. Zhou, X.; Zhu, Y.; Niu, Q.; Zeng, G.; Lai, C.; Liu, S.; Huang, D.; Qin, L.; Liu, X.; Li, B.; et al. New notion of biochar: A review on the mechanism of biochar applications in advannced oxidation processes. *Chem. Eng. J.* **2021**, *416*, 129027. [CrossRef]
46. Revathi, V.; Karthik, K. Microwave assisted CdO–ZnO–MgO nanocomposite and its photocatalytic and antibacterial studies. *J. Mater. Sci. Mater. Electron.* **2018**, *29*, 18519–18530. [CrossRef]
47. Panchal, P.; Paul, D.R.; Malik, R.; Sharma, A.; Meena, P.; Patidar, D.; Nehra, S. Phytoextract Mediated ZnO/MgO Nanocomposites for Photocatalytic and Antibacterial Activities. *J. Photochem. Photobiol. A Chem.* **2019**, *385*, 112049. [CrossRef]

48. Xiang, X.; Xie, L.; Li, Z.; Li, F. Ternary $MgO/ZnO/In_2O_3$ heterostructured photocatalysts derived from a layered precursor and visible-light-induced photocatalytic activity. *Chem. Eng. J.* **2013**, *221*, 222–229. [CrossRef]
49. Ansari, F.; Nasr, M.; Rawat, I.; Bux, F. Artificial neural network and techno-economic estimation with algae-based tertiary wastewater treatment. *J. Water Process Eng.* **2021**, *40*, 101761. [CrossRef]
50. León, M.; Silva, J.; Carrasco, S.; Barrientos, N. Design, Cost Estimation and Sensitivity Analysis for a Production Process of Activated Carbon from Waste Nutshells by Physical Activation. *Processes* **2020**, *8*, 945. [CrossRef]
51. Ansari, F.A.; Ravindran, B.; Gupta, S.K.; Nasr, M.; Rawat, I.; Bux, F. Techno-economic estimation of wastewater phycoremediation and environmental benefits using Scenedesmus obliquus microalgae. *J. Environ. Manag.* **2019**, *240*, 293–302. [CrossRef]
52. Mahamuni, N.N.; Adewuyi, Y.G. Advanced oxidation processes (AOPs) involving ultrasound for waste water treatment: A review with emphasis on cost estimation. *Ultrason. Sonochem.* **2010**, *17*, 990–1003. [CrossRef]
53. Hamdy, A.; Mostafa, M.K.; Nasr, M. Zero-valent iron nanoparticles for methylene blue removal from aqueous solutions and textile wastewater treatment, with cost estimation. *Water Sci. Technol.* **2018**, *78*, 367–378. [CrossRef] [PubMed]
54. Vishnuganth, M.; Remya, N.; Kumar, M.; Selvaraju, N. Photocatalytic degradation of carbofuran by TiO_2-coated activated carbon: Model for kinetic, electrical energy per order and economic analysis. *J. Environ. Manag.* **2016**, *181*, 201–207. [CrossRef] [PubMed]
55. Olya, M.E.; Akhi, Y.; Marandi, R. Improvement of efficiency and electrical energy consumption of AB74 degradation process using a novel cylindrical batch photochemical reactor. *Orient. J. Chem.* **2016**, *32*, 1295–1303. [CrossRef]
56. Ghadikolaee, M.R.; Cerro-Prada, E.; Pan, Z.; Korayem, A.H. Nanomaterials as Promising Additives for High-Performance 3D-Printed Concrete: A Critical Review. *Nanomaterials* **2023**, *13*, 1440. [CrossRef]
57. Tesfaye, M.; Gonfa, Y.; Tadesse, G.; Temesgen, T.; Periyasamy, S. Green synthesis of silver nanoparticles using Vernonia amygdalina plant extract and its antimicrobial activities. *Heliyon* **2023**, *9*, e17356. [CrossRef]
58. Lobregas, M.; Camacho, D. Green synthesis of copper-based nanoparticles using microbes. *Copp. Nanostructures Next-Gener. Agrochem. Sustain. Agroecosyst.* **2022**, *2022*, 17–44. [CrossRef]
59. Ajmal, A.; Majeed, I.; Malik, R.N.; Idriss, H.; Nadeem, M.A. Principles and mechanisms of photocatalytic dye degradation on TiO_2 based photocatalysts: A comparative overview. *RSC Adv.* **2014**, *4*, 37003–37026. [CrossRef]

water

MDPI

Article

Novel Fenton-like Catalyst HKUST-1(Cu)/MoS_2-3-C with Non-Equilibrium-State Surface for Selective Degradation of Phenolic Contaminants: Synergistic Effects of σ-Cu-Ligand and ≡Mo–$OOSO_3^-$ Complex

Xiaoze Yin [1,2,†], Huaqin Yin [1,2,†], Renjie Wang [1,2], Jinnan Wang [1,2,*] and Aimin Li [1,2]

[1] State Key Laboratory of Pollution Control and Resource Reuse, Nanjing 210023, China
[2] School of the Environment, Nanjing University, Nanjing 210023, China
* Correspondence: wjnnju@163.com
† These authors contributed equally to this work.

Abstract: Novel Fenton-like catalyst HKUST-1(Cu)/MoS_2-3-C with a non-equilibrium-state surface was constructed for selective degradation of phenolic contaminants. Electron-polarized distribution facilitated the formation of σ-Cu-ligand between electron-poor Cu centre and phenolic compounds, which not only enhanced radicals generation but also accelerated the Cu(I)/Cu(II) redox. Meanwhile, ≡Mo–$OOSO_3^-$ complexes formed by the electron-rich Mo centre and peroxymonosulfate (PMS), could directly oxidize phenolic contaminants with the generation of $SO_4^{\bullet-}$. The radical quenching experiments and EPR tests indicated that both $SO_4^{\bullet-}$ and •OH played a dominant role in the reaction. Additionally, O_2 could be reduced to $O_2^{\bullet-}$ by OVs and subsequently converted into $^1O_2$ over the Mo centre. DFT calculation, FT-IR, and in situ Raman spectra analysis results demonstrated that phenolic compounds and PMS were respectively adsorbed by electron-poor Cu centre and electron-rich Mo centre, favouring the electrons transfer from phenolic contaminants to Mo centre for PMS activation. With synergistic effects of σ-Cu-ligand and ≡Mo–$OOSO_3^-$ complexes, HKUST-1(Cu)/MoS_2-3-C achieved a high degradation rate of phenolic contaminants and utilization efficiency of PMS.

Keywords: fenton-like catalyst; non-equilibrium-state surface; σ-Cu-ligand; ≡Mo–$OOSO_3^-$ complex; phenolic contaminants degradation

1. Introduction

Cu-based catalysts attracted great interest in the selective degradation of phenolic contaminants due to the formation of σ-Cu-ligand between phenolic hydroxyl groups and Cu(II) [1]. Different from traditional Fenton-like reaction, σ-Cu-ligand could be ruptured by persulfate (PMS) with the generation of $SO_4^{\bullet-}$ and HO-adduct radicals, which were further oxidized to hydroxylation products by reduction of Cu(II) to Cu(I) [2]. However, the formation of σ-Cu-ligand in the system would be gradually decreased with the phenolic compounds degradation, which caused the reaction to follow the traditional Fenton-like reaction mechanism (Equations (1) and (2)). In such case, although PMS could act as the electron acceptor and electron donor with the redox of Cu(I)/Cu(II) [1], reduction of Cu(II) ($E^0_{Cu(II)/Cu(I)}$ = 0.15 V) by HSO_5^- ($E^0_{HSO5-/HSO4-}$ = 1.8 V) is thermodynamically unfavourable [3]. Thus, the reduction of Cu(II) to Cu(I) was very slow in the absence of reductive agents, which significantly restricted the Cu(I)/Cu(II) circulation [4]. Therefore, most Cu-based catalysts still faced the problems of the rate limitation step (Equation (2)) and low utilization of PMS.

$$Cu(I) + HSO_5^- \rightarrow Cu(II) + SO_4^{\bullet-} + OH^- \quad (1)$$

$$Cu(II) + HSO_5^- \rightarrow Cu(I) + SO_5^{\bullet-} + H^+ \quad (2)$$

To resolve the above problems, Cu-based catalysts with dual-reaction centres were constructed for the enhancement of catalytic activity and selective conversion of oxidants to radicals [5]. By doping different electro-negative metals, the non-uniform distribution of electron density of lattice oxygen was formed, which induced the formation of an electron-rich centre and an electron-poor centre. Due to the galvanic-like cell effect, electrons of organics could be captured by the electron-poor centre and subsequently transferred to the electron-rich centre for oxidant activation [6]. However, owing to high electronegativity, Cu (1.9 eV) generally acted as an electron-rich centre in dual reaction systems, which prevented the formation of σ-Cu-ligand with phenolic compounds because organics were preferentially accumulated around the electron-poor centre [7,8]. Previous work demonstrated that new ligands with strong electron-donating ability could act on MOFs to increase the charge density on the adjacent original ligand sites. This inspires us to construct an electron-poor Cu centre by doping higher electronegative metal into MOFs, which might facilitate the formation of σ-Cu-ligand for enhancement of PMS utilization and selective degradation of phenolic contaminants.

On the other hand, interface interaction between oxidant and catalyst also played an important role in the PMS activation process. Generally, as PMS was bounded on the catalyst surface, electrons were immediately transferred from transition metal oxides to PMS accompanied by O–O bond broken and generation of ROS (e.g., $SO_4^{\bullet-}$, •OH and $O_2^{\bullet-}$). Thus, a strong adsorption force between the catalyst and PMS could promote electron migration and ROS generation. As a commonly used co-catalyst, molybdenum disulfide (MoS_2) with a 2D layered structure could not only act as the electron bridge for electron transfer [9] but also form an intermediate metal–peroxy complex ($\equiv$Mo–$OOSO_3^-$) with PMS due to high adsorption energy. Notably, such $\equiv$Mo–$OOSO_3^-$ complex would decompose to other reactive species with the generation of radicals ($SO_4^{\bullet-}$, •OH) [10,11]. More importantly, Mo (2.16 eV) with higher electronegativity introduced into Cu-MOFs might induce the formation of a non-equilibrium-state surface, which can simultaneously enhance the phenolic compounds degradation around electron-poor Cu centre and PMS activation around the electron-rich Mo centre.

Herein, novel Fenton-like catalyst HKUST-1(Cu)/MoS_2-3-C with a non-equilibrium-state surface derived from Cu-MOFs was constructed for the highly efficient degradation of phenolic contaminants. MoS_2 with strong electron-accepting ability introduced into Cu-MOFs could act on the MOFs to decrease charge density on the adjacent original ligand sites, which induced the formation of electron-poor Cu centre and electron-rich Mo centre. In the catalytic reaction system, the electron-poor Cu centre facilitated the formation of σ-Cu-ligand complexes with phenolic compounds, while the electron-rich Mo centre favoured the formation of $\equiv$Mo–$OOSO_3^-$ complex with PMS. Meanwhile, electrons of organics could be captured by the electron-poor Cu centre and subsequently transferred to the electron-rich Mo centre for PMS activation, avoiding the accumulation of Cu(II)/Mo(VI) and PMS invalid decomposition. Combining the advantages mentioned above, HKUST-1(Cu)/MoS_2-3-C achieved a high degradation rate of phenolic contaminants and utilization efficiency of PMS.

2. Materials and Methods

2.1. Chemicals

The following agents were analytical grade, acquired from a commercial source:1,3,5-benzenetricarboxylic acid (H_3BTC, 98%, Aladdin Biochemical Technology Co., Ltd., Shanghai, China), copper(II) nitrate trihydrate ($Cu(NO_3)_2 \bullet 3H_2O$, 99%, Sinopharm Chemical Reagent Co., Ltd., Shanghai, China), ethanol($\geq$99.5%, Sinopharm Chemical Reagent Co., Ltd.), dimethylformamide (DMF, $\geq$99.5%, Sinopharm Chemical Reagent Co., Ltd.), sodium molybdatedihydrate ($Na_2MoO_4 \bullet 2H_2O$, 99%, Aladdin Biochemical Technology Co., Ltd.), thiuronium (CH_5N_2S, $\geq$98.0%, Aladdin Biochemical Technology Co., Ltd.), potassium peroxymonosulfate ($2KHSO_5 \bullet KHSO_4 \bullet K_2SO_4$, 99%, Macklin Biochemical Technology Co., Ltd., Shanghai, China), Bisphenol A (BPA, >99%, Aladdin Biochemical Technology Co., Ltd.),

Phenol (≥99.5%, Sinopharm Chemical Reagent Co., Ltd.), Nitrobenzene (≥99.5%, Aladdin Biochemical Technology Co., Ltd.), Benzoic acid (≥99.5%, Aladdin Biochemical Technology Co., Ltd.), Aniline (≥99.5%, Sinopharm Chemical Reagent Co., Ltd.), L-Histidine ($C_6H_9N_3O_2$, ≥99%, Aladdin Biochemical Technology Co., Ltd.), tertbutylalcohol (TBA, ≥99.5%, Sinopharm Chemical Reagent Co., Ltd.), methanol (≥99.5%, Sinopharm Chemical Reagent Co., Ltd.). 5,5-Dimethyl-1-pyrroline N-oxide (DMPO, 98%) was obtained from Sigma-Aldrich Co., Ltd., Shanghai, China. 2,2,6,6-Tetramethylpiperidine (TEMP, 98%) was purchased from Tokyo Chemical Industry Co., Ltd., Tokyo, Japan. Ultrapure water was used for all the experiments.

2.2. Preparation of Catalysts

The synthesis process of HKUST-1(Cu)/MoS_2-x-C is shown in Scheme 1. In addition, MoS_2 and MoS_2-C are also prepared for comparative studies (Text S1).

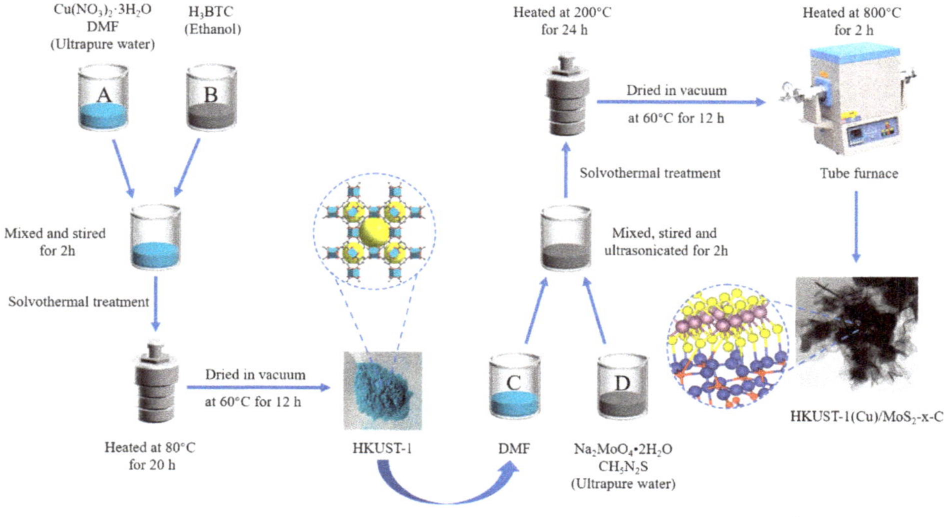

Scheme 1. Preparation process of HKUST-1(Cu)/MoS_2-x-C.

Synthesis of HKUST-1: In flask A, 5.0 g of $Cu(NO_3)_2 \bullet 3H_2O$ was dissolved in 60 mL of ultrapure water. In flask B, 1.36 g of H_3BTC was dissolved in 60 mL of ethanol. After mixing with 4 mL of DMF, $Cu(NO_3)_2$ solution was subsequently added to an ethanol solution of H_3BTC under continuous stirring for 2 h. Then the mixture was transferred to a stainless steel lined Teflon autoclave reactor. After being heated at 80 °C for 20 h, the obtained blue crystal HKUST-1(Cu) was washed with ultrapure water and ethanol successively and dried in a vacuum oven at 60 °C for 12 h before use.

Synthesis of HKUST-1(Cu)/MoS_2-x-C: In flask C, a certain amount of as-synthesized HKUST-1(Cu) was dispersed in 40 mL of DMF solution. In flask D, a certain amount of $Na_2MoO_4 \bullet 2H_2O$ and CH_5N_2S (the molar ratio of $Na_2MoO_4 \bullet 2H_2O$ to CH_5N_2S was 1:2) were dissolved in 40 mL of ultrapure water with continuously stirring for 30 min. Then the DMF solution of HKUST-1(Cu) was slowly added into the solution of $Na_2MoO_4 \bullet 2H_2O$ and CH_5N_2S with continuous stirring. After ultrasonic treatment for 2 h, the mixture was transferred to a stainless steel lined Teflon autoclave reactor and heated at 200 °C for 24 h. The black precipitate was collected by centrifugation and washed with ultrapure water and ethanol successively. After being dried in a vacuum oven at 60 °C for 12 h, the product was heated in a tube furnace at 800 °C for 2 h under an argon atmosphere. Finally, the black powder of HKUST-1(Cu)/MoS_2-x-C (x = 1, 2, 3 represented the molar ratio of Mo to Cu; C denoted calcination) was obtained.

2.3. Characterization

Surface morphology and lattice fringe of HKUST-1(Cu)/MoS_2-3-C were characterized by scanning electron microscopy (SEM, ZEISS Gemini 300, Tokyo, Japan) and transmission electron microscopy (TEM, FEI Talos F200X G2, Tokyo, Japan), respectively. To observe the crystal structure of the catalysts, X-ray diffraction (XRD) spectra were collected using an XRD-6000 X-ray diffractometer (Shimadzu, Kyoto, Japan) with a Cu K radiation (1.5406 Å) over the 2θ range of 10–90°. To identify the valence state of an element, X-ray photoelectron spectroscopy (XPS) of samples was analyzed by a PHI 5000 Versa Probe Instrument (Chigasaki, Japan) using monochromatic Al Kα radiation (225 W, 15 mA, 15 kV). During the deconvolution process of XPS curves, all binding energies were referenced to the C 1s peak at 284.8 eV.

FT-IR spectra (Thermo Scientific Nicolet iS20, Waltham, MA, USA) and ATR-FT-IR spectra were applied to analyze the formation of σ-Cu-ligand and $\equiv Mo–OOSO_3^-$ complex, respectively. Furthermore, in situ Raman spectra were conducted on Horiba LabRAM HR Evolution (Gloucestershire, London, UK) with a 532 nm excitation laser, which could demonstrate the interaction between PMS molecules and Mo. In addition, to evaluate the electron transfer ability of catalysts, electrochemical measurements were conducted by the electrochemical workstation (CHI760E, Shanghai, China) with a three-electrode system (80 mL 0.5 M Na_2SO_4 as the electrolyte). In addition, the reactive species produced in the catalytic oxidation system were captured by 5,5-dimethyl-1-pyrroline N-oxide (DMPO) and 2,2,6,6-tetramethyl-4-piperidone (TEMP). The formed adducts were analyzed using an EMX-10/12 Electron paramagnetic resonance spectrometer (Karlsruhe, Germany).

2.4. Catalytic Degradation Experiments

The catalytic performance of samples was evaluated by the degradation of phenolic compounds (BPA, phenol) and other aromatic compounds (Nitrobenzene, Benzoic acid, and Aniline). Typically, 50 mL aqueous solutions of a certain concentration of phenolic compound and catalyst were mixed in 100 mL beaker flasks. Before 1 mM PMS was added in, the suspensions were magnetically stirred for 30 min to reach absorption-desorption equilibrium between the catalyst and organics. Then, 1 mL aliquots were sampled at given time intervals and filtered through a Mixed Cellulose Ester (MCE) filter (0.22 μm) before analysis. Notably, to quench the residual oxidant, 50 μL MeOH was immediately added to the sample for termination reaction. The residual concentration of BPA in solution was measured by a 1200 series HPLC equipped with a UV–visible detector and C-18 column (Eclipse XBD-C18, Agilent Technologies, Santa Clara, CA, USA, 4.5 × 150 mm, 5 μm). The degradation intermediates of BPA were analyzed by HPLC-MS (Ultimate 3000 UHPLC-Q Exactive, Thermo Fisher, Waltham, MA, USA) with a scan range of m/z 50–750 in a negative mode. The capillary voltage and operating temperature were 2500 V and 300 °C, respectively. The leached metal ions in suspensions were measured by Inductively coupled plasma optical emission spectrometer (ICP-OES, Thermo Fisher iCAP PRO, Waltham, MA, USA). In addition, considering that inorganic anions and natural organic matter (NOM) widely existed both in natural water and wastewater, influences of anions (Cl^-, SO_4^{2-}, $NO_3^-$and CO_3^{2-}) and humic acid (HA) on catalytic degradation of BPA were also conducted.

3. Results and Discussion

3.1. Morphology and Structure of Catalysts

HKUST-1(Cu)/MoS_2-3-C shows a loosely amorphous structure with MoS_2 nanosheets embedded in HKUST-1(Cu) (Figure 1a). This structure makes the catalyst have an appropriate BET surface area (208.9509 m^2/g, Figure S5). The resolved lattice fringes of 0.21 nm and 0.25 nm are respectively ascribed to Cu (111) and Cu_2O (111) (Figure 1b), while the lattice fringe of 0.62 nm and 0.27 nm belong to the (002) and (100) plane of MoS_2. In this heterostructure, bonding among the interfaces of different components can enhance the rate of electron transfer [12]. EDS images indicate that Cu, Mo, O, S, and C are uniformly dispersed on the catalyst surface, which might favour the contact of active sites and reactants (Figure 1c). In addition, XRD spectra of HKUST-1(Cu)/MoS_2-3-C and other as-synthesized materials are shown in Figure 1d. For HKUST-1(Cu)-C, characteristic peaks at 38° and 45° respectively represent Cu_2O (PDF#78-2076) crystalline phases of (111) and (200), while the diffraction peaks at 46° and 79° are respectively ascribed to Cu (PDF#85-1326) crystalline phases of (111) and (220). Thus, both Cu_2O and metallic Cu phases were formed after carbonizing the Cu-MOF [13,14]. As for MoS_2-C, peaks at 14, 33, 39, and 58° are ascribed to MoS_2 (PDF#73-1508) phases of (002), (100), (103), and (110), respectively. Since both characteristic peaks of Cu_2O, Cu, and MoS_2 can be observed in the spectra of HKUST-1(Cu)/MoS_2-3-C, it can be concluded that MoS_2 is doped on HKUST-1(Cu) successfully [13,14].

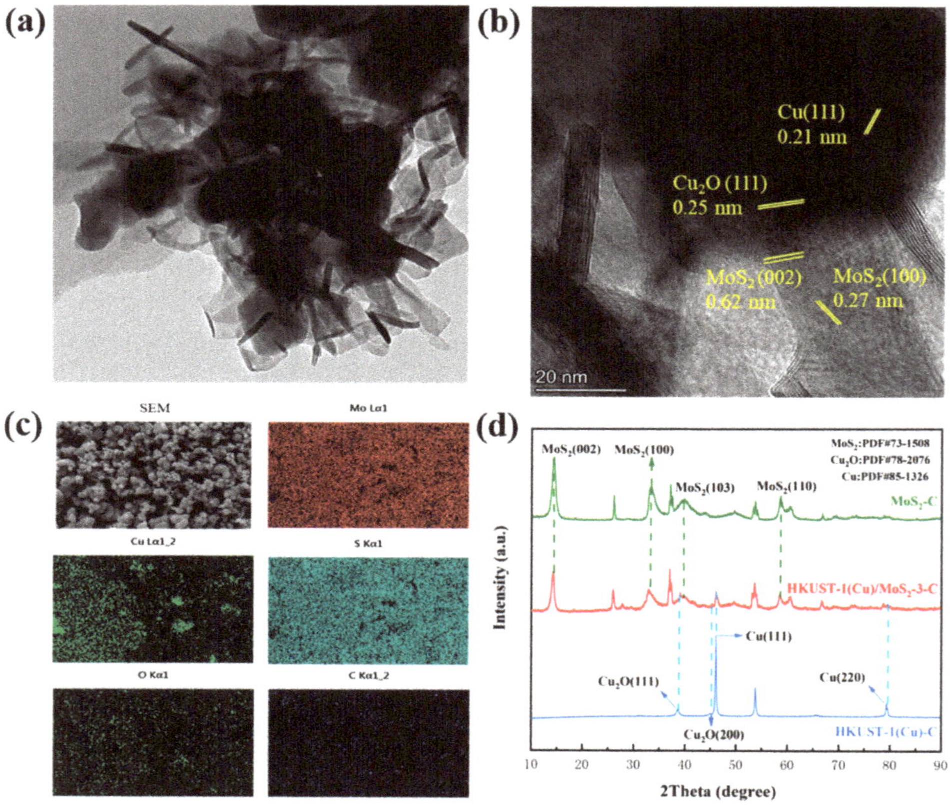

Figure 1. (**a**) TEM imagine of HKUST-1(Cu)/MoS_2-3-C; (**b**) HRTEM imagine of HKUST-1(Cu)/MoS_2-3-C; (**c**) SEM imagine of HKUST-1(Cu)/MoS_2-3-C and corresponding elemental mapping of Cu, O, Mo, S and C; (**d**) XRD spectra of several catalysts.

XPS spectra of HKUST-1(Cu)/MoS_2-3-C (Figure 2a) confirm the existence of C, O, Cu, Mo, and S, which agrees with the EDS analysis. In the spectra of C 1s (Figure 2b), the binding energy of 284.8, 285.6, and 286.4 eV are respectively ascribed to C–C, C–O, and C–O–C species [14], which indicates that organic ligands of MOF (1,3,5-benzenetricarboxylate) are successfully transformed into a carbon-based skeleton. As for the spectra of Cu 2p (Figure 2c), two main peaks of Cu $2p_{3/2}$ (932.5 eV) and Cu $2p_{1/2}$ (952.5 eV) are observed, together with their satellite peaks (943 eV and 963 eV). Due to the similar binding energy of Cu and Cu^+ (Cu_2O), these two characteristic peaks overlap. In order to identify the Cu element more accurately, a high-resolution Auger Cu LMM spectrum was recorded. In the spectrum of Cu LMM (Figure 2d), characteristic peaks located at 569.8 eV and 568.1 eV are respectively assigned to Cu(I) (Cu_2O) and Cu [15], while the peak at 572.7 eV corresponds to Cu(II) (CuO) [14,16].

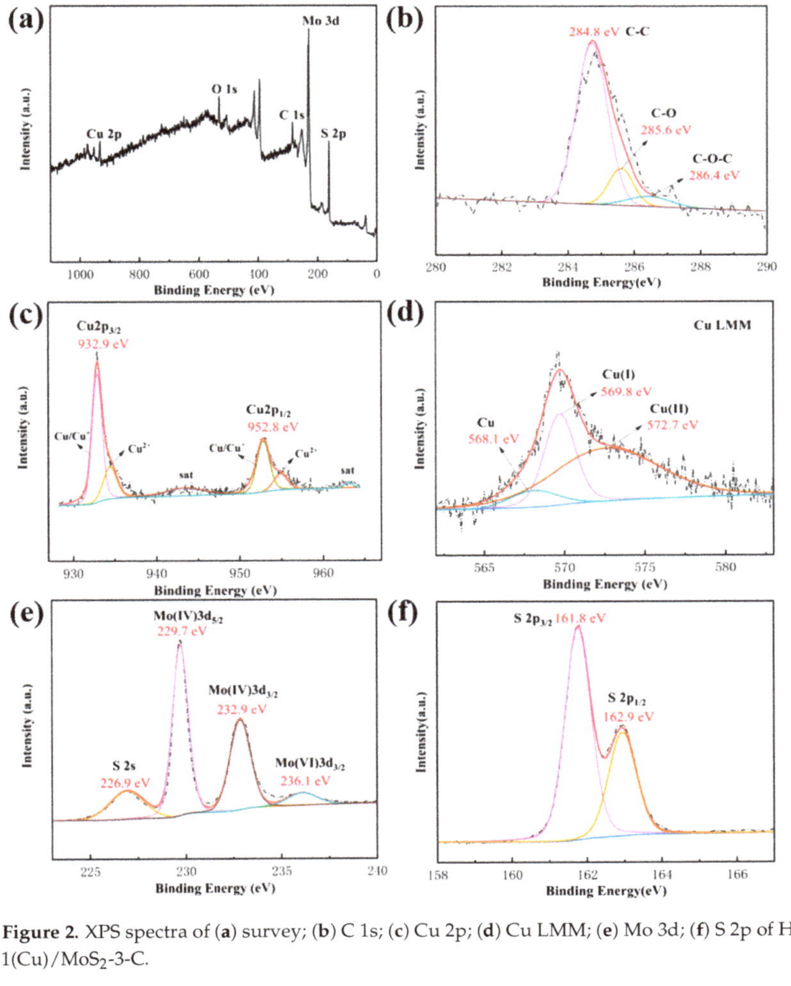

Figure 2. XPS spectra of (**a**) survey; (**b**) C 1s; (**c**) Cu 2p; (**d**) Cu LMM; (**e**) Mo 3d; (**f**) S 2p of HKUST-1(Cu)/MoS_2-3-C.

The proportion of Cu species is calculated by integration of the peak areas, which follows the order of Cu(II) (57.31%) > Cu(I) (35.09%) > Cu (7.60%). Herein, Cu(II) plays the dominant role in the formation of σ-Cu(II)-ligand, while Cu(I) can directly reduce oxidants with the generation of radicals [17]. Notably, binding energies of Cu $2p_{3/2}$ (932.9 eV) and Cu $2p_{1/2}$ (952.8 eV) in HKUST-1(Cu)/MoS_2-3-C are higher than those in HKUST-1(Cu)-C (932.6 eV and 952.4 eV) (Figure S2), suggesting the strong interaction between Mo and Cu in HKUST-1(Cu)/MoS_2-3-C [13,14]. It can be explained as follows: owing to higher electronegativity, Mo (2.16 eV) could capture the electrons from the Cu ligand, which induces the formation of a non-equilibrium-state surface with electron-poor Cu centre and electron-rich Mo centre in HKUST-1(Cu)/MoS_2-3-C. In addition, the peak at 236.1 eV corresponded to Mo $3d_{3/2}$ of Mo^{6+} might be ascribed to the partial oxidation of Mo^{4+} on the MoS_2 surface [18]. Furthermore, a slight shift to the higher binding energy of S 2p from 161.6 eV to 161.8 eV for S $2p_{3/2}$ and from 162.8 eV to 162.9 eV for S $2p_{1/2}$ (Figure 2e), suggesting the formation of S-C bonding in HKUST-1(Cu)/MoS_2-3-C [14,19]. Such S-C bonding not only strengthened the structural stability of catalysts but also accelerated the electron transfer between MoS_2 and HKUST-1(Cu)-C [20,21].

3.2. Catalytic Performance

Within 30 min, BPA degradation rates follow the order of HKUST-1(Cu)/MoS_2-3-C/PMS (100%) > HKUST-1(Cu)-C/PMS (59.7%) > MoS_2-C/PMS(57.6%) > PMS(5%) (Figure 3a), which suggests the highest catalytic activity of HKUST-1(Cu)/MoS_2-3-C. The pseudo-first-order equation is used to fit the BPA degradation curves (Figure 3c), with the rate constant (k) following the order of HKUST-1(Cu)/MoS_2-3-C (0.1404 min^{-1}) > HKUST-1(Cu)-C (0.0247 min^{-1}) > MoS_2-C (0.0184 min^{-1}). Additionally, the BPA degradation experiment over different catalysts (HKUST-1(Cu)/MoS_2-x-C) indicates that the optimal ratio (x) of Mo:Cu is 3 (Figure 3b). Such extraordinary performance might be explained as follows: high adsorption energies (E_{ads}) between Mo and PMS molecules [22] facilitated the formation of $\equiv$Mo–$OOSO_3^-$ complex, which favoured the BPA degradation and radicals generation [19,23,24]. On the other hand, Cu(II) induced the formation of σ-Cu-ligand with BPA, which promoted the BPA degradation and redox of Cu(II)/Cu(I) [1,25,26].

To evaluate the utilization of PMS (η) in different reaction systems, the ABTS method (detailed information provided in Text S4) is used to determine PMS concentration. The PMS utilization can be evaluated by using the stoichiometric efficiency method [27,28], which indicates that the introduction of MoS_2 into HKUST-1(Cu)-C obviously enhances PMS utilization. The corresponding PMS stoichiometry efficiency over different catalysts followed the order of HKUST-1(Cu)/MoS_2-3-C (0.14) > MoS_2-C (0.11) > HKUST-1(Cu)-C (0.09) (Table S1). Furthermore, catalytic degradations of various aromatic organics (Nitrobenzene, Benzoic acid, Phenol, and Aniline) over HKUST-1(Cu)/MoS_2-3-C demonstrate that functional groups show great influence on organics degradation. Being attributed to σ-Cu-ligand formed between HKUST-1(Cu)/MoS_2-3-C and phenolic compounds, the TOC removal rate of BPA and phenol are much higher than those aromatic organics without phenolic hydroxyl groups (Nitrobenzene, Benzoic acid, and Aniline) (Figure 3f). Thus, HKUST-1(Cu)/MoS_2-3-C can selectively degrade phenolic contaminants.

The influence of inorganic anions (Cl^-, SO_4^{2-}, $NO_3^-$and CO_3^{2-}) and humic acid (HA) on BPA degradation are conducted to evaluate the anti-inference ability. Previous work reported that inorganic anions could act as radicals quencher with the second-order kinetic rate constants following the order of k(•OH/Cl^-: 3~4.3 × 10^9 $M^{-1}s^{-1}$) > k(•OH/CO_3^{2-}: 3.2~4.2 × 10^8 $M^{-1}s^{-1}$) > k($SO_4^{\bullet-}$/Cl^-: 2.3~660 × 10^6 $M^{-1}s^{-1}$) > k($SO_4^{\bullet-}$/CO_3^{2-}: 1~6.1 × 10^6 $M^{-1}s^{-1}$) > k($SO_4^{\bullet-}$/NO_3^-: 5~210 × 10^4 $M^{-1}s^{-1}$) [19,29,30]. Herein, BPA removal is not decreased in the presence of three inorganic anions (Cl^-, SO_4^{2-}, NO_3^-) (Figure 3g), which might be explained by that σ-Cu-ligand can directly contribute to BPA degradation even without radicals. However, an obvious inhibition is observed as CO_3^{2-} is added in the reaction system, which can be explained by that σ-Cu-ligand can be hardly formed between ionized BPA and Cu centre in alkaline CO_3^{2-} solution (pH = 10.92). In addition, although HA shows a slight adverse effect on BPA degradation due to competition for radicals (Figure 3h) [31,32], the supramolecular structure of HA prevents the formation of σ-Cu-ligand with catalyst. Thus HKUST-1(Cu)-C/MoS_2-3-C exhibits strong anti-inference ability for co-existence of NOM. Addionally, the RhB degradation experiment also be implemented to verify the suitability of the catalyst for the degradation of dye (Figure S6), which proves the catalyst has the potential to be applied to various industrial wastewater treatment.

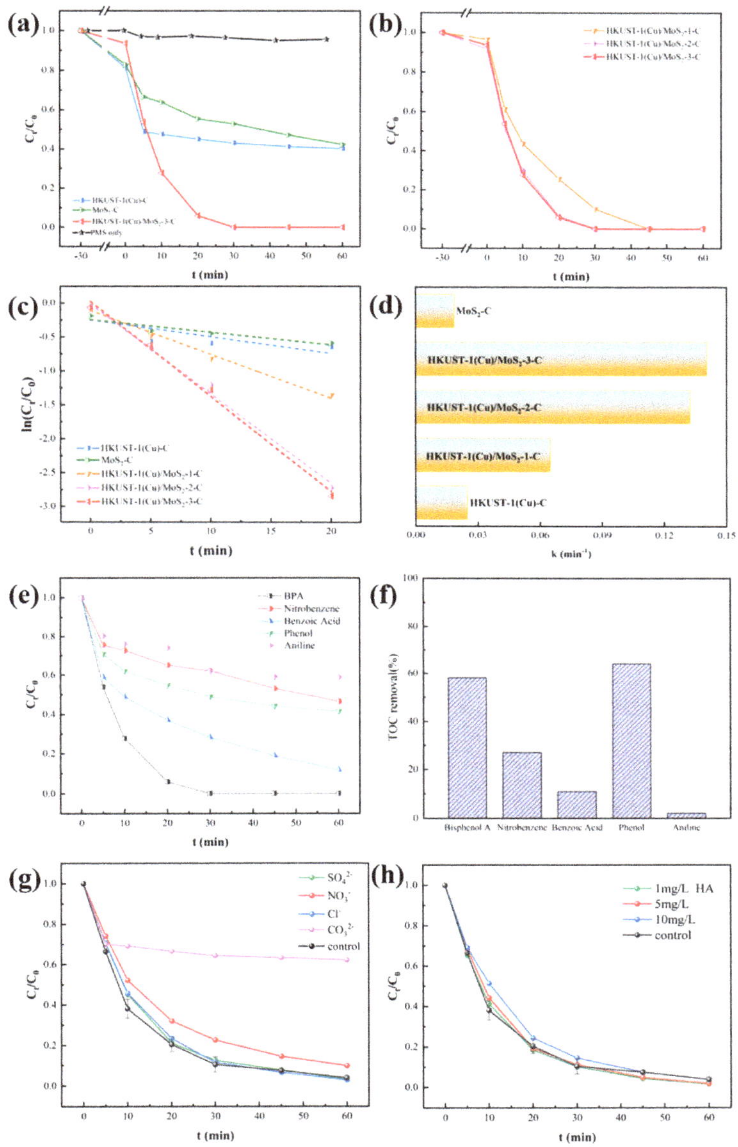

Figure 3. (**a**) Degradation of BPA in different catalysts systems; (**b**) Degradation of BPA in different HKUST-1(Cu)/MoS_2-x-C systems; (**c**) The kinetic degradation curves in different catalysts systems; (**d**) Pseudo-first-order rate constants k in different reaction systems; (**e**) Degradation of 20 mg/L BPA, Nitrobenzene, Benzoic Acid, Phenol and Aniline in HKUST-1(Cu)/MoS_2-3-C system; (**f**) TOC removal rate of 20 mg/L BPA, Nitrobenzene, Benzoic Acid, Phenol and Aniline in HKUST-1(Cu)/MoS_2-3-C system; (**g**) Effects of inorganic anions (SO_4^{2-}, NO_3^-, Cl^-, CO_3^{2-}) with the initial anion concentration of 1 mM on BPA degradation; (**h**) Effects of HA (1, 5, 10 mg/L) on BPA degradation. Reaction conditions: [BPA] = 20 mg/L, [PMS] = 1 mM, [Catalysts] = 0.3 g/L.

The reusability and stability of the catalyst are important factors related to the practical application. As expected, HKUST-1(Cu)-C/MoS_2-3-C exhibits satisfied stability. Even after five cycles, HKUST-1(Cu)-C/MoS_2-3-C could still maintain more than 80% degradation rate of BPA within 60 min (Figure 4a). The concentrations of Mo and Cu leached after the reaction are detected, which are much lower than the limitation of the standard proposed by EPA (Table S6). Additionally, the morphology and structure of HKUST-1(Cu)/MoS_2-3-C are not destroyed after successive experiments (Figure 4b), exhibiting superior stability. In addition, an obvious increase proportion of Cu(I) from 35.09% to 55.53% and Mo(IV) from 93.22% to 95.40% are observed after the reaction (Figure 4c–f), which suggests that PMS activation over HKUST-1(Cu)/MoS_2-3-C did not follow the electron transfer behavior of the classic Fenton reaction involving transition metal oxidized by oxidants. Thus, the catalytic process is dominated by the synergistic effect of non-equilibrium-state surface, σ-Cu-ligand, and ≡Mo–$OOSO_3^-$ that directly capture electrons from phenolic compounds for PMS activation.

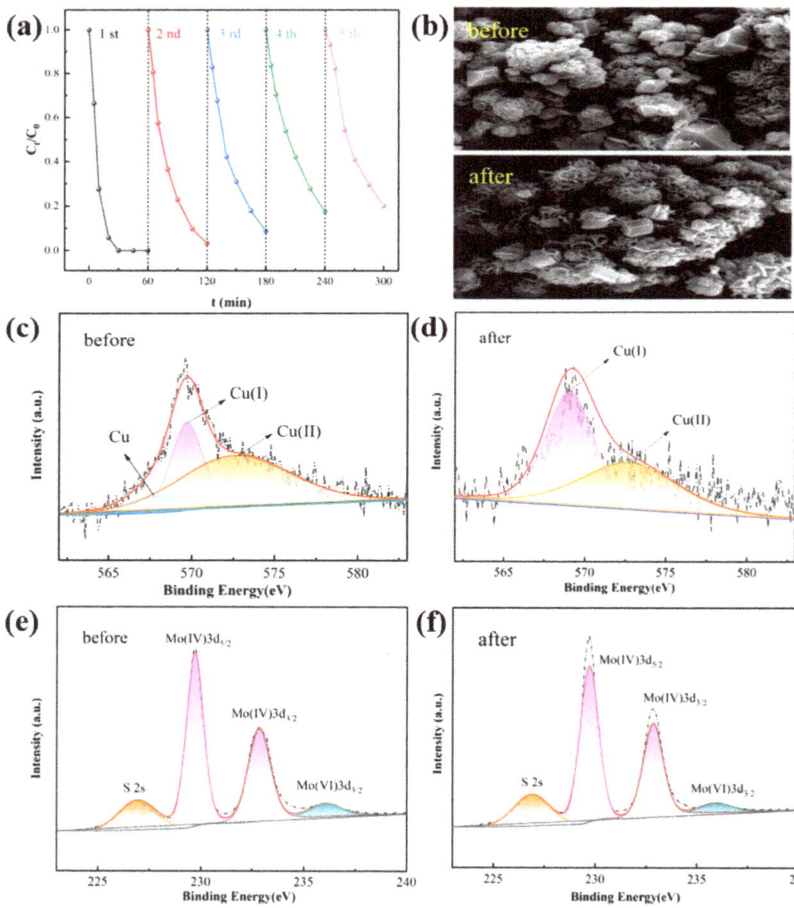

Figure 4. (**a**) Catalytic stability tests in five consecutive runs of HKUST-1(Cu)/MoS_2-3-C for BPA degradation; (**b**) SEM imagine of HKUST-1(Cu)/MoS_2-3-C before and after reaction; High-resolution XPS spectra of (**c**,**d**) Cu LMM; (**e**,**f**) Mo 3d for before and after reaction.

3.3. Catalytic Mechanism

The degradation intermediates of BPA over HKUST-1(Cu)/MoS_2-3-C are identified by ESI-MS analysis (Tables S2 and S3, Figures S7–S9), and the possible degradation pathways are proposed (Figure 5). On one hand, BPA was transformed into monohydroxylated BPA (m/z = 243.9) and 2,2-diphenylpropane (m/z = 195.1). Followed by the attack on alkyl carbon, these intermediates were decomposed to 4-hydroxyacetophenone (m/z = 135.0) and acetophenone (m/z = 117.9) respectively. On the other hand, the aromatic ring which was at the para-position of the phenolic hydroxyl group was initially hydroxylated to form 4-isopropylphenol (m/z = 135.1), hydroquinone (m/z = 109.0), and phenol (m/z = 93.0). With the proceeding of the ring-rupturing reaction, the above phenolic compounds were decomposed to maleic acid (m/z = 116.9), glycerol (m/z = 92.0), glycolic acid (m/z = 77.0), and carbonic acid (m/z = 62.0), and finally mineralized into CO_2 and H_2O.

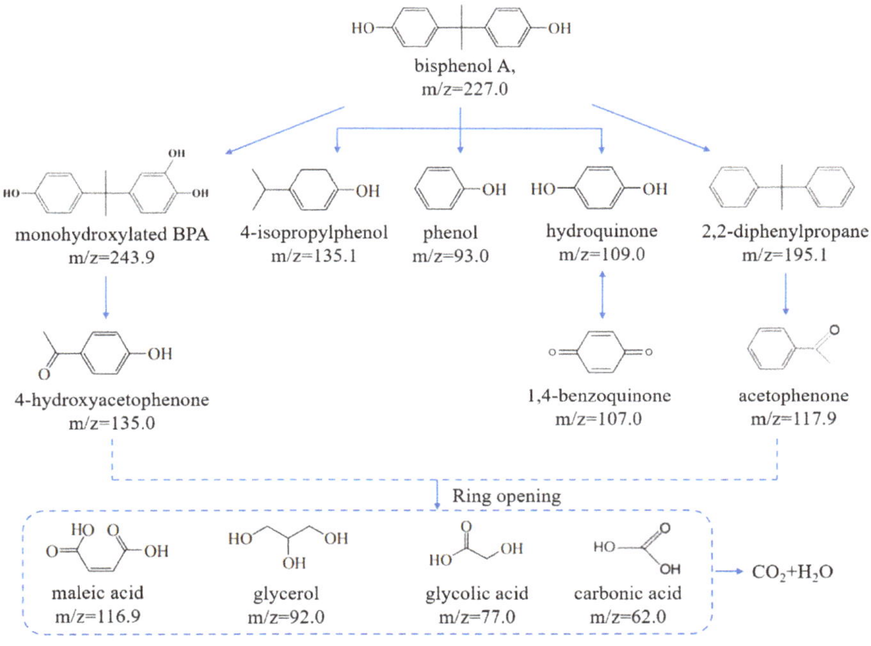

Figure 5. Possible degradation pathways of BPA in HKUST-1(Cu)/MoS_2-3-C+PMS system.

Aromatic organics preferred to be adsorbed and degraded on the surface of copper-based catalysts [1,6,26] due to the formation of complexes between Cu and phenolic hydroxyl groups [33,34]. In spectra of HKUST-1(Cu)/MoS_2-3-C (Figure 6a), the peak of the phenolic hydroxyl group shifts from 3331.7 cm^{-1} to 3450.1 cm^{-1} after BPA adsorption, which can be explained by that coordination sphere formed between phenolic hydroxyl groups of BPA and Cu(II) [35] causes the deprotonation of the phenolic hydroxyl group and the difference in surroundings. As PMS is added in, the characteristic peak of phenolic hydroxyl groups further shifts from 3450.1 cm^{-1} to 3451 cm^{-1}, suggesting the formation of new complexes between Cu(II) and degradation intermediates of BPA [1]. Meanwhile, vibration bands of aromatic rings are disappeared due to the destruction of aromatic rings. After the reaction for 60 min, the spectra of HKUST-1(Cu)/MoS_2-3-C were the same as a fresh catalyst, demonstrating that BPA and its intermediates were mineralized completely.

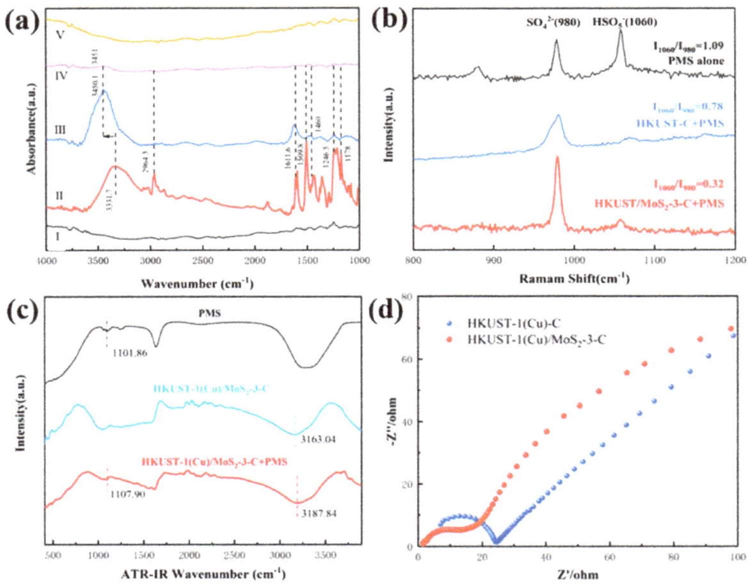

Figure 6. (**a**) FT-IR spectra for various samples [(I) fresh catalyst, (II) BPA, (III) catalyst absorbed BPA, (IV) catalyst in the reaction with BPA in the presence of PMS, (V) used catalyst]. Reaction conditions: [BPA] = 20 mg/L, [PMS] = 1 mM, [Catalysts] = 0.3 g/L; (**b**) In situ Raman spectra of PMS, HKUST-1(Cu)-C+PMS, HKUST-1(Cu)/MoS_2-3-C+PMS; (**c**) ATR-FTIR spectra for PMS solution alone, HKUST-1(Cu)/MoS_2-3-C in water, and HKUST-1(Cu)/MoS_2-3-C in PMS solution; (**d**) Nyquist plots of HKUST-1(Cu)-C and HKUST-1(Cu)/MoS_2-3-C electrodes.

In situ Raman spectra (Figure 6b) can demonstrate the adsorption affinity between PMS and catalytic sites. The band centred at 980 cm^{-1} and 1060 cm^{-1} are respectively ascribed to SO_4^{2-} and HSO_5^- [3,36,37], thus changes in the ratio of I_{1060}/I_{980} can represent PMS consumption rate which is related to the adsorption affinity for PMS on catalyst [22]. I_{1060}/I_{980} of HKUST-1(Cu)/MoS_2-3-C+PMS (0.32) is much lower than that of HKUST-1(Cu)-C+PMS (0.78), suggesting a higher utilization rate of PMS over HKUST-1(Cu)/MoS_2-3-C. In the composite of HKUST-1(Cu)/MoS_2-3-C, MoS_2 with superior adsorption affinity of PMS facilitated the HSO_5^- conversion to SO_4^{2-}, which can be further proved and discussed by DFT calculation. In addition, the ATR-FTIR test indicates that bands of S–O stretching vibration of HSO_5^- and SO_4^{2-} shift from 1101.86 cm^{-1} to 1107.90 cm^{-1} over HKUST-1(Cu)/MoS_2-3-C (Figure 6c), suggesting the formation of metastable ≡Mo–$OOSO_3^-$ complex [38]. In addition, the characteristic peak at 3163.04 cm^{-1} of HKUST-1(Cu)/MoS_2-3-C in solution was ascribed to the surface hydroxyl groups [3]. As PMS is added to the reaction system, the characteristic peak of hydroxyl groups on the catalyst surface shifts from 3163.04 to 3187.84 cm^{-1}, which suggests the electron transfer from hydroxyl groups to Mo [39]. Such phenomenon could be explained as follows: partial substitution of −OH by HSO_5^- on Mo strengthened the electron-withdrawing ability of Mo, which also confirms the formation of ≡Mo–$OOSO_3^-$ complex [3,10]. Additionally, the addition of MoS_2 into the catalyst enhances the conductivity for charge transfer during the reaction, which is shown in electrochemical impedance spectroscopy (EIS) (Figure 6d). The Nyquist plots semicircle radius of HKUST-1(Cu)/MoS_2-3-C was much smaller than that of HKUST-1(Cu)-C, demonstrating that MoS_2 effectively decreased electrical impedance.

Various radial scavengers are separately added in the reaction system to identify the major ROS (Figure 7a). As reported, methanol was commonly used to quench $SO_4^{\bullet-}$ ($k = 2.5 \times 10^7$ M^{-1} s^{-1}) and •OH ($k = 9.7 \times 10^8$ M^{-1} s^{-1}) [19]. TBA, L-histidine (L-his), and p-BQ were used to quench •OH (k = 3.8 ~ 7.6 × 10^8 M^{-1}•s^{-1}) [24], $^1O_2$ and $O_2^{\bullet-}$ [40], respectively. Considering that $SO_4^{\bullet-}$ and •OH were generally major ROS in the system of PMS activation [41], sufficient dosages of MeOH and TBA were added to the reaction system. Consequently, the BPA degradations are seriously inhibited in the presence of either p-BQ or L-his, which can be inferred that both $^1O_2$ and $O_2^{\bullet-}$ played the dominant role in the PMS activation process (Equations (3) and (4)) [42,43]. In addition, obvious decreases in BPA removal are observed in the presence of MeOH and TBA, suggesting the main ROS of $SO_4^{\bullet-}$ during the BPA degradation process. To directly identify the ROS, EPR tests are conducted using DMPO and TEMP as spin-trapping agents (Figure 7b–d). DMPO-•OH and DMPO-$SO_4^{\bullet-}$ signals observed in the system of (HKUST-1(Cu)/MoS_2-3-C+PMS+DMPO) indicate the generation of $SO_4^{\bullet-}$ and •OH. The signal intensity of DMPO-•OH is strengthened after the addition of BPA, suggesting that oxidation of σ-Cu-ligand contributed to radical generation. Additionally, the detected 1:1:1 triplet signal confirms the presence of $^1O_2$, which is consistent with the quenching experiment results (Figure 7a). Furthermore, the triplet signal is significantly strengthened as BPA is added, indicating that electrons transfer from BPA to PMS accelerates the $^1O_2$ production (Equations (9) and (10)) [44–47].

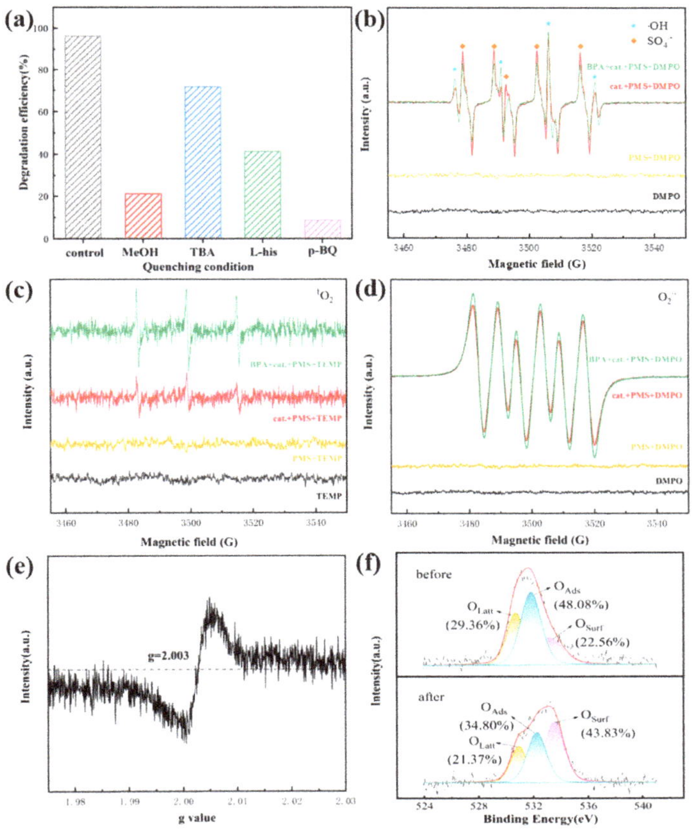

Figure 7. (**a**) Degradation efficiency of BPA under different quenching conditions. Quenching conditions: [MeOH] = 1 M, [TBA] = 1 M, [L-his] = 20 mM, [p-BQ] = 20 mM, [BPA] = 20 mg/L, [PMS] = 1 mM, [Catalysts] = 0.3 g/L. EPR detection for (**b**) •OH/$SO_4^{\bullet-}$; (**c**) $^1O_2$ and (**d**) $O_2^{\bullet-}$ in different reaction systems; (**e**) Solid EPR spectra of HKUST-1(Cu)/MoS_2-3-C; (**f**) High-resolution XPS spectra of O 1s before and after reaction.

Notably, although the addition of a high concentration of MeOH can completely block the $O_2^{\bullet-}$ production from the chain reactions of $SO_4^{\bullet-}$ and •OH (Equations (5)–(7)) [48], the removal rate of BPA can be still maintained at 21.15% (Figure 7a). In the present work, asymmetric signals at g = 2.003 ascribed to OVs unpaired electrons are trapped in Solid EPR (Figure 7e) [49,50], which suggested that O_2 might be directly reduced to $O_2^{\bullet-}$ by OVs over HKUST-1(Cu)/MoS_2-3-C (Equation (8)) [51,52]. By analyzing O 1s in XPS spectra before and after the reaction, three characteristic peaks at 530.7, 531.8, and 533.2 eV are identified, which are respectively assigned to lattice oxygen of metal oxides (O_{Latt}), absorbed oxygen (O_{Ads}) on OVs and surface absorbed water (O_{Surf}) (Figure 7f) [53,54]. The proportion of O_{Ads} decreased from 48.08% to 34.80% after the reaction, suggesting the consumption of OVs on HKUST-1(Cu)/MoS_2-3-C during the process of O_2 reduction to $O_2^{\bullet-}$. Meanwhile, BPA degradation kinetic constant under O_2 purging is higher than that under N_2 purging (Figure S4), which further proves that O_2 reduction to $O_2^{\bullet-}$ over OVs contributes to BPA degradation [42,49,55]

$$O_2^{\bullet-} + \equiv Mo(VI) \rightarrow + {}^1O_2 + \equiv Mo(IV) \tag{3}$$

$$2O_2^{\bullet-} + 2H_2O \rightarrow {}^1O_2 + H_2O_2 + 2OH^- \tag{4}$$

$$SO_4^{\bullet-} + H_2O \rightarrow \bullet OH + H^+ + SO_4^{2-} \tag{5}$$

$$HSO_5^- + H_2O \rightarrow H_2O_2 + HSO_4^- \tag{6}$$

$$H_2O_2 + \bullet OH \rightarrow H_2O + O_2^{\bullet-} + H^+ \tag{7}$$

$$O_2 + OVs \longrightarrow O_2^{\bullet-} \tag{8}$$

$$HSO_5^- \rightarrow SO_5^{\bullet-} + H^+ + e^- \tag{9}$$

$$2SO_5^{\bullet-} \rightarrow {}^1O_2 + 2SO_4^{2-} \tag{10}$$

Adsorption energy is calculated using the DFT analysis, which can demonstrate the interface interaction (Figure 8a–e). The adsorption energy of PMS on HKUST-1(Cu)/MoS_2-3-C ($E_{ads} = -2.728$ eV) is higher than that on HKUST-1(Cu)-C ($E_{ads} = -2.614$ eV) (Figure 8b,d), suggesting that PMS prefers to be adsorbed on MoS_2. Additionally, the O-O bond of PMS adsorbed on HKUST-1(Cu)/MoS_2-3-C is lengthened from 1.351 Å to 1.394 Å (Figure 8e), which induces the O-O bond is broken for the generation of ROS [22]. Meanwhile, due to the formation of a non-equilibrium-state surface, BPA prefers to be adsorbed on electron electron-poor Cu centre ($E_{ads} = -2.194$ eV) rather than electron-rich Mo centre ($E_{ads} = -0.869$ eV) to compose σ-Cu-ligand (Figure 8a,c). Thus, both PMS and BPA could be accumulated on HKUST-1(Cu)/MoS_2-3-C, which not only facilitated the PMS conversion to ROS but also promoted the BPA degradation. Furthermore, Bader charges can directly display the charge density between the interface of PMS, BPA, and catalytic sites during the reaction. Charge density between BPA and Cu(II) ($-0.451\,|e|$) is higher than that between BPA and MoS_2 ($-0.259\,|e|$), which indicates that the formation of σ-Cu-ligand strengthens the adsorption affinity of BPA. Furthermore, charge density between PMS and MoS_2 ($-0.787\,|e|$) is higher than that between PMS and HKUST-1(Cu)-C ($-0.622\,|e|$), also proves that PMS prefers to be adsorbed on the electron-rich Mo centre for activation.

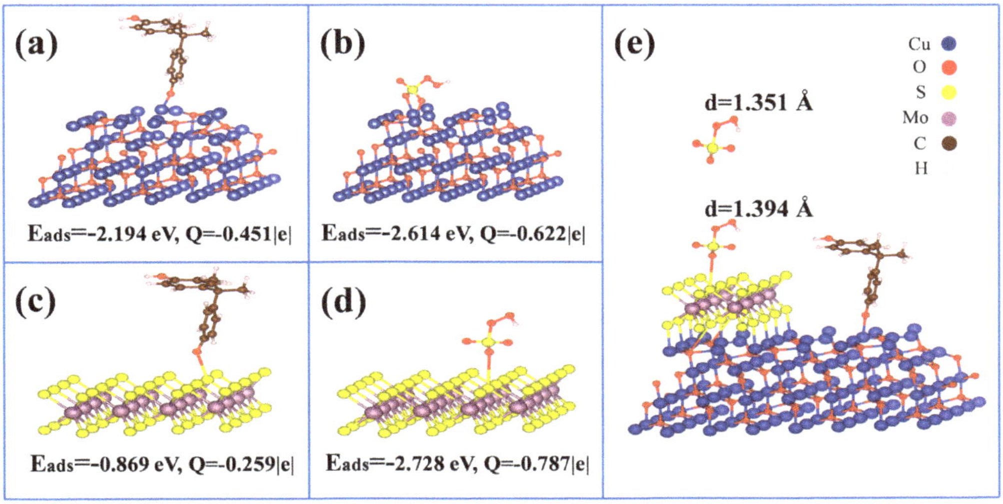

Figure 8. Optimized configurations of BPA/PMS absorbed on (**a**,**b**) HKUST-1(Cu)-C, (**c**,**d**) MoS_2; The E_{ads} and Q referred to the adsorption energy and Bader charge of PMS; (**e**) The optimized configuration of BPA/PMS absorbed on HKUST-1(Cu)/MoS_2-3-C.

Based on the above results, a possible catalytic mechanism is proposed (Figure 9). In the composite of HKUST-1(Cu)/MoS_2-3-C, a new ligand of MoS_2 with strong electron-accepting ability decreased the charge density of Cu, inducing the formation of a non-equilibrium-state surface with electron-poor Cu centre and electron-rich Mo centre in MOFs. During the initial reaction process, the electron-poor Cu centre facilitated the formation of σ-Cu-ligand with phenolic compounds (Equation (6)) [1]. Such σ-Cu-ligand would be ruptured under the attack of PMS, accompanied by the generation of $\bullet OH/SO_4^{\bullet -}$ and Cu(II) reduction to Cu(I) (Equation (7)). Notably, the non-equilibrium-state surface was broken by the formation of σ-Cu-ligand, which not only promoted the electrons (captured by electron-poor Cu centre) transfer to electron-rich Mo centre but also facilitated the formation of $\equiv Mo–OOSO_3^-$ complex (Equation (8)). Herein, $\equiv Mo–OOSO_3^-$ complex could oxidize organics with the generation of $SO_4^{\bullet -}$ (Equation (11)) [10,56,57]. Additionally, O_2 could be reduced to $O_2^{\bullet -}$ by OVs over HKUST-1(Cu)/MoS_2-3-C (Equation (10)) [49], also contributing to $^1O_2$ generation (Equation (11)) [43]. Thus, ROS were generated via four-electron transfer routes during the degradation process of phenolic compounds: (1) the first electron transfer route was from σ-Cu-ligand to PMS, with the generation of $SO_4^{\bullet -}/\bullet OH$ and Cu(II) reduction to Cu(I); (2) the second transfer route was from electron-rich Mo centre to PMS with the generation of $SO_4^{\bullet -}/\bullet OH$; (3) the third transfer route was from organics to $\equiv Mo–OOSO_3^-$ complexes, accompanying with organics oxidation and $SO_4^{\bullet -}$ generation; and (4) the fourth transfer route was from OVs to O_2 with the generation of $O_2^{\bullet -}$. Finally, BPA was degraded and mineralized by ROS (Equation (12)).

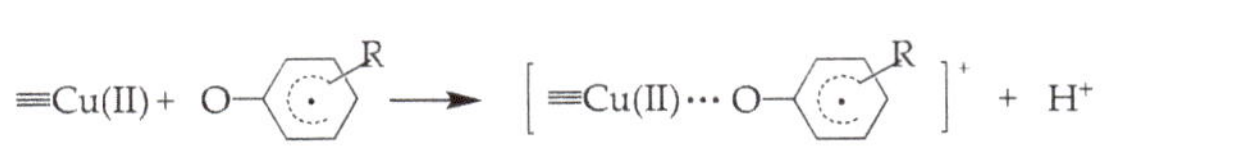

(11)

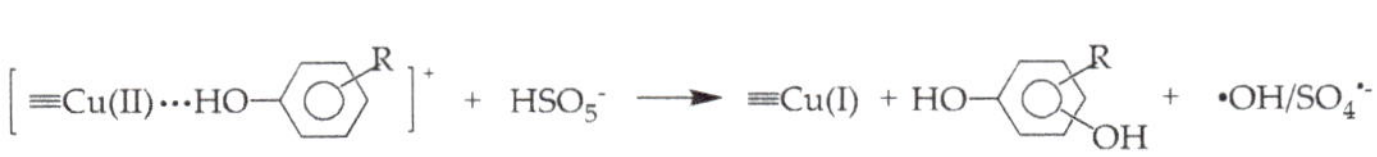

(12)

$$\equiv Mo(IV) + HSO_5^- \rightarrow \equiv Mo\text{-}OOSO_3^- + H^+ \quad (13)$$

$$\equiv Mo\text{-}OOSO_3^- + H^+ \rightarrow \equiv Mo\ (IV) + SO_4^{\bullet -} + OH^- \quad (14)$$

$$O_2 + e^- (OV_s) \rightarrow O_2^{\bullet -} \quad (15)$$

$$\equiv Mo(VI) + O_2^{\bullet -} \rightarrow \equiv Mo(IV) + {}^1O_2 \quad (16)$$

$$\bullet OH/SO_4^{\bullet -}/{}^1O_2/O_2^{\bullet -} + BPA \rightarrow CO_2 + H_2O \quad (17)$$

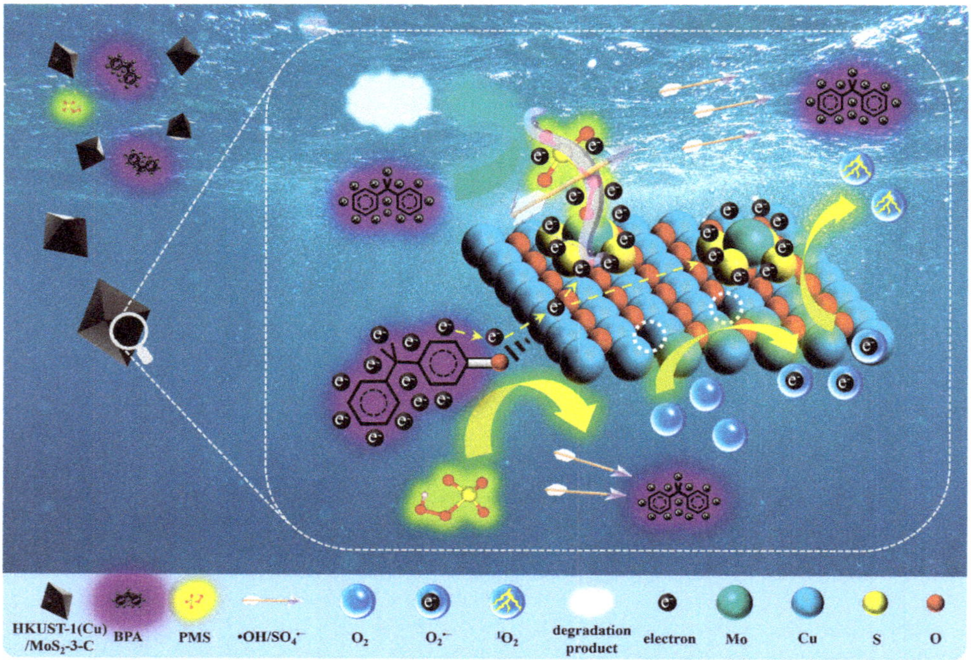

Figure 9. The possible mechanism of BPA degradation in HKUST-1(Cu)/MoS_2-3-C/PMS system.

4. Conclusions

In the present work, the PMS activation over novel catalyst HKUST-1(Cu)/MoS_2-3-C was investigated for selective degradation of phenolic compounds. A new ligand of MoS_2 with a strong electron-accepting ability decreased the charge density of Cu, resulting in the formation of non-equilibrium-state surface potential. σ-Cu-ligand formed between the electron-poor Cu centre and phenolic compounds not only contributed to radicals generation but also accelerated the redox of Cu(I)/Cu(II). Meanwhile, PMS preferred to be adsorbed on electron-rich Mo centre with the formation of $\equiv Mo–OOSO_3^-$ complex, which also enhanced phenolic compounds degradation and radicals production. In addition, the phenolic compounds and PMS were respectively adsorbed onto the electron-poor Cu centre and electron-rich Mo centre, promoting the electron transfer from phenolic compounds to PMS for PMS activation. Being attributed to the synergistic effect of σ-Cu-ligand and $\equiv Mo–OOSO_3^-$ complex, HKUST-1(Cu)/MoS_2-3-C achieved the high degradation rate of phenolic contaminants and utilization efficiency of PMS.

Supplementary Materials: The following supporting information can be downloaded at: https://www.mdpi.com/article/10.3390/w16010121/s1, Figure S1: TEM imagine of HKUST/MoS_2-3-C; Figure S2: High-resolution XPS spectra of Cu 2p for HKUST-1(Cu)-C; Figure S3: Degradation efficiency of BPA under different quenching conditions (p-BQ concentration = 5, 10, 15, 20 mM); Figure S4: Degradation efficiency and kinetics of BPA under nitrogen/oxygen purging; Figure S5: N_2 adsorption/desorption isotherms and pore-size distribution of HKUST-1(Cu)-C; Figure S6: Degradation of RhB in HKUST-1(Cu)/MoS_2-3-C system. Reaction conditions: [RhB] = 20 mg/L, [PMS] = 1 mM, [Catalysts] = 0.3 g/L;

Figure S7: ESI-MS spectra of BPA (sampling time: 10 min); Figure S8: ESI-MS spectra of BPA (sampling time: 30 min); Figure S9: ESI-MS spectra of BPA (sampling time: 60 min); Table S1: The effectiveness of PMS utilization for BPA (20 mg/L) degradation during the Fenton-like reaction; Table S2: The chemical formulas and fragments (m/z) of intermediate products during BPA; Table S3: The molecular structure and fragments (m/z) of intermediate products during BPA degradation; Table S4: Operation parameters of ESI-MS; Table S5: HPLC operating conditions for the detection of BPA; Table S6: The leached concentration of Mo and Cu ions after BPA degradation; Text S1: Synthesis of comparative catalysts; Text S2: Preparation of working electrode for EIS; Text S3: Raman analysis procedures for in situ characterization of the catalyst surface during catalytic PMS decomposition; Text S4: ABTS method for measuring PMS; Text S5: EPR analysis procedures for detecting radical signals.

Author Contributions: Conceptualization, X.Y. and J.W.; methodology, H.Y.; software, H.Y. and R.W.; validation, X.Y.; formal analysis, R.W.; investigation, X.Y. and H.Y.; data curation, H.Y.; writing—original draft preparation, X.Y. and H.Y.; writing—review and editing, J.W.; visualization, A.L.; supervision, J.W.; project administration, J.W. All authors have read and agreed to the published version of the manuscript.

Funding: This research was funded by the National Natural Science Foundation of China (No. 52070095), the Natural Science Foundation of Jiangsu Province (BK 20231407), and the social development project of Jiangsu Province (BE 2022771).

Data Availability Statement: The data are available from the corresponding author on reasonable request.

Conflicts of Interest: The authors declare no conflicts of interest.

References

1. Lyu, L.; Zhang, L.; Wang, Q.; Nie, Y.; Hu, C. Enhanced Fenton Catalytic Efficiency of gamma-Cu-Al_2O_3 by σ-Cu^{2+}-Ligand Complexes from Aromatic Pollutant Degradation. *Environ. Sci. Technol.* **2015**, *49*, 8639–8647. [CrossRef] [PubMed]
2. Salazar, R.; Brillas, E.; Sires, I. Finding the best Fe^{2+}/Cu^{2+} combination for the solar photoelectro-Fenton treatment of simulated wastewater containing the industrial textile dye Disperse Blue 3. *Appl. Catal. B Environ.* **2012**, *115*, 107–116. [CrossRef]
3. Zhang, T.; Zhu, H.; Croué, J.-P. Production of Sulfate Radical from Peroxymonosulfate Induced by a Magnetically Separable $CuFe_2O_4$ Spinel in Water: Efficiency, Stability, and Mechanism. *Environ. Sci. Technol.* **2013**, *47*, 2784–2791. [CrossRef] [PubMed]
4. Song, W.J.; Cheng, M.M.; Ma, J.H.; Ma, W.H.; Chen, C.C.; Zhao, J.C. Decomposition of hydrogen peroxide driven by photochemical cycling of iron species in clay. *Environ. Sci. Technol.* **2006**, *40*, 4782–4787. [CrossRef] [PubMed]
5. Lyu, L.; Yan, D.; Yu, G.; Cao, W.; Hu, C. Efficient Destruction of Pollutants in Water by a Dual-Reaction-Center Fenton-like Process over Carbon Nitride Compounds-Complexed Cu(II)-$CuAlO_2$. *Environ. Sci. Technol.* **2018**, *52*, 4294–4304. [CrossRef] [PubMed]
6. Lyu, L.; Zhang, L.; He, G.; He, H.; Hu, C. Selective H_2O_2 conversion to hydroxyl radicals in the electron-rich area of hydroxylated C-g-C_3N_4/CuCo–Al_2O_3. *J. Mater. Chem. A* **2017**, *5*, 7153–7164. [CrossRef]
7. Xu, S.; Zhu, H.; Cao, W.; Wen, Z.; Wang, J.; François-Xavier, C.P.; Wintgens, T. Cu-Al_2O_3-g-C_3N_4 and Cu-Al_2O_3-C-dots with dual-reaction centres for simultaneous enhancement of Fenton-like catalytic activity and selective H_2O_2 conversion to hydroxyl radicals. *Appl. Catal. B Environ.* **2018**, *234*, 223–233. [CrossRef]
8. Lyu, L.; Zhang, L.; Hu, C. Galvanic-like cells produced by negative charge nonuniformity of lattice oxygen on d-TiCuAl–SiO_2 nanospheres for enhancement of Fenton-catalytic efficiency. *Environ. Sci. Nano* **2016**, *3*, 1483–1492. [CrossRef]
9. Chen, J.; Wu, X.J.; Gong, Y.; Zhu, Y.; Yang, Z.; Li, B.; Lu, Q.; Yu, Y.; Han, S.; Zhang, Z.; et al. Edge Epitaxy of Two-Dimensional $MoSe_2$ and MoS_2 Nanosheets on One-Dimensional Nanowires. *J. Am. Chem. Soc.* **2017**, *139*, 8653–8660. [CrossRef]
10. Sun, Y.; Li, H.; Zhang, S.; Hua, M.; Qian, J.; Pan, B. Revisiting the Heterogeneous Peroxymonosulfate Activation by MoS_2: A Surface Mo–Peroxymonosulfate Complex as the Major Reactive Species. *ACS EST Water* **2022**, *2*, 376–384. [CrossRef]
11. Wang, Z.; Qiu, W.; Pang, S.-y.; Zhou, Y.; Gao, Y.; Guan, C.; Jiang, J. Further understanding the involvement of Fe(IV) in peroxydisulfate and peroxymonosulfate activation by Fe(II) for oxidative water treatment. *Chem. Eng. J.* **2019**, *371*, 842–847. [CrossRef]
12. Yang, L.; Chen, W.; Huang, J.; Tang, X.; Yang, R.; Zhang, H.; Tang, Z.; Gui, X. Resistance Switching and Failure Behavior of the MoO_x/Mo_2C Heterostructure. *ACS Appl. Mater. Interfaces* **2021**, *13*, 41857–41865. [CrossRef] [PubMed]
13. Wang, T.; Kim, H.-K.; Liu, Y.; Li, W.; Griffiths, J.T.; Wu, Y.; Laha, S.; Fong, K.D.; Podjaski, F.; Yun, C.; et al. Bottom-up Formation of Carbon-Based Structures with Multilevel Hierarchy from MOF–Guest Polyhedra. *J. Am. Chem. Soc.* **2018**, *140*, 6130–6136. [CrossRef] [PubMed]
14. Peng, D.; Zhang, Y.; Xu, G.; Tian, Y.; Ma, D.; Zhang, Y.; Qiu, P. Synthesis of Multilevel Structured MoS_2@Cu/Cu_2O@C Visible-Light-Driven Photocatalyst Derived from MOF–Guest Polyhedra for Cyclohexane Oxidation. *ACS Sustain. Chem. Eng.* **2020**, *8*, 6622–6633. [CrossRef]
15. Platzman, I.; Brener, R.; Haick, H.; Tannenbaum, R. Oxidation of Polycrystalline Copper Thin Films at Ambient Conditions. *J. Phys. Chem. C* **2008**, *112*, 1101–1108. [CrossRef]

16. Dubot, P.; Jousset, D.; Pinet, V.; Pellerin, F.; Langeron, J.P. Simulation of the LMM Auger spectra of copper. *Surf. Interface Anal.* **1988**, *12*, 99–104. [CrossRef]
17. Zhu, G.; Shi, C.; Jin, Y.; Ge, M. Enhanced degradation of polyvinyl alcohol over a Cu^0-carbon@γ-Al_2O_3 composite through heterogeneous Fenton-like reactions: Preparation, mechanism, and applications. *Appl. Catal. A Gen.* **2022**, *643*, 118721. [CrossRef]
18. Yang, Y.; Wang, Q.; Aleisa, R.; Zhao, T.; Ma, S.; Zhang, G.; Yao, T.; Yin, Y. MoS_2/FeS Nanocomposite Catalyst for Efficient Fenton Reaction. *ACS Appl. Mater. Interfaces* **2021**, *13*, 51829–51838. [CrossRef]
19. Lu, J.; Zhou, Y.; Zhou, Y. Efficiently activate peroxymonosulfate by Fe_3O_4@MoS_2 for rapid degradation of sulfonamides. *Chem. Eng. J.* **2021**, *422*, 130126. [CrossRef]
20. Nong, Y.; Zhang, M.; Li, Q.; Pan, Q.; Huang, Y.; Wang, H.; Zheng, F.; Li, Q. Carbon coated bimetallic sulfides Co_9S_8/ZnS heterostructures microrods as advanced anode materials for lithium ion batteries. *J. Taiwan Inst. Chem. Eng.* **2022**, *141*, 104601. [CrossRef]
21. Zhang, X.; Zhao, Y.; Zhang, C.; Wu, C.; Li, X.; Jin, M.; Lan, W.; Pan, X.; Zhou, J.; Xie, E. Cobalt sulfide embedded carbon nanofibers as a self-supporting template to improve lithium ion battery performances. *Electrochim. Acta* **2021**, *366*, 137351. [CrossRef]
22. Chen, Y.; Zhang, G.; Liu, H.; Qu, J. Confining Free Radicals in Close Vicinity to Contaminants Enables Ultrafast Fenton-like Processes in the Interspacing of MoS_2 Membranes. *Angew. Chem. Int. Ed.* **2019**, *58*, 8134–8138. [CrossRef] [PubMed]
23. Ai, K.; Ruan, C.; Shen, M.; Lu, L. MoS_2 Nanosheets with Widened Interlayer Spacing for High-Efficiency Removal of Mercury in Aquatic Systems. *Adv. Funct. Mater.* **2016**, *26*, 5542–5549. [CrossRef]
24. Bai, R.; Yan, W.; Xiao, Y.; Wang, S.; Tian, X.; Li, J.; Xiao, X.; Lu, X.; Zhao, F. Acceleration of peroxymonosulfate decomposition by a magnetic MoS_2/$CuFe_2O_4$ heterogeneous catalyst for rapid degradation of fluoxetine. *Chem. Eng. J.* **2020**, *397*, 125501. [CrossRef]
25. Zhu, Y.; Zhu, R.; Xi, Y.; Zhu, J.; Zhu, G.; He, H. Strategies for enhancing the heterogeneous Fenton catalytic reactivity: A review. *Appl. Catal. B Environ.* **2019**, *255*, 117739. [CrossRef]
26. Xie, Z.; Zhou, J.; Wang, J.; François-Xavier, C.P.; Wintgens, T. Novel Fenton-like catalyst γ-Cu-Al_2O_3-$Bi_{12}O_{15}C_{16}$ with electron-poor Cu centre and electron-rich Bi centre for enhancement of phenolic compounds degradation and H_2O_2 utilization: The synergistic effects of σ-Cu-ligand, dual-reaction centres and oxygen vacancies. *Appl. Catal. B Environ.* **2019**, *253*, 28–40. [CrossRef]
27. Zhou, X.; Luo, C.; Luo, M.; Wang, Q.; Wang, J.; Liao, Z.; Chen, Z.; Chen, Z. Understanding the synergetic effect from foreign metals in bimetallic oxides for PMS activation: A common strategy to increase the stoichiometric efficiency of oxidants. *Chem. Eng. J.* **2020**, *381*, 122587. [CrossRef]
28. Wang, Z.; Bush, R.T.; Sullivan, L.A.; Chen, C.; Liu, J. Selective oxidation of arsenite by peroxymonosulfate with high utilization efficiency of oxidant. *Environ. Sci. Technol.* **2014**, *48*, 3978–3985. [CrossRef]
29. Giannakis, S.; Lin, K.-Y.A.; Ghanbari, F. A review of the recent advances on the treatment of industrial wastewaters by Sulfate Radical-based Advanced Oxidation Processes (SR-AOPs). *Chem. Eng. J.* **2021**, *406*, 127083. [CrossRef]
30. Wacławek, S.; Lutze, H.V.; Grübel, K.; Padil, V.V.T.; Černík, M.; Dionysiou, D.D. Chemistry of persulfates in water and wastewater treatment: A review. *Chem. Eng. J.* **2017**, *330*, 44–62. [CrossRef]
31. Sun, P.; Liu, H.; Feng, M.; Zhai, Z.; Fang, Y.; Zhang, X.; Sharma, V.K. Strategic combination of N-doped graphene and g-C_3N_4: Efficient catalytic peroxymonosulfate-based oxidation of organic pollutants by non-radical-dominated processes. *Appl. Catal. B Environ.* **2020**, *272*, 119005. [CrossRef]
32. Lyu, Z.; Xu, M.; Wang, J.; Li, A.; François-Xavier Corvini, P. Hierarchical nano-vesicles with bimetal-encapsulated for peroxymonosulfate activation: Singlet oxygen-dominated oxidation process. *Chem. Eng. J.* **2022**, *433*, 133581. [CrossRef]
33. Oess, A.C.; Cheshire, M.V.; McPhail, D.B.; Stoll, S.; Alailia, M.E.; Vedy, J.C. Elucidation of phenol-Cu interaction mechanisms by potentiometry, ESR, UV absorption spectroscopy and molecular simulations. *Sci. Total Environ.* **1999**, *228*, 49–58. [CrossRef]
34. Han, J.; Du, Z.; Zou, W.; Li, H.; Zhang, C. In-situ improved phenol adsorption at ions-enrichment interface of porous adsorbent for simultaneous removal of copper ions and phenol. *Chem. Eng. J.* **2015**, *262*, 571–578. [CrossRef]
35. Mitić, Ž.; Nikolić, G.S.; Cakić, M.; Premović, P.; Ilić, L. FTIR spectroscopic characterization of Cu(II) coordination compounds with exopolysaccharide pullulan and its derivatives. *J. Mol. Struct.* **2009**, *924–926*, 264–273. [CrossRef]
36. Zuo, S.; Guan, Z.; Xia, D.; Yang, F.; Xu, H.; Huang, M.; Li, D. Polarized heterogeneous CuO-CN for peroxymonosulfate nonradical activation: An enhancement mechanism of mediated electron transfer. *Chem. Eng. J.* **2020**, *420*, 127619. [CrossRef]
37. Yu, L.; Zhang, G.; Liu, C.; Lan, H.; Liu, H.; Qu, J. Interface Stabilization of Undercoordinated Iron Centers on Manganese Oxides for Nature-Inspired Peroxide Activation. *ACS Catal.* **2018**, *8*, 1090–1096. [CrossRef]
38. Wang, Y.; Zhang, Z.; Yin, Z.; Liu, Z.; Liu, Y.; Yang, Z.; Yang, W. Adsorption and catalysis of peroxymonosulfate on carbocatalysts for phenol degradation: The role of pyrrolic-nitrogen. *Appl. Catal. B Environ.* **2022**, *319*, 121891. [CrossRef]
39. Zhang, T.; Li, C.; Ma, J.; Tian, H.; Qiang, Z. Surface hydroxyl groups of synthetic α-FeOOH in promoting OH generation from aqueous ozone: Property and activity relationship. *Appl. Catal. B Environ.* **2008**, *82*, 131–137. [CrossRef]
40. Sun, P.; Liu, H.; Feng, M.; Guo, L.; Zhai, Z.; Fang, Y.; Zhang, X.; Sharma, V.K. Nitrogen-sulfur co-doped industrial graphene as an efficient peroxymonosulfate activator: Singlet oxygen-dominated catalytic degradation of organic contaminants. *Appl. Catal. B Environ.* **2019**, *251*, 335–345. [CrossRef]
41. Chen, T.; Zhu, Z.; Zhang, H.; Qiu, Y.; Yin, D.; Zhao, G. Facile Construction of a Copper-Containing Covalent Bond for Peroxymonosulfate Activation: Efficient Redox Behavior of Copper Species via Electron Transfer Regulation. *ACS Appl. Mater. Interfaces* **2020**, *12*, 42790–42802. [CrossRef] [PubMed]

42. Dong, C.; Wang, Z.; Ye, Z.; He, J.; Zheng, Z.; Gong, X.; Zhang, J.; Lo, I.M.C. Superoxide radicals dominated visible light driven peroxymonosulfate activation using molybdenum selenide ($MoSe_2$) for boosting catalytic degradation of pharmaceuticals and personal care products. *Appl. Catal. B Environ.* **2021**, *296*, 120223. [CrossRef]
43. Yi, Q.; Ji, J.; Shen, B.; Dong, C.; Liu, J.; Zhang, J.; Xing, M. Singlet Oxygen Triggered by Superoxide Radicals in a Molybdenum Cocatalytic Fenton Reaction with Enhanced REDOX Activity in the Environment. *Environ. Sci. Technol.* **2019**, *53*, 9725–9733. [CrossRef] [PubMed]
44. Zhu, C.; Zhao, S.; Fan, Z.; Wu, H.; Liu, F.; Chen, Z.; Li, A. Confinement of CoP Nanoparticles in Nitrogen-Doped Yolk-Shell Porous Carbon Polyhedron for Ultrafast Catalytic Oxidation. *Adv. Funct. Mater.* **2020**, *30*, 2003947. [CrossRef]
45. Duan, P.; Qi, Y.; Feng, S.; Peng, X.; Wang, W.; Yue, Y.; Shang, Y.; Li, Y.; Gao, B.; Xu, X. Enhanced degradation of clothianidin in peroxymonosulfate/catalyst system via core-shell FeMn@N-C and phosphate surrounding. *Appl. Catal. B Environ.* **2020**, *267*, 118717. [CrossRef]
46. Huang, G.X.; Wang, C.Y.; Yang, C.W.; Guo, P.C.; Yu, H.Q. Degradation of Bisphenol A by Peroxymonosulfate Catalytically Activated with Mn1.8Fe1.2O4 Nanospheres: Synergism between Mn and Fe. *Environ. Sci. Technol.* **2017**, *51*, 12611–12618. [CrossRef] [PubMed]
47. Zhang, L.S.; Jiang, X.H.; Zhong, Z.A.; Tian, L.; Sun, Q.; Cui, Y.T.; Lu, X.; Zou, J.P.; Luo, S.L. Carbon Nitride Supported High-Loading Fe Single-Atom Catalyst for Activation of Peroxymonosulfate to Generate $^1O_2$ with 100% Selectivity. *Angew. Chem. Int. Ed. Engl.* **2021**, *60*, 21751–21755. [CrossRef] [PubMed]
48. Xie, Z.; Lyu, Z.; Wang, J.; Li, A.; Corvini, P.F.X. Ultrafine-Mn_2O_3@N-doped porous carbon hybrids derived from Mn-MOFs: Dual-reaction centre catalyst with singlet oxygen-dominant oxidation process. *Chem. Eng. J.* **2022**, *429*, 132299. [CrossRef]
49. Wu, L.; Sun, Z.; Zhen, Y.; Zhu, S.; Yang, C.; Lu, J.; Tian, Y.; Zhong, D.; Ma, J. Oxygen Vacancy-Induced Nonradical Degradation of Organics: Critical Trigger of Oxygen (O_2) in the Fe-Co LDH/Peroxymonosulfate System. *Environ. Sci. Technol.* **2021**, *55*, 15400–15411. [CrossRef]
50. Liu, L.; Liu, Q.; Wang, Y.; Huang, J.; Wang, W.; Duan, L.; Yang, X.; Yu, X.; Han, X.; Liu, N. Nonradical activation of peroxydisulfate promoted by oxygen vacancy-laden NiO for catalytic phenol oxidative polymerization. *Appl. Catal. B Environ.* **2019**, *254*, 166–173. [CrossRef]
51. Li, Z.; Yang, W.; Xie, L.; Li, Y.; Liu, Y.; Sun, Y.; Bu, Y.; Mi, X.; Zhan, S.; Hu, W. Prominent role of oxygen vacancy for superoxide radical and hydroxyl radical formation to promote electro-Fenton like reaction by W-doped CeO_2 composites. *Appl. Surf. Sci.* **2021**, *549*, 149262. [CrossRef]
52. Pan, Y.; Su, H.; Zhu, Y.; Vafaei Molamahmood, H.; Long, M. CaO_2 based Fenton-like reaction at neutral pH: Accelerated reduction of ferric species and production of superoxide radicals. *Water Res.* **2018**, *145*, 731–740. [CrossRef] [PubMed]
53. Zhu, J.; Zhu, Y.; Zhou, W. Cu-doped Ni-LDH with abundant oxygen vacancies for enhanced methyl 4-hydroxybenzoate degradation via peroxymonosulfate activation: Key role of superoxide radicals. *J. Colloid Interface Sci.* **2022**, *610*, 504–517. [CrossRef] [PubMed]
54. Lim, J.; Yang, Y.; Hoffmann, M.R. Activation of Peroxymonosulfate by Oxygen Vacancies-Enriched Cobalt-Doped Black TiO_2 Nanotubes for the Removal of Organic Pollutants. *Environ. Sci. Technol.* **2019**, *53*, 6972–6980. [CrossRef]
55. Qin, W.; Fang, G.; Wang, Y.; Zhou, D. Mechanistic understanding of polychlorinated biphenyls degradation by peroxymonosulfate activated with $CuFe_2O_4$ nanoparticles: Key role of superoxide radicals. *Chem. Eng. J.* **2018**, *348*, 526–534. [CrossRef]
56. Li, H.; Zhao, Z.; Qian, J.; Pan, B. Are Free Radicals the Primary Reactive Species in Co(II)-Mediated Activation of Peroxymonosulfate? New Evidence for the Role of the Co(II)-Peroxymonosulfate Complex. *Environ. Sci. Technol.* **2021**, *55*, 6397–6406. [CrossRef]
57. Li, X.; Zhao, Z.; Li, H.; Qian, J. Degradation of organic contaminants in the $CoFe_2O_4$/peroxymonosulfate process: The overlooked role of Co(II)-PMS complex. *Chem. Eng. J. Adv.* **2021**, *8*, 100143. [CrossRef]

molecules

MDPI

Article

In Situ Synthesis of $Bi_2S_3/BiFeO_3$ Nanoflower Hybrid Photocatalyst for Enhanced Photocatalytic Degradation of Organic Pollutants

Rentao Zhou [1], Xinman Tu [1,2,*], Peng Zheng [2], Li Zhang [2] and Zhenxing Zeng [3,*]

1 College of Environment Science and Engineering, Guilin University of Technology, Guilin 541004, China
2 Key Laboratory of Jiangxi Province for Persistent Pollutants Control and Resources Recycle, Nanchang Hangkong University, Nanchang 330063, China; 17770042612@163.com (P.Z.)
3 College of Environmental Sciences, Sichuan Agricultural University, Chengdu 611130, China
* Correspondence: tuxinman@126.com (X.T.); zxzeng@sicau.edu.cn (Z.Z.)

Abstract: Photocatalytic degradation of Malachite Green oxalate (MG) in a water body is of significant importance to our health protection, as it could cause various serious diseases. However the photocatalytic activity of most catalysts is still unsatisfactory, due to the poor reactive oxygen species production as a result of sluggish charge separation. Here, innovative nanoflower-shaped $Bi_2S_3/BiFeO_3$ heterojunctions are prepared via a facile sol–gel method, exhibiting an enhanced reactive oxygen species generation, which leads to the excellent photocatalytic performance toward MG degradation. We verify that interfacing $BiFeO_3$ with Bi_2S_3 could form a fine junction and offers a built-in field to speed up charge separation at the junction area; as a result, this shows much higher charge separation efficiency. By virtue of the aforementioned advantages, the as-prepared $Bi_2S_3/BiFeO_3$ heterojunctions exhibit excellent photocatalytic performance toward MG degradation, where more than 99% of MG is removed within 2 h of photocatalysis. The innovative design of nanoflower-like $Bi_2S_3/BiFeO_3$ heterojunctions may offer new viewpoints in designing highly efficient photocatalysts for environmentally related applications.

Keywords: photocatalysis; advanced oxidation processes; organic degradation; heterojunction

1. Introduction

In recent years, the world has been grappling with escalating energy crises and environmental pollution issues, particularly the pervasive presence of organic pollutants that significantly affect people's daily lives. As the organic pollutants in water bodies are mostly highly toxic and may cause various serious diseases, effectively eliminating harmful organic pollutants from the environment has become a paramount concern for the scientific community. As a safe and sustainable way for organic pollutant removal, photocatalytic degradation of organics following a radical oxidative manner in recent years has been intensively studied. The key to achieving highly efficient organic pollutant removal from a water body lies in the development of a cost-effective photocatalyst with high reactive radical generation efficiency [1,2]. Unfortunately, most widely used photocatalysts primarily respond to UV–visible irradiation, limiting their practicality under visible light. Consequently, the quest for highly efficient and stable visible-light-driven photocatalysts has emerged as a hot topic in the field of photocatalysis.

Recently, $BiFeO_3$ photocatalysts have garnered attention, due to their lower energy bandgap, excellent chemical stability at room temperature, and superior carrier transport properties. $BiFeO_3$ holds promise for catalytic applications under visible light. Research has shown that $BiFeO_3$, with a bandgap of approximately 2.2 eV [3,4], exhibits a favorable response to visible light and has been employed for the optical degradation of organic pollutants in water, including the degradation of organic dyes [5–13]. To further enhance the photocatalytic activity of $BiFeO_3$, researchers have explored various modification techniques. Common approaches include element doping [14–16], which can boost visible-light-driven photocatalytic activity but often fails to effectively suppress the recombination of photogenerated electron–hole pairs, resulting in limited solar energy utilization. Some scholars have also experimented with noble metal deposition methods [17,18]; however, the high cost of precious metals makes these approaches less practical for real-world applications. Additionally, these modifications require further optimization to achieve better responsiveness to the visible light spectrum. As a result, researchers have turned to the incorporation of various materials into $BiFeO_3$ to form composite heterojunction structures, such as CuO [19], TiO_2 [20,21], and graphene oxide (GO) [22,23], among others. These heterojunction structures have been effective in enhancing visible light photocatalytic activity by mitigating electron–hole-pair recombination. However, these heterojunctions still lack optimal reducibility and oxidizability for photogenerated electron–hole pairs.

Hence, the search for a new semiconductor material to create heterojunctions with stronger reducibility and oxidizability is imperative. Additionally, since $BiFeO_3$ nanoparticles possess a limited specific surface area, finding a substance capable of effectively increasing its surface area is crucial for enhancing photocatalytic reactions. Based on these considerations, our research aims to improve the reducibility and oxidizability of $BiFeO_3$ through the formation of heterojunction structures with other materials, while simultaneously increasing its surface area. Achieving this under economical, simple, and mild reaction conditions has been a primary objective in our field. Our investigation revealed that Bi_2S_3, with its narrow band gap (Eg = 1.38–1.71 eV) [24] and the ability to adjust heterojunction morphology, is an ideal material for combining with $BiFeO_3$. While previous studies have explored the potential of Bi_2S_3 as a visible-light-driven photocatalyst in conjunction with other photocatalysts, such as BiOCl [25,26], ZnO [27], TiO_2 [28], CdS [29], WO_3 [30], $BiFeO_3/\alpha\text{-}Fe_2O_3$ [31], $La_xBa_{1-x}Sr_yCd_{1-y}O_{3\pm\delta}$ [32], Bi_2WO_6 [33], $BiVO_4$ [34], and Bi_2O_3 [35], no reports exist on the $Bi_2S_3/BiFeO_3$ heterojunction photocatalysis.

In this study, we employed the in situ growth method to synthesize a novel $Bi_2S_3/BiFeO_3$ nanoflower structure, demonstrating enhanced photocatalytic performance under visible light irradiation. We employed MG as a model contaminant to evaluate the material's efficiency. Furthermore, we comprehensively investigated the structure, composition, morphology, and mechanism of this novel photocatalyst.

2. Results and Discussion

Figure 1 displays the X-ray diffraction (XRD) patterns of the $Bi_2S_3/BiFeO_3$ samples with varying molar mass ratios, ranging from 1:2 to 1:4, in addition to pure $BiFeO_3$, for comparison. The XRD patterns of pure $BiFeO_3$ align well with the JCPDS 20-0169 standard [36], indicating the presence of characteristic diffraction peaks associated with this compound. In the composite material, discernible diffraction peaks attributed to Bi_2S_3 are evident, indicating the successful incorporation of Bi_2S_3 into the structure. It is worth noting that the diffraction peak of Bi_2S_3 at $2\theta = 25.32$ exhibits a slight overlap with $BiFeO_3$ ($2\theta = 25.40$). A noteworthy observation is that the positions of the Bi_2S_3 diffraction peaks at 25.2 and 28.6 display minor shifts towards larger angles in the composite material. These shifts suggest that Bi_2S_3 has effectively integrated with $BiFeO_3$, leading to a reduction in the crystalline lattice spacing. Furthermore, it is important to acknowledge that the crystallinity of Bi_2S_3, synthesized via a hydrothermal method, is comparatively inferior to that of $BiFeO_3$. These XRD findings shed light on the successful synthesis of $Bi_2S_3/BiFeO_3$ composites and the structural modifications induced by their integration.

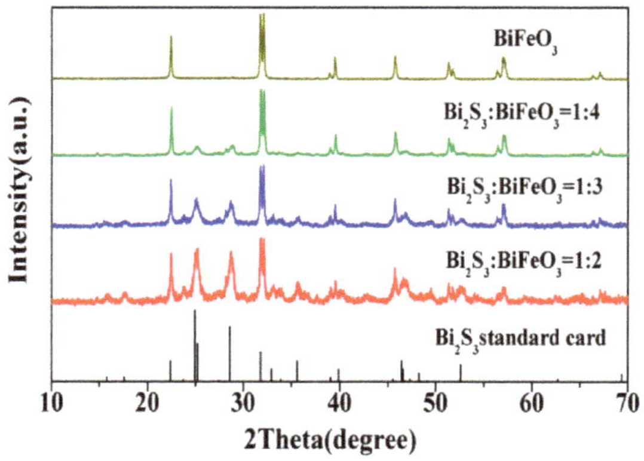

Figure 1. XRD patterns of $Bi_2S_3/BiFeO_3$ with different proportions.

The microcosmic morphologies and microstructure of both pure $BiFeO_3$ and $Bi_2S_3/BiFeO_3$ were meticulously examined using SEM, TEM, and HRTEM, revealing intriguing insights. SEM images of pure $BiFeO_3$, as shown in Figure 2a, depict an assembly of particles characterized by smooth borderlines. However, the introduction of sulfur for the ion exchange process brought about significant changes in the morphology of the composite materials. As evidenced in Figure 2d, there is a noticeable shift towards a nearly spherical flower-like shape. Detailed TEM analysis, displayed in Figure 2e, unequivocally indicates the growth of Bi_2S_3 on the surface of $BiFeO_3$ through ion exchange. HRTEM provided further elucidation, revealing that Bi_2S_3 exposes more planes corresponding to crystal planes (310) and (211), which correspond to lattice spacings of 0.3530 nm and 0.3120 nm, respectively. The formation mechanism of these changes can be understood as follows: Initially, the lattice of $BiFeO_3$ is pristine, with filled bonding orbitals and relatively few dangling bonds at the interface, rendering the entire $BiFeO_3$ molecule in a stable state. The introduction of sulfuric acid provides the necessary energy for the surface activation of $BiFeO_3$. This activation leads to damage to the original crystal lattice, exposing elemental bismuth and creating an abundance of dangling bonds on the surface of $BiFeO_3$. Consequently, surface energy increases significantly, enhancing surface activity. When L-cysteine is introduced, a critical chemical transformation occurs. The elemental sulfur in L-cysteine forms stable new chemical bonds with the elemental bismuth on the surface of $BiFeO_3$, resulting in the formation of $Bi_2S_3/BiFeO_3$ heterojunctions. Moreover, the EDX spectrum analysis corroborates these findings, as it reveals the presence of Bi, O, Fe, and S elements, further confirming the formation mechanism of the $Bi_2S_3/BiFeO_3$ heterojunction.

To gain further insight into the surface composition of the heterojunction material, X-ray photoelectron spectroscopy (XPS) was employed, providing compelling evidence for the coexistence of Bi_2S_3 and $BiFeO_3$, as well as the presence of Bi, Fe, O, and S elements (Figure 3). The high-resolution XPS spectra reveal overlapping peaks centered at 158.2 eV and 163.6 eV, as well as 159.2 eV and 164.5 eV. These peaks can be attributed to the binding energy of Bi^{3+} $4f_{7/2}$ and $4f_{5/2}$ in Bi_2S_3 [28] and the Bi-O bond in $BiFeO_3$ [37,38], respectively. The XPS spectrum of Fe(2p) is noteworthy, exhibiting six peaks at 709.9, 711.3, 715.5, 719.0, 723.4, and 725.0 eV. Notably, the Fe(2p) peaks at binding energies of 711.3/725.0 eV and 709.9/723.4 eV, along with a satellite signal peak at 719.0 eV and 715.5 eV, are characteristic of Fe^{3+} and Fe^{2+}. The Fe(2p) spectra primarily feature two maxima peaks corresponding to the electron levels of Fe $2p_{3/2}$ and Fe $2p_{1/2}$ [39,40]. It is worth noting that the peak of the high-spin Fe^{3+} ion is significantly broader, compared to the low-spin Fe^{2+} ions. This broadening can be attributed to electrostatic interactions and spin–orbit coupling among the 2p empty orbital and the unpaired 3d electrons of iron ions. In the XPS spectrum of O 1s, four distinct peaks were observed at 529.3, 530.8, 532.2, and 533.2 eV, corresponding to oxygen–metal bonds in the lattice O_2^- of $BiFeO_3$ [41], remaining OH^- at the material surface [42], adsorbed oxygen (O_{ads}), and surface-adsorbed H_2O [43]. The presence of a smaller-intensity peak at a binding energy of 225.4 eV in the S 2s spectrum can be attributed to the S-Bi bond of Bi_2S_3 [44]. In conclusion, the XPS analysis unequivocally confirms the successful growth of Bi_2S_3 nanomaterials on the surface of $BiFeO_3$, forming the $Bi_2S_3/BiFeO_3$ heterojunction.

Figure 2. SEM images of samples (**a**) $BiFeO_3$; (**b**) $Bi_2S_3/BiFeO_3$(1:4); (**c**) $Bi_2S_3/BiFeO_3$(1:3); (**d**) $Bi_2S_3/BiFeO_3$(1:2); (**e**) TEM of $Bi_2S_3/BiFeO_3$(1:3); (**f**) HRTEM of $Bi_2S_3/BiFeO_3$(1:3); (**g**) the selected area of the EDX; (**h**) the EDX spectrum of the $Bi_2S_3/BiFeO_3$(1:3).

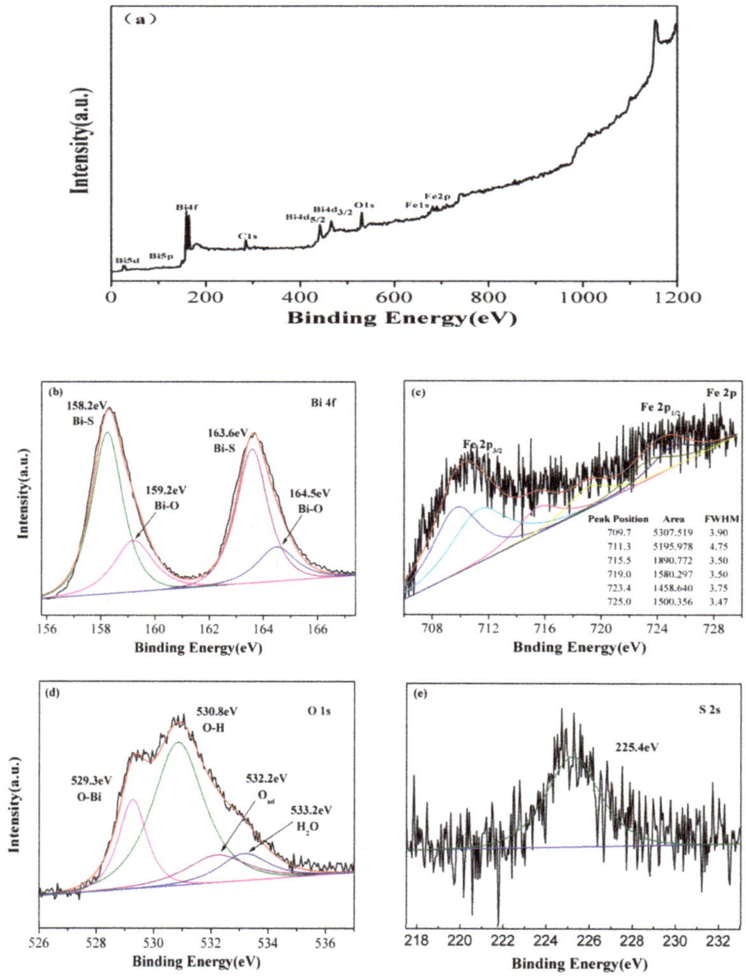

Figure 3. (**a**) Whole XPS spectra; (**b**) Bi(4f); (**c**) Fe(2p); (**d**) O(1s) and (**e**) S(2s) lines of the $Bi_2S_3/BiFeO_3$.

Photocurrent measurements were conducted for both pure $BiFeO_3$ and various molar ratios of $Bi_2S_3/BiFeO_3$, to probe electronic interactions, as depicted in Figure 4a. The composite materials exhibited markedly higher photocurrent responses compared to single $BiFeO_3$ during on–off cycles under visible-light irradiation. This enhancement can be directly attributed to the improved separation efficiency of photo-generated electrons within the heterojunction. Notably, when the molar mass ratio of $Bi_2S_3/BiFeO_3$ is 1:3, the photocurrent response surpassed that of the other ratios, signifying superior performance in charge separation. This enhanced photocurrent in the $Bi_2S_3/BiFeO_3$ heterojunction indicates heightened efficiency in separating electron–hole pairs, thereby enhancing the photocatalytic properties. To delve further into the charge separation process, electrochemical impedance spectroscopy (EIS) was employed. In EIS, a smaller arc radius signifies more effective separation of photo-generated electron–hole pairs and faster interface charge transfer.

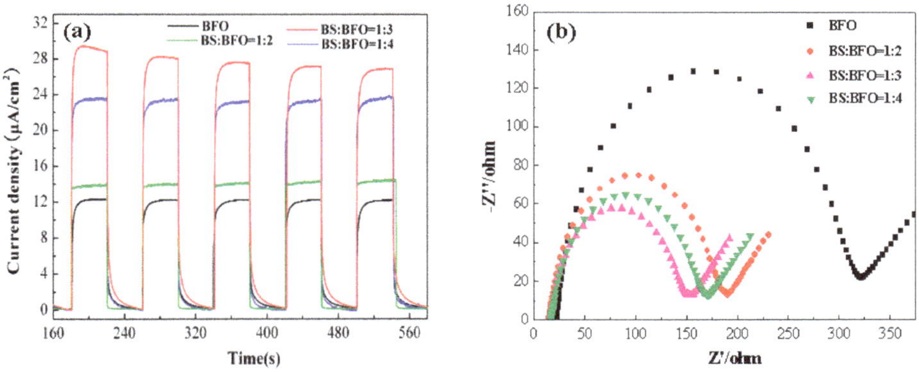

Figure 4. Photocurrents (**a**) $BiFeO_3$ and the proportions of Bi_2S_3 and $BiFeO_3$ ($[Na_2SO_4]$ = 0.5 M) and (**b**) electrochemical impedance spectroscopy (EIS) for $BiFeO_3$ and the proportions of Bi_2S_3 and $BiFeO_3$ ($[Na_2SO_4]$ = 0.5 M, $[K_4Fe(CN)_6{\bullet}3H_2O]$ = 2.5 mM, $[K_3[Fe(CN)_6]]$ = 2.5 mM).

Figure 4b illustrates the fact that the arc radius of the $Bi_2S_3/BiFeO_3$ heterojunction is notably smaller than that of single $BiFeO_3$. This difference is particularly pronounced when the ratio of $Bi_2S_3/BiFeO_3$ is 1:3, where the arc radius is minimized. These results underscore the significant role played by the $Bi_2S_3/BiFeO_3$ heterojunction in improving the separation of photogenerated electron–hole pairs and facilitating efficient charge transfer at the interface. Consequently, the utilization of light energy is greatly enhanced, contributing to the overall photocatalytic efficiency.

The UV-Vis DRS spectra of the $Bi_2S_3/BiFeO_3$ heterojunction photocatalysts, with varying proportions, along with pure $BiFeO_3$, are presented in Figure 5. These spectra offer valuable insights into the optical properties of the materials. The optical absorption spectra, as revealed by DRS, demonstrate distinct characteristics. The pure $BiFeO_3$ sample exhibits intrinsic semiconductor-like absorption, primarily in the orange region of the visible spectra. In contrast, the $Bi_2S_3/BiFeO_3$ heterojunction composite materials exhibit absorption across the entire visible region of the spectrum, indicating enhanced light absorption capabilities. Notably, pure $BiFeO_3$ shows an absorption edge at approximately 610 nm, corresponding to a band gap of 2.03 eV for $BiFeO_3$. In the case of the $Bi_2S_3/BiFeO_3$ heterojunction, the absorption edge shifts towards the visible light region with increased intensity, compared to $BiFeO_3$. This shift signifies the photosensitization effect of Bi_2S_3 on $BiFeO_3$, where Bi_2S_3 enhances the absorption of visible light by the composite material. These findings emphasize the significant role of Bi_2S_3 in expanding the spectral range of light absorption by the $BiFeO_3$ photocatalyst, thereby enhancing its photocatalytic potential.

The photocatalytic activity of the $Bi_2S_3/BiFeO_3$ heterojunction with varying proportions of Bi_2S_3 was investigated through the photodegradation of the organic pollutant MG, as illustrated in Figure 6a. In the absence of the catalyst, the rate of photodegradation of MG under visible light ($\lambda > 420$ nm) irradiation remained below 40% within 2 h. However, when the $Bi_2S_3/BiFeO_3$ heterojunction catalysts were introduced, notable improvements were observed. MG was completely photodegraded after 60 min, demonstrating the enhanced photocatalytic efficiency of these materials. The photocatalytic performance followed the following order: $Bi_2S_3/BiFeO_3$ (1:3) > $Bi_2S_3/BiFeO_3$ (1:2) > $Bi_2S_3/BiFeO_3$ (1:4) > $BiFeO_3$. In particular, $Bi_2S_3/BiFeO_3$ (1:3) exhibited the highest photocatalytic efficiency, with approximately 99% of MG molecules decomposed within 60 min under visible light. This result underscores the significance of the optimal proportion between $BiFeO_3$ and Bi_2S_3 for enhanced photocatalytic activity. Scanning electron microscopy revealed that $Bi_2S_3/BiFeO_3$ (1:3) possessed a more uniform morphology, further supporting the superior photocatalytic performance observed. Moreover, full-wavelength scanning in Figure 6b highlights the degradation process: as MG was degraded by the addition of the

$Bi_2S_3/BiFeO_3$ photocatalyst, the UV–visible absorption peak gradually weakened until it disappeared. During this weakening, there was a noticeable blue shift in the UV–visible absorption peak, indicating the generation of intermediate products in the optical degradation process, which, however, did not impede the degradation process. These findings underscore the substantial photocatalytic advantage of the $Bi_2S_3/BiFeO_3$ heterojunction, particularly when the proportion is optimized, and the importance of uniform particle surfaces in enhancing photocatalytic activity.

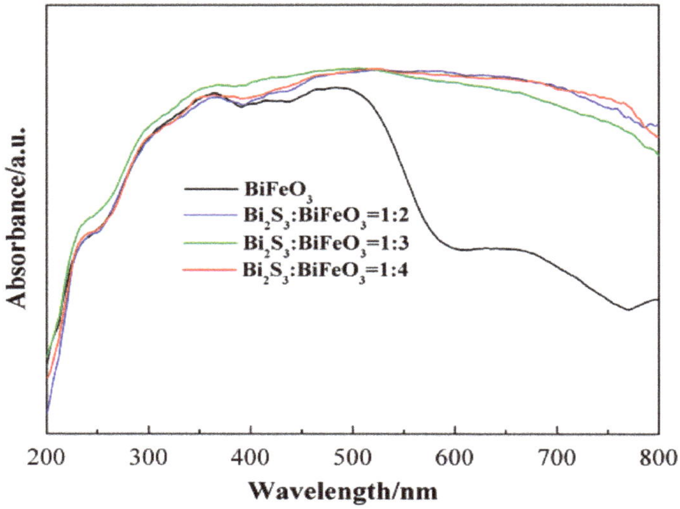

Figure 5. UV-vis diffuse reflection spectra of pure $BiFeO_3$ and the different proportions of $Bi_2S_3/BiFeO_3$.

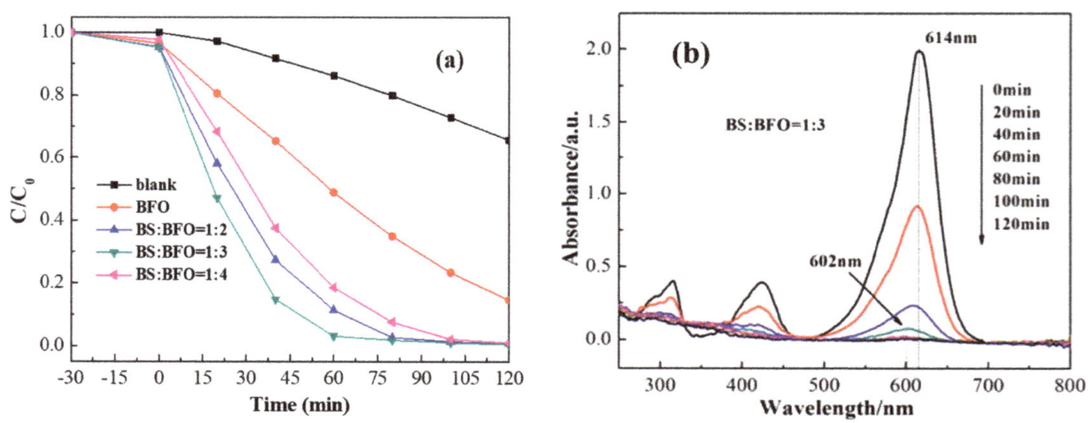

Figure 6. (**a**) Photocatalytic degradation of MG under visible-light irradiation using $Bi_2S_3/BiFeO_3$ heterojunctions as the catalyst; (**b**) the full-wavelength scanning of $Bi_2S_3/BiFeO_3$(1:3) heterojunction photocatalyst.

As shown in Figure 7a, Bi_2S_3/BiFeO (1:3) shows pH-dependent photocatalytic activity toward MG degradation, where the heterojunction catalyst exhibits excellent catalytic activity, with a pH around 6. An acidic or alkaline condition brings a negative effect toward

MG degradation, which may be attributed to the fact that the heterojunction catalyst is not stable enough against acid or alkaline corrosion. We also investigate the influence of catalyst dosage on the MG photo degradation, and the results are shown in Figure 7b. It is easy to find that, with the increase in the amount of catalyst used, the catalytic activity of heterojunctions toward MG degradation increases, which may be attributed to the increased catalyst amount, which could provide a more active side for oxygen activation to generate more reactive oxygen species. It should be noted here that an excess amount of catalyst would bring a light shading effect, which may bring negative effects toward pollutant removal. As can be seen from Figure 7b, limited activity improvement is found with the amount of catalyst increasing from 75 mg to 100 mg, and we believe that further increasing the amount of catalyst would not bring much improvement in catalytic activity. Thus, we then chose 100 mg of catalyst as the optimized condition to carry out other experiments.

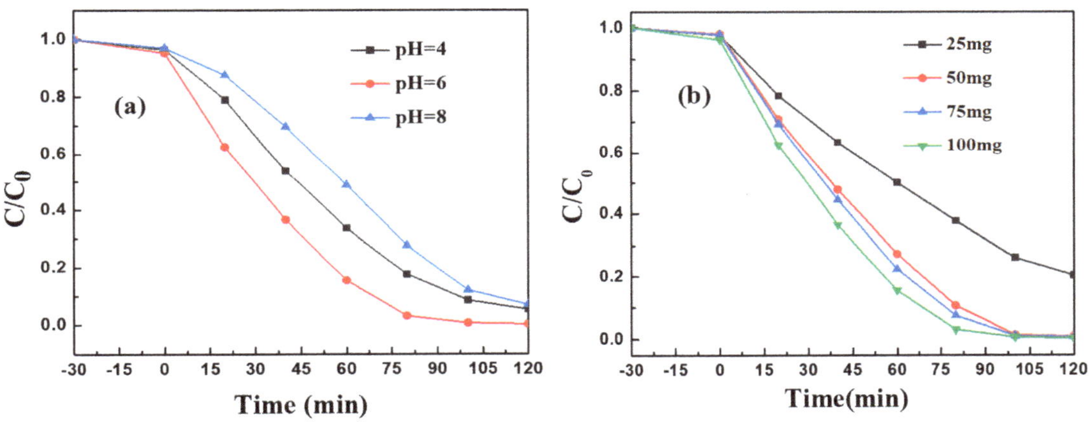

Figure 7. Photocatalytic performance of Bi_2S_3/BiFeO (1:3) toward MG degradation at (**a**) various pH conditions. (**b**) Effects of catalyst dosage on MG degradation.

To gain a deeper understanding of the photodegradation mechanism, various sacrificial agents were introduced to elucidate the roles of specific reactive groups in the reaction process. As depicted in Figure 8, when isopropanol (IPA) was introduced as a sacrificial agent to capture hydroxyl (•OH) radicals, the degradation of MG was almost negligible. This observation suggests that •OH radicals play a crucial role in the degradation process, as they are potent oxidants, capable of effectively oxidizing many organic pollutants. Conversely, when triethanolamine (TEOA) was added to capture the photo-generated holes (h^+), and 4-benzoquinone (BQ) was used to capture $•O_2^-$ radicals, the efficiency of MG degradation was significantly reduced. This reduction was particularly pronounced in the presence of TEOA. These results imply that, during the photodegradation reaction, $•O_2^-$ radicals and photo-generated holes (h^+) work in tandem to facilitate the degradation of organic pollutants. The •OH radical, being a robust oxidant capable of oxidizing a wide range of organic pollutants, is likely a key contributor to the degradation process. These findings align with the existing literature, further supporting the significance of •OH radicals in the photodegradation of organic pollutants.

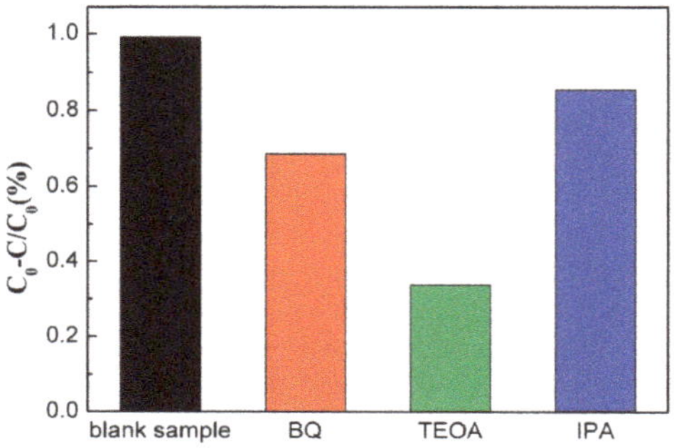

Figure 8. Effects of various scavengers on the photocatalytic efficiencies of $Bi_2S_3/BiFeO_3$ in MG degradation under visible-light irradiation.

To provide further authentication of the photocatalytic mechanisms at play, several key energy levels and oxidation–reduction potentials were considered (Figure 9). The band gap of $BiFeO_3$ was determined to be 2.05 eV, based on UV-Vis DRS spectra. Additionally, the conduction band of $BiFeO_3$ is at +0.21eV (E_{CB} = +0.21 eV) [45], and the valence band is at +2.26 eV (E_{VB} = +2.26 eV). In contrast, Bi_2S_3 has a conduction band at 0V (E_{CB} = 0 V) [46] and a valence band at +1.38 eV (E_{VB} = +1.38 eV) [47]. Notably, the conduction band of Bi_2S_3 is more negative than that of $BiFeO_3$. Based on the above energy-band structure analysis, the photogenerated electrons are transferred from the conduction band of Bi_2S_3 to the conduction band of $BiFeO_3$, due to the more negative conduction band of Bi_2S_3. This facilitates the generation of $\bullet O_2{}^-$ on the material surface by oxygen absorption, which can then directly oxidize the target contaminant MG. On the other hand, the oxidation–reduction potential of •OH is 2.38 V [48], significantly more positive than the valence band of $BiFeO_3$ (E_{VB} = +2.26 V) [49] and Bi_2S_3 (E_{VB} = +1.38 eV) [50]. Consequently, the photoexcited $Bi_2S_3/BiFeO_3$ generates holes that are unable to oxidize water to form hydroxyl radicals. Furthermore, the valence band energy of Bi_2S_3 is lower than that of $BiFeO_3$. Thus, the photogenerated holes in the valence band of $BiFeO_3$ are transferred to the valence band of Bi_2S_3. This energy transfer enables the direct oxidation of MG. Considering the properties of each substance in the reaction system, these insights suggest the procedures and mechanisms of the reaction [50].

$$Bi_2S_3 + hv \rightarrow h^+_{Bi_2S_3} + e^-_{Bi_2S_3} \quad (1)$$

$$BiFeO_3 + hv \rightarrow h^+_{BiFeO_3} + e^-_{BiFeO_3} \quad (2)$$

$$e^-_{Bi_2S_3} + BiFeO_3 \rightarrow e^-_{BiFeO_3} \quad (3)$$

$$e^-_{BiFeO_3} + O_2 \rightarrow \bullet O_2^- \quad (4)$$

$$\bullet O_2^- + MG \rightarrow products \quad (5)$$

$$h^+_{BiFeO_3} + Bi_2S_3 \rightarrow h^+_{Bi_2S_3} \quad (6)$$

$$h^+_{Bi_2S_3} + MG \rightarrow products \quad (7)$$

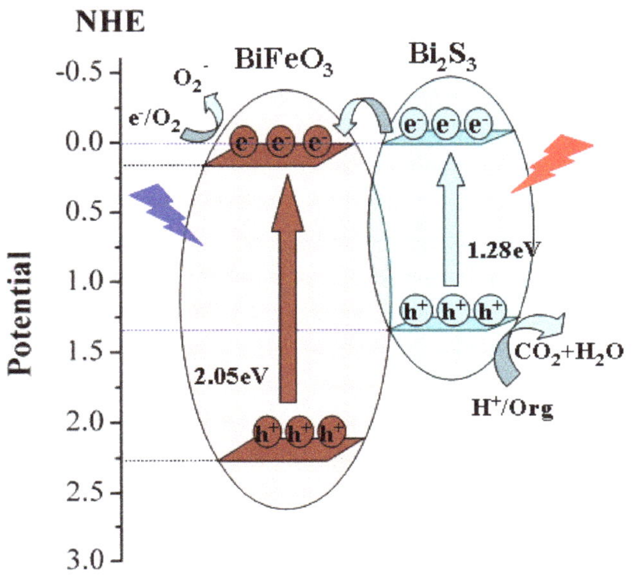

Figure 9. Schematic illustration of the $Bi_2S_3/BiFeO_3$ photocatalytic reaction process under visible-light irradiation.

3. Experimental

3.1. Material Synthesis

$BiFeO_3$ was prepared according to the reported method, with some modification [51]. Typically, 5 mM of bismuth nitrate pentahydrate ($Bi(NO_3)_3\bullet5H_2O$) and 5 mM of ferric nitrate ($Fe(NO_3)_3\bullet9H_2O$) were added into 35 mL of ethylene glycol monomethyl ether, to form a fine mixture. The solution was vigorously stirred until it reached a state of homogeneity, resulting in a transparent and stable solution at room temperature. Subsequently, the sample was left to dry for 24 h at 80 °C, yielding a gel-like powder. This powder was then carefully ground and subjected to heating in a muffle furnace at 550 °C for 4 h. After the heating process, the sample was allowed to cool to room temperature, and its final form was obtained through additional grinding.

Following this initial preparation, the next step involved the synthesis of $Bi_2S_3/BiFeO_3$ through an in situ growth process on the $BiFeO_3$ substrate under hydrothermal conditions, as depicted in Figure 10.

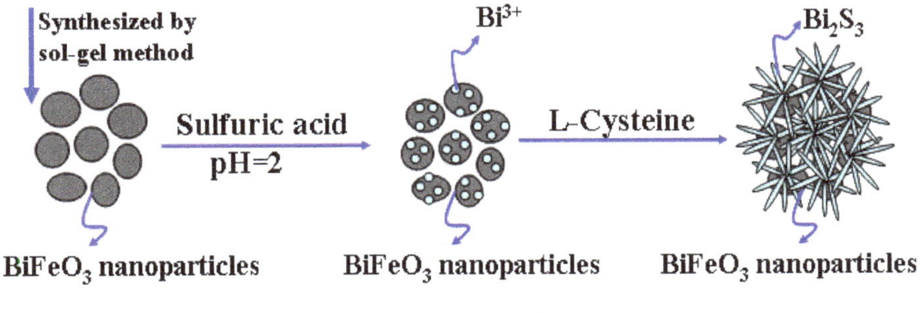

Figure 10. Schematic illustration of the $Bi_2S_3/BiFeO_3$ synthetic processes.

Commencing with three clean beakers, each received 35 mL of deionized water. Sequentially, L-cysteine was introduced at concentrations of 6 mM, 4 mM, and 3 mM into the respective beakers. Complete dissolution of L-cysteine was ensured. Sulfuric acid was employed to carefully adjust the pH level to 2, while maintaining precision in this process. With meticulous care, an equivalent quantity of $BiFeO_3$ (4 mM) was gradually added to each solution. Stirring continued until the $BiFeO_3$ was uniformly dispersed within the mix. This stirring process endured for an additional 30 min. The mixed solutions were then carefully transferred to a 50 mL autoclave. The autoclave was subjected to a hydrothermal reaction at 150 °C for a duration of 12 h. Following completion of the reaction, the contents underwent filtration, and the resulting material was thoroughly washed with deionized water and absolute ethanol. The $Bi_2S_3/BiFeO_3$ heterojunction materials obtained from the process were dried at 80 °C for 12 h under ambient air conditions. The final samples were categorized as $Bi_2S_3/BiFeO_3$ (1:2), $Bi_2S_3/BiFeO_3$ (1:3), and $Bi_2S_3/BiFeO_3$ (1:4), respectively, based on the ratios used during their preparation.

3.2. Characterization

The crystalline phase of the samples was investigated using X-ray diffraction (XRD, D8 Advance, Bruker, Mannheim, Germany), using graphite monochromatized Cu-Kα (λ = 1.5406 Å) radiation at 40 kv and 40 mA with a scan rate of 2 °/min between 10° and 70°. The UV–vis diffuse reflectance spectra (DRS) of the powders were collected using a U-3900/3900H UV-vis spectrophotometer (UV-vis DRS, U-3900/3900H, Hitachi High-Technologies, Tokyo, Japan) equipped with an integrating sphere attachment. The general morphologies and micro-area chemical analysis of the samples were obtained using scanning electron microscopy images (SEM, Sirion 200, FEI, Eindhoven, The Netherlands) and energy-dispersive X-ray spectroscopy (EDS) with a scanning voltage of 5.00 kV. Further microstructure and crystallinity of the sample was analyzed using transmission electron microscopy (TEM, Tecnai F20, FEI, Hillsboro, OR, USA) and high-resolution transmission electron microscopy (HRTEM) observation. The chemical states of as-prepared photocatalysts were tested using monochromated Al-Kα X-ray photoelectron spectroscopy (XPS, Axis Ultra DLD) at a residual gas pressure of less than 10^{-8} Pa.

Electrochemical measurements were performed on a CHI 660D electrochemical workstation (Shanghai Chenhua, Shanghai, China) by using a standard three-electrode system with a working electrode. (The working electrode was prepared by dip-coating on a 3 cm × 1 cm fluorine tin oxide (FTO) glass electrode, with the films drying under room condition.) A graphite electrode was used as the electrodes, and a standard calomel electrode in saturated KCl was used as a reference electrode. A total of 0.5 M sodium sulfate was used as the electrolyte solution for the photocurrent test, and 0.5 M sodium sulfate, 2.5 mM of potassium ferricyanide and potassium ferrocyanide were adopted as the supporting electrolyte for Ac impedance.

3.3. Photocatalytic Activity Evaluation

The effectiveness of the $Bi_2S_3/BiFeO_3$ heterojunction catalysts in photocatalysis was assessed through the degradation of Malachite Green oxalate (MG) under visible light conditions (λ > 420 nm). These photocatalytic reactions were conducted using simulated solar light with an intensity equivalent to 300 W, generated by a Xe-lamp system from Asahi Spectra Co., Ltd. (Torrance, CA, USA) The initial concentration of the Malachite Green solution was set at 20 mg/L. The photocatalytic procedure was performed in a homemade reactor with catalytic conditions quite similar to the reported method; typically, 0.1 g of the catalyst was dispersed into 100 mL of the Malachite Green oxalate (MG) solution [52]. To establish a balance between the adsorption and desorption of Malachite Green on the catalyst surface, the mixture was magnetically stirred for 30 min in the dark.

At specified irradiation time intervals, approximately 5 mL of the suspension was extracted every 10 min. These samples were subsequently centrifuged to obtain clear solutions, which were then analyzed using a UV–vis spectrophotometer (UV-vis DRS, U-

3900/3900H, Hitachi High-Technologies, Tokyo, Japan). The maximum absorbance of the solution was determined at 614 nm. To provide a basis for comparison, parallel experiments using pure $BiFeO_3$ were conducted alongside the $Bi_2S_3/BiFeO_3$ experiments, allowing for an evaluation of the relative performance of $Bi_2S_3/BiFeO_3$, compared to pure $BiFeO_3$.

4. Conclusions

In summary, we have shown the first instance of fabricating nanoflower-shaped $Bi_2S_3/BiFeO_3$ heterojunctions via a simple sol–gel process. The formed type II heterojunction is found capable of providing a powerful built-in field to accelerate electron–hole spatial separation, which leads to the enhanced reactive oxygen species generation. As a result, the as-prepared $Bi_2S_3/BiFeO_3$ heterojunctions exhibit excellent photocatalytic performance toward MG degradation, where more than 99% of MG is removed within 2 h of photocatalysis. The innovative design shown in this work may offer new viewpoints in designing highly efficient photocatalysts for environmental purification. For real water purification, however, we still need to immobilize the catalyst, so as to prevent catalyst loss. The immobilization of the as-prepared $Bi_2S_3/BiFeO_3$ on a light-weight carrier, for example loofah, to construct floating materials, is another possibility for its practical application.

Author Contributions: Conceptualization, L.Z; Methodology, R.Z. and P.Z.; Validation, L.Z.; Formal analysis, R.Z.; Investigation, R.Z. and P.Z.; Data curation, R.Z. and L.Z.; Writing—original draft, R.Z. and X.T.; Writing—review & editing, X.T. and Z.Z.; Supervision, X.T. and Z.Z.; Project administration, X.T.; Funding acquisition, X.T. All authors have read and agreed to the published version of the manuscript.

Funding: This work was financially supported by the National Natural Science Foundation of China (No. 52060019 and 22276087), the Key Project of Jiangxi Provincial Natural Science Foundation (No. 20202ACBL204015), the Key Research and Development Project of Jiangxi Province (No. 20212BBG73030).

Institutional Review Board Statement: Not applicable.

Informed Consent Statement: Not applicable.

Data Availability Statement: Data are contained within the article.

Conflicts of Interest: The authors declare no conflict of interest.

References

1. Zeng, Z.; Fan, Y.; Quan, X.; Yu, H.; Chen, S.; Zhang, S. Energy-transfer-mediated oxygen activation in carbonyl functionalized carbon nitride nanosheets for high-efficient photocatalytic water disinfection and organic pollutants degradation. *Water Res.* **2020**, *177*, 115798. [CrossRef]
2. Fung, C.-M.; Er, C.-C.; Tan, L.-L.; Mohamed, A.R.; Chai, S.-P. Red phosphorus: An up-and-coming photocatalyst on the horizon for sustainable energy development and environmental remediation. *Chem. Rev.* **2022**, *122*, 3879–3965. [CrossRef]
3. Li, S.; Lin, Y.H.; Zhang, B.P.; Wang, Y.; Nan, C.W. Controlled fabrication of $BiFeO_3$ uniform microcrystals and their magnetic and photocatalytic behaviors. *J. Phys. Chem. C* **2010**, *114*, 2903–2908. [CrossRef]
4. Gao, F.; Chen, X.Y.; Yin, K.B.; Dong, S.; Ren, Z.F.; Yuan, F.; Yu, T.; Zou, Z.G.; Liu, J.M. Visible-Light Photocatalytic Properties of Weak Magnetic $BiFeO_3$Nanoparticles. *Adv. Mater.* **2007**, *19*, 2889–2892. [CrossRef]
5. Yang, X.; Zhang, Y.; Xu, G.; Wei, X.; Ren, Z.; Shen, G.; Han, G. Phase and morphology evolution of bismuth ferrites via hydrothermal reaction route. *Mater. Res. Bull.* **2013**, *48*, 1694–1699. [CrossRef]
6. Wang, X.; Lin, Y.; Ding, X.; Jiang, J. Enhanced visible-light-response photocatalytic activity of bismuth ferrite nanoparticles. *J. Alloys Compd.* **2011**, *509*, 6585–6588. [CrossRef]
7. Zhang, J.; Gondal, M.A.; Wei, W.; Zhang, T.; Xu, Q.; Shen, K. Preparation of room temperature ferromagnetic $BiFeO_3$ and its application as an highly efficient magnetic separable adsorbent for removal of Rhodamine B from aqueous solution. *J. Alloys Compd.* **2012**, *530*, 107–110. [CrossRef]
8. Xian, T.; Yang, H.; Dai, J.F.; Wei, Z.Q.; Ma, J.Y.; Feng, W.J. Photocatalytic properties of $BiFeO_3$ nanoparticles with different sizes. *Mater. Lett.* **2011**, *65*, 1573–1575. [CrossRef]
9. He, J.; Guo, R.; Fang, L.; Dong, W.; Zheng, F.; Shen, M. Characterization and visible light photocatalytic mechanism of size-controlled $BiFeO_3$ nanoparticles. *Mater. Res. Bull.* **2013**, *48*, 3017–3024. [CrossRef]
10. Liu, Z.; Qi, Y.; Lu, C. High efficient ultraviolet photocatalytic activity of $BiFeO_3$ nanoparticles synthesized by a chemical coprecipitation process. *J. Mater.Sci. Mater. Electron.* **2010**, *21*, 380–384. [CrossRef]

11. Soltani, T.; Entezari, M.H. Photolysis and photocatalysis of methylene blue by ferrite bismuth nanoparticles under sunlight irradiation. *J. Mol. Catal. A Chem.* **2013**, *377*, 197–203. [CrossRef]
12. Mohan, S.; Subramanian, B. A strategy to fabricate bismuth ferrite ($BiFeO_3$) nanotubes from electro spun nanofibers and their solar light-driven photocatalytic properties. *RSC Adv.* **2013**, *3*, 23737–23744. [CrossRef]
13. Zhang, X.; Liu, H.; Zheng, B.; Lin, Y.; Liu, D.; Nan, C.W. Photocatalytic and magnetic behaviors observed in $BiFeO_3$ nanofibers by electrospinning. *J. Nanomater.* **2013**, *2013*, 917–948. [CrossRef]
14. Guo, R.; Fang, L.; Dong, W.; Zheng, F.; Shen, M. Enhanced photocatalytic activity and ferromagnetism in Gd Doped $BiFeO_3$ nanoparticles. *J. Phys. Chem. C* **2010**, *114*, 21390–21396. [CrossRef]
15. Zhang, Z.; Liu, H.; Lin, Y.; Wei, Y.; Nan, C.W.; Deng, X. Influence of La Doping onMagnetic and Optical Properties of Bismuth Ferrite Nanofibers. *J. Nanomater.* **2012**, *2012*, 238605. [CrossRef]
16. Feng, Y.N.; Wang, H.C.; Luo, Y.D.; Shen, Y.; Lin, Y.H. Ferromagnetic and photocatalytic behaviors observed in Ca-doped $BiFeO_3$ nanofibres. *J. Appl. Phys.* **2013**, *113*, 146101. [CrossRef]
17. Li, S.; Zhang, J.; Kibria, M.G.; Mi, Z.; Chaker, M.; Ma, D.; Nechache, R.; Rosei, F. Remarkably enhanced photocatalytic activity of laser ablated Au nanoparticle decorated $BiFeO_3$ nanowires under visible-light. *Chem. Commun.* **2013**, *49*, 5856–5858. [CrossRef]
18. Schultz, A.M.; Zhang, Y.; Salvador, P.A.; Rohrer, G.S. Effect of Crystal and Domain Orientation on the Visible-Light Photochemical Reduction of Ag on $BiFeO_3$. *ACS Appl. Mater. Inter.* **2011**, *3*, 1562–1567. [CrossRef]
19. Niu, F.; Chen, D.; Qin, L.; Zhang, N.; Wang, J.; Chen, Z.; Huang, Y. Facile Synthesis of Highly Efficient p–n Heterojunction $CuO/BiFeO_3$ Composite Photocatalysts with Enhanced Visible-Light Photocatalytic Activity. *ChemCatChem* **2015**, *7*, 3279–3289. [CrossRef]
20. Zhang, Y.; Schultz, A.M.; Salvador, P.A.; Rohrer, G.S. Spatially selective visible light photocatalytic activity of $TiO_2/BiFeO_3$ heterostructures. *J. Mater. Chem.* **2011**, *21*, 4168–4174. [CrossRef]
21. Zhu, A.; Zhao, Q.; Li, X.; Shi, Y. $BiFeO_3/TiO_2$ Nanotube Arrays Composite Electrode: Construction, Characterization, and Enhanced Photoelectrochemical Properties. *ACS Appl. Mater. Interfaces* **2014**, *6*, 671–679. [CrossRef] [PubMed]
22. Li, T.; Shen, J.; Li, N.; Ye, M. Hydrothermal preparation, characterization and enhanced properties of reduced grapheme-BiFeO3 nanocomposite. *Mater. Lett.* **2013**, *91*, 42–44. [CrossRef]
23. Li, Z.; Shen, Y.; Guan, Y.; Hu, Y.; Lin, Y.; Nan, C.W. Bandgap engineering and enhanced interface coupling of graphene–$BiFeO_3$ nanocomposites as efficient photocatalysts under visible light. *J. Mater. Chem. A* **2014**, *2*, 1967–1973. [CrossRef]
24. Pineda, E.; Nicho, M.E.; Nair, P.K.; Hu, H. Optoelectronic properties of chemically deposited Bi_2S_3 thin films and the photovoltaic performance of Bi_2S_3/P_3OT solar cells. *Sol. Energy* **2012**, *86*, 1017–1022. [CrossRef]
25. Cao, J.; Xu, B.; Lin, H.; Luo, B.; Chen, S. Novel Bi_2S_3-sensitized BiOCl with highly visible light photocatalytic activity for the removal of rhodamine B. *Catal. Commun.* **2012**, *26*, 204–208. [CrossRef]
26. Cheng, H.; Huang, B.; Qin, X.; Zhang, X.; Dai, Y. A controlled anion exchange strategy to synthesize Bi_2S_3 nanocrystals/BiOCl hybrid architectures with efficient visible light photoactivity. *Chem. Commun.* **2012**, *48*, 97–99. [CrossRef] [PubMed]
27. Chen, P.; Gu, L.; Cao, X. From single ZnO multipods to heterostructured ZnO/ZnS, ZnO/ZnSe, ZnO/Bi_2S_3 and ZnO/Cu_2S multipods: Controlled synthesis and tunable optical and photoelectrochemical properties. *CrystEngComm* **2010**, *12*, 3950–3958. [CrossRef]
28. Calva-Yanez, J.C.; Rincon, M.E.; Solís de la Fuente, M.; Alvarado-Tenorio, G. Structural and photoelectrochemical characterization of MWCNT-TiO_2 matrices sensitized with Bi_2S_3. *J. Solid State Electrochem.* **2013**, *17*, 2633–2641. [CrossRef]
29. Jana, A.; Bhattacharya, C.; Datta, J. Enhanced photoelectrochemical activity of electro-synthesized CdS-Bi_2S_3 composite films grown with self-designed cross-linked structure. *Electrochim. Acta* **2010**, *55*, 6553–6562. [CrossRef]
30. He, H.; Berglund, S.P.; Xiao, P.; Chemelewski, W.D.; Zhang, Y.; Mullins, C.B. Nanostructured Bi_2S_3/WO_3 heterojunction films exhibiting enhanced photoelectrochemical performance. *J. Mater. Chem. A* **2013**, *1*, 12826–12834. [CrossRef]
31. Ata, S.; Shaheen, I.; Aslam, H.; Mohsin, I.U.; Alwadai, N.; Al Huwayz, M.; Iqbal, M.; Younas, U. Barium and strontium doped La-based perovskite synthesis via sol-gel route and photocatalytic activity evaluation for methylene blue. *Results Phys.* **2023**, *45*, 106235. [CrossRef]
32. Jiao, S.; Zhao, Y.; Bi, M.; Bi, S.; Li, X.; Wang, B.; Li, C.; Dong, Y. Removal of Methylene Blue from Water by $BiFeO_3$/Carbon Fibre Nanocomposite and Its Photocatalytic Regeneration. *Catalysts* **2018**, *8*, 267. [CrossRef]
33. Tseng, W.J.; Lin, R.D. $BiFeO_3/\alpha\text{-}Fe_2O_3$ core/shell composite particles for fast and selective removal of methyl orange dye in water. *J. Colloid Interface Sci.* **2014**, *428*, 95–100. [CrossRef]
34. Qin, K.; Zhao, Q.; Yu, H.; Xia, X.; Li, J.; He, S.; Wei, L.; An, T. A review of bismuth-based photocatalysts for antibiotic degradation: Insight into the photocatalytic degradation performance, pathways and relevant mechanisms. *Environ. Res.* **2021**, *199*, 111360. [CrossRef] [PubMed]
35. Song, S.; Xing, Z.; Zhao, H.; Li, Z.; Zhou, W. Recent advances in bismuth-based photocatalysts: Environment and energy applications. *Green Energy Environ.* **2023**, *8*, 1232–1264. [CrossRef]
36. *JCPDS 20-0169*; Elements of X-ray Diffraction, B. D. Cullinity, 2nd ed., Addison-Wesley, Reading, MA, 1978. JCPDS (Joint Committee on Powder Diffraction Standards): Newtown Square, PA, USA, 2023.
37. Dai, Y.; Wang, Y.; Yao, J.; Wang, Q.; Liu, L.; Chu, W.; Wang, G. Phosgene-Free Synthesis of Phenyl Isocyanate by Catalytic Decomposition of Methyl N-Phenyl Carbamate over Bi_2O_3 Catalyst. *Catal. Lett.* **2008**, *123*, 307–316. [CrossRef]

38. Zeng, Z.; Ye, F.; Deng, S.; Fang, D.; Wang, X.; Bai, Y.; Xiao, H. Accelerated organic pollutants mineralization in interlayer confined single Pt atom photocatalyst for hydrogen recovery. *Chem. Eng. J.* **2022**, *444*, 136561. [CrossRef]
39. Lu, X.; Pu, F.; Xia, Y.; Huang, W.; Li, Z. Facile fabrication of porous thin films of Bi_2O_3/Bi_2S_3 nanocomposite semiconductors at gas/liquid interface and their photoelectrochemical performances. *Appl. Surface Sci.* **2014**, *299*, 131–135. [CrossRef]
40. Wang, N.; Zhu, L.; Wang, M.; Wang, D.; Tang, H. Sono-enhanced degradation of dye pollutants with the use of H_2O_2 activated by Fe_3O_4 magnetic nanoparticles as peroxidase mimetic. *Ultrason. Sonochem.* **2010**, *17*, 526–533. [CrossRef]
41. Jiang, J.; Zou, J.; Anjum, M.N.; Yan, J.; Huang, L.; Zhang, Y.; Chen, J. Synthesis and characterization of wafer-like $BiFeO_3$ with efficient catalytic activity. *Solid State Sci.* **2011**, *13*, 1779–1785. [CrossRef]
42. Zhong, Y.; Shi, J.; Li, K.; Guo, H.; Yan, L.; Zeng, Z. Silicotungstic acid as electron transfer mediator in Z-scheme g-C_3N_4/H4Si$W_{12}O_{40}$/Ag_3PO_4: Accelerated charge interface transportation for high-efficient photocatalysis. *Process Saf. Environ.* **2023**, *171*, 188–199. [CrossRef]
43. Xiao, H.; Zhang, Q.; Ahmad, M.; Dong, S.; Zhang, Y.; Fang, D.; Wang, X.; Peng, H.; Lei, Y.; Wu, G.; et al. Carbonate mediated hole transfer boosting the photocatalytic degradation of organic pollutants over carbon nitride nanosheets. *Sep. Purif. Technol.* **2023**, *306*, 122580. [CrossRef]
44. Marchon, B.; Carrazza, J.; Heinemann, H.; Somorjai, G.A. TPD and XPS studies of O_2, CO_2, and H_2O adsorption on clean polycrystalline graphite. *Carbon* **1988**, *26*, 507–514. [CrossRef]
45. Xu, W.; Fang, J.; Chen, Y.; Lu, S.; Zhou, G.; Zhu, X.; Fang, Z. Novel heterostructured $Bi_2S_3/Bi_2Sn_2O_7$ with highly visible light photocatalytic activity for the removal of rhodamine B. *Mater. Chem. Phys.* **2015**, *154*, 30–37. [CrossRef]
46. Humayun, M.; Zada, A.; Li, Z.; Xie, M.; Zhang, X.; Qu, Y.; Raziq, F.; Jing, L. Enhanced visible-light activities of porous $BiFeO_3$ by coupling with nanocrystalline TiO_2 and mechanism. *Appl. Catal. B Environ.* **2016**, *180*, 219–226. [CrossRef]
47. Li, X.; Ye, F.; Zhang, H.; Ahmad, M.; Zeng, Z.; Wang, S.; Wang, S.; Gao, D.; Zhang, Q. Ternary rGO decorated $W_{18}O_{49}$ @g-C3N4 composite as a full-spectrum-responded Z-scheme photocatalyst for efficient photocatalytic H_2O_2 production and water disinfection. *J. Environ. Chem. Eng.* **2023**, *11*, 110329. [CrossRef]
48. Kong, L.; Jiang, Z. Unusual reactivity of visible-light-responsive AgBr-BiOBr heterojunction photocatalysts. *J. Catal.* **2012**, *293*, 116–125. [CrossRef]
49. Bard, A.J.; Faulkner, L.R. *Electrochemical Methods: Fundamental and Application*; John Wiley & Sons: New York, NY, USA, 1980.
50. Li, X.; Zhu, J. Comparative study on the mechanism in photocatalytic degradation of different-type organic dyes on SnS_2 and CdS. *Appl. Catal. B Environ.* **2012**, *123–124*, 174–181. [CrossRef]
51. Xu, M.L.; Lu, M.; Qin, G.Y.; Wu, X.M.; Yu, T.; Zhang, L.N.; Li, K.; Cheng, X.; Lan, Y.Q. Piezo-Photocatalytic Synergy in $BiFeO_3$@COF Z-Scheme Heterostructures for High-Efficiency Overall Water Splitting. *Angew. Chem. Int. Ed.* **2022**, *61*, e202210700. [CrossRef]
52. Tang, J.; Wang, R.; Liu, M.; Zhang, Z.; Song, Y.; Xue, S.; Zhao, Z.; Dionysiou, D.D. Construction of novel Z-scheme Ag/$FeTiO_3$/Ag/$BiFeO_3$ photocatalyst with enhanced visible-light-driven photocatalytic performance for degradation of norfloxacin. *Chem. Eng. J.* **2018**, *351*, 1056–1066. [CrossRef]

catalysts

MDPI

Article

S-Scheme $WO_3/SnIn_4S_8$ Heterojunction for Water Purification: Enhanced Photocatalytic Performance and Mechanism

Pingfan Xu [1,2], Runqiu Zhang [1], Jiarong Gong [1], Yaofa Luo [1], Yihua Zhuang [1,*] and Peikun Zhang [1,*]

1 School of Advanced manufacturing, Fuzhou University, Jinjiang 362200, China; xupingfan@fzu.edu.cn (P.X.); zrq980916@163.com (R.Z.); jiarong_gong@163.com (J.G.); luoyaofa@fzu.edu.cn (Y.L.)

2 Key Laboratory of Ecological Environment and Information Atlas, Fujian Provincial University, Putian 351100, China

* Correspondence: yhzhuang@fzu.edu.cn (Y.Z.); pkzhang@fzu.edu.cn (P.Z.)

Abstract: Photocatalysis is a promising technology for removing micropollutants in water. However, developing efficient and stable catalysts remains a challenge. In this work, a novel step-scheme (S-scheme) heterojunction of $WO_3/SnIn_4S_8$ (WSI) was constructed through the combined process of in situ precipitation with hydrothermal synthesis to simultaneously realize photocatalytic degradation of bisphenol A(BPA) and reduction of Cr(VI) in contaminated water. Results showed that the WSI S-scheme heterojuction has a synergistic effect for the removal of BPA and Cr(VI). An optimum case of the WSI-12% heterojunction exhibited the highest photocatalytic efficiency in the degradation of BPA under visible light, which is ca. 2.5 and 3.8 times more than the pure WO_3 and SIS, respectively. The enhanced photocatalytic activity is attributed to the formation of the WSI S-scheme heterojunctions which facilitate the spatial separation of charge carriers and preserve strong photoredox ability. Further, the S-scheme mechanism of enhanced photocatalysis was examined by the radical-trapping experiment and ESR, and superoxide and hydroxyl radicals were determined to be the major reactive oxygen species responsible for BPA degradation and Cr(VI) reduction by WSI. This work provides a novel strategy for tailoring high-performance S-scheme heterojunctions and shows the promising application in purifying wastewater with complex pollutants.

Keywords: $WO_3/SnIn_4S_8$ heterostructure; bisphenol A; hexavalent chromium; photocatalytic degradation

1. Introduction

The increasing worldwide contamination of freshwater systems is a considerable environmental problem facing humanity in the twenty-first century. One cause of water quality degradation is micropollutants, including both inorganic and organic pollutants, such as toxic metal ions and endocrine-disrupting compounds (EDCs) in aquatic systems [1,2]. Of these micropollutants, hexavalent chromium (Cr(VI)) and bisphenol A (BPA) are the most common micropollutants found in industrial wastewater. Cr(VI) emitted from tanneries, electroplating and dyeing processes, is known to be toxic and carcinogenic and can cause health problems, such as liver damage, pulmonary congestions, vomiting, and severe diarrhea [3]. BPA, a raw material in the production of epoxy resin and polycarbonate plastic, may cause endocrine disruption in humans [4]. Although these pollutants are present at trace levels in effluents, their adverse effects on human health and aquatic ecosystems are extremely harmful, particularly when they are present in complex mixtures.

Wastewater treatment plants are usually designed to remove suspended particles and nutrient pollutants, while the overwhelming majority of micropollutants may be discharged directly into the environment or accumulated in sludge disposal [5,6]. Recent technologies including adsorption [7,8], membrane processes [9], electrochemical precipitation [10], and advanced oxidation processes [11], have been developed for the removal of micropollutants. Unfortunately, practical application of these technologies has suffered from several limitations, such as high energy consumption, causing secondary pollution, and a narrow pH range for application. Therefore, the provision of economically and environmentally sustainable wastewater treatment for simultaneous removal of organic compounds and toxic metal ions remains a great challenge for wastewater treatment.

Heterogeneous photocatalysis is recognized as a promising technology for the simultaneous removal of organic pollutants and heavy metals. Over the past few decades, various types of single photocatalysts, such as TiO_2, WO_3, g-C_3N_4 and MoS_2, have been used for environmental remediation due to their efficient photocatalytic performance, high stability and environmental safety [12–15]. However, single photocatalysts cannot ensure both efficient light harvesting and strong redox ability. Another issue associated with single photocatalysts is the rapid recombination of charge pairs. To prevent these challenges, diverse strategies have been developed to enhance the photocatalytic activity of photocatalysts, including doping [16], metal loading [17], type II heterojunctions [18,19] and S-scheme heterojunctions [20]. Among them, the emerging S-scheme heterojunction, by combining two types of semiconductors, has been demonstrated as a promising strategy for energy conversion and environmental remediation, due to the effective spatial separation of photogenerated electron–hole pairs through the band alignment [21].

Generally, three design principles for S-scheme heterojunctions have been proposed in the previous literature. (1) An S-scheme heterojunction is composed of an oxidation photocatalyst (OP) and a reduction photocatalyst (RP). (2) The conduction band (CB) edge and Fermi level of RP should be higher than that of OP. (3) The work function of RP should be lower than that of OP. When a RP and an OP come into contact, the electron with a higher Fermi level in the RP drifts to the OP until their Fermi levels reach equilibrium. The redistribution of charge leads to the band bending and the formation of a built-in electric field, whose direction points from the RP to the OP. Upon light irradiation, the photogenerated electrons in the CB of the OP cross the heterojunction interface and recombine with the photogenerated holes in the VB of the RP under the action of the internal field. Ultimately, the photogenerated electrons of the RP and photogenerated holes of the OP can be preserved. Therefore, S-scheme heterojunctions have efficient spatial separation of charge carriers and maintain the strong photo-redox ability of both photogenerated electrons and holes.

An efficient S-scheme heterojunction for water purification requires the CB potential of the RP to be more negative than the reduction potential of oxygen to superoxide radicals ($\cdot O_2^-/O_2 = -0.33$ eV vs. NHE), while the VB position of the OP must be more positive than the oxidization potential of water to hydroxyl radicals ($\cdot OH/H_2O = +2.4$ eV vs. NHE) [22,23]. Therefore, the accumulated holes and electrons are able to oxidize the H_2O to $\cdot OH$ and reduce the O_2 to $\cdot O_2^-$. In order to design an S-scheme heterojunction, tungsten oxide (WO_3) is selected to couple with tin indium sulfide ($SnIn_4S_8$). WO_3, which has a band-gap energy of approximately 2.7 eV, can absorb visible photons and has a positive VB potential (ca. +3.1 eV vs. NHE), which can be used as an ideal OP that allows photoexcited holes to oxidize water into hydroxyl radicals [24,25]. $SnIn_4S_8$ (SIS) is another efficient visible-light photocatalyst with a relatively narrow band gap energy of approximately 2.0 eV [26]. The VB potential of SIS is located at approximately +1.30 eV vs. NHE, which is close to the CB position (+0.26 eV vs. NHE) of WO_3. After contact, the photogenerated electrons in the CB of WO_3 are inclined to recombine with holes in the VB position of SIS due to the Coulomb attraction. On the other hand, the CB potential of SIS is positioned at approximately −0.7 eV vs. NHE, which is more negative than the standard redox potential of $\cdot O_2^-/O_2$. Therefore, the photogenerated electron–hole pairs of WO_3/SIS S-scheme heterojunction are spatially separated in the two semiconductors, maintaining a strong redox potential.

Herein, a novel WSI S-scheme heterojunction was first fabricated via the combined process of in situ precipitation with hydrothermal synthesis to realize the perfect integration of the two semiconductors, as depicted in Scheme 1. Yellow WO_3 nanoparticles were synthesized by alcoholysis of WCl_6 followed by calcination of the blue WO_3 power. The as-synthesized WO_3 were dispersed in water and metal ions, and In^{3+} and Sn^{4+} were deposited in advance on the WO_3 surface, where the metal ions can be adsorbed onto the surface through electrostatic interaction. Then, sodium diethyldithiocarbamate (DDTC-2Na) as sulfur source was added into the precursor solution to grow SIS nanosheets in situ onto the WO_3 surface, forming a direct S-scheme heterojunction. The crystal phase, morphology and optical property of the as-prepared samples were characterized by XRD, SEM, HRTEM, UV-vis DRS and PL spectrum. In addition, the simultaneous photocatalytic degradation of BPA and Cr(VI) reduction was conducted under visible light irradiation to evaluate the synergistic effect. Furthermore, the enhanced photocatalytic activity of the S-scheme mechanism was proposed based on the reactive oxygen species trapping experiments and the ESR characterizations. It was found that the WSI S-scheme heterojunction exhibits enhanced photocatalytic activity. This work provides a novel direct S-scheme system for efficient degradation of pollutant and potential application in both environmental protection and energy conversion.

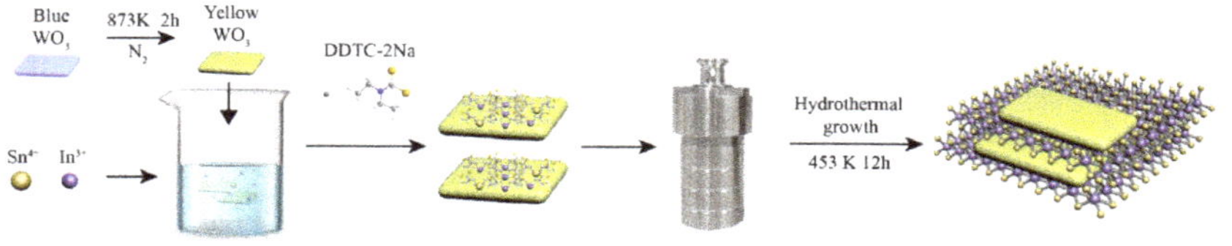

Scheme 1. Illustration of the fabrication of the WSI S-scheme heterojunction.

2. Results and Discussion

2.1. Structure and Morphology of WSI S-Scheme Heterojunction

The XRD patterns of the as-prepared samples showed a high degree of crystallinity, as shown in Figure 1. The distinct peaks of bare SIS at 14.3°, 23.4°, 27.5°, 33.3°, 43.7°, and 47.8° could be ascribed to (111), (220), (311), (400), (333), and (440) crystal planes of cubic SIS (JCPDS card # 42-1305, a = b = c = 10.7507 Å), which is consistent with a previous report on the crystal planes of cubic SIS [26]. In addition, the XRD patterns of the pure WO_3 are perfectly ascribed to the monoclinic phase of WO_3 (JCPDS card # 43-1035, a = 7.297 Å, b = 7.539 Å, c = 7.688 Å), which exhibits higher photocatalytic activity and is more stable than the orthorhombic and hexagonal phase [13]. Generally, the intensity of diffraction peaks is proportional to the amount of each phase present. In the present study, the characteristic peaks of WSI heterojunction are well consistent with the pattern of pure SIS. Compared with the peaks of SIS, the peak intensity of WO_3 is relatively weak owing to the low proportion of WO_3 in the heterojunction. No characteristic peaks of other impurities were observed, indicating that the WSI heterojunction was successfully prepared. Importantly, all diffraction peaks of SIS remain unchanged in their position and intensity after integration with WO_3 via hydrothermal synthesis, demonstrating that the introduction of WO_3 does not alter the crystal structure of SIS.

The morphologies of the as-prepared samples were observed by SEM and HRTEM. The SEM image shows that the as-synthesized WO_3 present a flat sheet-like structure (Figure 2a), which is similar to the previous report [27]. As shown in Figure 2b, SIS was grown in situ onto a WO_3 surface via hydrothermal synthesis to form a layered structure of WSI-12% heterojunction, indicating that a WO_3 nanosheet was embedded in SIS and the S-scheme heterojunction was successfully fabricated. The HRTEM image of the WSI-12% heterojunction was analyzed to further investigate the morphology and crystal structure. As shown in Figure 2c, the WO_3 nanosheet was wrapped uniformly through in situ synthesis of SIS, indicating that the direct S-scheme heterojunction was successfully synthesized. The intimate contact between WO_3 and SIS could promote electron transfer from WO_3 to SIS under visible light irradiation, which is the key factor for the successful construction of the S-scheme heterojunction. In addition, the lattice fringe of WO_3 and SIS was measured to further investigate the detailed information of the crystalline structure. As shown in Figure 2d, the lattice spacing of 0.52 nm, 0.384 nm, and 0.269 nm can be, respectively, assigned to the (110), (002), and (002) planes of WO_3, which agrees with the XRD analysis. For the SIS nanosheet, the obvious lattice spacing values were 0.267 nm and 0.62 nm, which match the distance between the (400) and (111) lattice planes of the cubic SIS structure, respectively. It should be noted that the WO_3 nanosheets were stuck in the middle of SIS, as shown in Figure 2e,f. The obvious layered structure indicates that the direct S-scheme heterostructure is successfully synthesized through a combined method of in situ precipitation with the hydrothermal synthesis.

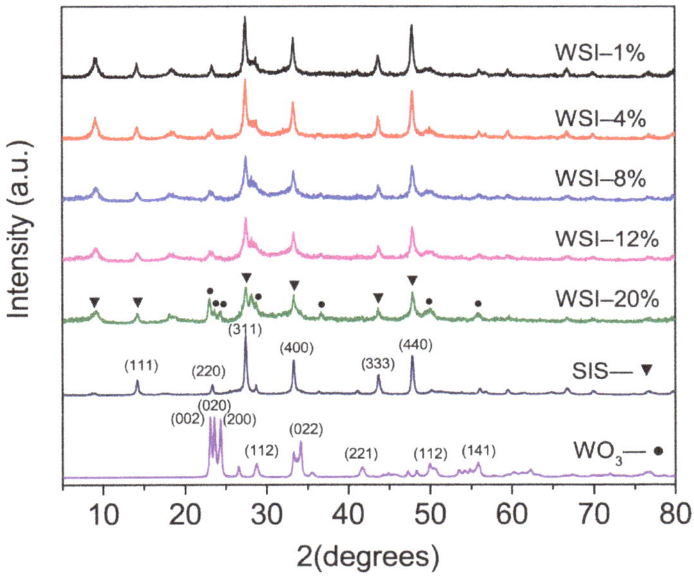

Figure 1. XRD patterns of SIS, WO_3, and WSI S-scheme heterojunction.

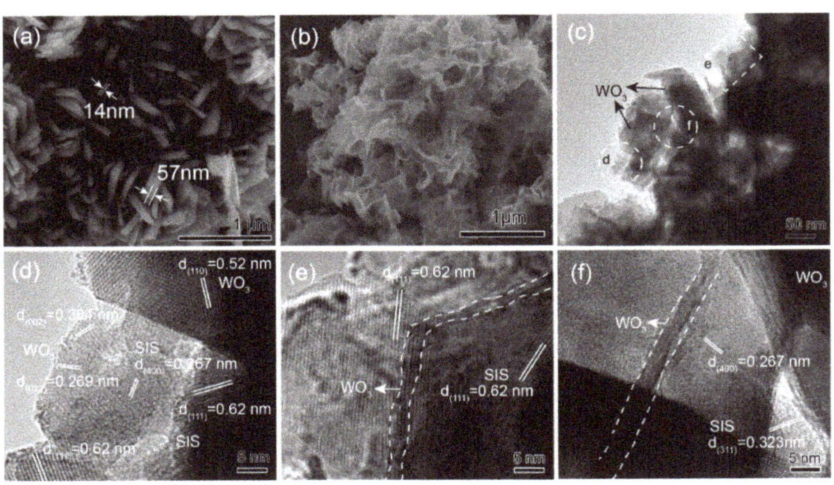

Figure 2. SEM images of (**a**) WO_3 and (**b**) WSI photocatalyst, (**c**–**f**) HRTEM image of WSI-12% photocatalyst.

The chemical states of SIS and the WSI-12% heterostructure were investigated by XPS analysis, as shown in Figure 3. The high-resolution Sn 3d and In 3d spectrum exhibit two peaks, 3d5/2 and 3d3/2, with spin-orbital splittings of 8.5 eV and 7.6 eV, respectively, indicating that the valence states in the composite are Sn^{4+} and In^{3+} [28,29]. The binding energy of Sn 3d in the WSI-12% heterostructure decreases compared with SIS, due to the electronegativity of oxygen atoms boned to Sn^{4+} after integration of WO_3. Similarly, the binding energy of In 3d in the WSI-12% heterostructure shifts from 444.5 eV to 444.4 eV under the influence of oxygen atoms in WO_3, as observed in $ZnIn_2S_4/TiO_2$, where the binding energy of In 3d is influenced by TiO_2 [30]. The S 2p spectrum shows a broad spectrum with a hump at higher binding energy, which is divided into two peaks, assigned to 2p3/2 and 2p1/2, respectively. The spin-orbital splitting of S 2p in the composites is 1.1 eV, indicating that the oxidation state of S is −2. Compared to S 2p in SIS and WSI-12%, the decrease of the binding energy is obvious under the influence of the W and O atoms, due to the increase in electron density of the S atom. Moreover, the peak area ratios and the spin-orbital splitting are 1.3 and 2.15 eV, respectively, indicating that the chemical state of W 4f in the composite is W^{6+} [31].

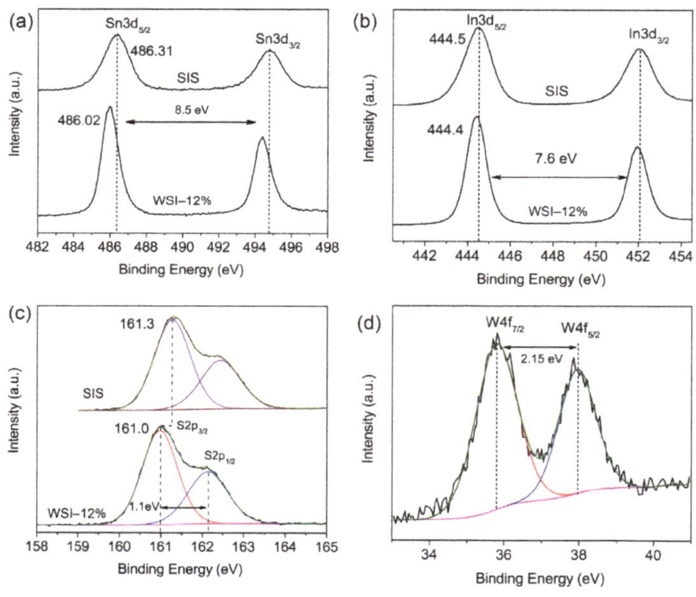

Figure 3. XPS spectrum of SIS and WSI-12% photocatalyst: (**a**) Sn 3d, (**b**) In 3d, (**c**) S 2p, (**d**) W 4f.

2.2. *Optical Properties and Charge Separation Efficiency of WSI S-Scheme Heterojunction*

The UV-visible diffuse reflectance spectra of pure SIS, WO_3 and the WSI heterojunction are shown in Figure 4a,b. Both the SIS and WO_3 semiconductors have strong absorption in the visible light because of their narrow band gaps. The visible light absorption property of WSI-12% is between that of SIS and WO_3, with a maximum absorption wavelength larger than 608 nm. The optical band gaps of SIS, WSI (1%, 4%, 8%, 12%, 20%) and WO_3 was estimated from the transmittance data using Tauc's plot ($\alpha h\upsilon \approx (h\upsilon\text{-}Eg)^{1/2}$) [32], which are 2.08, 2.17, 2.21, 2.26, 2.24, 2.29, and 2.83 eV, respectively. The band gap of the WSI heterojunction is located between that of SIS and WO_3, but the value does not increase proportionally with increasing WO_3 doping amount, indicating that the WSI heterojunction is not simply mixed.

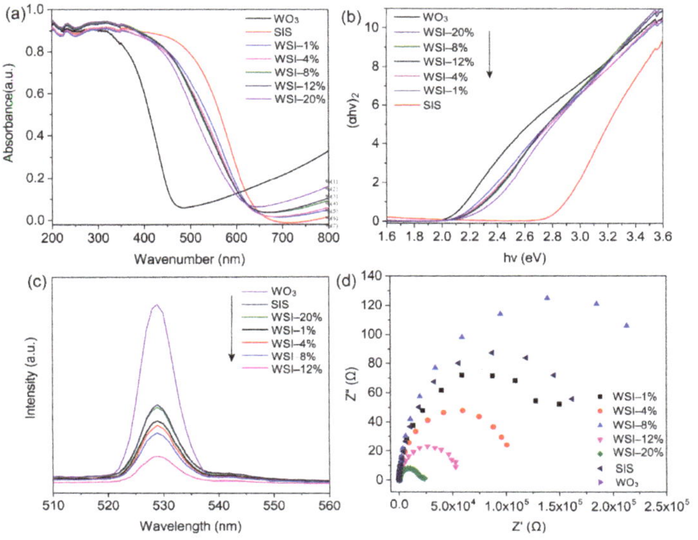

Figure 4. (**a**) DRS spectra of the as-prepared sample; (**b**) Plots of the $(\alpha hv)^2$ vs. photo energy (hv) for the as-prepared sample; (**c**) PL emission spectra (λ_{ex} = 350 nm) and (**d**) EIS spectra of the as-prepared sample. The direction of the arrows corresponds to the spectra of different samples.

The photocatalytic activity of a photocatalyst mainly depends on the separation of photogenerated carriers. The PL spectrum was utilized to investigate the separation efficiency of photogenerated electron–hole pairs. Generally, lower PL emission intensity indicates that the recombination rate of photogenerated electron–hole pairs is low, thus resulting in better photocatalytic activity [33]. As shown in Figure 4c, the PL intensity of WO_3 and SIS was stronger than that of the WSI heterojunction. The minimum PL intensity is ascribed to the WSI-12% heterojunction, suggesting that the construction of the S-scheme heterojunction facilitates the separation of photogenerated electron–hole pairs, which is expected to enhance photocatalytic activity.

Electrochemical Impedance Spectroscopy (EIS) of the photocatalysts was measured to further investigate the photogenerated electron transfer efficiency. It is reported that the smaller arc radius implies better conductivity and higher electron transfer efficiency [34]. As shown in Figure 4d, WSI-12% and WSI-20% have a relatively small arc radius compared with the bare WO_3 and SIS, indicating the highest conductivity and the fastest electron transfer efficiency are under visible light irradiation.

2.3. Photocatalytic Activity of WSI S-Scheme Heterojunction

The photocatalytic performance of the photocatalysts was evaluated by the degradation of BPA and reduction of Cr(VI), respectively, under visible light irradiation. As shown in Figure 5a, the photocatalytic efficiency of the WSI heterojunction is significantly higher than that of pure SIS and WO_3. Moreover, the photocatalytic degradation rate was calculated to evaluate the photocatalytic activity. The photocatalytic degradation rate fitted well with the pseudo-first-order kinetics. As shown in Table 1, the photocatalytic rate constant of BPA degradation by WSI-12% was estimated to be 0.2135 h^{-1}, which was 2.5 and 3.8 times that of WO_3 and SIS, respectively. For the reduction of Cr(VI), the apparent rate constant of WSI-12% was 6.5 and 1.5 times that of WO_3 and SIS, respectively. The enhanced photocatalytic activity can be attributed to the favorable band alignment between WO_3 and SIS, which forms a direct S-scheme heterojunction that greatly improves the separation of charge carriers in the two-phase interface. Hence, the construction of the S-scheme heterojunction shows great potential as an efficient photocatalyst for simultaneous oxidation of BPA and reduction of Cr(VI).

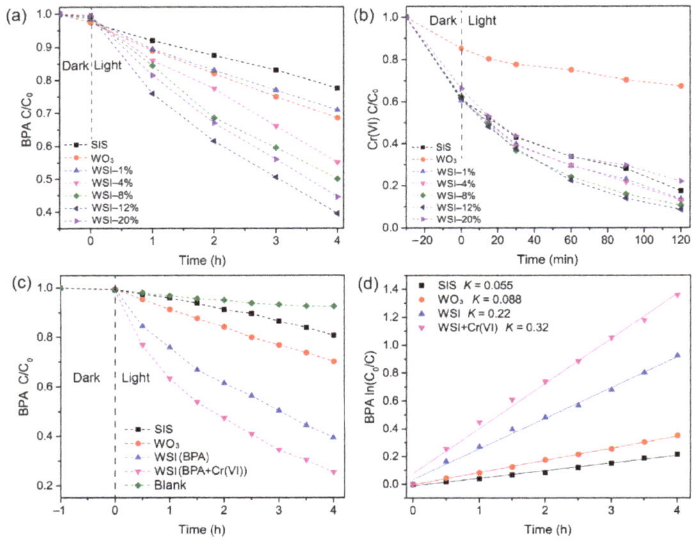

Figure 5. Photocatalytic performance of the as-prepared sample: (**a**) BPA oxidation, (**b**) Cr(VI) reduction; simultaneous photocatalytic oxidation of BPA and reduction of Cr(VI) over WO_3, SIS and WSI-12% heterojunction under visible irradiation, (**c**) photocatalytic efficiency, (**d**) photocatalytic degradation rate constant.

Table 1. Kinetic reaction rate constants for the degradation of BPA and reduction of Cr(VI).

Photocatalyst	Reaction Rate Constants (k)/min^{-1}	
	BPA	**Cr(VI)**
SIS	0.0557	0.6432
WO_3	0.0852	0.1494
WSI-1%	0.0780	0.7482
WSI-4%	0.1339	0.7728
WSI-8%	0.1614	0.8970
WSI-12%	0.2135	0.9834
WSI-20%	0.1853	0.5772

The simultaneous photocatalytic degradation of BPA and reduction of Cr(VI) with the WSI-12% heterojunction was studied to evaluate the S-scheme mechanism. Figure 5c,d show the BPA degradation efficiency and the corresponding rate constant in coexistence with Cr(VI). No significant degradation of BPA was detected in the dark without the photocatalyst, indicating that no reaction occurred between Cr(VI) and BPA. The degradation efficiency of BPA with and without Cr(VI) was 61.5% and 74.6%, respectively, which demonstrated that the coexistence of Cr(VI) could significantly enhance the BPA degradation efficiency. Moreover, the degradation process also fits well with the pseudo-first-order kinetics model when BPA is mixed with Cr(VI). As shown in Figure 5d, the degradation rate constant of BPA/Cr(VI) matrix is 1.5 times more than that observed with BPA alone. The photocatalytic reduction efficiency of Cr(VI) in the absence/presence of BPA is 85.9% and 96.9%, respectively, after 2 h of visible light irradiation. At the same time, the reduction rate constant of Cr(VI) in the presence of BPA is 1.75 times that of Cr(VI) alone. According to the basic principle of photocatalysis, the separation of photogenerated electrons and holes is accelerated in the presence of photogenerated electrons and hole traps. In this case, BPA and Cr(VI) were used as the electron donor and electron acceptor, respectively. Hence, the coexistence of Cr(VI) could accelerate the oxidation of BPA and vice versa. BPA was degraded by photogenerated holes and the corresponding hydroxyl radical produced by VB position of WO_3, and Cr(VI) as electron acceptor is directly reduced by photogenerated electrons produced by the CB of SIS, resulting in the efficient separation of photogenerated electron–hole pairs in the WSI heterojunction. As a result, there are synergy effects that mutually promote the degradation efficiency with the coexistence of BPA and Cr(VI).

2.4. Reaction Mechanism

Photocatalytic Degradation Intermediates and Pathway

Three-dimensional excitation emission matrix fluorescence spectroscopy (3D EEM) was used to identify the trace intermediates generated in the photocatalytic degradation of BPA. According to the previous report [35], the delocalized electrons in the conjugate system of benzene and its derivatives were easily excited to produce fluorescence signals. In this case, the degradation intermediates of BPA, including benzene, were determined by 3D EMMs to preliminarily estimate the pathway of photodegradation. As shown in Figure 6a, no obvious fluorescence spectrum peak was detected in ultrapure water, which served as a blank sample. All the fluorescence spectrum samples were subtracted to eliminate the Rayleigh scattering. As shown in Figure 6b, the three-dimensional EEM spectroscopy of a BPA sample displayed two weak fluorescence signals located at $\lambda_{ex}/\lambda_{em}$ = 273/305 nm and $\lambda_{ex}/\lambda_{em}$ = 226/305 nm [36]. The emission fluorescence peak of BPA disappeared under visible irradiation for one hour. Meanwhile, two new fluorescence signals located at $\lambda_{ex}/\lambda_{em}$ = 220/300 nm and $\lambda_{ex}/\lambda_{em}$ = 260/353 nm appeared, which were attributed to the phenol fluorescence signal and the low-molecular-weight organic matter or organic matter containing carboxyl groups [37]. After 2 h of irradiation, the above fluorescence peaks were significantly enhanced, indicating an increase in the concentration of intermediates. Meanwhile, the fluorescence peak located at $\lambda_{ex}/\lambda_{em}$ = 275/300 nm appeared, which was ascribed to isopropylated phenyl. The intensity of the fluorescence peak became weak after 3 h of photocatalytic degradation, indicating that the low-molecular-weight organic matter was mineralized. After 4 h of irradiation, all fluorescence signals expected for the signal phenol and low-molecular organic matter disappeared, indicating that BPA was decomposed into phenol and low-molecular organic matter.

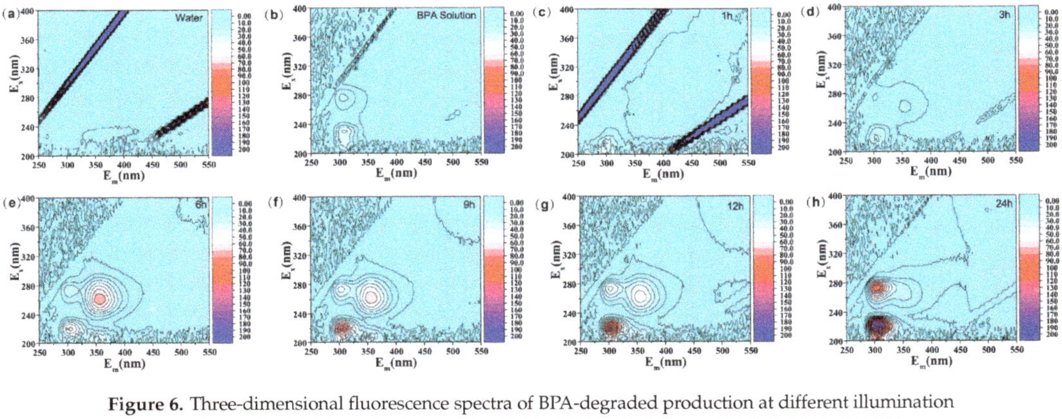

Figure 6. Three-dimensional fluorescence spectra of BPA-degraded production at different illumination times: (**a**) Blank–0 h; (**b**) Original solution–0 h; (**c**) 1 h; (**d**) 3 h; (**e**) 6 hours; (**f**) 9 h; (**g**) 12 h; (**h**) 24 h.

To further deepen our understanding of the photocatalytic degradation process, GC/MS was conducted to identify the degradation intermediates of BPA after 4 h of irradiation. Figure S2 displays the MS spectra of the intermediates observed during the photocatalytic oxidation reaction. Table S1 summarizes the main fragment ions (m/z) and abundance (%) of reaction products in comparison to the NIST data library. As shown in Figure S1 and Table S1, the chromatogram and mass spectrum of the BPA degraded intermediate were listed in the order according to the retention time. The intermediates were as follows: 4-hydroxy butyric acid(B), phenol(C), benzoquinone(D), 2-dipropylmalonic acid©, benzene acetaldehyde(F), hydroquinone(G), Penta-1,4-dien-3-one(H), 4-isopropenylphenol(I), 4-isopropylphenol(J) and p-Hydroxyacetophenone (K). Based on the detected intermediates and relevant literature reports, three possible degradation pathways of BPA were proposed, as illustrated in Figure 7 [38,39]. The first pathway is initiated by the cleavage of the C-C bond between phenyl groups and the central carbon atom and leads to the formation of a phenol radical and an isopropylphenol radical. In Pathway 1, hydroxyl radicals (·OH) attack the C-C bond between phenyl groups and the central carbon atom, leading to the formation of the phenol radical and isopropylphenol radical. Some of the phenol radicals can further oxidize to produce hydroquinone (G) and benzoquinone (D), while another portion of the phenol radicals is quenched to form phenol (C). The isopropylphenol radical undergoes additional transformation, resulting in the formation of 4-isopropenylphenol (J) and 4-isopropylphenol (I), with the latter being oxidized to p-Hydroxyacetophenone. The phenol products continue to oxidize, breaking the benzene ring and forming smaller molecules, such as 4-hydroxybutyric acid, Penta-1,4-dien-3-one, ultimately leading to their complete oxidation into CO_2 and H_2O. Another potential degradation pathway involves the direct oxidation and demethylation of BPA, resulting in the formation of phenol (C), benzene acetaldehyde (F), and p-Hydroxyacetophenone (K), which subsequently degrade into low-molecular-weight organic acids. The hydroxyl radical is the major active radical involved in decomposing the benzene ring due to its strong oxidation properties. In this work, we demonstrated that the redox potential of photogenerated holes was enhanced by constructing the WSI heterojunction and that BPA could be oxidized by ·OH and mineralized into low-molecular-weight organics.

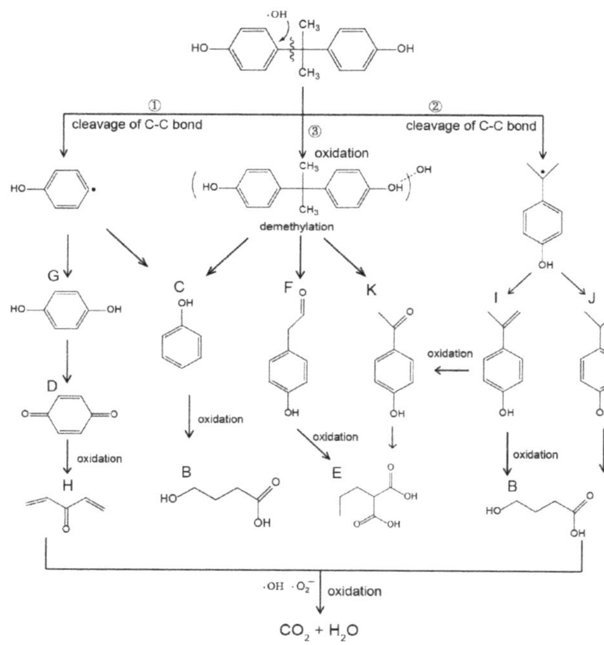

Figure 7. Possible degradation pathway of BPA over WSI-12% heterojunction.

2.5. S-Scheme Mechanism

The study investigated the reactive species of the photocatalytic reaction to explore the S-scheme mechanism of the WSI heterojunction. Initially, benzoquinone (BQ), isopropyl alcohol (IPA) and ammonium oxalate (AO) were selected to trap the radical of $\cdot O_2$-, h+ and $\cdot$OH, respectively, in the photocatalytic degradation of BPA. As shown in Figure 8a, the photocatalytic degradation efficiency of BPA was inhibited in the presence of traps. The degradation efficiency was 57.9%, 63.3% and 84.5% when adding the trap of BQ, IPA, and AO, respectively, while the degradation efficiency was 91.7% in the absence of traps, indicating that the major active species during BPA degradation were $\cdot O_2^-$ and $\cdot$OH. The minor inhibition of degraded BPA efficiency in the presence of AO suggested that the radical of h+ contributed little to the dye photodegradation. The role of the reactive species was as follows: $\cdot O_2^- > \cdot OH > h^+$.

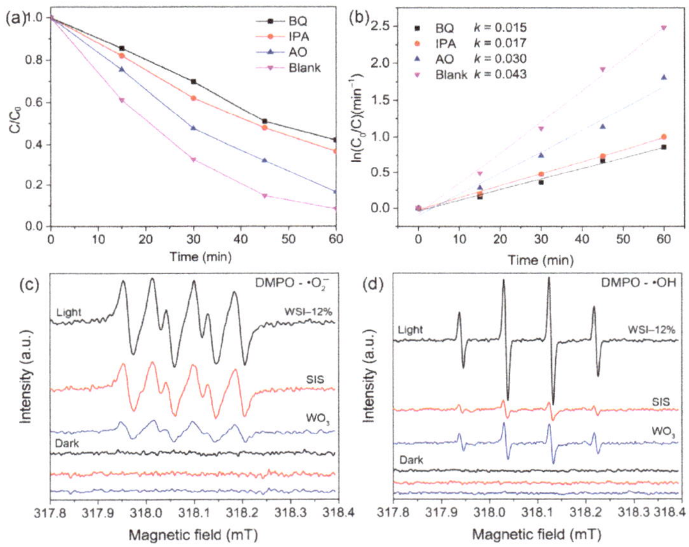

Figure 8. Effect of scavengers on the degradation of BPA over WSI-12%: (**a**) photocatalytic efficiency and (**b**) degradation rate, ESR signals of the (**c**) DMPO-$\cdot O_2^-$ and (**d**) DMPO-$\cdot$OH of WSI-12% photocatalyst.

To further investigate the concentration of reactive oxygen species in the photodegradation of organic compounds, the ESR spin-trap technique (with DMPO) was employed to measure the signals of DMPO-$\cdot O_2^-$ and DMPO-$\cdot$OH. The ESR signals of DMPO-$\cdot O_2^-$ and DMPO-$\cdot$OH were shown in Figure 8c,d. No obvious signals were detected in the dark reaction process. In contrast, the ESR signals of DMPO-$\cdot O_2^-$ and DMPO-$\cdot$OH appeared after 10 min of irradiation. Both the DMPO-$\cdot O_2^-$ and DMPO-$\cdot$OH signals were significantly enhanced in the WSI-12% heterojunction compared with the single-component photocatalyst, indicating that the concentrations of $\cdot O_2^-$ and $\cdot$OH radicals were higher in the WSI-12% reaction system. The ESR signals of DMPO-$\cdot$OH produced by WO_3 were relatively strong, which indicated that the $\cdot$OH radical played a more important role in the degradation of organic compounds. As for the SIS, the $\cdot O_2^-$ radical contributed significantly to the photocatalytic degradation based on the ESR signals of DMPO-$\cdot O_2^-$. According the ESR results, it is further confirmed that the efficient separation and transfer of photogenerated electron–hole pairs were improved by the construction of the heterojunction. In addition, the redox ability of the photogenerated hole and electron was enhanced, resulting in improved photocatalytic activity.

Based on the above results and analysis, a possible S-scheme mechanism of WSI-12% was proposed, as shown in Figure 9. The CB and VB redox potentials of WO_3 were +0.13 V vs. NHE and +2.96 V vs. NHE, respectively. The SIS photocatalyst, with a redox potential of −0.64 V vs. NHE (CB) and +1.44 V vs. NHE (VB), could couple with WO_3 to form the direct S-scheme system, due to the more negative CB potential. The redox potential between the VB of SIS and the CB of WO_3 is 1.31 V, which facilitates the transfer of the photogenerated electron to the VB position of SIS. Under visible light irradiation, the photogenerated electron of the WSI-12% S-scheme system transfers from the CB of WO_3 to the VB of SIS and recombines with the hole of SIS. Therefore, the electron on the CB position of SIS and the hole on the VB position of WO_3 are spatially separated, increasing the probability of reaction with reactants. The VB potential of WO_3 is larger than the oxidization potential of water to hydroxyl radicals ($\cdot OH/H_2O$ = 2.4 eV vs. NHE), while the CB potential of SIS is more negative than the reduction potential of oxygen to superoxide radicals ($\cdot O_2^-/O_2$ = −0.33 eV vs. NHE). The $\cdot OH$ and $\cdot O_2$- radical generated by the photo-introduced holes and electrons of the WSI S-scheme system can oxidize the refractory organics to low-molecular organics, and the photogenerated electrons can reduce the Cr(VI) to Cr(III). Hence, the S-scheme system can simultaneous oxidize organic and reduce heavy metal ions. The major reaction steps of the S-scheme mechanism under visible light irradiation are summarized in the following reactions:

$$\mathrm{WSI} \overset{h\nu}{\rightarrow} \mathrm{SIS}(e^-/h^+)/\mathrm{WO_3}(e^-/h^+) \quad (1)$$

$$\mathrm{SIS}(e^-/h^+)/\mathrm{WO_3}(e^-/h^+) \rightarrow \mathrm{SIS}(e^-) + \mathrm{WO_3}(h^+) \quad (2)$$

$$\mathrm{SIS}(e^-) + O_2 \rightarrow O_2^-\cdot \quad (3)$$

$$\mathrm{WO_3}(h^+) + H_2O \rightarrow \cdot OH \quad (4)$$

$$6\mathrm{SIS}(e^-) + Cr_2O_7^{2-} + 14H^+ \rightarrow 2Cr^{3+} + 7H_2O \quad (5)$$

$$\cdot O_2^-/\cdot OH + \mathrm{BPA} \rightarrow \cdots \rightarrow CO_2 + H_2O \quad (6)$$

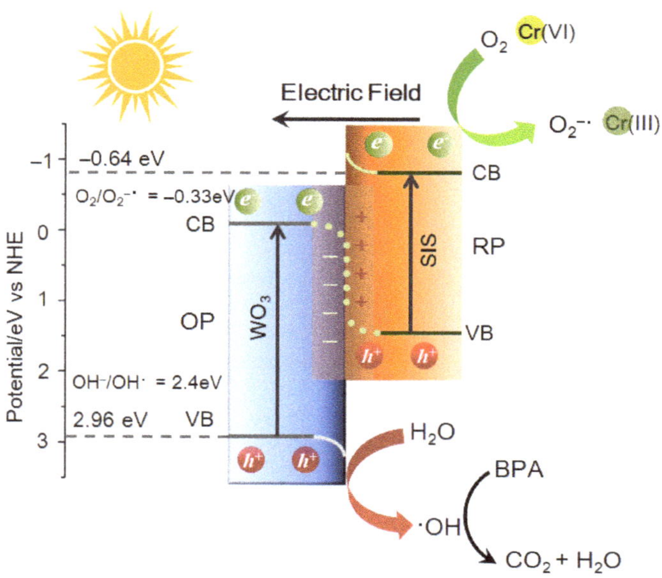

Figure 9. S-scheme transfer mechanism of photogenerated electrons under visible light irradiation.

3. Materials and Methods

3.1. Synthesis of WSI S-Scheme Heterojunction

Yellow WO_3 nanosheets were synthesized according to a previous report via a solvothermal alcoholysis process followed by calcination of the hierarchical structure of the as-prepared WO_3 [27]. Furthermore, a direct WSI S-scheme heterojunction was fabricated by combining the in situ precipitation and hydrothermal synthesis method. In brief, WO_3 nanosheets were dispersed in deionized water and ultrasonized for 10 min. Meanwhile, $SnCl_4{\cdot}5H_2O$ and $InCl_3{\cdot}4H_2O$ were dissolved in aqueous acetic acid solution, and the molar ratio of Sn^{4+} and In^{3+} was 1:4. The concentration of the Sn^{4+}/In^{3+} mixture was 0.3 mM. Then, the above solution was mixed with the WO_3 suspension, and 0.24 mM DDTC-2Na was dropwise added to the mixture. The suspension was vigorously stirred for 2 h and then introduced into a 50 mL Teflon-lined autoclave and maintained at 453 K for 12 h. After cooling down to room temperature, the yellow powder was collected by filtration and repeatedly washed by deionized water and ethanol, finally freeze-dried to obtain the WSI heterojunction. Bare WO_3 and SIS were also prepared under the same conditions. For the WSI heterojunction, the mass percentage of WO_3 to SIS ranged from 1% to 20%. The as-prepared composites were labeled as WSI-1%, WSI-4%, WSI-8%, WSI-12%, and WGS-20%, respectively.

3.2. Characterization of WSI S-Scheme Heterojunction

X-ray diffraction (XRD) patterns were obtained using a Miniflex600 XRD spectroscopy instrument (Rigaku, Tokyo, Japan) with Ni-filtered Cu Kα irradiation (λ = 1.5406 Å) at a scanning rate of 2°/min, ranging from 10° to 80°. The structures and morphologies of the samples were characterized using field emission scanning electron microscopy (FESEM, FEI, Nova Nano SEM450, Hillsboro, OR, USA) and a high-resolution transmission electronic micrograph (HRTEM, FEI, Tecnai G220, Hillsboro, OR, USA). The chemical bonding states of samples were investigated using X-ray photoelectron spectroscopy (XPS, ESCA Lab250, Waltham, MA, USA) equipped with twin anode Al Kα radiation, and the binding energies were calibrated by referencing the C 1s (284.6 eV) peak. The optical properties were measured by a UV-Vis spectrophotometer (Cary 5000, Agilent, Santa Clara, CA, USA) and a photoluminescence spectrum (PL, Cary Eclipse, Agilent). EIS measurements were conducted using the EIS system (EIS, CHI660, CH instruments, Beijing, China) in a frequency range of 0.1–10^5 Hz, utilizing an initial voltage of 0 V and amplitude of 5 mV. The photocatalytic degradation intermediates of BPA were identified using gas chromatography–mass spectrometry (GC/MS, 7890B/5977, Agilent, Santa Clara, CA, USA). Finally, the reactive oxygen species were verified using electron spin resonance (ESR, JES-FA200, JESO, Okinawa, Japan).

3.3. Photocatalytic Experiment

The photocatalytic experiments were performed in a double-walled quartz jacket filled with cool water under visible light irradiation (Figure S3). A 500W Xenon lamp coupled with a cutoff filter (λ > 420 nm) was used as the light source. The distance between Xe lamp and reactor was adjusted to 10 cm to achieve an incident-light intensity of 100 $mW{\cdot}cm^{-2}$ (equivalent to one sun). First, the photocatalytic activity of the WSI heterojunction was evaluated by photocatalytic degradation of BPA and reduction of Cr(VI), respectively. Subsequently, simultaneous photocatalytic oxidation of BPA and reduction of Cr(VI) was conducted to investigate the synergetic decontamination effects. Briefly, 20 mg of photocatalyst was dispersed in 50 mL aqueous solution with the initial concentration (C_0) of 20 $mg{\cdot}L^{-1}$ Cr(VI), 20 $mg{\cdot}L^{-1}$ BPA and a mixture of Cr(VI) and BPA. The pH of the solution was adjusted using diluted NaOH or HCl. Prior to irradiation, the suspension was magnetically stirred for 30 min in the dark to achieve an adsorption equilibrium between reactants and photocatalysts. During irradiation, 3 mL of suspension was sampled at specific time intervals and centrifuged to remove the photocatalysts. The concentration of Cr(VI) and BPA was measured using UV-Vis spectrophotometry and high-performance liquid chromatography (HPLC) with a UV detector. The photocatalytic efficiency was calculated according to C/C_0, where C_0 was the equilibrium concentration after adsorption and C was the concentration of Cr(VI) and BPA after the reaction.

4. Conclusions

In this work, a novel direct S-scheme photocatalyst was prepared by combining in situ precipitation with hydrothermal synthesis to achieve perfect integration of WO_3 and SIS. The construction of the S-scheme heterojunction promoted the separation of photogenerated charges while maintaining the strong redox ability, leading to its excellent photocatalytic activity towards the oxidation of BPA. Moreover, it exhibited outstanding performance for simultaneous photocatalytic degradation of BPA and reduction of Cr(VI). BPA was decomposed into low-molecular-weight organic compounds and finally mineralized into CO_2 and H_2O. Radical trap and ESR experiments indicated that the major active species in the process of BPA degradation were $\cdot O_2^-$ and $\cdot OH$. This study provides a novel strategy for designing and constructing a direct S-scheme photocatalyst for highly efficient decontamination of organic pollutants and heavy metal ions.

Supplementary Materials: The following supporting information can be downloaded at: https://www.mdpi.com/article/10.3390/catal13111450/s1, Figure S1: GC chromatogram of BPA after 4 h degradation; Figure S2: Mass spectrum of the intermediate products of BPA after 4 h degradation; Figure S3: Schematic diagram of photocatalytic reactor; Table S1: GC-MS data of the photodegradation products of BPA by WSI.

Author Contributions: P.X.: Conceptualization, Methodology, Writing—Original draft preparation. R.Z.: Investigation, Data Curation. J.G.: Investigation, Formal analysis. Y.L.: Investigation, Formal analysis. Y.Z.: Resources, Writing—Reviewing and Editing. P.Z.: Conceptualization, Writing—Reviewing and Editing. All authors have read and agreed to the published version of the manuscript.

Funding: This research was funded by Key Laboratory of Ecological Environment and Information Atlas (grant number ST21001), and Technology Innovation Center for Exploitation of Marine Biological Resources (TIRMBR202104). And The APC was funded by Technology Innovation Center for Exploitation of Marine Biological Resources.

Data Availability Statement: The data presented in this study are available in the article and its online supplementary information.

Conflicts of Interest: The authors declare no conflict of interest.

References

1. Petrie, B.; Barden, R.; Kasprzyk-Hordern, B. A review on emerging contaminants in wastewaters and the environment: Current knowledge, understudied areas and recommendations for future monitoring. *Water Res.* **2015**, *72*, 3–27. [CrossRef] [PubMed]
2. Sun, Y.; Xu, L.; Jin, P.K.; Bai, X.; Jin, X.; Shi, X. Simultaneous removal of colorless micropollutants and hexavalent chromium by pristine TiO_2 under visible light: An electron transfer mechanism. *Chem. Eng. J.* **2021**, *405*, 11. [CrossRef]
3. Fellahi, O.; Barras, A.; Fan, G.H.; Coffinier, Y.; Hadjersi, T.; Maamache, M.; Szunerits, S.; Boukherroub, R. Reduction of Cr(VI) to Cr(III) using silicon nanowire arrays under visible light irradiation. *J. Hazard. Mater.* **2016**, *304*, 441–447. [CrossRef] [PubMed]
4. Oh, W.D.; Lok, L.W.; Veksha, A.; Giannis, A.; Lim, T.T. Enhanced photocatalytic degradation of bisphenol A with Ag-decorated S-doped g-C_3N_4 under solar irradiation: Performance and mechanistic studies. *Chem. Eng. J.* **2018**, *333*, 739–749. [CrossRef]
5. Ben, W.W.; Zhu, B.; Yuan, X.J.; Zhang, Y.; Yang, M.; Qiang, Z.M. Occurrence, removal and risk of organic micropollutants in wastewater treatment plants across China: Comparison of wastewater treatment processes. *Water Res.* **2018**, *130*, 38–46. [CrossRef] [PubMed]
6. Guillossou, R.; Roux, J.L.; Mailler, R.; Vulliet, E.; Morlay, C.; Nauleau, F.; Gasperi, J.; Rocher, V. Organic micropollutants in a large wastewater treatment plant: What are the benefits of an advanced treatment by activated carbon adsorption in comparison to conventional treatment? *Pergamon* **2019**, *218*, 1050–1060. [CrossRef]
7. Park, H.G.; Kim, T.W.; Chae, M.Y.; Yoo, I.K. Activated carbon-containing alginate adsorbent for the simultaneous removal of heavy metals and toxic organics. *Process Biochem.* **2007**, *42*, 1371–1377. [CrossRef]
8. Dos Santos, G.E.D.; Lins, P.V.D.; Oliveira, L.; da Silva, E.O.; Anastopoulos, I.; Erto, A.; Giannakoudakis, D.A.; de Almeida, A.R.F.; Duarte, J.L.D.; Meili, L. Layered double hydroxides/biochar composites as adsorbents for water remediation applications: Recent trends and perspectives. *J. Clean Prod.* **2021**, *284*, 19. [CrossRef]

9. Pan, S.; Li, J.; Noonan, O.; Fang, X.; Wan, G.; Yu, C.; Wang, L. Dual-functional Ultrafiltration Membrane for Simultaneous Removal of Multiple Pollutants with High-performance. *Environ. Sci. Technol.* **2017**, *51*, 5098. [CrossRef]
10. Butkovskyi, A.; Jeremiasse, A.W.; Leal, L.H.; van der Zande, T.; Rijnaarts, H.; Zeeman, G. Electrochemical Conversion of Micropollutants in Gray Water. *Environ. Sci. Technol.* **2014**, *48*, 1893–1901. [CrossRef]
11. Song, X.L.; Ren, C.W.; Zhao, Q.Y.; Su, B.Q. Simultaneous removal of Cr(VI) and triclosan from aqueous solutions through Fe_3O_4 magnetic nanoscale-activated persulfate oxidation. *Chem. Eng. J.* **2020**, *381*, 11. [CrossRef]
12. Pelaez, M.; Nolan, N.T.; Pillai, S.C.; Seery, M.K.; Falaras, P.; Kontos, A.G.; Dunlop, P.S.M.; Hamilton, J.W.J.; Byrne, J.A.; O'Shea, K.; et al. A review on the visible light active titanium dioxide photocatalysts for environmental applications. *Appl. Catal. B-Environ.* **2012**, *125*, 331–349. [CrossRef]
13. Szilagyi, I.M.; Forizs, B.; Rosseler, O.; Szegedi, A.; Nemeth, P.; Kiraly, P.; Tarkanyi, G.; Vajna, B.; Varga-Josepovits, K.; Laszlo, K.; et al. WO3 photocatalysts: Influence of structure and composition. *J. Catal.* **2012**, *294*, 119–127. [CrossRef]
14. Wang, X.C.; Blechert, S.; Antonietti, M. Polymeric Graphitic Carbon Nitride for Heterogeneous Photocatalysis. *ACS Catal.* **2012**, *2*, 1596–1606. [CrossRef]
15. Wang, Z.Y.; Sim, A.; Urban, J.J.; Mi, B.X. Removal and Recovery of Heavy Metal Ions by Two-dimensional MoS2 Nanosheets: Performance and Mechanisms. *Environ. Sci. Technol.* **2018**, *52*, 9741–9748. [CrossRef]
16. Abdelraheem, W.H.M.; Patil, M.K.; Nadagouda, M.N.; Dionysiou, D.D. Hydrothermal synthesis of photoactive nitrogen- and boron- codoped TiO(2)nanoparticles for the treatment of bisphenol A in wastewater: Synthesis, photocatalytic activity, degradation byproducts and reaction pathways. *Appl. Catal. B-Environ.* **2019**, *241*, 598–611. [CrossRef]
17. Li, S.Y.; Zhang, M.; Qu, Z.H.; Cui, X.; Liu, Z.Y.; Piao, C.C.; Li, S.G.; Wang, J.; Song, Y.T. Fabrication of highly active Z-scheme Ag/g-C3N4-Ag-Ag_3PO_4(110) photocatalyst photocatalyst for visible light photocatalytic degradation of levofloxacin with simultaneous hydrogen production. *Chem. Eng. J.* **2020**, *382*, 13. [CrossRef]
18. Du, X.Y.; Bai, X.; Xu, L.; Yang, L.; Jin, P.K. Visible-light activation of persulfate by TiO_2/g-C_3N_4 photocatalyst toward efficient degradation of micropollutants. *Chem. Eng. J.* **2020**, *384*, 14. [CrossRef]
19. Low, J.X.; Dai, B.Z.; Tong, T.; Jiang, C.J.; Yu, J.G. In Situ Irradiated X-Ray Photoelectron Spectroscopy Investigation on a Direct Z-Scheme TiO_2/CdS Composite Film Photocatalyst. *Adv. Mater.* **2019**, *31*, 1802981. [CrossRef]
20. Fu, J.; Xu, Q.; Low, J.; Jiang, C.; Yu, J. Ultrathin 2D/2D WO_3/g-C_3N_4 step-scheme H_2-production photocatalyst. *Appl. Catal. B Environ.* **2019**, *243*, 556–565. [CrossRef]
21. Zhang, L.; Zhang, J.; Yu, H.; Yu, J. Emerging S-Scheme Photocatalyst. *Adv. Mater.* **2022**, *34*, e2107668. [CrossRef] [PubMed]
22. Li, X.; Xiong, J.; Gao, X.; Ma, J.; Chen, Z.; Kang, B.; Liu, J.; Li, H.; Feng, Z.; Huang, J. Novel BP/BiOBr S-scheme nano-heterojunction for enhanced visible-light photocatalytic tetracycline removal and oxygen evolution activity. *J. Hazard. Mater.* **2020**, *387*, 121690. [CrossRef]
23. Wang, W.; Zhao, W.; Zhang, H.; Dou, X.; Shi, H. 2D/2D step-scheme α-Fe_2O_3/Bi_2WO_6 photocatalyst with efficient charge transfer for enhanced photo-Fenton catalytic activity. *Chin. J. Catal.* **2021**, *42*, 97–106. [CrossRef]
24. Singh, T.; Mueller, R.; Singh, J.; Mathur, S. Tailoring surface states in WO_3 photoanodes for efficient photoelectrochemical water splitting. *Appl. Surf. Sci.* **2015**, *347*, 448–453. [CrossRef]
25. Subrahmanyam, A.; Karuppasamy, A. Optical and electrochromic properties of oxygen sputtered tungsten oxide (WO_3) thin films. *Sol. Energy Mater. Sol. Cells* **2007**, *91*, 266–274. [CrossRef]
26. Xu, P.; Huang, S.; Lv, Y.; Chen, Y.; Liu, M.; Fan, H. Surfactant-assisted hydrothermal synthesis of rGO/$SnIn_4S_8$ nanosheets and their application in complete removal of Cr(VI). *RSC Adv.* **2018**, *8*, 5749–5759. [CrossRef]
27. Xi, G.; Ye, J.; Ma, Q.; Su, N.; Bai, H.; Wang, C. In Situ Growth of Metal Particles on 3D Urchin-like WO_3 Nanostructures. *J. Am. Chem. Soc.* **2012**, *134*, 6508–6511. [CrossRef]
28. Prakash, A.; Dan, M.; Yu, S.; Wei, S.Q.; Li, Y.; Wang, F.; Zhou, Y. In2S3/CuS nanosheet composite: An excellent visible light photocatalyst for H_2 production from H_2S. *Sol. Energy Mater. Sol. Cells* **2018**, *180*, 205–212. [CrossRef]
29. Wang, C.P.; Zhang, Y.Y.; Li, Y.S.; Zhang, Y.; Dong, Y.T.; Li, D.; Zhang, J.M. Construction of uniform SnS_2/ZnS heterostructure nanosheets embedded in graphene for advanced lithium-ion batteries. *J. Alloys Compd.* **2020**, *820*, 7. [CrossRef]
30. Li, Q.; Xia, Y.; Yang, C.; Lv, K.L.; Lei, M.; Li, M. Building a direct Z-scheme heterojunction photocatalyst by ZnIn2S4 nanosheets and TiO_2 hollowspheres for highly-efficient artificial photosynthesis. *Chem. Eng. J.* **2018**, *349*, 287–296. [CrossRef]
31. Katherine, V.; Sebastián, M.; Joan Ramón, M.; Teresa, A. An insight on the role of La in mesoporous WO_3 for the photocatalytic conversion of methane into methanol. *Appl. Catal. B Environ. Int. J. Devoted Catal. Sci. Its Appl.* **2016**, *180*, 30–36.
32. Lee, G.Y.; Cho, E.C.; Lo, P.Y.; Zheng, J.H.; Huang, J.H.; Chen, Y.L.; Lee, K.C. Simultaneous formation of $Bi_2O_2(OH)(NO_3)$/BiOBr ultrathin hierarchical microspheres for effectively promoting visible-light-driven photocatalytic activity in environmental remediation. *Chemosphere* **2020**, *258*, 127384. [CrossRef]
33. An, L.; Park, Y.; Sohn, Y.; Onishi, H. Effect of Etching on Electron–Hole Recombination in Sr-Doped $NaTaO_3$ Photocatalysts. *J. Phys. Chem. C* **2015**, *119*, 28440–28447. [CrossRef]
34. Li, Y.B.; Jin, Z.L.; Wang, H.Y.; Zhang, Y.P.; Liu, H. Effect of electron-hole separation in MoO_3@Ni_2P hybrid nanocomposite as highly efficient metal-free photocatalyst for H_2 production. *J. Colloid Interface Sci.* **2019**, *537*, 629–639. [CrossRef]
35. Ren, H.; Li, J.; Ma, Z.; Wan, H.; Zheng, X.; Cheng, S. Use of three-dimensional excitation and emission matrix fluorescence spectroscopy for predicting the disinfection by-product formation potential of reclaimed water. *Water Res.* **2012**, *46*, 5765–5776.

36. Zhao, G.; Sun, Y.; Zhao, Y.; Wen, T.; Wang, X.; Chen, Z.; Sheng, G.; Chen, C.; Wang, X. Enhanced Photocatalytic Simultaneous Removals of Cr(VI) and Bisphenol A over Co(II)-Modified TiO_2. *Langmuir ACS J. Surf. Colloids* **2019**, *35*, 276–283. [CrossRef]
37. Maroto, A.; Kissingou, P.; Diascorn, A.; Benmansour, B.; Deschamps, L.; Stephan, L.; Giamarchi, C.P. Direct laser photo-induced fluorescence determination of bisphenol A. *Anal. Bioanal. Chem.* **2011**, *401*, 3011–3017. [CrossRef]
38. Gao, S.W.; Guo, C.S.; Lv, J.P.; Wang, Q.; Zhang, Y.; Hou, S.; Gao, J.F.; Xu, J. A novel 3D hollow magnetic Fe3O4/BiOI heterojunction with enhanced photocatalytic performance for bisphenol A degradation. *Chem. Eng. J.* **2017**, *307*, 1055–1065. [CrossRef]
39. Kanigaridou, Y.; Petala, A.; Frontistis, Z.; Antonopoulou, M.; Solakidou, M.; Konstantinou, I.; Deligiannakis, Y.; Mantzavinos, D.; Kondarides, D.I. Solar photocatalytic degradation of bisphenol A with $CuO_x/BiVO_4$: Insights into the unexpectedly favorable effect of bicarbonates. *Chem. Eng. J.* **2017**, *318*, 39–49. [CrossRef]

sustainability

Article

The Addition of Anthocyanin as a Sensitizer for TiO_2 Nanotubes in a Combined Process of Electrocoagulation and Photocatalysis for Methylene Blue Removal

Indar Kustiningsih [1,*], Hendrini Pujiastuti [1], Denni Kartika Sari [1], Agus Rochmat [1] and Slamet [2]

[1] Department of Chemical Engineering, Universitas Sultan Ageng Tirtayasa, Cilegon 42435, Indonesia; hendrini@untirta.ac.id (H.P.); denni.kartikasari@untirta.ac.id (D.K.S.); agus_rochmat@untirta.ac.id (A.R.)
[2] Department of Chemical Engineering, Universitas Indonesia, Depok 16424, Indonesia; slamet@che.ui.ac.id
* Correspondence: indar.kustiningsih@untirta.ac.id; Tel.: +62-85210723040

Abstract: Photocatalysis with TiO_2 semiconductors is one of several potential methods for removing Methylene Blue (MB) that is environmentally friendly, relatively cheap, and effective. The capability of TiO_2 photocatalysts for degrading MB can be improved by modifying the morphology of TiO_2 into nanotubes and adding anthocyanin sensitizers. The objective of this study was to investigate the effect of anthocyanin sensitizer addition for TiO_2 nanotubes on MB removal using a combined process of electrocoagulation and photocatalysis. TiO_2 nanotubes were prepared through an anodization method with a glycerol electrolyte containing NH_4F of 0.5% w/v and water of 25% v/v. The cathode and anode used in the electrocoagulation process were 316 stainless steel and aluminum, respectively. The characteristics of the resulting TiO_2 nanotubes were analyzed using SEM-EDX, UV–Vis DRS, and XRD analyses. The results showed that the electrocoagulation at a pH of 10 and a voltage of 50 volts resulted in an MB removal efficiency of 57.88%. In the photocatalysis process, sensitizer addition can increase the MB removal efficiency from 19.71% to 29.06%. Furthermore, a combined process of electrocoagulation and photocatalysis without and with sensitizer addition resulted in MB removal efficiencies of 59.66% and 64.30%, respectively.

Keywords: anthocyanin; electrocoagulation; methylene blue; nanotubes; photocatalysis; TiO_2

1. Introduction

Waste resulting from textile industries contains dyes with concentrations of 20–30 mg/L, which are hazardous and carcinogenic [1,2]. One of the dyes that are often used in industry is Methylene Blue (MB) ($C_{16}H_{18}N_3SCl$), which is a heterocyclic aromatic molecule [2,3]. The structure formula of MB can be seen in Figure 1. The MB has harmful effects such as vomiting, increased heart rate, cyanosis, digestive system irritation if inhaled, and skin irritation [3,4]. Because of the presence of aromatic amines in the MB structure, the MB is difficult to decompose [3,5].

Figure 1. The structure formula of Methylene Blue.

Commonly, MB can be treated through physical and/or chemical treatments, including filtration [6,7], coagulation and precipitation [8,9], adsorption [3,10], ozonation [11], reverse osmosis [12], ion exchange [13], and advanced oxidation [14,15]. However, these treatments need a high operating cost and result in a low pollutant removal efficiency, a high selectivity just for a certain dye group, and harmful intermediate compounds [16].

Photocatalysis is one of the promising methods for removing MB organic dyes in liquid waste because it can degrade MB to CO_2 and H_2O [17–21]. Titanium dioxide (TiO_2) is one of the semiconductors used in photocatalysis. TiO_2 has some advantages, including being stable, non-toxic, corrosion-resistant, abundant in nature, and relatively affordable [18,22,23]. On the other hand, TiO_2 photocatalysts have some disadvantages, including a low specific surface area, a high recombination rate, and a high band gap energy value [24]. The surface of TiO_2 can be improved by modifying its morphology into the shape of nanorods [24,25], nanowires [26,27], and nanotubes [23,26,28,29]. Nanotube morphology is often used because it has a larger surface area and is more effective in photocatalysis than other morphologies [26]. In water treatment, TiO_2 nanotubes can be used in particle or film form (titanium nanotubes arrays (TNTAs)), but the film form (TNTAs) is more attractive because it does not require costly solid–liquid separation [30].

In addition, the effectiveness of TiO_2 photocatalysts can also be improved by the addition of dye-sensitizers. The dye-sensitizer materials function to absorb visible light so electrons can be excited. With the addition of dye-sensitizers, TiO_2 becomes more responsive to visible light. Furthermore, dyes can easily adhere to the catalyst surface; thus, the electrons readily achieve an excited state by photon absorption in the visible light spectrum range [31–33]. Synthetic metal-based dyes can be used as sensitizers, but they can decompose quickly in the working solution [28,29]. Therefore, natural dyes are interesting alternative sensitizers in the photodegradation of aqueous organic contaminants. The natural dyes are harmless, abundant, and inexpensive and contain a hydroxyl carbonyl group to attach to the surface of TiO_2 [34,35]. Anthocyanin, carotenoid, chlorophyll, curcumin, and flavonoid are natural dyes that can be used as sensitizers [31,36–39]. Anthocyanin is a stable sensitizer that can absorb visible light with a high absorption coefficient. Moreover, it can be easily obtained because it is derived from plant sources, such as fruits, blossoms, and vegetables [32,40,41]. Figure 2 shows the structure formula of an anthocyanin molecule attached to TiO_2.

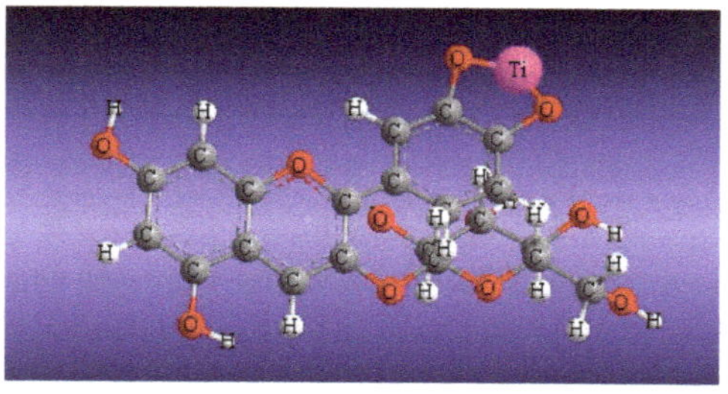

Figure 2. Structure formula of an anthocyanin molecule attached to TiO_2.

Anthocyanin and other natural dyes have been widely used in solar cell applications but are less used in water purification [40,42]. Some studies employed anthocyanins and other natural dyes as sensitizers for TiO_2 nanoparticles; however, there are still very few applications of these sensitizers for TiO_2 nanotubes [32,38,43–46].

To increase pollutant removal efficiency, photocatalysis can be combined with other methods, such as ozone [11,17,47], zeolite [48–50], and electrocoagulation [51–54]. The combination of photocatalysis and electrocoagulation treatments is often conducted in separated reactors [9,55,56], and a small number of researchers employ a single reactor for this purpose. In our previous studies, a combined process of electrocoagulation and photocatalysis in a single reactor for waste treatment was developed [51–53]. However, the modification of the TiO_2 nanotube arrays (TNTAs) through the sensitizer addition has not been conducted yet.

Based on the information above, the main question addressed by this study is how anthocyanins (as sensitizers) affect the characteristics of the resulting sensitized TNTAs and the performance of the sensitized TNTAs for removing MB through a combined process of electrocoagulation and photocatalysis using a single reactor. Therefore, the goal of this study was to investigate the effect of the addition of anthocyanin sensitizers for TNTAs applied in a combined process of electrocoagulation and photocatalysis, which was operated in a single reactor, for degrading MB. In addition, electrocoagulation alone and photocatalysis alone for degrading MB were also carried out under various operating conditions in this study. The authors hypothesized that the combined process of electrocoagulation and photocatalysis using sensitized TNTAs could result in a higher MB removal efficiency than the electrocoagulation alone, photocatalysis alone, and the combined process using unsensitized TNTAs. This study is original and has not been conducted by other authors yet. The novelty of this study is the utilization of anthocyanins as sensitizers for TNTAs used in a combined process of electrocoagulation and photocatalysis using a single reactor for removing the MB.

2. Materials and Methods

2.1. Material

This study used pro-analytic chemicals, such as ethanol, glycerol, HF, HNO_3, MB, and NH_4F obtained from Sigma-Aldrich (St. Louis, MO, USA). Meanwhile, pure water, aluminum as an anode, and 316 stainless steel as a cathode were collected from local stores.

2.2. Synthesis of TiO_2 Nanotubes (TNTAs)

Titanium plates with dimensions of length, width, and thickness of 8 cm, 4 cm, and 0.5 mm are cleaned using cloth to eliminate adherent dirt. Then, chemical polishing was performed using a mixture of HF (6 mL), HNO_3 (18 mL), and H_2O (36 mL). After that, the plates were cleaned using distilled water, dried at room temperature, and kept in a desiccator. The anodization process was carried out using a two-electrode configuration, in which titanium (Ti) and platinum (Pt) plates were anode and cathode, respectively, with an electrode distance of 1 cm. A DC power supply was utilized as a voltage source. The glycerol electrolyte solution contained NH_4F of 0.5% w/v and water of 25% v/v. Anodization was conducted at a voltage of 50 V for 2 h without heating. After anodizing, the plates were cleaned using distilled water and then dried at room temperature. The calcination process was carried out at 500 °C for 3 h. The scheme of synthesis of TNTAs is shown in Figure 3.

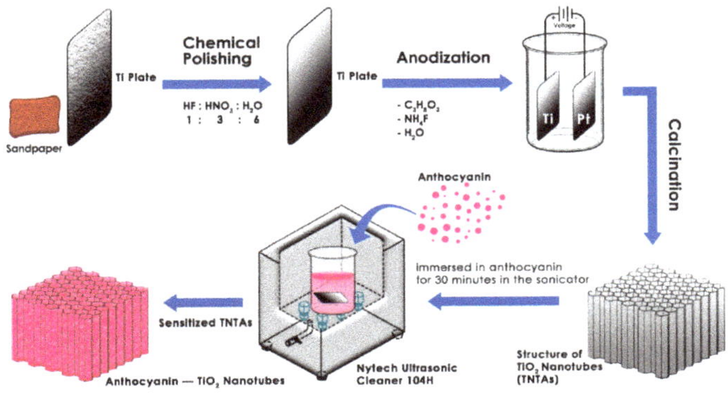

Figure 3. The scheme of synthesis of TNTAs and sensitized TNTAs.

2.3. Preparation of Sensitized TNTAs

Anthocyanin sensitizer was prepared by dissolving synthetic anthocyanin powder in ethanol to generate an anthocyanin solution with concentrations of 40, 80, and 160 ppm. Then, TiO_2 nanotube photocatalysts were immersed in an anthocyanin solution while ultrasonicated for 30 min (see Figure 3). Then, they were dried at room temperature.

2.4. Characterization of Catalyst

Characterization of catalyst was conducted to determine the physical and chemical properties of the resulting photocatalysts. In this study, the catalyst characterization methods included UV–Vis DRS, XRD, SEM-EDX, and FTIR.

2.5. Degradation of Methylene Blue by Electrocoagulation Process

The reactor was filled with 1000 mL of MB solution with a concentration of 10 ppm. A 316 stainless steel plat (length, width, and thickness of 10 cm, 2.5 cm, and 2 mm) as a cathode and two aluminum plates (length, width, and thickness of 10 cm, 2.5 cm, and 2 mm) as anodes were immersed to the solution with an electrode distance of 1 cm. The operating conditions were varied, including voltages (30 V, 40 V, and 50 V) and initial pHs (5, 7, and 10). A DC power supply was used as an electrical current. The electrocoagulation process was carried out for 120 min at room temperature. The liquid samples were taken at minutes 15, 30, 45, 60, 90, and 120. The MB concentrations in the liquid samples were analyzed using UV–Vis spectrophotometry at 663 nm [57].

2.6. Degradation of Methylene Blue by a Combined Process of Photocatalysis and Electrocoagulation

As much as 1000 mL of MB solution with a concentration of 10 ppm was placed in a reactor that was previously filled with sensitized TiO_2 nanotubes. A 316 stainless steel plate (length, width, and thickness of 10 cm, 2.5 cm, and 2 mm) as a cathode and two aluminum plates (length, width, and thickness of 10 cm, 2.5 cm, and 2 mm) as anodes were immersed to the solution with an electrode distance of 1 cm. Afterward, a 120 min irradiation with a 250-watt mercury lamp was conducted. The combined process was carried out for 120 min at room temperature. The liquid samples were taken at minutes 15, 30, 45, 60, 90, and 120. The MB concentrations in the liquid samples were analyzed using UV–Vis spectrophotometry at 663 nm.

2.7. Removal Efficiency of Methylene Blue

The removal efficiency of MB was calculated using Equation (1), where C_{MB_i} is the initial MB concentration (ppm), and C_{MB_f} is the final MB concentration (ppm).

$$MB\ removal\ efficiency = \frac{C_{MB_i} - C_{MB_f}}{C_{MB_i}} \times 100\% \quad (1)$$

3. Results and Discussion

3.1. Effect of Water Content in Glycerol Electrolyte on the Morphology of TiO_2 Nanotube Arrays (TNTAs)

The mechanism of TiO_2 nanotube formation can be explained by Equation (2). Equation (2) describes the anodic oxidation of Ti to generate TiO_2 [58].

$$Ti + 2H_2O \rightarrow TiO_2 + 4H^+ + 4e^- \quad (2)$$

The Ti surface is oxidized to generate a TiO_2 layer, which is then dissolved (electric field-assisted dissolution), producing a tube-like TiO_2 layer. The formation of pores (small pits) is due to dissolving TiO_2 in a solution containing H^+ and F^- ions (resulting from NH_4F), following Equation (3).

$$TiO_2 + 6F^- + 4H^+ \rightarrow (TiF_6)^{2-} + 2H_2O \quad (3)$$

According to Equation (3), Ti and TiO_2 are also chemically dissolved due to the availability of fluoride and the acidic environment. This reaction occurs across the entire porous surface.

Two tests were performed in this investigation, one with a water content of 2% v/v in glycerol, indicating a viscous electrolyte, and one with a water content of 25% v/v in glycerol, representing an aqueous electrolyte. Figure 4 shows the morphology of TNTAs resulting from the anodization process using viscous and aqueous electrolytes. Based on Figure 4, the difference in water content in the electrolyte solution had a significant impact on the morphology of the TNTAs. A higher water content in the electrolyte solution resulted in nanotubes with a larger diameter and a more regular shape. The electrolyte solution with a low water content hindered the mobility of ions in the solution, reducing the TNTA production reaction rate and then producing nanotubes with smaller diameters [58]. Table 1 shows the results of tube diameter measurement.

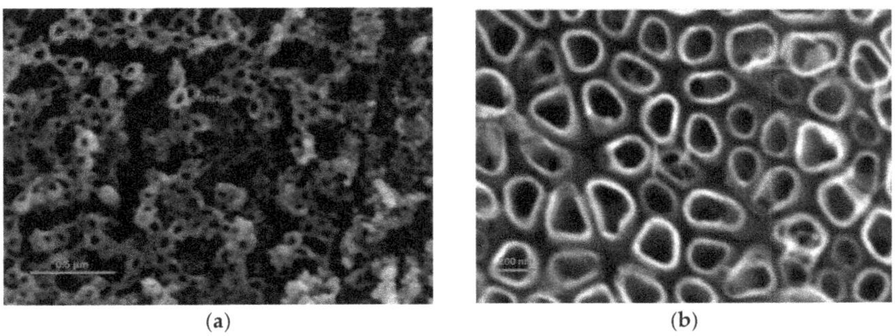

Figure 4. SEM results of photocatalyst morphology of TNTAs with water contents in glycerol of (**a**) 2% v/v and (**b**) 25% v/v.

Table 1. Comparison of nanotube diameter size at a variation of water content in glycerol used in the anodization process.

Water Content in Glycerol, v/v	Diameter of Tube (nm)
2%	24–29
25%	130–168

The presence of F^- ions can dissolve the resulting oxide layer, and F^- ions were more aggressive in aqueous conditions, resulting in shorter tubes with wider diameters [59]. TiO_2 nanotubes with a mean diameter of 55 nm were produced by anodization at a voltage of 50 volts using an electrolyte solution of ethylene glycol containing 0.5% w/v NH_4F and 2% v/v H_2O [28]. It means that the average diameter of TNTAs generated in this study using glycerol electrolyte solution with a water content of 2% v/v was less than that in the previous study using ethylene glycol electrolyte solution with a water content of 2% v/v [22]. The different values were because ethylene glycol had a lower viscosity (1.612×10^{-2} Pa·s) than glycerol (1.412 Pa·s), so the mobility of ions in the ethylene glycol solution was higher than that in the glycerol solution for the same water content.

3.2. *Effect of Water Content in Glycerol Electrolyte on Crystallinity of TNTAs*

The goal of calcination is to produce TiO_2 in the anatase crystalline phase. XRD testing was performed to determine the crystallinity of the photocatalysts. Figure 5 shows the XRD test results. Following the XRD result data, anatase crystals were formed either at a water content of 2% v/v or 25% v/v. Peaks of anatase appear at angles of 25.60°, 38.70°, 48.30°, 53.30°, 55.30°, 63.20°, 70.90°, 76.50°, and 82.50° at 2θ position. From Figure 5, it is clear that more frequent and higher peaks are anatase peaks. The peak altitude in the

XRD results reflects the crystallinity level. It means that the higher the anatase peak, the higher the anatase crystal content. The appearance of the anatase peak in Figure 5 indicates that the main component of the catalyst was the anatase crystal. The anatase content in catalysts at a water content of 25% *v*/*v* was higher than that at a water content of 2% *v*/*v*. Table 2 displays the size of anatase crystals at water contents of 2% *v*/*v* and 25% *v*/*v*.

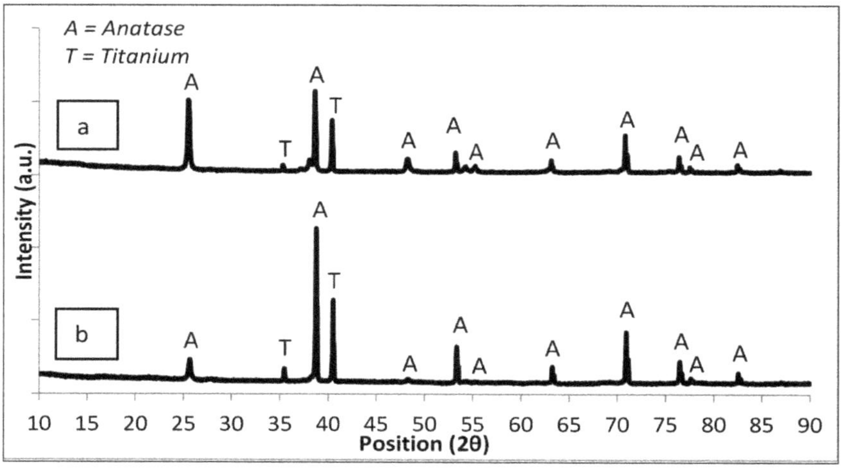

Figure 5. XRD analysis results of catalysts with water contents in glycerol of (**a**) 2% *v*/*v* and (**b**) 25% *v*/*v*.

Table 2. Comparison of anatase crystal size at a variation of water content in glycerol used in the anodization process.

Water Content in Glycerol, *v*/*v*	Anatase Crystal Size (nm)
2%	26.6
25%	23.6

The anodization process using glycerol electrolytes with a water content of 2% *v*/*v* resulted in anatase crystals with greater crystal sizes than those with a water content of 25% *v*/*v*. The resulting anatase crystal size tends to be set at tiny diameters between 10 and 30 nm, and anatase crystals are stabilized in larger-diameter nanotubes [2]. According to the results of catalyst characterization, the water content in the electrolyte solution affected the tube diameter size and the film layer thickness.

Based on the explanation above, the TNTAs resulting from the anodization process using electrolytes with a water content of 25% *v*/*v* were better than those with a water content of 2% *v*/*v*. Therefore, the electrolyte solution with a water content of 25% *v*/*v* was used in the investigation of the effect of sensitizer addition.

3.3. The Effect of Anthocyanin Sensitizer Addition on TNTA Morphology

Ti and O were the primary components of TNTAs, as seen in Figure 6. Other elements, including C, N, and F, were also identified because these elements were absorbed during the anodizing process. After the addition of anthocyanin sensitizer, the C element in TNTAs increased (Figure 6). Based on Figure 6b, TNTAs before sensitization contained Ti, C, and O elements of 67.68, 2.22, and 27.79%wt or 40.55, 5.32, and 49.85% atoms. Then, based on Figure 6d, TNTAs, after sensitization, contained Ti, C, and O elements of 67.96, 4.51, and 25.37%wt or 40.63, 10.74, and 45.39% atoms. This demonstrated that the increase in C element was caused by the diffusion of anthocyanins into TNTAs during the immersionultrasonication process. Zyoud et al. also reported that the C element in the catalyst increased after the addition of anthocyanin sensitizers [31].

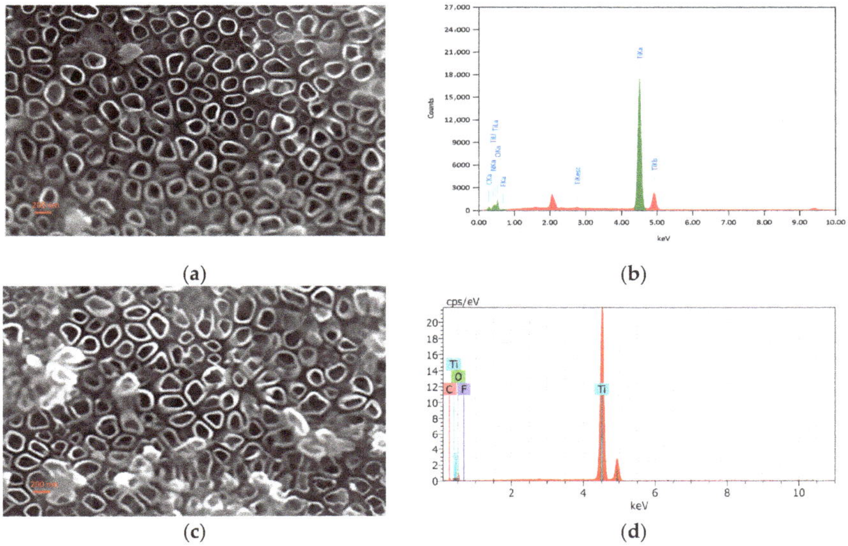

Figure 6. (**a**) SEM for TNTAs before sensitization, (**b**) EDX for TNTAs before sensitization, (**c**) SEM for TNTAs after sensitization, (**d**) EDX for TNTAs after sensitization. Electrolyte solution with a water content of 25% v/v and anthocyanin sensitizer addition of 160 ppm.

3.4. The Effect of Anthocyanin Sensitizer Addition on Crystallinity

The effect of adding anthocyanin sensitizers on the crystallinity of TNTAs was examined using XRD. The results of the XRD analysis are shown in Figure 7. For TNTAs without sensitizers, peaks at 2 θ at angles of 35.40° and 40.50° indicate diffraction from titanium (Ti). Meanwhile, peaks at 2 θ at angles of 25.60°, 38.70°, 48.30°, 53.30°, 55.30°, 63.20°, 70.90°, 76.50°, and 82.50° indicate diffraction from anatase. In detail, sensitized TNTAs had higher anatase peaks than unsensitized TNTAs, especially at an angle of 38.70°. This means that the sensitizer addition successfully increased the anatase crystal content in TNTAs.

The anatase phase crystal sizes without and with sensitizer additions are shown in Table 3. The addition of anthocyanin sensitizers caused the anatase phase crystal size to expand because there was an increase in temperature throughout the process of sensitizing TNTA catalysts with an anthocyanin sensitizer with ultrasound for 30 min. Following the addition of the sensitizer, the calcination effect caused the anatase crystal size of the catalyst to become larger.

3.5. The Effect of Anthocyanin Addition on the Band Gap Energy of TNTAs

The effect of anthocyanin addition on the band gap energy was examined using a UV–Vis DRS. Figure 8a,b depict the results of UV–Vis DRS analysis. Figure 8 depicts the results of UV–Vis DRS investigations to calculate the band gap energy using a Tauc plot, as well as extrapolating Tauc graphics on the linear part of the graph and cutting the x-axis. Figure 8 shows that unmodified TNTAs had a band gap energy of 3.25 eV, which was consistent with reports in the literature and confirmed that TiO_2 was only active when exposed to UV light [31,60]. The band gap energy for TNTAs + anthocyanin was reduced to 2.5 eV. When a photon is absorbed by the anthocyanin molecule, an electron is excited from its highest occupied molecular orbital (HOMO) to its lowest unoccupied molecular orbital (LUMO). Subsequently, a state of relaxation is achieved as a result of the process of electron loss to the conduction band (CB) of TNTAs [43,44,61].

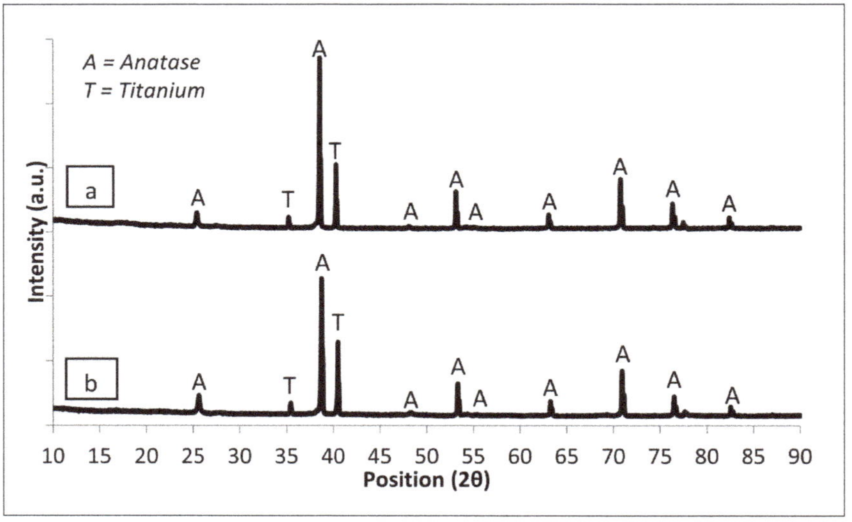

Figure 7. Results of XRD analysis of catalysts (**a**) with sensitizer addition and (**b**) without sensitizer addition. Electrolyte solution with a water content of 25% *v*/*v* and anthocyanin sensitizer addition of 160 ppm.

Table 3. Comparison of anatase crystal size as a result of the addition of anthocyanin sensitizer.

Catalyst	Anatase Crystal Size (nm)
TNTAs	23.6
TNTAs + anthocyanin 160 ppm	28.7

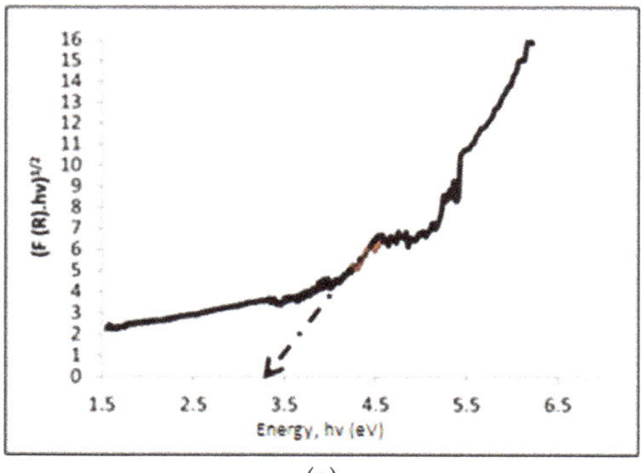

(a)

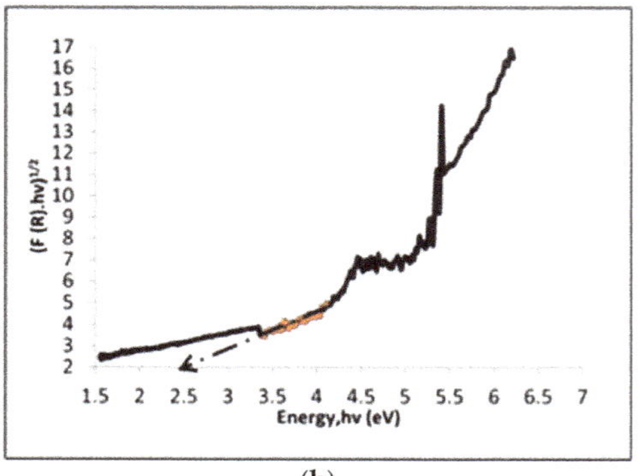

(b)

Figure 8. Tauc plot of Kubelka–Munk function transformation vs. band gap energy (eV) for (**a**) TNTAs and (**b**) for TNTAs + anthocyanin. Electrolyte solution with a water content of 25% *v*/*v* and anthocyanin addition of 160 ppm.

The UV–Vis DRS spectra in Figure 9 exhibited an increase in absorbance values in the wavelength range of 380–800 nm due to the addition of 160 ppm anthocyanin sensitizer.

This shows that catalysts with sensitizers can absorb photons more efficiently in the visible light spectrum.

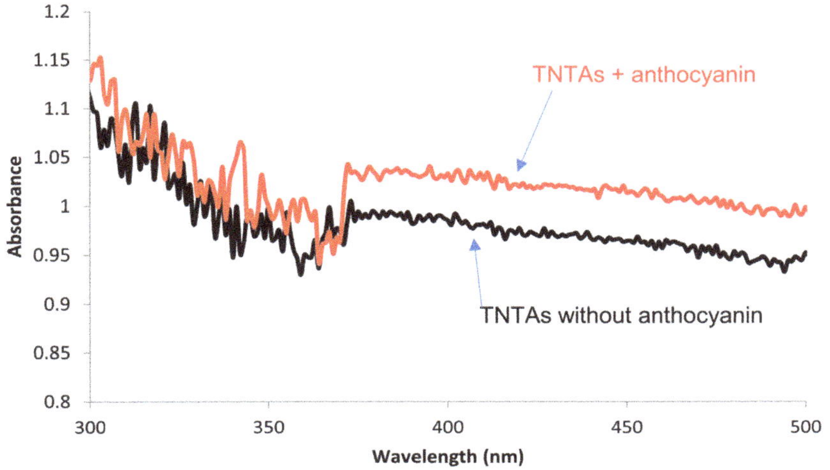

Figure 9. UV–Vis DRS spectra of catalysts with and without anthocyanin addition. Electrolyte solution with a water content of 25% v/v and anthocyanin sensitizer addition of 160 ppm.

3.6. Photocatalytic Process for Methylene Blue (MB) Degradation

3.6.1. The Effect of Initial pH

Figure 10 demonstrates that the rate of MB photodegradation increases when the initial pH increases from 5 to 10. After 120 min, initial pHs of 5, 7, and 10 resulted in MB removal efficiencies of 15.38, 17.47, and 19.71%, respectively. Hence, MB degradation was best achieved at an initial pH of 10, resulting in the highest MB removal efficiency. This was because the initial pH of 10 changed the charge on the surface of the TNTA photocatalysts to be negatively charged, so MB (which was a cationic dye) became more easily adsorbed, thereby increasing the photodegradation rate of MB. The adsorption rate of MB on the catalyst surface affected the variance in reaction rate.

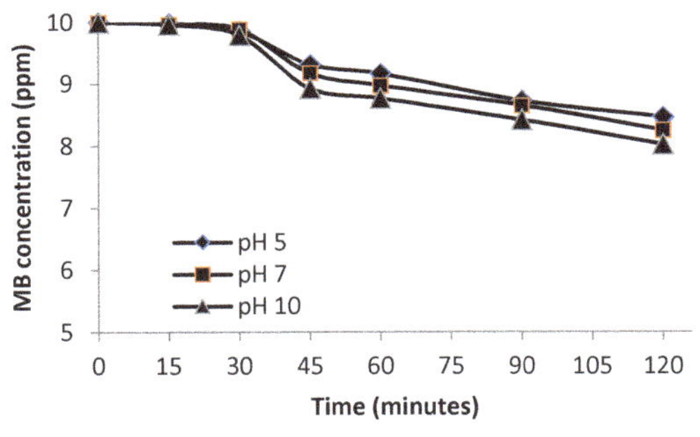

Figure 10. Effect of initial pH variation on the performance of TNTA photocatalyst for degrading MB. Electrolyte solution with a water content of 25% v/v in the anodization process and anthocyanin sensitizer addition of 160 ppm.

The findings of this study were consistent with the findings of a previous study by Dariani et al., who reported that an increase in initial pH from 3 to 11 can improve the MB removal efficiency through a photocatalysis process using 0.5 g of TiO_2 with a particle size of 200 nm and a process time of 5 h [62]. At high pH levels, the catalyst surfaces were dominated by negative charges, and there was a significant interaction between the catalyst surfaces and the dye cations, resulting in strong adsorption. On the other hand, at low pH, the catalytic surfaces and coloring molecules were positively charged. The surface of TiO_2 had a neutral charge at a pH of 6.8. In an acidic medium (pH < 6.8), the surface of TiO_2 had a positive charge; in a base medium (pH > 6.8), it had a negative charge [62].

3.6.2. The Effect of Sensitizer Addition for TNTAs on Photocatalytic Performance

The decrease in MB concentration during the photocatalysis process at various sensitizer additions is shown in Figure 11. An increase in sensitizer addition from 0 to 160 ppm successfully increased the rate of MB degradation. After 120 min, MB removal efficiencies at sensitizer additions of 0 (without sensitizer addition), 40, 80, and 160 ppm were 19.71, 19.94, 24.65, and 29.06%, respectively. The presence of sensitizers reduced the band gap energy, making the catalyst more receptive to photons and increasing the number of excited electrons. The addition of anthocyanin sensitizers did not affect the shape of the catalyst's nanotubes, so the area of the catalytic surface remained vast. The XRD data revealed that the addition of sensitizers increased the anatase crystal size. Anthocyanins had a hydroxyl (OH) group, so the anthocyanin-sensitized catalyst surface became hydrophilic. As a consequence, MB molecules were closer to the catalyst surface, and the tension between the catalyst surfaces and the MB solution became lower.

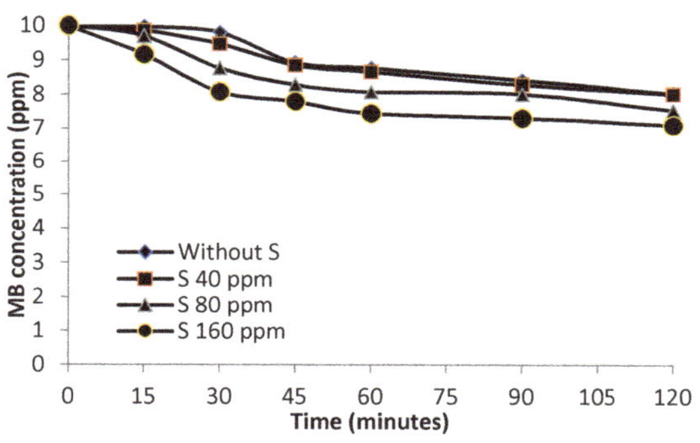

Figure 11. Effect of addition of anthocyanin sensitizers on the performance of TNTA photocatalysts for degrading MB. Electrolyte solution with a water content of 25% *v*/*v* in the anodization process and initial pH of photocatalysis process of 10.

The MB removal efficiency in the case without sensitizer addition was 19.71%. Furthermore, after the sensitizer addition of 40 ppm, the MB removal efficiency increased to 19.94%. It shows that the sensitizer addition of 40 ppm to TNTAs had no significant effect on the MB removal efficiency. However, the sensitizer addition of 160 ppm significantly increased the MB removal efficiency value, namely 29.06%.

According to Angulo et al., dye molecules can easily attach to the catalyst surface and acquire an excited state via photon absorption in the visible light spectrum region [33]. The injection of electrons from anthocyanin dye molecules into the conduction bands of TNTA photocatalysts increased the quantities of reactive oxygen species (ROS) and accelerated the breakdown of MB. Anthocyanin sensitizers absorbed visible light and allowed electrons to be activated. TNTA photocatalysts became more responsive to visible light when anthocyanin sensitizers were added [31,33]. Figure 12 depicts a scheme of MB photodegradation by TNTAs with anthocyanin sensitizers.

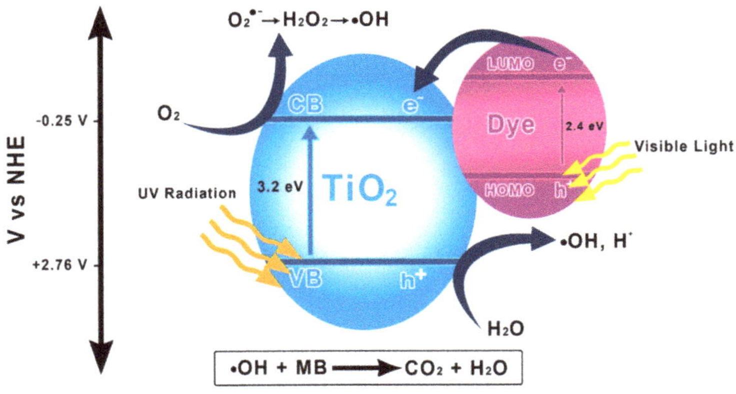

Figure 12. The scheme of MB photodegradation by TNTAs with anthocyanin sensitizers.

3.7. Electrocoagulation Process for Methylene Blue (MB) Degradation

3.7.1. Effect of Voltage

The effect of voltage on the electrocoagulation process for degrading MB is shown in Figure 13. Based on Figure 13, the higher the voltage was applied, the faster the MB degradation rate would be. After 120 min, the electrocoagulation process at voltages of 30, 40, and 50 volts resulted in MB removal efficiencies of 29.91, 43.35, and 51.85%, respectively. An increase in voltage was directly proportional to an increase in electrical current (I). The larger the electrical current was applied, the more the ions were generated at the anode and cathode, resulting in enhanced coagulant formation in the solution [63]. The more the coagulants were generated in the solution, the more the MB molecules could be adsorbed, enhancing the MB removal efficiency.

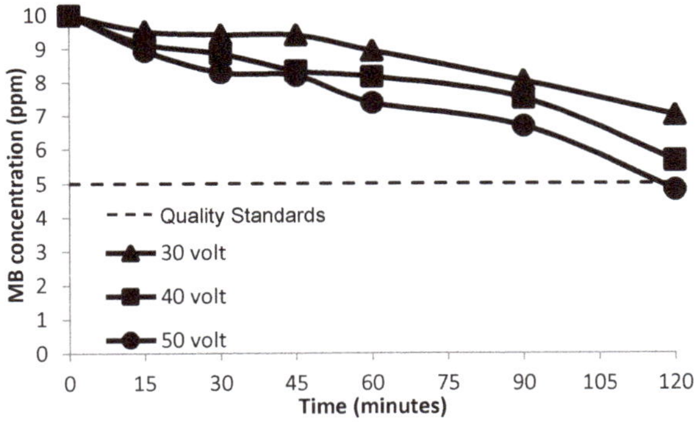

Figure 13. The profiles of MB concentrations during the electrocoagulation process at various voltages.

The waste quality standard (MB of 5 ppm) was reached in around 120 min at a voltage of 50 volts. The linear equations in Figure 14 can be used to approximate the reduction of the MB compound as a function of time during the electrocoagulation process. Through these linear equations, the waste quality standard for MB can be achieved after being processed through an electrocoagulation process at voltages of 30, 40, and 50 volts for 215 min, 153 min, and 120 min, respectively. Furthermore, the relationship between voltage and time needed to reach the waste quality standard can be predicted through the linear equations shown in Figure 15. Through the equations in Figures 14 and 15, the electrocoagulation process time required to reach the MB effluent quality standard at various voltages can be estimated within a voltage range of 30–50 V.

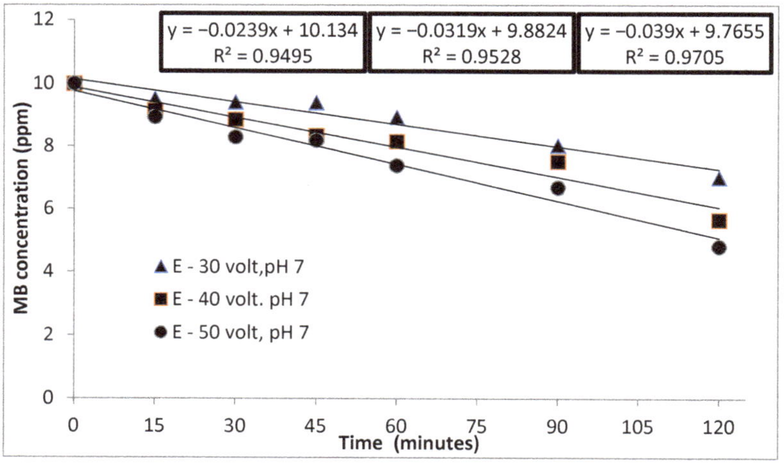

Figure 14. The relationship between MB concentration and electrocoagulation process time.

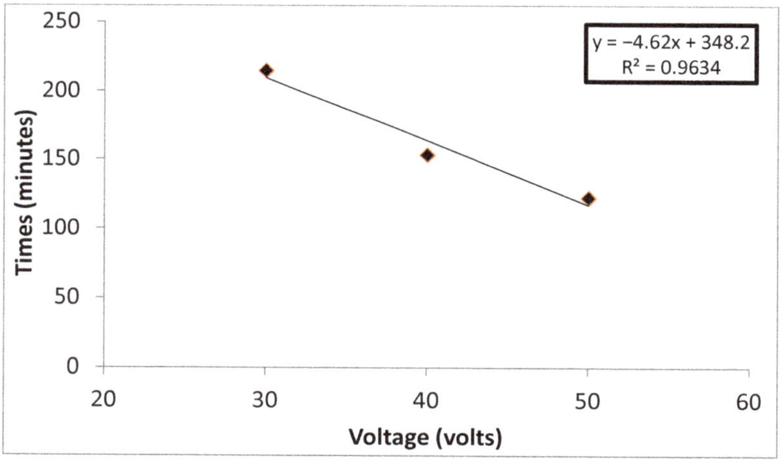

Figure 15. The relationship between voltage and time to achieve waste quality standards.

The degradation of MB during electrocoagulation can be modeled through the first-order kinetic model, following Equation (4).

$$-\frac{dC}{dt} = kC \tag{4}$$

The first-order kinetic model can be rearranged by integration to become Equations (5) and (6) with the boundary conditions of $C = C_0$ at $t = 0$ and $C = C_t$ at $t = t$

$$In\left(\frac{C_t}{C_0}\right) = -kt \tag{5}$$

$$lnC_t = -kt + lnC_0 \tag{6}$$

Equation (5) can be arranged into Equation (6).

$$C_t = C_0\exp(-kt) \tag{7}$$

where C_t is the concentration of MB at t time (ppm), C_0 is the initial concentration of MB (ppm), t is the electrocoagulation time (minutes), and k is the reaction rate constant (min^{-1}). The kinetic constants (k) can be obtained through optimization with the objective function of the Sum of Squared Error (SSE) using Microsoft Excel. The formula of SSE is shown in Equation (7).

$$SSE = \sum_{i=1}^{n} \left(C_{t\,model_i} - C_{t\,experiment_i} \right)^2 \tag{8}$$

where $C_{t\,model}$ is the concentration of MB at time t (ppm) obtained from modeling, and $C_{experiment}$ is the concentration of MB at time t (ppm) obtained from the experiment. The first-order kinetic model was successfully used to simulate the MB concentration profiles during the electrocoagulation process with SSE values of 0.522–0.725. Plotting between experimental data and modeled data is shown in Figure 16. Then, the reaction rate constants at 30, 40, and 50 V were 0.0025, 0.0040, and 0.0053 min^{-1}, respectively.

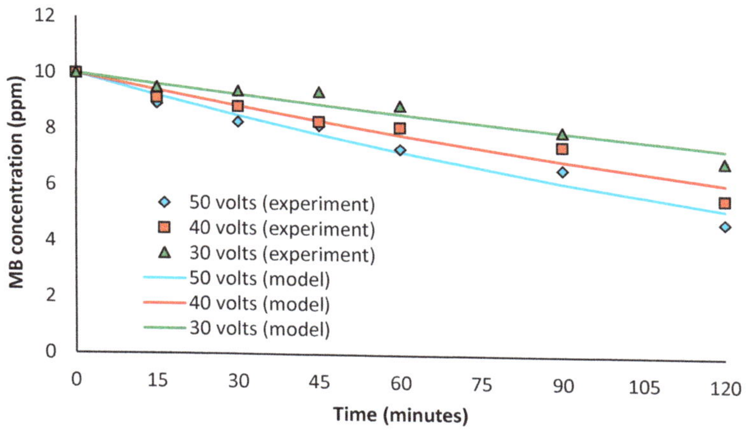

Figure 16. Plotting between experimental data and modeled data obtained from the first-order kinetic model.

3.7.2. The Effect of pH

In this stage, the electrocoagulation process was carried out at a voltage of 50 volts with a variation of initial pH of 5, 7, and 10. Figure 17 shows the profiles of MB concentration during the electrocoagulation process at various initial pHs. The electrocoagulation at initial pHs of 5, 7, and 10 resulted in MB removal efficiencies of 39.49, 51.85, and 57.88%, respectively. Hence, an increase in initial pH from 5 to 10 increased MB removal efficiency values. The highest MB removal efficiency of 57.88% was obtained at an initial pH of 10. According to Figure 17, after 2 h, the electrocoagulation process at initial pHs of 7 and 10 can decrease MB concentrations to meet the waste quality standard for MB. Meanwhile, the time to reach the waste quality standard at pH 5 can be predicted through linear regression of the relationship between MB concentration and time. As a result, for initial pHs of 5, 7, and 10, the time required to achieve the waste quality criteria was 147, 115, and 75 min, respectively. It can be concluded that an initial pH of 10 can lower the MB concentration to the quality standards in less time than initial pHs of 5 and 7. During electrocoagulation, Al^{3+} ions are generated at the anode, and OH^- ions are generated at the cathode. The Al^{3+} can react with OH^- to form various Al species. At high pH, the dominant Al species is $Al(OH)_3$, while at low pH, the dominant Al species is Al^{3+} ions. Hence, the $Al(OH)_3$ coagulant amount is higher at an initial pH of 10 than at initial pHs of 5 and 7. The $Al(OH)_3$ coagulants can adsorb MB. Thus, the MB was easier to remove at initial pH of 10 than at initial pHs of 5 and 7. As a consequence, the initial pH needed a shorter time to decrease the MB concentration from 10 ppm to 5 ppm (quality standard). In line with this study, Muttaqin et al. reported that an initial pH of 10 resulted in a higher $Al(OH)_3$ amount than initial pHs of 5 and 7, so an initial pH of 10 resulted in a higher MB removal efficiency than initial pHs of 5 and 7 [64].

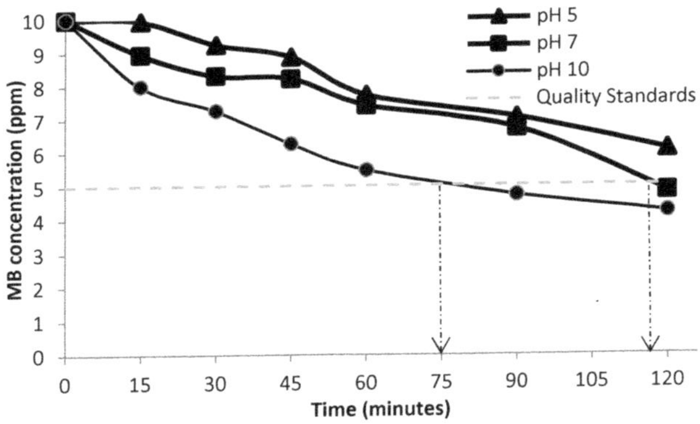

Figure 17. The profiles of MB concentrations during the electrocoagulation process at various initial pHs. Voltage of 50 Volts.

The rate of MB decolorization and the amount of sludge were affected by the initial pHs. As shown in Table 4, an increase in initial pH increased the sludge amount and the decolorization rate. At higher pH levels, $Al(OH)_3$ coagulations were generated in large amounts. The more the amount of $Al(OH)_3$ was generated in the solution, the higher the dye decolorization efficiency was obtained [42]. The solution pH has a significant impact on the electrocoagulation process performance [8].

Table 4. Sludge weights in the electrocoagulation process at various initial pHs.

Initial pH of Electrocoagulation	Sludge Weight (g)
5	2.42
7	2.83
10	3.85

3.8. A Combined Process of Electrocoagulation and Photocatalysis for Methylene Blue (MB) Degradation

At this stage, a combined process of electrocoagulation–photocatalysis without and with anthocyanin sensitizer addition for degrading MB compounds was performed. The profiles of MB concentrations during photocatalysis (P), electrocoagulation (E), and the combination process of E-P are shown in Figure 18. After 120 min, the electrocoagulation process achieved an MB removal efficiency of 57.88%, while the combined process of electrocoagulation and photocatalysis without sensitizer addition achieved a higher MB reduction efficiency of 59.66%. The presence of anthocyanin sensitizers in the combined process of electrocoagulation and photocatalysis can improve the MB removal efficiency from 59.66% to 64.30%. In this condition, waste quality standards can be reached within 48 min, whereas combining electrocoagulation and photocatalysis without a sensitizer takes 60 min (Figure 18). The presence of photocatalysis could assist in the degradation of MB dyes into non-hazardous compounds for the environment.

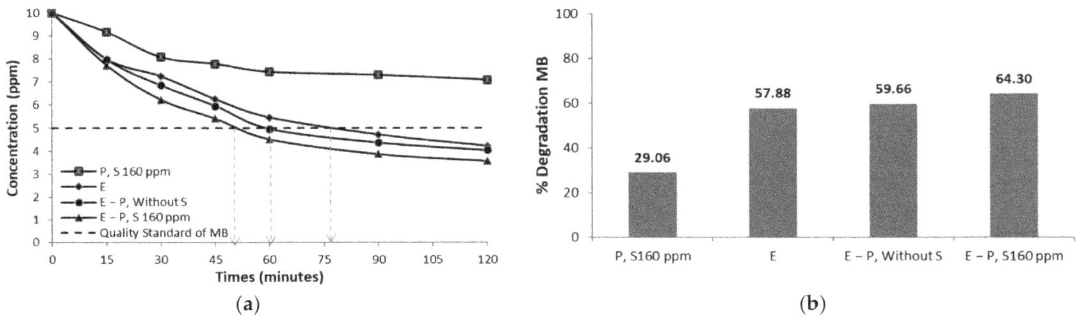

Figure 18. (**a**) Profiles of MB concentrations during the electrocoagulation, photocatalysis, and combined process of electrocoagulation and photocatalysis; (**b**) the total MB removal efficiencies after 120 min. Electrolyte solution with a water content of 25% v/v in the anodization process, anthocyanin sensitizer addition of 160 ppm, initial pH of 10, and voltage of 50 Volts. E = electrocoagulation, P = photocatalysis, E−P = combined process of electrocoagulation and photocatalysis. Blue arrows in (**a**) show the times that are needed by the processes to decrease the MB concentration from 10 ppm to 5 ppm (quality standard).

This was also reflected in the sludge's color of the combined process without sensitizer addition, which was more faded when compared to the sludge's color of the combined process without sensitizer addition. In other words, the presence of anthocyanin sensitizers in the combined process can improve its ability to degrade MB. The addition of anthocyanin sensitizers can improve photon absorption in the visible light spectrum, allowing TNTA photocatalysts to degrade MB dyes more effectively. The addition of sensitizers reduced the band gap energy, making the catalyst more receptive to photons and increasing the number of excited electrons. The addition of anthocyanin sensitizers did not change the morphology of the catalyst nanotubes, so the area of the catalytic surface remained vast.

The combined process of electrocoagulation and photocatalysis can decrease MB concentrations to waste quality standards in less time than the electrocoagulation alone and the photocatalysis alone. Figure 18 shows that the electrocoagulation alone needed around 75 min to reach waste quality criteria, while the combined process of electrocoagulation and photocatalysis with sensitizer addition needed only 48 min to reach waste quality criteria.

Table 5 provides a comparison of the results of this study with those of other studies. According to Table 5, the combined process in this study resulted in a higher MB removal per catalyst than that in a previous study [51] because this study used anthocyanin sensitizers, while a previous study did not use sensitizers [51]. However, another previous study [61] resulted in a higher MB removal per catalyst than this study, where the previous study [61] utilized the photocatalysis process with anthocyanin sensitizers, while this study utilized the combined process anthocyanin sensitizers. It was due to the difference in the ratio of MB/catalyst, namely 0.3125 mg/cm^2 in this study and 0.125 mg/cm^2 in the previous study [61]. Furthermore, the other previous studies (see Table 5) reported different pollutant removal efficiencies because of the differences in the types of methods, types and forms of photocatalysts, types of sensitizers, and types of pollutants.

Table 5. Comparison of the results of the current research with those from previous studies.

No	Methods	Photocatalyst	Power/Film	Sensitizer	Pollutant	Experiment	Results	References
1	E-P with a single reactor	TiO_2 nanotubes array	Film	Anthocyanin	Methylene Blue (MB)	Volume of 1000 mL, MB of 10 ppm, ratio of MB/catalyst of 0.3125 mg/cm^2	The removal efficiency of MB was 64.5%. The MB removal per catalyst was 2.016%/cm^2.	This study
2	E-P with a single reactor	Fe-TiO_2 nanotubes arrays	Film	-	Methylene Blue–ciprofloxacin (MB-CIP)	Volume of 1000 mL, 10 ppm MB-10 ppm CIP, film surface of 64 cm^2, ratio of MB/catalyst of 0.1562 mg/cm^2, ratio of CIP/catalyst of 0.1562 mg/cm^2	The removal efficiencies of CIP and MB were 90% and 100%, respectively. The CIP removal per catalyst was 1.406%/cm^2. The MB removal per catalyst was 1.5625%/cm^2.	[51]
3	E-P with separated reactors	Ag-TiO_2 P25	Powder	-	Tartrazine (TTZ)	Volume of 3 L, TTZ of 50 mg/L, 100 mg/L Ag-TiO_2 P25, ratio of TTZ/catalyst of 0.5 mg/mg	The mineralization of TTZ was 74%. The TTZ removal per catalyst was 0.2467%/mg.	[56]
4	E-P with separated reactor	Immobilized ZnO	Film	-	Petroleum	Initial COD of 1000 mg/L, ratio of petroleum/catalyst of 0.2711 mg/mg	The removal efficiency of COD was 94% after 60 min. The petroleum removal per catalyst was 0.0255%/mg	[9]
5	E-P with separated reactors	TiO_2 and ZnO	Powder	-	Olive washing wastewater	Initial pH of 6.9, voltage of 12.5 V, ZnO of 1 g/L	The removal efficiencies of COD and color were 88% and 100%, respectively.	[55]
6	E-P with a single reactor	ZnO	Powder	-	Malathion	Initial malathion concentration of 45 mg/L, ZNO of 1.6 mg/L, volume of 500 mL, ratio of malathion/catalyst of 28.125 mg/mg	The removal efficiency of malathion was 68.33%. The malathion removal per catalyst was 85.41%/mg.	[54]

Table 5. *Cont.*

No	Methods	Photocatalyst	Power/Film	Sensitizer	Pollutant	Experiment	Results	References
7	P	TiO_2 nanoparticles	Powder	Anthocyanin	Methylene Blue (MB)	Anthocyanin from blackberry, hibiscus, Urucum, Curcuma	The decolorization efficiency was 51–73%.	[45]
8	P	TiO_2 P25	Thin Film	Anthocyanin	Methylene Blue (MB)	Volume of 25 mL, MB of 10 ppm, surface area of 2 cm^2, ratio of MB/catalyst of 0.125 mg/cm^2	The decolorization efficiency was 25.6%. The MB removal per catalyst was 12.8%/cm^2.	[61]
9	P	TiO_2 nanoparticles	Powder	Chlorophyll	Methylene Blue (MB)	Volume of 100 mL, MB of 20 ppm, catalyst of 1 g/L, ratio of MB/catalyst of 20 mg/mg	The removal efficiency of MB was 85%. The MB removal per catalyst was 8.5%/mg.	[65]
10	P	TiO_2 aerogel	Powder	Anthocyanin	Cr (VI)	Volume of 50 mL, Cr(VI) concentration of 15 ppm, ratio of Cr(VI)/catalyst of 0.75 mg/mg	The removal efficiency of Cr(VI) was 100%. The Cr(VI) removal per catalyst was 2%/mg.	[46]

3.9. The Color and Weight of Sludges Resulting from the Electrocoagulation Process and the Combined Process of Electrocoagulation and Photocatalysis

The sludges from the electrocoagulation process and the combined process were analyzed. The sludge from the electrocoagulation process without MB was used as a control. Then, the control sludge was compared to the sludges from the electrocoagulation process, the combined process with and without sensitizer additions. Figure 19 shows the sludges of the electrocoagulation process and the combined processes.

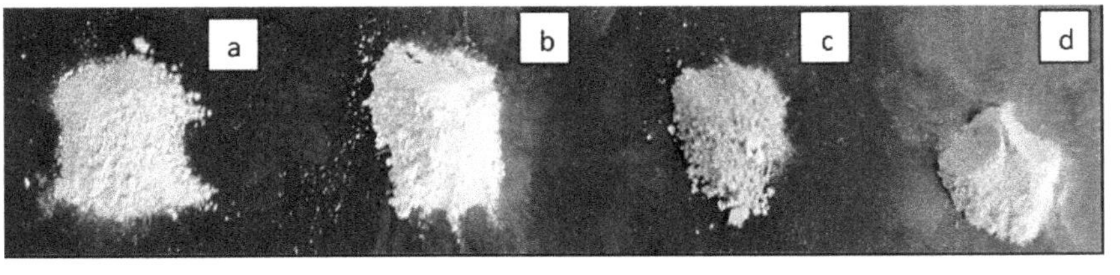

Figure 19. The result of sludges in the process of electrocoagulation and electrocoagulation–photocatalysis combination. (**a**) blank; (**b**) electrocoagulation; (**c**) combined process without sensitizer; (**d**) combined process with sensitizer.

Based on Figure 19, there was a difference in the color of sludges resulting from the electrocoagulation process and the combined process. The electrocoagulation process has three steps, namely electrolysis, coagulation, and flotation [42,66]. During this process, Al is oxidized to Al^{3+} at the anode, while water is reduced to H^+ and OH^- ions at the cathode. Then, Al^{3+} reacts with OH ions to form coagulants of $Al(OH)_3$, which can adsorb contaminants, while H_2 is produced through water electrolysis. The H_2 gas aids in the removal of contaminants through flotation [55,67,68]. Detailed reactions occurring during the electrocoagulation process are as follows:

$$\text{At Anode}: Al_{(s)} \rightarrow Al^{3+}_{(aq)} + 3e^- \tag{9}$$

$$\text{At Cathode}: 2H_2O_{(aq)} + 2e^- \rightarrow 2OH^-_{(aq)} + H_{2(g)} \tag{10}$$

$$\text{Chemical Reaction}: Al^{3+}_{(aq)} + 3OH^-_{(aq)} \rightarrow Al(OH)_{3(s)} \tag{11}$$

$$\text{Total Reaction}: 2Al_{(s)} + 6H_2O_{(aq)} \rightarrow 2Al(OH)_{3(s)} + 3H_{2(g)} \tag{12}$$

Equation (11) is applicable for alkaline medium [64]. Then, $Al(OH)_3$ can adsorb MB to form sludge (MB-$Al(OH)_3$), as shown in Equation (12).

$$Al(OH)_{3(s)} + MB \rightarrow (MB - Al(OH)_3)_{(s)} \tag{13}$$

Light and catalysts are required for the photocatalysis process to accelerate chemical reactions. When subjected to UV light with an energy larger than the band gap energy, these semiconductor materials create electron pairs $\left(e^-_{cb}\right)$ and holes, which can reduce and oxidize compounds. The reactions that occur in the photocatalytic process for waste degradation are as follows:

$$TiO_2 + hv \rightarrow TiO_2(h^+ + e^-_{cb}) \tag{14}$$

$$TiO_2(h^+) + H_2O \rightarrow OH\bullet + H^+ \tag{15}$$

$$TiO_2(e^-_{cb}) + O_2 \rightarrow O_2^-\bullet \tag{16}$$

$$OH\bullet \text{ or } O_2^-\bullet + MB \rightarrow CO_2 + H_2 \tag{17}$$

Therefore, the electrocoagulation process involves the adsorption of MB dye, but the combined process involves the degradation of MB dye by TNTA photocatalysts as well as the adsorption process. The MB removal mechanisms during a combined process of electrocoagulation and photocatalysis are shown in Figure 20 in detail.

The sludge formed from the electrocoagulation process was not blue compared to the sludge formed from the combined process of electrocoagulation–photocatalysis. Methylene Blue (MB) can turn into colorless leuco-Methylene Blue (LMB) due to a hydrogenation reaction, while the LMB can turn into MB due to an oxidation reaction [68]. A hydrogenation reaction is a reaction between hydrogen and another compound. During electrocoagulation, hydrogen is generated at the cathode. The presence of hydrogen might cause MB to change to LMB. Then, the coagulants might adsorb the LMB, thereby resulting in colorless sludge.

Meanwhile, during the combined process, the hydrogen resulted from the electrocoagulation, while the oxygen resulted from the photocatalysis process. The presence of hydrogen would change MB to LMB, and then the presence of oxygen would change LMB to MB again. Furthermore, the coagulants adsorbed MB, thereby resulting in blue sludge.

The sludges resulting from the combined process without and with sensitizers also appear to differ in color. When compared to the sludge from the electrocoagulation process without sensitizers, the sludge from the combined process with the sensitizers has a more fading color. The combined process with sensitizers degraded more MB than the combined process without sensitizers. As a consequence, the coagulants in the combined process with sensitizers adsorbed less MB than those in the combined process without sensitizers. Therefore, the sludge from the combined process with sensitizers has a more fading color.

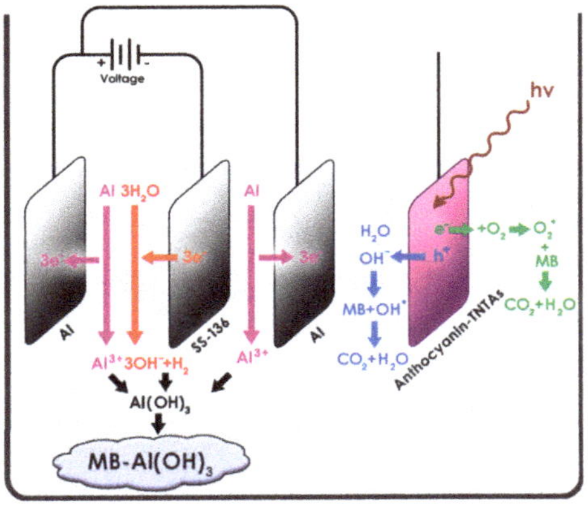

Figure 20. The MB removal mechanisms during a combined process of electrocoagulation and photocatalysis.

Table 6 shows the weight of the sludge produced in the electrocoagulation process and the combined process without and with sensitizer addition, indicating that there is a difference in the weight of the sludge produced. The sludge produced in the combined process was less than the sludge produced in the electrocoagulation alone. This was because, in the combined process, some OH^- ions became OH• radicals due to the photocatalysis process, so only a small amount of OH^- ions reacted with Al^{3+} ions to form $Al(OH)_3$ coagulants.

Table 6. Sludge weight in the electrocoagulation process and the combined process of electrocoagulation–photocatalysis.

Method	Sludge Weight (g)
Electrocoagulation	3.85
Combination without Sensitizer	2.15
Combination with Sensitizer	1.6

4. Conclusions

The water content in the glycerol electrolyte solution used in the anodization process significantly affected the diameter of nanotubes and the size of TNTAs. Specifically, a higher water content led to a bigger diameter while simultaneously causing a decrease in crystal size. The incorporation of anthocyanins into TNTAs did not alter the structural characteristics of the nanotubes. However, it led to an increase in the crystal size of the anatase phase from 23.6 nm to 28.7 nm and a decrease in the band gap energy from 3.2 eV to 2.5 eV. Utilization of the sensitized TNTAs in the combined process of electrocoagulation and photocatalysis can increase the removal efficiency from 59.66% to 64.30%. The results of this study confirmed the authors' hypothesis that the combined process of electrocoagulation and photocatalysis using sensitized TiO_2 nanotube photocatalysts can result in a higher MB removal efficiency than the electrocoagulation alone, photocatalysis alone, and the combined process using unsensitized TiO_2 nanotube photocatalysts. In general, the total cost of the photocatalysis process using sensitized TiO_2 nanotubes is higher than the total cost of the photocatalysis process using unsensitized TiO_2 nanotubes because there is the cost of synthesis of sensitized TiO_2 nanotubes in the former. Then, the MB removal efficiency in the former does not exceed 65%. Therefore, a feasibility study of the process is necessary to be conducted in the future.

This study has some limitations. This study used a wavelength of 663 nm to measure the MB concentration during electrocoagulation and photocatalysis at various pH conditions. However, in theory, the optimum wavelength of MB may be affected by pH levels. For the next study, this needs to be noted. Furthermore, the composition of sludges resulting from electrocoagulation, photocatalysis, and the combined process was not analyzed. Thus, this needs to be noted in the next study. The future research prospect might be the application of the combined process with sensitized TiO_2 nanotube photocatalysts for treating Rhodamine B, methyl orange, as well as real organic wastes. In addition, the economic analysis and feasibility study of the combined process are necessary to be conducted in the future. This study did not analyze the stability and recyclability of the sensitized TiO_2 nanotubes. Therefore, these analyses need to be carried out in the next research. Also, this study did not analyze the intermediate products resulting from the combined process. Hence, identification and characterization of these intermediates need to be conducted in the future to know detailed mechanisms during the combined process.

Author Contributions: Conceptualization, I.K., S. and A.R.; methodology, I.K. and A.R.; software, H.P. and I.K.; validation, S.; formal analysis, H.P. and D.K.S.; investigation, H.P. and D.K.S.; resources H.P. and I.K.; data curation, H.P., I.K. and S.; writing—original draft preparation, H.P. and I.K.; writing—review and editing, S.; visualization, I.K.; supervision, I.K. and S.; project administration, D.K.S.; funding acquisition, I.K. All authors have read and agreed to the published version of the manuscript.

Funding: This research was funded by the Directorate General of Higher Education, Ministry of National Education Indonesia, through HIBAH FUNDAMENTAL 2023, grant number B/609/UN43.9/PT.00.03.

Informed Consent Statement: Not applicable.

Data Availability Statement: Data will be made available on request.

Acknowledgments: The main financial support from the Directorate General of Higher Education, Ministry of National Education Indonesia, through HIBAH FUNDAMENTAL 2023, is gratefully acknowledged.

Conflicts of Interest: The authors declare no conflict of interest.

References

1. Karim, A.V.; Shriwastav, A. Degradation of Ciprofloxacin Using Photo, Sono, and Sonophotocatalytic Oxidation with Visible Light and Low-Frequency Ultrasound: Degradation Kinetics and Pathways. *Chem. Eng. J.* **2020**, *392*, 124853. [CrossRef]
2. Vasiraja, N.; Saravana Sathiya Prabhahar, R.; Joshua, A. Preparation and Physio–Chemical Characterisation of Activated Carbon Derived from Prosopis Juliflora Stem for the Removal of Methylene Blue Dye and Heavy Metal Containing Textile Industry Effluent. *J. Clean. Prod.* **2023**, *397*, 136579. [CrossRef]
3. Marahel, F.; Mombeni Goodajdar, B.; Niknam, L.; Faridnia, M.; Pournamdari, E.; Mohammad Doost, S. Ultrasonic Assisted Adsorption of Methylene Blue Dye and Neural Network Model for Adsorption of Methylene Blue Dye by Synthesised Mn-Doped PbS Nanoparticles. *Int. J. Environ. Anal. Chem.* **2023**, *103*, 3059–3080. [CrossRef]
4. Mustafa, F.H.A.; Gad ElRab, E.K.M.; Kamel, R.M.; Elshaarawy, R.F.M. Cost-Effective Removal of Toxic Methylene Blue Dye from Textile Effluents by New Integrated Crosslinked Chitosan/Aspartic Acid Hydrogels. *Int. J. Biol. Macromol.* **2023**, *248*, 125986. [CrossRef]
5. Zhao, Q.; Zhang, L.; Wang, X.; Jia, X.; Xu, P.; Zhao, M.; Dai, R. Simultaneous Efficient Adsorption and Photocatalytic Degradation of Methylene Blue over Iron(III)-Based Metal–Organic Frameworks: A Comparative Study. *Transit. Met. Chem.* **2019**, *44*, 789–797. [CrossRef]
6. Li, Q.; Li, Y.; Ma, X.; Du, Q.; Sui, K.; Wang, D.; Wang, C.; Li, H.; Xia, Y. Filtration and Adsorption Properties of Porous Calcium Alginate Membrane for Methylene Blue Removal from Water. *Chem. Eng. J.* **2017**, *316*, 623–630. [CrossRef]
7. Pasichnyk, M.; Gaálová, J.; Minarik, P.; Václavíková, M.; Melnyk, I. Development of Polyester Filters with Polymer Nanocomposite Active Layer for Effective Dye Filtration. *Sci. Rep.* **2022**, *12*, 973. [CrossRef]
8. Ihaddaden, S.; Aberkane, D.; Boukerroui, A.; Robert, D. Removal of Methylene Blue (Basic Dye) by Coagulation-Flocculation with Biomaterials (Bentonite and Opuntia Ficus Indica). *J. Water Process Eng.* **2022**, *49*, 102952. [CrossRef]
9. Keramati, M.; Ayati, B. Petroleum Wastewater Treatment Using a Combination of Electrocoagulation and Photocatalytic Process with Immobilized ZnO Nanoparticles on Concrete Surface. *Process Saf. Environ. Prot.* **2019**, *126*, 356–365. [CrossRef]
10. Meili, L.; Lins, P.V.S.; Costa, M.T.; Almeida, R.L.; Abud, A.K.S.; Soletti, J.I.; Dotto, G.L.; Tanabe, E.H.; Sellaoui, L.; Carvalho, S.H.V.; et al. Adsorption of Methylene Blue on Agroindustrial Wastes: Experimental Investigation and Phenomenological Modelling. *Prog. Biophys. Mol. Biol.* **2019**, *141*, 60–71. [CrossRef]
11. Babar, M.; Munir, H.M.S.; Nawaz, A.; Ramzan, N.; Azhar, U.; Sagir, M.; Tahir, M.S.; Ikhlaq, A.; Mohammad Azmin, S.N.h.; Mubashir, M.; et al. Comparative Study of Ozonation and Ozonation Catalyzed by Fe-Loaded Biochar as Catalyst to Remove Methylene Blue from Aqueous Solution. *Chemosphere* **2022**, *307*, 135738. [CrossRef] [PubMed]

12. Li, H.; Lin, Y.; Luo, Y.; Yu, P.; Hou, L. Relating Organic Fouling of Reverse Osmosis Membranes to Adsorption during the Reclamation of Secondary Effluents Containing Methylene Blue and Rhodamine B. *J. Hazard. Mater.* **2011**, *192*, 490–499. [CrossRef] [PubMed]
13. El-Moselhy, M.M.; Kamal, S.M. Selective Removal and Preconcentration of Methylene Blue from Polluted Water Using Cation Exchange Polymeric Material. *Groundw. Sustain. Dev.* **2018**, *6*, 6–13. [CrossRef]
14. Kumar, M.S.; Sonawane, S.H.; Pandit, A.B. Degradation of Methylene Blue Dye in Aqueous Solution Using Hydrodynamic Cavitation Based Hybrid Advanced Oxidation Processes. *Chem. Eng. Process. Process Intensif.* **2017**, *122*, 288–295. [CrossRef]
15. Mohammed, H.A.; Khaleefa, S.A.; Basheer, M.I. Photolysis of methylene blue dye using an advanced oxidation process (ultraviolet light and hydrogen peroxide). *J. Eng. Sustain. Dev.* **2022**, *25*, 59–67. [CrossRef]
16. Huseynov, K.; Ramazanov, M.; Garib, M. Photocatalytic degradation of organic pollutants in air by application of titanium dioxide nanoparticles. *Chem. Eng. Trans.* **2017**, *60*, 241–246. [CrossRef]
17. Sonmez, B.; Baser, E.; Gel, O.Y. Photodecolourization of Methylene Blue by Fe- and Cd-Incorporated Titania-Supported Zeolite Clinoptilolite. *Microporous Mesoporous Mater.* **2022**, *340*, 112001. [CrossRef]
18. Xing, Z.; Zhang, J.; Cui, J.; Yin, J.; Zhao, T.; Kuang, J.; Xiu, Z.; Wan, N.; Zhou, W. Recent Advances in Floating TiO_2-Based Photocatalysts for Environmental Application. *Appl. Catal. B Environ.* **2018**, *225*, 452–467. [CrossRef]
19. Tan, H.; Zhang, Y.; Li, B.; Yang, H.; Hou, H.; Huang, Q. Preparation of TiO_2-Coated Glass Flat Membrane and Its Photocatalytic Degradation of Methylene Blue. *Ceram. Int.* **2023**, *49*, 17236–17244. [CrossRef]
20. Yang, R.; Zhu, Z.; Hu, C.; Zhong, S.; Zhang, L.; Liu, B.; Wang, W. One-Step Preparation (3D/2D/2D) BiVO4/FeVO4@rGO Heterojunction Composite Photocatalyst for the Removal of Tetracycline and Hexavalent Chromium Ions in Water. *Chem. Eng. J.* **2020**, *390*, 124522. [CrossRef]
21. Yang, R.; Zhong, S.; Zhang, L.; Liu, B. PW12/CN@Bi_2WO_6 Composite Photocatalyst Prepared Based on Organic-Inorganic Hybrid System for Removing Pollutants in Water. *Sep. Purif. Technol.* **2020**, *235*, 116270. [CrossRef]
22. Abdullah, H.; Khan, M.M.R.; Ong, H.R.; Yaakob, Z. Modified TiO_2 Photocatalyst for CO_2 Photocatalytic Reduction: An Overview. *J. CO_2 Util.* **2017**, *22*, 15–32. [CrossRef]
23. Roy, P.; Berger, S.; Schmuki, P. TiO_2 Nanotubes: Synthesis and Applications. *Angew. Chem. Int. Ed.* **2011**, *50*, 2904–2939. [CrossRef] [PubMed]
24. Singh, J.; Juneja, S.; Soni, R.K.; Bhattacharya, J. Sunlight Mediated Enhanced Photocatalytic Activity of TiO_2 Nanoparticles Functionalized CuO-Cu_2O Nanorods for Removal of Methylene Blue and Oxytetracycline Hydrochloride. *J. Colloid Interface Sci.* **2021**, *590*, 60–71. [CrossRef] [PubMed]
25. You, M.; Kim, T.G.; Sung, Y.-M. Synthesis of Cu-Doped TiO_2 Nanorods with Various Aspect Ratios and Dopant Concentrations. *Cryst. Growth Des.* **2010**, *10*, 983–987. [CrossRef]
26. Khan, M.A.M.; Ansari, A.A.; Choudhary, P.; Ahmed, J.; Kumar, S.; Hussain, S. Reduced Graphene Oxide Supported Ag-Loaded Brookite TiO2 Nanowires: Enhanced Photocatalytic Degradation Performance and Electrochemical Energy Storage Applications. *Diam. Relat. Mater.* **2023**, *139*, 110397. [CrossRef]
27. Safajou, H.; Khojasteh, H.; Salavati-Niasari, M.; Mortazavi-Derazkola, S. Enhanced Photocatalytic Degradation of Dyes over Graphene/Pd/TiO_2 Nanocomposites: TiO_2 Nanowires versus TiO_2 Nanoparticles. *J. Colloid Interface Sci.* **2017**, *498*, 423–432. [CrossRef]
28. Abdullah, M.; Kamarudin, S.K. Titanium Dioxide Nanotubes (TNT) in Energy and Environmental Applications: An Overview. *Renew. Sustain. Energy Rev.* **2017**, *76*, 212–225. [CrossRef]
29. Wtulich, M.; Szkoda, M.; Gajowiec, G.; Jurak, K.; Trykowski, G.; Lisowska-Oleksiak, A. Hydrothermal Modification of TiO_2 Nanotubes in Water and Alkali Metal Electrolytes ($LiNO_3$, $NaNO_3$, KNO_3) — Direct Evidence for Photocatalytic Activity Enhancement. *Electrochim. Acta* **2022**, *426*, 140802. [CrossRef]
30. Feng, Y.; Rijnaarts, H.H.M.; Yntema, D.; Gong, Z.; Dionysiou, D.D.; Cao, Z.; Miao, S.; Chen, Y.; Ye, Y.; Wang, Y. Applications of Anodized TiO_2 Nanotube Arrays on the Removal of Aqueous Contaminants of Emerging Concern: A Review. *Water Res.* **2020**, *186*, 116327. [CrossRef]
31. Zyoud, A.H.; Saleh, F.; Helal, M.H.; Shawahna, R.; Hilal, H.S. Anthocyanin-Sensitized TiO_2 Nanoparticles for Phenazopyridine Photodegradation under Solar Simulated Light. *J. Nanomater.* **2018**, *2018*, 2789616. [CrossRef]
32. Diaz-Uribe, C.; Vallejo, W.; Camargo, G.; Muñoz-Acevedo, A.; Quiñones, C.; Schott, E.; Zarate, X. Potential Use of an Anthocyanin-Rich Extract from Berries of Vaccinium Meridionale Swartz as Sensitizer for TiO_2 Thin Films—An Experimental and Theoretical Study. *J. Photochem. Photobiol. A Chem.* **2019**, *384*, 112050. [CrossRef]
33. Diaz-Angulo, J.; Lara-Ramos, J.; Mueses, M.; Hernández-Ramírez, A.; Li Puma, G.; Machuca-Martínez, F. Enhancement of the Oxidative Removal of Diclofenac and of the TiO_2 Rate of Photon Absorption in Dye-Sensitized Solar Pilot Scale CPC Photocatalytic Reactors. *Chem. Eng. J.* **2020**, *381*, 122520. [CrossRef]
34. Khalil, M.; Gunlazuardi, J.; Ivandini, T.A.; Umar, A. Photocatalytic Conversion of CO_2 Using Earth-Abundant Catalysts: A Review on Mechanism and Catalytic Performance. *Renew. Sustain. Energy Rev.* **2019**, *113*, 109246. [CrossRef]
35. Rekha, M.; Kowsalya, M.; Ananth, S.; Vivek, P.; Jauhar, R.O.M.U. Current–Voltage Characteristics of New Organic Natural Dye Extracted from Terminalia Chebula for Dye-Sensitized Solar Cell Applications. *J. Opt.* **2019**, *48*, 104–112. [CrossRef]
36. Pujiastuti, H.; Kustiningsih, I.; Slamet, S. Improvement of the Efficiency of TiO_2 Photocatalysts with Natural Dye Sensitizers Anthocyanin for the Degradation of Methylene Blue. *J. Rekayasa Kim. Lingkung.* **2021**, *16*, 84–99. [CrossRef]

37. Degradation of Ciprofloxacin Antibiotic Waste Using TiO2 Nanotube with Addition of Anthocyanin Dye-Sensitizer in Photocatalysis Process: Review | Fidarohman | Jurnal Rekayasa Kimia & Lingkungan. Available online: https://jurnal.usk.ac.id/RKL/article/view/28520 (accessed on 5 October 2023).
38. Díaz-Uribe, C.; Vallejo, W.; Campos, K.; Solano, W.; Andrade, J.; Muñoz-Acevedo, A.; Schott, E.; Zarate, X. Improvement of the Photocatalytic Activity of TiO_2 Using Colombian Caribbean Species (Syzygium Cumini) as Natural Sensitizers: Experimental and Theoretical Studies. *Dyes Pigments* **2018**, *150*, 370–376. [CrossRef]
39. Chandel, M.; Thakur, M.; Sharma, A.; Pathania, D.; Kumar, A.; Singh, L. Chlorophyll Sensitized $(BiO)_2CO_3/CdWO_4/rGO$ Nano-Hybrid Assembly for Solar Assisted Photo-Degradation of Chlorzoxazone. *Chemosphere* **2022**, *305*, 135472. [CrossRef]
40. Elseman, A.M. *Solar Cells: Theory, Materials and Recent Advances*; BoD—Books on Demand: Norderstedt, Germany, 2021; ISBN 978-1-83881-016-0.
41. Mohammed, H.A.; Khan, R.A. Anthocyanins: Traditional Uses, Structural and Functional Variations, Approaches to Increase Yields and Products' Quality, Hepatoprotection, Liver Longevity, and Commercial Products. *Int. J. Mol. Sci.* **2022**, *23*, 2149. [CrossRef]
42. Daneshvar, N.; Ashassi Sorkhabi, H.; Kasiri, M.B. Decolorization of Dye Solution Containing Acid Red 14 by Electrocoagulation with a Comparative Investigation of Different Electrode Connections. *J. Hazard. Mater.* **2004**, *112*, 55–62. [CrossRef]
43. Giang, N.T.H.; Thinh, N.T.; Hai, N.D.; Loc, P.T.; Thu, T.N.A.; Loan, N.H.P.; Quang, D.M.; Anh, L.D.; Truong An, V.N.T.; Phong, M.T.; et al. Application of TiO_2 Nanoparticles with Natural Chlorophyll as the Catalyst for Visible Light Photocatalytic Degradation of Methyl Orange and Antibacterial. *Inorg. Chem. Commun.* **2023**, *150*, 110513. [CrossRef]
44. Zyoud, A.; Hilal, H. Investigation of Curcumin as Sensitizer for Anatase TiO_2 Nanoparticles in Photodegradation of of Phenazopyridine with Visible Light. *PhytoChem BioSub J.* **2014**, *8*, 127–137.
45. Goulart, S.; Jaramillo Nieves, L.J.; Dal Bó, A.G.; Bernardin, A.M. Sensitization of TiO_2 Nanoparticles with Natural Dyes Extracts for Photocatalytic Activity under Visible Light. *Dyes Pigments* **2020**, *182*, 108654. [CrossRef]
46. Yang, H.; Jiang, L.; Li, Y.; Li, G.; Yang, Y.; He, J.; Wang, J.; Yan, Z. Highly Efficient Red Cabbage Anthocyanin Inserted TiO_2 Aerogel Nanocomposites for Photocatalytic Reduction of Cr(VI) under Visible Light. *Nanomaterials* **2018**, *8*, 937. [CrossRef]
47. Mecha, A.C.; Chollom, M.N. Photocatalytic Ozonation of Wastewater: A Review. *Environ. Chem. Lett.* **2020**, *18*, 1491–1507. [CrossRef]
48. Mahamud, M.; Taddesse, A.M.; Bogale, Y.; Bezu, Z. Zeolite Supported $CdS/TiO_2/CeO_2$ Composite: Synthesis, Characterization and Photocatalytic Activity for Methylene Blue Dye Degradation. *Mater. Res. Bull.* **2023**, *161*, 112176. [CrossRef]
49. Gong, P.; Li, B.; Kong, X.; Shakeel, M.; Liu, J.; Zuo, S. Hybriding Hierarchical Zeolite with Pt Nanoparticles and Graphene: Ternary Nanocomposites for Efficient Visible-Light Photocatalytic Degradation of Methylene Blue. *Microporous Mesoporous Mater.* **2018**, *260*, 180–189. [CrossRef]
50. Girma, S.; Taddesse, A.M.; Bogale, Y.; Bezu, Z. Zeolite-Supported g-C3N4/ZnO/CeO_2 Nanocomposite: Synthesis, Characterization and Photocatalytic Activity Study for Methylene Blue Dye Degradation. *J. Photochem. Photobiol. A Chem.* **2023**, *444*, 114963. [CrossRef]
51. Muttaqin, R.; Pratiwi, R.; Ratnawati; Dewi, E.L.; Ibadurrohman, M.; Slamet. Degradation of Methylene Blue-Ciprofloxacin and Hydrogen Production Simultaneously Using Combination of Electrocoagulation and Photocatalytic Process with Fe-TiNTAs. *Int. J. Hydrogen Energy* **2022**, *47*, 18272–18284. [CrossRef]
52. Slamet, S.; Kurniawan, R. Degradation of Tartrazine and Hydrogen Production Simultaneously with Combination of Photocatalysis-Electrocoagulation. *AIP Conf. Proc.* **2018**, *2024*, 020064. [CrossRef]
53. Slamet, N.S.; Shobri, A.; Anindria, F.A.; Mauricio, R.; Tafsili, M.A.B.; Slamet, S. Treatment of Batik Industry Waste with a Combination of Electrocoagulation and Photocatalysis. *Chem. Eng.* **2018**, *9*, 936–943. [CrossRef]
54. Rangel-Peraza, J.G.; Prado, M.A.R.; Amabilis-Sosa, L.E.; Bustos-Terrones, Y.A.; Ramírez-Pereda, B. Malathion Removal through Peroxi-Electrocoagulation and Photocatalytic Treatments. Optimization by Statistical Analysis. *Int. J. Electrochem. Sci.* **2020**, *15*, 8253–8264. [CrossRef]
55. Ates, H.; Dizge, N.; Yatmaz, H.C. Combined Process of Electrocoagulation and Photocatalytic Degradation for the Treatment of Olive Washing Wastewater. *Water Sci. Technol.* **2016**, *75*, 141–154. [CrossRef] [PubMed]
56. Santos, L.M.; Amorim, K.P.D.; Andrade, L.S.; Batista, P.S.; Trovó, A.G.; Machado, A.E.H. Dye Degradation Enhanced by Coupling Electrochemical Process and Heterogeneous Photocatalysis. *J. Braz. Chem. Soc.* **2015**, *26*, 1817–1823. [CrossRef]
57. Blanckenberg, A.; Malgas-Enus, R. Raspberry-like Gold-Decorated Silica (SSx–AMPS–Au) Nanoparticles for the Reductive Discoloration of Dyes. *SN Appl. Sci.* **2019**, 787. [CrossRef]
58. Ratnawati; Gunlazuardi, J.; Slamet. Development of Titania Nanotube Arrays: The Roles of Water Content and Annealing Atmosphere. *Mater. Chem. Phys.* **2015**, *160*, 111–118. [CrossRef]
59. Regonini, D.; Bowen, C.R.; Jaroenworaluck, A.; Stevens, R. A Review of Growth Mechanism, Structure and Crystallinity of Anodized TiO_2 Nanotubes. *Mater. Sci. Eng. R Rep.* **2013**, *74*, 377–406. [CrossRef]
60. Madhusudan Reddy, K.; Manorama, S.V.; Ramachandra Reddy, A. Bandgap Studies on Anatase Titanium Dioxide Nanoparticles. *Mater. Chem. Phys.* **2003**, *78*, 239–245. [CrossRef]
61. Diaz-Uribe, C.; Vallejo, W.; Romero, E.; Villareal, M.; Padilla, M.; Hazbun, N.; Muñoz-Acevedo, A.; Schott, E.; Zarate, X. TiO_2 Thin Films Sensitization with Natural Dyes Extracted from Bactris Guineensis for Photocatalytic Applications: Experimental and DFT Study. *J. Saudi Chem. Soc.* **2020**, *24*, 407–416. [CrossRef]

62. Dariani, R.S.; Esmaeili, A.; Mortezaali, A.; Dehghanpour, S. Photocatalytic Reaction and Degradation of Methylene Blue on TiO_2 Nano-Sized Particles. *Optik* **2016**, *127*, 7143–7154. [CrossRef]
63. Liu, C.; Neale, Z.G.; Cao, G. Understanding Electrochemical Potentials of Cathode Materials in Rechargeable Batteries. *Mater. Today* **2016**, *19*, 109–123. [CrossRef]
64. Muttaqin, R.; Ratnawati, R.; Slamet, S. Batch Electrocoagulation System Using Aluminum and Stainless Steel 316 Plates for Hospital Wastewater Treatment. *IOP Conf. Ser. Earth Environ. Sci.* **2022**, *963*, 012056. [CrossRef]
65. Krishnan, S.; Shriwastav, A. Application of TiO_2 Nanoparticles Sensitized with Natural Chlorophyll Pigments as Catalyst for Visible Light Photocatalytic Degradation of Methylene Blue. *J. Environ. Chem. Eng.* **2021**, *9*, 104699. [CrossRef]
66. Liu, N.; Wu, Y. Removal of Methylene Blue by Electrocoagulation: A Study of the Effect of Operational Parameters and Mechanism. *Ionics* **2019**, *25*, 3953–3960. [CrossRef]
67. Holt, P.; Barton, G.; Mitchell, C. Electrocoagulation as a Wastewater Treatment. In Proceedings of the Third Annual Australian Environmental Engineering Research Event, Castlemaine, VIC, Australia, 23–26 November 1999.
68. Jegadeesan, C.; Somanathan, A.; Jeyakumar, R.B.; Godvin Sharmila, V. Combination of Electrocoagulation with Solar Photo Fenton Process for Treatment of Landfill Leachate. *Environ. Technol.* **2023**, *44*, 4441–4459. [CrossRef]

water

MDPI

Article

Study of the Process of Electrochemical Oxidation of Active Pharmaceutical Substances on the Example of Nitrofurazone ((2*E*)-2-[(5-Nitro-2-furyl)methylene]hydrazine Carboxamide)

Vitalyi Vladimirovich Kuznetsov [1,2], Natalya Andreevna Ivantsova [1], Evgenii Nikolaevich Kuzin [1,*], Andrey Vladimirovich Pirogov [3], Yaroslav Olegovich Mezhuev [1], Elena Alexeevna Filatova [1] and Yulia Michailovna Averina [1]

1 Mendeleev University of Chemical Technology of Russia, 125047 Moscow, Russia; vitkuzn1@mail.ru (V.V.K.); mezhuev.i.o@muctr.ru (Y.O.M.); filatova.e.a@muctr.ru (E.A.F.)
2 Frumkin Institute of Physical Chemistry and Electrochemistry, 119071 Moscow, Russia
3 Lomonosov Moscow State University M. V., 119234 Moscow, Russia; pirogov@analyt.chem.msu.ru
* Correspondence: kuzin.e.n@muctr.ru

Abstract: The effective mineralization of nitrofurazone (10–100 mg L^{-1}) was performed in aqueous solutions in the presence of chloride ions by electrochemical treatment. The destruction of the organic pollutant molecules was due to their interaction with active oxygen- and chlorine-containing species forming at the inert anode (Pt/Ti or BDD) during electrolysis. Measurements of nitrofurazone concentration, chemical oxygen demand (COD) and total organic carbon (TOC) were used to estimate the removal efficiency of the pollutant. Both the pollutant oxidation rate and the degree of its mineralization were higher for the BDD anode due to the higher anode potentials on it in the course of electrolysis, which provides a high rate of active oxidizer species generation. As a result, practically full nitrofurazone molecule destruction (>99%) was achieved in 30 min at an anodic current density of 0.1 A cm^{-2}, a volume current density of 1.33 A L^{-1} and pH 2 using BDD anodes. On the other hand, the nitrafurazone degradation efficiency was about 95% for Pt/Ti anodes under the same conditions. Additionally, byproducts of nitrofurazone electrooxidation were investigated by means of liquid chromato-mass-spectrometry (LC/MS). It was found that the initial decolorization of nitrofurazone solution, which occurs during the first 5 min of electrolysis, is due to the formation of a dichloro derivative of nitrofurazone, which causes the destruction of the π−conjugated bond system. Further electrolysis resulted in the almost complete destruction of the dichloro derivative within 30 min of electrochemical treatment.

Keywords: nitrofurazone; indirect electrochemical oxidation; mineralization; wastewater treatment

1. Introduction

The degradation of surface waters is one of the key issues of the modern world, and it has attracted attention worldwide. Population growth, an increase in industrial production, as well as the development and implementation of new materials and technologies have led to the complicated composition of wastewater. As a result, current wastewater treatment methods are ineffective. For this reason, the development of innovative, highly effective methods of wastewater treatment is relevant. Specific pollutants, the neutralization of which is undoubtedly necessary, are active pharmaceutical ingredients (APIs). Increasing the use of antibiotics in agriculture results in a significant increase in API emissions into the environment and, in particular, in the hydrosphere [1–3]. The bioaccumulation of APIs, the inhibition of vital activities of microorganisms, the formation of resistant and highly toxic complexes, biocenoses degradation and the emergence of resistant strains of microorganisms should be mentioned among the negative consequences of the entry of APIs into the environment [4–7]. The contamination of domestic wastewater by APIs is another consequence of their use, which can lead to the complete inactivation of deep biological wastewater treatment and, therefore, the discharge of wastewater with a high content of nutrients into various water bodies, which causes eutrophication. According to various sources, antibiotics and their metabolites are already found in both natural [8,9] and domestic wastewater [10]. Solving this problem is urgent, not only for countries with significant sources of fresh water [11], but also for countries that produce drinking water through the desalination of sea or ocean water [12–18]. Antibiotics and their metabolites are just the tip of the iceberg. The chemical contamination of wastewater by other forms of APIs is, essentially, not controlled. Nonsteroidal anti-inflammatory drugs, hormonal drugs (contraceptives), antihistamines, antibacterials (with the exception of antibiotics), antivirals and painkillers, unlike antibiotics, are not subject to strict control in many countries. For this reason, they are used far more often. Their entry into the environment can exceed the volumes of antibiotics entry by an order of magnitude [19,20]. Since the health effects of these pollutants have not yet been fully studied, they can be hazardous for both humans and nature.

Many studies devoted to the issues of the purification of both natural waters and wastewater from various kinds of pharmaceutical substances have been recently published [21–26]. They describe the treatment of model or real wastewater containing less than 1 mg L^{-1} APIs. Advanced oxidation processes (AOPs) are traditionally used for the destruction of API molecules. The combination of catalytic photo-oxidation processes and oxidation with Fenton's reagent and ozone makes it possible to destruct microamounts of APIs with high efficiency and the formation of a minimal amount of organic metabolites. The high efficiency of AOPs is due to the formation of highly active oxidizers, namely, the hydroxyl radical, hydroperoxyl radical, ozone and other species, which are significantly superior in their oxidative power to most individual oxidative agents. Data on the redox potentials of some oxidizers generated in AOPs are given in Table 1 [27–29].

Table 1. Standard oxidation potentials of some oxidizers in aqueous solutions.

Oxidant	$OH^{\bullet}$	O_3	H_2O_2	Hydroperoxyl Radicals	Cl_2	O_2
Standard oxidation potential $E^{\circ}_{298.15}$/V vs. SHE	2.80	2.07	1.77	~1.70	1.36	1.23

Despite all of the advantages described above, the use of AOPs is only reasonable for trace amounts of APIs in waste or natural waters. The treatment of more concentrated wastewater (c(APIs) > 10 mg L^{-1}) becomes either insufficiently effective or too expensive [30,31]. The use of supercritical water oxidation (SCWO) is relevant in these cases. Although SCWO has demonstrated high efficiency in wastewater purification, its application requires high energy consumption and only becomes economically feasible if there is a significant amount of organic compounds in the treated water, which are a source of thermal energy for the process.

Untreated pharmaceutical industry wastewater contains relatively high (>5 mg L^{-1}) concentrations of nitrofurazone. The use of AOPs associated with treatment with ozone or Fenton's reagent is not effective in this case, since it will lead to a large consumption of the reagents. In addition, the turbidity of real wastewater can be rather high, which makes photochemical methods inefficient. Electrooxidation is one of the methods for AOPs, which can be used alone or in combination with other methods for wastewater treatment. It is suitable for the local purification of industrial wastewater containing high concentrations of APIs. Electrooxidation can provide a high degree of mineralization of pollutants, while the residue amounts can be further oxidized using advanced oxidative processes [32–35]. Electrochemical treatment has recently been successfully applied to the degradation of paracetamol [36] and tetracycline [37] in aqueous solutions. The efficiency of electrooxidation for the decomposition of such difficult-to-oxidize compounds, such as non-ionic surfactants based on ethers, was shown in [38]. The mechanism of electrooxidation is mainly divided into direct and indirect mechanisms [39]. Direct oxidation means that the organic molecule is oxidized on the anode by direct electron transfer. The processes of direct oxidation are very important in electroorganic synthesis; however, their role in electrodestruction is limited. Indirect oxidation implies the generation of active species at the anode (Table 1), which further react with pollutant molecules. The full mineralization of organic molecules during the processes of indirect oxidation may be due to the high oxidation power of species generated during electrolysis.

Nitrofurazone ((2*E*)-2-[(5-Nitro-2-furyl)methylene]hydrazine carboxamide) (Figure 1) is antibiotically active against both Gram-positive and Gram-negative bacteria. Its chemical formula is given below:

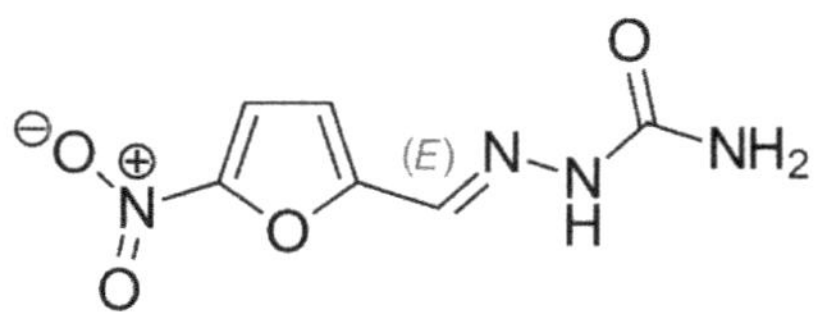

Figure 1. Chemical formula of nitrofurazone.

The use of nitrofurazone in the treatment of animal diseases, in particular, in animal husbandry, is still widespread. Moreover, it is still very popular as a topical solution for the treatment of tonsillitis in Russia and some other countries. Therefore, the neutralization of wastewater containing nitrofurazone is an urgent task, which should be solved in the near future taking into account environmental management requirements [40–42]. Since synthetic pollutants are not removed by conventional wastewater treatment plants, the costs of neutralizing them can be high [40]. The rational choice of electrode material and electrolysis conditions is needed for cost reduction [41]. In this study, nitrofurazone was chosen as a model active pharmaceutical substance to evaluate the possibilities and limitations of electrochemical methods for the destruction of small API molecules. Nitrofurazone is a heterocyclic compound containing electron-withdrawing groups, which hinders its oxidation. It should be noted that data on the indirect electrochemical oxidation of nitrofurazone, the degree of its mineralization, as well as the products of its electrochemical transformations, are lacking in the literature. A quite complex photoelectrochemical method for nitrofurazone removal was proposed in [43], in which about 80% of pollutant removal was achieved within 10 h.

Electrochemical treatment methods can be applied for the mineralization of nitrofurazone [44]; however, a number of issues need to be solved before using them. Firstly, the electrochemical oxidation of organic pollutants of complex chemical nature can be incomplete, which results in the accumulation of various metabolites in aqueous solutions under the treatment. The incomplete oxidation products can be highly toxic, and their accumulation in wastewater is completely unacceptable. Therefore, the identification of metabolites is necessary for the evaluation of the possibility of using electrochemical methods. Secondly, it is necessary to determine the time for wastewater treatment, as well as to choose the optimal solution composition and electrolysis conditions. Based on the above, the aims of the present study were: (1) to estimate the efficiency of the use of electrochemical oxidation for wastewater containing relatively high concentrations (10–100 mg L^{-1}) of nitrofurazone; (2) to find a degree of organic pollutant mineralization in the course of electrochemical treatment; (3) to identify metabolites of nitrofurazone electrooxidation and, therefore, to estimate the safety of electrochemical methods for wastewater treatment.

2. Material and Reagents

Aqueous solutions containing 10–100 mg L^{-1} nitrafuranzone were used for electrooxidation. To prepare them, the required amount of Avexima© (Russian Federation), a drug based on nitrofurazone, was dissolved in distilled water. According to the manufacturer, one pill contains 20 mg nitrofurazone, 100 mg sodium chloride and excipients: tartaric acid, sodium bicarbonate, sodium carbonate, macrogol and medium molecular weight povidone. The drug was crushed and ground in an agate mortar. After that, the necessary amount of the powder (in terms of nitrofurazone) was weighed using an analytical balance (CAS CAUW-120D, South Korea) and dissolved in 0.75 L of deionized water. The solution was stirred using a magnetic stirrer with slight heating of the solution ($t \approx 50$ °C). After that, the solution was filtered from insoluble impurities using the Büchner funnel with the Millipore membrane (0.45 μm pore size, 47 mm diameter). The resulting solution was transparent. The concentration of nitrofurazone in the prepared solutions was determined by its specific reaction with a 10 wt.% solution of decarbonized sodium hydroxide resulting in the appearance of an orange color. The optical density of solutions ($\lambda = 450$ nm) was measured on a photoelectric colorimeter DR 2800 (HACH, Ames, IA, USA).

The concentration of chloride ions in solutions under electrochemical treatment varied within 0.001–0.010 M, which was five times higher than the concentration of nitrofurazone and corresponded to the level of chloride ions in natural waters. The oxidation of chloride ions on the inert anode, followed by the interaction of oxidation products with water molecules, leads to the formation of highly-active chlorine-containing species, e.g., HClO, $OCl^{\bullet}$ in the solution under electrochemical treatment. Since an activation barrier of redox reactions involving them is low, this facilitates the destruction of organic pollutant molecules. However, the interaction of chlorine-containing species with organic molecules can lead to the formation of toxic chlorine derivatives. Therefore, it is necessary to control their formation and avoid their accumulation in treated aqueous solutions.

Preliminary experiments on the electrochemical oxidation of nitrofurazone were carried out in cells with both separated and undivided cathode and anode compartments. As was later shown, the effective oxidation of nitrofurazone is only possible in cells with separated spaces (see below); it occurs in the anolyte. A cation exchange membrane (Nafion®) was used (Figure 2) in the experiments on nitrofurazone electrooxidation. It allows for stabilizing the acidity of the anolite. Since the hydronium ions transfer number is large enough, H^+-ions form during electrolysis at the anode transfer through the membrane without causing excessive acidification of the solution in the anode compartment of the electrochemical cell.

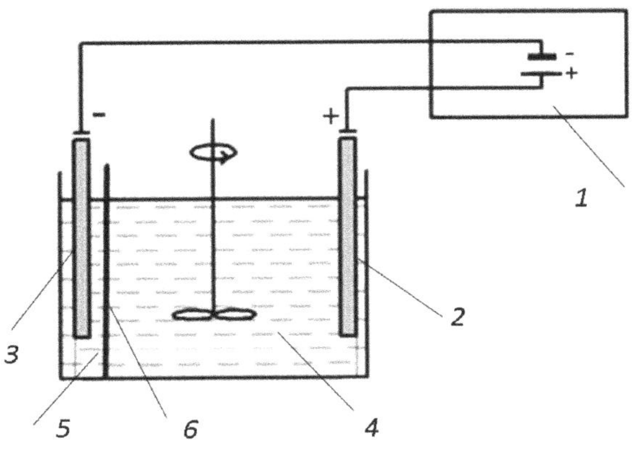

Figure 2. Electrochemical reactor. 1-DC source, 2-anode, 3-cathode, 4-anodic compartment ($V = 150$ mL), 5-cathodic compartment ($V = 50$ mL), 6-cation exchange membrane.

The volume of the anolyte was 150 mL. Sulfuric acid was added to the solution under electrochemical treatment in such a way that its concentration was ~0.01 M in order to increase its electrical conductivity and facilitate the oxidation of the organic pollutant. The cathodic compartment of the electrolyzer ($V = 50$ mL) was filled with 0.01 M H_2SO_4 (pH $\approx$ 2.0).

The electrooxidation of nitrafurazone was carried out on platinized titanium or boron-doped diamond (BDD) anodes with S_{geom} = 2 cm^2. A counter electrode was a mesh of platinized titanium. Electrolysis was performed in galvanostatic mode using a DC source DAZHENG PS-305D (China). The solution was stirred with a magnetic stirrer. In the course of the preliminary experiments, it was found that the rapid decoloration of the solution under electrochemical treatment, and, hence, the oxidation of nitrofurazone, occurs when the current passing through the cell is more than 0.15 A. For this reason, I = 0.2 A was chosen for the electrochemical destruction of a pollutant. This value corresponds to the anodic current density of 0.1 A cm^{-2} and the volume current density per the volume of the anolyte of 1.33 A L^{-1}.

The efficiency of the nitrofurazone removal was calculated using the formula:

$$\text{Removal efficiency } (\%) = \frac{C_0 - C_i}{C_0} \cdot 100\%,$$

where C_0 (g L^{-1}) is the initial concentration of the pollutant, C_i (g L^{-1}) is the concentration of the pollutant after electrochemical treatment.

The rate of pollutant degradation can be expressed by the formula [42]:

$$-\frac{dC}{dt} = k \cdot C^{\alpha} \cdot C(\mathrm{Ox})^{\beta},$$

where C (g L^{-1}) is the concentration of the pollutant, t (min) is the treatment time, k is the Arrhenius constant in the relevant units, C(Ox) (mol L^{-1}) is the concentration of active oxidizing species in the solution, which is approximately constant in steady state electrolysis. Under conditions of constant C(Ox), one can write:

$$-\frac{dC}{dt} = k' \cdot C^{\alpha}$$

If the reaction obeys a first order equation (α = 1):

$$-\frac{dC}{dt} = k' \cdot C, \text{ and}$$

$$C = C_0 \cdot e^{-k't}$$

where k' is the rate constant (min^{-1}). Therefore, the dependence of ln C vs. t is a straight line in the case of the first-order equation, which makes it possible to determine the rate constant.

A digital potentiostat IPC-Pro MF (Volta, Russian Federation) was used for voltammetric measurements. The potential of the working electrode was set and measured versus the Ag/AgCl reference electrode in saturated KCl solution. In the present paper, all electrode potentials are given against this electrode unless otherwise stated. The processing of experimental data was performed using Origin 8® software (Cheshire, CT, USA).

To estimate the degree of organic pollutant destruction, the chemical oxygen demand (COD), total organic carbon content (TOC) and the open-circuit potential of Pt/Pt electrode (ORP) were measured. In addition, since formaldehyde and monobasic carboxylic acids can potentially be formed during electrooxidation, it is also necessary to determine their concentrations in the solution under electrochemical treatment. The color reaction with phenylhydrazine and potassium ferricyanide in an alkaline medium was used for the determination of formaldehyde, and the reaction with ammonium metavanadate was used for carboxylic acid detection.

The identification of organic metabolites that can potentially accumulate in aqueous solutions of nitrofurazone in the course of electrochemical treatment was performed using the HPLC-MS technique. The analysis was carried out on a hybrid mass-spectrometer for tandem mass spectrometry QTrap 3200 AB Sciex (Canada). The device was equipped with two ion sources: Turbo Spraytm (electrospray ionization) and Heated Nebulizertm (chemical ionization at atmospheric pressure). The separation of organic substances was performed using Acclaim RSLC. The sample was loaded in solvent A (0.5% aqueous solution of formic acid) followed by gradient elution with solvent B (acetonitrile) over 39 min according to the following program: 0.00–2 min—20% B, 2–31 min—20–100% B, 31–35 min—100% B, 35.1–39 min—20% B. A full survey mass scan was performed in the interval of an m/z range of 50–1070.

The use of the HPLC-MS technique was due to the complex chemical nature of the mixtures of species formed during electrolysis. Intermediates formed in the course of electrochemical treatment have similar chemical structures, and it is necessary to separate them before, for instance, NMR analysis. This was beyond the scope of this study.

3. Results and Discussion

3.1. Electrochemical Behavior of Nitrofurazone

According to the literature data, nitrofurazone is electrochemically active in aqueous solutions [45,46] due to easily reduced nitro groups in its molecule. Nitrofurazone reduction waves were used for its quantitative determination by means of electrochemical sensors [46]. The mechanism of nitrafurazone electroreduction is quite complex, includes several electrochemical stages and finally leads to the formation of a hydroxylamine (RNHOH) derivative. The route of nitrofurazone electrochemical reduction depends on both the electrode material and the pH of the solution, and can generally be represented as follows [46]:

$$R\text{–}NO_2 + e \rightleftarrows R\text{–}NO_2^{\bullet -}$$

$$R\text{–}NO_2^{\bullet -} + 2\,H^+ + e \rightarrow R\text{–}NO + H_2O$$

$$R\text{–}NO + 2\,H^+ + 2e \rightarrow R\text{–}NHOH.$$

Nitro radicals forming at the first stage of nitrafurazone electroreduction react with oxygen molecules under aerobic conditions, resulting in the formation of suboxide radicals, which can later be transformed into hydrogen peroxide. It is very likely that the antimicrobial activity of nitrofurazone solutions is due to these reactions.

Voltammetric measurements showed two reduction waves of nitrofurazone in acidic solution at the Pt/Ti electrode (Figure 3). The redox transitions are irreversible because both the electrode potentials and the electrical charges corresponding to them depend on the scan rate.

There were no responses associated with the electrochemical transformation of nutrofurazone in the anodic region of potentials. The destruction of nitrofurazone molecules at high anodic potentials is due to their interaction with highly reactive oxygen- ($O_2^{\bullet -}$, $HO_2^{\bullet -}$, etc.) or chlorine- ($ClO^{\bullet}$, HClO) containing species generated at the anode [47]. From a practical point of view, anodic oxidation is preferable for wastewater treatment since it can lead to the complete destruction of organic molecules, which avoids the accumulation of toxic metabolites in wastewater. The cathodic and anodic compartments of the electrochemical cell should be separated to prevent the useless consumption of the electrical charge in the reduction of nitrofurazone at the cathode and the subsequent oxidation-formed products at the anode. In addition, the interaction between reactive intermediates of the cathode and anode reaction results in the formation of unpredictable products during the electrochemical treatment, the potential danger of which is unknown. The separation of the cathode and anode compartments of the electrolytic cell avoids the mixing of cathodic and anodic reaction products.

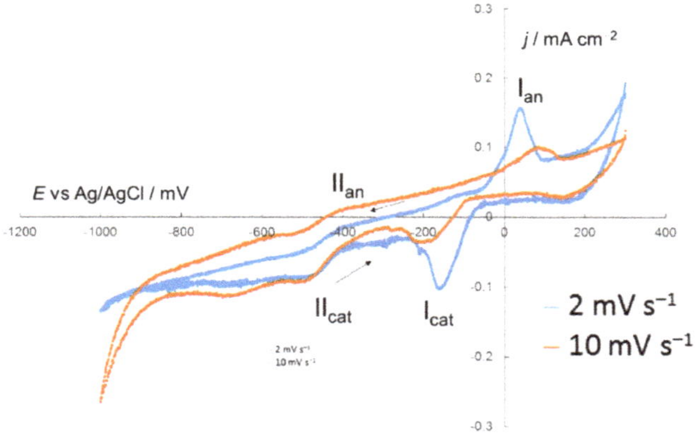

Figure 3. Cyclic voltammograms in solution of nitrafurazone (26.7 mg L^{-1}) obtained by dissolving Avexima® in 0.01 M H_2SO_4. Numbers I and II correspond to the redox transitions of nitrafurazone in the cathodic region of potentials (cat—redox transitons on the cathodic scans of CVs, an—redox transitions on the anodic scans of CVs). See comments in the text.

The use of a cation-exchange membrane prevents the acidification of anolytes due to the transport of hydronium ions formed during water molecule oxidation through them toward the cathode. The solution containing nitrofurazone should be placed into the anodic compartment of the electrolyzer (see above). All subsequent experiments were performed under such conditions.

3.2. Electrooxidation of Nitrafurazone

According to the optical density of solutions measurements (λ = 450 nm), nitrafurazone is effectively oxidized at a platinized titanium anode (Figure 4a). Decoloration of the solution occurs in the first 10 min of electrolysis. The color of nitrofurazone is due to the π–π^* electron transition in the π-conjugated system of the molecule. Electrooxidation leads to the destruction of the system of π-conjugated bonds, which results in the disappearance of the absorption maximum at 320–400 nm in the UV-Visible spectrum (Supplementary Figure S1).

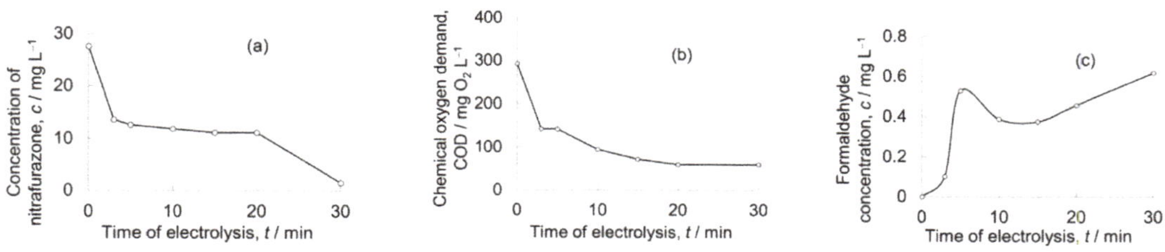

Figure 4. Nitrofurazone concentration (**a**), COD (**b**), and formaldehyde concentration (**c**) in the course of electrolysis using Pt/Ti anodes. See comments in the text.

The degree of nitrafurazone destruction reaches ca. 95% for 30 min of electrolysis. The values of chemical oxygen demand (COD) and total organic carbon (TOC) characterizing water quality symbatically diminish with a decrease in the concentration of nitrafurazone (Figure 4b).

The rate of nitrofuranone oxidation slows down 10–15 min after the start of electrolysis using Pt/Ti anodes (Figure 4a,b). It is rather difficult to exactly explain this effect at this stage of the study. It can be assumed that various intermediates formed during nitrofurazone oxidation are adsorbed on the Pt surface, affecting the kinetics of oxygen evolution. As a result, the concentration of active species in the bulk of the solution decreases, causing a decrease in the nitrofurazone oxidation rate. The COD is rather high after 10–15 min of electrolysis. Later, the rate of nitrofurazone oxidation increases again (Figure 4a). This is probably due to the oxidation of inhibing species on the electrode surface. A decrease in COD to ~60 mg O_2 L^{-1} occurs after 30 min of electrochemical treatment. In addition, monocarboxylic acids are not detected in the solution after electrolysis. These results might be considered satisfactory.

However, the residual COD of about 60 mg O_2 L^{-1} after 30 min of electrolysis can be considered as being high enough. Similarly, in solutions containing 13.3 mg L^{-1} of nitrofurazone, TOC reduces from 4.85 mg L^{-1} to 3.17 mg L^{-1} in 15 min of electrochemical treatment (Table 2). Further electrolysis leads to a decrease in TOC value, though only to 2.48 mg L^{-1}.

Table 2. The TOC values after electrolysis using Pt/Ti and BDD anodes. The initial concentration of nitrofurazone is 13.3 mg L^{-1}.

Time of Electrolysis, min	TOC, mg L^{-1}	
	Pt/Ti	BDD
0	4.85	4.85
15	3.17	1.87
30	2.48	0.85

Therefore, the mineralization of nitrofurazone is not quite complete after electrochemical treatment using Pt/Ti anodes. In addition, formaldehyde is formed during electrochemical treatment (Figure 4c). This result can be considered as unfavorable due to the high toxicity of formaldehyde.

A more complete destruction of nitrofurazone after electrochemical treatment is highly desirable. This issue can be solved using the anode with a higher positive electrode potential during electrolysis. A more positive value of electrode potential promotes the formation of highly active oxygen- and chlorine-containing species, which will then destroy organic pollutants molecules. Boron-doped diamond (BDD) is an appropriate electrode material for this purpose since its oxygen evolution potential is 2.06 V (j = 0.1 A cm^{-2}). This value is 210 mV more positive than that of the Pt/Ti anode (1.85 V, Figure 5).

The difference in electrode potentials at which oxygen evolution occurs on Pt/Ti and BDD anodes is due to the different mechanisms of this reaction in catalytic and non-catalytic electrodes. The evolution of oxygen on platinum proceeds through various oxygen-containing species (Pt–OH_{ads} and others) adsorbed on the electrode surface. In contrast, outer sphere electron transfer takes place in the case of the BDD anode. However, as voltammetric experiments have shown, nitrofurazone does not participate in electrochemical reactions in the anodic potential region. The oxidation of nitrofurazone molecules is due to its interaction with active oxygen- and chorine-containing species accumulated in the solution in the vicinity of the anode. The difference in both the concentration and chemical nature of these species should lead to different routes of nitrofurazone oxidation. It is highly likely that the difference in oxidation rates is explained this way.

Experiments on the electrochemical oxidation of nitrofurazone have shown that the efficiency of the BDD anode is much higher compared to the Pt/Ti anode. The concentration of nitrofurazone dropped to zero in the first half an hour of electrolysis; therefore, the degree of its destruction was about 100% (Figure 6a). This conclusion was confirmed by the residual COD (18.0 mg O_2 L^{-1}) and TOC (<1 mg L^{-1}) values (Tables 2 and 3).

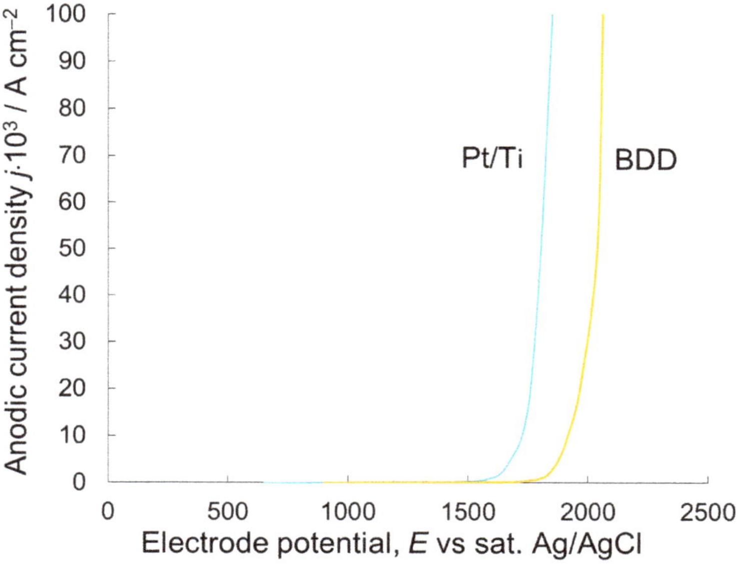

Figure 5. Potentiodynamic (v = 1 mV s^{-1}) oxygen evolution polarization curves on Pt/Ti and BDD electrodes in 0.5 M H_2SO_4.

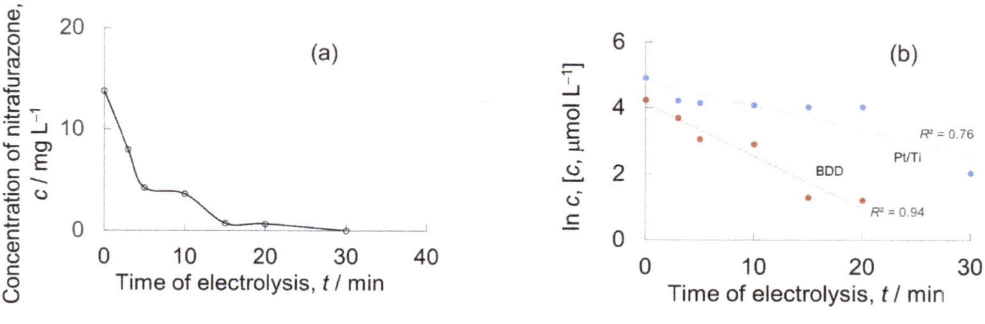

Figure 6. Changes in nitrofurazone concentration (**a**) in the course of electrolysis using BDD anodes and kinetic curves in first-order equation coordinates (**b**).

Table 3. The changes in COD and formaldehyde concentration during electrolysis with BDD anodes. The initial concentration of nitrofurazone is 13.3 mg L^{-1}

Time of Electrolysis, min	COD, mg O_2 L^{-1}	Formaldehyde Concentration, mg L^{-1}
0	140	<0.1
15	45	<0.1
30	18	<0.1

It is notable that formaldehyde is not formed in the solution during electrolysis, which makes the purification process safer than in the case of using Pt/Ti anodes (Table 3). Reducing the concentration of nitrofurazone to practically zero occurred for the investigated initial concentration in the interval 10–100 mg L^{-1} during 30 min of electrolysis. The optimal concentration of chloride ions should be considered in the range of 0.002–0.005 mol

L^{-1}, obtained after dissolving the Avexima tablet in distilled water and corresponding to its concentration in tap water. The rate of nitrofurazone mineralization slows down when the concentration of chloride ions in the solution is less than 0.002 mol L^{-1}. On the other hand, chlorine-worsening working conditions occur when the concentration of chloride ions is more than 0.005 mol L^{-1}.

The value of the redox potential (ORP) of nitrofurazone solution is equal to 0.435 V (versus standard hydrogen electrode, s.h.e, pH 2.0) before electrochemical treatment. It increases to 0.90 V (s.h.e, pH 2.0) after 30 min of electrolysis. The highly positive value of ORP indicates that the solution is well oxidized.

It should be noted that, as shown by the ICP-MS analysis of solutions after electrolysis, no degradation of electrode materials (c(Pt) < 0.1 ppb, c(Ti) < 0.5 ppb) was detected.

The kinetics of nitrofurazone electrooxidation using the BDD anode can be described by the first-order equation (R^2 = 0.94, Figure 6b). The rate constant is determined to be 0.16 min^{-1} for BDD. The data for the Pt/Ti anode do not linearize well in the coordinates of the first-order equation (R^2 = 0.76). Apparently, this is due to the complicated kinetics of nitrofurazone oxidation during electrolysis using this anode. Nevertheless, the rate of nitrofurazone oxidation is 2–2.5 times higher for the BDD anode. Therefore, one can conclude that BDD is a more preferable anode material both in terms of efficiency and oxidation rate. Energy consuption for the complete neutralization of nitrofurazone is estimated to be equal to 70 W·h mg^{-1} (BDD anode).

The better efficiency of the BDD anode compared to the Pt/Ti anode is due to the different mechanisms of oxygen evolution on them (see above). High positive electrode potentials at the BDD anode during oxygen evolution promote the formation of highly active oxygen-containing species in the volume of the treated solution. This results in the effective destruction of organic molecules.

The removal rate of nitrofurazone by electrolysis was higher compared to a water-splitting bio-photoelectrochemical cell, where about 80% of NFZ removal was achieved within 10 h of treatment [43]. An enhanced coupling photocatalysis and biodegradation method for nitrofurazone removal was proposed in [48]. In a cited article, the concentration of nitrofurazone decreased from 5 to 1 mg L^{-1} in 30 min of treatment; i.e., 80% removal of the pollutant was achieved. From this point of view, the results obtained for electrolysis with the BDD anode can be considered promising.

3.3. Identification of Products

The identification of metabolites of organic pollutants oxidation is necessary as they may be potentially hazardous. In the present study, this was carried out using Liquid Chromatography Tandem Mass Spectrometry (LC-MS-MS). Mass spectra were recorded in both positive and negative ions. As shown by mass spectrometry experiments, the nature and composition of intermediates formed at Pt/Ti and BDD anodes in the course of nitrofurazone oxidation are practically the same. The difference between these two anodes lies in the rate of accumulation and the consumption of intermediates.

Since nitrofurazone is an electron acceptor, the spectrum of nitrofurazone was observed in negative ions at a retention time of 0.3–0.5 min (Figure 7a,b).

The line at m/z = 198 corresponds to the ions formed upon ionization of a nitrofurazone molecule. In addition, associates with higher molecular weight are present in the mass spectra. Lines at m/z 269–272 can be assigned to ions formed upon the ionization of associates of nitrofurazone molecules with semicarbazide. It is known that semicarbazide is accumulated in the solutions of nitrafurazone as a product of its spontaneous decomposition [49]. Associates between nitrofurazone and semicarbazide molecules are formed due to hydrogen bonds between nitro and amino groups. The peaks with long retention times (5.2 min, 5.6 min, 8.6 min) seem to be associated with more complex associates, since they correspond to large m/z values in the mass spectrum.

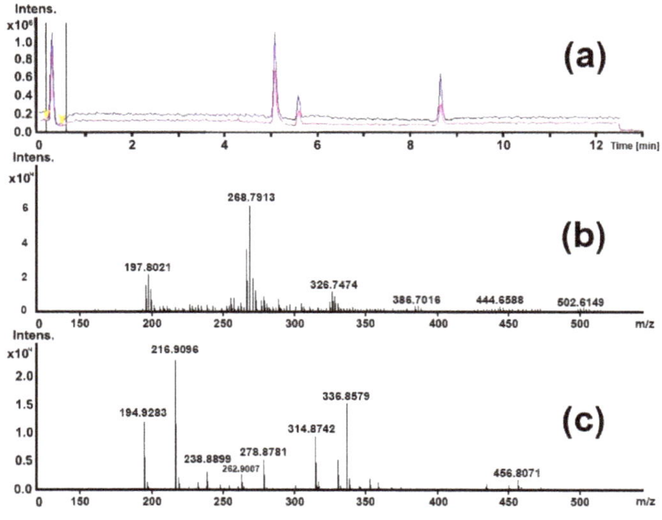

Figure 7. Chromatogram (**a**) and mass spectrum (t_R = 0.3–0.5 min) (**b**) in the initial solution of nitrofurazone in negative ions. The figure (**c**) presents the mass spectrum (t_R = 0.3–0.5 min) after electrochemical treatment for 30 min, BDD anode.

Lines of nitrofurazone disappeared in the mass spectrum after 30 min of electrolysis using BDD anodes (Figure 7c), which corresponds to the results of its concentration determination. The mass spectrum of the solution after electrolysis only contained residual lines of low intensity.

The nature of the intermediates formed during electrochemical treatment is better seen in the mass spectra recorded in positive ions.

The peak at t_R = 3.9 min present in the chromatogram was recorded in positive ions already in the initial solution of nitrofurazone (Figure 8). Its intensity increased during the first minutes of electrochemical treatment, reaching a maximum after ca. 5 min (Pt/Ti anode) and after ca. 2 min (BDD anode) after the start of electrolysis. After that, the intensity of this peak decreased and finally became negligible after 30 min of current passing.

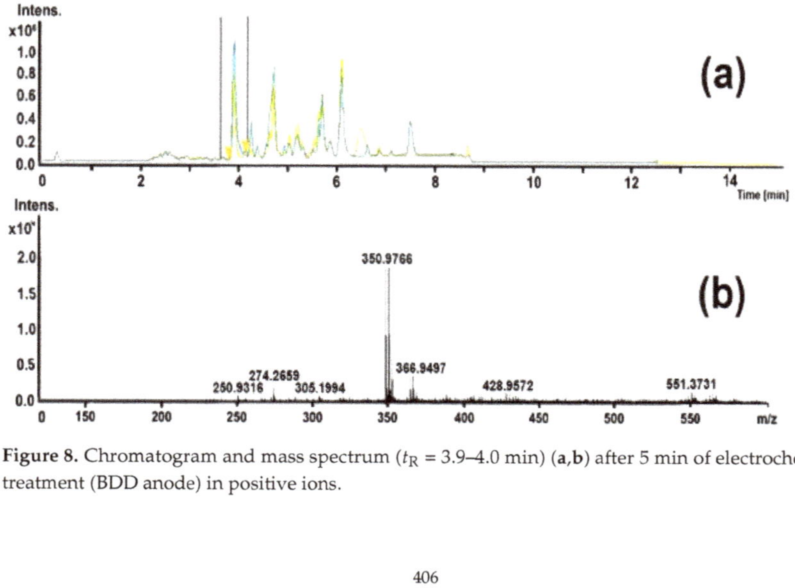

Figure 8. Chromatogram and mass spectrum (t_R = 3.9–4.0 min) (**a**,**b**) after 5 min of electrochemical treatment (BDD anode) in positive ions.

The ion with $m/z = 349$ dominates in the mass spectrum of the peak with $t_R = 3.9–4.0$ min; there are also ions with $m/z = 351$ and 353. A difference in m/z value by two units indicates the presence of chlorine atoms in the molecule. The mass of the ion corresponds to the protonated form of 5-nitro-2-furaldehyde azine $[M–H]^+$ with two chlorine atoms. The structure of the discussed molecule and the pathway of its formation can be represented as follows (Figure 9):

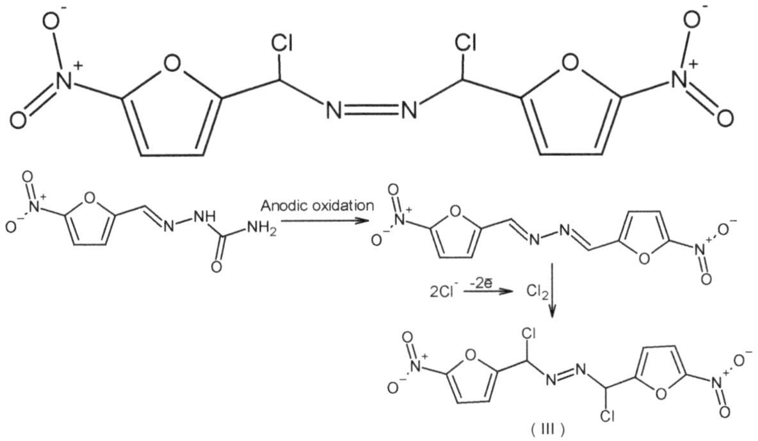

Figure 9. Chemical formula of the product of the addition of chlorine to the azine structure and the route of its formation. See comments in the text.

According to [50], 5-nitro-2-furaldehyde azine is formed upon the chlorination of aqueous solutions of nitrofurazone. The initial decoloration of nitrofurazone solution is probably due to the destruction of the system of conjugated π-bonds in its molecule. The chlorination of azines can occur by the type 1–4 addition in the azine fragment [51]. The mass spectrum also contains ions with $m/z = 365$, 367, and 369, which differs from the previous three by 16. It is possible that this is the product of the addition of water and the elimination of two hydrogen atoms.

The discussed intermediates accumulate in the first minutes of electrochemical treatment; however, they are destroyed by further electrolysis.

The chromatogram of the solution after 5 min of electrooxidation (Figure 8) also contains peaks ($t_R = 4.8$, 5.7, 6.1 min) corresponding to other intermediates formed during nitrofurazone electrooxidation. There are no lines different by 2 units of m/z in their mass spectra; therefore, their molecules do not contain chlorine atoms. The proposed structures of oxidation products are given in Table 4.

Oxidative polycondensation reactions are responsible for the formation of oxygen-containing intermediates of nitrofurazone. One of the possible reaction paths is given in Figure 10.

The potential toxicity of the resulting nitrofurazone oxidation intermediates was previously not known. It seems that it is advisable to apply an approach based on the use of neural networks and that published in [52] in order to evaluate it. Nevertheless, it is very important that these intermediates undergo further oxidation during electrolysis with BDD anodes, which leads to the complete mineralization of organic pollutants.

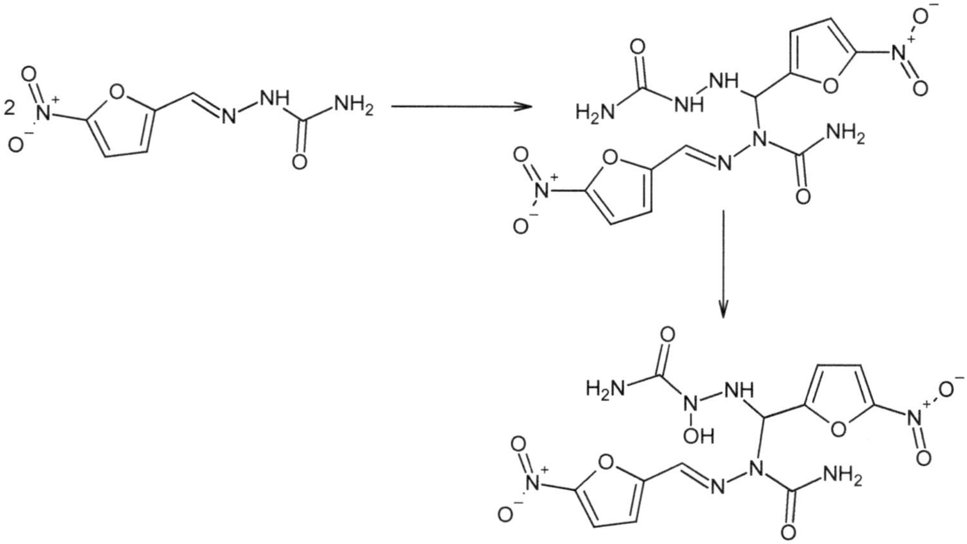

m /z 412.27

Figure 10. One of the possible reaction paths of nitrofurazone oxidative polycondensation.

Table 4. The proposed products of nitrofurazone oxidation (BDD anode, 5 min).

Retention Time, t_R, min	Dominant Peak, *m/z*	Proposed Structure
3.9	351, 353	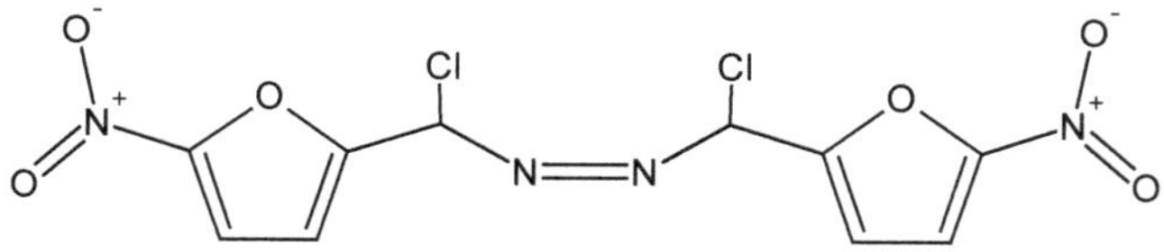
4.8	265	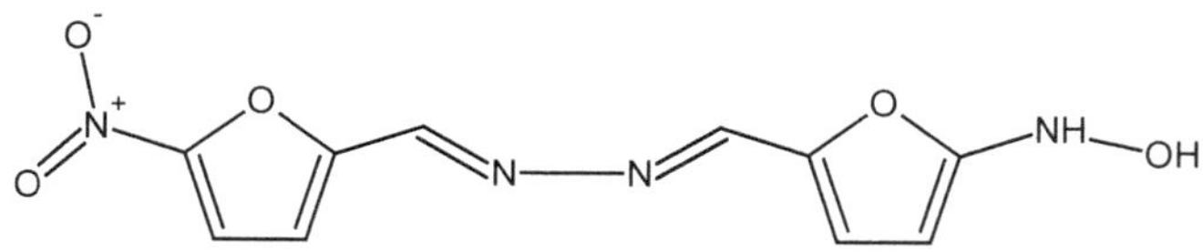
5.7	413	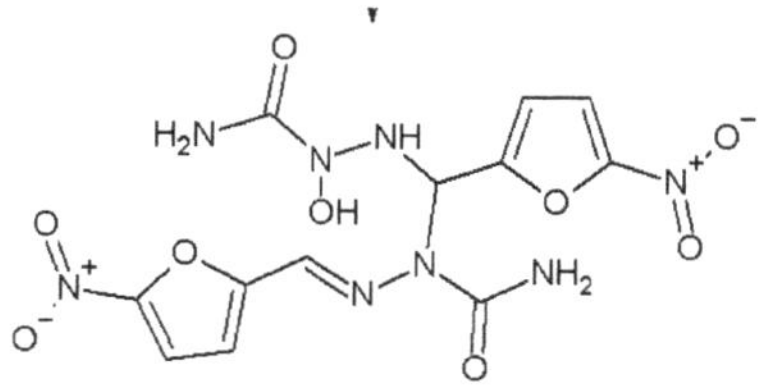

Table 4. *Cont.*

Retention Time, t_R, min	Dominant Peak, *m/z*	Proposed Structure
6.1	569	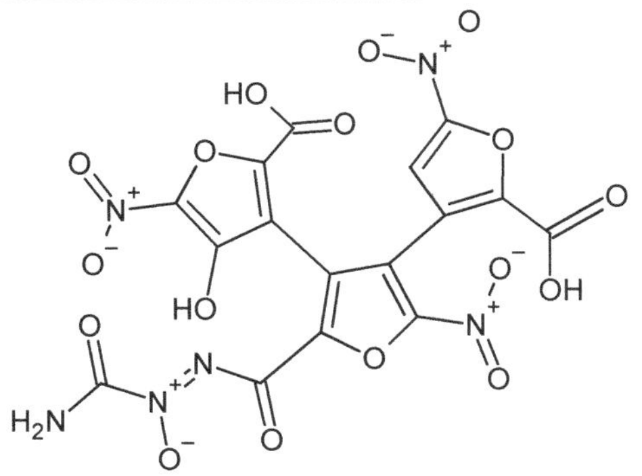

4. Conclusions

The electrochemical oxidative degradation of nitrofurazone in aqueous solutions in the presence of chloride ions was performed. Solutions under the electrochemical treatment contained 10–100 mg L^{-1} of nitrofurazone. Electrolysis was carried out under galvanostatic conditions at an anodic current density of 0.1 A cm^{-2} and a volume current density of 1.33 A L^{-1} for 30 min.

Since nitrofurazone reduces in the cathodic region of potentials, cathodic and anodic compartments of the electrochemical cell should be separated in order to exclude the cost of electricity on the reduction of nitrofurazone at the cathode and the subsequent oxidation of formed products at the anode. The separation of cathode and anode compartments also prevents the accumulation of nitrofurazone reduction products of various chemical natures in the solution under electrochemical treatment.

In principle, the electrochemical destruction of nitrofurazone is possible on both Pt/Ti and BDD anodes; however, the process is more efficient and safe when using BDD anodes. The preference for using BDD anodes is due to their higher electrode potential during electrolysis, which leads to a higher rate of active species generation.

The addition of the products of chlorine to 5-nitro-2-furaldehyde azine was formed in the solution under electrochemical treatment in the first minutes (0–15 min) of electrolysis. Almost complete mineralization of organic pollutants occurred after ca. 30 min of electrolysis using BDD anodes. Therefore, electrochemical treatment for 30 min can be recommended for the mineralization of nitrofurazone.

Supplementary Materials: The following supporting information can be downloaded at: https://www.mdpi.com/article/10.3390/w15193370/s1.

Author Contributions: V.V.K. Conceptualization, Methodology, Writing, N.A.I.—intermediate analytical control, selection of experimental conditions, Data Processing, E.N.K.—literature review, writing, experiment, A.V.P.—analytical control, description of the mechanism of destruction, Y.O.M.—interpretation of the results obtained, E.A.F.—analytical control of oxidation products. Writing, Y.M.A.—experiment, Data Processing. All authors have read and agreed to the published version of the manuscript.

Funding: This work was supported by the Russian Science Foundation grant no. 23-23-00067.

Institutional Review Board Statement: Not applicable.

Informed Consent Statement: Not applicable.

Data Availability Statement: The data can be found within the manuscript.

Acknowledgments: This work was supported by the Russian Science Foundation grant no. 23-23-00067.

Conflicts of Interest: The authors declare that they have no known competing financial interest or personal relationship that could have appeared to influence the work reported in this paper.

References

1. Wang, Q.; Wang, P.; Yang, Q. Occurrence and diversity of antibiotic resistance in untreated hospital wastewater. *Sci. Total Environ.* **2018**, *621*, 990–999. [CrossRef] [PubMed]
2. Aali, R.; Nikaeen, M.; Khanahmad, H.; Hassanzadeh, A. Monitoring and comparison of antibiotic resistant bacteria and their resistance genes in municipal and hospital wastewaters. *Int. J. Prev. Med.* **2014**, *5*, 887–894. [PubMed]
3. Adator, E.H.; Narvaez-Bravo, C.; Zaheer, R.; Cook, S.R.; Tymensen, L.; Hannon, S.J.; Booker, C.W.; Church, D.; Read, R.R.; McAllister, T.A. A one health comparative assessment of antimicrobial resistance in generic and Escherichia coli from beef production, sewage and clinical settings. *Microorganisms* **2020**, *8*, 885. [CrossRef] [PubMed]
4. Aubertheau, E.; Stalder, T.; Mondamert, L.; Ploy, M.-C.; Dagot, C.; Labanowski, J. Impact of wastewater treatment plant discharge on the contamination of river biofilms by pharmaceuticals and antibiotic resistance. *Sci. Total Environ.* **2017**, *579*, 1387–1398. [CrossRef]
5. Thomas, K.V.; Langford, K.H. Point Sources of Human Pharmaceuticals into the Aquatic Environment. In *Green and Sustainable Pharmacy*; Kümmerer, K., Hempel, M., Eds.; Springer: Berlin/Heidelberg, Germany, 2010; pp. 211–223.
6. Kotwani, A.; Joshi, J.; Kaloni, D. Pharmaceutical effluent: A critical link in the interconnected ecosystem promoting antimicrobial resistance. *Environ. Sci. Pollut. Res. Int.* **2021**, *28*, 32111–32124. [CrossRef] [PubMed]
7. Kabanov, M.A.; Ivantsova, N.A.; Kuzin, E.N.; Murzina, E.D.; Korobov, A.Y. Evaluation of the influence of drug complex formation on the efficiency of water conditioning with reagents for tetracycline. *Pharm. Chem. J.* **2022**, *55*, 1245–1249. [CrossRef]
8. Li, C.; Xu, Y.; Song, W. Pollution Characteristics and Risk Assessment of Typical Antibiotics and Persistent Organic Pollutants in Reservoir Water Sources. *Water* **2023**, *15*, 259. [CrossRef]
9. Wang, P.; He, Y.L.; Huang, C.H. Reactions of tetracycline antibiotics with chlorine dioxide and free chlorine. *Water Res.* **2011**, *45*, 1838–1846. [CrossRef]
10. Gu, C.; Karthikeyan, K.G. Interaction of tetracycline with aluminum and iron hydrous oxides. *J. Environ. Sci. Technol.* **2005**, *39*, 2660–2667. [CrossRef]
11. Ng, K.; Alygizakis, N.A.; Thomaidis, N.S.; Slobodnik, J. Wide-Scope Target and Suspect Screening of Antibiotics in Effluent Wastewater from Wastewater Treatment Plants in Europe. *Antibiotics* **2023**, *12*, 100. [CrossRef]
12. Pereira, A.; Silva, L.J.G.; Laranjeiro, C.S.M.; Meisel, L.M.; Lino, C.M.; Pena, A. Human pharmaceuticals in Portuguese rivers: The impact of water scarcity in the environmental risk. *Sci. Total Environ.* **2017**, *609*, 1182–1191. [CrossRef] [PubMed]
13. Kim, H.-Y.; Lee, I.-S.; Oh, J.-E. Human and veterinary pharmaceuticals in the marine environment including fish farms in Korea. *Sci. Total Environ.* **2017**, *579*, 940–949. [CrossRef] [PubMed]
14. Gevao, B.; Uddin, S.; Dupont, S. Baseline concentrations of pharmaceuticals in Kuwait's coastal marine environment. *Mar. Pollut. Bull.* **2021**, *173*, 113040. [CrossRef]
15. Younes, H.A.; Mahmoud, H.M.; Abdelrahman, M.M.; Nassar, H.F. Seasonal occurrence, removal efficiency and associated ecological risk assessment of three antibiotics in a municipal wastewater treatment plant in Egypt. *Environ. Nanotechnol. Monit. Manag.* **2019**, *12*, 100239. [CrossRef]
16. Qiao, M.; Ying, G.-G.; Singer, A.C.; Zhu, Y.-G. Review of antibiotic resistance in China and its environment. *Environ. Int.* **2018**, *110*, 160–172. [CrossRef]
17. Pan, M.; Chu, L.M. Occurrence of antibiotics and antibiotic resistance genes in soils from wastewater irrigation areas in the Pearl River Delta region, southern China. *Sci. Total Environ.* **2018**, *624*, 145–152. [CrossRef]
18. Tiwari, A.; Kurittu, P.; Al-Mustapha, A.I.; Heljanko, V.; Johansson, V.; Thakali, O.; Mishra, S.K.; Lehto, K.M.; Lipponen, A.; Oikarinen, S.; et al. WastPan Study Group; A. Heikinheimo, Wastewater surveillance of antibiotic-resistant bacterial pathogens: A systematic review. *Front. Microbiol.* **2022**, *13*, 977106. [CrossRef]
19. Desai, M.; Njoku, A.; Nimo-Sefah, L. Comparing Environmental Policies to Reduce Pharmaceutical Pollution and Address Disparities. *Int. J. Environ. Res. Public Health* **2022**, *19*, 8292. [CrossRef]
20. Kusturica, M.P.; Jevtic, M.; Ristovski, J.T. Minimizing the environmental impact of unused pharmaceuticals: Review focused on prevention. *Front. Environ. Sci.* **2022**, *10*, 1077974. [CrossRef]
21. de Ilurdoz, M.S.; Sadhwani, J.J.; Reboso, J.V. Antibiotic removal processes from water & wastewater for the protection of the aquatic environment—A review. *J. Water Process Eng.* **2022**, *45*, 102474.
22. Wang, J.; Wang, S. Removal of pharmaceuticals and personal care products (PPCPs) from wastewater: A review. *J. Environ. Manag.* **2016**, *182*, 620–640. [CrossRef] [PubMed]
23. Zhang, D.; Gersberg, R.M.; WJ, N.; SK, T. Removal of pharmaceuticals and personal care products in aquatic plant-based systems: A review. *Environ. Pollut.* **2014**, *184*, 620–639. [CrossRef] [PubMed]
24. Loganathan, P.; Vigneswaran, S.; Kandasamy, J.; Cuprys, A.K.; Maletskyi, Z.; Ratnaweera, H. Treatment Trends and Combined Methods in Removing Pharmaceuticals and Personal Care Products from Wastewater—A review. *Membranes* **2023**, *13*, 158. [CrossRef] [PubMed]

25. Kanakaraju, D.; Glass, B.D.; Oelgemöller, M. Advanced oxidation process-mediated removal of pharmaceuticals from water: A review. *J. Environ. Manag.* **2018**, *219*, 189–207. [CrossRef]
26. Thiruven Katachari, R.; Vigneswaran, S.; Moon, I.S. A review on UV/TiO_2 photocatalytic oxidation process. *Korean J. Chem. Eng.* **2008**, *25*, 64–72. [CrossRef]
27. Sharma, S.; Ruparelia, J.P.; Patel, M.L. A General Review on Advanced Oxidation Processes for Waste Water Treatment. In Proceedings of the Nirma University International Conference, Ahmedabad, India, 8–10 December 2011.
28. Jiao, J.; Li, Y.; Song, Q.; Wang, L.; Luo, T.; Gao, C.; Liu, L.; Yang, S. Removal of Pharmaceuticals and Personal Care Products (PPCPs) by Free Radicals in Advanced Oxidation Processes. *Materials* **2022**, *15*, 8152. [CrossRef]
29. Emzhina, V.V.; Kuzin, E.N.; Babusenko, E.S.; Krutchinina, N.E. Photodegradation of tetracycline in presence of H_2O_2 and metal oxide based catalysts. *J. Water Process Eng.* **2021**, *39*, 101696. [CrossRef]
30. Mylapilli, S.V.P.; Reddy, S.N. Sub and supercritical water oxidation of pharmaceutical wastewater. *J. Environ. Chem. Eng.* **2019**, *7*, 103165. [CrossRef]
31. Johnston, J.B.; Hannah, R.E.; Cunningham, V.L.; Daggy, B.P.; Sturm, F.J.; Kelly, R.M. Destruction of Pharmaceutical and Biopharmaceutical Wastes by the Modar Supercritical Water Oxidation Process. *Nat. Biotechnol.* **1988**, *6*, 1423–1427. [CrossRef]
32. Trellu, C.; Chaplin, B.P.; Coetsier, C.; Esmilaire, R.; Cerneaux, S.; Causserand, C.; Cretin, M. Electro-oxidation of organic pollutants by reactive electrochemical membranes. *Chemosphere* **2018**, *208*, 159–175. [CrossRef]
33. Lan, Y.; Coetsier, C.; Causserand, C.; Serrano, K. An experimental and modelling study of the electrochemical oxidation of pharmaceuticals using a boron-doped diamond anode. *Chem. Eng. J.* **2018**, *333*, 486–494. [CrossRef]
34. Yang, Y.; Ramos, N.C.; Clark, J.A.; Hillhouse, H.W. Electrochemical oxidation of pharmaceuticals in synthetic fresh human urine: Using selective radical quenchers to reveal the dominant degradation pathways and the scavenging effects of individual urine constituents. *Water Res.* **2022**, *221*, 118722. [CrossRef]
35. da Silva, S.W.; Welter, J.B.; Albornoz, L.L.; Heberle, A.N.; Ferreira, J.Z.; Bernardes, A.M. Advanced Electrochemical Oxidation Processes in the Treatment of Pharmaceutical Containing Water and Wastewater: A Review. *Curr. Pollut. Rep.* **2021**, *7*, 146–159. [CrossRef]
36. Karaoğlu, G.; Öztürk, D.; Akyol, A.; Kara, S. PCT degradation with electrooxidation (EOx) and ultrasound (US) hybrid process using different type electrodes: BDD, Ti/PbO_2 and Ti/Pt. *Sep. Purif. Technol.* **2023**, *311*, 123313. [CrossRef]
37. Köktaş, Y.; Gökkuş, Ö.; Kariper, İ.A.; Othmani, A. Tetracycline removal from aqueous solution by electrooxidation using ruthenium-coated graphite anode. *Chemosphere* **2023**, *315*, 137758. [CrossRef] [PubMed]
38. Kuznetsov, V.V.; Kapustin, E.S.; Pirogov, A.V.; Kurdin, K.A.; Filatova, E.A.; Kolesnikov, V.A. An effective electrochemical destruction of non-ionic surfactants on bismuth-modified lead dioxide anodes for wastewater pretreatment. *Solid. State Electrochem.* **2020**, *24*, 173–183. [CrossRef]
39. Fu, R.; Zhang, P.-S.; Jiang, Y.-X.; Sun, L.; Sun, X.-H. Wastewater treatment by anodic oxidation in electrochemical advanced oxidation process: Advance in mechanism, direct and indirect oxidation detection methods. *Chemosphere* **2023**, *311*, 136993. [CrossRef]
40. Bellver-Domingo, A.; Fuentes, R.; Hernández-Sancho, F. Shadow prices of emerging pollutants in wastewater treatment plants: Quantification of environmental externalities. *J. Environ. Manag.* **2017**, *203*, 439–447. [CrossRef]
41. Symes, D.; Al-Duri, B.; Bujalski, W.; Dhir, A. Cost-effective design of the alkaline electrolyser for enhanced electrochemical performance and reduced electrode degradation. *Int. J. Low-Carbon Technol.* **2015**, *10*, 452–459. [CrossRef]
42. Aviezer, Y.; Lahav, O. Removal of contaminants of emerging concern from secondary-effluent reverse osmosis retentates by continuous supercritical water oxidation parametric study and conceptual design. *J. Hazard. Mater.* **2022**, *437*, 129379. [CrossRef]
43. Hou, Y.; Yuan, G.; Wang, S.; Yu, Z.; Qin, S.; Tu, L.; Yan, Y.; Chen, X.; Zhu, H.; Tang, Y. Nitrofurazone degradation in the self-biased bio-photoelectrochemical system: g-C_3N_4/CdS photocathode characterization, degradation performance, mechanism and pathways. *J. Hazard. Mater.* **2020**, *384*, 121438. [CrossRef] [PubMed]
44. Shmychkova, O.; Zahorulko, S.; Girenko, D.; Luk, T.; Dmitrikova, L.; Velichenko, A. Material Selection and Optimization of Conditions for Electrooxidation of Nitrofurazone: A Comparative Study of Tin and Lead Dioxides. *J. Electrochem. Soc.* **2021**, *168*, 086507. [CrossRef]
45. Brito, C.L.; Ferreira, E.I.; La-Scalea, M.A. Application of multi-walled carbon nanotubes functionalized with hemin to evaluate the electrochemical behavior of nitrofurazone in aqueous media. *Electrochim. Acta* **2023**, *459*, 142486. [CrossRef]
46. Rahi, A.; Sattarahmady, N.; Vais, R.D.; Heli, H. Sonoelectrodeposition of gold nanorods at a gold surface—Application for electrocatalytic reduction and determination of nitrofurazone. *Sens. Actuators B* **2015**, *210*, 96–102. [CrossRef]
47. Jiang, Y.; Zhao, H.; Liang, J.; Yue, L.; Li, T.; Luo, Y.; Liu, Q.; Lu, S.; Asiri, A.M.; Gong, Z.; et al. Anodic oxidation for the degradation of organic pollutants: Anode materials, operating conditions and mechanisms. A mini review. *Electrochem. Commun.* **2021**, *123*, 106912. [CrossRef]
48. Smułek, W.; Bielan, Z.; Zgoła-Grześkowiak, A.; Zielińska-Jurek, A.; Kaczorek, E. Nitrofurazone Removal from Water Enhanced by Coupling Photocatalysis and Biodegradation. *Int. J. Mol. Sci.* **2021**, *22*, 2186. [CrossRef]
49. Li, G.; Tang, C.; Wang, Y.; Yang, J.; Wu, H.; Chen, G.; Kong, X.; Kong, W.; Liu, S.; You, J. A Rapid and Sensitive Method for Semicarbazide Screening in Foodstuffs by HPLC with Fluorescence Detection. *Food Anal. Methods* **2015**, *8*, 1804–1811. [CrossRef]
50. Nakamura, H.; Kawakami, T.; Niino, T.; Takahashi, Y.; Onodera, S. Chemical fate and changes in mutagenic activity of antibiotics nitrofurazone and furazolidone during aqueous chlorination. *J. Toxicol. Sci.* **2008**, *33*, 621–629. [CrossRef]

51. Al-Masoudi, N.A.; Al-Soud, Y.A.; Ali, I.A.I. Synthesis of 1,2,4-triazole C-nucleosides from hydrazonyl chlorides and nitriles. *Nucleosides Nucleotides Nucleic Acids* **2007**, *26*, 37–43. [CrossRef]
52. Cremer, J.; Sandonas, L.M.; Tkatchenko, A.; Clevert, D.-A.; Fabritiis, C.D. Equivariant Graph Neural Networks for Toxicity Prediction. *Chem. Res. Toxicol.* **2023**. [CrossRef]

nanomaterials

MDPI

Review

Metal-Based Electrocatalysts for Selective Electrochemical Nitrogen Reduction to Ammonia

Yi-Zhen Zhang [1,2], Peng-Hui Li [1], Yi-Nuo Ren [1], Yun He [3], Cheng-Xu Zhang [4], Jue Hu [4,*], Xiao-Qiang Cao [1,*] and Michael K. H. Leung [2,*]

1 College of Safety and Environmental Engineering, Shandong University of Science and Technology, Qingdao 266590, China; yizhzhang@sdust.edu.cn (Y.-Z.Z.)

2 Ability R&D Energy Research Centre, School of Energy and Environment, City University of Hong Kong, Hong Kong, China

3 School of Chemical and Environmental Engineering, Wuhan Polytechnic University, Wuhan 430024, China

4 Faculty of Metallurgical and Energy Engineering, Kunming University of Science and Technology, Kunming 650093, China

* Correspondence: hujue@kmust.edu.cn (J.H.); caoxiaoqiang@sdust.edu.cn (X.-Q.C.); mkh.leung@cityu.edu.hk (M.K.H.L.)

Abstract: Ammonia (NH_3) plays a significant role in the manufacture of fertilizers, nitrogen-containing chemical production, and hydrogen storage. The electrochemical nitrogen reduction reaction

(e-NRR) is an attractive prospect for achieving clean and sustainable NH_3 production, under mild conditions driven by renewable energy. The sluggish cleavage of N≡N bonds and poor selectivity of e-NRR are the primary challenges for e-NRR, over the competitive hydrogen evolution reaction (HER). The rational design of e-NRR electrocatalysts is of vital significance and should be based on a thorough understanding of the structure–activity relationship and mechanism. Among the various explored e-NRR catalysts, metal-based electrocatalysts have drawn increasing attention due to their remarkable performances. This review highlighted the recent progress and developments in metal-based electrocatalysts for e-NRR. Different kinds of metal-based electrocatalysts used in NH_3 synthesis (including noble-metal-based catalysts, non-noble-metal-based catalysts, and metal compound catalysts) were introduced. The theoretical screening and the experimental practice of rational metal-based electrocatalyst design with different strategies were systematically summarized. Additionally, the structure–function relationship to improve the NH_3 yield was evaluated. Finally, current challenges and perspectives of this burgeoning area were provided. The objective of this review is to provide a comprehensive understanding of metal-based e-NRR electrocatalysts with a focus on enhancing their efficiency in the future.

Keywords: electrochemical nitrogen reduction reaction; NH_3 synthesis; noble-metal-based catalysts; non-noble-metal-based catalysts; metal compound catalysts

1. Introduction

With the ever-increasing global population, pressing worldwide environmental concerns and the challenges in energy supply and conservation, the search for sustainable and eco-friendly energy pathways is strongly demanded to secure our energy future [1,2]. Elemental nitrogen (N_2) is indispensable to human society and the planet's ecosystem. Although nitrogen is abundant in the atmosphere (~78% by volume), it cannot be directly utilized by humans. However, nitrogen fixation can convert atmospheric N_2 to ammonia (NH_3), which is a more active nitrogen-containing alternative than N_2. Traditionally, NH_3 serves as a raw material for the synthesis of fertilizers to sustain the rising global population. NH_3 is also extensively applied to produce explosives, plastics, resins, pharmaceuticals, and many other chemical compounds for industrial use. Currently, NH_3 has received considerable attention as a promising carbon-free energy carrier, due to its high hydrogen content (17.65%) and energy density (4.3 kW h kg^{-1}), as well as its easy storage in liquids for transportation (9–10 bar) [3,4]. Compared to C-containing fuels, N-containing fuels do not emit CO_2 upon final decomposition.

Benefiting from one of the most significant scientific inventions in the early 20th century, the Haber–Bosch process ($N_2 + 3H_2 \rightleftharpoons 2NH_3$, $\Delta_f H^\circ = -45.940$ kJ mol^{-1}, $\Delta_f G^\circ = -16.407$ kJ mol^{-1}) is a huge leap towards the mass production of NH_3 [5,6]. To activate the strong N≡N bonds (46.1 kJ mol^{-1}), the Haber–Bosch process requires high temperature (300–500 °C) and high pressure (150–300 atm) with heterogeneous iron-based catalysts. Here, the H_2 required depends on the carbon-intensive steam reforming of methane, with the input of energy derived from fossil fuels [7]. Consequently, NH_3 production accounts for 1–2% of the global energy consumption each year and over the 2% of world's natural gas, giving rise to 3% of energy-related CO_2 emissions. In this regard, alternative approaches of NH_3 synthesis need to be developed with relatively low energy consumption, low pollutant production and mild operating conditions. In 2016, the US Department of Energy (DOE) launched the Renewable Energy to Fuels through Utilization of Energy-dense Liquids (REFUEL) program, including NH_3 as the Carbon-Neutral Liquid Fuels (CNLFs).

In nature, biological N_2 fixation occurs through multiple proton- and electron-transfer steps relying on the partnership of reductase and nitrogenase enzymes in certain bacteria. Notably, nitrogenase enzymes operate under mild conditions with a significant energy input by the hydrolysis of adenosine triphosphate (ATP) molecules ($N_2 + 6H^+ + nMg–ATP + 6e^-$ (enzyme) $\rightarrow 2NH_3 + nMg–ADP + nPi$) [8]. Electrochemical catalytic reactions including the hydrogen evolution reaction (HER), oxygen evolution reaction (OER), oxygen reduction reaction (ORR), and carbon dioxide reduction reaction (CO_2RR) have rapidly developed and achieved excellent results over the past few years [9–14]. In addition, nitrate-containing wastewater streams could serve as a nitrogen source via the electrochemical reduction of nitrate into ammonia [15–18]. Inspired by the biological nitrogen-fixation process, the emerging electrochemical nitrogen-reduction reaction (e-NRR) is promising for achieving NH_3 production, directly from N_2 and water under mild conditions with the assistance of renewable energy.

Given that steam reforming for hydrogen production accounts for approximately 75% of the energy consumption in the Haber–Bosch process (Figure 1a), the steam-reforming unit with an electrocatalysis unit is a highly effective strategy [19]. Additionally, only two major blocks were observed in this electrocatalytic strategy (Figure 1b). If renewable electricity (from solar energy, wind, etc.) is available, its use in the electrocatalysis strategy will become nearly 100% renewable. In the NH_3 economy, electrochemical NH_3 synthesis and NH_3-powered fuel cells are two critical technologies [20]. The water and nitrogen in air are used as the only reactants to produce NH_3. The generated NH_3 can be distributed to users, including farms, NH_3/H_2 refuelling stations and residents. The NH_3-based infrastructure provides a promising way to solve the challenges related to the spatiotemporal fluctuations and the mismatch between supply and demand of electricity. Typically, the detailed cathodic and anodic reactions for e-NRR can be expressed as shown below (Equations (1)–(5)), under different pH aqueous electrolytes [21].

$$\text{Anodic reaction (acidic condition)}: 3H_2O \rightarrow \frac{3}{2}O_2 + 6H^+ + 6e^- \quad (1)$$

$$\text{Cathodic reaction (acidic condition)}: N_2 + 6H^+ + 6e^- \rightarrow 2NH_3 \quad (2)$$

$$\text{Anodic reaction (basic condition)}: 6OH^- \rightarrow \frac{3}{2}O_2 + 3H_2O + 6e^- \quad (3)$$

$$\text{Cathodic reaction (basic condition)}: N_2 + 6H_2O + 6e^- \rightarrow 2NH_3 + 6OH^- \quad (4)$$

$$\text{Overall reaction: } 2N_2+6H_2O \rightarrow 4NH_3+3O_2 \quad (5)$$

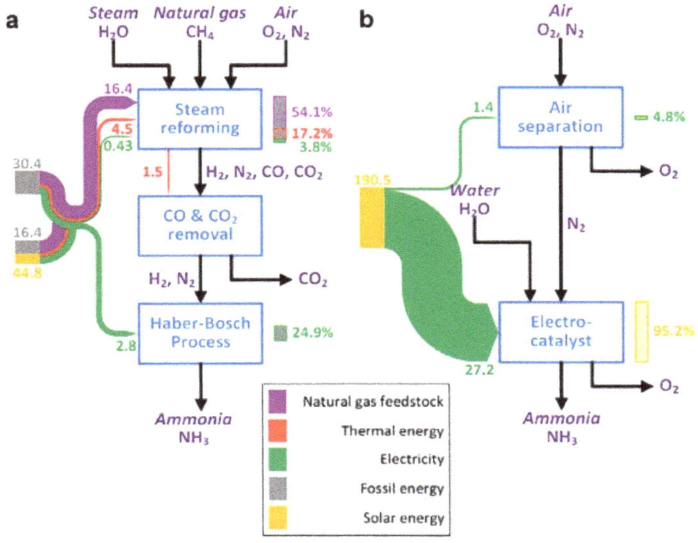

Figure 1. Energy efficiency of NH_3-synthesis strategies for the (**a**) Haber–Bosch strategy and (**b**) electrocatalytic strategy [19]. Copyright 2018, Elsevier.

Tremendous efforts have been devoted towards the development of e-NRR since 2016, aimed at promoting NH_3 yield and Faradaic efficiency (FE). However, e-NRR activity is hindered by the poor selectivity of e-NRR and poor activity of current e-NRR electrocatalytic designs. The poor selectivity of e-NRR arises from the competing HER. And the poor catalytic activity is mainly due to the weak affinity of N_2 to the catalyst surface, which hinders the activation of N_2 and the corresponding e-NRR efficiency. Accordingly, it is necessary to explore suitable electrocatalysts to overcome these limitations and improve catalytic activity towards e-NRR. Recently, extensive research has focused on designing a suitable electrocatalyst for efficient e-NRR, including noble-metal-based materials, non-noble-metal-based materials, and metal-free materials [2,3,22–24].

There are some reviews published elsewhere highlighting the progress of e-NRR electrocatalysts [3,25–28]. More specifically, Wen et al. [29] summarized the recent progress in low-dimensional nanomaterials with various structures and mentioned the relationship between this structure and e-NRR activities from both theoretical and experimental perspectives. Liu et al. [22] systematically outlined the latest development in novel electrocatalysts, including noble-metal-based catalysts, single-metal-atom catalysts, non-noble metal and their compounds, as well as metal-free catalysts, with various strategies to enhance the e-NRR activities through surface control, defect engineering, and hybridization. From the view of defect engineering, structural manipulation, crystallographic tailoring, and interface regulation, Shi et al. [2] comprehensively summarized the recent development of heterogeneous e-NRR catalysts, together with the catalytic mechanisms, current issues, and critical challenges.

In this review, we begin with the configurations and fundamentals of e-NRR under ambient conditions. Subsequently, the developed e-NRR metal-based catalysts based on noble-metal-based catalysts, non-noble-metal-based catalysts and metal compound catalysts were summarized, from experimental and theoretical perspectives, discussing the structure–function relationship. Finally, current challenges and perspectives of this burgeoning area are provided.

2. Configurations of Electrochemical-Reactor for e-NRR

The electrochemical reactor is important for performing e-NRR. Generally, the configuration of such reactors can be divided into four categories, namely, back-to-back cell, proton-exchange membrane (PEM)-type cell, single-chamber cell, and H-type cell (Figure 2) [30].

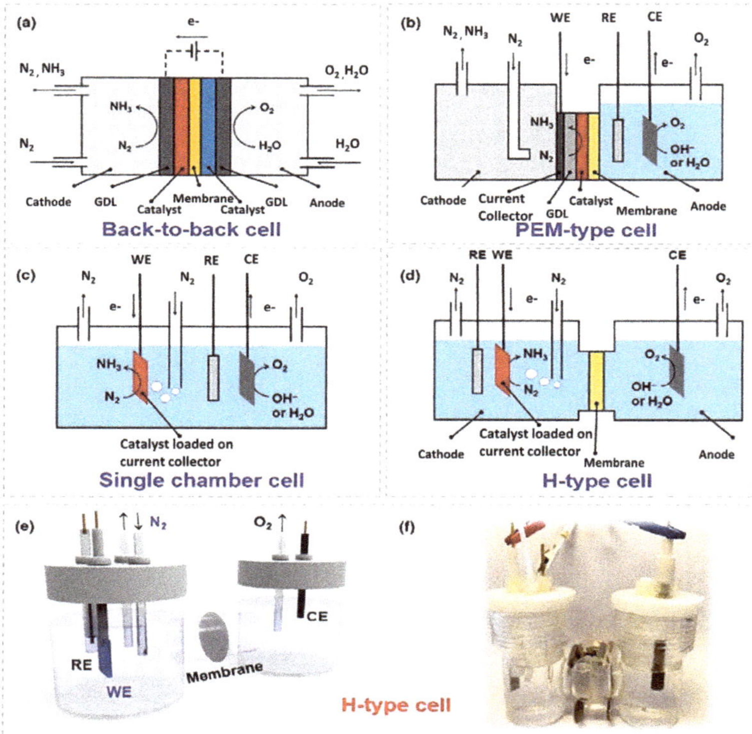

Figure 2. Schematics of different electrochemical-reactor configurations for e-NRR, including (**a**) back-to-back cell, (**b**) proton-exchange membrane (PEM)-type cell, (**c**) single-chamber cell, and (**d**) H-type cell. (**e**,**f**) Schematic and photograph of an H-type cell, including working electrode (WE), reference electrode (RE), counter electrode (CE), and membrane [5]. Copyright 2019, Elsevier.

In the back-to-back cell (Figure 2a), two gas-diffusion electrodes (anode and cathode) are usually separated by a dense membrane, and the anode and cathode are supplied with H_2O and N_2 gas, respectively. Among the various types of polymer membranes, perfluorosulfonic acid proton-exchange membranes such as Nafion membranes are widely used [31]. Apart from the use of solid-state electrolytes, liquid electrolytes can also be utilized in back-to-back cells [7].

In the PEM-type cell (Figure 2b), the cathodic chamber is fed with nitrogen gas, and the synthesized NH_3 gas is directly dissolved in the acidic electrolyte. Different from the back-to-back cell, the anodic chamber in the PEM-type cell is filled with liquid electrolyte, where water is electrolyzed to supply protons for the cathodic reaction [32]. In the back-to-back and PEM-type cells, no direct contact exists between the cathode and electrolyte. Consequently, they possess a favorable advantage for suppressing the HER by limiting the proton supply. However, these two types of cells are not ideal for e-NRR measurements. Several shortcomings including the complex preparation process and the effect from the use of an anion-exchange membrane lead to underestimated NH_3 production.

In the single-chamber cell (Figure 2c), the anode and cathode are placed in the same chamber without any separator. Nitrogen gas is continuously bubbled into the electrolyte and is subsequently reduced into NH_3 at the cathode under an applied potential. Simultaneously, OERs occur at the anode. One disadvantage of the single-chamber cell is that the NH_3 produced from the cathodic reaction may be oxidized at the anode, leading to inaccurate NH_3 determination. Given that the cathodic and anodic reactions are not separated, gases consisting of NH_3, hydrogen, and oxygen are produced and discharged from the reactor together at the same time. The dominant HER could also suppress e-NRR at the same region of potential in an aqueous electrolyte-based cell. Consequently, the NH_3 production rate and FE are limited.

The H-type cell is the most widely studied reactor configuration. The anode and cathode chambers are separated by a membrane to prevent the mixing of products (Figure 2d–f). The working and reference electrodes are located on the same side, attributed to the accurate measurement of the applied potentials and significant decrease in resistance between the working and reference electrodes. In the cathode chamber, nitrogen gas is purged into the electrolyte and is reduced into NH_3. The anodic reaction is mainly the oxidation of water molecules, also known as the OER. The anodic chamber was sealed without any gas purge in most reported work. Compared with the single-chamber cell, the products in each chamber can be separated in a double-chamber reactor, thereby preventing further oxidation and a mixture of gaseous products. Moreover, different electrolytes can be separated into two chambers so that the cathodic reactions are controlled independently with little influence from the anode. Accordingly, instrumental errors in measurements of e-NRR activity are minimized in the H-type cell compared with other reactors. However, the contribution of NH_3 dissolved in or leached from the ion-exchange membrane should be carefully handled.

3. Fundamental Comprehension on e-NRR

3.1. Adsorption of Nitrogen onto the Catalyst Surface

The process of e-NRR includes N_2 adsorption onto the active sites, activation of $N{\equiv}N$ bonds, and a final hydrogenation process. However, the high efficiency of e-NRR is hindered by the difficulty of N_2 adsorption and activation. Thus, a comprehensive understanding on the dominant reaction pathways is important to design highly efficient electrocatalysts. In the following subsection, an outline of the fundamental comprehension of e-NRR is discussed, with a focus on key steps and dominant reaction pathways.

The first step is the chemical adsorption of N_2 onto the catalyst surface. A large amount of N_2 adsorption sites can be provided by e-NRR catalysts with large surface areas. Accordingly, catalysts with relatively large specific surface areas are promising for the enhancement of e-NRR performance, such as porous-structure materials. The qualitative trends of catalytic activities on different metal surfaces have been summarized by Skúlason and co-workers through theoretical calculations [33]. They assumed that the activation-energy scales with the free energy differed in each elementary step of e-NRR, for the range of flat and stepped transition metals. The key results of this study were illustrated as a volcano plot (Figure 3a), in which the theoretical limiting potentials (U) on different metal surfaces were plotted versus their adsorption energy of N atom (ΔE_{N^*}). This plot also showed the relatively limited region in white shading (N-binding), where binding N-adatoms were able to compete with the H-adatoms on the metal surface. With regard to minimizing the parallel HER process, Mo, Fe, Rh, and Ru on top of the volcano diagrams were bound to be the most active surfaces for e-NRR. Metals (Rh, Ru, Ir, Co, Ni, and Pt) on the right legs of the volcano plot were prone to adsorb H-adatoms instead of N-adatoms. In addition, more negative potentials were required for these metals to activate N_2, resulting in HER that overwhelms e-NRR. Several flat metal surfaces of early transition metals such as Sc, Y, Ti, and Zr, on the left side of the volcano plots, tended to bind N-adatoms more strongly than H-adatoms. However, it remains unclear whether these early transition metals are effective e-NRR electrocatalysts, owing to easy oxidation. As a result, the authors encouraged experimental studies by using some of these metals. Beyond pure metal binding, some nanostructured catalysts with metal–nitrogen bonds have also been found capable of adsorbing N_2, such as heteroatom-doped carbons (e.g., N/B-doped porous carbon) and metal/nonmetal nitrides (e.g., Mo_2N, C_3N_4) [34]. Therefore, active sites on catalysts can be engineered to preferentially adsorb nitrogen species for e-NRR.

Moreover, the adsorbed amount of N_2 near the active sites possibly affects e-NRR activity. The solubility of N_2 gas in water-based electrolytes is about 0.66 mmol/L. Suryanto et al. [35] reported that the use of hydrophobic fluorinated aprotic electrolyte effectively enhanced N_2 solubility, which could significantly improve the FE of e-NRR. Their results also indicated that the availability of protons was effectively limited and thus suppressed the competing HER. Therefore, the increase in N_2 solubility near the active sites may be a promising strategy to achieve high e-NRR performance.

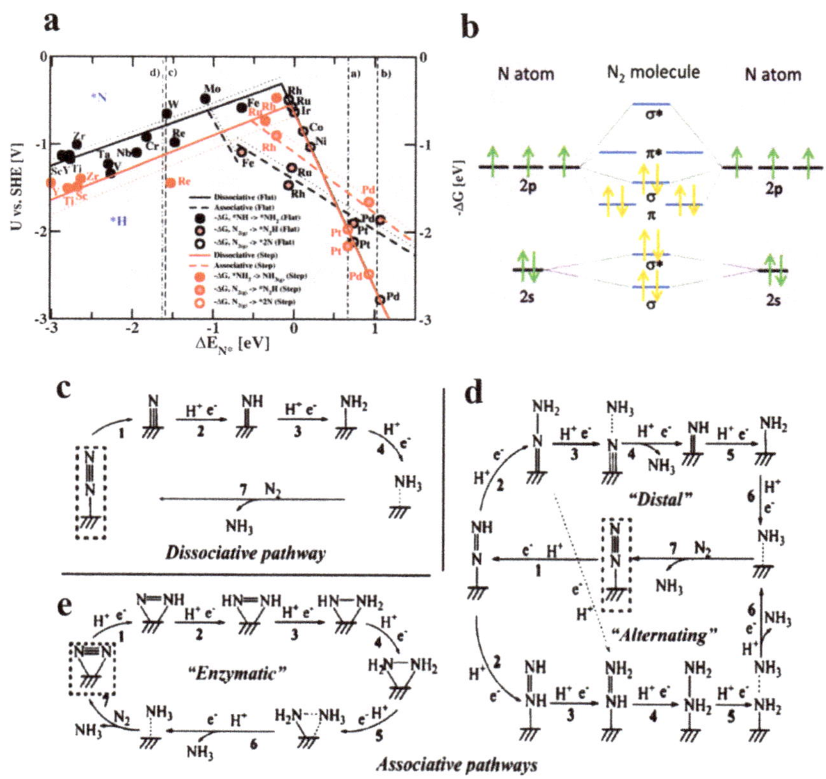

Figure 3. (**a**) Volcano plot of e-NRR for the flat (black lines) and stepped (red lines) transition metal, via dissociative (solid lines) and associative (dotted lines) mechanisms (the redox-potential-limiting step for each metal is highlighted with circles) (an asterisk, *, denotes the adsorption site; the vertical lines (a, b, c, and d) separate different parts and display which species are most strongly bound to the surface) [33], copyright 2012, Royal Society of Chemistry; (**b**) diagrams of N atomic orbitals and their hybridization as N_2 molecular orbitals [36], copyright 2018, Wiley-VCH; schematic diagram of nitrogen reduction pathway on heterogeneous catalysts [37] ((**c**) dissociative pathway, associative pathways including (**d**) distal and alternating pathway, and (**e**) enzymatic pathway), copyright 2016, American Chemical Society.

3.2. Catalytic Activation of N_2

Due to the high dissociation energy of N≡N bonds (941 kJ mol^{-1}) and first-bond cleavage energy (410 kJ mol^{-1}), N_2 is kinetically inert under mild reaction conditions. Activation of N_2 occurs following the chemical adsorption of N_2, and is usually considered as one of the rate-limiting steps in e-NRR. The change in the electron density and electron density distribution of the N_2 molecule can trigger its activation, during adsorption on the catalyst surface. In Figure 3b, an N atom has five valence electrons outside its nucleus, arranged in the configuration $2s^22p^3$ [36]. After bonding, the hybridization of the atomic orbitals is divided into four bonding orbitals and four anti-bonding orbitals. Furthermore, there is a large gap (10.82 eV) between the highest occupied molecular orbital (HOMO) with the lowest unoccupied molecular orbital (LUMO), as well as high ionization energy (15.58 eV) that blocks electron transfer [34]. Consequently, N_2 activation is extremely difficult to achieve under ambient conditions. The first strategy to accelerate N_2 activation is the improvement of the electron donation and back-donation effect between catalysts and the adsorbed N_2. Appropriate strong binding on metals, positively charged carbons or vacancies may promote electron transferring from the electrocatalyst matrix to the dinitrogen molecule, therefore accelerating the triple bond activation. Oxygen vacancies in oxides, nitrogen vacancies in nitrides, and surface defects on metals have been further investigated. In another way, it was reported that N_2 could be activated using the lithium (Li)-mediation assistant method [38]. Li can react directly with N_2 and dissociate to form Li_3N. Subsequently, NH_3 was transformed after the hydrolysis of Li_3N. However, the procedures of this strategy are relatively complicated.

3.3. The Hydrogenation Pathway of N_2 to NH_3

Generally speaking, the current reaction mechanisms of the e-NRR on the heterogeneous catalysts are mainly divided into two kinds: the dissociative pathway (Figure 3c), and the associative pathway (Figure 3d,e) [37]. In the dissociative pathway, the triple bonds of an adsorbed nitrogen are firstly cleaved before the hydrogenation reaction, and two independent N atoms subsequently go through a catalytic hydrogenation reaction [39]. The traditional Haber–Bosch process for industrial NH_3 production mainly follows this mechanism, in which extraordinarily high energy input is required, whereas the process of e-NRR tends to undergo the associative pathway under ambient conditions [40]. In this case, according to the different types of hydrogenation sequences, the associative pathway can be carried out in three possible pathways: the distal, alternating, and enzymatic pathways [30,36]. It is assumed that only one N atom is fixed to the active site, namely, end-on adsorption [41]. In the distal pathway, the N-atom distant from the adsorption site is preferentially hydrogenated continuously. After the release of the first NH_3, the other N atom bound to the catalyst surface begins to form the second NH_3, through the hydrogenation process. In contrast, the alternating pathway is to hydrogenate two N atoms in turn, with two NH_3 molecules generated simultaneously. Instead of an end-on adsorption mode, the enzymatic pathway exhibits side-on adsorption, in which two N- atoms are both adsorbed on the active sites. Additionally, the hydrogenation process involved in the enzymatic path is similar to the alternating pathway. The reduction of nitrogen to NH_3 undergoes these possible mechanisms, resulting in different intermediates, such as diazene (N_2H_2), NH_3, and hydrazine (N_2H_4).

However, apart from the difficulty of nitrogen activation, the e-NRR in aqueous solution is limited by the competition of HER [42]. Several processes at the electrode–electrolyte interface occur concurrently, involving the diffusion and adsorption of reactant species, transfer of electrons and protons, as well as desorption of species, where e-NRR and HER share some reaction species for basically electro-hydrogenation reactions [43,44]. Moreover, the standard equilibrium potential of HER (E^0 = 0 V, vs. RHE) is similar to that of e-NRR (E^0 = 0.092 V, vs. RHE), but HER has much faster reaction kinetics [36]. As a result, e-NRR typically suffers from a low reaction rate and low selectivity (FE) for NH_3 production. The competition between HER and e-NRR can be controlled by optimizing the electrolyte and potential, the local availability of protons and N_2 molecules near the catalyst, and revealing the relationships between structure and activity for rational catalyst design.

4. Advances in Metal Catalysts Design for e-NRR

In recent years, significant efforts have been devoted to design and fabricate an efficient electrocatalyst for NH_3 production [45]. In view of the different compositions and characteristics, the various e-NRR metal catalysts can be classified into metal-based materials and metal compound materials. The recent progress of reported e-NRR metal catalysts is summarized and discussed, with a particular emphasis on their e-NRR performance and catalytic reaction.

4.1. Metal-Based Catalysts

4.1.1. Noble-Metal-Based Catalysts (Ru, Rh, Pt, Au, and Pd)

Noble metal catalysts have been proved as promising electrocatalysts in plenty of electrochemical reactions (such as HER, OER, and ORR), due to their marvelous conductivity, active polycrystalline surfaces, and appropriate adsorption of various reactants. Recently, noble metal catalysts such as Pt, Au, Ag, Ru, and Rh, have been explored for e-NRR to NH_3 synthesis.

Au electrocatalysts have been studied as the most promising noble catalysts for e-NRR [46], by controlling the morphology-dependent effect and metal–support synergetic effect. The former involves the creation of additional active sites by controlling morphology, crystal facets orientation, and crystallinity. As shown in Figure 4, Yan et al. [47] synthesized tetra-hexahedral Au nanorods (Au THH NRs) as heterogeneous electrocatalysts and characterized e-NRR activity in an N_2-saturated 0.10 M KOH solution. The measured angle implied that the bevels on the THH Au NR were high-index (730) planes, composed of the (210) and (310) sub-facets (Figure 4a,b). A large number of active sites can be provided to capture and activate N_2, due to the exposed high-index (210) and (310) facets. The Au THH NRs endowed a highest NH_3 production rate of 1.648 $\mu g\ h^{-1}\ cm^{-2}$ and maximum FE of 4.02% at −0.20 V vs. reversible hydrogen electrode (RHE). The density functional theory (DFT) calculations predicted that the e-NRR process preferably follows the alternating pathway, with the rate-determining step of N_2 dissociation for both Au (210) and Au (310) (Figure 4c). Wang et al. [48] reported a rapid approach for the fabrication of flower-like Au microstructures (Au flowers). In a comparative experiment, the e-NRR performance of Au flowers (NH_3 rate: 25.57 $\mu g\ h^{-1}\ mg_{cat.}^{-1}$, FE: 6.05%) outperformed the Au sphere counterpart. It indicated that Au flowers (particle size: ~900 nm) with highly dendritic structures could provide abundant electrocatalytically active sites, and therefore, promote e-NRR activity. The use of hollow gold nanocages (Au HNCs) as an effective electrocatalyst was also evaluated for e-NRR in 0.50 M $LiClO_4$ [49]. The highest FE of Au HNCs (30.2%) was achieved at −0.40 V vs. RHE, while the maximum NH_3 production (3.90 $\mu g\ h^{-1}\ cm^{-2}$) was obtained at −0.50 V vs. RHE. In contrast experiments, the e-NRR activity of Au HNCs was much better than other Au nanoparticles of various shapes (i.e., Au nanorods, Au nanospheres, and Au nanocubes) in the same conditions, resulting from the increased surface area and confinement effects.

Unlike the morphology-dependent effect, the metal–support synergetic effect contributes to improving the intrinsic activity of active centers. Therefore, the combination of the metal with support as composite may present a new route for reducing the usage of noble metal. For example, Zhao et al. [50] reported a nano-gold catalyst supported on a boron organic polymer (Au/M-BOP) as electrocatalyst for electrochemical reduction from N_2 to NH_3. Yan and co-workers [51] studied the Au/TiO_2 catalyst as a heterogeneous catalyst for e-NRR, synthesized using Au sub-nanoclusters (~0.50 nm) embedded in commercial TiO_2 support. Unexpectedly, the obtained Au/TiO_2 endowed the e-NRR with a high yield (NH_3: 21.4 $\mu g\ h^{-1}\ mg^{-1}$, FE: 8.11%) at −0.20 V vs. RHE. Moreover, it should be noted that the apparent catalytic activity decreased after tuning particle size of Au species dispersed on TiO_2 ranging from nanometer down to sub-nanometer sizes. This work also indicated that the isolated precious metal onto oxide supports provided a well-defined system. The proposed pathway for the NH_3 synthesis using Au/TiO_2 catalyst was shown in Figure 4d, displaying a distal hydriding pathway.

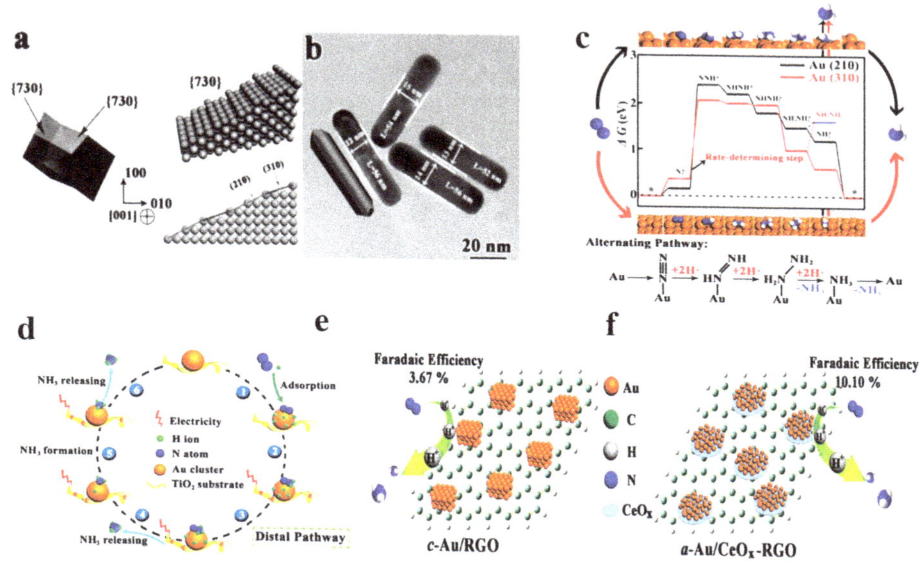

Figure 4. (**a**) Geometric models of Au THH NR and exposed facet [47]; (**b**) TEM images of Au THH NRs [47]; (**c**) free energy diagram and alternating hydriding pathway for e-NRR on Au (210) and Au (310) at equilibrium potential (* represents the adsorption site) [47], copyright 2016, WILEY-VCH Verlag GmbH & Co. KGaA, Weinheim; (**d**) proposed pathway for the NH_3 synthesis using Au/TiO_2 catalyst [51], copyright 2017, WILEY-VCH Verlag GmbH & Co. KGaA, Weinheim; schematic illustration of e-NRR by catalysts of (**e**) c-Au/RGO and (**f**) a-Au/CeOx–RGO [52], copyright 2017, WILEY-VCH Verlag GmbH & Co. KGaA, Weinheim.

This group continued to explore the effectiveness of Au and proposed CeO_x-induced amorphous Au nanoparticles on reduced graphite oxide (a-Au/CeO_x-rGO) as e-NRR electrocatalysts [52]. As shown in Figure 4e,f, it was found that the CeO_x played an important role in transferring the crystalline Au NPs into the amorphous ones. Compared with its crystalline counterpart, a-Au/CeO_x-rGO achieved a higher 10.10% FE with an NH_3 yield of 8.3 μg h^{-1} $mg_{cat.}{}^{-1}$ at −0.20 V vs. RHE, because of their higher concentration of active sites and more structural distortion. In another study, Wang and co-workers [53] reported the Au/N-doped nano-porous graphitic carbon membrane (NCM) electrocatalyst. The synergistic effect between NCM and Au promoted the N_2 adsorption and thereby improved the conversion of N_2 to NH_3.

The precious metal Ru is also a hot research subject in the electroreduction of N_2. Similarly, the crystal structure and the particle size of Ru also have a great influence on e-NRR activity. Wang et al. [54] studied Ru nanoparticles as e-NRR electrocatalysts in 0.01 M HCl aqueous solution. The maximum yield rate of 5.50 mg h^{-1} m^{-2} was achieved at −0.10 V vs. RHE, whereas the highest FE was 5.40% at 0.10 V vs. RHE. The DFT calculations indicated that the efficient e-NRR activity at the low overpotential was attributed to instantaneous N_2 adsorption on Ru (001) surfaces and the spontaneous hydrogenation process by a dissociative mechanism. In another study, isolating Ru single atoms in N-doped porous carbon as electrocatalyst greatly promoted N_2-to-NH_3 conversion (Figure 5a,b), affording an NH_3 formation rate of 3.665 mg h^{-1} $mg_{Ru}{}^{-1}$ at −0.21 V vs. RHE [55]. It was found that the addition of ZrO_2 can effectively suppress the competitive

HER, reaching a high FE of 21% at a low overpotential. From calculation results, the e-NRR mainly occurred at Ru sites with O vacancies, which was permitted through the stabilization of *NNH (low overpotential), destabilization of *H (high e-NRR/HER selectivity), and enchantment of N_2 adsorption (to initiate the e-NRR process).

In addition to the above catalysts, Rh, Ag, and Pd have also been studied for e-NRR, due to their strong adsorption energy and low overpotentials [33]. Surfactant-free atomically ultrathin Rh nanosheets (Rh NSs) were synthesized and used as an effective e-NRR in a 0.10 M KOH solution [54]. Benefiting from their unique ultrathin two-dimensional (2D) structure with abundant surface and modified electronic structure, Rh NSs exhibited an excellent e-NRR performance with a high NH_3 yield rate (23.88 μg h^{-1} $mg_{cat.}^{-1}$) and selectivity (no N_2H_4 generation) at −0.20 V vs. RHE. But, the FE at the same potential was only 0.217%, due to the dominant HER process. Yin et al. [56] reported Ag triangular nanoplates (Ag TPs) as e-NRR catalysts with efficient activity of NH_3 generation. The e-NRR activity of Ag TPs was much more efficient than circular Ag nanoparticles, owing to the more anchored atoms at sharp edges and corners on Ag TPs. Single Ag sites with the Ag-N_4 coordination (SA-Ag/NC) were synthesized massively by targeting the admolecules (Figure 5c), confirming that abundant Ag SAs exist in the carbon matrix by TEM and HAADF-STEM (Figure 5d–f). SA-Ag/NC achieved a record-high NH_3 yield rate (270.9 μg h^{-1} $mg_{cat.}^{-1}$ or 69.4 mg h^{-1} mg Ag^{-1}) and a desirable FE (21.9%) in HCl aqueous solution [57]. Through 20 consecutive cycle tests, the stability of SA-Ag/NC was maintained. Furthermore, to eliminate or quantify the sources of contamination, a rigorous reduction experiment was recommended by the isotopic labeling experiment using $^{15}N_2$, reliably confirming the ammonia production only from the N_2 source [4,58]. As expected, through the isotopic labeling experiment, the NH_3 generation was verified from the gaseous N_2 over SA-Ag/NC during the e-NRR process. Based on first principles calculations (Figure 5g–j), the emergence of vertical end-on *N_2 and oblique end-on *NNH admolecules on single metal sites in succession were energetically favorable for e-NRR.

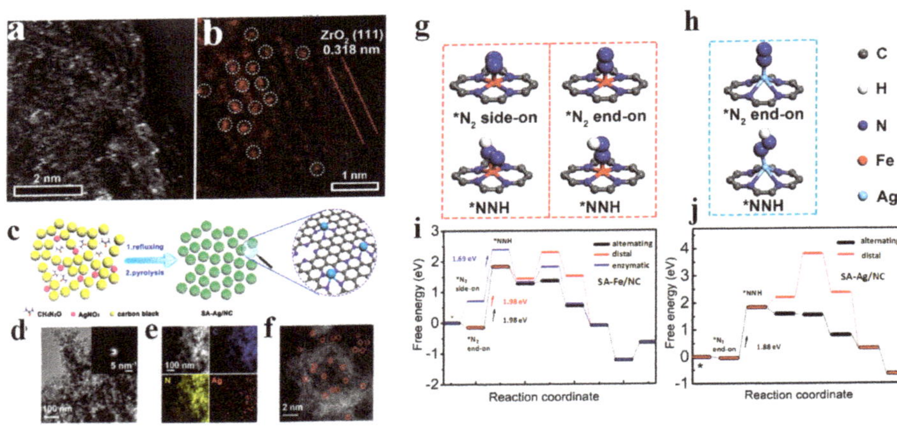

Figure 5. (**a**,**b**) HAADF-STEM image of the Ru@ZrO_2/NC [55], copyright 2018, Elsevier Inc. (**c–f**) Synthesis and structural characterization of SA-Ag/NC [57]: (**c**) schematic illustration; (**d**) TEM image; (**e**) HAADF-STEM images and corresponding element maps; (**f**) aberration-corrected HAADF-STEM image); (**g–j**) DFT computation results of e-NRR on SA-Fe/NC and SA-Ag/NC [57], copyright 2020, American Chemical Society.

From the point of view of morphology control, crystallographic tailoring, structural manipulation, and defect engineering, the noble metal catalysts were summarized, aiming to enhance the e-NRR activity. Combined with discussing the relationship between the structure and e-NRR activity based on experimental and theoretical results, they are expected to provide a reference for the rational design of e-NRR electrocatalysts in a targeted manner.

4.1.2. Non-Noble-Metal-Based Catalysts

There is a growing desire to explore resource-rich metal in the earth, in order to reduce cost and improve the applicability of e-NRR technology. Due to the important role of the FeMo cofactor in biological nitrogen fixation and Fe-based catalysts in Haber–Bosch technology, Mo- and Fe-based electrocatalysts have been explored toward e-NRR.

To explore the effect of crystal phase orientations for Mo catalysts, Yang et al. [59] prepared four kinds of Mo-based nanofilms with different facet orientations and surface morphology. Mo (110) plane can adsorb N adatoms more strongly than H adatoms, while Mo (211) dominantly follows competitive HER. As a result, Mo (110) was more efficient with FE of 0.72% at a low overpotential of −0.49 V vs. RHE. This study showed that morphology control was also a feasible way to improve the catalytic activity of pure non-noble metal toward e-NRR. A study of DFT simulation predicted that single Mo atom fixed on a defective boron nitride (BN) monolayer could be potentially used as a N_2 fixation electrocatalyst, where dispersed Mo atoms bonded to N atoms contributed to activate N_2 molecules, selectively stabilize N_2H^*, or destabilize NH_2^* during e-NRR [60]. Based on this study, Han et al. [61] reported single Mo atoms anchored onto N-doped porous carbon (SA-Mo/NPC) as e-NRR electrocatalysts. Benefiting from the optimized abundance of active sites and 3D hierarchically porous carbon frameworks, SA-Mo/NPC achieved a high NH_3 yield rate (34.0 ± 3.6 μg h^{-1} $mg_{cat.}{}^{-1}$) and a high FE (14.6 ± 1.6%) in 0.10 M KOH electrolyte at −0.30 V vs. RHE. Similarly, efficient e-NRR activity and durability were also obtained by SA-Mo/NPC in 0.10 M HCl acid electrolyte. The authors also concluded that Mo–N sites of atomically dispersed Mo atoms bonding to N were the catalytic active sites. As shown in Figure 6a,b, the stabilized single Mo atoms anchored on holey N-doped graphene (Mo/HNG), with a continuous porous skeleton and plenty of edges containing N-coordination sites, were synthesized through a potassium salt-assisted activation process [62]. As plotted in Figure 6c, at −0.05 V vs. RHE, Mo/HNG exhibited an exceptional FE of 50.2% for NH_3 production (partial reduction current density: 17.0 μA cm^{-2}) and a NH_3 production yield rate of 3.6 μg h^{-1} $mg_{cat}{}^{-1}$. During continuous electrolysis (20,000 s), Mo/HNG still maintained over 50% FE (−0.05 V vs. RHE), with only 0.0125% of Mo on the electrode dissolved (ICP-MS test), exhibiting good stability. The isotopic labeling experiments were measured, respectively, using abundant natural $^{14}N_2$ and $^{15}N_2$ as feed gas [63]. As shown in Figure 6d, the NH_4^+ splitting patterns in 1H nuclear magnetic resonance were consistent with the corresponding resultant electrolyte using isotopic $^{14}N_2$ or $^{15}N_2$ source, presenting a specific double peak for $^{15}NH_4^+$ and three peaks for $^{14}NH_4^+$ [64]. Through theoretical calculations, it is unveiled that the edge coordinated Mo atoms and the existence of vacancies on holey graphene jointly contribute to the intriguing e-NRR activity.

As one of the most earth-abundant metals, Fe-based catalysts have also shown great potential as excellent e-NRR electrocatalysts. For instance, Wang et al. [65] theoretically proposed the catalytic mechanisms of single Fe atom embedded N-doped graphene for e-NRR. The results indicated that the magnetic moment of the Fe atom increased with the increase in coordination of the neighboring N atom, resulting in a lower overpotential of N_2 reduction. In experiment, Wang et al. [66] recently used a single-atom catalyst (iron on N-doped carbon, Fe_{SA}-N-C) as an e-NRR electrocatalyst, enabling a dramatically enhanced FE. Here, the DFT calculations suggested that the Fe_{SA}-N-C structure could effectively attract the access of N_2 molecules with a small energy barrier, which benefits preferential N_2 adsorption instead of H adsorption. The isotope-labeling experiments and control experiments indicated that the generated NH_3 entirely comes from the e-NRR process catalyzed by Fe_{SA}-N-C. Careful characterization and consecutive recycling electrolysis were preformed, suggesting its excellent stability. In another study, an Fe-N/C-carbon nanotube catalyst (Fe-N/C-CNTs) was designed, through carbonizing a metal–organic framework and carbon-nanotube-based composite [67], with built-in Fe−N_3 sites. The corresponding synthesis process was shown in Figure 6e. In 0.10 M KOH electrolyte, the optimal NH_3 formation rate was 34.83 μg h^{-1} $mg_{cat.}^{-1}$ with an FE of 9.28% at −0.20 V vs. RHE. The favorable e-NRR activity was attributed to Fe−N_3 species as active sites. The theoretical results further revealed that the e-NRR reaction proceeded preferentially via the distal pathway.

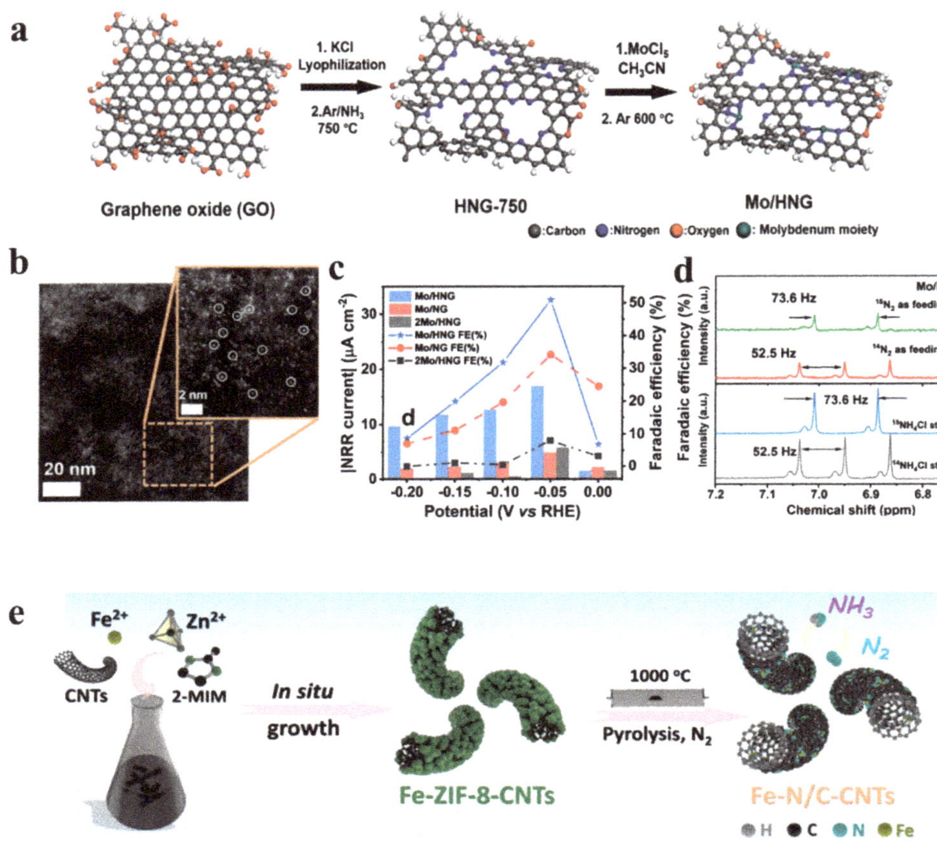

Figure 6. (**a**) Schematic illustration of the synthetic process for the Mo/HNG catalyst [62], (**b**) atomic resolution HAADF-STEM images of Mo-HNG and magnified area with circled individual Mo atoms anchored on the carbon matrix at N-rich edges [62], (**c**) NH_3 yields, FEs and partial current densities for e-NRR on Mo/HNG, Mo/NG, 2Mo/HNG catalysts determined from chronoamperometric measurements [62], (**d**) 1H NMR spectra of the resultant electrolyte obtained from the e-NRR measurement of Mo/HNG, respectively, using $^{14}N_2$ or $^{15}N_2$ as the isotopic nitrogen source at −0.05 V [62], copyright 2022, Wiley-VCH GmbH. (**e**) Schematic illustration of the synthesis of Fe−N/C−CNTs [67], copyright 2018, American Chemical Society.

Recently, Leung's group was dedicated to non-noble bimetals on nitrogen-doped carbons, selecting from either side of the theoretical volcano plot for the e-NRR. Dispersed Mo-Co bimetallic nanoparticles immobilized on N-doped porous carbon (Mo-Co/NC) were developed and exhibited the enhanced activity and selectivity of e-NRR electrocatalysis with an ammonia yield, in comparison to single-metallic Co/NC [68]. Additionally, to overcome the sluggish kinetics of the proton-coupled electron transfer on the single-atom site, Leung et al. [69] synthesized atomically dispersed Co-Mo pairs anchored on N-doped carbon frameworks (Co-Mo-SA/NC) through calcinating Co–Mo-doped zinc-based zeolite imidazole framework precursors. Revealed by Bader charge analysis and charge density difference analysis, 0.35 e^- and 0.30 e^- were, respectively, transferred to N1 on the Mo-end and N_2 on the Co-end from Co–Mo active sites; simultaneously, Co and Mo atoms with two occupied d orbitals possess the capability to donate their electrons to the empty p* orbital of N, ultimately forming triple bonds. Nevertheless, the N_2 on a single active site follows the electron acceptance–donation concept, resulting in a significant increase in energy required for the initial activation of N_2. Consequently, the Co–Mo-SA/NC catalyst achieves outstanding e-NRR performance in 0.1 M Na_2SO_4 solution with 37.73 μg h^{-1} $mg_{cat.}{}^{-1}$ and a desirable FE of 23.18% at −0.1 V vs. RHE, which are twofold higher than those of the isolated single-atom Co (Co-SA/NC) or Mo (Mo-SA/NC) catalyst.

Apart from the aforementioned Fe- and Mo-based catalyst, other transition metals have also emerged as electrocatalysts for N_2 fixation [70–73]. For example, Co single-atom-embedded N-doped porous carbon (CSA/NPC) was synthesized as an electrocatalyst for e-NRR [74]. At a low overpotential of −0.20 V, CSA/NPC presented a high NH_3 yield rate of 0.86 mmol cm^{-2} h^{-1} and a FE of 10.50%, attributed to the positive effects of Co single atoms, N-doping, and porous structure. In another work, Wang et al. [75] reported atomically dispersed Ni sites on a carbon framework with nitrogen-vacancy (Ni_x-N-C) as an effective non-noble-metal electrocatalyst for the e-NRR, synthesized from a Ni-doped ZIF-8 precursor. Compared with Ni clusters supported on the N-doped carbon framework, significant e-NRR activity was observed on Ni_x-N-C with an NH_3 production rate of 115 μg cm^{-2} h^{-1} at −0.80 V (vs. RHE) and FE of 20% at −0.60 V (vs. RHE) in $LiClO_4$ solution. From calculation results, Ni-N_x sites were responsible for the experimentally observed activity and the potential determining step was the hydrogenation during the e-NRR. Cu as a common and low-cost metal has also been studied for e-NRR by Zang and co-workers [76]. In this experiment, a Cu single atom on a porous N-doped carbon network (NC-Cu SA) was studied for catalytic performance toward e-NRR in both alkaline and acidic solutions. The NC-Cu SA exhibited a high NH_3 yield rate and FE, specifically ~53.30 μg h^{-1} $mg_{cat.}{}^{-1}$ and 13.80% under 0.10 M KOH, ~49.30 μg h^{-1} $mg_{cat.}{}^{-1}$ and 11.70% under 0.10 M HCl. Similarly, the experimental analysis and DFT calculations indicated that the local Cu−N_2 coordination was identified as the efficient sites and responsible for the outstanding e-NRR performance.

Up to now, non-noble-metal-based materials have been reported as efficient e-NRR electrocatalysts, most investigations mainly focused on their composites including nitrides, carbides and oxides. More discussion about metal compounds will be summarized in the following section.

4.2. Metal Compound Catalysts

4.2.1. Metal Sulfide and Metal Nitride Catalysts

In recent years, a range of metal sulfides and metal nitrides have been taken into consideration for e-NRR. Although the intrinsic catalytic activity of MoS_2 for water reduction suppressed the e-NRR process, MoS_2 is still considered and utilized as an electrocatalyst to catalyze the N_2 reduction reaction. Sun's group theoretically predicted and experimentally confirmed that MoS_2 as an active e-NRR electrocatalyst achieved a high NH_3 yield rate (8.08×10^{-11} mol s^{-1} cm^{-1}) and FE (1.17%) at −0.50 V vs. RHE [77]. Impressively, this study further indicated that MoS_2 was still active for e-NRR, where a strong HER occurs. Soon thereafter, they found that defect-rich MoS_2 (DR MoS_2) nanoflower could greatly boost electrocatalytic N_2 reduction to NH_3 [39]. Compared with the defect-free counterpart, a high FE (8.34%) and high NH_3 yield (29.28 μg h^{-1} $mg_{cat.}{}^{-1}$) were obtained at −0.40 V vs. RHE. DFT calculations revealed that the potential-determining step was $^*NH_2 \rightarrow {}^*NH_3$, and the barrier of DR MoS_2 (0.60 eV) was lower than the barrier of the defect-free catalyst (0.68 eV). In another theoretical study, Fe-doped MoS_2 through an associative distal pathway revealed that the presence of a vicinal Fe atom enabled highly selective chemisorption of N_2, which was conducive to the efficient activation of the N≡N bonds [78]. This investigation provides some new ideas for designing active metal sulfides for the electrochemical synthesis of NH_3.

Under the comprehensive theoretical investigation on a range of transition metal nitrides (TMN) for e-NRR, metal nitrides are believed to offer the potential advantages for N_2 fixation [79]. Skúlason et al. [80] studied the possibility of nitrogen activation for electrochemical NH_3 formation on a range of (111) TMN surfaces (ScN, TiN, VN, CrN, MnN, YN, ZrN, NbN, MoN, HfN, TaN, WN, and ReN). It was found that VN, CrN and MnN were the most promising candidates, which were expected to catalyze e-NRR at the relatively low onset potential (from −0.80 V to −0.50 V vs. SHE). However, the possibility of poisoning toward MnN and WN was found in an electrochemical environment. Only NbN with the (111) plane can be regenerated itself and can activate N_2 to NH_3, with active and stable activity.

To date, only Mo- and V-based nitrides have been experimentally studied and proved to enable efficient catalytic activity toward e-NRR. Based on the theoretical investigations, vanadium nitride nanosheet [81], vanadium nitride nanowire array [82], and vanadium nitride nanoparticles [83] have been fabricated and tested for e-NRR activity. Comparatively, vanadium nitride (VN) nanoparticles exhibited better catalytic performance for e-NRR, with an NH_3 production rate and FE of 3.30×10^{-10} mol s^{-1} cm^{-2} and 6.0%, respectively [83]. According to a combination of ex situ and operando characterizations, multiple vanadium oxide, oxynitride and nitride species were present on the surface. Among them, $VN_{0.7}O_{0.45}$ was identified as the active phase in the e-NRR, and the conversion of $VN_{0.7}O_{0.45}$ to VN phase was proposed as the deactivation pathway.

Except for vanadium nitride, concerns regarding molybdenum nitride have been recently discussed. Li et al. [84] theoretically studied the 2D layered molybdenum nitride nanosheets (MoN_2) as NH_3 synthesis catalysts at room temperature. According to calculations, MoN_2 exhibited excellent performance for adsorption and activation of N_2 molecules, but large energy input was requested to regenerate the MoN_2 surface. However, the e-NRR performance can be remarkably promoted after Fe-doping, with ΔG_{max} = 0.47 eV for the rate-determining step. The conclusion about Fe-doping agreed with the recent report regarding Fe-doped MoS_2 [78]. Experimentally, the MoN nanosheet array on a carbon cloth (MoN NA/CC) was explored as a high-performance catalyst towards e-NRR in 0.10 M HCl under ambient conditions [81]. This catalyst achieved an NH_3 yield of 3.01×10^{-10} mol s^{-1} cm^{-2} and an FE of 1.15% at −0.30 V vs. RHE. Moreover, N_2H_4 was not detected, and therefore, MoN NA/CC showed excellent selectivity to NH_3. The potential-determining step of this catalyst was the second protonation of the surface N, confirmed by DFT calculations. In another study, this group reported a Mo_2N nanorod as an efficient electrocatalyst to electrochemically convert N_2 to NH_3 [85]. Mo_2N nanorods were prepared by nitriding of the precursor MoO_2 in an NH_3 atmosphere. Compared with MoO_2, the NH_3 yield of Mo_2N was much higher. When tested in 0.10 M HCl, Mo_2N could enhance the FE to 4.5% at an applied potential of −0.30 V vs. RHE, which was higher than MoN NA/CC in the previous report. DFT calculations also confirmed that the free energy barrier of the potential-determining step for the Mo_2N catalyst was dramatically lower than MoO_2. Based on the studies above, the FE in e-NRR still needs to be further improved in the future.

4.2.2. Metal Carbide Catalysts

The metal carbides are an interesting class of catalysts. According to the d orbital theory, transition metal carbides with unoccupied d orbitals should have good adsorption ability for electron-enriched reactants [86,87]. In order to investigate the viability of using molybdenum carbide as an e-NRR electrocatalyst, a computational study was conducted by Matanovic and co-workers [88]. The comparison between two competing reactions (HER and NRR) revealed that MoC (111) was the only surface that suppressed the adsorption of H-atoms at low overpotentials, among various crystallographic surfaces. Additionally, the e-NRR in MoC (111) surface could take place at small negative potentials of −0.30 V vs. SHE, and followed an associative reaction pathway. The authors also illustrated that introducing carbon vacancies could mitigate hydrogen evolution and H-adatom accumulation. Recently, molybdenum carbide nanodots embedded in ultrathin carbon nanosheets (Mo_2C/C) were designed by molten salt synthesis, and used as a catalyst candidate for e-NRR [89]. The obtained Mo_2C/C nanosheets exhibited efficient e-NRR catalytic activity with an NH_3 production rate of 11.3 μg h^{-1} mg^{-1} and FE of 7.8%. Based on the experiments and DFT calculations, the catalytic active center of Mo_2C nanodots was favorable for adsorbing N_2, and their unique electronic structure was feasible for N_2 activation and hydrogenation. MoS_2, MoO_3, MoN and Mo_2N have been reported as e-NRR electrocatalysts with relatively lower FE of 1.17%, 1.9%, 1.15%, and 4.5% [77,81,85,90], respectively. To continuously enhance the performance, Sun's group reported Mo_2C nanorod as a catalyst for electrocatalytic N_2 reduction to NH_3 production [90]. At the applied potential of −0.30 V vs. RHE, such a catalyst achieved a high FE of 8.13% and NH_3 yield rate of 95.10 μg h^{-1} $mg_{cat.}{}^{-1}$ in 0.10 M HCl electrolyte. To date, metal carbide nanocomposites such as the e-NRR catalyst are rarely reported.

MXenes, a group of 2D layers of transition metal carbides, are promising catalysts for e-NRR [91,92]. A large number of studies have focused on theoretical calculations. In 2019, Luo et al. [93] firstly reported that the MXene ($Ti_3C_2T_x$) nanosheets attached to a vertically aligned metal host could achieve a high NH_3 FE (5.78%) at an ultralow overpotential of −0.10 V vs. RHE. From the combined experimental and theoretic results, a greater number of exposed edge sites and a metal host with poor HER activity were responsible for higher e-NRR activity. In another work, a $Ti_3C_2T_x$ MXene nanosheet was used as both a conductive and Ti source toward the in situ hydrothermal growth of TiO_2 nanoparticles [94]. The combination of TiO_2 and $Ti_3C_2T_x$ led to a synergistically active Ti-based nanohybrid catalyst with enhanced activity. As a result, such a $TiO_2/Ti_3C_2T_x$ hybrid catalyst exhibited an NH_3 yield of 26.32 μg h^{-1} $mg_{cat.}{}^{-1}$ with an FE of 8.42% in 0.10 M HCl electrolyte (−0.60 V vs. RHE). It is universally known that 3D porous MXene-based aerogel architectures are beneficial for rapid mass diffusion, higher exposure of electrochemically active sites, and faster mass diffusion and charge/electron transport. Herein, Li et al. [95] designed a functional 3D MXene-based composite heterojunction aerogel (MS@S-MAs) for e-NRR, fabricating metal sulfide nanoparticles confined in 3D S-doped MXene sheets (Figure 7a), via divalent metal-ion-induced assembly following the thermal sulfidation method. Remarkably, CoS@S-MAs gave the best reactivity among metal sulfide nanoparticles (M = Co, Fe, Cu, Ni), showing an NH_3 yield rate and a FE of 12.4 μg h^{-1} $mg_{cat.}{}^{-1}$ and 27.05% at the lower potential of −0.15 V vs. RHE in Na_2SO_4 solution. Additionally, CoS@S-MAs, after 50 h of e-NRR, displayed a slight loss in FE and NH_3 yield rate (Figure 7b), indicating the excellent long-term stability of the catalyst. This study offers a new prospect for 3D porous aerogel materials for application in e-NRR metal oxide catalysts.

Metal oxides have been widely applied in chemical research and also exhibited great potential in e-NRR. To find the viability of electrocatalysts for catalyzing NH_3 formation electrochemically at ambient conditions, 11 types of transition metal dioxides (NbO_2, TaO_2, RuO_2, ReO_2, TiO_2, OsO_2, RhO_2, MnO_2, CrO_2, IrO_2, and PtO_2) in the rutile structure were investigated by DFT calculations on their (110) lattice planes [96]. The predicted onset potentials as a function of the binding energy of NNH were given in Figure 7c, with only two potential-determining steps. Among 11 types of transition metal dioxides, ReO_2, TaO_2, and OsO_2, required an overpotential similar to, or lower than, the overpotential required for reducing nitrogen through nitrogensase, but that is believed to be approximately 0.63 V.

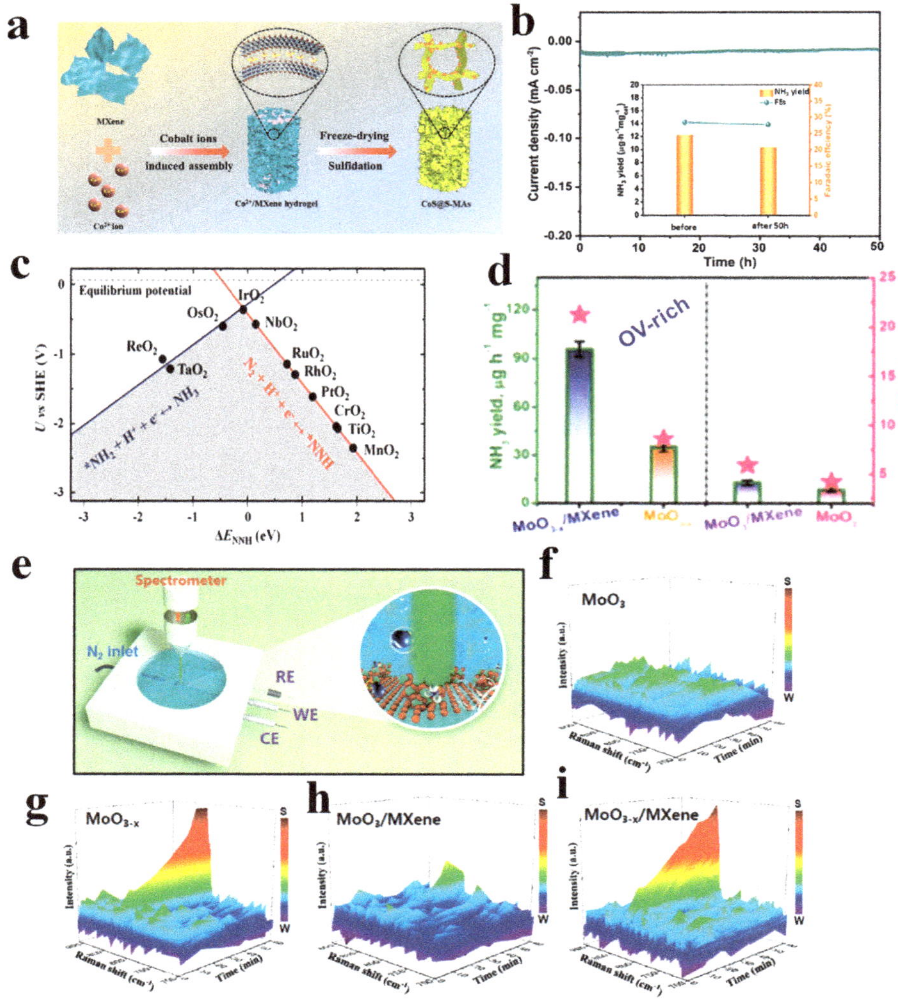

Figure 7. (**a**) Schematic illustration of the synthesis of CoS@S-MAs. (**b**) Chronoamperometry curve of CoS@S-MAs for 50 h electrolysis (−0.15 V vs. RHE), and corresponding NH_3 yield and FEs before and after 50 h (inset), Copyright 2021, Wiley-VCH GmbH. (**c**) Volcano plot of plotting the predicted onset potentials for e-NRR on the (110) facet of transition-metal dioxides against the binding energy of NNH, ΔE_{NNH}, as the descriptor of catalytic activity [96]. Copyright 2017, American Chemical Society; (**d**) NH_3 yields/FEs at −0.4 V of MoO_3, $MoO_{3\text{-}x}$, MoO_3/MXene and $MoO_{3\text{-}x}$/MXene (pink star represents FE, dotted line separates OV-rich materials and OV-poor materials), (**e**) schematic of tailor-made electrolytic cell, (**f**–**i**) 3D plots of the time-dependent in situ Raman spectroscopy of different catalysts for e-NRR process at −0.4 V [97], Copyright 2021, Wiley-VCH GmbH.

Moreover, the (110) facets of ReO_2 and TaO_2 were found to favor NNH adsorption over H adsorption, whereas IrO_2 and NbO_2 surfaces might be poisoned by adsorbed hydrogen atoms. Huang et al. [98] experimentally verified the potential of NbO_2 nanoparticles as an efficient e-NRR electrocatalyst. Compared to Nb_2O_5 with a similar crystal structure but a different linkage style, the Nb^{4+} cation of NbO_2 enabled effective N_2 adsorption by proving empty d-orbitals and subsequent activation by back donation. Consequently, the NbO_2 nanoparticles presented both an efficient NH_3 production rate (11.60 μg h^{-1} $mg_{cat.}{}^{-1}$) at −0.65 V and FE (32%) at −0.60 V, significantly higher than those of Nb_2O_5 nanoparticles under similar conditions.

Due to the industrial application of Fe-based catalysts in the Haber–Bosh process, Fe-based oxide materials were also considered as an efficient candidate in the field of e-NRR. Ever since nano-Fe_2O_3 was reported as an e-NRR electrocatalyst by Licht and co-workers, Fe-based oxides have attracted wide attention. Later, a complementary theoretical study demonstrated the chemical formation process of NH_3 on two kinds of hematite (γ-Fe_2O_3) surfaces. Compared with single-iron (Fe–O_3–Fe–), double-iron (Fe–Fe–O_3–) needed a smaller applied bias for proton transfer, owing to the two available reactive Fe sites on this surface [99]. Kong et al. [7] firstly investigated the e-NRR activity of nanosized γ-Fe_2O_3 electrocatalysts at low temperature (<65 °C), in basic aqueous solution and in the membrane electrode assembly (MEA)-based reactors, respectively. Compared with the half-cell, the e-NRR activity in MEA-based reactors was observed with a dramatical increase to 55.90 nmol h^{-1} mg^{-1}. The enhanced catalytic performance may be attributed to the efficient utilization of γ-Fe_2O_3 after it is coated on the porous carbon paper. Furthermore, the Fe_2O_3-CNT and oxygen-vacancy-enriched-Fe_2O_3/CNT catalysts were also reported [100,101]. In addition to Fe_2O_3, Fe_3O_4 was also reported to be catalytically active for e-NRR. A spinel Fe_3O_4 nanorod on a Ti mesh (Fe_3O_4/Ti) was fabricated as a catalyst for electrochemical N_2 conversion to NH_3, with long-term electrochemical durability [102]. Hu et al. [103] investigated the Fe-based materials for electrocatalytic NH_3 production and revealed the effect of different chemical states of Fe on the e-NRR activity. The Fe/Fe_3O_4 catalyst was fabricated via in situ growth on the Fe foil. In particular, the activity and selectivity of Fe/Fe_3O_4 were superior to those of Fe, Fe_3O_4 and Fe_2O_3 nanoparticles. It has been concluded that the e-NRR catalytic performance was related to Fe/Fe oxide ratio.

Much attention has also been focused on developing Mo-based oxides as high-performance e-NRR electrocatalysts. Sun et al. [90] discovered that MoO_3 nanosheets exhibited remarkable e-NRR activity with excellent selectivity in 0.10 M HCl electrolyte (NH_3 yield: 29.43 μg h^{-1} $mg_{cat.}{}^{-1}$ and FE: 1.9%). It was found that the outermost Mo atoms served as the active sites for effective N_2 adsorption, by DFT calculations. To further tailor the performance of Mo oxides, a hybrid catalyst of MoO_2 on reduced graphene oxide (MoO_2/RGO) was fabricated to catalyze the e-NRR. In 0.10 M Na_2SO_4 electrolyte, an enhanced e-NRR performance was obtained, with an NH_3 yield of 37.4 μg h^{-1} mg^{-1} and FE of 6.6% at the potential of −0.35 V (vs. RHE) [104]. Relative to MoO_2 alone, MoO_2/RGO hybrid promoted the electronic interactions with *N_2H, and enabled the donation of more electrons from the active Mo sites to *N_2H, leading to the enhanced e-NRR activity. Based on the vacancy and heterostructure engineering, O-vacancy-rich MoO_{3-x} anchored on $Ti_3C_2T_x$-MXene (MoO_{3-x}/MXene) was explored, as a highly efficient and selective e-NRR electrocatalyst, obtaining an exceptional e-NRR performance with an NH_3 yield of 95.8 μg h^{-1} mg^{-1} at −0.4 V and a FE of 22.3% at −0.3 V [97]. MoO_{3-x}/MXene produce steady NH_3 yields and FEs during consecutive seven cycles of electrolysis, while just a very small change compared to the initial one. In Figure 7d, OV-rich MoO_3 and MoO_3/MXene achieved higher e-NRR activities with respect to their corresponding OV-rich MoO_3 and MoO_3/MXene, indicating the critical role of OVs for substantially improving e-NRR performance. Through in situ Raman spectroscopy adopted in a tailor-made electrolytic cell (Figure 7e), the 3D plots for the time-dependent Raman spectra traces of various catalysts at −0.4 V were shown in Figure 7f–i, to track the changes in surface chemical bonds of considerable catalysts. Together with molecular dynamics simulations and DFT computations, the synergistic effects of OVs and MXene on the e-NRR of MoO_{3-x}/MXene were confirmed.

SnO_2, known for its low cost and high chemical stability, was initially developed by Zhang et al. [105] as an e-NRR electrocatalyst in the form of cubic sub-micron SnO_2 particles loaded on carbon cloth (SnO_2/CC). However, to enhance conductivity and active sites of such catalyst, Chu et al. [106] developed a novel fluorine-doped SnO_2 mesoporous nanosheets on carbon cloth (F-SnO_2/CC) as an e-NRR electrocatalyst. From the calculations, F-doping contributed to readily enhance the conductivity and increase the positive charge density on active Sn sites, resulting in reduced reaction energy barriers and enhanced e-NRR activities. This group also investigated the e-NRR performance of supporting the ultrasmall SnO_2 QDs on RGO [106]. Similarly, the experimental and theoretical results confirmed that coupling SnO_2 QDs and RGO could readily tailor the electronic structure of SnO_2, leading to fascinating e-NRR activity.

As one of the classical semiconductors, TiO_2-based materials have been firstly investigated as efficient photocatalysts in the photo-reduction of N_2 to NH_3. Lately, Sun et al. [107] explored the TiO_2 nanosheets array on the Ti plate (TiO_2/Ti) for electrochemical N_2 conversion to NH_3. When measured in 0.10 M Na_2SO_4, TiO_2/Ti achieved a high NH_3 yield of 9.16×10^{-11} mol s^{-1} cm^{-2} with an FE of 2.50% at −0.70 V vs. RHE, due to the enhancement of adsorption and activation of N_2 by in situ-generated oxygen vacancies. To further enhance electronic conductivity, a TiO_2 nanoparticle-reduced graphene oxide hybrid (TiO_2-rGO) was fabricated as an e-NRR electrocatalyst, by Sun's group [108]. The FE of TiO_2-rGO was enhanced to 3.30% at −0.90 V vs. RHE.

Inspired by the enhanced activity of nitrogenases with Mn^{2+}, Wang et al. [109] reported MnO particles on Ti mesh (MnO/TM) as a robust e-NRR catalyst. In 0.10 M Na_2SO_4 electrolyte, such catalyst achieved a high FE up to 8.02% and a large NH_3 production of 1.11×10^{-10} mol s^{-1} cm^{-2} at −0.39 V (vs. RHE). Theoretical calculations further revealed that the MnO(200) surface preferentially adsorbed N atoms instead of H atoms, and the potential-determining step was $*N_2 \rightarrow *N_2H$ transformation. Additionally, a spinel $LiMn_2O_4$ nanofiber could act as a noble-metal-free electrocatalyst for NH_3 synthesis with an excellent FE of 7.44% [110], much higher than that of the previous Mn_3O_4 nanocube (3.00%) [111] and Mn_3O_4 nano-particles-reduced graphene oxide (3.52%) [112]. Besides, Cr_2O_3 nanofiber was fabricated as a non-noble-metal e-NRR electrocatalyst [113]. This catalyst achieved an efficient performance in both FE and NH_3 formation, with favorable electrochemical durability.

5. Conclusions and Outlook

In conclusion, we discuss the recent advances of metal-based e-NRR electrocatalysts using the structure–function relationship, concluding that noble-metal-based catalysts, non-noble-metal-based catalysts and metal compound catalysts provide a fundamental basis for rational electrocatalyst design. Additionally, the challenges and prospects for e-NRR were proposed. Although the encouraging progress on e-NRR electrocatalysts has been achieved with favorable performance, the reported studies of e-NRR still have a long distance to go in contrast with the industrial Haber–Bosch process, from the point of view of industrialization and commercialization. Research in e-NRR still faces several key challenges in the near future.

(1) Selectivity of catalysts is a much larger issue for improving FE, due to the competitive reactions. The hydrogen and hydrazine simultaneously generated during the ammonia production resulted in a relatively low selectivity towards e-NRR [114–116]. The designed catalysts are required to have a much stronger binding energy of *N compared to the *H. In addition, the strategy of enhancing the solubility of N_2 in the electrolyte also needs to be developed [117–120].

(2) In-depth studies of the e-NRR mechanism are still limited and plain. Most of the research only simulated the possible reaction pathways and the energy barriers using theoretical calculations. Most reported theoretical studies for identifying the research direction toward electrocatalyst design were performed on appropriate and simplified models. However, the real-time operation of e-NRR is always in combination with different reaction conditions and parameters (pH, environmental electrolyte, voltage, environmental temperature, and pressure, etc.) that ought to be considered in further calculations [3,28,121].

(3) The stability of the catalyst is as important as catalyst activity and selectivity. After longtime electrolysis operation, the electrocatalyst may undergo decomposition and deactivation [4,122]. Therefore, the electrocatalysts should be designed with a stable structure. Additionally, the prolonged periods for the stability tests are recommended to screen active electrocatalysts for e-NRR [28,123].

(4) The relationship between structure and activity for e-NRR is of significance to provide a guideline on the rational design of novel catalysts. Despite the great efforts on developing advanced materials for e-NRR, it remains challenging to reveal the relationship between the structure and activity under the reaction conditions [124,125]. In situ analytical techniques and theoretical experiments are highly desirable and beneficial in providing evidence of catalyst surface reconstruction and generation of key intermediates under real-time reaction conditions, as well as in achieving a comprehensive understanding of the kinetic mechanism [25,124,126,127].

Funding: The authors acknowledge the support from the Natural Science Foundation of Shandong Province Youth Project (ZR2021QE067), Research Grants Council of the Hong Kong Special Administrative Region (CityU 11206520), Shenzhen Knowledge Innovation Program (Basic Research, JCYJ20190808181205752), Development Plan of Youth Innovation Team in Shandong Province (2022KJ213), the Outstanding Young Talents Project of Shandong University of Science and Technology (SKR21-3-A-011) and Innovation and Entrepreneurship Training Program for College Students of Shandong University of Science and Technology (X202210424006).

Data Availability Statement: Not applicable.

Conflicts of Interest: The authors declare no conflict of interest.

References

1. Wang, Y.; Wang, J.; Li, H.; Li, Y.; Li, J.; Wei, K.; Peng, F.; Gao, F. Nitrogen Reduction Reaction: Heteronuclear Double-Atom Electrocatalysts. *Small Struct.* **2023**, *4*, 2200306. [CrossRef]
2. Shi, L.; Yin, Y.; Wang, S.; Sun, H. Rational Catalyst Design for N_2 Reduction under Ambient Conditions: Strategies toward Enhanced Conversion Efficiency. *ACS Catal.* **2020**, *10*, 6870–6899. [CrossRef]
3. Yang, B.; Ding, W.; Zhang, H.; Zhang, S. Recent progress in electrochemical synthesis of ammonia from nitrogen: Strategies to improve the catalytic activity and selectivity. *Energy Environ. Sci.* **2021**, *14*, 672–687. [CrossRef]
4. Suryanto, B.H.R.; Du, H.-L.; Wang, D.; Chen, J.; Simonov, A.N.; MacFarlane, D.R. Challenges and prospects in the catalysis of electroreduction of nitrogen to ammonia. *Nat. Catal.* **2019**, *2*, 290–296. [CrossRef]
5. Wan, Y.; Xu, J.; Lv, R. Heterogeneous electrocatalysts design for nitrogen reduction reaction under ambient conditions. *Mater. Today* **2019**, *27*, 69–90. [CrossRef]
6. Tang, C.; Qiao, S.Z. How to explore ambient electrocatalytic nitrogen reduction reliably and insightfully. *Chem. Soc. Rev.* **2019**, *48*, 3166–3180. [CrossRef] [PubMed]
7. Kong, J.; Lim, A.; Yoon, C.; Jang, J.H.; Ham, H.C.; Han, J.; Nam, S.; Kim, D.; Sung, Y.-E.; Choi, J.; et al. Electrochemical Synthesis of NH_3 at Low Temperature and Atmospheric Pressure Using a γ-Fe_2O_3 Catalyst. *ACS Sustain. Chem. Eng.* **2017**, *5*, 10986–10995. [CrossRef]
8. Ithisuphalap, K.; Zhang, H.; Guo, L.; Yang, Q.; Yang, H.; Wu, G. Photocatalysis and Photoelectrocatalysis Methods of Nitrogen Reduction for Sustainable Ammonia Synthesis. *Small Methods* **2019**, *3*, 1800352. [CrossRef]
9. Cheng, S.; Liu, Y.; Zheng, H.; Pan, Y.; Luo, J.; Cao, J.; Liu, F. FeCo-ZIF derived carbon-encapsulated metal alloy as efficient cathode material for heterogeneous electro-Fenton reaction in 3-electron ORR pathway: Enhanced performance in alkaline condition. *Sep. Purif. Technol.* **2023**, *325*, 124545. [CrossRef]
10. Hou, M.; Zhang, Y.; Jiao, X.; Ding, N.; Jiao, Y.; Pan, Y.; Xue, J.; Zhang, Y. Polyphenol-modified zero-valent iron prepared using ball milling technology for hexavalent chromium removal: Kinetics and mechanisms. *Sep. Purif. Technol.* **2023**, *326*, 124874. [CrossRef]
11. Li, X.; Song, H.; Zhang, G.; Zou, W.; Cao, Z.; Pan, Y.; Zhang, G.; Zhou, M. Enhanced organic pollutant removal in saline wastewater by a tripolyphosphate-Fe^0/H_2O_2 system: Key role of tripolyphosphate and reactive oxygen species generation. *J. Hazard. Mater.* **2023**, *457*, 131821. [CrossRef] [PubMed]
12. Lin, L.; Miao, N.; Wallace, G.G.; Chen, J.; Allwood, D.A. Engineering Carbon Materials for Electrochemical Oxygen Reduction Reactions. *Adv. Energy Mater.* **2021**, *11*, 2100695. [CrossRef]
13. Kim, C.; Dionigi, F.; Beermann, V.; Wang, X.; Möller, T.; Strasser, P. Alloy Nanocatalysts for the Electrochemical Oxygen Reduction (ORR) and the Direct Electrochemical Carbon Dioxide Reduction Reaction (CO_2RR). *Adv. Mater.* **2019**, *31*, 1805617. [CrossRef] [PubMed]

14. Eon Jun, S.; Choi, S.; Kim, J.; Kwon, K.C.; Park, S.H.; Jang, H.W. Non-noble metal single atom catalysts for electrochemical energy conversion reactions. *Chin. J. Catal.* **2023**, *50*, 195–214. [CrossRef]
15. Theerthagiri, J.; Park, J.; Das, H.T.; Rahamathulla, N.; Cardoso, E.S.F.; Murthy, A.P.; Maia, G.; Vo, D.V.N.; Choi, M.Y. Electrocatalytic conversion of nitrate waste into ammonia: A review. *Environ. Chem. Lett.* **2022**, *20*, 2929–2949. [CrossRef]
16. van Langevelde, P.H.; Katsounaros, I.; Koper, M.T.M. Electrocatalytic Nitrate Reduction for Sustainable Ammonia Production. *Joule* **2021**, *5*, 290–294. [CrossRef]
17. Flores, K.; Cerrón-Calle, G.A.; Valdes, C.; Atrashkevich, A.; Castillo, A.; Morales, H.; Parsons, J.G.; Garcia-Segura, S.; Gardea-Torresdey, J.L. Outlining Key Perspectives for the Advancement of Electrocatalytic Remediation of Nitrate from Polluted Waters. *ACS EST Eng.* **2022**, *2*, 746–768. [CrossRef]
18. Liu, M.J.; Miller, D.M.; Tarpeh, W.A. Reactive Separation of Ammonia from Wastewater Nitrate via Molecular Electrocatalysis. *Environ. Sci. Technol. Lett.* **2023**, *10*, 458–463. [CrossRef]
19. Wang, L.; Xia, M.; Wang, H.; Huang, K.; Qian, C.; Maravelias, C.T.; Ozin, G.A. Greening Ammonia toward the Solar Ammonia Refinery. *Joule* **2018**, *2*, 1055–1074. [CrossRef]
20. Jiao, L.; Wang, Y.; Jiang, H.-L.; Xu, Q. Metal–Organic Frameworks as Platforms for Catalytic Applications. *Adv. Mater.* **2018**, *30*, 1703663. [CrossRef]
21. Hou, J.; Yang, M.; Zhang, J. Recent advances in catalysts, electrolytes and electrode engineering for the nitrogen reduction reaction under ambient conditions. *Nanoscale* **2020**, *12*, 6900–6920. [CrossRef] [PubMed]
22. Liu, D.; Chen, M.; Du, X.; Ai, H.; Lo, K.H.; Wang, S.; Chen, S.; Xing, G.; Wang, X.; Pan, H. Development of Electrocatalysts for Efficient Nitrogen Reduction Reaction under Ambient Condition. *Adv. Funct. Mater.* **2020**, *31*, 2008983. [CrossRef]
23. Pang, Y.; Su, C.; Xu, L.; Shao, Z. When nitrogen reduction meets single-atom catalysts. *Prog. Mater. Sci.* **2023**, *132*, 101044. [CrossRef]
24. Xue, C.; Zhou, X.; Li, X.; Yang, N.; Xin, X.; Wang, Y.; Zhang, W.; Wu, J.; Liu, W.; Huo, F. Rational Synthesis and Regulation of Hollow Structural Materials for Electrocatalytic Nitrogen Reduction Reaction. *Adv. Sci.* **2022**, *9*, 2104183. [CrossRef] [PubMed]
25. Chebrolu, V.T.; Jang, D.; Rani, G.M.; Lim, C.; Yong, K.; Kim, W.B. Overview of emerging catalytic materials for electrochemical green ammonia synthesis and process. *Carbon Energy* **2023**, *e361*, 1–52. [CrossRef]
26. Guo, X.; Wan, X.; Shui, J. Molybdenum-based materials for electrocatalytic nitrogen reduction reaction. *Cell Rep. Phys. Sci.* **2021**, *2*, 100447. [CrossRef]
27. Zhang, G.; Li, Y.; He, C.; Ren, X.; Zhang, P.; Mi, H. Recent Progress in 2D Catalysts for Photocatalytic and Electrocatalytic Artificial Nitrogen Reduction to Ammonia. *Adv. Energy Mater.* **2021**, *11*, 2003294. [CrossRef]
28. Huang, Z.; Rafiq, M.; Woldu, A.R.; Tong, Q.-X.; Astruc, D.; Hu, L. Recent progress in electrocatalytic nitrogen reduction to ammonia (NRR). *Coord. Chem. Rev.* **2023**, *478*, 214981. [CrossRef]
29. Wen, J.; Zuo, L.; Sun, H.; Wu, X.; Huang, T.; Liu, Z.; Wang, J.; Liu, L.; Wu, Y.; Liu, X.; et al. Nanomaterials for the electrochemical nitrogen reduction reaction under ambient conditions. *Nanoscale Adv.* **2021**, *3*, 5525–5541. [CrossRef]
30. Cui, X.; Tang, C.; Zhang, Q. A Review of Electrocatalytic Reduction of Dinitrogen to Ammonia under Ambient Conditions. *Adv. Energy Mater.* **2018**, *8*, 1800369. [CrossRef]
31. Yao, Y.; Wang, J.; Shahid, U.B.; Gu, M.; Wang, H.; Li, H.; Shao, M. Electrochemical Synthesis of Ammonia from Nitrogen Under Mild Conditions: Current Status and Challenges. *Electrochem. Energy Rev.* **2020**, *3*, 239–270. [CrossRef]
32. Wang, J.; Yu, L.; Hu, L.; Chen, G.; Xin, H.; Feng, X. Ambient ammonia synthesis via palladium-catalyzed electrohydrogenation of dinitrogen at low overpotential. *Nat. Commun.* **2018**, *9*, 1795. [CrossRef] [PubMed]
33. Skulason, E.; Bligaard, T.; Gudmundsdottir, S.; Studt, F.; Rossmeisl, J.; Abild-Pedersen, F.; Vegge, T.; Jonsson, H.; Norskov, J.K. A theoretical evaluation of possible transition metal electro-catalysts for N_2 reduction. *Phys. Chem. Chem. Phys.* **2012**, *14*, 1235–1245. [CrossRef] [PubMed]
34. Wang, S.; Ichihara, F.; Pang, H.; Chen, H.; Ye, J. Nitrogen Fixation Reaction Derived from Nanostructured Catalytic Materials. *Adv. Funct. Mater.* **2018**, *28*, 1803309. [CrossRef]
35. Suryanto, B.H.R.; Kang, C.S.M.; Wang, D.; Xiao, C.; Zhou, F.; Azofra, L.M.; Cavallo, L.; Zhang, X.; MacFarlane, D.R. Rational Electrode–Electrolyte Design for Efficient Ammonia Electrosynthesis under Ambient Conditions. *ACS Energy Lett.* **2018**, *3*, 1219–1224. [CrossRef]
36. Zhang, L.; Chen, G.-F.; Ding, L.-X.; Wang, H. Advanced Non-metallic Catalysts for Electrochemical Nitrogen Reduction under Ambient Conditions. *Chem.—A Eur. J.* **2019**, *25*, 12464–12485. [CrossRef] [PubMed]
37. Shi, R.; Zhao, Y.; Waterhouse, G.I.N.; Zhang, S.; Zhang, T. Defect Engineering in Photocatalytic Nitrogen Fixation. *ACS Catal.* **2019**, *9*, 9739–9750. [CrossRef]
38. Qin, Q.; Oschatz, M. Overcoming Chemical Inertness under Ambient Conditions: A Critical View on Recent Developments in Ammonia Synthesis via Electrochemical N_2 Reduction by Asking Five Questions. *ChemElectroChem* **2020**, *7*, 878–889. [CrossRef]
39. Wang, J.; Chen, S.; Li, Z.; Li, G.; Liu, X. Recent Advances in Electrochemical Synthesis of Ammonia through Nitrogen Reduction under Ambient Conditions. *ChemElectroChem* **2020**, *7*, 1067–1079. [CrossRef]
40. Li, X.-F.; Li, Q.-K.; Cheng, J.; Liu, L.; Yan, Q.; Wu, Y.; Zhang, X.-H.; Wang, Z.-Y.; Qiu, Q.; Luo, Y. Conversion of Dinitrogen to Ammonia by FeN_3-Embedded Graphene. *J. Am. Chem. Soc.* **2016**, *138*, 8706–8709. [CrossRef]
41. Chen, X.; Guo, Y.; Du, X.; Zeng, Y.; Chu, J.; Gong, C.; Huang, J.; Fan, C.; Wang, X.; Xiong, J. Atomic Structure Modification for Electrochemical Nitrogen Reduction to Ammonia. *Adv. Energy Mater.* **2019**, *10*, 1903172. [CrossRef]

42. Xie, M.; Dai, F.; Guo, H.; Du, P.; Xu, X.; Liu, J.; Zhang, Z.; Lu, X. Improving Electrocatalytic Nitrogen Reduction Selectivity and Yield by Suppressing Hydrogen Evolution Reaction via Electronic Metal–Support Interaction. *Adv. Energy Mater.* **2023**, *13*, 2203032. [CrossRef]
43. Hu, L.; Xing, Z.; Feng, X. Understanding the Electrocatalytic Interface for Ambient Ammonia Synthesis. *ACS Energy Lett.* **2020**, *5*, 430–436. [CrossRef]
44. Kani, N.C.; Prajapati, A.; Collins, B.A.; Goodpaster, J.D.; Singh, M.R. Competing Effects of pH, Cation Identity, H_2O Saturation, and N_2 Concentration on the Activity and Selectivity of Electrochemical Reduction of N_2 to NH_3 on Electrodeposited Cu at Ambient Conditions. *ACS Catal.* **2020**, *10*, 14592–14603. [CrossRef]
45. Iqbal, M.S.; Yao, Z.-B.; Ruan, Y.-K.; Iftikhar, R.; Hao, L.-D.; Robertson, A.W.; Imran, S.M.; Sun, Z.-Y. Single-atom catalysts for electrochemical N_2 reduction to NH_3. *Rare Met.* **2023**, *42*, 1075–1097. [CrossRef]
46. Nazemi, M.; El-Sayed, M.A. Electrochemical Synthesis of Ammonia from N_2 and H_2O under Ambient Conditions Using Pore-Size-Controlled Hollow Gold Nanocatalysts with Tunable Plasmonic Properties. *J. Phys. Chem. Lett.* **2018**, *9*, 5160–5166. [CrossRef]
47. Bao, D.; Zhang, Q.; Meng, F.-L.; Zhong, H.-X.; Shi, M.-M.; Zhang, Y.; Yan, J.-M.; Jiang, Q.; Zhang, X.-B. Electrochemical Reduction of N_2 under Ambient Conditions for Artificial N_2 Fixation and Renewable Energy Storage Using N_2/NH_3 Cycle. *Adv. Mater.* **2017**, *29*, 1604799. [CrossRef]
48. Wang, Z.; Li, Y.; Yu, H.; Xu, Y.; Xue, H.; Li, X.; Wang, H.; Wang, L. Ambient Electrochemical Synthesis of Ammonia from Nitrogen and Water Catalyzed by Flower-Like Gold Microstructures. *ChemSusChem* **2018**, *11*, 3480–3485. [CrossRef] [PubMed]
49. Nazemi, M.; Panikkanvalappil, S.R.; El-Sayed, M.A. Enhancing the rate of electrochemical nitrogen reduction reaction for ammonia synthesis under ambient conditions using hollow gold nanocages. *Nano Energy* **2018**, *49*, 316–323. [CrossRef]
50. Zhao, X.; Yao, C.; Chen, H.; Fu, Y.; Xiang, C.; He, S.; Zhou, X.; Zhang, H. In situ nano Au triggered by a metal boron organic polymer: Efficient electrochemical N2 fixation to NH3 under ambient conditions. *J. Mater. Chem. A* **2019**, *7*, 20945–20951. [CrossRef]
51. Shi, M.M.; Bao, D.; Wulan, B.R.; Li, Y.H.; Zhang, Y.F.; Yan, J.M.; Jiang, Q. Au Sub-Nanoclusters on TiO_2 toward Highly Efficient and Selective Electrocatalyst for N_2 Conversion to NH_3 at Ambient Conditions. *Adv. Mater.* **2017**, *29*, 1606550. [CrossRef] [PubMed]
52. Li, S.J.; Bao, D.; Shi, M.M.; Wulan, B.R.; Yan, J.M.; Jiang, Q. Amorphizing of Au Nanoparticles by CeO_x-RGO Hybrid Support towards Highly Efficient Electrocatalyst for N_2 Reduction under Ambient Conditions. *Adv. Mater.* **2017**, *29*, 1700001. [CrossRef]
53. Wang, H.; Wang, L.; Wang, Q.; Ye, S.; Sun, W.; Shao, Y.; Jiang, Z.; Qiao, Q.; Zhu, Y.; Song, P.; et al. Ambient Electrosynthesis of Ammonia: Electrode Porosity and Composition Engineering. *Angew. Chem. Int. Ed.* **2018**, *57*, 12360–12364. [CrossRef]
54. Wang, D.; Azofra, L.M.; Harb, M.; Cavallo, L.; Zhang, X.; Suryanto, B.H.R.; MacFarlane, D.R. Energy-Efficient Nitrogen Reduction to Ammonia at Low Overpotential in Aqueous Electrolyte under Ambient Conditions. *ChemSusChem* **2018**, *11*, 3416–3422. [CrossRef] [PubMed]
55. Tao, H.; Choi, C.; Ding, L.-X.; Jiang, Z.; Han, Z.; Jia, M.; Fan, Q.; Gao, Y.; Wang, H.; Robertson, A.W.; et al. Nitrogen Fixation by Ru Single-Atom Electrocatalytic Reduction. *Chem* **2019**, *5*, 204–214. [CrossRef]
56. Gao, W.Y.; Hao, Y.C.; Su, X.; Chen, L.W.; Bu, T.A.; Zhang, N.; Yu, Z.L.; Zhu, Z.; Yin, A.X. Morphology-dependent electrocatalytic nitrogen reduction on Ag triangular nanoplates. *Chem. Commun.* **2019**, *55*, 10705–10708. [CrossRef] [PubMed]
57. Chen, Y.; Guo, R.; Peng, X.; Wang, X.; Liu, X.; Ren, J.; He, J.; Zhuo, L.; Sun, J.; Liu, Y.; et al. Highly Productive Electrosynthesis of Ammonia by Admolecule-Targeting Single Ag Sites. *ACS Nano* **2020**, *14*, 6938–6946. [CrossRef]
58. Andersen, S.Z.; Colic, V.; Yang, S.; Schwalbe, J.A.; Nielander, A.C.; McEnaney, J.M.; Enemark-Rasmussen, K.; Baker, J.G.; Singh, A.R.; Rohr, B.A.; et al. A rigorous electrochemical ammonia synthesis protocol with quantitative isotope measurements. *Nature* **2019**, *570*, 504–508. [CrossRef] [PubMed]
59. Yang, D.; Chen, T.; Wang, Z. Electrochemical reduction of aqueous nitrogen (N_2) at a low overpotential on (110)-oriented Mo nanofilm. *J. Mater. Chem. A* **2017**, *5*, 18967–18971. [CrossRef]
60. Zhao, J.; Chen, Z. Single Mo Atom Supported on Defective Boron Nitride Monolayer as an Efficient Electrocatalyst for Nitrogen Fixation: A Computational Study. *J. Am. Chem. Soc.* **2017**, *139*, 12480–12487. [CrossRef] [PubMed]
61. Han, L.; Liu, X.; Chen, J.; Lin, R.; Liu, H.; Lü, F.; Bak, S.; Liang, Z.; Zhao, S.; Stavitski, E.; et al. Atomically Dispersed Molybdenum Catalysts for Efficient Ambient Nitrogen Fixation. *Angew. Chem. Int. Ed.* **2019**, *58*, 2321–2325. [CrossRef] [PubMed]
62. Zhang, C.; Wang, Z.; Lei, J.; Ma, L.; Yakobson, B.I.; Tour, J.M. Atomic Molybdenum for Synthesis of Ammonia with 50% Faradic Efficiency. *Small* **2022**, *18*, 2106327. [CrossRef] [PubMed]
63. Greenlee, L.F.; Renner, J.N.; Foster, S.L. The Use of Controls for Consistent and Accurate Measurements of Electrocatalytic Ammonia Synthesis from Dinitrogen. *ACS Catal.* **2018**, *8*, 7820–7827. [CrossRef]
64. Nielander, A.C.; McEnaney, J.M.; Schwalbe, J.A.; Baker, J.G.; Blair, S.J.; Wang, L.; Pelton, J.G.; Andersen, S.Z.; Enemark-Rasmussen, K.; Colic, V.; et al. A Versatile Method for Ammonia Detection in a Range of Relevant Electrolytes via Direct Nuclear Magnetic Resonance Techniques. *ACS Catal.* **2019**, *9*, 5797–5802. [CrossRef]
65. Guo, X.; Huang, S. Tuning nitrogen reduction reaction activity via controllable Fe magnetic moment: A computational study of single Fe atom supported on defective graphene. *Electrochim. Acta* **2018**, *284*, 392–399. [CrossRef]
66. Wang, M.; Liu, S.; Qian, T.; Liu, J.; Zhou, J.; Ji, H.; Xiong, J.; Zhong, J.; Yan, C. Over 56.55% Faradaic efficiency of ambient ammonia synthesis enabled by positively shifting the reaction potential. *Nat. Commun.* **2019**, *10*, 341. [CrossRef] [PubMed]

67. Wang, Y.; Cui, X.; Zhao, J.; Jia, G.; Gu, L.; Zhang, Q.; Meng, L.; Shi, Z.; Zheng, L.; Wang, C.; et al. Rational Design of Fe–N/C Hybrid for Enhanced Nitrogen Reduction Electrocatalysis under Ambient Conditions in Aqueous Solution. *ACS Catal.* **2018**, *9*, 336–344. [CrossRef]
68. Zhang, Y.; Hu, J.; Zhang, C.; Liu, Y.; Xu, M.; Xue, Y.; Liu, L.; Leung, M.K.H. Bimetallic Mo–Co nanoparticles anchored on nitrogen-doped carbon for enhanced electrochemical nitrogen fixation. *J. Mater. Chem. A* **2020**, *8*, 9091–9098. [CrossRef]
69. Li, X.; Liu, J.; Zhang, Y.; Leung, M.K.H. Changing charge transfer mode with cobalt–molybdenum bimetallic atomic pairs for enhanced nitrogen fixation. *J. Mater. Chem. A* **2022**, *10*, 15595–15604. [CrossRef]
70. Talukdar, B.; Kuo, T.C.; Sneed, B.T.; Lyu, L.M.; Lin, H.M.; Chuang, Y.C.; Cheng, M.J.; Kuo, C.H. Enhancement of NH_3 Production in Electrochemical N_2 Reduction by the Cu-Rich Inner Surfaces of Beveled CuAu Nanoboxes. *ACS Appl. Mater. Interfaces* **2021**, *13*, 51839–51848. [CrossRef]
71. Maia, V.A.; Santos, C.M.G.; Azeredo, N.F.B.; Zambiazi, P.J.; Antolini, E.; Neto, A.O.; de Souza, R.F.B. Conversion of nitrogen to ammonia using a Cu/C electrocatalyst in a polymeric electrolyte reactor. *Electrochem. Commun.* **2023**, *146*, 107421. [CrossRef]
72. Mukherjee, J.; Adalder, A.; Mukherjee, N.; Ghorai, U.K. Solvothermal synthesis of α–CuPc nanostructures for electrochemical nitrogen fixation under ambient conditions. *Catal. Today* **2023**, *423*, 113905. [CrossRef]
73. Singh Verma, T.; Paramita Samal, P.; Selvaraj, K.; Krishnamurty, S. Can Li Atoms Anchored on Boron- and Nitrogen-Doped Graphene Catalyze Dinitrogen Molecules to Ammonia? A DFT Study. *ChemPhysChem* **2023**, *24*, e202200750. [CrossRef] [PubMed]
74. Liu, Y.; Su, Y.; Quan, X.; Fan, X.; Chen, S.; Yu, H.; Zhao, H.; Zhang, Y.; Zhao, J. Facile Ammonia Synthesis from Electrocatalytic N2 Reduction under Ambient Conditions on N-Doped Porous Carbon. *ACS Catal.* **2018**, *8*, 1186–1191. [CrossRef]
75. Mukherjee, S.; Yang, X.; Shan, W.; Samarakoon, W.; Karakalos, S.; Cullen, D.A.; More, K.; Wang, M.; Feng, Z.; Wang, G.; et al. Atomically Dispersed Single Ni Site Catalysts for Nitrogen Reduction toward Electrochemical Ammonia Synthesis Using N_2 and H_2O. *Small Methods* **2020**, *4*, 1900821. [CrossRef]
76. Zang, W.; Yang, T.; Zou, H.; Xi, S.; Zhang, H.; Liu, X.; Kou, Z.; Du, Y.; Feng, Y.P.; Shen, L.; et al. Copper Single Atoms Anchored in Porous Nitrogen-Doped Carbon as Efficient pH-Universal Catalysts for the Nitrogen Reduction Reaction. *ACS Catal.* **2019**, *9*, 10166–10173. [CrossRef]
77. Zhang, L.; Ji, X.; Ren, X.; Ma, Y.; Shi, X.; Tian, Z.; Asiri, A.M.; Chen, L.; Tang, B.; Sun, X. Electrochemical Ammonia Synthesis via Nitrogen Reduction Reaction on a MoS_2 Catalyst: Theoretical and Experimental Studies. *Adv. Mater.* **2018**, *30*, 1800191. [CrossRef]
78. Li, X.; Li, T.; Ma, Y.; Wei, Q.; Qiu, W.; Guo, H.; Shi, X.; Zhang, P.; Asiri, A.M.; Chen, L.; et al. Boosted Electrocatalytic N_2 Reduction to NH_3 by Defect-Rich MoS_2 Nanoflower. *Adv. Energy Mater.* **2018**, *8*, 1801357. [CrossRef]
79. Abghoui, Y.; Skúlason, E. Onset potentials for different reaction mechanisms of nitrogen activation to ammonia on transition metal nitride electro-catalysts. *Catal. Today* **2017**, *286*, 69–77. [CrossRef]
80. Abghoui, Y.; Garden, A.L.; Hlynsson, V.F.; Bjorgvinsdottir, S.; Olafsdottir, H.; Skulason, E. Enabling electrochemical reduction of nitrogen to ammonia at ambient conditions through rational catalyst design. *Phys. Chem. Chem. Phys.* **2015**, *17*, 4909–4918. [CrossRef] [PubMed]
81. Zhang, L.; Ji, X.; Ren, X.; Luo, Y.; Shi, X.; Asiri, A.M.; Zheng, B.; Sun, X. Efficient Electrochemical N_2 Reduction to NH_3 on MoN Nanosheets Array under Ambient Conditions. *ACS Sustain. Chem. Eng.* **2018**, *6*, 9550–9554. [CrossRef]
82. Zhang, X.; Kong, R.M.; Du, H.; Xia, L.; Qu, F. Highly efficient electrochemical ammonia synthesis via nitrogen reduction reactions on a VN nanowire array under ambient conditions. *Chem. Commun.* **2018**, *54*, 5323–5325. [CrossRef] [PubMed]
83. Yang, X.; Nash, J.; Anibal, J.; Dunwell, M.; Kattel, S.; Stavitski, E.; Attenkofer, K.; Chen, J.G.; Yan, Y.; Xu, B. Mechanistic Insights into Electrochemical Nitrogen Reduction Reaction on Vanadium Nitride Nanoparticles. *J. Am. Chem. Soc.* **2018**, *140*, 13387–13391. [CrossRef] [PubMed]
84. Li, Q.; He, L.; Sun, C.; Zhang, X. Computational Study of MoN_2 Monolayer as Electrochemical Catalysts for Nitrogen Reduction. *J. Phys. Chem. C* **2017**, *121*, 27563–27568. [CrossRef]
85. Ren, X.; Cui, G.; Chen, L.; Xie, F.; Wei, Q.; Tian, Z.; Sun, X. Electrochemical N_2 fixation to NH_3 under ambient conditions: Mo_2N nanorod as a highly efficient and selective catalyst. *Chem. Commun.* **2018**, *54*, 8474–8477. [CrossRef] [PubMed]
86. Michalsky, R.; Zhang, Y.-J.; Medford, A.J.; Peterson, A.A. Departures from the Adsorption Energy Scaling Relations for Metal Carbide Catalysts. *J. Phys. Chem. C* **2014**, *118*, 13026–13034. [CrossRef]
87. Yu, G.; Guo, H.; Liu, S.; Chen, L.; Alshehri, A.A.; Alzahrani, K.A.; Hao, F.; Li, T. Cr_3C_2 Nanoparticle-Embedded Carbon Nanofiber for Artificial Synthesis of NH_3 through N_2 Fixation under Ambient Conditions. *ACS Appl. Mater. Interfaces* **2019**, *11*, 35764–35769. [CrossRef] [PubMed]
88. Matanovic, I.; Garzon, F.H. Nitrogen electroreduction and hydrogen evolution on cubic molybdenum carbide: A density functional study. *Phys. Chem. Chem. Phys.* **2018**, *20*, 14679–14687. [CrossRef]
89. Cheng, H.; Ding, L.X.; Chen, G.F.; Zhang, L.; Xue, J.; Wang, H. Molybdenum Carbide Nanodots Enable Efficient Electrocatalytic Nitrogen Fixation under Ambient Conditions. *Adv. Mater.* **2018**, *30*, 1803694. [CrossRef]
90. Han, J.; Ji, X.; Ren, X.; Cui, G.; Li, L.; Xie, F.; Wang, H.; Li, B.; Sun, X. MoO_3 nanosheets for efficient electrocatalytic N_2 fixation to NH_3. *J. Mater. Chem. A* **2018**, *6*, 12974–12977. [CrossRef]
91. Gouveia, J.D.; Morales-García, Á.; Viñes, F.; Gomes, J.R.B.; Illas, F. Facile Heterogeneously Catalyzed Nitrogen Fixation by MXenes. *ACS Catal.* **2020**, *10*, 5049–5056. [CrossRef]
92. Khalil, I.E.; Xue, C.; Liu, W.; Li, X.; Shen, Y.; Li, S.; Zhang, W.; Huo, F. The Role of Defects in Metal–Organic Frameworks for Nitrogen Reduction Reaction: When Defects Switch to Features. *Adv. Funct. Mater.* **2021**, *31*, 2010052. [CrossRef]

93. Luo, Y.; Chen, G.-F.; Ding, L.; Chen, X.; Ding, L.-X.; Wang, H. Efficient Electrocatalytic N_2 Fixation with MXene under Ambient Conditions. *Joule* **2019**, *3*, 279–289. [CrossRef]
94. Zhang, J.; Yang, L.; Wang, H.; Zhu, G.; Wen, H.; Feng, H.; Sun, X.; Guan, X.; Wen, J.; Yao, Y. In Situ Hydrothermal Growth of TiO_2 Nanoparticles on a Conductive $Ti_3C_2T_x$ MXene Nanosheet: A Synergistically Active Ti-Based Nanohybrid Electrocatalyst for Enhanced N_2 Reduction to NH_3 at Ambient Conditions. *Inorg. Chem.* **2019**, *58*, 5414–5418. [CrossRef] [PubMed]
95. Li, Q.; Song, T.; Wang, Z.; Wang, X.; Zhou, X.; Wang, Q.; Yang, Y. A General Strategy toward Metal Sulfide Nanoparticles Confined in a Sulfur-Doped Ti_3C_2Tx MXene 3D Porous Aerogel for Efficient Ambient N_2 Electroreduction. *Small* **2021**, *17*, e2103305. [CrossRef]
96. Höskuldsson, Á.B.; Abghoui, Y.; Gunnarsdóttir, A.B.; Skúlason, E. Computational Screening of Rutile Oxides for Electrochemical Ammonia Formation. *ACS Sustain. Chem. Eng.* **2017**, *5*, 10327–10333. [CrossRef]
97. Chu, K.; Luo, Y.; Shen, P.; Li, X.; Li, Q.; Guo, Y. Unveiling the Synergy of O-Vacancy and Heterostructure over MoO_{3-x}/MXene for N_2 Electroreduction to NH_3. *Adv. Energy Mater.* **2021**, *12*, 2103022. [CrossRef]
98. Huang, L.; Wu, J.; Han, P.; Al-Enizi, A.M.; Almutairi, T.M.; Zhang, L.; Zheng, G. NbO_2 Electrocatalyst Toward 32% Faradaic Efficiency for N_2 Fixation. *Small Methods* **2018**, *3*, 1800386. [CrossRef]
99. Nguyen, M.T.; Seriani, N.; Gebauer, R. Nitrogen electrochemically reduced to ammonia with hematite: Density-functional insights. *Phys. Chem. Chem. Phys.* **2015**, *17*, 14317–14322. [CrossRef]
100. Cui, X.; Tang, C.; Liu, X.M.; Wang, C.; Ma, W.; Zhang, Q. Highly Selective Electrochemical Reduction of Dinitrogen to Ammonia at Ambient Temperature and Pressure over Iron Oxide Catalysts. *Chemistry* **2018**, *24*, 18494–18501. [CrossRef]
101. Chen, S.; Perathoner, S.; Ampelli, C.; Mebrahtu, C.; Su, D.; Centi, G. Electrocatalytic Synthesis of Ammonia at Room Temperature and Atmospheric Pressure from Water and Nitrogen on a Carbon-Nanotube-Based Electrocatalyst. *Angew. Chem. Int. Ed.* **2017**, *56*, 2699–2703. [CrossRef] [PubMed]
102. Liu, Q.; Zhang, X.; Zhang, B.; Luo, Y.; Cui, G.; Xie, F.; Sun, X. Ambient N_2 fixation to NH_3 electrocatalyzed by a spinel Fe_3O_4 nanorod. *Nanoscale* **2018**, *10*, 14386–14389. [CrossRef]
103. Hu, L.; Khaniya, A.; Wang, J.; Chen, G.; Kaden, W.E.; Feng, X. Ambient Electrochemical Ammonia Synthesis with High Selectivity on Fe/Fe Oxide Catalyst. *ACS Catal.* **2018**, *8*, 9312–9319. [CrossRef]
104. Wang, J.; Liu, Y.-p.; Zhang, H.; Huang, D.-j.; Chu, K. Ambient electrocatalytic nitrogen reduction on a MoO_2/graphene hybrid: Experimental and DFT studies. *Catal. Sci. Technol.* **2019**, *9*, 4248–4254. [CrossRef]
105. Zhang, L.; Ren, X.; Luo, Y.; Shi, X.; Asiri, A.M.; Li, T.; Sun, X. Ambient NH_3 synthesis via electrochemical reduction of N_2 over cubic sub-micron SnO_2 particles. *Chem. Commun.* **2018**, *54*, 12966–12969. [CrossRef] [PubMed]
106. Liu, Y.P.; Li, Y.B.; Zhang, H.; Chu, K. Boosted Electrocatalytic N_2 Reduction on Fluorine-Doped SnO_2 Mesoporous Nanosheets. *Inorg. Chem.* **2019**, *58*, 10424–10431. [CrossRef] [PubMed]
107. Zhang, R.; Ren, X.; Shi, X.; Xie, F.; Zheng, B.; Guo, X.; Sun, X. Enabling Effective Electrocatalytic N_2 Conversion to NH_3 by the TiO_2 Nanosheets Array under Ambient Conditions. *ACS Appl. Mater. Interfaces* **2018**, *10*, 28251–28255. [CrossRef] [PubMed]
108. Kong, W.; Gong, F.; Zhou, Q.; Yu, G.; Ji, L.; Sun, X.; Asiri, A.M.; Wang, T.; Luo, Y.; Xu, Y. An MnO_2–$Ti_3C_2T_x$ MXene nanohybrid: An efficient and durable electrocatalyst toward artificial N_2 fixation to NH_3 under ambient conditions. *J. Mater. Chem. A* **2019**, *7*, 18823–18827. [CrossRef]
109. Wang, Z.; Gong, F.; Zhang, L.; Wang, R.; Ji, L.; Liu, Q.; Luo, Y.; Guo, H.; Li, Y.; Gao, P.; et al. Electrocatalytic Hydrogenation of N_2 to NH_3 by MnO: Experimental and Theoretical Investigations. *Adv. Sci.* **2019**, *6*, 1801182. [CrossRef] [PubMed]
110. Li, C.; Yu, J.; Yang, L.; Zhao, J.; Kong, W.; Wang, T.; Asiri, A.M.; Li, Q.; Sun, X. Spinel $LiMn_2O_4$ Nanofiber: An Efficient Electrocatalyst for N_2 Reduction to NH_3 under Ambient Conditions. *Inorg. Chem.* **2019**, *58*, 9597–9601. [CrossRef]
111. Wu, X.; Xia, L.; Wang, Y.; Lu, W.; Liu, Q.; Shi, X.; Sun, X. Mn_3O_4 Nanocube: An Efficient Electrocatalyst Toward Artificial N_2 Fixation to NH_3. *Small* **2018**, *14*, e1803111. [CrossRef] [PubMed]
112. Huang, H.; Gong, F.; Wang, Y.; Wang, H.; Wu, X.; Lu, W.; Zhao, R.; Chen, H.; Shi, X.; Asiri, A.M.; et al. Mn_3O_4 nanoparticles@reduced graphene oxide composite: An efficient electrocatalyst for artificial N_2 fixation to NH_3 at ambient conditions. *Nano Res.* **2019**, *12*, 1093–1098. [CrossRef]
113. Du, H.; Guo, X.; Kong, R.M.; Qu, F. Cr_2O_3 nanofiber: A high-performance electrocatalyst toward artificial N_2 fixation to NH_3 under ambient conditions. *Chem. Commun.* **2018**, *54*, 12848–12851. [CrossRef]
114. He, Y.; Liu, S.; Wang, M.; Cheng, Q.; Qian, T.; Yan, C. Deciphering engineering principle of three-phase interface for advanced gas-involved electrochemical reactions. *J. Energy Chem.* **2023**, *80*, 302–323. [CrossRef]
115. Shi, Z.; Yang, W.; Gu, Y.; Liao, T.; Sun, Z. Metal-Nitrogen-Doped Carbon Materials as Highly Efficient Catalysts: Progress and Rational Design. *Adv. Sci.* **2020**, *7*, 2001069. [CrossRef]
116. Zeng, Y.; Priest, C.; Wang, G.; Wu, G. Restoring the Nitrogen Cycle by Electrochemical Reduction of Nitrate: Progress and Prospects. *Small Methods* **2020**, *4*, 2000672. [CrossRef]
117. Lin, S.; Zhang, X.; Chen, L.; Zhang, Q.; Ma, L.; Liu, J. A review on catalysts for electrocatalytic and photocatalytic reduction of N_2 to ammonia. *Green Chem.* **2022**, *24*, 9003–9026. [CrossRef]
118. Rehman, F.; Delowar Hossain, M.; Tyagi, A.; Lu, D.; Yuan, B.; Luo, Z. Engineering electrocatalyst for low-temperature N_2 reduction to ammonia. *Mater. Today* **2021**, *44*, 136–167. [CrossRef]
119. Pang, Y.; Su, C.; Jia, G.; Xu, L.; Shao, Z. Emerging two-dimensional nanomaterials for electrochemical nitrogen reduction. *Chem. Soc. Rev.* **2021**, *50*, 12744–12787. [CrossRef]

120. Wang, T.; Guo, Z.; Zhang, X.; Li, Q.; Yu, A.; Wu, C.; Sun, C. Recent progress of iron-based electrocatalysts for nitrogen reduction reaction. *J. Mater. Sci. Technol.* **2023**, *140*, 121–134. [CrossRef]
121. Zhao, X.; Hu, G.; Chen, G.F.; Zhang, H.; Zhang, S.; Wang, H. Comprehensive Understanding of the Thriving Ambient Electrochemical Nitrogen Reduction Reaction. *Adv. Mater.* **2021**, *33*, 2007650. [CrossRef] [PubMed]
122. Niu, L.; An, L.; Wang, X.; Sun, Z. Effect on electrochemical reduction of nitrogen to ammonia under ambient conditions: Challenges and opportunities for chemical fuels. *J. Energy Chem.* **2021**, *61*, 304–318. [CrossRef]
123. Gu, H.; Chen, W.; Li, X. Atomically dispersed metal catalysts for the electrochemical nitrogen reduction reaction. *J. Mater. Chem. A* **2022**, *10*, 22331–22353. [CrossRef]
124. Qiang, S.; Wu, F.; Yu, J.; Liu, Y.T.; Ding, B. Complementary Design in Multicomponent Electrocatalysts for Electrochemical Nitrogen Reduction: Beyond the Leverage in Activity and Selectivity. *Angew. Chem. Int. Ed.* **2023**, *62*, 202217265. [CrossRef] [PubMed]
125. Lv, X.W.; Liu, X.L.; Suo, Y.J.; Liu, Y.P.; Yuan, Z.Y. Identifying the Dominant Role of Pyridinic-N-Mo Bonding in Synergistic Electrocatalysis for Ambient Nitrogen Reduction. *ACS Nano* **2021**, *15*, 12109–12118. [CrossRef] [PubMed]
126. Yang, C.; Zhu, Y.; Liu, J.; Qin, Y.; Wang, H.; Liu, H.; Chen, Y.; Zhang, Z.; Hu, W. Defect engineering for electrochemical nitrogen reduction reaction to ammonia. *Nano Energy* **2020**, *77*, 105126. [CrossRef]
127. Min, B.; Gao, Q.; Yan, Z.; Han, X.; Hosmer, K.; Campbell, A.; Zhu, H. Powering the Remediation of the Nitrogen Cycle: Progress and Perspectives of Electrochemical Nitrate Reduction. *Ind. Eng. Chem. Res.* **2021**, *60*, 14635–14650. [CrossRef]

separations

MDPI

Article

ZIF-67(Co)-Loaded Filter Paper for In Situ Catalytic Degradation of Bisphenol A in Water

Zhimin Cai [1], Yutao Luo [1] and Lu Gan [2,3,*]

1 College of Economics and Management, Nanjing Forestry University, Nanjing 210037, China
2 College of Materials Science and Engineering, Nanjing Forestry University, Nanjing 210037, China
3 Jiangsu Co-Innovation Center of Efficient Processing and Utilization of Forest Resources, International Innovation Center for Forest Chemicals and Materials, College of Materials Science and Engineering, Nanjing Forestry University, Nanjing 210037, China
* Correspondence: ganlu@njfu.edu.cn

Abstract: Herein, we loaded cobalt-based zeolite imidazolate frameworks, ZIF-67 (Co), onto commercial filter paper to prepare catalytic filter paper (ZFP) for the in situ degradation of bisphenol A (BPA) in water by activating peroxymonosulfate. The results showed that ZIF-67 (Co) was densely and uniformly distributed on the surface of the filter paper. The prepared ZFP could effectively degrade BPA in situ through a gravity-driven filtration process. Specifically, when the flow rate of the BPA solution passing through ZFP was lower than 10 mL/min, 0.02 mM of BPA could be completely degraded by ZFP. Furthermore, ZFP showed promising water matrix adaptability, which could provide promising BPA degradation efficiency in a wide pH range or in the existence of multiple anions. The scavenging tests demonstrated that both sulfate radical and hydroxyl radical were generated for BPA degradation, in which hydroxyl radical was the dominant active species. The ZFP also exhibited promising long-time use stability with a high mineralization rate. This study provides a novel method to prepare high-efficient catalyst paper for the in situ organic pollutant removal in water media via a prompt filtration process.

Keywords: peroxymonosulfate; degradation; catalytic filter paper; zeolite imidazolate frameworks

1. Introduction

In recent decades, water contamination by organic pollutants has caused enormous threats to both the ecological environment and human health [1–3]. Among all the organic pollutants, endocrine-disrupting chemicals (EDCs) have received tremendous concern since EDCs can disrupt the endocrine system of human beings seriously if they are excessively accumulated in the internal organs [4]. Bisphenol A (BPA), a classic EDC that is commonly used in the chemical synthesis field, has been frequently detected in various water bodies, with concentrations exceeding the threshold of many national standards [5,6]. However, traditional wastewater treatment methods, including activated sludge process, adsorption, and flocculation processes, are not capable of separating BPA from water completely [7,8]. Thus, it is a great necessity to develop ways to remove BPA effectively.

Advanced oxidation processes (AOPs), including the photocatalytic process, the Fenton process, etc., are regarded as significant approaches for the removal of organic pollutants in water, in which sulfate radical-based (SO_4^-) AOPs are frequently selected since SO_4^- can degrade almost all the organic pollutants with a wide pH application range and long half-life period [9]. In general, SO_4^- can be generated by activating persulfate, including peroxymonosulfate (PMS) and/or peroxydisulfate (PDS), through various means [10]. Compared with heat, ultraviolet, ultrasound, and electrochemical approaches, transition-metal-based catalysts can promptly activate persulfate under ambient conditions with a high SO_4^- generation rate, in which the cobalt catalyst/PMS system is regarded as one of the most promising systems for $SO_4^{\cdot-}$ generation and organic pollutant degradation [11]. Recently, cobalt-based metal–organic frameworks (MOFs) have been of particular interest as PMS-activating catalysts since MOFs have many promising characteristics such as controllable porosity, multiple functionalities, and high catalytic activity [12]. Compared with other Co-MOFs, zeolite imidazolate frameworks (ZIFs) such as ZIF-67(Co) are frequently selected since they have desirable chemical stability and synthesis simplicity [13]. Moreover, ZIF-67(Co) has been proven to be an efficient PMS activation agent for organic pollutant degradation [14].

Another concern of using the cobalt catalyst/PMS system for organic pollutant degradation is that the catalysts cannot readily be separated from the reaction solution after the degradation process is finished. It has been widely accepted that membrane filtration is an effective and easy-to-operate method for the purification of wastewater [15]. Thus, in the present study, ZIF-67(Co) is prepared and loaded on commercial filter paper to prepare catalytic filter paper. Since the catalyst was loaded in the filter matrix, the filter paper could immobilize the ZIF-67(Co) and had the potential to in situ degrade the BPA in the water when the BPA solution passed through the catalytic paper. Furthermore, the used catalytic paper could be readily recycled and reused, which avoided the introduction of secondary pollution to the water body like other powdery catalysts did. The impact of many parameters on the degradation performance of the catalytic paper is systematically studied. The active species in the reaction system are also elucidated.

2. Materials and Methods

2.1. Chemicals

Cobalt nitrate hexahydrate ($Co(NO_3)_2{\cdot}6H_2O$) and 2-methylimidazole (≥99%) were purchased from Aladdin Chemical Co., Ltd. (Shanghai, China). BPA (≥99%) and PMS (Oxone: $KHSO_5$, $0.5KHSO_4$, $0.5K_2SO_4$) were purchased from Sigma-Aldrich Company (USA). Commercial qualitative filter paper (d = 5 cm) was purchased from Double Ring (China). Other reagents were of analytical grade and used as received without purification. Distilled water (H_2O) was obtained by the Hi-Tech water purification instrument.

2.2. Preparation of Catalytic Paper

ZIF-67(Co)-loaded catalytic paper (ZFP) was prepared by a precipitation method described as follows. Firstly, one piece of filter paper was immersed in 50 mL of methanol. Afterward, 500 mg of $Co(NO_3)_2{\cdot}6H_2O$ was added and dissolved. Then, 20 mL methanol solution containing 660 mg 2-methylimidazole was dropped into the above mixed solution gradually under slow magnetic stirring. The mixture was aged for 24 h, and ZIF-67(Co)-loaded filter paper with ~700 mg ZIF-67(Co) loading amount was obtained after being washed twice with methanol and then dried in an oven at 60 °C for 12 h.

2.3. Characterizations

The morphology of the ZFP was observed by a scanning electron microscope (SEM, JEOL SEM 6490, Japan). An X-ray diffraction diffractometer (XRD, Rigaku Smartlab, Japan) was employed to analyze the crystalline structure of the ZFP. The thermogravimetric analysis of the composites was conducted with a TGA instrument (Mettler Toledo; TGA/DSC 1 Simultaneous Thermal Analyzer) from room temperature to 700 °C and 5 °C/min under N_2 atmosphere. The concentration of the BPA was analyzed by high-performance liquid chromatography (HPLC, Dionex Ultimate 3000) equipped with a Thermo C18 column (5 μm particle size, 250 × 4.6 mm). The concentrations of dissolved Co ions in the solutions were also determined with atomic absorption spectroscopy (AAS, AA800, PerkinElmer). The total organic carbon (TOC) was measured by a MutiN/C 2100 analyzer.

2.4. Catalytic Degradation of BPA in ZFP/PMS System

The in situ degradation of BPA by ZFP was carried out by a gravity-induced filtration process, as illustrated in Figure 1. Specifically, the ZFP was put in a sand core filter (d = 5 cm) with a vacuum pump. Typically, 20 mL of BPA solution with a certain concentration and 0.3 mL of PMS (1.5 mM) solution were added into the filter, and the flow rate was adjusted from 2 mL/min to 20 mL/min (10 min to 1 min for 20 mL BPA of solution passing through the ZFP) by the vacuum pump with controllable negative pressure. A 1.0 mL volume of BPA solution passing through the ZFP was collected, filtered with a 0.22 μm polytetrafluoroethylene membrane, and immediately quenched with 0.5 mL methanol before analysis. The initial pH value of the solution was measured after PMS and catalyst were added and was adjusted by NaOH (1 M) and H_2SO_4 (1 M). Methanol (MeOH), isobutanol (TBA), and L-histidine were selected as scavengers for scavenging tests. The long-time use stability of the ZFP was tested by repeating the above BPA test for 5 consecutive runs, in which the flow rate was adjusted to be 5 mL/min. All the degradation tests were triplicated, and the data were the average values with standard deviations.

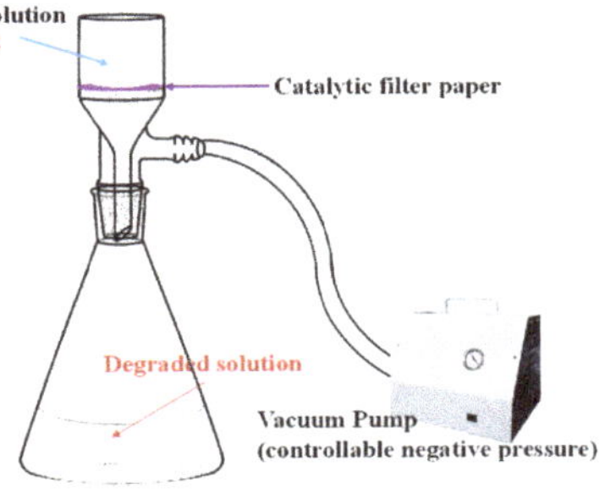

Figure 1. Schematic illustration of the BPA degradation by ZFP.

3. Results and Discussion

The morphology of the prepared ZFP was investigated in terms of SEM, with the results shown in Figure 2. As illustrated in Figure 2a, pristine ZIF-67(Co) had a polyhedral structure with uniform particle distributions. As for the ZFP shown in Figure 2b, it was seen that ZIF-67(Co) particles were densely and uniformly loaded on the surface of the filter paper, in which the ZIF-67(Co) also exhibited its pristine polyhedral morphology with smaller particle sizes. This might be because the commercial paper had a cellulose structure with abundant hydroxyl groups that could interact with ZIF-67(Co), resulting in a restraining effect on the growth of the ZIF-67(Co) on the filter paper surface. From the EDS mapping images shown in Figure 2b,c, it was seen that the Co element was uniformly distributed within the ZFP. Additionally, the EDS elemental analysis indicated that C, N, O, and Co occupied 58.96 wt%, 6.61 wt% 6.46 wt%, and 27.97 wt% in the ZFP, indicating that there are enough active sites available for PMS activation.

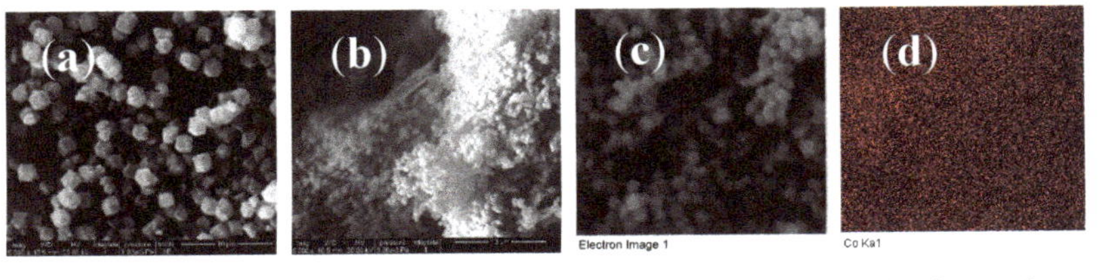

Figure 2. SEM images of (**a**) ZIF-67 and (**b**) ZFP (**c**) selected area and (**d**) EDS Co elemental mapping in ZFP.

The crystalline structure of the prepared ZFP was afterward studied using XRD, with the results shown in Figure 3a. It was seen that the unloaded filter paper had a main characteristic peak at 22.1°, which was associated with the (002) crystalline plane of the Cellulose I structure [16,17]. This meant the filter paper was composed of natural cellulose without post-treatment. At the same time, the prepared ZIF-67(Co) had multiple shape peaks, which were all assigned to the characteristic peaks of the cobalt-based ZIF structure [18]. After ZIF-67(Co) was loaded on the filter paper, the resulting ZFP exhibited a similar XRD pattern with ZIF-67(Co), and the pattern of filter paper was not observed. This was probably because the sharp peaks of large amounts of ZIF-67(Co) particles shielded the broad diffraction peaks of the filter paper with comparatively weak intensity. The TGA of the ZFP conducted under N_2 atmosphere is shown in Figure 3b. As illustrated, the ZFP began to decompose when the temperature increased to ~300 °C, which was ascribed to the decomposition of the lignocellulose component in the filter paper and the organic ligand in ZIF-67(Co) [19]. When the temperature increased to 500 °C, the weight loss gradually ceased, indicating that the organic compounds and polymers were converted to carbonaceous materials. Figure 3c shows the N_2 adsorption/desorption curves for the pristine filter paper and the ZFP. It was found that since ZIF-67(Co) was porous material with rich porosity, the loading of ZIF-67(Co) significantly increased the surface area of the pristine filter paper.

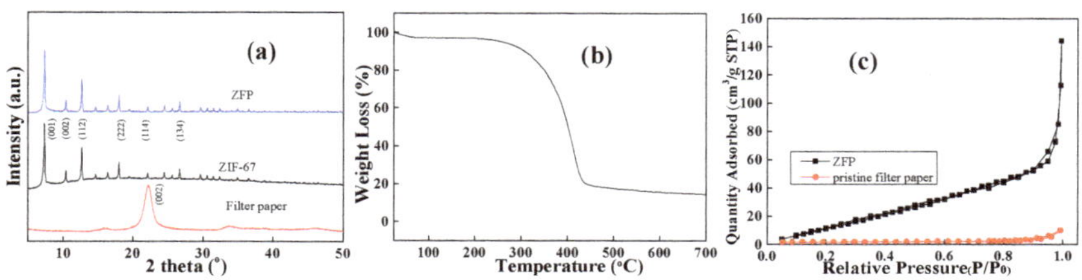

Figure 3. (**a**) XRD patterns of the prepared samples; (**b**) TGA and (**c**) N_2 adsorption/desorption curves of ZFP.

The catalytic degradation performance of the ZFP was then investigated by degrading BPA in water. Figure 4a shows the in situ BPA removal efficiency after the BPA solution with different flow rates passed through the ZFP. It is shown from the inset figure of Figure 4a that 0.2 g/L of pristine ZIF-67(Co) could completely degrade 0.02 mM BPA within 15 min. When ZIF-67(Co) was loaded on the filter paper, the resulting ZFP also exhibited high BPA removal efficiency. Furthermore, the lower flow rate of the solution facilitated the BPA degradation. When the flow rate of BPA decreased to lower than 10 mL/min, 0.02 mM of the BPA in the solution could be completely degraded after the solution passed through the ZFP. This was because ZIF-67(Co) had a higher possibility to contact with PMS and generate more active species at a lower flow rate.

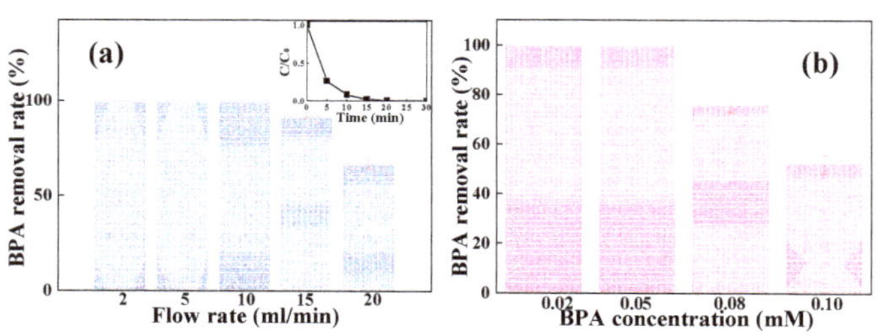

Figure 4. Impact of (**a**) solution flow rate (inset: catalytic performance of ZIF-67(Co) (0.2 g/L) toward BPA (0.02 mM) degradation) and (**b**) BPA concentration on the catalytic performance of ZFP.

The impact of BPA concentration on the performance of the ZFP was then investigated, in which the solution flow rate was fixed at 5 mL/min. As illustrated in Figure 4b, increased BPA concentration inhibited the performance of the ZFP since active species with a fixed generation speed were not capable of degrading the increased number of BPA molecules in the solution. It was still investigated that the ZFP could completely degrade the BPA in the solution when the concentration of the BPA was lower than 0.05 mM, indicating the promising in situ catalytic capability of the ZFP.

The in situ BPA degradation performance of the ZFP at different water conditions was then investigated, with the results shown in Figure 5, in which the solution flow rate was fixed at 5 mL/min, and the BPA concentration was set at 0.02 mM. Figure 5a shows the impact of the solution pH value on the performance of the ZFP. It was seen that the BPA degradation efficiency fluctuated with the change in the solution pH, which indicated that active species mainly dominated the BPA degradation in the reaction system. Meanwhile, the acidic condition was beneficial for the BPA degradation, which reflected the nature of the Fenton-like reactions [20]. When the pH value increased to basic conditions, the activity of the ZFP increased, and the BPA degradation efficiency reached a peak at the pH value of 9, since the generated SO_4^- could contact with more hydroxyl ions (OH^-) to produce hydroxyl radicals (OH) [21]. When further increasing the pH value to 11, excessive OH^- would inhibit the contact between the ZFP and the PMS, resulting in the decline of the ZFP activity [22].

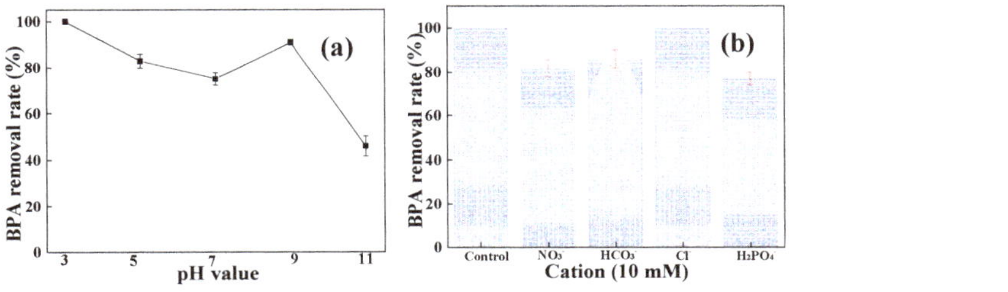

Figure 5. Impact of (**a**) pH value and (**b**) anions on the catalytic performance of ZFP ($[BPA]_0$ = 0.02 mM, flow rate = 5 mL/min).

Figure 5b shows the impact of different anions on the performance of the ZFP. As illustrated, all the introduced anions obviously inhibited the activity of the ZFP except Cl^-. This was because Cl^- could react with OH in the solution and generate abundant chlorine-based active species such as Cl and $ClOH^-$, which also took part in the BPA degradation [23]. Otherwise, the foreign anions would either occupy the active sites on the ZFP or compete with the BPA for active species consumptions, which both had a negative impact on the ZFP performance toward the BPA degradation [24].

The generated active species that dominated the BPA degradation in the ZFP/PMS system were analyzed through scavenging tests. Table 1 summarizes the BPA degradation rate in the ZFP/PMS system at the existence of various scavengers. It was first observed that L-histidine did not show an inhibition effect on the ZFP performance, indicating that non-radical active species were not generated in the reaction system. Meanwhile, both methanol and t-butyl alcohol had a significant quenching effect on the BPA degradation [25]. This indicated that the BPA degradation was dominated by the radical species. Methanol was an effective quencher for both OH ($k_{\bullet OH}$ = ~1.9×10^9 $M^{-1}s^{-1}$) and SO_4^- ($k_{SO4\bullet-}$ = ~2.6×10^7 $M^{-1}s^{-1}$), whereas tert-butyl alcohol had a higher quenching activity to OH ($k_{\bullet OH}$ = ~6.2×10^8 $M^{-1}s^{-1}$) than SO_4^- ($k_{SO4\bullet-}$ =~5.2×10^5 $M^{-1}s^{-1}$) [26]. This meant both OH and SO_4^- were generated in the reaction system for BPA degradation, in which OH was the dominant radical.

Table 1. Impact of different scavengers on the performance of SZP ($[BPA]_0$ = 0.02 mM, flow rate = 5 mL/min).

	Scavenger			
	Control	L-Histidine	Methanol	T-Butyl Alcohol
Degradation Rate (%)	100	100	0	34

The long-time use stability of the ZFP was then investigated through continuous BPA degradation cycles, with the results shown in Figure 6a. As shown, the ZFP provided promising recyclability, which could completely remove the BPA in the reaction system for five consecutive cycles. Based on the AAS results, the concentrations of the leached Co2+ for five cycles were all lower than 50 µg/L, indicating the promising stability of the ZFP. Furthermore, the XRD pattern of the used ZFP (Figure 6b) indicated that ZIF-67(Co) in the ZFP still maintained its crystalline structure after recycling. This meant the prepared catalytic filter paper had wide application potential for the in situ removal of organic pollutants in water.

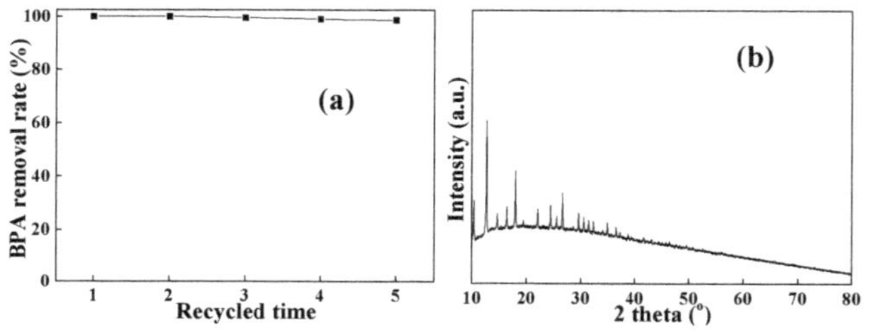

Figure 6. (**a**) Recyclability of ZFP ($[BPA]_0$ = 0.02 mM, flow rate = 5 mL/min), (**b**) XRD pattern of recycled ZFP.

The TOC removal rate of BPA in the ZFP/PMS system was also investigated, with the results shown in Figure 7. It was shown that when BPA passed through the ZFP once, only 18% of the TOC was removed, which indicated that the degraded BPA molecules were only converted into degradation organic intermediates. Instead, when the degraded solution passed through the ZFP multiple times, the TOC removal rate would increase with the increase in the filtration times. When the BPA solution passed through the ZFP five times, more than 60% of the TOC could be removed. This was because the degradation intermediates could be further degraded by the radical active species to inorganic molecules such as CO_2 and H_2O if the pollutant and radical had a longer contact time. The TOC results indicated that the ZFP was a promising catalyst that could completely mineralize inorganic pollutants without introducing toxic degradation intermediates into the reaction solution.

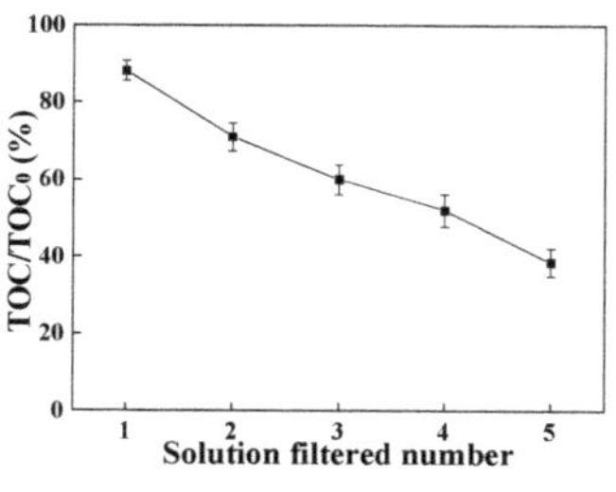

Figure 7. TOC removal of BPA in ZFP/PMS system ($[BPA]_0$ = 0.02 mM, flow rate = 5 mL/min).

4. Conclusions

To conclude, ZFP was prepared in this study by loading ZIF-67(Co) onto the surface of commercial filter paper. The ZFP/PMS system exhibited promising in situ BPA degradation performance as a gravity-driven filter. A 0.02 mM amount of BPA could be completely degraded after passing through ZFP at the rate of 10 mL/min or lower. The ZFP also exhibited promising water matrix adaptability, which exhibited high BPA degradation efficiency at a wide pH application range and at the existence of various foreign anions. Additionally, the ZFP also provided excellent long-time use stability with a high TOC removal rate. Compared with previously reported studies, which frequently utilized powdery catalysts for PMS activation, this study provides a new catalytic filter for the convenient removal of organic pollutants in water without introducing secondary pollution to the water media.

Author Contributions: Writing—original draft preparation, methodology, investigation, Z.C.; formal analysis, data curation, Y.L.; writing—review and editing, funding acquisition, project administration, L.G. All authors have read and agreed to the published version of the manuscript.

Funding: This research was funded by Natural Science Foundation of Jiangsu Province, China (BK20201385).

Data Availability Statement: Data are available upon request.

Acknowledgments: The Advanced Analysis and Testing Center of Nanjing Forestry University is acknowledged.

Conflicts of Interest: The authors declare no conflict of interest.

References

1. You, W.; Liu, L.; Xu, J.; Jin, T.; Fu, L.; Pan, Y. Effect of Anions and Cations on Tartrazine Removal by the Zero-Valent Iron/Peroxymonosulfate Process: Efficiency and Major Radicals. *Catalysts* **2022**, *12*, 1114. [CrossRef]
2. Pan, Y.; Qin, R.; Hou, M.; Xue, J.; Zhou, M.; Xu, L.; Zhang, Y. The interactions of polyphenols with Fe and their application in Fenton/Fenton-like reactions. *Sep. Purif. Technol.* **2022**, *300*, 121831. [CrossRef]
3. Yu, L.; Cheng, W.-x.; Wang, Q. The enhancement on biohydrogen production by the driving forces from extracellular iron oxide respiration. *Bioresour. Technol.* **2022**, *361*, 127679. [CrossRef]
4. Shi, J.; Dai, B.; Fang, X.; Xu, L.; Wu, Y.; Lu, H.; Cui, J.; Han, S.; Gan, L. Waste preserved wood derived biochar catalyst for promoted peroxymonosulfate activation towards bisphenol A degradation with low metal ion release: The insight into the mechanisms. *Sci. Total Environ.* **2022**, *813*, 152673. [CrossRef]
5. Xu, L.; Qi, L.; Han, Y.; Lu, W.; Han, J.; Qiao, W.; Mei, X.; Pan, Y.; Song, K.; Ling, C.; et al. Improvement of Fe^{2+}/peroxymonosulfate oxidation of organic pollutants by promoting Fe^{2+} regeneration with visible light driven g-C_3N_4 photocatalysis. *Chem. Eng. J.* **2022**, *430*, 132828. [CrossRef]
6. Lu, H.; Gan, L. Catalytic Degradation of Bisphenol A in Water by Poplar Wood Powder Waste Derived Biochar via Peroxymonosulfate Activation. *Catalysts* **2022**, *12*, 1164. [CrossRef]
7. Wang, Z.; Jia, H.; Liu, Z.; Peng, Z.; Dai, Y.; Zhang, C.; Guo, X.; Wang, T.; Zhu, L. Greatly enhanced oxidative activity of δ-MnO_2 to degrade organic pollutants driven by dominantly exposed {−111} facets. *J. Hazard. Mater.* **2021**, *413*, 125285. [CrossRef]
8. Wang, Z.; Ju, C.; Zhang, R.; Hua, J.; Chen, R.; Liu, G.; Yin, K.; Yu, L. Acceleration of the bio-reduction of methyl orange by a magnetic and extracellular polymeric substance nanocomposite. *J. Hazard. Mater.* **2021**, *420*, 126576. [CrossRef]
9. Liu, C.; Liu, S.; Liu, L.; Tian, X.; Liu, L.; Xia, Y.; Liang, X.; Wang, Y.; Song, Z.; Zhang, Y.; et al. Novel carbon based Fe-Co oxides derived from Prussian blue analogues activating peroxymonosulfate: Refractory drugs degradation without metal leaching. *Chem. Eng. J.* **2020**, *379*, 122274. [CrossRef]
10. Ling, C.; Wu, S.; Dong, T.L.; Dong, H.F.; Wang, Z.X.; Pan, Y.W.; Han, J.G. Sulfadiazine removal by peroxymonosulfate activation with sulfide-modified microscale zero-valent iron: Major radicals, the role of sulfur species, and particle size effect. *J. Hazard. Mater.* **2022**, *423*, 127082. [CrossRef]
11. Liang, L.; Yue, X.; Dong, S.; Feng, J.; Sun, J.; Pan, Y.; Zhou, M. New insights into the effect of adsorption on catalysis in the metal-free persulfate activation process for removing organic pollutants. *Sep. Purif. Technol.* **2021**, *272*, 118923. [CrossRef]
12. Zhang, M.; Xiao, C.; Zhang, C.; Qi, J.; Wang, C.; Sun, X.; Wang, L.; Xu, Q.; Li, J. Large-Scale Synthesis of Biomass@MOF-Derived Porous Carbon/Cobalt Nanofiber for Environmental Remediation by Advanced Oxidation Processes. *ACS EST Eng.* **2021**, *1*, 249–260. [CrossRef]
13. Wu, Z.L.; Wang, Y.P.; Xiong, Z.K.; Ao, Z.M.; Pu, S.Y.; Yao, G.; Lai, B. Core-shell magnetic Fe_3O_4@Zn/Co-ZIFs to activate peroxymonosulfate for highly efficient degradation of carbamazepine. *Appl. Catal. B-Environ.* **2020**, *277*, 119136. [CrossRef]

14. Fang, X.; Gan, L.; Wang, L.; Gong, H.; Xu, L.; Wu, Y.; Lu, H.; Han, S.; Cui, J.; Xia, C. Enhanced degradation of bisphenol A by mixed ZIF derived CoZn oxide encapsulated N-doped carbon via peroxymonosulfate activation: The importance of N doping amount. *J. Hazard. Mater.* **2021**, *419*, 126363. [CrossRef]
15. Wu, Y.; Xu, L.; Shi, J.; Cui, J.; Han, S.; Xia, C.; Gan, L. Cobalt ferrite/cellulose membrane inserted catalytic syringe filter for facile in-situ filtration/degradation of emerging organic pollutants in water via activating peroxymonosulfate. *Mater. Design* **2022**, *220*, 110817. [CrossRef]
16. Zhao, Z.; Wu, D.; Huang, C.; Zhang, M.; Umemura, K.; Yong, Q. Utilization of enzymatic hydrolysate from corn stover as a precursor to synthesize an eco-friendly adhesive for plywood II: Investigation of appropriate manufacturing conditions, curing behavior, and adhesion mechanism. *J. Wood Sci.* **2020**, *66*, 85. [CrossRef]
17. Zhu, S.; Kumar Biswas, S.; Qiu, Z.; Yue, Y.; Fu, Q.; Jiang, F.; Han, J. Transparent wood-based functional materials via a top-down approach. *Prog. Mater. Sci.* **2023**, *123*, 101025. [CrossRef]
18. Zhuang, G.; Gao, Y.; Zhou, X.; Tao, X.; Luo, J.; Gao, Y.; Yan, Y.; Gao, P.; Zhong, X.; Wang, J. ZIF-67/COF-derived highly dispersed Co_3O_4/N-doped porous carbon with excellent performance for oxygen evolution reaction and Li-ion batteries. *Chem. Eng. J.* **2017**, *330*, 1255–1264. [CrossRef]
19. Pan, Y.; Bu, Z.; Li, J.; Wang, W.; Wu, G.; Zhang, Y. Sulfamethazine removal by peracetic acid activation with sulfide-modified zero-valent iron: Efficiency, the role of sulfur species, and mechanisms. *Sep. Purif. Technol.* **2021**, *277*, 119402. [CrossRef]
20. Yang, X.; Biswas, S.K.; Han, J.; Tanpichai, S.; Li, M.-C.; Chen, C.; Zhu, S.; Das, A.K.; Yano, H. Surface and Interface Engineering for Nanocellulosic Advanced Materials. *Adv. Mater.* **2021**, *33*, 2002264. [CrossRef]
21. Bu, Z.; Li, X.; Xue, Y.; Ye, J.; Zhang, J.; Pan, Y. Hydroxylamine enhanced treatment of highly salty wastewater in $Fe0/H_2O_2$ system: Efficiency and mechanism study. *Sep. Purif. Technol.* **2021**, *271*, 118847. [CrossRef]
22. Fang, X.; Wu, Y.; Xu, L.; Gan, L. Fast removal of bisphenol A by coconut shell biochar incorporated α-MnO_2 composites via peroxymonosulfate activation. *J. Water Process Eng.* **2022**, *49*, 103071. [CrossRef]
23. Dong, X.; Ren, B.; Sun, Z.; Li, C.; Zhang, X.; Kong, M.; Zheng, S.; Dionysiou, D.D. Monodispersed $CuFe_2O_4$ nanoparticles anchored on natural kaolinite as highly efficient peroxymonosulfate catalyst for bisphenol A degradation. *Appl. Catal. B-Environ.* **2019**, *253*, 206–217. [CrossRef]
24. Wang, A.; Ni, J.; Wang, W.; Liu, D.; Zhu, Q.; Xue, B.; Chang, C.-C.; Ma, J.; Zhao, Y. MOF Derived Co−Fe nitrogen doped graphite carbon@crosslinked magnetic chitosan Micro−nanoreactor for environmental applications: Synergy enhancement effect of adsorption−PMS activation. *Appl. Catal. B-Environ.* **2022**, *319*, 121926. [CrossRef]
25. Gan, L.; Zhong, Q.; Geng, A.; Wang, L.; Song, C.; Han, S.; Cui, J.; Xu, L. Cellulose derived carbon nanofiber: A promising biochar support to enhance the catalytic performance of $CoFe_2O_4$ in activating peroxymonosulfate for recycled dimethyl phthalate degradation. *Sci. Total Environ.* **2019**, *694*, 133705. [CrossRef]
26. Zong, Y.; Guan, X.; Xu, J.; Feng, Y.; Mao, Y.; Xu, L.; Chu, H.; Wu, D. Unraveling the Overlooked Involvement of High-Valent Cobalt-Oxo Species Generated from the Cobalt(II)-Activated Peroxymonosulfate Process. *Environ. Sci. Technol.* **2020**, *54*, 16231–16239. [CrossRef]

catalysts MDPI

Article

Effect of Anions and Cations on Tartrazine Removal by the Zero-Valent Iron/Peroxymonosulfate Process: Efficiency and Major Radicals

Wenqi You [1], Liang Liu [2], Junjie Xu [3], Tao Jin [4], Lichun Fu [5,6,*] and Yuwei Pan [1,*]

[1] College of Biology and the Environment, Nanjing Forestry University, Nanjing 210037, China
[2] CCCC First Harbor Consultants Co., Ltd., Tianjin 300222, China
[3] Zhejiang Environmental Science Reseach Institute Co., Ltd., Hangzhou 311121, China
[4] China Construction Eco-Environmental Group Co., Ltd., Beijing 100037, China
[5] School of Iron and Steel, Soochow University, Suzhou 215000, China
[6] School of the Environment, Nanjing University, Nanjing 210023, China
* Correspondence: lcfu@suda.edu.cn (L.F.); panyw@njfu.edu.cn (Y.P.)

Abstract: Zero-valent iron/peroxymonosulfate (Fe^0/PMS) has been considered as a promising approach for wastewater treatment. Anions and cations are widely present in wastewater and have significant effects on the performance of the Fe^0/PMS system for wastewater treatment. Thus, in the present study, tartrazine was selected as the target model; $SO_4{}^{2-}$, $NO_3{}^-$, $HCO_3{}^-$, and Cl^- were selected as representative anions and Ca^{2+}, Cu^{2+}, Mg^{2+}, and Mn^{2+} were chosen as representative cations. The effect of these anions and cations on tartrazine removal and major radicals in the Fe^0/PMS were systematically investigated. The presence of a certain concentration of $SO_4{}^{2-}$ and Cl^- had positive, $NO_3{}^-$ had negative, and $HCO_3{}^-$ had negligible effects on tartrazine removal in the Fe^0/PMS system. $SO_4{}^{2-}$ and $HCO_3{}^-$ had a small effect on the contribution proportion of reduction, $SO_4{}^{\bullet-}$ and $^\bullet OH$; a certain concentration of Cl^- could enhance the contribution proportion of $^\bullet OH$; and $NO_3{}^-$ would decrease the contribution proportion of $SO_4{}^{\bullet-}$ and $^\bullet OH$. A certain concentration of each of Ca^{2+}, Cu^{2+}, Mg^{2+}, and Mn^{2+} could enhance the tartrazine removal in the Fe^0/PMS system. Ca^{2+}, Cu^{2+}, and Mg^{2+} had no effect of the contribution of reduction, $SO_4{}^{\bullet-}$ and $^\bullet OH$, while a certain concentration of Mn^{2+} could enhance the contribution proportion of $SO_4{}^{\bullet-}$. These results can provide some references for the Fe^0/PMS system to treat actual wastewater containing anions and cations.

Keywords: zero-valent iron; peroxymonosulfate; anions; cations

1. Introduction

Recently, with the rapid development of the economy, water pollution has had a great adverse impact on human health and ecosystems [1]. The pollutants of water mainly include antibiotic pollution, phenol pollution, and dye pollution, among others [2–4]. Advanced oxidation processes (AOPs) have been declared as efficient techniques for wastewater treatment owing to the production of reactive species such as sulfate radical ($SO_4{}^{\bullet-}$) and hydroxyl radical ($^\bullet OH$) [5–10].

Peroxymonosulfate (PMS) has been seen as the emerging oxidant because PMS's O-O bond length is short, which is more easily broken [11]. PMS can be catalyzed by metal ions, ultraviolet, ultrasound, and carbon-based materials producing $SO_4{}^{\bullet-}$, $^\bullet OH$, and other reactive species [12–16]. Among these catalytic methods, metal ions can be conducted simply and occurred in conventional environments. Among various mental ions (Co^{2+}, Mn^{2+}, Fe^{2+}, and so on), Fe^{2+} is non-toxic, abundant, and cheap. Reactive species can be produced in the Fe^{2+}/PMS system via Equations (1)–(3) [17–20]. However, the Fe^{2+}/PMS process's efficiency is limited by the low recycle efficiency of Fe^{2+}.

$$Fe^{2+} + HSO_5^- \rightarrow Fe^{3+} + SO_4{}^{\bullet-} + OH^- \quad (1)$$

$$Fe^{2+} + HSO_5^- \rightarrow Fe^{3+} + SO_4^{2-} + {}^{\bullet}OH \quad (2)$$

$$SO_4^{\bullet-} + H_2O \rightarrow SO_4^{2-} + {}^{\bullet}OH + H^+ \quad (3)$$

Nowadays, zero-valent iron (Fe^0) can be used to substitute Fe^{2+} because Fe^{2+} can be generated from the corrosion of Fe^0 via Equation (4) [21]. Fe^0 can also activate PMS directly via Equation (5), producing reactive species and Fe^{2+}. Meanwhile, the produced Fe^{3+} via Equation (1) could be reduced to Fe^{2+} via Equation (6), which can further increase the efficiency [22]. Moreover, Fe^0 loses electrons and undergoes a reduction reaction, which degrades pollutants [23].

$$Fe^0 + O_2 + 2H^+ \rightarrow Fe^{2+} + H_2O_2 \quad (4)$$

$$2Fe^0 + 3HSO_5^- \rightarrow 2Fe^{2+} + 2SO_4^{\bullet-} + 3OH^- + SO_4^{2-} \quad (5)$$

$$2Fe^{3+} + Fe^0 \rightarrow 3Fe^{2+} \quad (6)$$

Besides, anions and cations such as SO_4^{2-}, HCO_3^-, Cl^-, Ca^{2+}, and Mg^{2+} are generally present in wastewater. These ions have a significantly effect on AOP's efficiency for pollutants' removal. Ions can enhance the corrosion rate of Fe^0, but radicals could react with ions, forming less reactive species [24]. Therefore, the effect of both anions and cations on the Fe^0/PMS process's efficiency for pollutants' removal and the role of various radicals ($SO_4^{\bullet-}$, ${}^{\bullet}OH$, and so on) have not been systematically investigated and need further investigation. It is possible that these results might help to promote the application of the Fe^0/PMS process.

In the present study, tartrazine was chosen as the model pollutant; SO_4^{2-}, NO_3^-, Cl^-, and HCO_3^- were chosen as anion representatives; and Ca^{2+}, Cu^{2+}, Mg^{2+}, and Mn^{2+} were chosen as anion representatives. Then, the effect of these anions and cations on efficiency of Fe^0/PMS process for tartrazine removal was investigated. Meanwhile, the effect of these anions and cations on the role of $SO_4^{\bullet-}$ and ${}^{\bullet}OH$ and reduction in the Fe^0/PMS system for tartrazine removal was also systematically studied.

2. Materials and Methods

2.1. Chemicals

Fe^0 powder was bought from Shanghai Jinshan smelter (Shanghai, China). Tartrazine was obtained from Shanghai Macklin Biochemical Technology Co., Ltd., (Shanghai, China). Sodium sulfate (Na_2SO_4), sodium chloride (NaCl), sodium nitrate ($NaNO_3$), and sodium bicarbonate ($NaHCO_3$) were supplied by Aladdin China Company (Shanghai, China). Tert-butyl alcohol (TBA) was bought from Sinopharm Chemical Reagent Co., Ltd., (Shanghai, China).

2.2. Batch Experiments

All degradation experiments were performed in a 500 mL system using glass beakers with a 50 mg/L tartrazine solution under a constant stirring rate of 350 rpm. After adding 4 mM PMS and various ions, the initial pH would be adjusted to 3 using 1 M/0.1 M NaOH and HCl solutions. Then, 0.8 g/L Fe^0 was added into the working solution and stirred with a mechanical stirrer to trigger the degradation reaction. Samples (1 mL) withdrawn at scheduled time intervals by a pipette were immediately quenched by rapidly adding 0.5 mL MeOH, and filtered through 0.22 μm membrane filters for further analysis. Quenching experiments were performed in order to distinguish the reactive species by adding 500 mM TBA into the solution before initiating the reaction.

2.3. Analytical Methods

Tartrazine concentration was detected by a spectrophotometer (VI-1501, Tianjin Gangdong Sci & Tech Development Co., Ltd., Tianjin, China) at 428 nm.

Tartrazine removal efficiency ($\eta/\%$) and value of k were calculated by Equations (7) and (8), respectively.

$$\eta = \left(1 - \frac{C_t}{C_0}\right) * 100\% \tag{7}$$

$$k = ln\frac{C_t}{C_0} \tag{8}$$

where C_0 is the initial tartrazine concentration and C_t is the tartrazine concentration at reaction time t.

3. Results and Discussion

3.1. Effect of SO_4^{2-} on Tartrazine Removal and Major Radicals

It can be seen from Figure 1a,b that the removal efficiency of tartrazine within 60 min by the Fe^0/PMS process was 43.1% and the value of k was 0.0084 min^{-1} in the absence of any ion. The tartrazine removal efficiency and value of k were 56.5%, 0.0146 min^{-1}; 64.7%, 0.0206 min^{-1}; 64.1%, 0.0195 min^{-1}; 57.5%, 0.0159 min^{-1}; and 53.5%, 0.0132 min^{-1} in the presence of 1, 2, 5, 10, and 20 mM SO_4^{2-}, respectively. The removal efficiency and value k increased with a low concentration of SO_4^{2-}, but decreased with a large concentration of SO_4^{2-}. To maintain the local charge balance, Fe^{2+} would move together with negatively charged anions, thus the presence of SO_4^{2-} could also enhance the Fe^0 corrosion rate, thereby enhancing the Fe^{2+} generation rate [24]. From this point, the tartrazine removal efficiency and value of k would increase in the presence of SO_4^{2-}. However, the large concentration of SO_4^{2-} induced the decrease in the tartrazine removal efficiency and value of k, which might be because the oxidation reduction potential (ORP) of $SO_4^{\bullet-}/SO_4^{2-}$ would decrease in the presence of SO_4^{2-}, thereby significantly decreasing the PMS activation efficiency [25]. The results were consistent with the results using Fe^0/PMS treating Rhodamine B [25].

Tartrazine could be removed by reduction and oxidized by $SO_4^{\bullet-}$ and $^{\bullet}OH$ in the Fe^0/PMS system. Therefore, the contributions of the reduction of $SO_4^{\bullet-}$ and $^{\bullet}OH$ for tartrazine removal in the Fe^0/PMS system in the absence and presence of SO_4^{2-} were calculated by Equations (9)–(11), and the results are shown in Figure 1c–f. $k_{reduction}$, $k_{\bullet OH}$, and $k_{SO4\bullet-}$ for tartrazine removal in the Fe^0/PMS system were all enhanced in the presence of SO_4^{2-}, suggesting that the presence of SO_4^{2-} could increase the generation of $SO_4^{\bullet-}$ and $^{\bullet}OH$. Reduction and oxidized by $SO_4^{\bullet-}$ and $^{\bullet}OH$ exhibited 17.8%, 23.1%, and 59.1% contributions for tartrazine removal in the Fe^0/PMS system in the absence of ion. Further, reduction and oxidization by $SO_4^{\bullet-}$ and $^{\bullet}OH$ exhibited 15.0%, 20.1%, and 64.9% contribution for tartrazine removal in the Fe^0/PMS system in the presence of 5 mM SO_4^{2-}. In the presence of 20 mM SO_4^{2-}, reduction and oxidization by $SO_4^{\bullet-}$ and $^{\bullet}OH$ exhibited 15.4%, 29.6%, and 55.0% contribution for tartrazine removal in the Fe^0/PMS system. Radical scavenging experiments suggested that the presence of SO_4^{2-} had a negligible effect on the contribution role of reduction and oxidization by $SO_4^{\bullet-}$ and $^{\bullet}OH$ in removing tartrazine, which might be because SO_4^{2-} would increase the number of electrons lost from Fe^0 and enhanced the reduction reaction. The increased Fe^0 corrosion rate would also enhance the Fe^{2+} production rate, thereby enhancing PMS activation for $SO_4^{\bullet-}$ and $^{\bullet}OH$ production.

$$k(\text{reduction}) = k\left(Fe^0/TBA\right) \tag{9}$$

$$k(^{\bullet}\text{OH}) = k\left(Fe^0/PMS\right) - k\left(Fe^0/PMS/TBA\right) \tag{10}$$

$$k\left(SO_4{}^{\bullet-}\right) = k\left(Fe^0/PMS/TBA\right) - k\left(Fe^0/TBA\right) \tag{11}$$

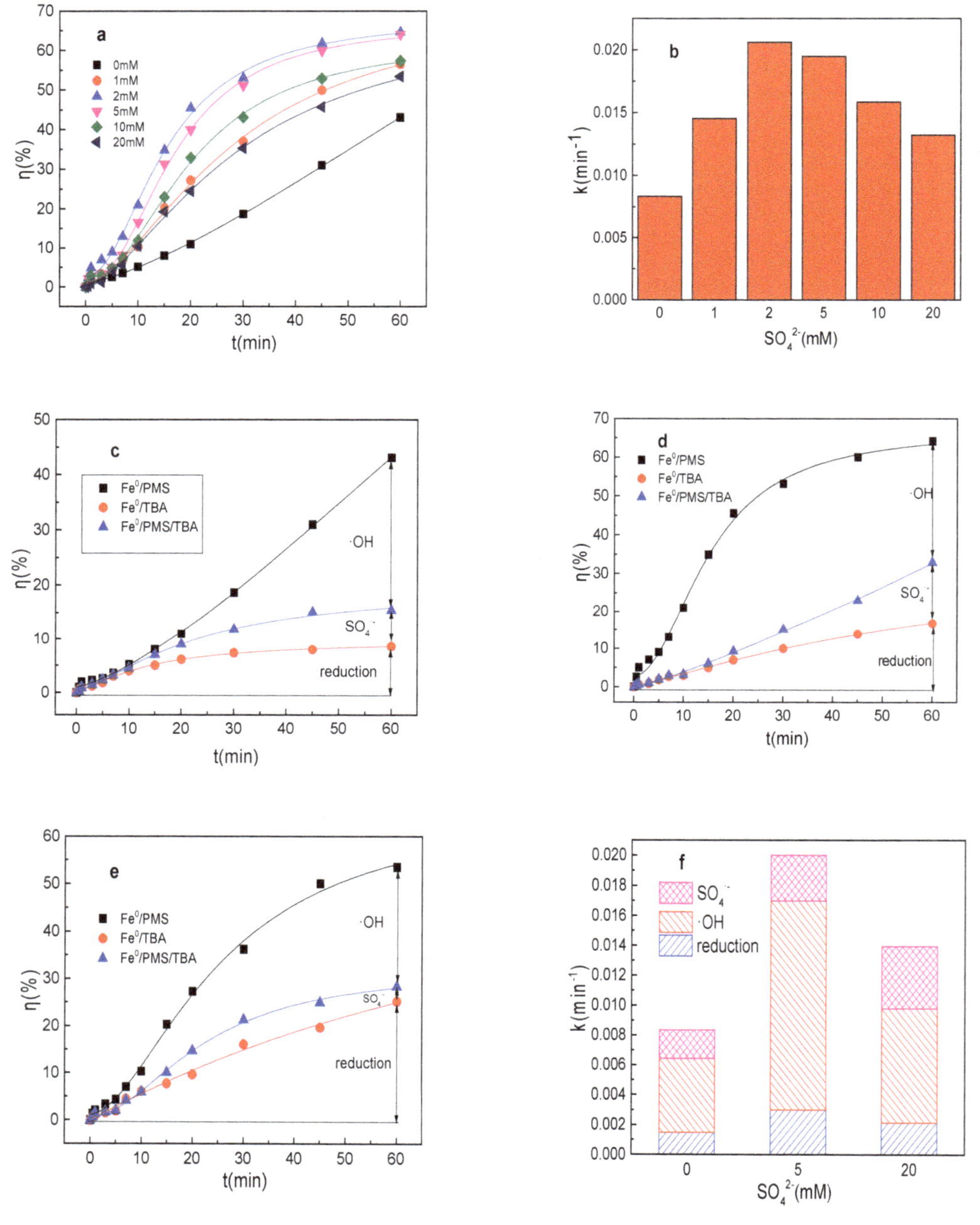

Figure 1. (**a**) Tartrazine removal efficiency by the Fe^0/PMS process in the presence of different $SO_4{}^{2-}$ concentrations; (**b**) value of k in the presence of different $SO_4{}^{2-}$ concentrations; (**c–e**) tartrazine removal under different inhibitors in the presence of 0, 5, and 20 mM $SO_4{}^{2-}$; and (**f**) contributions of $SO_4{}^{\bullet-}$ and $^{\bullet}OH$ and reduction for tartrazine removal. Condition: tartrazine 50 mg/L, Fe^0 0.8 g/L, pH 3, PMS 4 mM, TBA 500 mM.

3.2. Effect of NO_3^- on Tartrazine Removal and Major Radicals

As shown in Figure 2a,b, the tartrazine removal efficiency and value of *k* were obviously decreased in the presence of NO_3^- in the Fe^0/PMS system and the results were not in accordance with the results of SO_4^{2-}. The tartrazine removal efficiency and value of *k* were only 26.4%, 0.0053 min^{-1}; 23.3%, 0.0045 min^{-1}; 20.3%, 0.0041 min^{-1}; 19.3%, 0.0037 min^{-1}; and 16.5%, 0.0032 min^{-1} in the presence of 1, 2, 5, 10, and 20 mM NO_3^-, respectively. NO_3^-, like SO_4^{2-}, is also an anion that can improve the corrosion of Fe^0. The tartrazine removal efficiency and value of *k* were obviously decreased in the presence of NO_3^- in the Fe^0/PMS system, which might because of the following reasons: (1) Fe^0 could directly react with NO_3^- to form iron oxides and affected the formation of Fe^{2+} via Equation (12) [26]; (2) NO_3^- could adsorb on Fe^0 surface while iron powder eroded and many cavities were produced, which could also affect the Fe^{2+} generation [26]; and (3) $SO_4^{\bullet-}$ and $^{\bullet}OH$ could react with NO_3^- to form less reactive species via Equations (13) and (14) [27]. According to literature reports, the effect of NO_3^- on the pollutant treatment efficiency of the PMS-based system was not obvious because of reason (3) [27]. Therefore, the effect of NO_3^- on the Fe^0 corrosion rate was the main reason for the decrease in the treatment efficiency in the Fe^0/PMS system.

$$10Fe^0 + 6NO_3^- + 3H_2O \rightarrow 5Fe_2O_3 + 6OH^- + 3N_2 \quad (12)$$

$$SO_4^{\bullet-} + NO_3^- \rightarrow SO_4^{2-} + NO_3^{\bullet} \quad (13)$$

$$^{\bullet}OH + NO_3^- \rightarrow OH^- + NO_3^{\bullet} \quad (14)$$

Radical scavenging experiments in the presence of NO_3^- in the Fe^0/PMS system were also performed. As shown in Figure 2c–e, reduction of $SO_4^{\bullet-}$ and $^{\bullet}OH$ also all contributed to tartrazine removal in the presence of NO_3^- in the Fe^0/PMS system. $k_{reduction}$, $k_{\bullet OH}$, and $k_{SO4\bullet-}$ for tartrazine removal in the Fe^0/PMS system were all decreased in the presence of NO_3^-, suggesting Fe^{2+} released by the reduction reaction was related to the PMS activation efficiency. Reduction and oxidized by $SO_4^{\bullet-}$ and $^{\bullet}OH$ exhibited 26.2, 35.2%; 24.7%, 33.5%; and 49.1%, 31.3% contribution for tartrazine removal the Fe^0/PMS system in the presence of 5 and 20 mM NO_3^-. Oxidation contribution for tartrazine removal decreased in the presence of NO_3^-, which might be because the presence of NO_3^- inhibited the generation of Fe^{2+}.

3.3. Effect of Cl^- on Tartrazine Removal and Major Radicals

As shown in Figure 3a,b, the tartrazine removal efficiency and value of *k* were significantly increased in the presence of Cl^- in the Fe^0/PMS system. The tartrazine removal efficiency and value of *k* increased from to 43.1%, 0.0084 min^{-1} to 75.4%, 0.024 min^{-1}; 76.1%, 0.025 min^{-1}; 77.2%, 0.026 min^{-1}; 81.4%, 0.029 min^{-1}; and 87.0%, 0.035 min^{-1} in the presence of 1, 2, 5, 10, and 20 mM Cl^-, respectively. Although Cl^- could react with $SO_4^{\bullet-}$ and $^{\bullet}OH$ to generate less reactive radicals via Equations (15) and (16) [27], the presence of Cl^- could significantly enhance the Fe^{2+} generation, and then improved the production of $SO_4^{\bullet-}$ and $^{\bullet}OH$. Likewise, according to the literature reports, the reaction between Cl^- and $SO_4^{\bullet-}$ and $^{\bullet}OH$ had no obvious effect on pollutants' removal in PMS-based systems [27]. Furthermore, the results also demonstrated that the positive effects of Cl^- on tartrazine removal in the Fe^0/PMS system were significantly stronger than the side effects.

According to the radical scavenging experiment results in Figure 3c–e, $k_{reduction}$, $k_{\bullet OH}$, and $k_{SO4\bullet-}$ for tartrazine removal in the Fe^0/PMS system were all enhanced with the increase in the concentration of Cl^-. Reduction and oxidization by $SO_4^{\bullet-}$ and $^{\bullet}OH$ exhibited 34.6%, 17.1%; 11.6%, 11.4%; and 53.8%, 71.6% contribution to tartrazine removal in the Fe^0/PMS system in the presence of 5 and 20 mM Cl^-. The proportion of $^{\bullet}OH$ contribution was significantly enhanced in the presence of Cl^-, which might be because the oxidizing ability of $^{\bullet}OH$ was stronger than that of $SO_4^{\bullet-}$ and the enhanced Fe^{2+} generation could improve both $SO_4^{\bullet-}$ and $^{\bullet}OH$, thus the contribution of $^{\bullet}OH$ would be significantly enhanced.

$$SO_4^{\bullet-} + Cl^- \rightarrow SO_4^{2-} + Cl^{\bullet} \quad (15)$$

$$^{\bullet}OH + Cl^- \rightarrow OH^- + Cl^{\bullet} \quad (16)$$

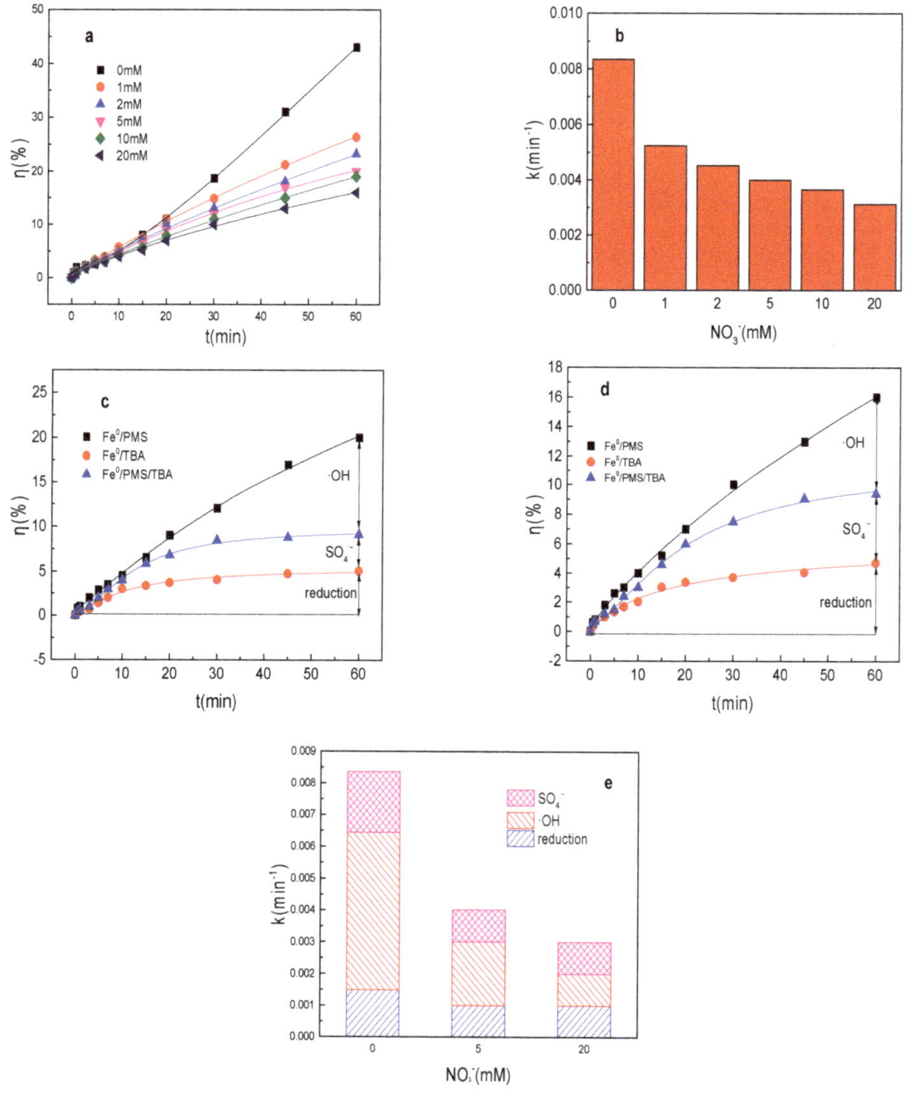

Figure 2. (**a**) Tartrazine removal efficiency by the Fe^0/PMS process in the presence of different NO_3^- concentrations; (**b**) value of k in the presence of different NO_3^- concentrations; (**c**,**d**) tartrazine removal under different inhibitors in the presence of 5 and 20 mM NO_3^-; and (**e**) contributions of $SO_4^{\bullet-}$ and $^{\bullet}OH$ and reduction to tartrazine removal. Condition: tartrazine 50 mg/L, Fe^0 0.8 g/L, pH 3, PMS 4 mM, TBA 500 mM.

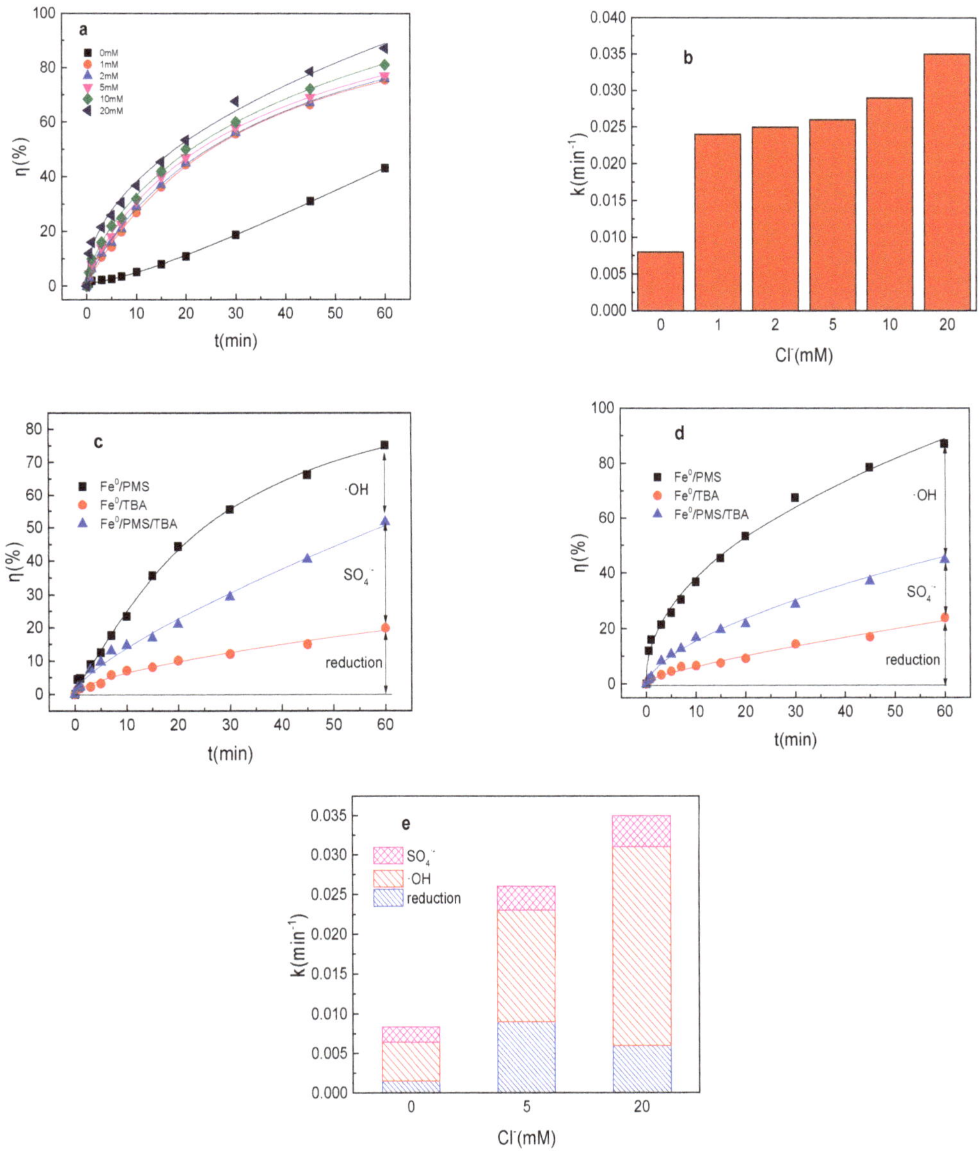

Figure 3. (**a**) Tartrazine removal efficiency by the Fe^0/PMS process in the presence of different Cl^- concentrations; (**b**) value of *k* in the presence of different Cl^- concentrations; (**c**,**d**) tartrazine removal under different inhibitors in the presence of 5 and 20 mM Cl^-; and (**e**) contributions of $SO_4^{\bullet-}$ and $^\bullet OH$ and reduction for tartrazine removal. Condition: tartrazine 50 mg/L, Fe^0 0.8 g/L, pH 3, PMS 4 mM, TBA 500 mM.

3.4. Effect of HCO_3^- on Tartrazine Removal and Major Radicals

As depicted in Figure 4a,b, the tartrazine removal efficiency and value of *k* were almost unchanged in the presence of HCO_3^- in the Fe^0/PMS system. HCO_3^- was also able to enhance the Fe^{2+} generation via Equation (17) [28]. Meanwhile, HCO_3^- might inhibit the release of Fe^{2+} from Fe^0 owing to the generation of an insulating film and iron compounds on the Fe^0 surface. Meanwhile, HCO_3^- could react with $SO_4^{\bullet-}$ and $^{\bullet}OH$ to generate less reactive radicals via Equations (18) and (19). Moreover, HCO_3^- could consume both H^+ and OH^- by Equations (20)–(21), and thus maintain a stable pH during the reaction, which further inhibited Fe^0 corrosion and PMS activation [28]. In general, HCO_3^- had a small effect on the removal of pollutants by the Fe^0/PMS system. According to the literature reports, HCO_3^- had a significant effect on pollutants' removal in PMS-based systems [27], thus the promotion of Fe^0 corrosion by HCO_3^- had a very positive effect on pollutants' removal efficiency in the Fe^0/PMS system. According to the radical scavenging experiment results in Figure 4c–e, HCO_3^- also had a small effect on $k_{reduction}$, $k_{\bullet OH}$, and $k_{SO4\bullet-}$ and the contribution proportion of reduction of $SO_4^{\bullet-}$ and $^{\bullet}OH$ for tartrazine removal in the Fe^0/PMS system.

$$Fe^0 + 2HCO_3^- \rightarrow 2CO_3^{2-} + Fe^{2+} + H_2 \quad (17)$$

$$SO_4^{\bullet-} + HCO_3^- \rightarrow SO_4^{2-} + HCO_3^{\bullet} \quad (18)$$

$$^{\bullet}OH + HCO_3^- \rightarrow OH^- + HCO_3^{\bullet} \quad (19)$$

$$H^+ + HCO_3^- \rightarrow H_2O + CO_2 \quad (20)$$

$$OH^- + HCO_3^- \rightarrow H_2O + CO_3^{2-} \quad (21)$$

3.5. Effect of Ca^{2+} on Tartrazine Removal and Major Radicals

The effect of Ca^{2+} on tartrazine removal and major radicals in Fe^0/PMS was also investigated and the results are shown in Figure 5a–e. Ca^{2+} had no obvious effect on tartrazine removal and value of k, and tartrazine removal and value of k could be enhanced from 43.1% and 0.0084 min^{-1} to 49.8% and 0.0114 min^{-1}, respectively, while Ca^{2+} increased from 0 to 50 mg/L. According to the previous studies, Ca^{2+} could not efficiently activate PMS for pollutants' removal. The adsorbance of Ca^{2+} on the Fe^{0}' surface was conducive to enhancing the PMS activation, and Ca^{2+}' bridging of negatively charged pollutants also favored degradation, possibly because of the enrichment of the catalyst by the substrate. Thus, a certain amount of Ca^{2+} could enhance the tartrazine removal in the Fe^0/PMS system. Meanwhile, 50 mg/L Ca^{2+} also could enhance $k_{reduction}$, $k_{\bullet OH}$, and $k_{SO4\bullet-}$, while it had no obvious effect on the contribution proportion of reduction of $SO_4^{\bullet-}$ and $^{\bullet}OH$ for tartrazine removal in the Fe^0/PMS system.

3.6. Effect of Cu^{2+} on Tartrazine Removal and Major Radicals

As shown in Figure 6a,b, the tartrazine removal and value of k increased from 43.1%, 0.0084 min^{-1} to 49.2%, 0.010 min^{-1}, 60.6%, 0.015 min^{-1}; 64.3%, 0.016 min^{-1}; 66.9%, 0.018 min^{-1}; and 70.8%, 0.020 min^{-1}, respectively, while the Cu^{2+} concentration increased from 0 to 1, 5, 10, 20, and 50 mg/L in the Fe^0/PMS system. The tartrazine removal efficiency increased with the addition of Cu^{2+}, which might be mainly because PMS could be also activated by Cu^{2+} and Cu^{2+} could also promote the corrosion of Fe^0. Meanwhile, Cu^{2+} could also inhibit $SO_4^{\bullet-}$ and $^{\bullet}OH$ via Equations (22) and (23) [25]. In general, the presence of Cu^{2+} could enhance the tartrazine removal in the Fe^0/PMS system. It could found from Figure 6c–e that $k_{reduction}$, $k_{\bullet OH}$, and $k_{SO4\bullet-}$ were all enhanced in the presence of Cu^{2+}. Reduction and oxidization by $SO_4^{\bullet-}$ and $^{\bullet}OH$ exhibited 12.4%, 21.4%; 29.7%, 38.4%; and 57.9%, 40.2% contributions for tartrazine removal in the Fe^0/PMS system in the presence of 10 and 50 mg/L Cu^{2+}. Cu^{2+} had no obvious effect on the contribution of reduction of $SO_4^{\bullet-}$ and $^{\bullet}OH$ on tartrazine removal, which might be because Cu^{2+} could affect not only the PMS activation, but also the corrosion of Fe^0.

$$SO_4^{\bullet-} + Cu^{2+} \rightarrow SO_4^{2-} + Cu^{3+} \quad (22)$$

$$^{\bullet}OH + Cu^{2+} \rightarrow OH^- + Cu^{3+} \quad (23)$$

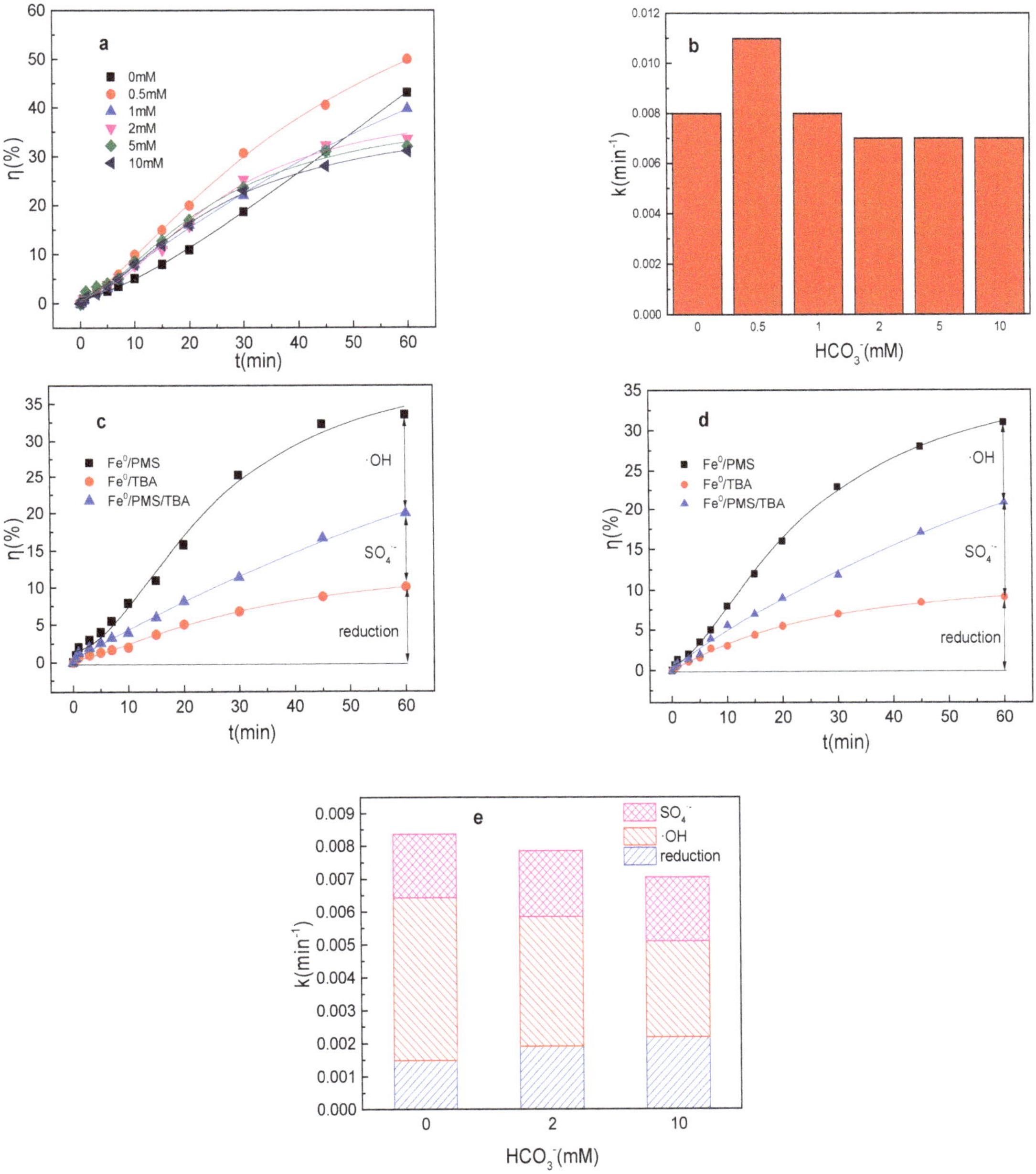

Figure 4. (**a**) Tartrazine removal efficiency by the Fe^0/PMS process in the presence of different HCO_3^- concentrations; (**b**) value of k in the presence of different HCO_3^- concentrations; (**c**,**d**) tartrazine removal under different inhibitors in the presence of 5 and 20 mM HCO_3^-; and (**e**) contributions of $SO_4^{\bullet-}$ and $^{\bullet}OH$ and reduction for tartrazine removal. Condition: tartrazine 50 mg/L, Fe^0 0.8 g/L, pH 3, PMS 4 mM, TBA 500 mM.

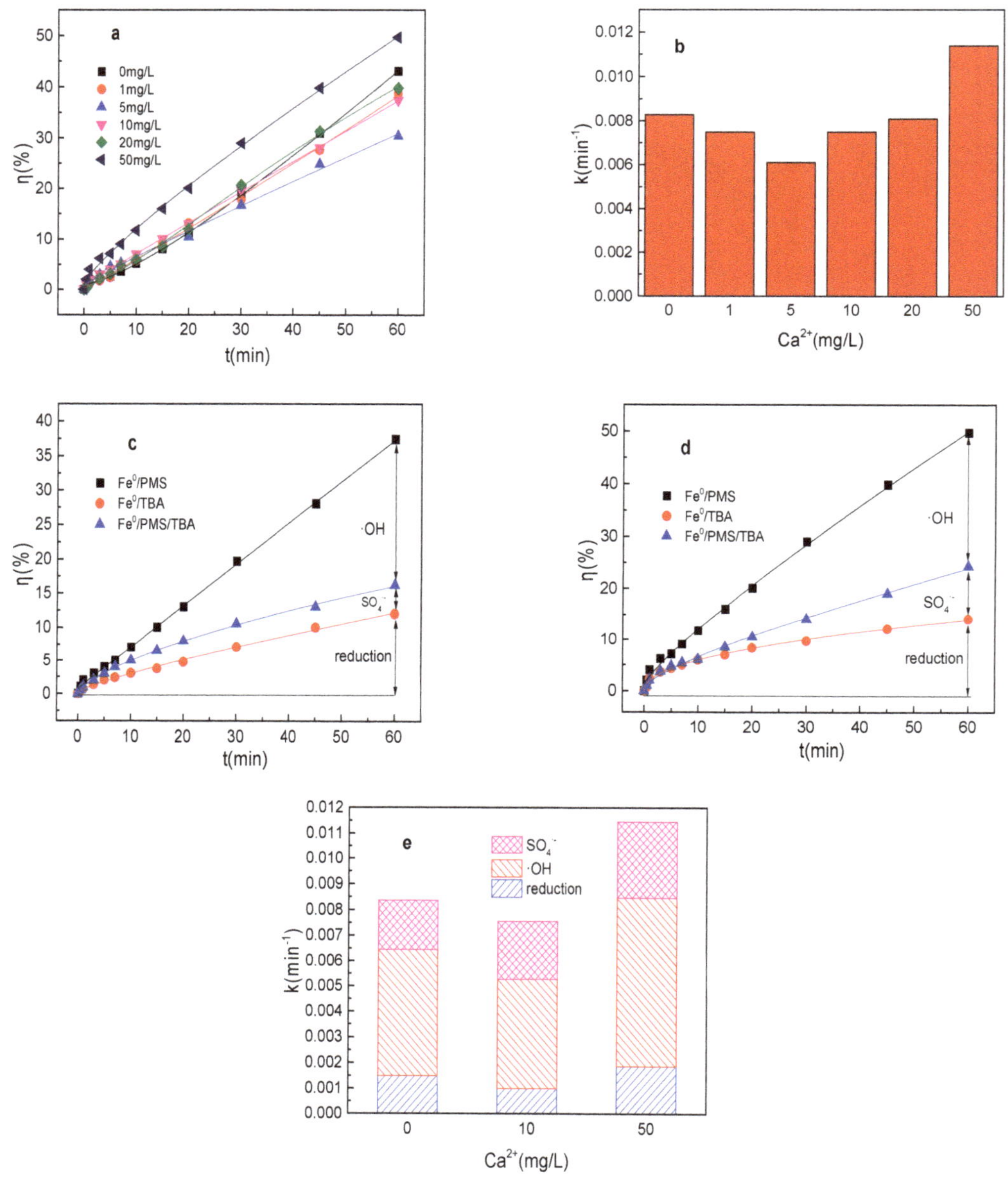

Figure 5. (**a**) Tartrazine removal efficiency by the Fe^0/PMS process in the presence of different Ca^{2+} concentrations; (**b**) value of *k* in the presence of different Ca^{2+} concentrations; (**c**,**d**) tartrazine removal under different inhibitors in the presence of 5 and 20 mM Ca^{2+}; and (**e**) contributions of $SO_4{}^{\bullet-}$ and $^{\bullet}OH$ and reduction for tartrazine removal. Condition: tartrazine 50 mg/L, Fe^0 0.8 g/L, pH 3, PMS 4 mM, TBA 500 mM.

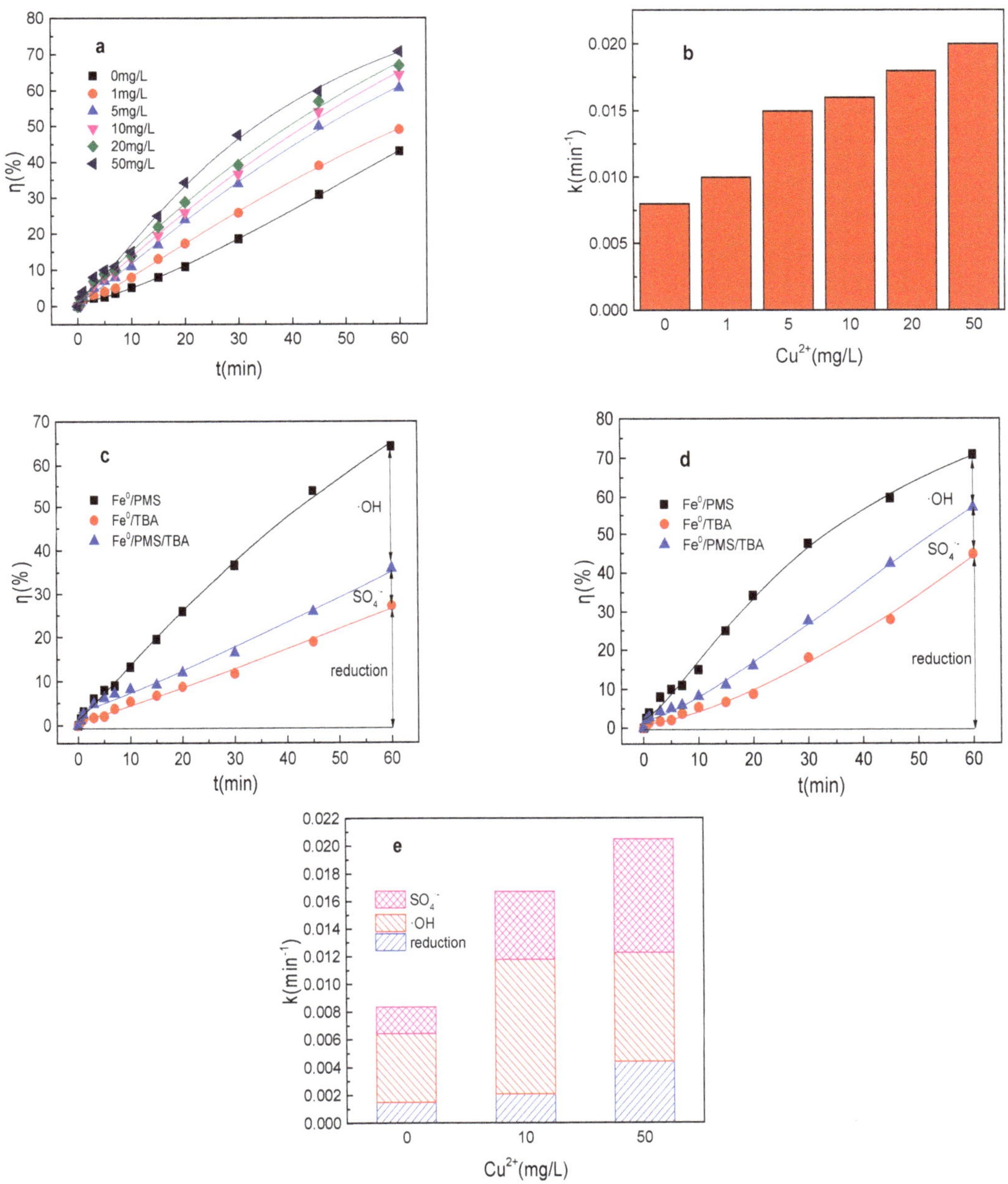

Figure 6. (**a**) Tartrazine removal efficiency by the Fe^0/PMS process in the presence of different Cu^{2+} concentrations; (**b**) value of *k* in the presence of different Cu^{2+} concentrations; (**c**,**d**) tartrazine removal under different inhibitors in the presence of 5 and 20 mM Cu^{2+}; and (**e**) contributions of $SO_4^{\bullet-}$ and $^{\bullet}OH$ and reduction for tartrazine removal. Condition: tartrazine 50 mg/L, Fe^0 0.8 g/L, pH 3, PMS 4 mM, TBA 500 mM.

3.7. Effect of Mg^{2+} on Tartrazine Removal and Major Radicals

As shown in Figure 7a,b, tartrazine removal efficiency increased with the increase in Mg^{2+} concentration, and tartrazine removal and the value of k were 41.0%, 0.0086 min^{-1}; 43.4%, 0.0091 min^{-1}; 52.1%, 0.012 min^{-1}; 77.1%, 0.029 min^{-1}; and 68.4%, 0.020 min^{-1} in the presence of 1, 5, 10, 20, and 50 mg/L Mg^{2+} in the Fe^0/PMS system, respectively. Mg^{2+} was similar to Cu^{2+}, which could activate PMS and enhance the corrosion of Fe^0, and thus could improve the Fe^0/PMS system for tartrazine removal. The removal efficiency decreased at 50 mg/L Mg^{2+} in the Fe^0/PMS system, which might be because of the reaction between Mg^{2+} and $SO_4^{\bullet-}$ and $^{\bullet}OH$ via Equations (24) and (25) [25]. According to the radical scavenging experiment results from Figure 7c–e, $k_{reduction}$, $k_{\bullet OH}$, and $k_{SO4\bullet-}$ were all enhanced in the presence of Mg^{2+}, which also confirmed that the presence of Mg^{2+} could enhance the corrosion of Fe^0 and PMS activation. Moreover, Mg^{2+} also had no obvious effect of the contribution of reduction of $SO_4^{\bullet-}$ and $^{\bullet}OH$.

$$SO_4^{\bullet-} + Mg^{2+} \rightarrow SO_4^{2-} + Mg^{3+} \tag{24}$$

$$^{\bullet}OH + Mg^{2+} \rightarrow OH^- + Mg^{3+} \tag{25}$$

3.8. Effect of Mn^{2+} on Tartrazine Removal and Major Radicals

It could found from Figure 8a,b that Mn^{2+} could also significantly enhance the Fe^0/PMS system for tartrazine removal. Tartrazine removal and value of k increased from 43.1%, 0.0084 min^{-1} to 46.1%, 0.0099 min^{-1}; 47.6%, 0.010 min^{-1}; 53.1%, 0.013 min^{-1}; 58.4%, 0.015 min^{-1}; and 61.2%, 0.019 min^{-1}, while Mn^{2+} concentration increased from 0 to 1, 5, 10, 20, and 50 mg/L in the Fe^0/PMS system, respectively. As previous literatures reported that Mn^{2+} was a good transition metal for PMS activation via Equations (26) and (27), it thus induced the increase in tartrazine removal efficiency in the Fe^0/PMS system [29]. Meanwhile, the effect of Mn^{2+} on the Fe^0/PMS system improvement for tartrazine removal was lower than that of Mg^{2+}, which might be because Mn^{3+} and Mn^{2+} could react with Fe^{2+} via Equations (28) and (29), thus affecting its performance [29]. As depicted in Figure 8c–e, both $k_{SO4\bullet-}$ and the contribution of $SO_4^{\bullet-}$ were significantly enhanced in the presence of Mn^{2+}. The results also indicate that Mn^{2+} might affect the activation ability of Fe^{2+} and become the dominant activated metal.

$$Mn^{2+} + HSO_5^- \rightarrow Mn^{3+} + SO_4^{\bullet-} + OH^- \tag{26}$$

$$Mn^{3+} + HSO_5^- \rightarrow Mn^{4+} + SO_4^{\bullet-} + OH^- \tag{27}$$

$$Mn^{3+} + Fe^{2+} \rightarrow Fe^{3+} + Mn^{2+} \tag{28}$$

$$Mn^{4+} + Fe^{2+} \rightarrow Fe^{3+} + Mn^{3+} \tag{29}$$

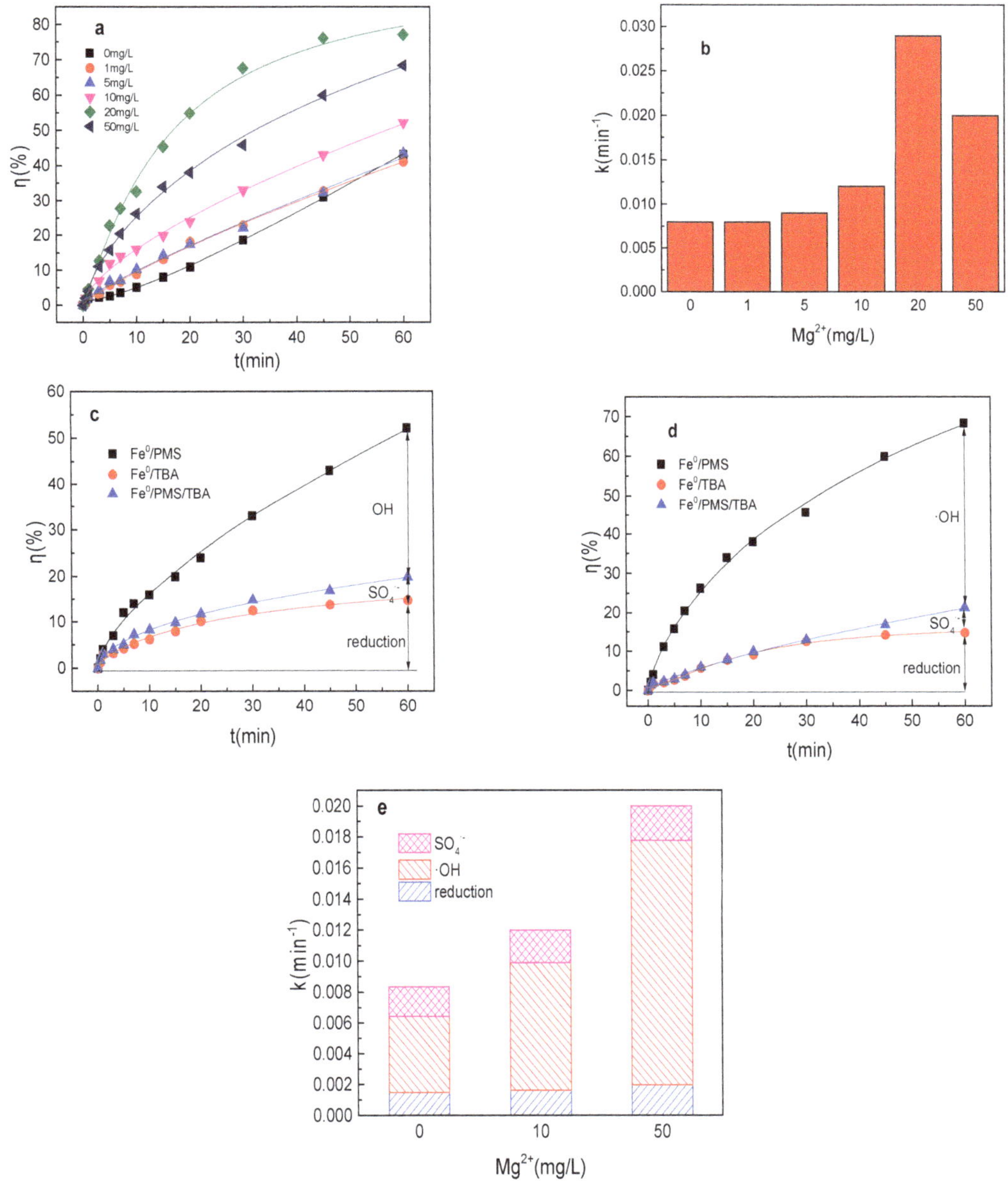

Figure 7. (**a**) Tartrazine removal efficiency by the Fe^0/PMS process in the presence of different Mg^{2+} concentrations; (**b**) value of *k* in the presence of different Mg^{2+} concentrations; (**c**,**d**) tartrazine removal under different inhibitors in the presence of 5 and 20 mM Mg^{2+}; and (**e**) contributions of $SO_4^{\bullet-}$ and $^{\bullet}OH$ and reduction for tartrazine removal. Condition: tartrazine 50 mg/L, Fe^0 0.8 g/L, pH 3, PMS 4 mM, TBA 500 mM.

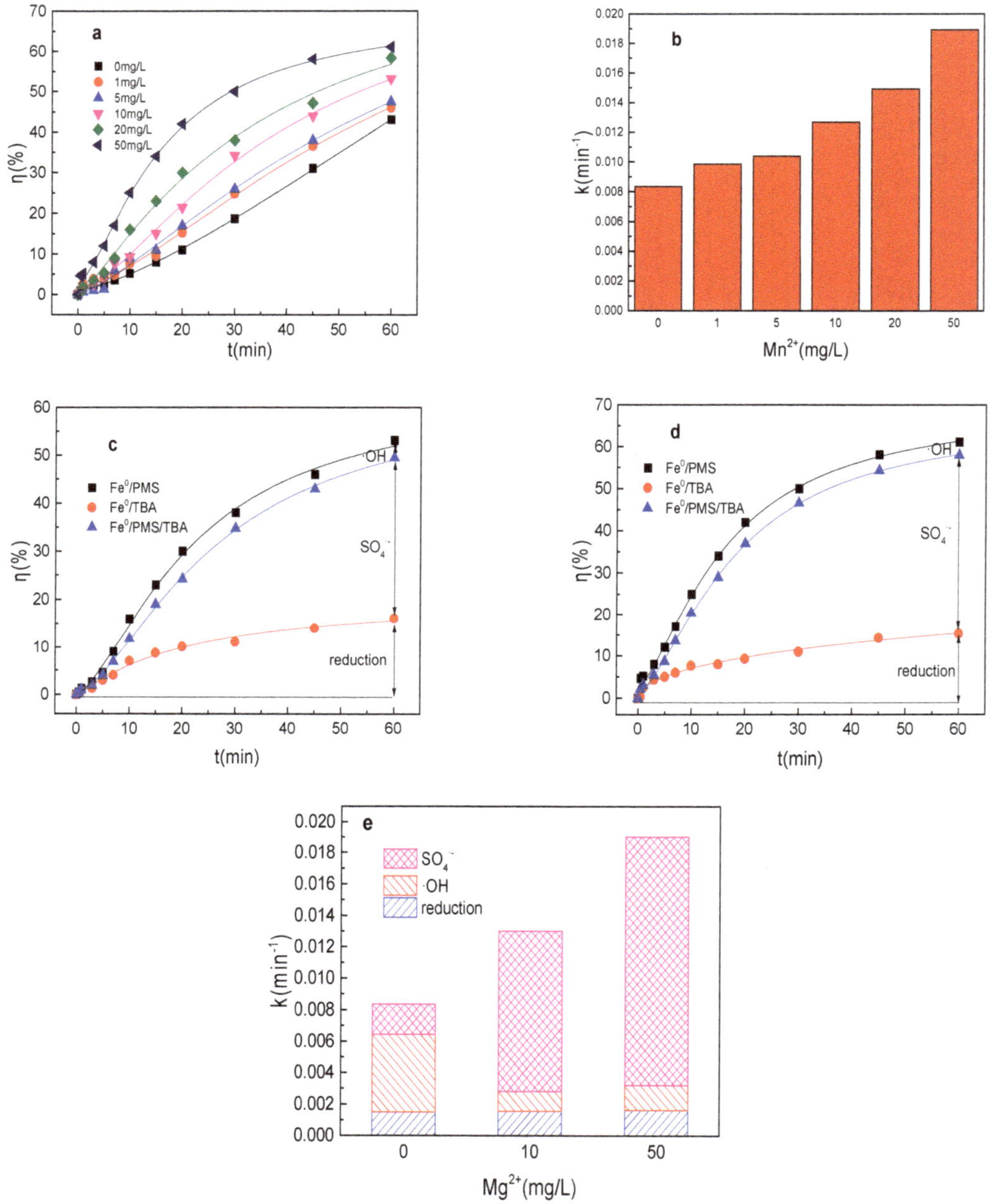

Figure 8. (**a**) Tartrazine removal efficiency by the Fe^0/PMS process in the presence of different Mn^{2+} concentrations; (**b**) value of k in the presence of different Mn^{2+} concentrations; (**c**,**d**) tartrazine removal under different inhibitors in the presence of 5 and 20 mM Mn^{2+}; and (**e**) contributions of $SO_4^{\bullet-}$ and $^{\bullet}OH$ and reduction for tartrazine removal. Condition: tartrazine 50 mg/L, Fe^0 0.8 g/L, pH 3, PMS 4 mM, TBA 500 mM.

4. Conclusions

In this study, the effect of anions (SO_4^{2-}, NO_3^-, HCO_3^-, and Cl^-) and cations (Ca^{2+}, Cu^{2+}, Mg^{2+}, and Mn^{2+}) on tartrazine removal and major radicals in Fe^0/PMS were systematically investigated. SO_4^{2-}, Cl^- Ca^{2+}, Cu^{2+}, Mg^{2+}, and Mn^{2+} could enhance tartrazine removal; NO_3^- could inhibit tartrazine removal; and HCO_3^- had no obvious effect on tartrazine removal. SO_4^{2-}, HCO_3^-, Ca^{2+}, Cu^{2+}, and Mg^{2+} had little effect on the contribution proportion of reduction of $SO_4^{\bullet-}$ and $^{\bullet}OH$; NO_3^- would decrease the contribution proportion of $SO_4^{\bullet-}$ and $^{\bullet}OH$, a certain concentration Cl^- could enhance the contribution proportion of $^{\bullet}OH$; and a certain concentration of Mn^{2+} could enhance the contribution proportion of $SO_4^{\bullet-}$. These dates can provide some references for the Fe^0/PMS system to treat actual wastewater containing anions and cations.

Author Contributions: Conceptualization, Y.P.; data curation, W.Y.; formal analysis, T.J.; funding acquisition, L.F. and Y.P.; supervision, L.F. and Y.P.; writing—review and editing, W.Y., L.L. and J.X. All authors have read and agreed to the published version of the manuscript.

Funding: This work was supported by Student Innovation Training Program (2021NFUSPITP0758) and China Postdoctoral Science Foundation Project (2020M681552). This study was also financially supported by Priority Academic Program Development of Jiangsu Higher Education Institutions (PAPD).

Data Availability Statement: Data available on request due to restrictions e.g. privacy or ethical.

Conflicts of Interest: The authors declare no conflict of interest.

References

1. Martinez, L.B.; Diza, C.B.; Morelos, C.S. Synergy of electrochemical and ozonation processes in industrial wastewater treatment. *Chem. Eng. J.* **2014**, *165*, 71–77. [CrossRef]
2. Qazi, U.Y.; Iftikhar, R.; Ikhlaq, A.; Riaz, I.; Jaleel, R.; Nusrat, R.; Javaid, R. Application of Fe RGO for the removal of dyes by catalytic ozonation process. *Environ. Sci. Pollut. Res.* **2022**, *22*, 1–13. [CrossRef] [PubMed]
3. Con, Y.Q.; Li, Z.; Zhang, Y.; Wang, Q.; Xu, Q. Synthesis of α-Fe_2O_3/TiO_2 nanotube arrays for photoelectro-Fenton degradation of phenol. *Chem. Eng. J.* **2012**, *191*, 356–363.
4. Chang, M.-C.; Shu, H.-Y.; Yu, H.-H. An integrated technique using zero-valent iron and UV/H_2O_2 sequential process for complete decolorization and mineralization of C.I. Acid Black 24 wastewater. *J. Hazard. Mater.* **2006**, *138*, 574–581. [CrossRef] [PubMed]
5. Pan, Y.; Zhang, Y.; Zhou, M.; Cai, J.; Tian, Y. Enhanced removal of emerging contaminants using persulfate activated by UV and pre-magnetized Fe^0. *Chem. Eng. J.* **2019**, *361*, 908–918. [CrossRef]
6. Masood, Z.; Ikhlaq, A.; Akram, A.; Qazi, U.Y.; Rizvi, O.S.; Javaid, R.; Alazmi, A.; Madkour, M.; Qi, F. Application of Nanocatalysts in Advanced Oxidation Processes for Wastewater Purification: Challenges and Future Prospects. *Catalysts* **2022**, *12*, 741. [CrossRef]
7. Qazi, U.Y.; Javaid, R.; Ikhlaq, A.; Al-Sodani, K.A.A.; Rizvi, O.S.; Alazmi, A.; Asiri, A.M.; Ibn Shamsah, S.M. Synergistically Improved Catalytic Ozonation Process Using Iron-Loaded Activated Carbons for the Removal of Arsenic in Drinking Water. *Water* **2022**, *14*, 2406. [CrossRef]
8. Fang, X.Y.; Wu, Y.; Xu, L.J.; Gan, L. Fast removal of bisphenol A by coconut shell biochar incorporated α-MnO_2 composites via peroxymonosulfate activation. *J. Water Process Eng.* **2022**, *49*, 103071. [CrossRef]
9. Qin, W.; Fang, G.; Wang, Y.; Zhou, D. Mechanistic understanding of polychlorinated biphenyls degradation by peroxymonosulfate activated with $CuFe_2O_4$ nanoparticles: Key role of superoxide radicals. *Chem. Eng. J.* **2018**, *348*, 526–534. [CrossRef]
10. Segura, Y.; Martínez, F.; Melero, J.A.; Molina, R.; Chand, R.; Bremner, D.H. Enhancement of the advanced Fenton process (Fe^0/H_2O_2) by ultrasound for the mineralization of phenol. *Appl. Catal. B Environ.* **2012**, *113–114*, 100–106. [CrossRef]
11. Ling, C.; Wu, S.; Dong, T.; Dong, H.; Wang, Z.; Pan, Y.; Han, J. Sulfadiazine removal by peroxymonosulfate activation with sulfide-modified microscale zero-valent iron: Major radicals, the role of sulfur species, and particle size effect. *J. Hazard. Mater.* **2022**, *423*, 127082. [CrossRef]
12. Wang, J.; Hasaer, B.; Yang, M.; Liu, R.; Hu, C.; Liu, H.; Qu, J. Anaerobically-digested sludge disintegration by transition metal ions-activated peroxymonosulfate (PMS): Comparison between Co^{2+}, Cu^{2+}, Fe^{2+} and Mn^{2+}. *Sci. Total Environ.* **2020**, *713*, 136530. [CrossRef]
13. Zhang, H.; Ji, Q.Q.; Lai, L.D.; Yao, G.; Lai, B. Degradation of p-nitrophenol (PNP) in aqueous solution by mFe/Cu-air-PS system. *Chin. Chem. Lett.* **2019**, *30*, 1129–1132. [CrossRef]
14. Huang, W.; Xiao, S.; Zhong, H.; Yan, M.; Yang, X. Activation of persulfates by carbonaceous materials: A review. *Chem. Eng. J.* **2021**, *418*, 129297. [CrossRef]
15. Giannakis, S.; Lin, K.-Y.A.; Ghanbari, F. A review of the recent advances on the treatment of industrial wastewaters by Sulfate Radical-based Advanced Oxidation Processes (SR-AOPs). *Chem. Eng. J.* **2021**, *406*, 127083. [CrossRef]

16. Kohantorabi, M.; Moussavi, G.; Giannakis, S. A review of the innovations in metal and carbon-based catalysts explored for heterogeneous peroxymonosulfate (PMS) activation, with focus on radical vs. non-radical degradation pathways of organic contaminants. *Chem. Eng. J.* **2021**, *411*, 127957. [CrossRef]
17. Li, Z.-Y.; Liu, Y.-L.; He, P.-N.; Zhang, X.; Wang, L.; Gu, H.-T.; Zhang, H.-C.; Ma, J. Further understanding the role of hydroxylamine in transformation of reactive species in Fe(II)/peroxydisulfate system. *Chem. Eng. J.* **2021**, *418*, 129464. [CrossRef]
18. Li, X.; Liu, X.; Lin, C.; Zhou, Z.; He, M.; Ouyang, W. Catalytic oxidation of contaminants by Fe^0 activated peroxymonosulfate process: Fe(IV) involvement, degradation intermediates and toxicity evaluation. *Chem. Eng. J.* **2020**, *382*, 123013. [CrossRef]
19. Cao, J.; Lai, L.; Lai, B.; Yao, G.; Chen, X.; Song, L. Degradation of tetracycline by peroxymonosulfate activated with zero-valent iron: Performance, intermediates, toxicity and mechanism. *Chem. Eng. J.* **2019**, *364*, 45–56. [CrossRef]
20. Li, W.; Zhang, Y.; Zhao, P.; Zhou, P.; Liu, Y.; Cheng, X.; Wang, J.; Yang, B.; Guo, H. Enhanced kinetic performance of peroxymonosulfate/ZVI system with the addition of copper ions: Reactivity, mechanism, and degradation pathways. *J. Hazard. Mater.* **2020**, *393*, 122399. [CrossRef] [PubMed]
21. Pan, Y.; Wang, Q.; Zhou, M.; Cai, J.; Tian, Y.; Zhang, Y. Kinetic and mechanism study of UV/pre-magnetized-Fe^0/oxalate for removing sulfamethazine. *J. Hazard. Mater.* **2020**, *398*, 122931. [CrossRef] [PubMed]
22. Guo, W.; Zhao, Q.; Du, J.; Wang, H.; Li, X.; Ren, N. Enhanced removal of sulfadiazine by sulfidated ZVI activated persulfate process: Performance, mechanisms and degradation pathways. *Chem. Eng. J.* **2020**, *388*, 124303. [CrossRef]
23. Pan, Y.W.; Zhou, M.H.; Li, X.; Xu, L.T.; Tang, Z.X.; Liu, M.M. Novel Fenton-like process (pre-magnetized Fe^0/H_2O_2) for efficient degradation of organic pollutants. *Sep. Purif. Technol.* **2016**, *169*, 83–92. [CrossRef]
24. Pan, Y.W.; Zhou, M.H.; Cai, J.J.; Li, X.; Wang, W.; Li, B.; Sheng, X.J.; Tang, Z.X. Significant enhancement in treatment of salty wastewater by pre-magnetization Fe^0/H_2O_2 process. *Chem. Eng. J.* **2018**, *339*, 411–423. [CrossRef]
25. Pan, Y.W.; Zhou, M.H.; Zhang, Y.; Cai, J.J.; Li, B.; Sheng, X.J. Enhanced degradation of Rhodamine B by pre-magnetized Fe^0/PS process: Parameters optimization, mechanism and interferences of ions. *Sep. Purif. Technol.* **2018**, *203*, 66–74. [CrossRef]
26. Fan, X.; Guan, X.; Ma, J.; Ai, H. Kinetics and corrosion products of aqueous nitrate reduction by iron powder without reaction conditions control. *J. Environ. Sci.* **2009**, *21*, 1028–1035. [CrossRef]
27. Long, X.; Xiong, Z.; Huang, R.; Yu, Y.; Zhou, P.; Zhang, H.; Yao, G.; Lai, B. Sustainable Fe(III)/Fe(II) cycles triggered by co-catalyst of weak electrical current in Fe(III)/peroxymonosulfate system: Collaboration of radical and non-radical mechanisms. *Appl. Catal. B Environ.* **2022**, *327*, 121716. [CrossRef]
28. Pan, Y.W.; Zhang, Y.; Zhou, M.H.; Cai, J.J.; Tian, Y.S. Enhanced removal of antibiotics from secondary wastewater effluents by novel UV/pre-magnetized Fe^0/H_2O_2 process. *Water Res.* **2019**, *153*, 144–159. [CrossRef]
29. Yang, Q.; Yang, X.; Yan, Y.; Sun, C.; Wu, H.; He, J.; Wang, D. Heterogeneous activation of peroxymonosulfate by different ferromanganese oxides for tetracycline degradation: Structure dependence and catalytic mechanism. *Chem. Eng. J.* **2018**, *348*, 263–270. [CrossRef]

MDPI AG
Grosspeteranlage 5
4052 Basel
Switzerland
Tel.: +41 61 683 77 34

MDPI Books Editorial Office
E-mail: books@mdpi.com
www.mdpi.com/books

www.ingramcontent.com/pod-product-compliance
Lightning Source LLC
Chambersburg PA
CBHW061507220926
51850CB00004B/190
9783725878130